Goods and Services of Marine Bivalves

Just the pearl II, by Frank van Driel, fine art photography (www.frankvandriel.com), with painted oyster shells of www.zeeuwsblauw.nl

Aad C. Smaal • Joao G. Ferreira • Jon Grant
Jens K. Petersen • Øivind Strand
Editors

Goods and Services of Marine Bivalves

Springer Open

Editors
Aad C. Smaal
Wageningen Marine Research and
Aquaculture and Fisheries group
Wageningen University and Research
Yerseke, The Netherlands

Jon Grant
Department of Oceanography
Dalhousie University
Halifax, Nova Scotia, Canada

Øivind Strand
Institute of Marine Research
Bergen, Norway

Joao G. Ferreira
Universidade Nova de Lisboa
Monte de Caparica, Portugal

Jens K. Petersen
Technical University of Denmark
Nykøbing Mors, Denmark

ISBN 978-3-319-96775-2 ISBN 978-3-319-96776-9 (eBook)
https://doi.org/10.1007/978-3-319-96776-9

Library of Congress Control Number: 2018951896

This Springer imprint is published by the registered company Springer Nature Switzerland AG
The registered company address is: Gewerbestrasse 11, 6330 Cham, Switzerland

Foreword

Bivalves are key to the development, functioning, and sustainability of coastal environments. Molluscs have long been revered for the beauty of their shells, culinary attributes, and as the basis for many successful aquaculture ventures. Long overdue, however, is wider recognition and understanding of their extraordinary abilities to shape, control, and improve their environments. As highly efficient filter feeders, bivalves facilitate benthic-pelagic coupling, influence sediment processes, provide structure, and contribute to habit diversity and biodiversity. While the term 'ecosystem services' is relatively new, the role of molluscs in performing those services has been recognised for centuries. Only in recent decades, however, have these attributes been studied, quantified, modelled, and put forth as integral to ecosystem development, maintenance, and sustainability.

In recent years, there have been two areas of major advancement in understanding how these bivalves 'make a living' – function at the molecular level and the part played by bivalves in the ecology of coastal seas. The development of advanced models to capture the complex integrative nature of the functions of bivalves has provided both theorists and practitioners with the means to understand these interactions. To wit, much of the advancement in these arenas has been through the contributions of the editors of this volume.

To summarise and explain complex systems and concepts associated with bivalves, it seems only fitting that the information is presented here by an equally integrated and diverse group of experts. Just as aggregations of individual bivalves increase their collective ability to influence their surroundings, so the current book brings together a stellar group of editors and authors of varied backgrounds who place bivalves in a well-deserved and prominent position as ecosystem engineers and providers of ecosystem services. Integration of the individual efforts of these scientists, their collaborators, and contributors to this volume has moved the importance of mussels, oysters, and other bivalves to new levels of understanding and acceptance.

As the field moves forward, their efforts will serve as a template for new investigators, as a valuable resource for managers, and as a launch pad for as-yet undefined and integrated studies. It is a dynamic future ahead.

Groton, CT, USA Sandra E. Shumway

Foreword

In 2050 – when the world population will have grown to almost ten billion people – the increase in income and the demand for more and better food will mean that food production needs to increase by 50% compared to its present. In many areas, but not everywhere, the available land for food production is decreasing due to competition with urbanisation and other uses, nutrient depletion, soil degradation, water scarcity, and climate change. Given the fact that the largest part of the world's population lives in coastal areas, there is great potential for marine ecosystems to contribute to the production of food. The Blue Growth Agenda provides a strategy to explore these resources to contribute to the production of high-quality and attractive food products as well as the production of feed, bioactive compounds, energy, and other valuable products.

Marine bivalves like oysters, clams, and mussels have been cultivated for ages and are recognised as a sustainable low food chain resource that acquires feed from natural resource in their environment. They provide a rich source for human nutrition and an associated economic value for local communities. Total bivalve aquaculuture and fisheries production amounted 16 million tons in 2015 with a landing value of 26 million US dollars.

Besides human nutrition, they provide food for birds and benthos and a habitat for a large number of species; they regulate water quality and sequester carbon and nitrogen. As eco-engineers, epibenthic bivalve beds are used for coastal defence and nature conservation. They also produce significant amounts of shell material that has many applications. These functions can be defined as ecological goods and services. This concept provides a framework for description and analysis of the role of bivalves in the ecosystem and a basis for addressing a wide range of topics, benefits, and controversies related to the use of bivalves for production, habitat restoration, water quality, and coastal management.

The book presents comprehensive reviews and analyses of the goods and services of bivalve shellfish. How they are defined, what determines the ecological functions that are the basis for the goods and services, what controversies in the use of goods and services exist, and what is needed for sustainable exploitation of bivalves from the perspective of the various stakeholders.

The reviews and analysis are based on case studies that exemplify the concept and show the strengths and weaknesses of the current applications. The multi-authored reviews cover ecological, economic, and social aspects of bivalve goods and services.

The transdisciplinary approaches as applied in this book represent a major strength in modern science. This approach is the core of the programmes of Wageningen University and Research, where various disciplines are integrated in order to achieve solutions. The international cooperation as exemplified in this book contributes to exploring the potential of the marine bivalves, to improve quality of life.

CEO Wageningen University and Research
Wageningen, The Netherlands

Louise O. Fresco

Preface

Marine bivalves have been a resource for human nutrition since prehistoric times. Their easy access and high nutritional quality have favoured their use throughout human history. Bivalve aquaculture and wild catch have shown a steady increase from 5 to 16 million tons per year over the period 1995–2015. Bivalve aquaculture nowadays dominates over wild catch almost ninefold, and this figure still increases. Bivalves are low food chain filter feeders. For their aquaculture, they rely on feed from their natural environment; hence, it is a non-fed extensive aquaculture.

The interactions with the environment are manifold. Main issues deal with competition with other filter feeders, overstocking, accumulation of biodeposits on the bottom, introduction of invasive species with bivalve transplantations, impacts of biotoxins for the consumer, and bivalve diseases. As impacts of bivalve aquaculture have gained much attention in literature, in this book, we focus on the goods and services of the bivalves.

In addition to aquaculture for production, both wild and cultivated bivalves have a suite of functions in the ecosystem. Through their filtration capacity, they clear water from particles, and under certain conditions, this increases the transparency of the water column. Better light penetration stimulates the production of phytoplankton if sufficient inorganic nutrients are available. Direct ammonia excretion and mineralisation of biodeposits, produced by the bivalves, act as a source of inorganic nutrients. So the uptake of phytoplankton by the bivalves gives a positive feedback on the growth of phytoplankton through increase in both light and nutrient availability. This is an example of a service of the bivalves to the ecosystem. This service can also be used to reduce the excess of nutrients in eutrophic conditions. Through uptake and assimilation of phytoplankton, the bivalves accumulate nutrients in their tissue, and harvesting of the product removes the accumulated nutrients from the ecosystem. Hence, the bivalves play a role in water quality management.

These examples brought the initiators of this book to the idea that the goods and services of marine bivalves cover a broad suite of bivalve characteristics that are worthwhile to be better explored. During a workshop in 2016, held in Celleno, Italy, a core group of almost 20 participants discussed the various topics that contribute to a more complete picture of the goods and services, as well as the controversies and limitations of the approach. It was concluded that the goods and services concept is a good basis for a comprehensive review of the functions of marine bivalves. Moreover, we realised that the more functions we addressed, even more ideas on further use of the bivalves emerged.

So, the initiators brought together a group of ca 100 authors and co-authors that are experts in the respective goods and services of the marine bivalves, in order to produce this book. We limited ourselves to the marine bivalves as a lot of knowledge is available from bivalve aquaculture. We also did not focus on adverse impacts of bivalve aquaculture on the environment as a lot of excellent literature is available on these issues.

The aim of the book is to review the knowledge of the various functions of natural and cultivated bivalves with relevance for human use, direct or indirect. This should deliver a better understanding of the bivalves and their various options for making better use of them.

This approach is relevant for anybody that deals with marine bivalves. Bivalve shellfish farmers can get a better understanding of the role the animals play in the ecosystem and for society; this may gain interest in combining different services to make use of the multiple potentials the bivalves have. This also holds for people that deal with shellfish restoration and conservation, as some of the reviews clearly show

Participants of the workshop on Bivalve Goods and Services, June 2016, Il Convento, Celleno, Italy, from left to right: Henrice Jansen, Cedric Bacher, Roberto Pastres, Camille Saurel, Luca van Duren, Ramon Filgueira, Peter Cranford, Pauline Kamermans, Jon Grant, Tom Ysebaert, Jacob Capelle, Jeroen Wijsman, Tore Strohmeier, Øivind Strand, Jens Petersen, Aad C. Smaal and in front Joao Ferreira; not on the photo Boze Hancock, Alessandra Roncarati that there is synergy in the combination of functions. This aspect is particularly relevant for policy advisors that need to prepare decisions on spatial planning and competing claims. As nowadays bivalve reefs are used for coastal defence, the book is also relevant for coastal engineers. The section on cultural services may inspire foodies as well as gardeners to start growing their own bivalves, as a sea garden or as a social community event. The goods and services concept is now further developed, in this case for the marine bivalves, and this contributes to further scientific knowledge that is relevant for students and scientists.

The book is set up for the reader with different chapters that can be read standalone as scientific papers. All chapters have been subject to peer reviews.

We are grateful for the help of many people. In particular, the referees for their constructive comments on the different chapters: Dr Andrea Alfaro, Dr Martin Baptist, Dr Jeff Barrell, Dr Bas Borsje, Dr Carrie Byron, Dr Matthieu Carre, Dr Loren Coen, Dr Luc Coumeau, Dr Steve Cross, Dr Jan Drent, Dr Ramon Filgueira, Dr Gef Flimlin, Dr Tom Gill, Dr Ing-Marie Gren, Dr Boze Hancock, Dr Vivian Husa, Dr John Icely, Dr Fred Jean, Dr Nigel Keely, Dr Lotte Kluger, Dr Thomas Landry, Dr Claire Lazareth, Dr Marie Maar, Dr Stein Mortensen, Dr Yngvar Olsen, Dr Christopher Pearce, Dr Theo Prins, Dr Julie Rose, Dr Matt Service, Dr Sandy Shumway, Dr Cosimo Solidoro, Dr ir Nathalie Steins, Dr Tore Strohmeier, Dr Jon Svendsen, Dr Mette Termansen, Dr Brenda Walles, Dr Gary Wickfors, and Dr Tom Ysebaert.

We are grateful to the colleagues of the Yellow Sea Fisheries Institute in Qingdao, China, for the Chinese translations of the abstracts.

We also thank Wageningen Marine Research, the Netherlands, for sponsoring the workshop. Special thanks to the Institute of Marine Research, Norway, the University of Applied Science Vlissingen, the Netherlands; Wageningen Marine Research, the Netherlands; and DTU Aqua, Denmark, and many of the authors institutions to facilitate the open access availability of the book. We thank Alexandrine Cheronet and Judith Terpos from Springer Nature for their help in publishing the book.

Yerseke, The Netherlands — Aad C. Smaal
Monte de Caparica, Portugal — Joao G. Ferreira
Halifax, NS, Canada — Jon Grant
Nykøbing Mors, Denmark — Jens K. Petersen
Bergen, Norway — Øivind Strand

General Introduction

The application in an ecological context of the economic and sociocultural concept of goods and services has been developed as a response to environmental degradation and the need to pay more attention to ecosystem functions and biodiversity in international policy. Loss of natural values due to human activities was recognised already long ago as a drawback not only for environmental quality but also for economic and social welfare. In the traditional economic theory, these were defined as (negative) external effects. In the course of the twentieth century, research started to quantify environmental impacts in economic terms, to include impacts in market decisions. This turned out to be complicated because environmental impacts were difficult to quantify and it was criticised because of market imperfections. It was recognised that more attention needed to be given to ecosystem functions in order to link economy and ecology (de Groot 1987). Ecosystem functions can be considered as the basis for the goods and services the ecosystems deliver to society. These ecosystem functions can be defined as 'the capacity of natural processes and components to provide goods and services that satisfy human needs, directly or indirectly' (de Groot et al. 2002). In this definition, ecosystem functions are explicitly coupled to human needs, rather than internal ecological processes, implying that 'ecosystem functions provide the goods and services that are valued by humans' (Fig. 1).

Meanwhile methodology has further been developed to express the goods and services in monetary values (Costanza et al. 1997; Pimentel and Wilson 1997).

The concept of ecosystem functions has been used as a basis for policy development. In the Convention on Biological Diversity (CBD), agreed upon at the Earth Summit in Rio, 1992, the ecosystem approach was adopted as a basis for international policy. It stands for a holistic approach in environmental policy, including environmental, economic, and social impacts of developments on the short and long terms. At the Johannesburg World Summit, 2002, the ecosystem approach was endorsed as a basis for the CBD. So the ecosystem approach stands for the ecosystem functions as a basis for ecosystem goods and services. As stated by Beaumont et al. (2007) the ecosystem goods and services concept provides a method to ensure the integration of environmental, economic, and social demands and pressures.

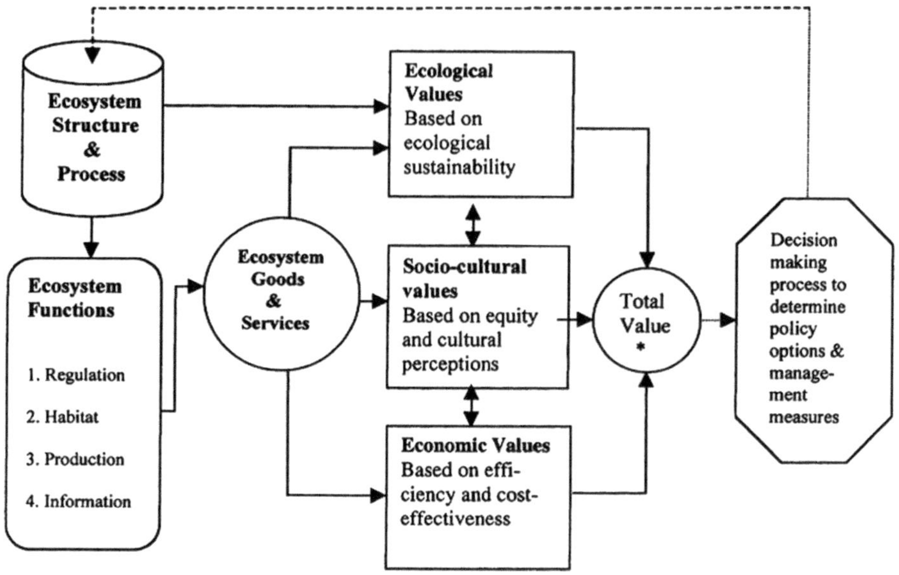

Fig. 1 Framework for the integrated assessment and valuation of ecosystem functions, goods, and services. (de Groot et al. 2002)

Goods and services are defined as 'the direct and indirect benefits people obtain from ecosystems' (Beaumont et al. 2007).

Assessing ecological processes and resources in terms of the goods and services translates the complexity of the environment into a series of functions. The concept has been further developed in the framework of the Millennium Ecosystem Assessment (MEA 2005). In the MEA approach, ecosystem goods and services are divided into provisioning, regulating, supportive, and cultural services, where supportive stands for habitats and genetic diversity. Many studies have been carried out on quantification of the ecosystem goods and services in the project The Economics of Ecosystems and Biodiversity (TEEB 2010). It is a global initiative focused on 'making nature's values visible'. Its principal objective is to mainstream the values of biodiversity and ecosystem services into decision-making at all levels. It aims to achieve this goal by following a structured approach to valuation that helps decision-makers recognise the wide range of benefits provided by ecosystems and biodiversity, demonstrate their values in economic terms, and, where appropriate, suggest how to capture those values in decision-making (www.teebweb.org).

The ecosystem goods and services concept is promoted as a basis for decision-making that now has a methodology to include not only an integrated approach to human impacts on the environment but also to evaluate the services that ecosystems provide for human use. This can be considered as a paradigm shift in environmental management. From a focus on adverse impacts, now ecosystem functions and their benefits for society can be analysed, quantified, and evaluated in more detail. This is of particular relevance for bivalve aquaculture. Farming of bivalves is an extensive type of aquaculture as the natural environment generally provides feed, seed, and space. Bivalve farming makes use of nature but also depends on nature. The close link between bivalve culture and nature has posed questions about possible negative impacts. In fact, these questions are dominant topics in many public debates all over the world. It is about impacts on habitats, landscape, sediment, carrying capacity, and other users, resulting in competing claims. Yet the ecological role of bivalves in the ecosystem provides a suite of goods and services to society. This has not yet been addressed in scientific literature in a comprehensive way. Reviews are available on specific ecosystem functions that exemplify the relevance of the concept (Coen et al. 2011; Ferreira and Bricker 2015; Petersen et al. 2015). Yet many questions remain to be addressed. A part of these deals with the discussion on the goods and services concept in broader sense, such as the debate about valorisation in monetary units (see TEEB 2010).

The aim of this book is to review and analyse the goods and services of bivalve shellfish. Given the debate about the different types of goods and services and their content (Haines-Young and Potschin 2017), we included bivalve habitats in the section on regulation and did not address a separate section on supportive functions. So, the papers have been ordered as provisioning, regulating, and cultural services, and there is a separate section on the assessment of services.

Wageningen Marine Research and Aquaculture and Fisheries group, Wageningen University and Research
Yerseke, The Netherlands
aad.smaal@wur.nl

Aad C. Smaal

References

Beaumont NJ, Austen MC, Atkins JP, Burdon D, Degraer S, Dentinho TP, Derous S, Holm P, Horton T, van Ierland E, Marboe AH, Starkey DJ, Townsend M, Zarzycki T (2007) Identification, definition and quantification of goods and services provided by marine biodiversity: implications for the ecosystem approach. Mar Pollut Bull 54:253–265

Coen LD, Dumbauld BR, Judge ML (2011) Expanding shellfish aquaculture: a review of the ecological services provided by and impacts of native and cultured bivalves in shellfish-dominated ecosystems. In: Shumway SE (ed) Shellfish aquaculture and the environment. Wiley-Blackwell Sussex, United Kingdom, pp 239–296

Costanza R, d'Arge R, de Groot RS, Farber S, Grasso M, Hannon B, Limburg K, Naeem S, O'Neill RV, Paruelo J, Raskin RG, Sutton P, van den Belt M (1997) The value of the world's ecosystem services and natural capital. Nature 387:253–260

De Groot RS (1987) Environmental functions as a unifying concept for ecology and economics. Environmentalist 7(2):105–109

De Groot RS, Wilson MA, Boumans RMJ (1992) A typology for the classification, description and valuation of ecosystem functions, goods and services. Ecol Econ 41:393–408

Ferreira JG, Bricker SB (2016) Goods and services of extensive aquaculture: shellfish culture and nutrient trading. Aquac Int 24(3):803–825

Haines-Young R, Potschin M (2017) Common international classification of ecosystem services (CICES). V5.1 Guidance on the application of the revised structure. Fabis Consulting Ltd, Nottingham.

Millennium Ecosystem Assessment (2005) Ecosystems and human well-being: synthesis. Island Press, Washington, DC

Petersen JK, Saurel C, Nielsen P, Timmermann K (2016) The use of shellfish for eutrophication control. Aquacult Internat 24(3):857–878

Pimentel D, Wilson C (1997) Economic and environmental benefits of biodiversity. Bioscience 47(11):747–758

TEEB (2010) The economics of ecosystems and biodiversity ecological and economic foundations. In: Pushpam K (ed). Earthscan, London/Washington, DC

Contents

Contributors

Stephan Abel Oyster Recovery Partnership, Annapolis, MD, USA

Cedric Bacher IFREMER, Centre de Bretagne, DYNECO-LEBCO, Plouzané, France

Anne Birch The Nature Conservancy, Florida Chapter, Maitland, FL, USA

Damian C. Brady School of Marine Sciences, Darling Marine Center, University of Maine, Walpole, ME, USA

Annette Breckwoldt Alfred Wegener Institut Helmholtz Center for Polar and Marine Research, Bremerhaven, Germany

Suzanne B. Bricker NOAA – National Ocean Service, NCCOS, Silver Spring, MD, USA

Bela H. Buck Alfred Wegener Institut Helmholtz Center for Polar and Marine Research, Bremerhaven, Germany

Meghan Burchell Department of Archaeology, Faculty of Humanities & Social Sciences, Memorial University, St. John's, NL, Canada

Paul G. Butler College of Life and Environmental Sciences, University of Exeter, Penryn Campus, Cornwall, UK

Jacob J. Capelle Wageningen UR, Wageningen Marine Research, Yerseke, The Netherlands

Laurent Chauvaud IUEM-UBO, UMR CNRS 6539, Technopôle Brest-Iroise, Plouzané, France

Johan A. Craeymeersch Wageningen UR, Wageningen Marine Research, Yerseke, The Netherlands

Peter J. Cranford St. Andrews Biological Station, St. Andrews, NB, Canada

Philippe Cugier IFREMER, Centre de Bretagne, DYNECO-LEBCO, Plouzané, France

Bryan DeAngelis The Nature Conservancy, North America Oceans and Coasts Program, URI Graduate School of Oceanography, Narragansett, RI, USA

Jeff DeQuattro The Nature Conservancy, Gulf of Mexico Program, Mobile, AL, USA

Paul Dinnel Skagit County Marine Resources Committee, Mount Vernon, WA, USA

Stefania Domeneghetti Department of Biology, University of Padova, Padova, Italy

Meirong Du Yellow Sea Fisheries Research Institute, Chinese Academy of Fishery Sciences, Qingdao, China

Peter F. Duncan University of the Sunshine Coast, Maroochydore DC, QLD, Australia

Jianguang Fang Yellow Sea Fisheries Research Institute, Chinese Academy of Fishery Sciences, Qingdao, China

Jinghui Fang Yellow Sea Fisheries Research Institute, Chinese Academy of Fishery Sciences, Qingdao, China

Fan Lin Yellow Sea Fisheries Research Institute, Chinese Academy of Fishery Sciences, Qingdao, China

Joao G. Ferreira DCEA, FCT, New University of Lisbon, Monte de Caparica, Portugal

Ramon Filgueira Marine Affairs Program, Dalhousie University, Halifax, NS, Canada

Institute of Marine Research, Bergen, Norway

Katia Frangoudes Univ Brest, IFREMER, CNRS, UMR 6308, AMURE, IUEM, Plouzané, France

Pedro S. Freitas Divisão de Geologia e Georecursos Marinhos, Instituto Português do Mar e da Atmosfera (IPMA), Lisbon, Portugal

Aline Gangnery IFREMER, Station de Port en Bessin, Port en Bessin, France

Marco Gerdol Department of Life Sciences, University of Trieste, Trieste, Italy

Arne Ghys Engelstraat, Deerlijk, Belgium

Jon Grant Department of Oceanography, Dalhousie University, Halifax, NS, Canada

Boze Hancock The Nature Conservancy, URI Graduate School of Oceanography, Narragansett, RI, USA

Judy Haner The Nature Conservancy, Coastal Programs Office, Mobile, AL, USA

Berit Hasler Environmental Social Science, Department of Environmental Science, Aarhus University, Roskilde, Denmark

Marianne Holmer Department of Biology, University of Southern Denmark, Odense, Denmark

Henrice M. Jansen Institute of Marine Research (IMR), Bergen, Norway

Wageningen UR – Wageningen Marine Research (WMR), Yerseke, The Netherlands

Zengjie Jiang Yellow Sea Fisheries Research Institute, Chinese Academy of Fishery Sciences, Qingdao, China

Pauline Kamermans Wageningen UR - Wageningen Marine Research, Yerseke, The Netherlands

Gesche Krause Alfred Wegener Institut Helmholtz Center for Polar and Marine Research, Bremerhaven, Germany

Jiaqi Li Yellow Sea Fisheries Research Institute, Chinese Academy of Fishery Sciences, Qingdao, China

Ting Li Key Laboratory of South China Sea Fishery Resources Exploitation & Utilization, Ministry of Agriculture, Guangzhou, People's Republic of China

Peter Malinowski Billion Oyster Project, New York Harbor Foundation, Brooklyn, NY, USA

Yuze Mao Yellow Sea Fisheries Research Institute, Chinese Academy of Fishery Sciences, Qingdao, China

Laboratory for Marine Ecology and Environmental Science, Qingdao National Laboratory for Marine Science and Technology, Qingdao, China

Remi Mongruel IFREMER, Centre de Bretagne, Ifremer, UMR 6308, AMURE, Plouzané, France

Carter R. Newell Maine Shellfish R+D, Damariscotta, ME, USA

Alberto Pallavicini Department of Life Sciences, University of Trieste, Trieste, Italy

Roberto Pastres Dipartimento di Scienze Ambientali, Informatica e Statistica, Mestre, VE, Italy

Betsy Peabody Puget Sound Restoration Fund, Bainbridge Island, WA, USA

Jens Kjerulf Petersen Danish Shellfish Centre, Institute of Aquatic Resources, Danish Technical University, DK, Nykoebing Mors, Denmark

John Richardson Blue Hill Hydraulics, Blue Hill, ME, USA

Shawn M. C. Robinson Department of Fisheries and Oceans Biological Station, Fisheries and Oceans Canada, St. Andrews Biological Station, St. Andrews, NB, Canada

Alessandra Roncarati URDIS Centre, University of Camerino, San Benedetto del Tronto, AP, Italy

Umberto Rosani Department of Biology, University of Padova, Padova, Italy

Camille Saurel Danish Shellfish Centre, DTU Aqua, Nykøbing M, Denmark

Nidhi Sharma Regional Centre for Biotechnology, NCR Biotech Science Cluster, Faridabad, Haryana (NCR Delhi), India

Aad C. Smaal Wageningen UR – Wageningen Marine Research (WMR), Yerseke, The Netherlands

Department of Aquaculture and Fisheries, Wageningen University, Wageningen, The Netherlands

Paul C. Southgate Faculty of Science, Health, Education and Engineering, University of the Sunshine Coast, Sippy Downs, QLD, Australia

Øivind Strand Institute of Marine Research, Bergen, Norway

Tore Strohmeier Institute of Marine Research (IMR), Bergen, Norway

Daniel Patrick Taylor Danish Shellfish Centre, DTU Aqua, Nykøbing M, Denmark

Mette Termansen Section for Environment and Natural Resources, Department of Food and Resource Economics, University of Copenhagen, Copenhagen, Denmark

Kim Tetrault Cornell Cooperative Extension, Suffolk County, Riverhead, NY, USA

Karin Troost Wageningen UR, Wageningen Marine Research, Yerseke, The Netherlands

Wouter van Broekhoven Wageningen UR – Wageningen Marine Research (WMR), Yerseke, The Netherlands

Department of Aquaculture and Fisheries, Wageningen University, Wageningen, The Netherlands

Luca A. van Duren Deltares, Delft, The Netherlands

Paola Venier Department of Biology, University of Padova, Padova, Italy

Marc C. Verdegem Department of Aquaculture and Fisheries, Wageningen University, Wageningen, The Netherlands

Brenda Walles Wageningen UR, Wageningen Marine Research, Yerseke, The Netherlands

Jeroen W. M. Wijsman Wageningen UR, Wageningen Marine Research, Yerseke, The Netherlands

Tom Ysebaert Wageningen UR, Wageningen Marine Research, Yerseke, The Netherlands

NIOZ Yerseke, Royal Netherlands Institute for Sea Research and Utrecht University, Yerseke, The Netherlands

Changbo Zhu Key Laboratory of South China Sea Fishery Resources Exploitation & Utilization, Ministry of Agriculture, Guangzhou, People's Republic of China

South China Sea Fisheries Research Institute, CAFS, Guangzhou, China

Philine zu Ermgassen Changing Oceans Group, School of Geosciences, University of Edinburgh, Grant Institute, Edinburgh, UK

The original version of this book was revised. The correction to this chapter is available at https://doi.org/10.1007/978-3-319-96776-9_28

Part I
Provisioning Services

Chapter 1
Introduction to Provisioning Services

Jon Grant and Øivind Strand

Abstract Food provisioning is a prominent feature of marine bivalve production, applicable worldwide since ancient times. Easy accessibility of this food source and high nutritional value make bivalves a possible driver in human evolution. In this section bivalve meat production is addressed, as well as other provisioning services including pearls and bio-active compounds. In both bivalve aquaculture and fisheries, harvest and production for meat provisioning must be balanced against carrying capacity and its implications for other services including water quality maintenance and habitat structure. Provisioning of meat through aquaculture can be improved via hatchery and breeding advances, a necessity in the changing ocean climate.

Keywords Human health · Evolution · Production

Provisioning of bivalves as food is perhaps the 'original' ecosystem commodity derived from the ocean, going back to the earliest humans. Indeed, the 'Aquatic Ape Hypothesis' links us directly with an ocean origin and dependence on bivalves (Morgan 1982). The specific consequences of fatty acid intake through bivalve consumption are thought to be critical in the evolution of the human brain (Crawford 2002). Bivalves remain tremendously popular as seafood, procured by hand in shallow water and cooked with the simplest of methods. Their position low in the food chain with no addition of feed and medicine makes bivalve aquaculture eminently future-proof. Interestingly, there is a caste system among bivalves, with oysters perceived as having more cachet than lowly mussels. The concept of white tablecloth dining goes hand in hand with oysters on the menu in France, although perhaps the southern US tradition of an oyster with hot sauce between crackers provides an alternative model. Regardless, bivalves are one of the few seafoods that are

J. Grant (✉)
Department of Oceanography, Dalhousie University, Halifax, NS, Canada
e-mail: jon.grant@dal.ca

Ø. Strand
Institute of Marine Research, Bergen, Norway
e-mail: oivind.strand@imr.no

A. C. Smaal et al. (eds.), *Goods and Services of Marine Bivalves*,
https://doi.org/10.1007/978-3-319-96776-9_1

purchased and sometimes eaten live, and thus embody fresh seafood. Their legendary reputation, particularly of oysters, is further enhanced with promises of aphrodisiac properties.

Bivalve fisheries have a similar long tradition with the prominence of regions such as the Limfjorden in Denmark, Zeeland Delta in the Netherlands and Chesapeake Bay in the USA, famed for mussels and oysters respectively. Some aspects of the fishery have become controversial due to fishing methods including dredging for scallops and suction dredging for clams. Removal of bivalve populations through fisheries has consequences for the provisioning of protein, but also for the removal of their many other services, a major theme of this book.

As with fisheries, bivalve aquaculture was developed initially for its provisioning potential. However, as detailed in other chapters, cultured bivalves provide a myriad of different services such as mitigation of eutrophication, and there are reasons to grow them besides food. An interesting aspect of suspended bivalve aquaculture is the way that it expands the habitat of the cultured species well beyond its natural benthic occurrence. A variety of production models have been developed at farm scale and beyond to predict biomass outcomes of farming. They have subsequently been extended to economic returns on farm yields. The deployment of these models, verified through individual growth rates and production statistics, has contributed to the success of bivalve culture worldwide. Moreover, integration of carrying capacity into these models is a means of forecasting maximum production before growth becomes self-limiting through food depletion.

An important caveat to bivalve production is the health benefit of low fat, high protein meat, rich in marine lipids and minerals. Bivalves do not receive the same attention as finfish regarding health consciousness in the media, but bivalve tissue is well known for its food value.

Like other marine products, bivalves provide a wide array of natural products based on both meat and shell. Joining the host of other marine organisms yielding potential therapeutants, bivalves contain both anti-microbial and anti-cancer candidates among other compounds. Beyond the value of soft tissue, in the tropical oyster *Pinctada maxima* pearl culture is far more valuable than oyster meat. Other uses for shell range from paving material to mother of pearl for inlays in furniture and musical instruments.

Although juveniles for many species of cultured bivalves are obtained from wild spat, the potential for improvements in growth rate and disease resistance via selective breeding are well known. Systemic bivalve diseases, perhaps best known in the Eastern oyster *Crassostrea virginica*, have decimated wild populations, and resistant stocks are an important tool in recovery. Triploidy is an important approach to introductions of alternative species. In cold waters, growth rates of cultured animals are slow to the detriment of profitability, and hatchery production is being established even for species with abundant spatfall. The necessity of breeding for potential climate resistance has become urgent with the impact of ocean acidification on early life history stages.

In this section, authors take a diverse view of these topics, and provide an account of the state of the art in the many direct beneficial uses of the Bivalvia.

References

Crawford MA (2002) Cerebral evolution. Nutr Health 16:29–34
Morgan E (1982) The aquatic ape. Souvenir, London

Chapter 2
Global Production of Marine Bivalves. Trends and Challenges

J. W. M. Wijsman, K. Troost, J. Fang, and A. Roncarati

Abstract The global production of marine bivalves for human consumption is more than 15 million tonnes per year (average period 2010–2015), which is about 14% of the total marine production in the world. Most of the marine bivalve production (89%) comes from aquaculture and only 11% comes from the wild fishery. Asia, especially China, is by far the largest producer of marine bivalves, accounting for 85% of the world production and responsible for the production growth. In other continents, the production is stabilizing or decreasing (Europe) the last decades. In order to stimulate growth, sustainability (Planet, Profit, People) of the aquaculture activities is a key issue. Environmental (Planet) aspects for sustainable aquaculture include the fishery on seed resources, carrying capacity, invasive species and organic loading. Food safety issues due to environmental contaminants and biotoxines should be minimized to increase the reliability of marine bivalves as a healthy food source and to stimulate market demands. Properly designed monitoring programs are important tools to accomplish sustainable growth of marine bivalve production.

Abstract in Chinese 在2010~2015年间，海水双壳贝类的年产量超过1500万吨，约占同时段全球海洋渔业总量的14%。其中约89%的贝类产量来自于水产养殖，野生采捕量仅占11%左右。亚洲（尤其是中国）是迄今为止最大的海水双壳贝类生产地，约占世界总产量的85%，同时也是全球双壳贝类生产的主要增长点。相比之下，在其他大陆（如欧洲）等地双壳贝类的产量在过去几十年均

J. W. M. Wijsman (✉) · K. Troost
Wageningen UR, Wageningen Marine Research, Yerseke, The Netherlands
e-mail: jeroen.wijsman@wur.nl; karin.troost@wur.nl

J. Fang
Yellow Sea Fisheries Research Institute, Chinese Academy of Fishery Sciences, Qingdao, China
e-mail: fangjg@ysfri.ac.cn

A. Roncarati
URDIS Centre, University of Camerino, San Benedetto del Tronto, AP, Italy
e-mail: alessandra.roncarati@unicam.it

A. C. Smaal et al. (eds.), *Goods and Services of Marine Bivalves*,
https://doi.org/10.1007/978-3-319-96776-9_2

保持稳定或呈下降趋势。经济效益和可持续性是驱动养殖产量持续稳定增长的关键因素。在可持续性方面,应重点关注养殖方式、养殖容量、入侵物种以及富营养化等因素,并将由环境污染物和生物毒素引起的食品安全问题最小化,这是提高海水双壳贝类作为食品可靠性并刺激市场需求的必要条件。合理的环境监测计划是保障双壳贝类产业可持续发展的重要手段。

Keywords Bivalves · Oysters · Mussels · China · Europe · Stock assessment · Sustainability

关键词 双壳贝类 · 牡蛎 · 贻贝 · 中国 · 欧洲 · 资源评估 · 可持续性

2.1 Introduction

Food production has been recognised as one of the most direct provisioning ecosystem functions of marine environments (Costanza et al. 1997). Food production of marine ecosystems comprises various types of organisms of which macroalgae, fish, crustaceans and molluscs are the most important. The increase of marine food production has been recognised as an important solution to fulfil the increasing protein demands of the growing world population in the future (Naylor et al. 2000). The total global food production of marine ecosystems in the period 2009 to 2014 was 104.3 million tonnes per year and consisted of wild capture (80.4 million tonnes per year) and marine aquaculture (23.9 million tonnes per year) (FAO 2016a, b). Marine bivalves account for about 14% of the global marine production (tonnes) in this period. Most of the marine bivalve production (89%) comes from aquaculture, with a total economic value of 20.6 billion US$ per year. Only 11% of the marine bivalve production comes from the wild fishery. However, the seed resources that form the basis for aquaculture production are often fished or collected from natural stocks as well. Due to decreasing seed resources and environmental issues with the seed fishery, more and more of the seed resources for marine bivalve aquaculture are produced within land-based hatcheries. The direct capture production of marine bivalves remained relatively constant since the 1970's (1.78 million tonnes per year), but the aquaculture production of marine bivalves increased from 1.18 million tonnes per year in the period 1970–1974 to 13.47 million tonnes per year in the period 2010–2015.

The total market value of marine bivalves is about 23 billion US$ per year (2010–2015), however, the full economic value is much higher due to the economic benefits from secondary products and services (e.g. shucking and packaging houses, transport, manufacture of prepared products and retail sales) (Schug et al. 2009). The value of the production in terms of US$ kg^{-1} is depending on the market demands and the supply of the specific species.

Marine bivalves are appreciated by consumers due to their nutritional benefits as well as their taste. Bivalves are healthy sources of energy and protein, rich in vitamins (A and D) and essential minerals (iodine, selenium calcium), low in fat and a good source of omega-3 fatty acids with well-established health benefits (Orban et al. 2002; Schug et al. 2009; EFSA 2014). Selenium for example is an essential trace element that is required by the human body for proper functioning of the thyroid gland, and may help protect against free radical damage of the tissue. Most of the dietary human intake of selenium occurs via plants (Brazil nuts) and seafood (Ariard et al. 1993; Kristan et al. 2015). There is evidence that selenium deficiency may be related to a variety of degenerative diseases (Reilly 1998). However, it is also known that there is also a narrow concentration window between essentiality and toxicity of selenium for humans (Kristan et al. 2015). The unavoidable presence of environmental contaminants, such as mercury and biotoxins in bivalves could also result in a risk to the health of consumers (Sadhu et al. 2015; Visciano et al. 2016). Regular monitoring programs, therefore, are essential to prevent food safety issues.

Marine bivalves are also a sustainable type of food production. As herbivores, they are low in the trophic chain. The trophic position of marine bivalves like mussels, oysters, clams and cockles is 2 (herbivores), while the average trophic position of the total marine capture fishery is 3.1 (Duarte et al. 2009).

In contrast to the intensive fish aquaculture, bivalve aquaculture is an extensive form of aquaculture while the bivalves feed on algae that occur naturally in the ecosystem and no additives such as vitamins and antibiotics are added. The production relies merely on the natural productivity of marine phytoplankton, either in the form of living algae or as detritus, transported to the bivalves by water flow e.g., currents and tidal exchange. Bivalves can enhance primary production by increased nutrient recycling (Prins and Smaal 1994). At high stocking densities, however, the bivalves can result in overgrazing and thereby reduce primary production (Smaal et al. 2013b; Filgueira et al. 2015). Management by farmers is an important factor whereas the farmers will try to maximise their profits within their aquaculture sites. This is done by growing the bivalves at specific locations where the conditions for growth and survival are maximized (Capelle 2017). Numerous management activities are possible among which active removal of predators (Calderwood et al. 2016) and thinning-out and sorting the bivalves to optimise growth efficiency and shape. The moment of harvesting is also decided by the farmers, based on the quality of the bivalves but also on market prices.

Since aquaculture of marine bivalves takes place in natural environments, it often results in conflicts with other functions such as nature conservation, recreation, economic development, etc. Also the fishery on marine bivalves might result in conflicts since natural stocks that are an important food source for fish and birds are removed from the system (Ens et al. 2004; Ens 2006). Moreover, the fishery with dredges is a bottom disturbing activity that might impact the seafloor integrity. Also aquaculture often depends on the wild fishery for the seed resources (Smaal and Lucas 2000).

For aquaculture purposes, bivalves and associated organisms are often translocated between sites and ecosystems which has resulted in introduction and spreading of (invasive) exotic species (Minchin and Gollasch 2002; Wolff 2005). Proper management of bivalve transports are important to reduce environmental impact.

In this paper an overview is given of the trends in global production of marine bivalves based on FAO data. The production figures for different continents are discussed and compared with each other. As case studies, the trends and developments in China – by far the largest producer of marine bivalves – and Europe are presented. In China, the production of marine bivalves is still increasing tremendously due to the increasing protein demand of the growing population. In Europe, however, the total production is decreasing the last decades due to various reasons such as competing claims on space, diseases and carrying capacity issues. For both case studies an overview is presented of the trends and developments of production, import and export and legislation. Finally, in this paper, special attention is paid to stock assessment of marine bivalves since this provides essential information for sustainable management of natural stocks in order to reduce environmental impact of the fishery on marine bivalves. This is based on a case study of the stock assessment for natural bivalve species in the Wadden Sea, The Netherlands.

2.2 Global Trends

In the FAO Global Fishery and Aquaculture Statistics database a total 79 marine bivalve species are listed as cultured and 93 species are listed as captured species. They can be grouped into four major groups: clams, oysters, mussels and cockles. Clams and oysters are the major species groups that contribute 38% and 33%, respectively, to the global production. Scallops account for 17% and mussels for 13% of the global production. The global production of marine bivalves is more than 15 million tonnes per year (data FishStat FAO 2010–2015) (Fig. 2.1). More than 85% of the total marine bivalve production in comes from Asia (Fig. 2.2). As a

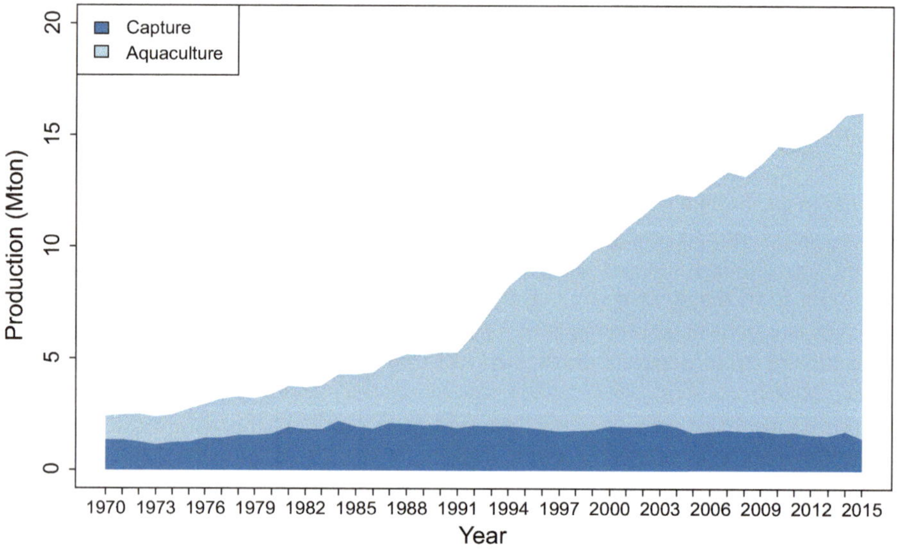

Fig. 2.1 Evolution of the total global production (million tonnes per year) of marine bivalves by the fishery and aquaculture. (Data from FAO FishStat (1970–2015))

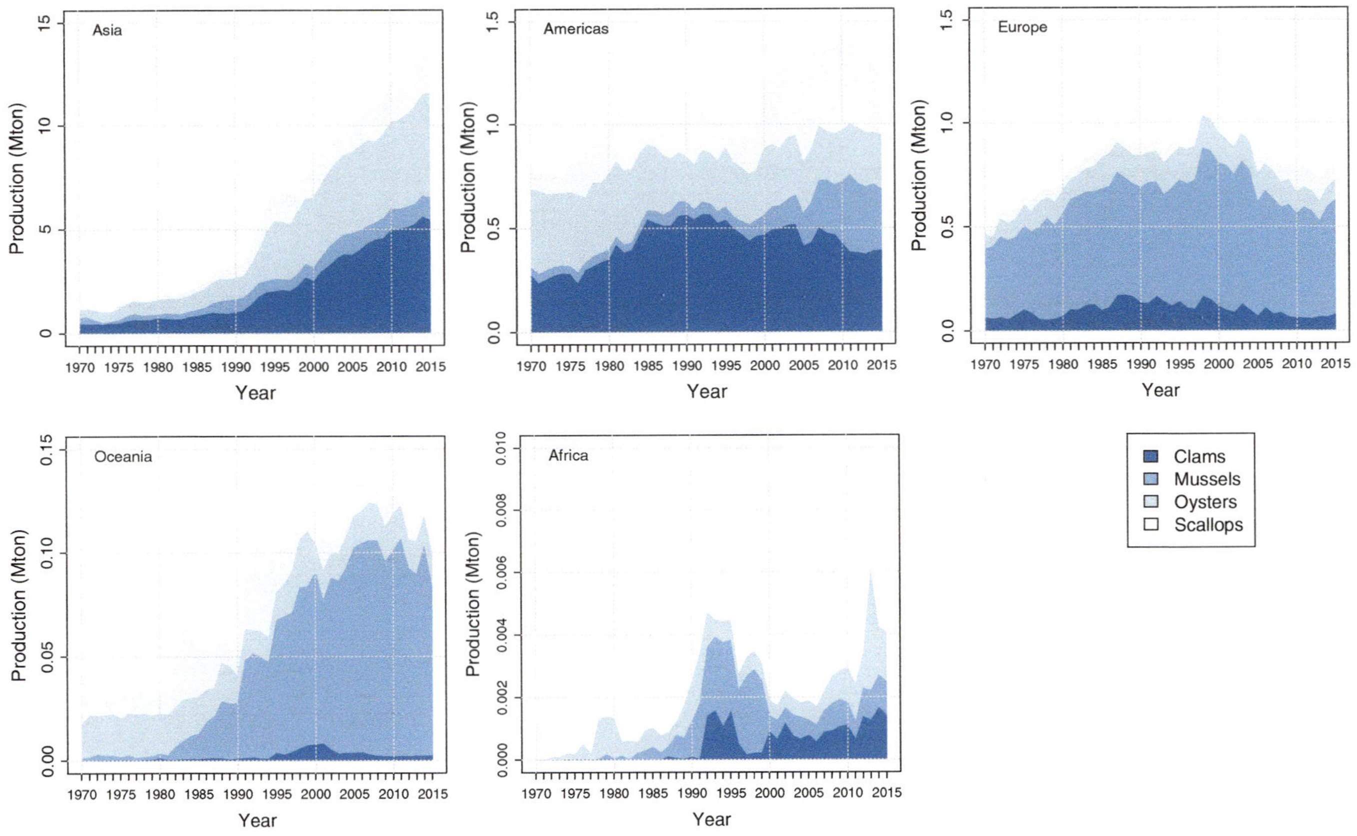

Fig. 2.2 Evolution of the total production (million tonnes per year) of marine bivalves by the fishery and aquaculture together for the different continents from 1970 to 2015. Marine bivalves are grouped as clams, mussels, oysters and scallops. (Data from FAO FishStat (1970–2015))

result the production in Asia, specifically China, largely dominates the patterns and trends in the world production.

The total production of marine bivalves is the result of a complex interaction between the market demand and the production capacity of the system. If the market demand increases, this will be a trigger to increase production. However, the production will be limited by the carrying capacity of the system. There are different types of carrying capacity that could potentially limit the production: physical, production, ecological and social carrying capacity (Inglis et al. 2000; Gibbs 2009; Smaal and Van Duren 2019).

The bivalve production in Asia is increasing on average with 0.42 million tonnes per year since 1990. The majority of the production in Asia comes from clams (5.4 million tonnes in 2015) and oysters (ca 5.1 million tonnes in 2015). The production of scallops and mussels in 2015 was 2.3 and 1.1 million tonnes, respectively. Production in Asia is dominated by the production in China (more than 90% of the marine bivalve production in Asia). Other marine bivalve producing countries of importance in Asia are Japan (0.75 million tonnes per year), Republic of Korea (0.4 million tonnes per year) and Thailand (0.23 million tonnes per year). The major reason for the increase in marine bivalve production in China is the increased demand for proteins from the growing population and the increased standard of living in China. As a result, social and ecological carrying capacity are no major issues yet. Spatial and production carrying capacity limitations might be occurring locally since the availability of suitable productive sites can sometimes be limiting. The wild fishery on marine bivalves in China is not specifically documented in the Fishstat database. Japan is the most important country in Asia in terms of the fishery on marine bivalves, mainly scallops, with an average yearly production of 0.38 million tonnes in the period 2010–2015. In Indonesia the fishery on blood cockles produce on average about 74 thousand tonnes per year (2010–2015).

North and south America is responsible for 9% of the global marine bivalve production. Most of the aquaculture production is in Chile (mussels and scallops), Peru (scallops), the United States (American and Pacific cupped oysters, hard clams) and Canada (mussels). The wild fishery is mainly practiced in the United Stated of America on scallops, hard clams and surf clams, with a mean total production of about 510 thousand tonnes per year (2010–2015). Also in Canada there is a wild fishery (ca 92 thousand tonnes per year) mainly on Atlantic deep-sea scallops. The total production in north and south America increased from about 1 million tonnes per year in the period 1995–2000 to about 1.3 million tonnes per year in the period 2010–2015. This increase is mainly due to the increase of aquaculture production. Clams used to be the most important species but the production is slightly decreasing since 1988. This is mainly due to a decrease in wild catches of clams in the United States from about 450 thousand tonnes per year in 1985 to a total production of 250 thousand tonnes per year at present (2010–2015). From 2000 the mussel-, but also the scallop production is increasing in the Americas. The increase in mussel production is mainly due to an increase in the aquaculture production in Chile with a tenfold increase in this century from 23 thousand tonnes in 2000 to a current production of about 244 thousand tonnes per year (2010–2015). In the United States of America, the wild fishery on oysters decreased from 200 thousand tonnes in the early 70's of the last century to a production of about 59 thousand tonnes per year in the period 2010–2015. The aquaculture production of eastern oysters increased from about 106 thousand tonnes per year in the period 1995–1999 to a total production of 142 thousand tonnes per year at present (2010–2015).

In Europe, responsible for 5.5% of the world production of marine bivalves, the production has decreased since 1998. This decrease is mainly due to a decrease in mussel production by aquaculture activities from about 600 thousand tonnes per year in 1998 to about 465 thousand tonnes per year in the period 2010 to 2015. The production of bottom culture mussels in the Netherlands is responsible for part of this reduction since the production in the Netherlands decreased from 113 thousand tonnes in 1998 to 46 thousand tonnes per year in the period 2010–2015. The production is limited by a reduction in physical space due to competing claims with nature conservation and occasional recruitment failures. Production of oysters, clams and scallops in Europe is much lower than the mussel production. The oyster production decreased from 150 thousand tonnes in 1998 to about 94 thousand tonnes per year (average 2010–2015), with the largest production in France (ca 78 thousand tonnes per year). In Ireland, however, the production of oysters is increasing. Almost 25% of the marine bivalve production in Europe, yearly about 205 thousand tonnes per year, comes from the fishery. The highest capture production is in the UK (scallops and cockles), Denmark (blue mussels), France (scallops) and Italy (venus clams).

The production in Africa and Oceania is less than 1% of the world production. In Oceania mussels, mainly produced in New Zealand, are by far the most important bivalve species, with a total production of about 94 thousand tonnes per year (2010–2015). In Australia there is additionally some production of flat and cupped oysters. The fishery on marine bivalves is very limited in Oceania. In Africa, there is some fisheries (ca 2 thousand tonnes per year) on carpet shells and cupped oysters in Tunisia and Senegal. Mussels are cultured in South Africa with a total production of 800 tonnes per year. The low production in Africa is low due to the limited market demands. The local community has no tradition in consuming bivalves, since it is often difficult to keep the healthy sanitary conditions.

2.3 China

2.3.1 Aquaculture Production in China

Aquaculture production of China is the highest in the world (61.5 million tonnes in 2015). The total output of marine aquaculture in China in 2015 was 29.5 million tonnes and consists of marine bivalve production of 12.4 million tonnes, macroalgae production of 13.8 million tonnes, fish production around 1.6 million

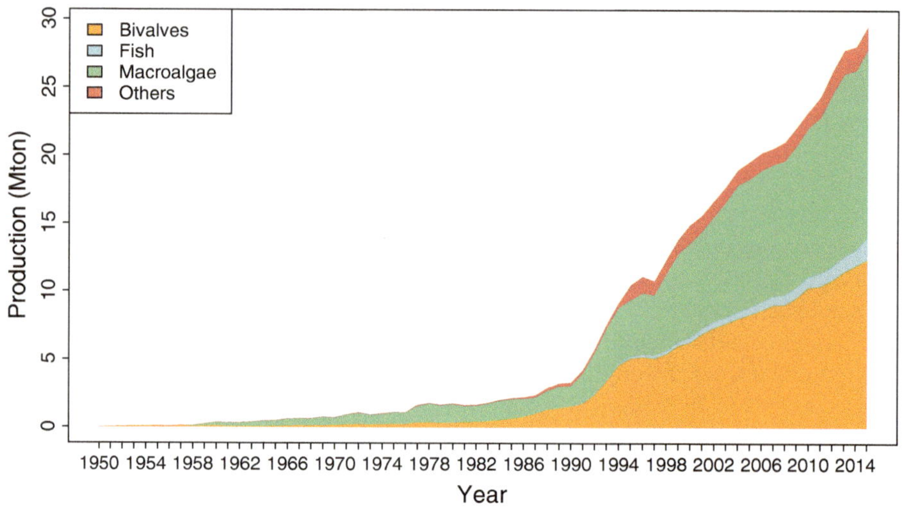

Fig. 2.3 Changes in mariculture production (million tonnes per year) in China. (Data from FAO FishStat (1950–2015))

tonnes[1] and other organisms (e.g. molluscs, crustaceans, echinoderms) about 1.7 million tonnes (FAO FishStat). Marine bivalves represented 42% of the total mariculture production in China in 2015. The production increased from an average of 51 thousand tonnes per year in 1950–1959 to 335 thousand tonnes per year in 1975–1979, 7.3 million tonnes per year in 2000–2004 to 12.4 million tonnes in 2015 (Fig. 2.3). Besides marine bivalves, macroalgae are also responsible for the enormous growth in marine aquaculture production in China since 1990 (Fig. 2.3). The major shellfish cultured in China include 8 categories (oysters, clams, scallops, mussels, razor clams, cockles, sea snails and abalones) and 48 species (Tang et al. 2016), among which oysters, clams and scallops yield more than 1 million tonnes annually, and the production of mussels and razor clams fall between 0.5 to 1 million tonnes each year.

2.3.2 *Trends and Developments*

Bivalve aquaculture has a long history in China, the record of oyster farming can be traced back to 2400 years ago, in the ancient book "Pisciculture" written by Fan Li, a famous politician, strategist, Taoist and Economist. In the 1950s and 1960s of the twentieth century, the main species of Chinese bivalve culture were oyster and mussel. The major farming methods were tideland cultivation and natural sea area nursing (Liu 1959).

[1] In the China Fishery statistical yearbook a production of 1.3 million tonnes fish is reported for 2015.

In the beginning of 1970s, the technologies for seed production of mussel in hatcheries and natural sea seed collection made great progress, which promoted the rapid development of mussel culture industry. In 1977, the national mussel farming area was more than 2000 ha, and the annual production exceeded 60,000 tonnes, about 200 times and 75 times respectively compared to those in 1970. In late 1970s, the success in artificial breeding of cockles *Tegillarca granosa*, and *Sinonovacula constricta*, clams *Ruditapes philippinarum* and *Cyclina sinensis* laid the foundation for development of the large-scale culture of these species. In the early 1980s, the breakthrough of artificial breeding in hatcheries and natural sea seed collection of *Chlamys farreri*, had led to the rapidly development of the scallop culture at industrial level. Particularly, the introduction of bay scallop *Argopecten irradians* from Atlantic coast in 1982 brought a prosperous stage for Chinese scallop aquaculture development.

New Eco-farming aquaculture modes such as integrated aquaculture of shellfish and seaweed in shallow-sea, and pond farming of shrimp-shellfish, has contributed greatly to the development of modern Chinese marine aquaculture. In recent years, China has carried out research on varieties of shellfish selective breeding. Until to 2015, 18 new varieties of shellfish were determined by genetic and selective breeding, including oysters, scallops, hard clams, abalone, pearl oyster and manila clam, which had been certificated by the national new variety committee in China. Shellfish farming methods now include maritime longline culture (northern China) and raft culture (southern China), mud flat farming, bottom sow farming, and pond culture. Integrated aquaculture of shellfish-fish, shellfish-shrimp and shellfish-seaweed has become the new trend for mariculture development in China.

From 2005 to 2014, the bivalve culture production maintained an overall growth. During these 10 years, production of scallops, clams, oysters and mussels increased by 80.4%, 40.8%, and 30.0% and 19.3%, respectively. Shellfish prices showed overall rise during the last 10 years with inter-annual fluctuations. In 2015, the domestic shellfish wholesale price data shows that, the average price of live oysters was increased from 0.87 US$/kg to 0.98 US$/kg, an increase of approximately 12.1%. The average price of live razor clam, from the same period last year, increased from 3.99 US$/kg to 4.09 US$/kg, an increase of slightly 1.9%. Scallop adductor muscle average price, reduced by 7.8% from 3.50 US$/kg to 3.23 US$/kg in the same period last year; the average price of fresh clams decreased from 1.16 US$/kg to 1.13 US$/kg compare to the same period last year, down by 2.6%.

2.3.3 *Import and Export*

In 2014, scallops, oysters and mussels were the major imported and exported molluscs, with the net import and export being 33.3 thousand tonnes and 32.1 thousand tonnes, respectively. The scallops, oysters and mussels import were 29.0 thousand tonnes, 2.6 thousand tonnes and 1.6 thousand tonnes, respectively, and the export of these 3 bivalve species were 29.2 thousand tonnes, 1.3 thousand tonnes and

1.5 thousand tonnes for each. The annual import and export volume were 135.7 and 453.0 million US dollars respectively. From 2008 to 2014, China imported shellfish mainly from the United States, Japan, North Korea, South Korea, France and New Zealand, and the shellfish exported went to United States, South Korea, Hong Kong, Macao and Australia. In 2014, the Chinese imports of oysters, scallops and mussels were mainly from France, Japan and North Korea, while the export of these species went to Hong Kong, the United States and South Korea. Data from China Customs show that from January to October 2015, China's shellfish export amount and revenue was 219 thousand tonnes (1.87% increase compared to the same period in 2014) and 1.38 billion US$ (1.11% decrease compared to the same period in 2014).

2.3.4 *Legislation*

The impact of marine bivalve culture to the environment is expected to be relatively small. This is mainly due to the filtering capacity, removing particles from the water column. Moreover, no additives (food, antibiotics, etc.) are added to the system. Nevertheless, there are many laws and regulations related to mariculture in China (Table 2.1). Besides the state-level management, protection and zoning regulations, there are also provincial level laws and regulations on natural resources exploitation and development. For instance, "Marine Functional Zoning of Shandong Province" has clearly clarified the scope and area that can be applied for aquaculture. Since 2007, the Ministry of Agriculture Fisheries Bureau executed the functional zoning for shellfish mariculture in Guangdong Province and 11 other areas. Reference from the relevant provisions of the EU, the Ministry announced the "Requirements for shellfish mariculture regional zoning", which defined the 3 categories of shellfish products according to the content of *Escherichia coli* (MPN/100 g) in meat and juice in the shellfish. For category one, the *Escherichia coli* content should be no more

Table 2.1 Relevant legislation concerning marine shellfish production in China

"Law on the Administration of Sea Area Use of the People's Republic of China", 2002
"Law on Marine Environmental Protection of the People's Republic of China", 2000
"Fisheries Law of the People's Republic of China"
"Standard for Seawater Quality of the People's Republic of China" (GB3097-1997), 1997
"National Marine Functional Zoning (2011–2020)", 2012
"Regulations of Marine Environmental Protection of Shandong Province", 2004
"Regulations on the Administration of Sea Area Use in Shandong Province" 2004
"Marine Functional Zoning of Shandong Province", 2012
"Requirements for Shellfish Mariculture Regional Zoning"
"Law of Quality and Safety of Agricultural Products of People's Republic of China"
"Provisions on the Administration of Aquaculture Quality"
"Provisional Regulations on Supervision and Management of Shellfish Production Environment"
"Provisions on the Hygiene Management of Exporting Shellfish"

than 230 *E. coli*/100 g, bivalves can be put into the market directly; the second category refers to the *Escherichia coli* content can be greater than 230 *E. coli*/100 g and no more than 4600 *E. coli*/100 g, which can be put into the market directly without raw food permit. Bivalves with *Escherichia coli* content more than 4600 *E. coli*/100 g and no more than 46,000 *E. coli*/100 g are in the third category, which need to be kept depurated until reached the standard in the second category before sales.

2.4 Europe

2.4.1 Aquaculture Production in Europe

In Europe, aquaculture production has remained relatively constant in the last years. In 2015, the total output of European aquaculture was 3.0 million tonnes, of which the majority (2.4 million tonnes) was marine production (FAO FishStat). The marine aquaculture production was represented almost exclusively by fish production (about 1.8 million tonnes) and bivalve production (about 598 thousand tonnes) (FEAP 2016; FAO 2017). Culture of other marine organisms like macroalgae and crustaceans is negligible in Europe (Fig. 2.4). The most important species (freshwater and marine) reared in Europe in 2015 are Atlantic salmon (1.6 million tonnes per year), mussels (497 thousand tonnes per year), rainbow trout (290 thousand tonnes per year), common carp (154 thousand tonnes per year), Pacific cupped oyster (89 thousand tonnes per year), gilthead sea bream (79 thousand tonnes per year) and European sea bass (68 thousand tonnes per year) (FAO 2017). Among the EU

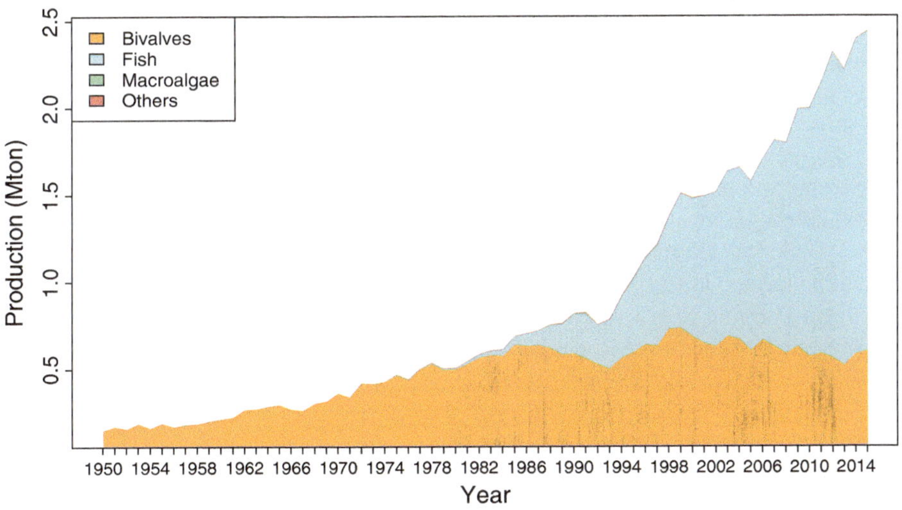

Fig. 2.4 Changes in mariculture production (million tonnes per year) in Europe. Macroalgae and others are hardly visible. (Data from FAO FishStat (1950–2015))

Member States, the largest producers of marine aquaculture products are Norway (1.4 million tonnes, mainly Atlantic salmon), Spain (271 thousand tonnes per year), United Kingdom (196 thousand tonnes per year), France (161 thousand tonnes per year) and Greece (103 thousand tonnes per year). With regard to the aquaculture of marine bivalves in the different countries, Mediterranean mussels accounted for 83.0% of the marine aquaculture in Spain whereas in France, the largest volumes were produced by Pacific cupped oyster (46.6%), blue mussel (37.9%) and Mediterranean mussel (8.8%). The growth of marine Aquaculture production in Europe is mainly caused by the increase in fish culture (Atlantic salmon) since 1985–1990 (Fig. 2.4). The production of marine bivalves by European aquaculture is decreasing from an average production of 661 thousand tonnes per year in the period 1995–1999 to an average of 560 thousand tonnes per year in the period 2010–2014.

2.4.2 Trends and Developments

In Europe, bivalve aquaculture has ancient origins, both for oysters and mussels. Images on archaeological findings (pots) date back the oyster farm to Roman times, between the second and first century BC. In Italy, in the lakes Lucrino and Fusaro (Campania Region) flat oysters were reared for the Roman nobles consumption. In Spain, in the fourth century BC, the natives used to leave bivalve molluscs in large deposits denominated ‘concheiros’; the first findings of bivalve culture were discovered near to the Roman villages in the first century A.D. In France, the mussel culture was practised in the intertidal zone since the thirteenth century using wooden stakes called “bouchots”. This technique spread widely along the French Atlantic coastline over the nineteenth century, while Northern European countries (the Netherlands, Ireland and the United Kingdom) developed bottom culture plots where juveniles were spread over the plots in shallow water, generally in bays or in sheltered areas on the ground. The French “bouchot” system, currently still in use, consists of ropes carrying young mussels placed on vertical poles and then, as the mussels grow, they move onto the pole where they will grow until they reach their commercial size. In the Middle Ages, oyster culture was widespread in the Sea of Taranto (Puglia Region, Italy). Under the kingdom of Ferdinand IV of Bourbon, around 1764, oysters continued to be farmed in the Fusaro Lake. In the sixteenth century, in Spain, people coming from Portugal began to gather mussels, clams, and cockles in the ria of Arosa (NW Spain). At the turn of the nineteenth century, flat oyster culture was well developed especially in the Bay of Arcachon (France), reaching 15–20,000 tonnes per year between 1908 and 1912. In 1979 the disease caused by the exotic parasite *Bonamia ostreae* broke down the productions (Buestel et al. 2009). Between 1971 and 1973, after the depletion of the Portuguese oyster (*Crassostrea angulata*) decimated by several successive diseases, several hundred tonnes of *Crassostrea gigas* were imported from Canada and the species became established and an abundant spat was settled in Marennes-Oleron (France). In the Mediterranean, flat oysters were cultured until 1950 when high mortalities strongly reduced productions due to a disease caused by the protozoan, *Marteilia refringens.*

In the Netherlands, due to numerous conflicts in the nineteenth century among fishermen for the open access to fish blue mussels (*Mytilus edulis*) and flat oysters (*Ostrea edulis*) in the delta region in the southwest of the Netherlands, the rearing system changed and mussel and oyster fishermen could rent exclusive access rights to plots in the sea. This plot system facilitated the beginning of bottom culture of blue mussel and flat oyster because only the person who rented the plot draw the benefits from the harvest. In 1952, plots to grow mussels could also be leased in the shallow Wadden Sea in the north of the Netherlands and this led to the development of a second region where blue mussel was cultured.

Mussels became important in Spain when farmers started culturing them in the beginning of the twentieth century. In longline systems, mussels are cultured on ropes that remain suspended in the water from a long line composed of buoys, whereas oysters are introduced in trays or "poches", attached to the rope. The long lines can be semi-submerged, submerged or buoyant depending on the farming environment. "Bateas mussel rafts" are largely employed in Spain. Rafts are composed of a solid structure from which the mussels hang in the water. In the bottom mussels system, predominantly used in the Netherlands, Germany, Ireland and the UK, large flat boats equipped with 2 to 4 dredges, fish juvenile mussels from natural beds which then are relayed in sheltered areas for further growth until they reach the commercial size. Currently, the Pacific cupped oyster is the most widely reared oyster species in Europe thanks to its fast growth, adaptability to different settings and improving breeding lines in the hatchery. Since 2008, high mortalities have been recorded in many European countries due to herpesviruses affecting larvae, spats and juveniles of cupped oysters highlighting the emergence of a global problem involving not only the European countries, but also New Zealand and Australia. Concerning clams (*Ruditapes decussatus, Tapes philippinarum*), the farming began in the 1980s, when harvesting wild stocks by hand or by dredging was discouraged in order to protect resources. Currently, clam farming depends mainly on natural recruitment and reproduction in hatchery. Spat is grown in nursery areas or tanks and seeded in shallow areas managed by fishermen's cooperatives.

2.4.3 Import and Export

Data from EUMOFA Report 2016, based on the elaboration of Eurostat data, show that in 2015, EU imports of mussels totalled 200,000 tonnes, the lowest volume in the past 6 years, 10,000 tonnes less than the average import volume from 2010–2014. France, the EU's largest market, recorded stable imports in 2015 when compared with the 2010–2014 average, while Italy, the second largest importer, demonstrated a remarkable increase in import volumes (+28%) compared with the average volume imported in the 2010–2014 period. Portugal recorded a significant growth of the import as well. Otherwise, import to all other EU markets declined rather sharply: the Netherlands (−49%), the UK and Germany (−19% each), and Spain and Belgium (−10% each). This reduction in imports can be explained by the economic crisis as well as the increase in prices (average price from US$ 10.30 per kg in 2010 to US$ 15.40 per kg in 2015). European bivalve export amounted at 20,000 tonnes (+9% respect to 2014) and 172 million US$ (+24% respect to 2014). EU self-sufficiency for this commodity fell to 61%. The EU consumption of mussels registered a slightly fluctuating trend from 2005 to 2014, with the apparent consumption moving from 1.36 kg per capita in 2005 to 1.27 kg per capita in 2014. Chile and New Zealand are the two main suppliers of mussels to Europe, providing the market with frozen and conserved products. Intra-EU trade is well developed with a value around half the total value of the EU supply. There are major trade flows from Spain, the Netherlands and Denmark (wild mussels in the case of Denmark) to Belgium, France and Italy. The European consumption of scallop in 2014 was almost at the same level as in 2005. Its peak of 0.63 kg per capita was registered in 2010, and a 4% decrease was recorded between 2013 and 2014, due to the reduction in catches in the United Kingdom and France of 11% and 29%, respectively. Since 2005, consumption of clam has remained stable at an average of 0.35 kg per capita (EUFOMA 2016).

2.4.4 Legislation, Environmental Issues

In the European Union, in 1979, the "Shellfish Water Directive 79/923/EEC" concerning the quality of shellfish waters to protect populations from the harmful consequences resulting from the discharge of polluting substances into the sea, was enacted. This legislation has laid down and updated official controls for monitoring bivalve production and relaying areas (Table 2.2). The authorities, based on faecal indicator organisms (*E. coli*), determine the classification of a production area and the treatment required in growing areas during the production cycle and for the end-product. The classification marks three classes: Class A (≤230 *E. coli*/100 g), molluscs can be harvested for direct human consumption; Class B (90% of samples must be ≤4600 *E. coli*/100 g; all samples must be less than 46,000 E. coli/100 g), molluscs can be sold for human consumption after purification in an approved plant, or after re-laying in an approved Class A re-laying area, or after an EC-approved heat treatment process; Class C (≤46,000 *E. coli*/100 g), molluscs can be sold for human consumption only after re-laying for at least 2 months in an approved re-laying area followed, where necessary, by treatment in a purification centre, or after an EC-approved heat treatment process. The European Food Safety Authority Panel on Biological Hazards has reviewed the hazards and has also determined the need to restrict shellfish harvesting from areas contaminated with faecal pollution. Molluscs must not be subject to production or collected in prohibited areas. In 2010, the EU Commission Regulation was enacted to identify the presence of OsHV-1 μvar associated with the massive mortality in oysters in order to reduce the spread

Table 2.2 Relevant legislation concerning marine shellfish production in Europe

"Regulation (EC) NO 178 of the European Parliament and of the Council laying down the general principles and requirements of food law, establishing the European Food Safety Authority and laying down procedures in matters of food safety", 2002
"Commission Decision establishing special health checks for the harvesting and processing of certain bivalve molluscs with a level of amnesic shellfish poison (ASP) exceeding the limit laid down by Council Directive 91/492/EEC", 2002
"Regulation (EC) No 852 of the European Parliament and of the Council on the hygiene of foodstuffs", 2004
"Regulation (EC) No 853 of the European Parliament and of the Council laying down specific hygiene rules for on the hygiene of foodstuffs", 2004
"Regulation (EC) No 854 of the European Parliament and of the Council laying down specific rules for the organisation of official controls on products of animal origin intended for human consumption", 2004
"Commission Regulation (EC) No 2073 on microbiological criteria for foodstuffs", 2005
"Commission Regulation (EC) No 2074 laying down implementing measures for certain products under Regulation (EC) No 853/2004 of the European Parliament and of the Council and for the organisation of official controls under Regulation (EC) No 854/2004 of the European Parliament and of the Council and Regulation (EC) No 882/2004 of the European Parliament and of the Council, derogating from Regulation (EC) No 852/2004 of the European Parliament and of the Council and amending Regulations (EC) No 853/2004 and (EC) No 854/2004", 2005
"Commission Regulation (EC) No 1664 amending Regulation (EC) No 2074/2005 as regards implementing measures for certain products of animal origin intended for human consumption and repealing certain implementing measures", 2006

of the virus to uninfected regions. According to the regulation, disease control measures must be implemented. This includes the establishment of containment areas and the restriction of movement from these areas if OsHV-1 μvar is identified.

2.5 Stock Assessment

Culture of some marine bivalve species is dependent on fishery on wild stocks. Seed is for instance collected using spat collectors, or fished in the natural environment. Culture of such species (e.g. blue mussels (*Mytilus edulis*) and Pacific cupped oysters *Crassostrea gigas*) is therefore dependent on the availability of natural stocks. The natural stocks of most bivalve species show large fluctuations from year to year, depending on the success of natural spatfall. Moreover, the spatial heterogeneity is high because many species occur locally within dense beds. Stock assessments are of key importance for fisheries regulation and management and provide essential information for impact assessment studies.

We illustrate the role of stock assessment with the case study of blue mussels in the Dutch part of the Wadden Sea. The Netherlands is, after Spain and France the third producer of mussels in Europe, with a total production of about 63 million kg per year (1990–2015). In contrast to the suspended culture in Spain, in the Netherlands the mussels are mainly cultured on-bottom at designated culture plots that are located in the Wadden Sea and in the Oosterschelde. The mussels are cultured by about 50 companies, operating 60 vessels (Capelle 2017). The mussel culture depends largely on natural seed resources. Mussel seed is dredged from naturally occurring subtidal mussel beds in the Wadden Sea in Autumn and Spring. From there they are translocated to the culture plots in the Wadden Sea and the Oosterschelde where they are kept for 1–3 years until they reach consumption size. The mussels are harvested mainly in Summer and Autumn and sold at the auction in Yerseke. From there they are processed and distributed over Europe (mainly Belgium, France, the Netherlands, Germany).

Since 1992 the natural mussel stock in the subtidal areas of the Wadden Sea is assessed annually from two different surveys. A quantitative survey in early Spring and a qualitative survey in Autumn (Van Stralen et al. 2016, 2017). In Autumn, the mussel seed fishery is exclusively allowed in subtidal areas that are designated as being unstable due to starfish (*Asterias rubens*) predation and exposure to unfavourable hydrodynamic conditions. In other words, in areas where the seed beds are likely to disappear before, or in the course of, the following winter. Designation of areas as stable or instable was made based on survey results since 1992 and expert judgement of fishermen and fisheries inspectors (Smaal et al. 2014). To determine the amount of mussels to be fished during the autumn fisheries in these instable areas, an estimate of the total stock of seed mussels is made in late summer or early autumn. A qualitative assessment of starfish abundance gives insight in the likelihood of particular beds disappearing before winter, which is used in the fisheries plan to identify beds to be fished first. In early spring a second stock assessment is carried out, with the primary purpose to prepare the fisheries permit for the spring fisheries, and with the secondary purpose to be able to assess effects of changes in the fisheries policy and management. Where the autumn assessment is a qualitative survey, the spring assessment is set up as a quantitative survey in which not only mussels but all species of bivalves, starfish and crabs are recorded. This dataset gives insight in distribution patterns of mussels and other bivalves, fishery impacts, as well as the main benthic predators, and is therefore of key importance in studies on effects of fisheries and changes in fisheries and nature policy (Smaal et al. 2013a).

The autumn assessment is carried out with a mussel dredge. Historical information as well as observations by fisheries inspectors and fishermen is used to determine the areas with a high encounter probability. Using the mussel dredge, operated by a commercial mussel fisheries vessel, the bed contours and kilograms per square meter are estimated. The total seed mussel stock, as well as the exploitable stock size in areas open to the fishery in autumn is estimated based on the dredge data and expert judgement.

The spring assessment is carried out with a suction dredge. For stations with a water depth over 10 meters a towed bottom dredge is used. Both sampling gears fish along a track with known length (ca 150 m) and surface area. The sampling locations are distributed along a stratified regular sampling grid where the distance between stations is smaller in areas with a high encounter probability. The encounter probability is estimated based on the autumn survey, the autumn fishery (gps-data of the fishing vessels), historical information and observations by fisheries inspectors and fishermen. During the Spring survey (March–April) 400–600 locations are sampled within a period of 3–4 weeks. The samples are sieved over a mesh of 5 mm, and all species of shellfish, crabs and starfish are counted and weighed per station (total wet weight). The total stock is calculated as the sum of all stations: biomass (wet weight) per square meter per station multiplied by the surface area the sampling station is representative for (which is determined by the stratum).

The amount of wild sublittoral mussels in the western Wadden Sea (Spring survey) is presented in Fig. 2.5 (bars). The lines indicate the total amount that has been harvested for grow-out on mussel culture plots in spring and autumn. As can be seen from this figure, in some years more seed has been fished than found during the spring survey. This is due to a new recruitment during the summer months, after the spring survey and spring fisheries and before autumn fisheries of the same year.

Due to competing claims with shellfish-eating birds, one of the nature conservation goals in the Wadden Sea, a transition from bottom fisheries to seed collection using suspended seed collectors (SMCs) has taken place within the mussel culture since 2010. According to an agreement between the mussel producers' organization, NGO's and the Dutch government, a gradually increasing portion of the stable areas are closed for fishing. The area available for SMCs is proportionally increased. The total harvest of the SMCs in the Wadden Sea increased from 1.3 Mkg in 2009 to 15.2 Mkg in 2016 (Capelle and Van Stralen 2017). The SMCs resulted in a more

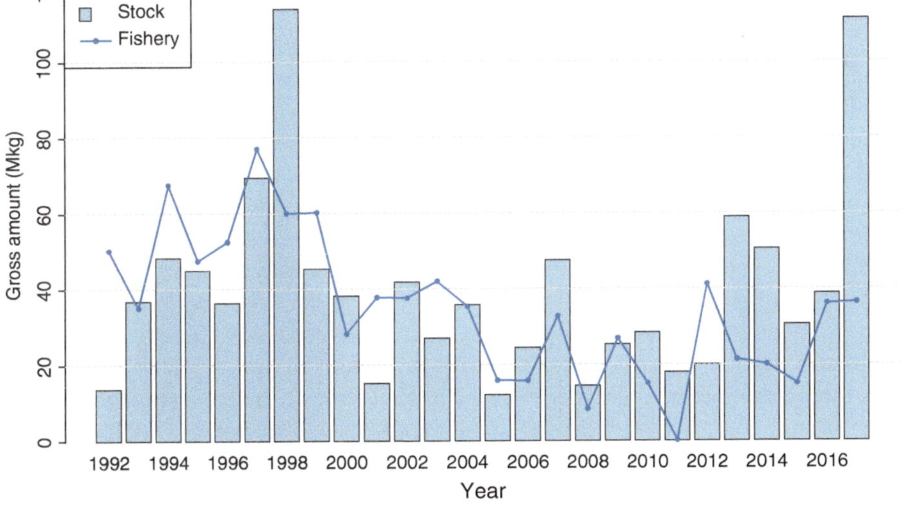

Fig. 2.5 Total wild stock of mussels in the sublitoral part of the western Wadden Sea (spring survey) and the total gross amount seed fished (Spring and Autumn) of that year. To calculate the Gross amount 40% debris and associated fauna is assumed for mussel seed and 25% for adult mussels. It is also assumed that the seed will gain 20% in weight between survey and fishery

stable supply of mussel seed for the mussel farmers, making them less depending on the fluctuations in natural spatfall on the bottom. Data from the stock assessments are an essential tool for the (evaluation of the) management decisions.

2.6 Conclusions

Food production is an important provisioning ecosystem function of marine bivalves. The global production is growing, although this growth is mainly caused by the increase in aquaculture production in Asia, in particular China. The bivalve farming has already become a considerable scale industry in China and has provided high quality proteins for humans. The production in North and South America, however, is stabilizing since 2000 and the production in Europe is decreasing.

It is expected that the global production of marine bivalves, particularly in Asia, will continue to grow in the future in order to fulfil part of the protein demand of the growing world population, especially since bivalves are a sustainable form of protein production. The expected growth in production of marine bivalves will come mainly from an increase in aquaculture production since it can be foreseen that the production from wild catches is relatively limited and will probably only decrease in the future. Sustainability (People, Profit, Planet) is an important factor for a further increase in marine bivalve production. Bivalve aquaculture is depending to a large extent, if not completely, on natural ecosystems, which are in many cases nature conservation areas. Removal of seed resources and microalgae as food source for the bivalves can, in some areas, result in competing claims with other ecosystem values.

Stock assessments are of primary importance in determining sustainable seed supply. The case study Wadden Sea shows that annual monitoring of bivalve stocks, resulting in long-term time series, is important for the year-to-year management of bivalve stocks since it gives insight in the population dynamics as well as potential ecological impacts of fisheries and aquaculture targeted on marine bivalves. This can also be applied to other regions where aquaculture is depending on wild stocks.

With increasing emphasis on sustainability, the balance between aquaculture development and ecology/environment has become a new requirement and challenge in both research and commercial aspects. The development of a sustainable bivalve aquaculture will promote employment in the coastal fishing zones supporting diversification in areas linked to changes in the fisheries sector. It could be of great socio-economic importance because it would allow the recovery and enhancement of traditional activities related to the region. New opportunities for local management of commercial fishing may open up to guarantee the characteristics of the product being of great interest to the consumer.

Acknowledgements Dr. Yuze Mao from YSFRI is acknowledged for providing valuable information on aquaculture production in China. The authors are grateful to the referees Dr. J. Grant and Dr. R. Filgueira for their valuable comments on the manuscript.

References

Ariard JC, Berthet B, Boutaghou S (1993) Seasonal selenium variations in mussels and oysters from a French marine farm. J Food Compos Anal 6:370–380

Buestel D, Ropert M, Prou J, Goulletquer P (2009) History, status, and future of oyster culture in France. J Shellfish Res 28:813–820

Calderwood J, O'Connor NE, Roberts D (2016) Efficiency of starfish mopping in reducing predation on cultivated benthic mussels (*Mytilus edulis* Linnaeus). Aquaculture 452:88–96

Capelle JJ (2017) Production efficiency of mussel bottom culture. PhD thesis, Wageningen University

Capelle JJ, Van Stralen MR (2017) Invang van mosselzaad in MZI's. Resultaten 2016. Wageningen Marine Research, Report number: C044/17, 30 pages

Costanza R, D'Arge R, De Groot R, Farber S, Grasso M, Hannon B, Limburg K, Naeem S, O'Neill RV, Paruelo J, Raskin RG, Sutton P, Van Den Belt M (1997) The value of the world's ecosystem services and natural capital. Nature 387:253–260

Duarte CM, Holmer M, Olsen Y, Soto D, Marba N, Guiu J, Black KD, Karakassis I (2009) Will the oceans help feed humanity? Bioscience 59:967–976

EFSA (2014) Scientific opinion on health benefits of seafood (fish and shellfish) consumption in relation to health risks associated with exposure to methylmercury. EFSA J 12(7):80

Ens BJ (2006) The conflict between shellfisheries and migratory waterbirds in the Dutch Wadden Sea. Waterbirds around the world. In: Broere GC, Galbraith CA, Stroud DA (eds) Waterbirds around the world. The Stationery Office, Edinburth, pp 806–811

Ens BJ, Smaal AC, De Vlas J (2004) The effects of shellfish fishery on the ecosystems of the Dutch Wadden Sea and Oosterschelde. Final report of thesecond phase of the scientific evaluation of the Dutch shellfish fishery policy (EVA II), Report number: C056/04

EUFOMA (2016) Highlights the EU in the world EU market, supply consumption, trade EU landings aquaculture processing. www.eumofa.eu

FAO (2016a) The state of world fisheries and aquaculture. Contributing to food security and nutrition for all. Food and Agriculture Organisation of the United Nations, Report, 200 pages

FAO (2016b) FAO yearbook. Fishery and Aquaculture Statistics. 2014. Food and Agriculture Organisation of the United Nations, Report, 76 pages

FAO (2017) FAO yearbook. Fishery and Aquaculture Statistics. 2015. Food and Agriculture Organisation of the United Nations Report, 78 pages

FEAP (2016) European aquaculture production report 2007–2015. FEAP Secretariat, Report, 46 pages

Filgueira R, Comeau LA, Guyondet T, Mckindsey CW, Byron CJ (2015) Modelling carrying capacity of bivalve aquaculture: a review of definitions and methods. Encyclopedia of Sustainability Science and Technology

Gibbs MT (2009) Implementation barriers to establishing a sustainable coastal aquaculture sector. Mar Policy 33:83–89

Inglis GJ, Hayden BJ, Ross AH (2000) An overview of factors affecting the carrying capacity of coastal embayments for mussel culture. NIWA, Report number: Client Report: CHC00/69, 38 pages

Kristan U, Planišek P, Benedik L, Falnoga I, Stibilj V (2015) Polonium-210 and selenium in tissues and tissue extracts of the mussel *Mytilus galloprovincialis* (Gulf of Trieste). Chemosphere 119:231–241

Liu CG (1959) The earliest literature of aquaculture worldwide- scientific value of "Pisciculture" by Fan Li (written in Chinese). China Fish 22:45–46

Minchin D, Gollasch S (2002) Vectors – how exotics get around. In: Leppäkoski E, Gollasch S, Olenin S (eds) Invasive aquatic species of Europe. Distribution, impacts and management. Kluwer Academic Publishers, Dordrecht, pp 183–192

Naylor RL, Goldburg RJ, Primavera JH, Kautsky N, Beveridge MCM, Clays J, Folke C, Lubchenco J, Mooney H, Troell M (2000) Effects of aquaculture on world fish supplies. Nature 405:1017–1024

Orban E, Di Lena G, Nevigato T, Casini I, Marzetti A, Caprioni R (2002) Seasonal changes in meat content, condition index and chemical composition of mussels (*Mytilus galloprovincialis*) cultured in two different Italian sites. Food Chem 77:57–65

Prins TC, Smaal AC (1994) The role of the blue mussel *Mytilus edulis* in the cycling of nutrients in the Oosterschelde estuary (The Netherlands). Hydrobiologia 282(283):413–429

Reilly C (1998) Selenium: a new entrant into the functional food arena. Trends Food Sci Technol 9:114–118

Sadhu AK, Kim JP, Furrell H, Bostock B (2015) Methyl mercury concentrations in edible fish and shellfish from Dunedin, and other regions around the South Island, New Zealand. Mar Pollut Bull 101:386–390

Schug DM, Baxter K, Wellman K (2009) Valuation of ecosystem services from shellfish restoration, enhancement and management: a review of the literature. Northern Economics Inc., Report, 58 pages

Smaal AC, Lucas L (2000) Regulation and monitoring of marine aquaculture in The Netherlands. J Appl Ichtyol 16:187–191

Smaal AC, van Duren L (2019) Bivalve aquaculture carrying capacity: concepts and assessment tools. In: Smaal AC (ed) Goods and services of marine bivalves. Springer, Dordrecht, pp 451–483

Smaal AC, Craeymeersch J, Drent J, Jansen JM, Glorius S, Van Stralen MR (2013a) Effecten van mosselzaadvisserij op sublitorale natuurwaarden in de westelijke Waddenzee: samenvattend eindrapport. Wageningen IMARES, Report number: C006/13, 162 pages

Smaal AC, Schellekens T, Van Stralen MR, Kromkamp JC (2013b) Decrease of the carrying capacity of the Oosterschelde estuary (SW Delta, NL) for bivalve filter feeders due to overgrazing? Aquaculture 404-405:28–34

Smaal AC, Brinkman AG, Schellekens T, Aguera A, Van Stralen MR (2014) Ontwikkeling en stabiliteit van sublitorale mosselbanken, samenvattend eindrapport. IMARES, Report

Tang Q, Han D, Mao Y, Zhang W, Shan X (2016) Species composition, non-fed rate and trophic level of Chinese aquaculture. J Fish Sci China 23:729–758

Van Stralen MR, Troost K, Van den Ende D (2016) Inventarisatie van het sublitorale wilde mosselbestand in de Oosterschelde en Voordelta in de zomer en najaar van 2016. MarinX, Report number: 2016.169, 9 pages

Van Stralen M, Van den Ende D, Troost K (2017) Inventarisatie van het sublitorale wilde mosselbestand in de westelijke Waddenzee in het voorjaar van 2017. MarinX, Report number: 2017.175, 25 pages

Visciano P, Schirone M, Berti M, Milandri A, Tofalo R, Suzzi G (2016) Marine biotoxins: occurrence, toxicity, regulatory limits and reference methods. Front Microbiol 7:1–10

Wolff WJ (2005) Non-indigenous marine and estuarine species in the Netherlands. Zoologische mededelingen 79:1–116

Chapter 3
Provisioning of Mussel Seed and Its Efficient Use in Culture

P. Kamermans and J. J. Capelle

Abstract Mussel culture largely depends on seed and feed from the natural environment. This paper focusses on seed provisioning and efficient use of these resources in mussel production. Approaches and technologies for seed supply and efficient use of seed in mussel production are described for the different culture techniques. This includes potential interactions and conflicts with the natural environment. Three methods are used to provide seed: wild harvest, use of suspended collectors and hatchery production. Harvest of wild seed from seaweed (in New Zealand) or natural beds is still a major source for culture in some areas, costs are low but provisioning is often unreliable. Most research concerning spat collection deals with comparison of different types of suspended collectors, settlement cues and problems with biofouling. Hatchery seed is more expensive, but hatcheries provide the opportunity for selective breeding and triploid production giving the product an added value. The challenge is to bring hatchery production costs more in line with the actual sale value of mussel seed. Monitoring genetic diversity can give insight in whether collector seed or hatchery seed growth and survival is negatively affected by reduced diversity. Grow-out occurs in bottom culture, bouchot culture and off-bottom longline and raft culture. In bottom-culture, the focus is on developing better seeding techniques, predator control and optimizing culture practices such as timing of relay, substrate use and harvest. For bouchot culture, technical developments are directed to mechanical methods to increase efficiency in size grading, restocking, harvesting and processing. Innovation in growing-out techniques for longline and raft culture are directed towards the investigation of optimal stocking densities, and on material type and configuration of farms. Production efficiency increases from bottom culture to bouchot culture, to rope and raft culture and are related to the sources of mortality and differences in growth rate. Growth rate of mussels is higher in off bottom culture than in on bottom culture and higher when submerged than in intertidal. Mussels from the *Perna* genus are found to have a higher growth rate but a lower production efficiency than mussels from the *Mytilus* genus. Efficient use of seed in mussel culture should aim at a reduction of mussel

P. Kamermans (✉) · J. J. Capelle
Wageningen UR – Wageningen Marine Research, Yerseke, The Netherlands
e-mail: pauline.kamermans@wur.nl

A. C. Smaal et al. (eds.), *Goods and Services of Marine Bivalves*,
https://doi.org/10.1007/978-3-319-96776-9_3

losses and an increase in growth rates. Important tools are adjusting seeding densities in relation to system design, reducing seeding stress, predator control and applying thinning out or relay.

Keywords Mussels · Seed · Culture · Efficiency

3.1 Mussel Aquaculture Production

Mussels are found in large quantities in coastal areas all around the world. Mussels, often organized in patches or in beds, are easily collected and have been an important protein source (an ecosystem good) for mankind since prehistoric times (Erlandson 1988). Mussels are commonly cultured, all that is needed is protection against dislodgement, by using sheltered sites or attachment substrate and protection against predation, supply of oxygen and seston, which is sufficient in most coastal environments. Mussel culture is carried out according to a variety of techniques, often developed in the course of centuries and adapted to the local culture environment. Mussel culture is based on seed and nourishment from the natural environment. This paper focusses on seed provisioning and efficient use of this resource in mussel production.

Global mussel culture mainly concerns two genera (*Mytilus* and *Perna*) and 9 species (*Mytilus edulis*, *Mytilus galloprovincialis*, *Mytilus californianus*, *Mytilus platensis (also called M. chilensis)*, *Mytilus unguiculatus*, *Mytilus planulatus*, *Perna canaliculus*, *Perna perna* and *Perna viridis*). In addition, a small production of *Aulacomya atra* and *Choromytilus chorus* takes place in Chile and Argentina. Mussel production comprises around 1.8 million tonnes with a value of 2.7 billion US dollars (average of 2010–2015, FAO statistics). In 2015, the largest production took place in Asia (1.05 million tonnes), followed by Europe (0.50 million tonnes), the Americas (0.25 million tonnes), Oceania and Africa (0.08 million tonnes) (FAO statistics, www.fao.org). The main mussel producing countries are China in Asia, Spain in Europe, Chile in the Americas, New Zealand in Oceania and South Africa in Africa (Table 3.1). Production in China, Chile and New Zealand started in the seventies of the last century and showed a rapid increase (Fig. 3.1). This levelled off for New Zealand around 2005 and continues to increase in China. In Chile production declined fast around 2011, mainly due to problems with toxic algae (Reguera et al. 2014).

3.2 Culture Techniques and Innovations

Mussels culture is based on recently settled individuals called spat, or juveniles called seed. This resource is collected in different ways depending on the local circumstances and grow-out methods. In general, three methods are used to harvest spat or seed: wild harvest, use of suspended collectors and hatchery production

Table 3.1 Mussel aquaculture production (tonnes) in 2015 per species and per country (FAO statistics, www.fao.org)

Species	Country	Tonnes
Mytilus e dulis	France	61,000
	Netherlands	54,100
	Canada	22,725
	United Kingdom	20,112
	Ireland	16,015
	Germany	10,875
	Norway	2731
	United States of America	1788
	Sweden	1525
	Denmark	1229
	Iceland	140
	Senegal	16
	Namibia	10
	Argentina	6
	Argentina	6
	St. Pierre and Miquelon	3
Mytilus galloprovincialis	China	845,038
	Spain	225,308
	Italy	63,700
	Greece	18,628
	France	14,100
	Bulgaria	3373
	Portugal	1315
	South Africa	950
	Croatia	746
	Slovenia	573
	Albania	295
	Russian Federation	207
	Montenegro	189
	Ukraine	70
	Romania	35
	Turkey	3
Mytilus californianus	Mexico	270
Mytilus platensis	Chile	208,707
	Argentina	6
Mytilus plan u latus	Australia	3679
Mytilus unguiculatus	Korea, Republic of	53,536
Perna perna	Brazil	18,364
	Venezuela, Boliv Rep of	1

(continued)

Table 3.1 (continued)

Species	Country	Tonnes
Perna viridis	Thailand	118,775
	Philippines	15,949
	India	8700
	Malaysia	1673
	Cambodia	1500
	Singapore	906
Perna canaliculus	New Zealand	76,811
Aulacomya atra	Chile	1068
	Argentina	4
Choromytilus chorus	Chile	1581

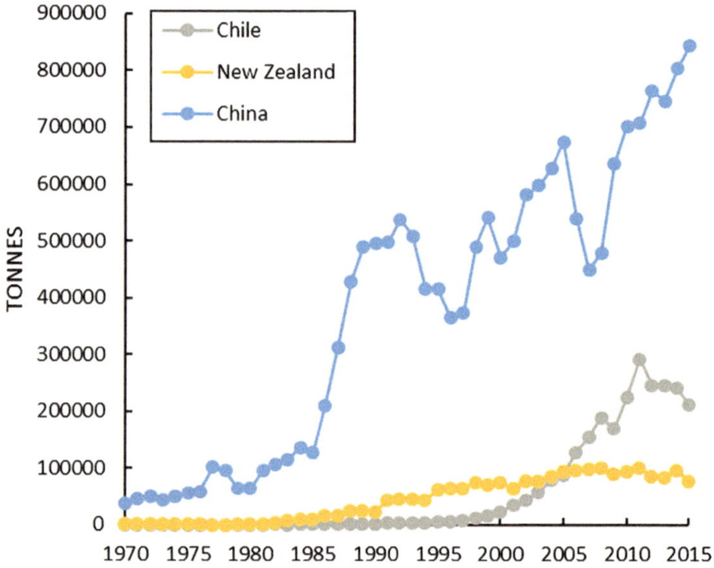

Fig. 3.1 Mussel aquaculture production (tonnes) in Chile, New Zealand and China (FAO statistics, www.fao.org)

(Fig. 3.2). The majority of the grow-out occurs in bottom culture, bouchot culture and off-bottom longline and raft culture (Fig. 3.2).

Each technique to acquire seed has different costs. In general, the least labour-intensive method (wild harvest or fishing) has the lowest cost. Fished seed is mostly used in low-effort grow-out such as bottom culture. However, dredging for seed can result in overexploitation. In New Zealand, this made the industry look for alternatives (Jeffs et al. 1999). Longline and raft culture use collected seed. The system to collect seed is usually the same as what is used for grow-out to make it cost efficient. The most expensive method to acquire seed is hatchery production (Kamermans et al. 2013). This is currently only used in longline culture.

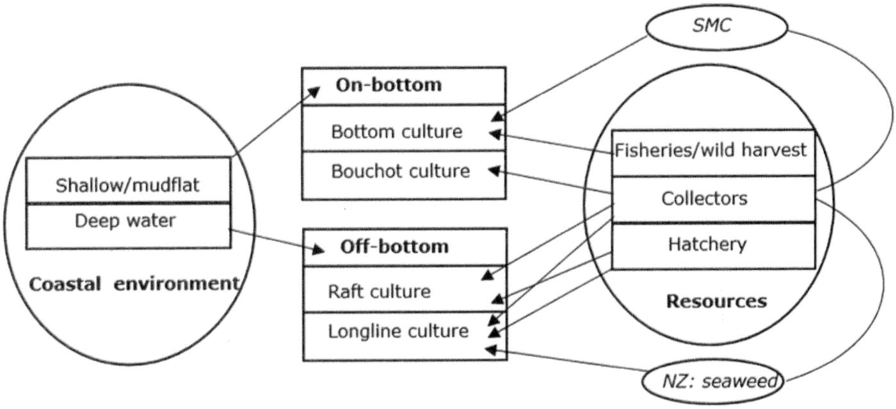

Fig. 3.2 Overview of culture techniques used for mussel production at different environments and for different resources (SMC = Seed Mussel Collectors, NZ = New Zealand)

3.2.1 *Bottom Culture*

Mussel bottom culture is typically practised on shallow mudflats in areas where there are extensive naturally occurring mussel seed beds (Fig. 3.3d). In the Netherlands, Germany, UK and Ireland, seed fished from natural beds is the main source for bottom culture (Kamermans and Smaal 2002).

Mussel seed from wild beds are relayed on bottom plots (lease sites) where the mussels are maintained until harvest. Bottom culture is an extensive culture where the mussels are still, to a large extent, subjected to, and dependent on the environment. The Netherlands are the centre of the bottom culture industry in Europe. In the 1970s most of the hand labour was mechanized leading to bulk production of mussels, limited by external factors such as seed availability and culture area. From the 2000s onwards, system innovation took place resulting in the deployment of seed mussel collectors (SMCs, Fig. 3.2). The first tests with seed mussel collectors started in 2000 (Kamermans et al. 2002) and the method showed a rapid development. In 2016 the total yield was about 20,000 tonnes (Capelle 2017). The main drivers for system innovation through SMCs were: (i) to safeguard a steady supply of seed, (ii) to become more sustainable by reducing bottom dredging, and (iii) pressure from green NGOs.

Mussel farmers in the Netherlands are in a transition process from fishing seed from natural beds to harvesting seed with collectors. A stepwise approach is taken: every 2 years a decision on reduction of seed fishing and expansion of the area reserved for seed collection is made based on the annual yield of the collectors. The shift from fishing to using collectors results in a higher mussel biomass in the system, because areas with natural beds are no longer fished and spat survival is enhanced on the collectors. However, competition for food (phytoplankton) between the extra mussel biomass and natural bivalve populations may result in overgrazing

Fig. 3.3 Mussel culture methods: (**a**) seed fishery, (**b**) seed mussels collectors, (**c**) hatchery production, (**d**) bottom culture, all in The Netherlands, (**e**) Bouchot culture in France (https://reporterre.net/Les-moules-du-Mont-Saint-Michel-etouffent-la-baie-magnifique), (**f**) raft culture in Spain and (**g**) longline culture in New Zealand. Source of pictures: Jacob Capelle (**a** and **d**), Aad Smaal (**e**) and Pauline Kamermans (**b**, **c**, **f**, and **g**)

and possibly affect the production capacity. This can have consequences for the yields of cultured bivalves and for organisms that depend on bivalve stocks for their food such as birds. A recent study used time-series data analysis and model calculations to estimate effects on production capacity (Kamermans et al. 2014). In addition, different indicators, such as meat content and growth rates of bivalves for assessment of changes in production capacity for bivalve shellfish were investigated. Kamermans et al. (2014) concluded that when all reserved space for SMCs is exploited at the envisioned end of the transition, expected effects on total bivalve biomass production will be less than or proportional to the increase in biomass of seed from SMC, depending on the area. In some areas, survival of wild, unfished beds is quite limited, due to predation.

The development of new technology that came with SMCs, increased the costs for the resource and will require innovations in other forms, notably an increase in production efficiency (Capelle 2017). Several research projects have been initiated to investigate this topic. Focus is on developing better seeding techniques (Capelle et al. 2014, 2016), predator control using starfish mops (Calderwood et al. 2016) or crab pots (Calderwood et al. 2015) and optimizing culture practices such as timing of relay, substrate use (Christensen et al. 2015) and harvest (Newell et al. 1998; Ferreira et al. 2007; Newell 2007).

3.2.2 *Bouchot Culture*

Bouchot culture (pole culture) is conducted exclusively in France, in areas with flat intertidal mudflats and a relatively large tidal range (Fig. 3.3e). In bouchot culture, mussel seed is collected on ropes, that are placed in horizontal racks in the water column when larvae are present. The ropes are then wound around poles in the intertidal zone for grow-out. Bouchot culture dates back to the thirteenth century and the principles and methods remain largely unchanged. Technical developments are very much restricted to mechanical methods to increase harvest efficiency. Amphibious vehicles are used to harvest the bouchots by means of a cylinder that can be lowered over the poles and scrapes off the mussels (Prou and Goulletquer 2002). Processing, size grading and restocking is also mechanized. Spatial conflicts on bivalve culture with other users is limiting the expansion of bouchot culture in France and has stimulated the development of longline cultures (Prou and Goulletquer 2002).

3.2.3 *Raft and Longline Culture*

In bays with deep waters and bays with rocky shores, rafts and longlines are more commonly used for the grow-out of mussels (Fig. 3.3f, g). Originally developed in the Mediterranean, large-scale raft culture is conducted primarily in Spain, and in more recent times, extensively at the northwest coast of Spain, where local upwelling results in a high food availability (Figueiras et al. 2002). In raft culture, mussels are grown on ropes hanging from rafts. In rope or longline culture mussels are grown on ropes attached to floating buoys at the water surface or submerged buoys. Longline culture is globally the most used culture method for mussels. Countries where high biomasses of mussels are produced on longlines are New Zealand, China, Italy and Chile. Culture practices can be summarized as (1) obtaining seed, (2) stocking and growing on rope, (3) restocking after thinning out and outgrow to consumption size. Major issues in off-bottom culture is resource requirement, density dependent growth and losses and biofouling. Self-thinning occurs when biomass increased through growth and food or space becomes limited (Fréchette and Lefaivre 1995; Guiñez 2005).

Seed for off-bottom culture is obtained mainly with seed collectors. However, when natural settlement is scarce other methods are used. For example, in New Zealand the spat for long-line culture is collected on Ninety Mile Beach in the far North of the North Island, where seaweed covered with recently settled natural spat washes upon a beach. Spat density varies from 200 to 2 million per kg of macroalgae. It is then transported to the culture areas in Coromandel on the North Island and Marlborough Sounds on the South Island (Jeffs et al. 1999).

Most research concerning spat collection deals with settlement cues, comparison of different types of collectors, and problems with biofouling. Understanding the impact of temperature on the rate of larval development is key to predicting the timing of settlement and optimizing mussel seed collection (Filgueira et al. 2015; Jacobs et al. 2014). However, other factors, such as food availability and quality, are important too (Bos et al. 2006; Philippart et al. 2012). Settlement is significantly higher on rough compared to smooth surfaces (Gribben et al. 2011). The most efficient type of SMC has a large surface area, and there is also thought to be a negative relationship between growth and density (e.g. Çelik et al. 2016). Identification and quantification of the presence of mussel larvae is important for optimising the use of suspended seed collectors. With this information timing of deployment can be optimised. Abalde et al. (2003) used mouse monoclonal antibodies to identify *M. galloprovincialis* larvae. The recent development of another identification method involving molecular tools can speed up processing of samples (Ranjith Kumar et al. 2015). After settlement, mussels can show gregarious behaviour on the collector ropes which is influenced by temperature or food availability (e.g. Aghzar et al. 2012). Failure of the collectors, other than insufficient availability of larvae, is mainly due to biofouling. For example, in Canada, the vase tunicate *Ciona intestinalis* reduces mussel production (Ramsay et al. 2008).

Recently, the focus of research on spat collectors extends towards interactions and conflicts with the natural environment. For example, carrying capacity (see box 1) and genetic diversity are a concern. Larraín et al. (2015) showed that blue mussels in southern Chile, raised from wild-caught seed obtained from relatively few collection sites, have lower genetic diversity than in other countries, and limited genetic differentiation among locations. Transplants of seed from other areas can result in mortality due to adaptation problems (Kautsky et al. 1990). Mussel seed has a high adaptive capacity (Widdows et al. 1984; Stirling and Okumuş 1994), but this varies among sources (Tremblay et al. 2011). Thus, adaptation capacity depends on the genetic composition of the stock and local environmental conditions.

Hatchery production of mussels (Fig. 3.3) is not as common as hatchery production of oysters and clams. One of the reasons why hatchery production of mussel seed is less developed for mussels than for other bivalves is that demand for the industry has been limited until now and that very large-scale production is required to make hatchery seed competitive with wild seed. However, commercial hatcheries that produce mussel spat are present (Kamermans et al. 2013). Optimisation of hatchery production is an ongoing process. For example, a recent study by Gui et al. (2016) showed that gill filaments in small *Perna canaliculus* are not fully developed and capture particles between 15–25 μm, while the filaments in bigger mussels are able to capture bacteria-sized particles around 2 μm. This type of information can be used to select the best algal diet for each life stage.

Generally, mussel hatcheries are only feasible when the price of the product allows it and when alternative sources of seed are scarce or unreliable. A pre-feasibility study for the installation of a Chilean mussel seed hatchery showed that seed production in a hatchery was not profitable due to both the low price of Chilean mussels in national and international markets and the high cost of production, mainly associated to the production of microalgae as feed for the larvae (Carrasco 2015). Seed from hatcheries is more expensive, but hatcheries provide the opportunity for selective breeding. Researchers in New Zealand have developed a selective breeding programme for the Greenshell™ mussel (*Perna canaliculus*) (Camara and Symonds, 2014). Innovative tools, such as cryopreservation that enables genetic material from selected stock to be stored, are being developed (Gale et al. 2014; Wang et al. 2014). Another advantage of hatchery production is the ability to produce triploids. Recently spawned mussels cannot be sold due to insufficient meat. Triploids are non-maturing mussels which have the advantage that they can be sold year-round. Two EU projects (BLUE SEED and REPROSEED) looked into hatchery production for mussels in Europe, including triploid production and the use of recirculation systems (Kamermans et al. 2013; Blanco and Kamermans 2015). Recently, a new project was started in Scalloway, Shetland, to test the commercial feasibility of producing mussel spat.

Kamermans et al. (2013) identified some areas where changes could be made to bring hatchery production costs more into line with the potential sale value of mussel seed: (i) use low-tech algal culture; (ii) restrict activities to the natural season and take seed into the field at the smallest size possible; (iii) scale up culture volumes during this restricted period of activity. In addition, production of higher added-value products, such as triploids or selective breeding for specific traits, is needed. Otherwise, the production of seed by hatchery techniques will be not be profitable in most cases compared with the cost of obtaining the wild counterparts.

Grow-out with hatchery seed is uncertain when it comes to the origin of the harvested strain. This can be the initially seeded hatchery material or wild recruits. Díaz-Puente et al. (2016) used multiplexed microsatellites to trace back the individual origin of a batch of harvested mussels and showed that 98.3% of the adult harvest came from the original hatchery full-sib family while only 1.7% of the mussels were recruited from the wild. A microsatellite genetic analysis of *M. edulis* on the west coast of Canada showed significant reduced genetic diversity in cultured populations compared to the wild population (Gurney-Smith et al. 2017). According to the authors, this is partially due to small effective breeding groups during hatchery propagation, creating genetic drift over successive generations. These results indicate the need for pedigree programs. The European network GENIMPACT evaluated genetic impact of aquaculture activities on native populations. Beaumont et al. (2006) concluded for mussels that it is essential to precisely characterize the true distributions of *M. edulis*, *M. galloprovincialis* and their hybrids in all European regions, but especially where mussel aquaculture takes place. Based on such a survey, a series of sites should be identified that are to be genetically monitored on a regular basis to identify any changes in species composition over time. As far as we are aware such monitoring has not started yet. Effects of climate change, such as ocean acidification, may have a serious impact on larval production. A recent study by Waldbusser et al. (2015) showed that larval shell development and growth in *Mytilus galloprovincialis* are dependent on aragonite saturation state, and not on carbon dioxide partial pressure or pH. With increasing acidification the aragonite saturation state decreases resulting in malformations and reduced growth of D-larvae. Hatcheries have the possibility for chemical manipulation of the seawater in larval tanks.

Innovation in grow-out techniques for longline and raft culture are mainly directed towards the investigation of optimal stocking densities and farm configuration. A few examples are: growing mussels without the need for thinning (Pérez-Camacho et al. 2013), using size grading (Cubillo et al. 2012), stocking as a function of food availability (Fréchette and Bacher 1998; Grant et al. 2008; Cranford et al. 2008; Strohmeier et al. 2005), and investigating the effect of spacing of mussel ropes (Drapeau et al. 2006; Aure et al. 2007). Effect of the culture structures on food provisioning to the mussels, can reduce mussel quality when scaling up (Rosland et al. 2011). Innovation in raft design is directed to deal with harsh environmental conditions, that results for example in submerged raft designs (Wang et al. 2015) and in optimizing food availability by raft design and orientation (Newell and Richardson 2014).

Biofouling on mussels grown on ropes or nets reduces mussel growth and quality (Sievers et al. 2013). In Canada up to 50% mortality was observed under heavy tunicate fouling (Locke and Carman 2009). Biofouling organisms that are causing major problems are ascidians, especially *Ciona intestinalis,* but may also consist of conspecific mussels or other species of mussels, for instance in New Zealand *M. galloprovincialis* is causing large fouling problems on the more valuable *P. canaliculus*. Forrest and Atalah (2017) used a 4-year dataset to calculate that *M. galloprovincialis* cover caused a 5 to 10% decrease in annual yield of *P. canaliculus*. Woods et al. (2012) reported an average of 54% biofouling organisms of the total rope biomass after 6 months. The reseeding of ropes reduced the amount of biofouling to 15% of the total rope biomass 6 months later. Innovations to reduce fouling are directed at reducing settlement. This can be done for instance by occupying 100% of the rope with mussels, or by manual removal of fouling or by using anti-foulants (Fitridge et al. 2012).

Space restrictions in the coastal zone and developments such as off-shore wind-farms, have speeded up developments towards off-shore mussel farms (Buck et al. 2004; Plew et al. 2005; Brenner et al. 2007; Ferreira et al. 2009; Van den Burg et al. 2017). However, off-shore conditions are much more challenging, also from a regulatory perspective (Corbin et al. 2017) and is an important driver of innovation in system design such as on the mooring of the systems (Ögmundarson et al. 2011), material use (Buck 2007) float design and food availability (Stevens et al. 2008).

3.3 Efficient Use

Culture efficiency is defined as how many units of end product (marked sized mussels) are harvested from one unit of resource (mussel seed). The index of culture efficiency is the average physical product APP (Ferreira et al. 2007), the Harvest to Seed Ratio (Newell 2007) or the relative biomass production (RBP) (Capelle et al. 2016). Efficient use is defined as by what means mussels growers can maximize their culture efficiency. Culture efficiency is biologically defined by the dynamics of growth and survival between resource and end product. There are several stages in the mussel culture cycle where management measures are or can be taken to improve growth and survival. These are: at seeding or stocking of seed, at relaying or thinning out and by predator control.

Survival of cultured mussels is dependent on the environment and on stress experienced in culture. In bottom mussel culture, large losses were found associated with seed handling (Calderwood et al. 2014; Capelle et al. 2016). Mussels are gregarious, but high mussel densities will increase competition and may result in substantial losses, that are witnessed in bouchot culture (Soletchnik et al. 2013), rope culture (Fréchette and Bacher 1998; Lauzon-Guay et al. 2005), but also in bottom culture (Capelle et al. 2014). In rope culture mussel losses can peak as a result of secondary settlement, when mussels that were initially attached (primary settlement), detach from the ropes in search for a different attachment substrate (South et al. 2017).

3.3.1 Stocking Density

Stocking mussels at optimal densities will enhance the culture efficiency. High mussel densities will increase competition and might result in substantial losses in bouchot culture (Soletchnik et al. 2013) and rope culture (Fréchette and Bacher 1998; Lauzon-Guay et al. 2005). Stocking in lower densities typically increases efficiency in rope culture (Cubillo et al. 2012), as well as in bottom culture (Capelle et al. 2016). Mussel size at stocking is an important parameter that effects culture efficiency: smaller mussels show higher losses (Lauzon-Guay et al. 2005), but have a higher biomass production potential (Petraitis 1995). However, stocking in low densities will expose more substrate for other species to settle on and enhances biofouling (South et al. 2017; Cubillo et al. 2015). Furthermore, when costs are considered higher biomass production at higher densities might compensate a reduction in quality and survival (Pérez-Camacho et al. 2013; Capelle et al. 2017). In several reports, mussel losses were attributed to seed handling. In bottom culture these losses are density dependent and can be reduced by applying a more homogeneous seeding pattern (Capelle et al. 2014) and by limiting the handling time (Calderwood et al. 2014). In rope culture, losses of 54% were observed within 1 month after stocking (South et al. 2017).

3.3.2 Relaying and Thinning Out

Selecting the best site, with high food availability, may substantially increase culture productivity in mussel bottom culture (Herman et al. 1999; Ferreira et al. 2007). Feeding rates may increase up to a flow velocity of 0.8 m s^{-1} (Widdows et al. 2002); at a certain threshold, mussels may be dislodged, and as such, mussel farmers need to optimize production within this range. In bottom mussel farming, relaying is common practice. Mussels are often kept on sheltered plots over winter and relayed to plots with good growing conditions in spring. Mussels might also be relayed from intertidal plots to deeper plots, to stimulate survival and growth (Beadman et al. 2003). Mussels that are transplanted between areas may require physiological adaptations. Especially in the size of the gills that are used to capture particles and in the size of the labial palps that are used to sort particles into edible and not edible (Bayne 2004). In areas with high turbidity, gills are small and labial palps are large (Theisen 1982). In mussels, an adaptation in the gill-to-palp ratio was observed after transplantation to sites with different turbidity values (Essink and Bos 1985; Payne et al. 1995). After a transplantation experiment between two systems in southern England, it took 2 months for the mussels to adapt the gill-to-palp ratio to the new environment (Widdows et al. 1984).

Ropes or nets have limited attachment area, hence mussels will start to fall off when mussel densities are too high. Self-thinning occurs when mussel biomass increases and space or food becomes limiting, causing a reduction in growth and survival (Alunno-Bruscia et al. 2000; Guiñez et al. 2005). Manual thinning out on ropes in raft culture in Galicia Spain occurs after 4–7 months of growing when the mussels reach 4–5 cm (Cubillo et al. 2012). In the thinning process mussels are detached from the ropes and re-socked in a lower density around a new rope. During the thinning process size grading can take place that will result in a more uniform mussel size at harvest and in less mussel discards (Pérez Camacho et al. 1991). The thinning process in Spain was associated with mussel losses (Pérez Camacho et al. 1991, 2013).

3.3.3 Predator Control

Mussels are not only providing goods for human consumption, but also for a range of other species, some of which depend on them as a food source. Several management measures to prevent predation in *bouchot culture* are described by Dardignac-Corbeil (1975): (1) Crabs (*Carcinus meanas, Maja brachydactyla*) which predate on the bouchot mussels, can be prevented by placing a sheet around the bouchots. (2) Predation by birds (e.g. gulls or molluscivorous ducks) on mussels on bouchots can be reduced by using nylon threads to prevent the birds landing. (3) When starfish and mollusc drilling snails (*Nucella lapillus*) are present in high densities and predation levels are high they need to be manually removed.

Predation may exert a top down limitation on production. Especially, in *bottom culture*, because mussel plots are accessible for benthic predators as well as for fish and birds. Intertidal mussels are preyed upon by shore crabs and birds (oystercatchers, herring gulls), while subtidal mussels are preyed upon by shore crabs, sea stars and molluscivorous (diving) ducks. The number of sea stars on culture plots is reduced by freshwater treatment and there is a selective fishery on sea stars with sea star mops (Netherlands, United Kingdom, Germany, and Ireland) and purse-seines (Denmark Petersen et al. 2016). Freshwater treatment is applied before seeding when mussels are in the vessels' hold; the process consists of the joint exposure of mussels and associated sea stars to freshwater for several hours Mussels will keep their shells shut, while sea stars are unable to protect themselves against osmotic stress and will not survive. Sea star mops are made of fuzzy rope entwined around small chains that are towed over the mussel plots, which ensnares the sea stars thereby enabling removal. The efficiency of sea star removal by mops was estimated in a case study in Belfast Lough in Northern Ireland. The results show a large variation in the catch efficiency (4–78%), while the mean sea star reduction applying this method was 27% (Calderwood et al. 2016).

When Davies et al. (1980) tested the effect of exclusion of shore crabs in newly formed intertidal mussel beds on a scale of 800 m^2; they found that exclusion of shore crabs resulted in a 400–500% increase in yield over a period of 2 years. Experiments have been conducted on selective crab fisheries in a comparative study on culture plots in the Wadden Sea, but no differences in survival between culture plots where crabs were removed vs. where no crab fishery took place could be found (Kamermans et al. 2010). Therefore, exclusion of shore crabs seems to be more effective than a selective fishery.

Rope or net culture of mussels have the advantage above bottom culture that benthic predators cannot reach the mussels directly. Predation by mobile predators on mussels in raft or longline culture are therefore limited to molluscivorous birds and fishes. However, predators with pelagic larvae can settle between the mussels. Sea stars commonly settle in long-line farms and marine flatworms (*Turbellaria* or *Plathyhelminthes*) can infest the mussels and cause substantial losses (Galleni et al. 1980; Robledo et al. 1994). Ducks such as eider ducks that primarily feed on mussels can cause extensive damage to longline mussel cultures (Dunthorn 1971;

Žydelis et al. 2009). In Maine (USA) mussels are protected by nets placed around the mussel rafts (Newell and Richardson 2014). Mussel ropes and nets are very attractive for a range of fish species (Šegvić-Bubić et al. 2011). In the Mediterranean, sea breams are considered a pest that is very difficult to handle and may require nets as physical barriers (Prou and Goulletquer 2002).

3.3.4 Other Loss Factors

Sometimes environmental events result in mussel losses and the only option mussel growers have are mitigation measures. Environmental factors such as harmful algal blooms (HABs) (Peperzak and Poelman 2008) or diseases and parasites (mainly limited to *Myticola intestinalis* in *Mytilidea* (Bower et al. 1994) and *Bucephalus sp.* in *Perna* (da Silva et al. 2002), on bouchot mussels heat stress might increase losses up to 70% (Soletchnik et al. 2013). Ice scour is a catastrophic event for intertidal mussel populations (Donker et al. 2015) However, not all mussel losses can be explained. In recent years, abnormal high mussel losses were observed at mussel production sites in the Atlantic coast in France (2014–2016) and at the Oosterschelde estuary in the Netherlands (2016). Mussel meat at sites with abnormal mortality rates contained higher densities of granulomas, inflammatory inclusions at the Atlantic coast in France, suggesting that the mussels experienced stress (Robert and Soletchnik 2016). In a follow-up study, climatic events tied to climate change that affected abiotic conditions, but also algal compositions and timing of blooms were linked to higher mortality events, although a conclusion is still lacking (Travers et al. 2016; Soletchnik et al. 2017). Elevation of atmosphere and sea surface temperatures resulted in shifts of the geographical distribution of mussels to colder areas (Berge et al. 2005) and catastrophic summer mortalities at intertidal sites due to heating stress (Jones et al. 2010).

3.3.5 Differences in Efficiency Between Species and Culture Methods

Reported culture efficiencies are shown in Table 3.2, expressed as Relative Biomass Production (RBP): the biomass of harvestable product from one biomass unit of seed. It appears from this table that bottom culture is the least efficient, which can be explained by the high density dependent losses, predation pressure and dislodgement vulnerability for the mussels in this type of culture. Major improvements are expected in reducing handling stress and density dependent losses (Capelle et al. 2017). Production efficiencies of mussels from the *Perna* species are around 5 kg of harvestable product from 1 kg of seed, despite having the largest growth rates. It seems that survival rates for *Perna* mussels are lower than for other rope or raft

Table 3.2 Differences in mussel culture efficiency including growth and mortality, expressed as Biomass Production Rate (RBP) between countries, systems and species

Country	System	Species	Typical Growth rate	Typical Mortality rate	Typical annual mussel production	Yield (RBP)	Source
Spain, Galicia	Raft	*M galloprovinciales*	87.6 mg/day	33–36% seed	150 kg m^{-2}	27.6	Figueras (1990), Pérez Camacho et al. (1991) and Figueiras et al. (2002)
			0.13 mm/day	Thinning: 14–15%			
				Grow out: 19–20%			
Italy, PO Delta	Rope	*M. galloprovinciales*	0.12 mm/day	59–82%	–	–	Ceccherelli and Barboni (1983)
UK, Menai Strait	Bottom, intertidal	*M. edulis*	0.04 mm/day	82–90%	12.8 kg m^{-2}	0.89–1.45	Dare and Edwards (1976)
India, Goa	Raft	*P. viridis*	0.27 mm/day	–	48 kg m^{-2}	–	Qasim et al. (1977)
India, Vizhinjam	Raft	*P. viridis*	0.10 mm/day	–	15 kg m^{-2}	5–11	Appukuttan et al. (1980)
Ireland, Killany	Rope	*M. edulis*	–	4–48%	5 kg m^{-1} ~ 250 kg m^{-2}		Rodhouse et al. (1985)
Loughs, Northern Ireland, UK	Bottom	*M. edulis*	20 mg/day, 0.06 mm/day	70% (prod. cycle, 26 months)	0.15–2.9 kg m^{-2}	1–4	Ferreira et al. (2007)
France	Bouchot – naturally settled	*M. edulis*	6 mg/day		6.07 kg m^{-1}	–	Boromthanarat and Deslous-Paoli (1988)
Marennes-Oleron	Bouchot Transplants		1.1 mg/day		10.78 kg m^{-1}	2.9	
Agadir, Maroc	Rope	*M. galloprovincialis*	–	87–88 % (prod cycle, 12 months)	9.1 kg, m^{-1}	2.8–5.7	Idhalla et al. (2017)

(continued)

Table 3.2 (continued)

Country	System	Species	Typical Growth rate	Typical Mortality rate	Typical annual mussel production	Yield (RBP)	Source
NZ	Rope	*P. perna*	–	86% (prod cycle, 12 months)	16.4 kg, m^{-1}	3.9–6.5	
NZ	Rope	*Perna canaliculus*	0.20 mm/day	54% (1 month), 81% (5 moths)	2.03–3.91 kg, m^{-1}	4.5–6.9	South et al. (2017) and Jeffs et al. (1999)
NL	Bottom	*M. edulis*	0.04 mm/day	–	–	–	Unpublished data
Oosterschelde			20 mg/day				
Wadden Sea	Bottom	*M. edulis*	0.07 mm/day	42% (1 month)	4.9 kg m^{-2}	1.0–1.9	Capelle et al. (2016)
			30 mg/day	77% (cycle)			
France	Bouchot	*M. edulis*	0.023 mm/day	–	–	–	Garen et al. (2004)
Pertuis Breton	Longline		0.034 mm/day				
	Bottom		0.016 mm/day				
	Intertidal						

grown mussel families, and are in fact comparable with mussel bottom culture. Note that RBPs of *Perna* mussels are higher than for mussel bottom culture, caused by faster growth rates of *Perna* mussels. It is reported that detachment from ropes is a major problem during the grow out of *Perna* mussels (South et al. 2017; Petes et al. 2007). Bouchot culture is slightly more efficient than bottom culture but less efficient than rope culture. This can be explained by the low growth rates which are experienced in this type of intertidal culture, and the fact that bouchot mussels are more vulnerable to benthic predators than rope cultured mussels. Raft culture of *M. galloprovincialis* in Spain is a very effective culture. High yields are reached because the culture starts with small seeds which increase in weight tenfold when they are thinned out and the mussel seed is re-socked in a lower density over three new ropes (Pérez Camacho et al. 1991).

3.4 Conclusions

The starting material for mussel culture is wild harvest of seed, use of SMC or hatchery production. Fished seed is mostly used in bottom culture, while longline and raft culture predominantly use seed collectors. Hatchery seed is only used in longline culture. Most research concerning spat collection deals with comparisons of different types of seed collectors, settlement cues and problems with biofouling. Optimising the timing of deployment of the collectors and the timing of harvest can increase the yield of seed collectors. Hatchery seed is more expensive, but hatcheries provide the opportunity for selective breeding and triploid production giving the product an added value. The challenge is to bring hatchery production costs more in line with the potential sale value of mussel seed. Monitoring can give insight in whether genetic diversity of collector seed or hatchery seed is negatively affected.

Efficiency in use of mussel seed shows large differences between species, regions and culture techniques. Survival rates seem higher for mussels from the *Mytilus* genus, than for mussels from the *Perna* genus. Several key processes were identified that can explain these differences. Losses differ because of different predation pressures or because of differences between substrate and the relationship between food, space and density. Other sources of losses can be related to anomalous, environmental events, such as storms or heat stress. Losses due to such events might become more common in the near future, for example, with the effects of climate change. Growth rates differ between species and between production systems. In general, mussels form the *Perna* genus display higher growth rates than mussels from the *Mytilus* genus. Rope and raft culture is more efficient in terms of yield than bouchot, while bouchot seems a little more efficient than bottom culture.

For bottom culture, seed from SMCs has gradually become an important seed source complementary to seed from wild harvest. However, seed is more expensive from SMCs than from wild harvest and several research programs were carried out towards methods to increase efficient use. Technical developments in off-bottom culture mainly concern optimizing system designs and are particularly innovative in the way in which they relate system design to optimal feeding rates and dealing with harsh hydrodynamic conditions. Spatial conflicts in traditional culture areas may provoke the development of off-shore culture implying risk of exposure to hydrodynamic stress.

Acknowledgements We would like to thank Nigel Keeley and Tore Strohmeier for their constructive review of the manuscript.

References

Abalde SL, Fuentes J, González-Fernández Á (2003) Identification of Mytilus galloprovincialis larvae from the Galician rıas by mouse monoclonal antibodies. Aquaculture 219:545–559

Aghzar A, Talbaoui M, Benajiba M, Presa P (2012) Influence of depth and diameter of rope collectors on settlement density of Mytilus galloprovincialis spat in Baie de M'diq (Alboran Sea). Mar Freshw Behav Physiol 45:51–61

Alunno-Bruscia M, Petraitis PS, Bourget E, Fréchette M (2000) Body size-density relationship for Mytilus edulis in an experimental food-regulated situation. Oikos 90:28–42

Appukuttan K, Nair TP, Joseph M, Thomas K (1980) Culture of brown mussel at Vizhinjam. CMFRI Bull 29:30–32

Aure J, Strohmeier T, Strand Ø (2007) Modelling current speed and carrying capacity in long-line blue mussel (Mytilus edulis) farms. Aquac Res 38:304–312

Aypa SM (1990) Mussel culture. Fisheries and Aquaculture Department, FAO, Rome

Bayne BL (2004) Phenotypic flexibility and physiological tradeoffs in the feeding and growth of marine bivalve molluscs. Integr Comp Biol 44:425–432

Beadman HA, Caldow RWG, Kaiser MJ, Willows RI (2003) How to toughen up your mussels: using mussel shell morphological plasticity to reduce predation losses. Mar Biol 142:487–494

Beaumont A, Gjedrem T, Moran P (2006) Genetic effects of domestication, culture and breeding of fish and shellfish, and their impacts on wild populations. Blue mussel – Mytilus edulis and Mediterranean mussel – M. galloprovincialis. In: Svåsand TCD, García-Vázquez E, Verspoor E (eds) Evaluation of genetic impact of aquaculture activities on native populations: a European network, pp 83–90. GENIMPACT final report (EU contract n. RICA-CT-2005-022802). http://genimpact.imr.no/

Berge J, Johnsen G, Nilsen F, Gulliksen B, Slagstad D (2005) Ocean temperature oscillations enable reappearance of blue mussels Mytilus edulis in Svalbard after a 1000 year absence. Mar Ecol Prog Ser 303:167–175

Blanco Garcia A, Kamermans P (2015) Optimization of blue mussel (Mytilus edulis) seed culture using recirculation aquaculture systems. Aquac Res 46:977–986

Boromthanarat S, Deslous-Paoli JM (1988) Production of Mytilus edulis L. reared on bouchots in the Bay of Marennes-Orleon: comparison between two methods of culture. Aquaculture 72:255–263

Bos OG, Hendriks IE, Strasser M, Dolmer P, Kamermans P (2006) Estimation of food limitation of bivalve larvae in coastal waters of North-Western Europe. J Sea Res 55:191–206

Bower SM, McGladdery SE, Price IM (1994) Synopsis of infectious diseases and parasites of commercially exploited shellfish. Annu Rev Fish Dis 4:1–199

Brenner M, Buck BH, Köhler A (2007) New concept combines offshore wind farms, mussel cultivation. Glob Aquacult Advocate 10(1):79–81

Buck BH (2007) Experimental trials on the feasibility of offshore seed production of the mussel Mytilus edulis in the German Bight: installation, technical requirements and environmental conditions. Helgol Mar Res 61:87–101

Buck BH, Krause G, Rosenthal H (2004) Multifunctional use, environmental regulations and the prospect of offshore co-management: potential for and constraints to extensive open ocean aquaculture development within wind farms in Germany. Ocean Coast Manag 47:95–122
Calderwood J, O'Connor NE, Sigwart J, Roberts D (2014) Determining optimal duration of seed translocation periods for benthic mussel (Mytilus edulis) cultivation using physiological and behavioural measures of stress. Aquaculture 434:288–295
Calderwood J, O'Connor NE, Roberts D (2016) Efficiency of starfish mopping in reducing predation on cultivated benthic mussels (Mytilus edulis Linnaeus). Aquaculture 452:88–96
Camara M, Symonds J (2014) Genetic improvement of New Zealand aquaculture species: programmes, progress and prospects. N Z J Mar Freshw Res 48:466–491
Capelle JJ (2017) Production efficiency of mussel bottom culture. Wageningen University, Wageningen, p 240
Capelle JJ, Wijsman JWM, Schellekens T, van Stralen MR, Herman PMJ, Smaal AC (2014) Spatial organisation and biomass development after relaying of mussel seed. J Sea Res 85:395–403
Capelle JJ, Wijsman JWM, van Stralen MR, Herman PMJ, Smaal AC (2016) Effect of seeding density on biomass production in mussel bottom culture. J Sea Res 85:395–403
Capelle JJ, van Stralen MR, Wijsman JWM, Herman PMJ, Smaal AC (2017) Population dynamics of subtital mussels (Mytilus edulis) and the impact of cultivation. Aquac Environ Interact 9:155–168
Carrasco AV (2015) Pre-feasibility study for the installation of a Chilean Mussel Mytilus chilensis (Hupé, 1854) seed hatchery in the lakes region, Chiles. Fish Aquac J 3:102
Ceccherelli VU, Barboni A (1983) Growth, survival and yield of Mytilus galloprovincialis Lamk. on fixed suspended culture in a bay of the Po River Delta. Aquaculture 34:101–114
Çelik MY, Karayücel S, Karayücel İ, Eyüboğlu B, Öztürk R (2016) Settlement and growth of the mussels (Mytilus galloprovincialis, Lamarck, 1819) on different collectors suspended from an offshore submerged longline system in the Black Sea. Aquac Res 47:3765–3776
Christensen HT (2012) Area-intensive bottom culture production of blue mussels, Mytilus edulis (L.), PhD dissertation, DTU aqua (Danmarks Tekniske Universitet), Denmark
Christensen HT, Dolmer P, Hansen BW, Holmer M, Kristensen LD, Poulsen LK, Stenberg C, Albertsen CM, Støttrup JG (2015) Aggregation and attachment responses of blue mussels, *Mytilus edulis* – impact of substrate composition, time scale and source of mussel seed. Aquaculture 435:245–251
Corbin JS, Holmyard J, Lindell S (2017) Regulation and permitting of standalone and co-located open ocean aquaculture facilities. In: Aquaculture perspective of multi-use sites in the open ocean. Springer, Dordrecht, pp 187–229
Cranford PJ, Li W, Strand Ø, Strohmeier T (2008) Phytoplankton depletion by mussel aquaculture: high resolution mapping, ecosystem modeling and potential indicators of ecological carrying capacity. Ecological carrying capacity in shellfish aquaculture. ICES CM, pp 1–5
Cubillo AM, Peteiro LG, Fernández-Reiriz MJ, Labarta U (2012) Influence of stocking density on growth of mussels (Mytilus galloprovincialis) in suspended culture. Aquaculture 342-343:103–111
Cubillo AM, Fuentes-Santos I, Labarta U (2015) Interaction between stocking density and settlement on population dynamics in suspended mussel culture. J Sea Res 95:84–94
da Silva PM, ARM M, Barracco MA (2002) Effects of Bucephalus sp. (Trematoda: Bucephalidae) on Perna perna mussels from a culture station in Ratones Grande Island, Brazil. J Invertebr Pathol 79:154–162
Dardignac-Corbeil M-J (1975) La culture des moules sur bouchots. Science et Pêche 244:1–10
Dare PJ, Edwards DB (1976) Experiments on the survival, growth and yield of relaid seed mussels (Mytilus edulis L.) in the Menai Straits, North Wales. J Cons Int Explor Mer 37:16–28
Davies GP, Dare PJ, Edwards DB (1980) Fenced enclosures for the protection of seed mussels (Mytilus edulis L.) from predation by shore crabs (Carcinus maenas L.). Fisheries research technical report, 56
Díaz-Puente B, Miñambres M, Rosón G, Aghzar A, Presa P (2016) Genetic decoupling of spat origin from hatchery to harvest of Mytilus galloprovincialis cultured in suspension. Aquaculture 460:124–135

Donker JJA, van der Vegt M, Hoekstra P (2015) Erosion of an intertidal mussel bed by ice- and wave-action. Cont Shelf Res 106:60–69
Drapeau A, Comeau LA, Landry T, Stryhn H, Davidson J (2006) Association between longline design and mussel productivity in Prince Edward Island, Canada. Aquaculture 261:879–889
Dunthorn A (1971) The predation of cultivated mussels by eiders. Bird Study 18:107–112
Erlandson JM (1988) The role of shellfish in prehistoric economies: a protein perspective. Am Antiq 53:102–109
Essink K, Bos AH (1985) Growth of three bivalve molluscs transplanted along the axis of the Ems estuary. Neth J Sea Res 19:45–51
Ferreira JG, Hawkins AJS, Monteiro P, Service M, Moore H, Edwards A, Gowen R, Lourenco P, Mellor A, Nunes JP, Pascoe PL, Ramos L, Sequeira A, Simas T, Strong J (2007) SMILE – sustainable mariculture in northern Irish lough ecosystems – Assesment of carrying capacity for environmental sustainable shelfish culture in Carlingford Lough, Strangford Lough, Belfast Lough, Larne Lough and Lough Foyle. IMAR – Institute of Marine Research, p 100
Ferreira J, Sequeira A, Hawkins A, Newton A, Nickell T, Pastres R, Forte J, Bodoy A, Bricker S (2009) Analysis of coastal and offshore aquaculture: application of the FARM model to multiple systems and shellfish species. Aquaculture 289:32–41
Figueiras F, Labarta U, Reiriz MF (2002) Coastal upwelling, primary production and mussel growth in the Rías Baixas of Galicia. In: Sustainable increase of marine harvesting: fundamental mechanisms and new concepts. Springer, Dordrecht, pp 121–131
Figueras A (1990) Mussel culture in Spain. Mar Behav Physiol 16:177–207
Filgueira R, Brown MS, Comeau LA, Grant J (2015) Predicting the timing of the pediveliger stage of Mytilus edulis based on ocean temperature. J Molluscan Stud 81:269–273
Fitridge I, Dempster T, Guenther J, de Nys R (2012) The impact and control of biofouling in marine aquaculture: a review. Biofouling 28:649–669
Forrest BM, Atalah J (2017) Significant impact from blue mussel Mytilus galloprovincialis biofouling on aquaculture production of green-lipped mussels in New Zealand. Aquac Environ Interact 9:115–126
Fréchette M, Bacher C (1998) A modelling study of optimal stocking density of mussel populations kept in experimental tanks. J Exp Mar Biol Ecol 219:241–255
Fréchette M, Lefaivre D (1995) On self-thinning in animals. Oikos 73:425–428
Gale SL, Burritt DJ, Tervit HR, Adams SL, McGowan LT (2014) An investigation of oxidative stress and antioxidant biomarkers during Greenshell mussel (Perna canaliculus) oocyte cryopreservation. Theriogenology 82:779–789
Galleni L, Tongiorgi P, Ferrero E, Salghetti U (1980) Stylochus mediterraneus (Turbellaria: Polycladida), predator on the mussel Mytilus galloprovincialis. Mar Biol 55:317–326
Garen P, Robert S, Bougrier S (2004) Comparison of growth of mussel, Mytilus edulis, on longline, pole and bottom culture sites in the Pertuis Breton, France. Aquaculture 232:511–524
Grant J, Bacher C, Cranford PJ, Guyondet T, Carreau M (2008) A spatially explicit ecosystem model of seston depletion in dense mussel culture. J Mar Syst 73:155–168
Gribben PE, Jeffs AG, de Nys R, Steinberg PD (2011) Relative importance of natural cues and substrate morphology for settlement of the New Zealand Greenshell™ mussel, Perna canaliculus. Aquaculture 319:240–246
Gui Y, Zamora L, Dunphy B, Jeffs A (2016) Understanding the ontogenetic changes in particle processing of the greenshell™ mussel, Perna canaliculus, in order to improve hatchery feeding practices. Aquaculture 452:120–127
Guiñez R (2005) A review on self-thinning in mussels. Revista de Biologia Marina y Oceanografia 40:1–6
Guiñez R, Petraitis PS, Castilla JC (2005) Layering, the effective density of mussels and mass-density boundary curves. Oikos 110:186–190
Gurney-Smith HJ, Wade AJ, Abbott CL (2017) Species composition and genetic diversity of farmed mussels in British Columbia, Canada. Aquaculture 466:33–40
Herman P, Middelburg J, Van de Koppel J, Heip C (1999) Ecology of estuarine macrobenthos. Adv Ecol Res 29:195–240

Idhalla M, Nhhala H, Kassila J, Ait Chattou EM, Orbi A, Moukrim A (2017) Comparative production of two mussel species (Perna perna and Mytilus galloprovincialis) reared on an offshore submerged longline system in Agadir, Morocco. Int J Sci Eng Res 8:1
Jacobs P, Beauchemin C, Riegman R (2014) Growth of juvenile blue mussels (Mytilus edulis) on suspended collectors in the Dutch Wadden Sea. J Sea Res 85:365–371
Jeffs AG, Holland RC, Hooker SH, Hayden BJ (1999) Overview and bibliography of research on the greenshell mussel, Perna canaliculus, from New Zealand waters. J Shellfish Res 18:347–360
Jones SJ, Lima FP, Wethey DS (2010) Rising environmental temperatures and biogeography: poleward range contraction of the blue mussel, Mytilus edulis L., in the western Atlantic. J Biogeogr 37:2243–2259
Kamermans P, Smaal AC (2002) Mussel culture and cockle fisheries in The Netherlands: finfing a balance between economy and ecology. J Shellfish Res 21:509–517
Kamermans P, Brummelhuis E, Smaal A (2002) Use of spat collectors to enchange supply of seed for bottom culture of blue mussels (Mytilus edulis) in the Netherlands. World Aquacult 33:12–15
Kamermans P, de Jong ML, van Hoppe M (2010) PRODUS 1 d: rendement MZI zaad op percelen: effect van wegvissen van krabben – perceelproef 2009. IMARES, Yerseke, Report C075/10
Kamermans P, Galley T, Boudry P, Fuentes J, McCombie H, Batista F, Blanco A, Dominguez L, Cornette F, Pincot L (2013) Blue mussel hatchery technology in Europe. In: Advances in aquaculture hatchery technology. Elsevier, New York, pp 339–373
Kamermans P, Smit CJ, Wijsman JWM, Smaal AC (2014) Meerjarige effect-en productiemetingen aan MZI's in de Westelijke Waddenzee, Oosterschelde en Voordelta: samenvattend eindrapport. IMARES. Report C191/13
Kautsky N, Johannesson K, Tedengren M (1990) Genotypic and phenotypic differences between Baltic and North Sea populations of Mytilus edulis evaluated through reciprocal transplantations. I. Growth and morphology. Mar Ecol Prog Ser 59:203–210
Larraín MA, Díaz NF, Lamas C, Uribe C, Jilberto F, Araneda C (2015) Heterologous microsatellite-based genetic diversity in blue mussel (Mytilus chilensis) and differentiation among localities in southern Chile. Lat Am J Aquat Res 43:998
Lauzon-Guay J-S, Dionne M, Barbeau MA, Hamilton DJ (2005) Effects of seed size and density on growth, tissue-to-shell ratio and survival of cultivated mussels (Mytilus edulis) in Prince Edward Island, Canada. Aquaculture 250:652–665
Locke A, Carman M (2009) Ecological interactions between the vase tunicate (Ciona intestinalis) and the farmed blue mussel (Mytilus edulis) in Nova Scotia, Canada. Aquat Invasions 4:177–187
Newell CR (2007) Case study 1 – factors which influence mussel production on bottom leases. In: SMILE – Sustainable Mariculture in northern Irish Lough Ecosystems – Assesment of carrying capacity for environmental sustainable Shelfish culture in Carlingford lough, Strangford lough, Belfast lough, Larne lough and lough Foyle. IMAR – Institute of Marine Research, p 100
Newell CR, Richardson J (2014) The effects of ambient and aquaculture structure hydrodynamics on the food supply and demand of mussel rafts. J Shellfish Res 33:257–272
Newell CR, Campbell DE, Gallagher SM (1998) Development of the mussel aquaculture lease site model MUSMOD©: a field program to calibrate model formulations. J Exp Mar Biol Ecol 219:143–169
Ögmundarson Ó, Holmyard J, Þórðarson G, Sigurðsson F, Gunnlaugsdóttir H (2011) Offshore aquaculture farming – report from the initial feasibility study and market requirements for the innovations from the project. Icelandic Food and Biotech, Reykjavík
Payne BS, Miller AC, Jin L (1995) Palp to gill area ratio of bivalves: a sensitive indicator of elevated suspended solids. Regul Rivers Res Manag 11:193–200
Peperzak L, Poelman M (2008) Mass mussel mortality in The Netherlands after a bloom of Phaeocystis globosa (prymnesiophyceae). J Sea Res 60:220–222
Pérez Camacho A, González R, Fuentes J (1991) Mussel culture in Galicia (N.W. Spain). Aquaculture 94:263–278

Pérez-Camacho A, Labarta U, Vinseiro V, Fernández-Reiriz MJ (2013) Mussel production management: raft culture without thinning-out. Aquaculture 406:172–179
Petersen AJK, Gislason H, Fitridge I, Saurel C, Degel H, Nielsen CF (2016) Fiskeri efter søstjerner i Limfjorden. Fagligt grundlag for en forvaltningsplan. Institut for Akvatiske Ressourcer, Danmarks Tekniske Universitet
Petes LE, Menge BA, Murphy GD (2007) Environmental stress decreases survival, growth, and reproduction in New Zealand mussels. J Exp Mar Biol Ecol 351:83–91
Petraitis PS (1995) The role of growth in maintaining spatial dominance by mussels (Mytilus edulis). Ecology 76:1337–1346
Philippart CJ, Amaral A, Asmus R, van Bleijswijk J, Bremner J, Buchholz F, Cabanellas-Reboredo M, Catarino D, Cattrijsse A, Charles F (2012) Spatial synchronies in the seasonal occurrence of larvae of oysters (Crassostrea gigas) and mussels (Mytilus edulis/galloprovincialis) in European coastal waters. Estuar Coast Shelf Sci 108:52–63
Plew DR, Stevens CL, Spigel RH, Hartstein ND (2005) Hydrodynamic implications of large offshore mussel farms. IEEE J Ocean Eng 30:95–108
Prou J, Goulletquer P (2002) The French mussel industry: present status and perspectives. Bull Aquac Assoc Can 102:17–23
Qasim S, Parulekar A, Harkantra S, Ansari Z, Nair A (1977) Aquaculture of green mussel Mytilus viridis L.: cultivation on ropes from floating rafts. Indian J Mar Sci 6:15–25
Ramsay A, Davidson J, Landry T, Stryhn H (2008) The effect of mussel seed density on tunicate settlement and growth for the cultured mussel, Mytilus edulis. Aquaculture 275:194–200
Ranjith Kumar R, Vijayan K, Thomas P, Mohamed K, Gopalakrishnan A (2015) Identification of brown mussel (Perna indica) larvae using molecular tool. Indian J Fish 62:128–131
Reguera B, Riobó P, Rodríguez F, Díaz PA, Pizarro G, Paz B, Franco JM, Blanco J (2014) Dinophysis toxins: causative organisms, distribution and fate in shellfish. Mar Drugs 12:394–461
Robert S, Soletchnik P (2016) Réseau national d'observation de la moule bleue, MYTILOBS/Campagne 2015
Robledo J, Caceres-Martinez J, Sluys R, Figueras A (1994) The parasitic turbellarian Urastoma cyprinae (Platyhelminthes: Urastomidae) from blue mussel Mytilus galloprovincialis in Spain: occurrence and pathology. Dis Aquat Org 18:203–210
Rodhouse PG, Roden CM, Hensey MP, Ryan TH (1985) Production of mussels, Mytilus edulis, in suspended culture and estimates of carbon and nitrogen flow: Killary Harbour, Ireland. J Mar Biol Assoc U K 65:55–68
Rosland R, Bacher C, Strand Ø, Aure J, Strohmeier T (2011) Modelling growth variability in longline mussel farms as a function of stocking density and farm design. J Sea Res 66:318–330
Šegvić-Bubić T, Grubišić L, Karaman N, Tičina V, Jelavić KM, Katavić I (2011) Damages on mussel farms potentially caused by fish predation – self service on the ropes? Aquaculture 319:497–504
Sievers M, Fitridge I, Dempster T, Keough MJ (2013) Biofouling leads to reduced shell growth and flesh weight in the cultured mussel Mytilus galloprovincialis. Biofouling 29:97–107
Smaal AC (2002) European mussel cultivation along the Atlantic coast: production status, problems and perspectives. Hydrobiologia 484:89–98
Soletchnik P, Robert S, Le Moine O (2013) Suivi expérimental de la croissance de la moule, Mytilus edulis, sur les bouchots des Pertuis Charentais entre 2000 et 2010. Etude des performances de croissance en liens avec l'environnement des élevages. http://archimer.ifremer.fr/doc/00120/23097/
Soletchnik P, Le Moine O, Polsenaere P (2017) Evolution de l'environnement hydroclimatique du bassin de Marennes-Oléron dans le contexte du changement global. http://archimer.ifremer.fr/doc/00387/49815/
South PM, Floerl O, Jeffs AG (2017) Differential effects of adult mussels on the retention and fine-scale distribution of juvenile seed mussels and biofouling organisms in long-line aquaculture. Aquac Environ Interact 9:239–256
Stevens C, Plew D, Hartstein N, Fredriksson D (2008) The physics of open-water shellfish aquaculture. Aquac Eng 38:145–160

Stirling H, Okumuş İ (1994) Growth, mortality and shell morphology of cultivated mussel (Mytilus edulis) stocks cross-planted between two Scottish Sea lochs. Mar Biol 119:115–123

Strohmeier T, Aure J, Duinker A, Castberg T, Svardal A, Strand Ø (2005) Flow reduction, seston depletion, meat content and distribution of diarrhetic shellfish toxins in a long-line blue mussel (Mytilus edulis) farm. J Shellfish Res 24:15–23

Theisen BF (1982) Variation in size of gills, labial palps, and adductor muscle in Mytilus edulis L. (Bivalvia) from Danish waters. Ophelia 21:49–63

Travers M-A, Pepin J-F, Soletchnik P, Guesdon S, Le Moine O (2016) Mortalités de moules bleues dans les Pertuis Charentais: description et facteurs liés–MORBLEU. http://archimer.ifremer.fr/doc/00324/43539/

Tremblay R, Landry T, Leblanc N, Pernet F, Barkhouse C, Sévigny J-M (2011) Physiological and biochemical indicators of mussel seed quality in relation to temperatures. Aquat Living Resour 24:273–282

van den Burg S, Kamermans P, Blanch M, Pletsas D, Poelman M, Soma K, Dalton G (2017) Business case for mussel aquaculture in offshore wind farms in the North Sea. Mar Policy 85:1–7

Waldbusser GG, Hales B, Langdon CJ, Haley BA, Schrader P, Brunner EL, Gray MW, Miller CA, Gimenez I (2015) Saturation-state sensitivity of marine bivalve larvae to ocean acidification. Nat Clim Chang 5:273–280

Wang H, Li X, Wang M, Clarke S, Gluis M (2014) The development of oocyte cryopreservation techniques in blue mussels Mytilus galloprovincialis. Fish Sci 80:1257–1267

Wang X-x, Swift MR, Dewhurst T, Tsukrov I, Celikkol B, Newell C (2015) Dynamics of submersible mussel rafts in waves and current. China Ocean Eng 29:431–444

Widdows J, Donkin P, Salkeld PN, Cleary JJ, Lowe DM, Evans SV, Thompson PE (1984) Relative importance of environmental factors in determining physiological differences between two populations of mussels (Mytilus edulis). Mar Ecol Prog Ser 17:33–47

Widdows J, Lucas JS, Brinsley MD, Salkeld PN, Staff FJ (2002) Investigation of the effects of current velocity on mussel feeding and mussel bed stability using an annular flume. Helgol Mar Res 56:3–12

Woods CM, Floerl O, Hayden BJ (2012) Biofouling on Greenshell™ mussel (Perna canaliculus) farms: a preliminary assessment and potential implications for sustainable aquaculture practices. Aquac Int 20:537–557

Žydelis R, Esler D, Kirk M, Sean Boyd W (2009) Effects of off-bottom shellfish aquaculture on winter habitat use by molluscivorous sea ducks. Aquat Conserv Mar Freshwat Ecosyst 19:34–42

Chapter 4
Bivalve Production in China

Yuze Mao, Fan Lin, Jianguang Fang, Jinghui Fang, Jiaqi Li, and Meirong Du

Abstract Bivalve is the main species of mariculture in China. In 2015, bivalve production was about 12.4 million tonnes, accounting for more than 66% of China's total mariculture production. The first record of shellfish culture in China, about oyster culture, can be tracked back to 2000 years ago. The large-scale aquaculture started in the 1950s with the breakthrough in seed breeding techniques for *Tegillarca granosa* and *Ruditapes philippinarum*. Subsequently, with the promotion of seed breeding and artificial seed collection for mussels, scallops and oysters, the bivalve aquaculture industry has rapidly developed. In the twenty-first century, the scale of bivalve farming is constantly expanding, with increasing culture species and yield.

The length of the coastline of China is about 18,000 km comprising 11 coastal provinces (Liaoning, Hebei, Tianjin, Shandong, Jiangsu, Shanghai, Zhejiang, Fujian, Guangdong, Guangxi and Hainan provinces), all suitable for bivalve culture. Due to the significant difference in climate, the distribution of bivalve species is obviously regional. The major culture methods in China are longline culture (major species oysters, scallops, mussels, etc.) and bottom culture (clams). In this paper, we will describe the process of the longline cultured bivalve (Pacific oyster *Crassostrea gigas* and thick shell mussel *Mytilus coruscus*), and the bottom cultured ones (Manila clam *Ruditapes philippinarum* and cockle clam *Tegillarca granosa*).

Y. Mao
Yellow Sea Fisheries Research Institute, Chinese Academy of Fishery Sciences, Qingdao, China

Laboratory for Marine Ecology and Environmental Science, Qingdao National Laboratory for Marine Science and Technology, Qingdao, China
e-mail: maoyz@ysfri.ac.cn

F. Lin · J. Fang (✉) · J. Fang · J. Li · M. Du
Yellow Sea Fisheries Research Institute, Chinese Academy of Fishery Sciences, Qingdao, China
e-mail: linfan@ysfri.ac.cn; fangjg@ysfri.ac.cn; fangjh@ysfri.ac.cn; lijq@ysfri.ac.cn; dumr@ysfri.ac.cn

A. C. Smaal et al. (eds.), *Goods and Services of Marine Bivalves*,
https://doi.org/10.1007/978-3-319-96776-9_4

Abstract in Chinese 贝类是中国海水养殖的主要种类，2015年贝类养殖产量约为1360万吨，占中国海水养殖总产量的72.4%，其中双壳贝类产量约为1240万吨，占贝类总产量的90%以上。中国的贝类养殖历史悠久，距今2000多年前就有牡蛎养殖的记载，但规模化养殖始于20世纪70年代，这主要得益于贝类的苗种繁育技术得到了提高；随后，贻贝、扇贝、牡蛎等多种贝类的苗种繁育和人工采苗技术的建立，推动了贝类产业的迅速发展。进入21世纪，贝类养殖规模不断扩大，养殖种类不断增加，养殖产量大幅度提高。中国海岸线长度为1.8万公里，从北到南跨越辽宁、河北、天津、山东、江苏、上海、浙江、福建、广东、广西和海南11个省(直辖市)，大多数海域都适合贝类养殖，因气候差异显著，贝类分布具有明显地域性。中国贝类养殖方式主要包括筏式养殖和底播养殖，前者主要养殖种类包括牡蛎、扇贝、贻贝等；后者主要养殖种类包括菲律宾蛤仔、毛蚶、文蛤、虾夷扇贝等。本文分别以长牡蛎、厚壳贻贝为代表介绍了筏式养殖贝类的苗种生产和养殖过程；以菲律宾蛤仔和泥蚶为例介绍了滩涂贝类的苗种繁育和养殖过程。

Keywords Longline culture · Bottom culture · Seed breeding · Production process

关键词 筏式养殖 • 底播养殖 • 苗种繁育 • 生产过程

4.1 Overview of the Bivalve Production

The historical evidence of bivalve culture in China can be traced back to 2000 years ago, but the large-scale mariculture of bivalves was extensively practiced since the 1950s. The annual production of mariculture in China was about 10,000 t in 1950 and oyster was the major culture species then.

In the following 20 years, mussel and kelp had joined the oysters to make up the most cultured species in China. However, the bivalve seeding mainly comes from wild breeding in this period. After the 1950s, Chinese government and scientists paid great efforts on artificial breeding and natural collection of clam seeds such as cockle *Tegillarca granosa*, razor clam *Sinonovacula constricta*, clams *Ruditapes philippinarum* and *Cyclina sinensis*. In the 1970s, the mussel farming industry grew rapidly according to the persistent exploration of mussel hatchery and wild seed collection techniques. The farming area for mussels exceeded 2000 hectares and the annual production approached 60 kt in 1977, marking the rise of the Chinese shallow sea bivalve culture industry.

In the early 1980s, when the artificial breeding of scallops became mature and applicable in hatcheries (especially for the imported species bay scallop *Argopecten irradians*), together with the wild seed collection and improved longline culture technologies, scallop mariculture has greatly expanded. With the development of feeding eco-physiology, bivalve aquaculture industry gradually stepped into a new era in the fields of natural seed collection, seed breeding in hatchery, and variety of culture methods such as longline, sea ranching and pond culture.

Since 1990, mariculture (main categories: molluscs (bivalves and gastropods), algae, crustaceans, fish and others) in China has experienced a stage with flourishing

Table 4.1 The production of molluscs from 2006–2015 in China in tons * 10000 (10^4)

Year / Species	2006	2007	2008	2009	2010	2011	2012	2013	2014	2015
Total	1113.6	993.8	1008.1	1053.0	1108.2	1154.4	1208.4	1272.8	1316.6	1358.4
Oyster	389.2	350.9	335.4	350.4	364.3	375.6	394.9	421.9	435.2	457.3
Clam	301.9	295.7	305.8	319.2	353.9	361.3	373.5	385.4	396.7	400.9
Scallop	114.9	116.5	113.7	127.7	140.7	130.6	142.0	160.8	164.9	178.5
Mussel	74.6	44.9	48.0	63.7	70.2	70.7	76.4	74.7	80.6	84.5
Razor clam	67.9	66.7	74.2	68.4	71.4	74.5	72.0	72.1	78.7	79.4
Cockle	31.6	28.0	29.0	27.7	31.0	29.3	27.8	33.7	35.3	36.4
Sea snail	24.9	25.9	22.5	20.4	20.8	20.3	21.4	21.3	23.3	24.3
Abalone	2.2	2.5	3.3	4.2	5.7	7.7	9.1	11.0	11.5	12.8
Pen shell	1.8	1.2	1.1	1.5	3.1	3.0	1.5	1.7	1.8	1.8
Others	104.6	61.5	75.0	69.8	47.1	81.2	89.7	90.2	88.5	82.3

development (FAO 2014; Bureau of Fisheries, Ministry of Agriculture 2016). Until 2015, the cultured mollusc production is about 13.6 Mt., accounting for 72.4% of the total mariculture production (Bureau of Fisheries, Ministry of Agriculture 2016), and is about 4.4 times of that in 1995 (3.1 Mt), 48.9 times of that in 1975 (277,538 tonnes) The annual production of cultured bivalves in 2015 is around 12.4 Mt., accounting for about 91.2% of the total annual mollusc yield. Table 4.1 showed the annual production of cultured molluscs from 2006 to 2015.

Nowadays, the bivalves cultured in China has rose from around 10 species to approximately 70 since the 1960s (Tang et al. 2016), and among them two species, bay scallop (*A. irradians*) and Yesso scallop (*Patinopecten yessoensis*), were successfully introduced and applied in commercial scale production. The most productive bivalves include oysters (*Crassostrea gigas*, *C. rivularis* and *C. plicatula*); scallops (*Chlamys farreri*, *P. yessoensis*, *A. irradians* and *C. nobilis*); clams (*Meretrix meretrix*, *Ruditapes philippinarum* and *Mactra veneriformis*), razor clams (*Sinonovacula constricta* and *Solen grandis*), cockles (*Scaphaributica subcrenata*, *Scapharca broughtonii* and *Tegillarca granosa*), mussels (*Mytilus galloprovincialis*, *M. coruscus* and *Perna viridis*) and etc.

4.1.1 Production Distribution

China's coastline is about 18,000 km, crossing the tropics, subtropical and temperate zones, Different climatic zones and eco-environment provide varieties of survival and reproduction condition for various bivalve species (Fig. 4.1).

The major cultured bivalves and gastropods include scallops, abalones, mussels, sea snails and manila clam in Liaoning province along the North Yellow Sea coast. The sea ranching and longline culture of Japanese scallop *P. yessoensis* and mudflat

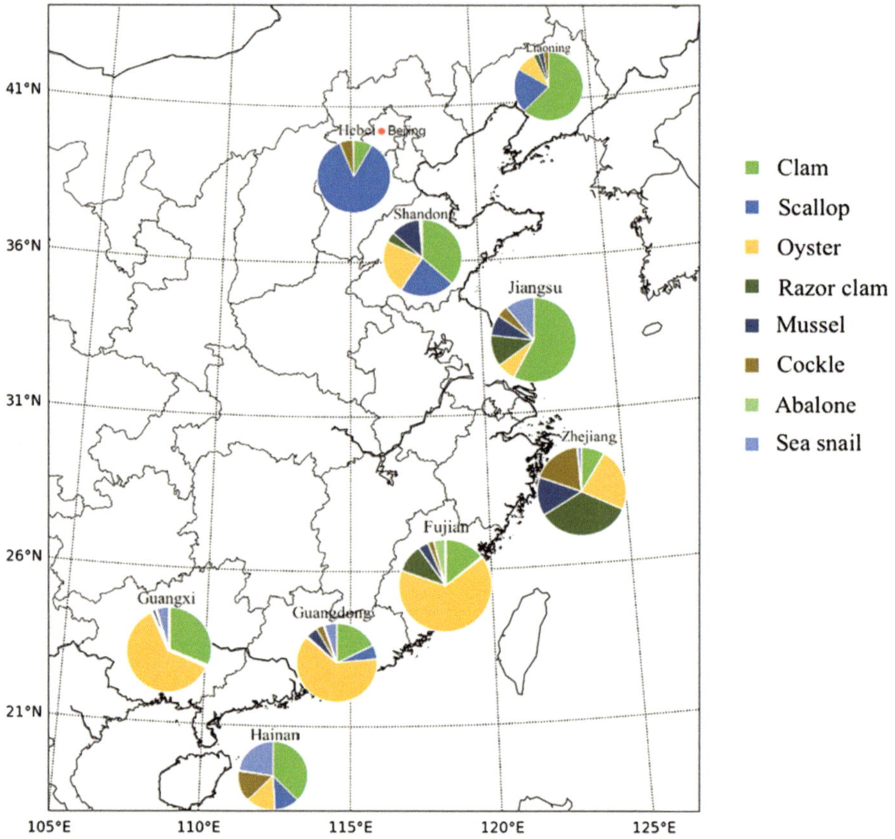

Fig. 4.1 Major culture shellfish species and production percentage in coastal provinces in China

culture of *R. philippinarum* are the major farming methods for aquaculture industries here.

Bohai Bay, Liaodong Bay and Laizhou Bay are the major aquaculture areas along the Bohai Sea coast with the major cultured species being mudflat shellfish such as clams, razor clams, conchs, oyster and cockle. Changdao Islands, located crossing the boundary of the Yellow Sea and Bohai Sea, are the high yield area for abalone (*Haliotis discus hannai*), Scallop (*C. farreri* and *P. yessoensis*) and cockle (*S. broughtonii*).

High diversity of mariculture species has been well practiced in Shandong Peninsula with various culture methods such as bottom, pond and longline culture. Mostly popular cultivated species includes abalone (*Holiotis discuss hannai*), scallop (*C. farreri*), Pacific oysters, manila clam (*R. philippinarum*), snails (*Bullacta exarata*), razor clam (*S. constricta*) and cockle (*S. broughtonii*). The seaweed-bivalve polyculture and Integrated Multi-Trophic Aquaculture (IMTA) of seaweed, bivalves, fish and sea cucumber have been conducted in Sanggou Bay for decades, leading the development of eco-farming in the world.

Haizhou Bay is located between southern Shandong peninsula and north of Jiangsu province, which is productive in blue mussel (*M. galloprovincialis*) and mudflat bivalves; the intertidal bottom clam culture and shallow sea longline culture are the major culture modes.

Culture species in Zhejiang and Fujian provinces include clams, Fujian oyster (*C. angulata*), abalone, mussel, conch and others. Pond culture of clams in Zhejiang province is well known in China, even in the world. The abalone culture has become popular in Fujian province in the last decade, which also promoted the culture of seaweed as feed for abalone. Meanwhile, the seed breeding of Manila clam in ponds gradually became one of the most important industries for local communes in Fujian province. In 2013, the total seed production of Manila clam (with shell length about 1 cm) in Fujian province was 7952 tons, which has fulfilled more than 80% of the seed demand for Manila clam farming in China. Moreover, the Manila clam is also farmed with shrimp, fish and crab in pond IMTA systems; this mode has been well practiced at commercial scale in the above two provinces.

Guangdong and Guangxi provinces are located along the coast of South China Sea. Major culture species here are Hong Kong oyster (*C. hongkongensis*), pearl oyster (*Pinctada martensii*), scallops (*C. nobilis*) and clams. The mariculture of pearl oyster is the traditional industry but recently has suffered a depression; the causes are supposed to be the stress from both climate change and human activities. The production of seawater pearl oyster has dropped from 38.6 tons in 2000 to 3.6 tons in 2015 (Zhu et al. 2019).

Hainan Island is located between tropic and subtropic zones, and the major cultured species include scallops (*C. nobilis*), sea snails (*Babylonia areolata*), green mussel (*P. viridis*), pearl oyster (*P. martensii*), oyster (*C. hongkongensis*) and others.

4.2 Bivalve Seed Production

Bivalve breeding technology is the basis of large-scale bivalve farming in China. After the 1950s, China has conducted artificial breeding on mud flat species such as *T. granosa*, *S. constricta*, *R. Philippinarum* and *Cyclina sinensis*, and successfully established the artificial breeding techniques. In the 1970s, the indoor hatchery technology and wild seed collection of mussel had been well practiced. In the early 1980s, artificial breeding, wild seed collection and longline culture technology of scallop gradually matured. Especially in 1982, the introduction of the bay scallop (*A. irradians*) greatly promoted China's scallop aquaculture, and contributed to the formation of several latest culture modes such as the alternative culture of seaweed and scallop and the polyculture of scallop, seaweed and shrimp. New culture modes and technology effectively promoted the development of China's marine aquaculture industry, and formed China's third wave of large scale mariculture activity. At present, there are two major seed production methods of bivalves in China, one is factory hatchery and the other is eco-hatchery in earth pond. The following introduces these two methods with representative species, Pacific oyster (*C. gigas,* factory) and Manila clam (R. *philippinarum*, earth pond), respectively.

4.2.1 Artificial Breeding of Pacific Oysters

The oyster is a worldwide commercial bivalve with diverse species, wide distribution and high adaptability. China's oyster farming can be traced back 2400 years to the ancient book "*Pisciculture*" written by the famous politician, strategist, Taoist and Economist FAN Li (Liu 1959). The production of oysters has ranked first position in a variety of cultivated bivalves in China. Natural seed collection was the major means for seed production before the 1980s, and was then replaced by the artificial seeding technique. In Shandong and Liaoning province the reproductive season continues from May to August, while in Zhejiang coastal region the reproductive time is in June and July (Gao et al. 1982; Wang and Wang 2008).

Taking the Pacific oyster (*C. gigas*) as an example, the seed breeding process was introduced below:

Bivalve farming covers the life cycle from larvae to adult, mainly including seed production and commercial size production (Fig. 4.2). Seed production is of vital

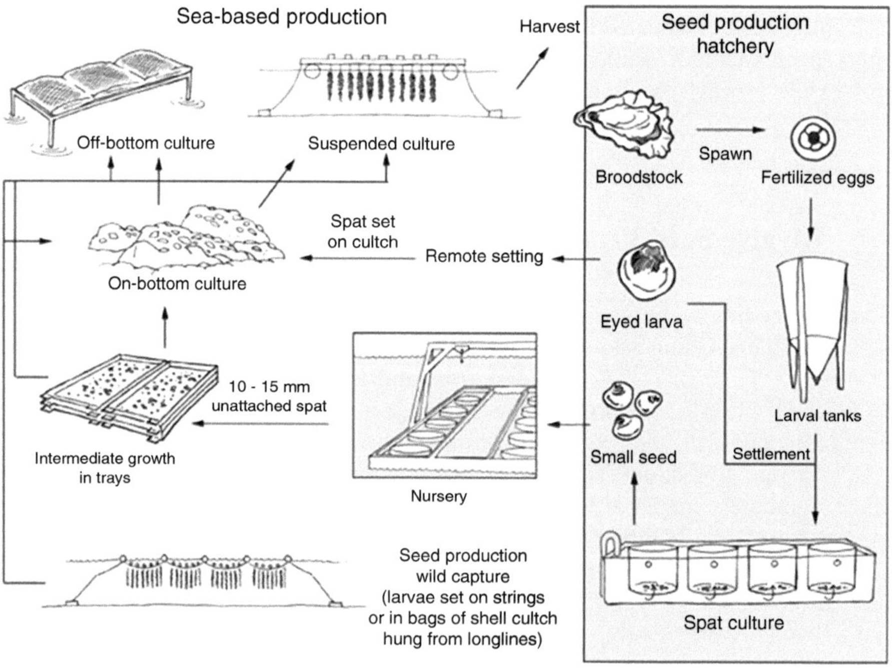

Fig. 4.2 Process and method of oyster farming (http://www.fao.org/fishery/culturedspecies/*Crassostrea_gigas*/en)

importance for the sustainable development of bivalve farming. The major methods are seed production in hatchery, semi-artificial seed breeding in ponds, wild seed collection and intermediate nursery in ponds or shallow seas.

4.2.1.1 Choice and Conditioning of Broodstock

The choice of broodstock oysters is of vital importance because high quality germplasm is fundamental to producing excellent offspring (Sui et al. 1997). Each year in March–April, oysters with shell length greater than 12 cm will be selected from phytoplankton-rich waters as broodstock and moved to the hatchery. After surface attachments and creatures have been removed, broodstock oysters will be transferred to the indoor tank of the hatchery for conditioning with a density of 35–50 ind./m^3. During the conditioning period, microalgae, such as diatoms (*Phaeodactylum tricornutum, Nitzschia closterium* and *Chaetoceros muelleri*), Chrysophyta (*Isochrysis galbana* and *Dicrateria zhanjiangensis*), Chlorophyta (*Chlorella vulgaris* and *Platymonas hegolandica*) are the major feed for broodstock oysters with feeding density about 200,000 cell/ml in 24 h. The conditioning water temperature is gradually increased from the beginning by 1 °C per day, and maintained stable at around 20 °C until ready for spawning (Yang et al. 1995).

4.2.1.2 Hatching and Larval Rearing

When the broodstock oysters' gonads mature, stimulation for spawning can be conducted. After drying 6–10 h in shade, the broodstock is put into the floating cages and placed into hatching tanks prepared for spawning. When the spawning egg density reaches 30–50 cells/ml, the spawning broodstock will be moved to another tank for continuous spawning. Generally, with the water temperature about 20 °C, in 24 h or so, fertilized eggs can be hatched to veliger larvae (D-larvae).

4.2.1.3 Larval Rearing

Larval rearing refers to the process of the veliger larvae (D-larvae) growth to spat. It takes about 7–9 days for veliger larvae grow to the umbo larvae, then about 19–22 days to grow into post larvae (eyespot larvae) and finally about 21–26 days to finish metamorphosis and transform into spat. The larvae are cultured at a temperature of 23–24 °C, and about 40%, 50% and 80% of water need be replaced daily for D-larvae, umbo larvae and eyespot larvae stage respectively. Usually every 7–8 days the tank will be refreshed. After the last refresh, the substrate will be placed in the tank for seed settlement. Chrysophyta are the best starting feed for D-larvae. When shell length of the larvae reached greater than 130 μm, high concentration Chlorophyta (*Platymonas* spp.) can be added to the feed. Feed density is better at 20,000 cell/ml for D larvae stage, 30,000–40,000 cell/ml for the umbo

Fig. 4.3 Substrate used in artificial breeding of oyster. (Photo from Mao)

larvae stage and 50,000 cell/ml for eyespot larvae stage per day. The best substrate for oyster larvae is scallop shells. When the larvae shell length grows to more than 280 μm and eyespots emerge gradually, then substrates are gradually placed into the settling tank with a density about 5000–6000 shells/m^3 (Fig. 4.3). When larvae attach to substrate and finished metamorphosis, the amount of feeding needs to be increased based on the feeding status of the spat. The feed is made up of diatom (*P. tricornutum, N. closterium*), Chlorophyta (*Platymonas* spp., *Chlorella* spp.) and Chrysophyta (*I. galbana*). Fifteen days after the settlement, spats can be moved to outdoor ponds or the sea for nursery.

*4.2.2 Artificial Breeding of Manila Clam (*R. philippinarum*) in Ponds*

R. philippinarum, which belongs to *Veneridae*, *Ruditapes*, is a species widely distributed in coastal areas of China and is highly productive in Shandong, Liaoning, Zhejiang and Fujian coastal areas. Annual production of *R. philippinarum* (about 3.2 Mt) accounts for about 62.7% of mudflat bivalve production and about 24.3% of total bivalve production in 2014 (Yan 2014).

The reproductive season of Manila clams varies in regions: from June to August along Liaoning coast, in late May and late September along Shandong coast, and in late September to November along Fujian coastal region. Appropriate reproduction water temperature is at 20 °C. With a fecundity of about 2–6 million per clam per year, Manila clam spawns 3–4 times during the reproductive season, and most spawning activity happens during the high tide and in the evening (Yan 2005; Zhang et al. 2006). The process and methods of seed production of Manila clam in ponds are as follows: Seed production pond is usually built in the intertidal zone near the shore, with no flood or storm threats and sufficient water exchange. The most suitable region is in the sheltered area with sandy muddy sediment.

4.2.2.1 Construction of the Seed Production Pond

Specified area for seed production ponds are varied from 1–100 ha and rectangle ponds are recommended. With water depth of 1.5–2 m, the pond wall height should be at least 1 m above the maximum tide level. A gate is used to control the water exchange. The number, size and location of gates should be determined according to the topography, area, flow direction, water flux and other related aspects. Generally, inlet gate and outlet gate should be built in pairs, the size of the gate should be able to fill or drain the pond in one day during spring tides (Fig. 4.4).

The bottom of the pond should be flattened with a longitudinal ditch about 0.5 m deep in the middle of the pond. A thin layer of fine sand with particle size about 1–2 mm should be laid on the pond bottom for spats and juveniles. A broodstock support frame is designed for stimulating spawning of broodstock, which is built close to the inside of inlet gate. The frame needs to be covered with netting in order to support and prevent the broodstock clam from escaping into the pond. Size of the support frame is varied with the pond size.

4.2.2.2 Preparatory Work Before Seed Production

Pond Cleaning: Clean the mud, stones and other debris, seaweed (*Ulva* spp.) and other attachments in the pond.

Drying: Drain the water in the pond, flatten the pond bottom, disinfect and bleach to improve substrate condition for helping spat settlement.

Cultivation algae: After disinfection and 7–10 days before the nursery stage, 30–50 cm of seawater filtered with nylon screen (ca. 50–5 μm) should be filled in the pond. About 0.5–1 ppm urea, 0.25–0.5 ppm superphosphate and 0.1 ppm silicate are added into the pond to promote the growth of phytoplankton.

4.2.2.3 Spawning

Gonadal status need to be identified before cleaning and temporarily reared.

Broodstock shellfish are usually to be placed in the support frame for temporary rearing at a density of 300–600 kg/ha. The procedure of stimulating spawning including drying the broodstock in the shade for 4–8 h and then exchange the pond water with flow rate of 35 cm/s for 2–3 h. When the broodstock spawned, continuing inflow is necessary for well fertilization and evenly distributed for fertilized eggs. The suitable D-larvae density is 3–4 ind./ml.

Fig. 4.4 Inlet gate with channel higher than water level; and outlet gate with channel lower than water level. (From Mao)

4.2.2.4 Larval Rearing

Water supplement: about 10–20 cm water filtered with the screen will flow into the pond daily during high tide in order to promote larvae growing and stabilize the temperature and salinity of pond water.

Fertilization: The density of phytoplankton in the pond should be maintained at the concentration of 20,000–40,000 cell/ml for meeting the feeding demand of larvae. To maintain such density, fertilizers should be applied according to the variation of phytoplankton concentration. About 0.5–1 ppm urea, 0.5 ppm superphosphate and other nutrients should be added every 1–2 days. During D-larvae period, feed density should be around 15,000 cell/ml, and increase to 30,000 cell/ml during umbo larvae period. If the microalgae density is high enough, fertilize is not necessary.

4.2.2.5 Spat and Juvenile Cultivation

Juvenile cultivation is a key stage of shellfish culture. The growth from spat to juvenile (about 1 cm in length) is an important stage in shellfish lifecycle. If right after the settlement, the filter organ gill, water pipe and shell are not well formed for the spat, mortality will be high and the survival rate is about 10%. At this point, measures such as suitable substrate selection, water quality management and sufficient food supplement should be applied (Wang and Wang 2008).

4.3 Shellfish Longline Farming

Longline farming is one of the most common bivalve culture methods in China (Fig. 4.5), the annual production of longline cultured bivalves exceeded 4 Mt. in recent years. There are many species of longline cultured bivalves and gastropods, such as scallops, oysters, mussels and abalone. Longline farming has a variety of types according to different regions. Aquaculture technology has developed through years, and currently the poly-culture of bivalves and seaweed (i.e. IMTA) has become the latest eco-farming mode. With this method, bivalves and algae mutually benefit, which achieves a double-win result with both ecological and economic benefits.

Below, approaches to longline culture of two important taxa, oysters and mussels, are discussed.

4.3.1 Oyster Farming

Oyster farming methods are mainly shallow sea longline farming in northern China and mudflat farming in southern China. China's oyster production ranked first in the total production of shellfish and has been on rise continuously. Oyster yield reached 4.6 million tons in 2015, accounting for 33.7% of total bivalves output (Bureau of Fisheries, Ministry of Agriculture). The major Chinese longline cultured oysters are *C. gigas*, *C. rivularis*, *C. plicatula*, *O. denselamellosa*, and *C. angulata*. Among

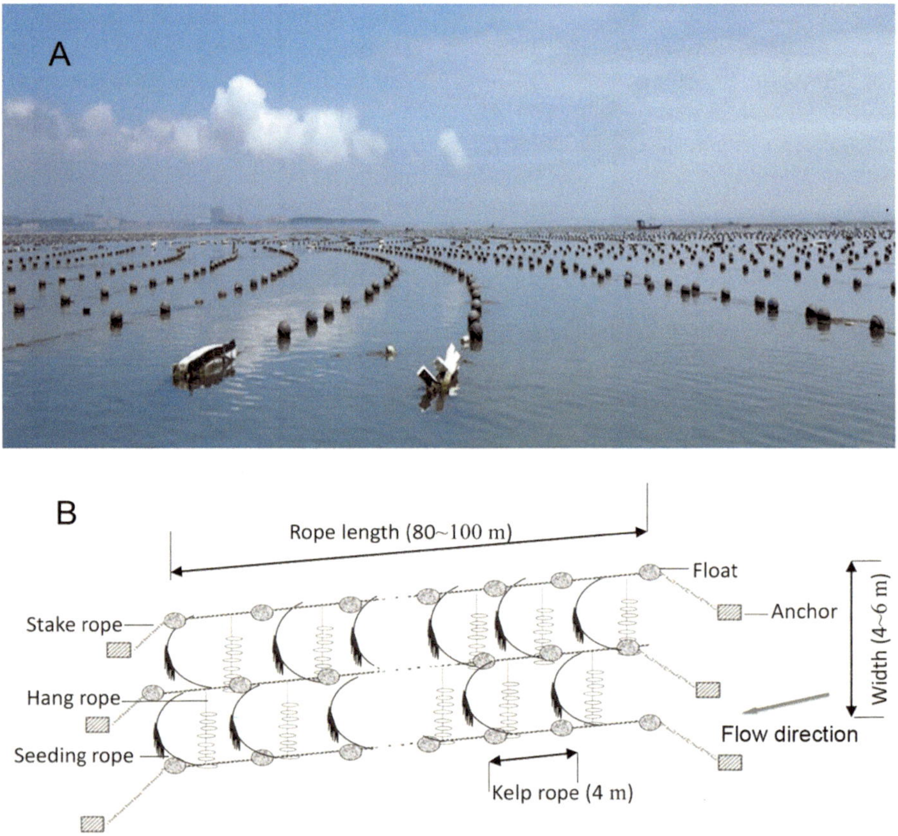

Fig. 4.5 Longline culture in China (above); Schematic diagram of longline culture in China (below)

them, Pacific oyster (*C. gigas*) is mainly cultured in northern China, while the other three species are mainly cultured in southern China. Below, we introduce the longline culture technology with the Pacific oyster as an example.

4.3.1.1 Pacific Oyster (*C. gigas*)

Pacific oysters have the advantages of high growth rate and high yield. The successful artificial breeding of oysters has provided abundant seeds for large-scale culture.

The main process of Pacific oyster longline culture is as follows:

4.3.1.2 Area Selection

Sea area for Pacific oyster longline culture should be relatively calm, water depth at low tide greater than 4 m, and water temperature above the freezing point in winter and less than 30 °C in summer. Flow rate of about 0.3–0.5 m/s is appropriate for oyster longline culture. The amount of phytoplankton in the sea area is generally no less than 40,000 cells/L. Additionally the culture area should be far from where mussels, sea squirts and other competitive species exist, and away from pollution (Li 2006).

4.3.1.3 Facility Set Up

The direction of the longline stake rope should follow the current, and polyethylene rope with a diameter of 2.4 cm is used as the stake raft rope. The raft rope length is about 150 m in total, about 80–100 m of which is used for cultivation. There are about 25 m at each end attaching to the fixed pile. The space between two consecutive longlines is 7 m wide. Float number is gradually increased according to the oyster growth. Polyethylene rope with 0.4 cm diameter and 3.0 m long is used for hanging oyster, hanging space between each rope is about 1 m. When oyster is cultured in a cage, the hanging space should be 1.2–1.5 m (Fig. 4.5).

4.3.1.4 Density and Scale

Oyster farming is mainly conducted in sheltered waters. The farming density and farming scale is planned based on carrying capacity according to local environmental parameters (Fang et al. 1996). To prevent over farming, a better way is to implement shellfish-algae polyculture.

4.3.1.5 Harvest

The harvest of Pacific oysters is slightly different according to the situation in north China. Some aquaculture areas have been harvested in November–December, some areas with sufficient food supply and non-frozen winter usually harvest from March–June in the following year (Lian and Mao 2010). At present, the harvest of oysters is in a traditional and high labor cost way with manual operation. Nowadays, researchers and enterprises are developing relevant mechanized harvesting means to reduce the labor costs.

4.3.2 *Mussel Farming*

Longline culture is the major mussel culture mode in China. Mussel species include *Mytilus galloprovincialis*, *M. coruscus* and *Perna viridis*. *M. galloprovincialis* and *M. coruscus* are mainly distributed in the Yellow Sea and the East China Sea, while *P. viridis* is only found in the South China Sea. At present, the majority of mussel seed comes from natural sea area collection, and a small amount from artificial breeding. Below, the culture method of thick shell mussel is described as an example.

4.3.2.1 Thick Shell Mussel (*Mytilus coruscus*)

M. coruscus are the representative mussel species as their higher market price. The raft culture of *M. coruscus* is introduced as an example at a typical area in Shengsi, Zhejiang.

The main process of Thick shell mussel longline culture is as follows:

4.3.2.2 Area Selection

Mussels are usually cultured in sheltered areas with sand and mud sediment. Sufficient water exchange, abundant natural phytoplankton and detritus, and a water depth between 5–20 m at low tide are preferred conditions. *M. galloprovincialis* and *M. coruscus* can survive in a condition with salinity between 18–34 psu and temperature between 0–29 °C (temporary frozen period in winter).

4.3.2.3 Facility Set Up

Longline raft set up for mussel is similar to that of Pacific oyster. The major differences are in the longline and float distances. The raft rope length is about 63 m in total, raft is set every 17 m along the rope. There are about 25 m at each end attaching to the fixed pile. The space between two consecutive longlines is 7 m wide. Length of the longline is about 65 meter, the cultured thick shell mussels are attached to a rope hanging every 4 m on the longline and is attached to a float; Distance between each longline is about 17 m and 7 longlines forms a culture unit, with an area about 6667 m^2, culturing a total of 105 ropes of thick shell mussel (Fig. 4.6).

4.3.2.4 Nursery Facility

During the breeding, the thick shell mussel seed will be sorted several times. When seed leaves the nursery, the seed will be put into net bags for intermediate cultivation. After the seed grows to about 0.5–1.0 cm in length, farmers will conduct the second resocking into polyethylene mesh.

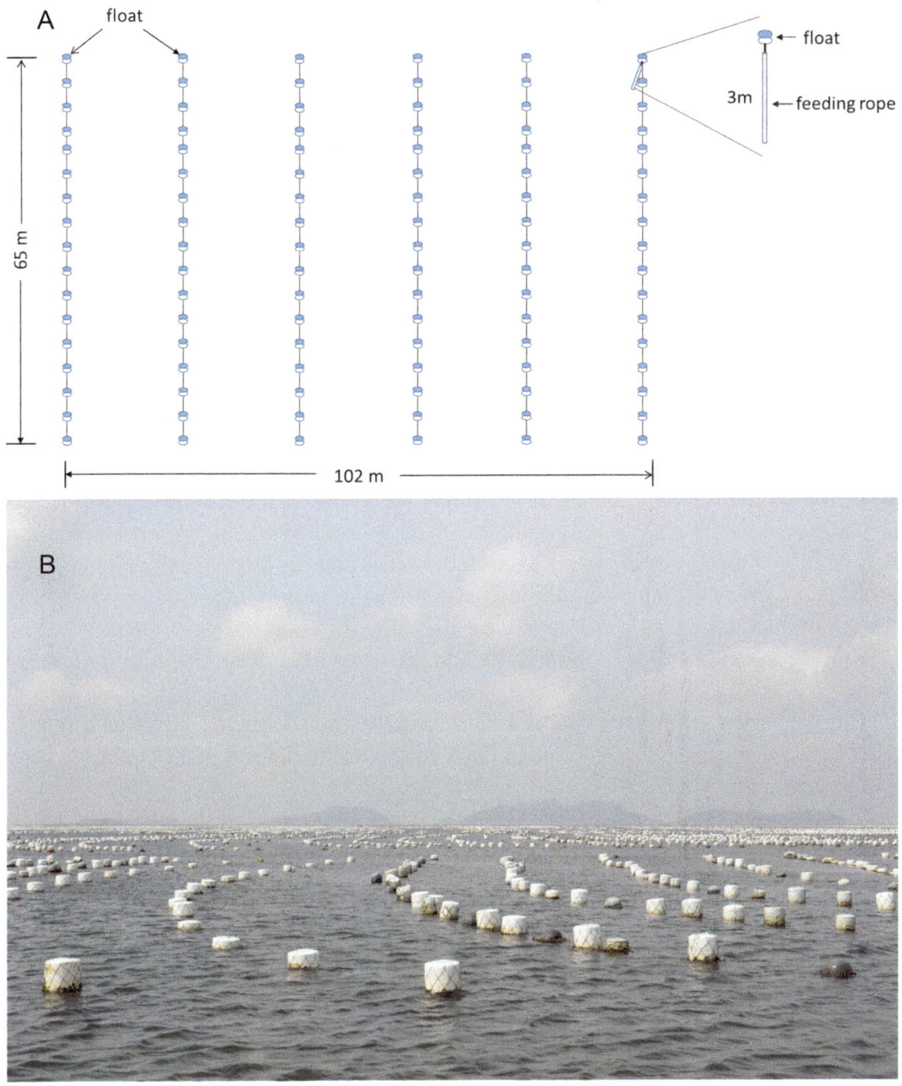

Fig. 4.6 Facility of thick shell mussel longline culture (**a**, Schematic; **b**, Field photograph)

Packing mesh and tying rope: Seed packing mesh should be woven with polyethylene. The length of the package mesh should be 30–50 cm longer than the breeding rope. The initial package is 30 cm wide and 0.5–1.0 cm aperture mesh, and latter package is 40 cm wide and 1.5–2.5 cm aperture mesh.

Normally, rubber or polyethylene ropes are used as the nursery and culture ropes for mussel farming in China, while for thick shell mussel, hemp rope is usually applied as the culture rope, with the diameter around 2.0 cm and length between 2.5–3.0 m (Fig. 4.7a). At present, the widespread use of a self-dissolving material is conducive to the growth of mussels with less labor and financial cost. The sock-type

Fig. 4.7 Culture ropes and bagging the juvenile mussel (**a** the Rope; **b** bagging juvenile mussel; **c** shematic of the bags)

bag will automatically break when the mussels attach to the breeding rope. The sock-type bag is usually made with paper-like material, the width of the bag is about 20–25 cm and the length is a bit shorter than the culture rope. The juvenile mussels are filtered through a sieve with a diameter of 2–3 cm to ensure they are separated sufficiently. When putting juvenile mussel into the bag, farmers usually place 30–50 individuals every 10 cm along the net, when all the individuals are placed, rolling the net and attach it to the culture rope with some thin hemp ropes to make the bag (Fig. 4.7b, c), each culture rope will be attached to a float during the culture period (Fig. 4.6).

Hanging rope: Generally, polyethylene rope with a diameter of about 5 mm is used for hanging mussels. To prevent the rotation of the device, paired hanging rope can be applied.

4.3.2.5 Harvest of Mussels

Mussels are usually harvested twice a year, during early summer and early winter, when the shell height reached about 6–7 cm. Now the thick shell mussels are harvested in a semi-mechanized way (Fig. 4.8). Harvested mussels are usually sorted and cleared for further processing.

4.4 Bivalve Bottom Culture

Bivalve bottom culture is another major method in China. Mudflat bivalves, such as clam, razor clam and arc shell, are the main bottom culture species. In 2015, 3 species mentioned before produced more than 5 Mt. of shellfish, accounting for more than 1/3 of the total shellfish production. Mudflat shellfish farming has developed rapidly in recent years, the price showed a trend of continuous increasing, and has become major species in Chinese market.

Fig. 4.8 Harvesting of the thick shell mussel

General methods to bottom culture of two important taxa, Manila clam and cockle clam, are discussed below.

4.4.1 Manila Clam Farming

4.4.1.1 Mudflat Modification

The mudflat is usually modified to suitable condition before the clam culture. By applying bottom plowing and sediment drying to clean the dead shells and loosen the sediment, the humus in the sand is decomposed and then washed away by seawater.

4.4.1.2 Seed Source

Major sources include: Indoor artificial seed production, natural seeds collected from Shandong and Liaoning province, artificial seed produced in ponds from Fujian Putian city.

4.4.1.3 Sowing

Seeds are sown in spring or autumn. Neap tide and sunny windless days are preferred for the dry release of seeds. Seed size from 3–10 mm can be sown; sowing density is controlled at 1000–2000 ind./m^2 and distributed uniformly (Mitchell et al. 1992; Cigarría and Fernández 1998; Zhou et al. 1998). Seed around 10 mm have better survival rate, growth rate and yield.

4.4.1.4 Subtidal Zone Culture (Water Depth Within 20 m)

Clams cultured in the subtidal zone have no exposure time, and are less affected by high temperature, freezing and flood, which prolongs the effective feeding time. The growth rate and relative fatness of clams are found increased by 67.9% and 26.9%, respectively, compared to the clams cultured in mudflats, and the survival rate was above 80%. In the subtidal zone, shell growth rate of cultured Manila clams is more than 4.0 mm/month and the culture cycle is shorted by 6 to 12 months compare to the mudflat culture, which improves both product quality and commodity value.

4.4.1.5 Predators

The major predators affecting Manila clam culture are shrimp and crabs, snails (*Rapana venosa* and *Glossaulax didyma*), sea anemones, starfish, fish (*Acanthogobius* spp.), and birds (seagulls and sea ducks). In China, artificial catchment of predators is the major way to protect the clams from being predated. The coverage of plastic protective nets in the heavily damaged areas has greatly increased the survival rate of cultured clams abroad (Cigarría and Fernández 2000; Spencer et al. 1992).

4.4.1.6 Harvest

Manila clams are harvested throughout the year. In the intertidal zone, manual capture is the major method and smaller individuals will be left for continued growth. In the subtidal zone the clams will be captured with motor boats.

4.4.2 *Cockle Clam* (T. granosa) *Farming*

T. granosa bottom culture can be divided into two ways: field farming and pond farming.

4.4.2.1 Field Farming

Field farming is popular in Guangdong and Fujian coastal regions, and refers to the farming on non-water retained flat sediment. A common choice for farming sites is on the inner soft mudflat in the intertidal zone. Such method benefits from construction convenience and sufficient water exchange, which is suitable for large-scale farming. Field farming area can reach up to 50–60 ha. Selected areas are divided into several square or rectangular zones according to topography and marked with bamboo or sticks. Shallow channels are constructed between each square for water outlet and prevention of clam seed escape. In some regions, the farm area is surrounded by nets to protect clams from predators.

4.4.2.2 Pond Farming

Clam ponds are usually built in the mid tidal flushing area; in the low tide area the pond walls are frequently eroded by tidal flow and in the high tide area water exchange is insufficient (Mojica and Nelson 1993). Tidal cycles are used for water level control in the pond, which leads to the comprehensive utilization of the tidal zone. In addition, feeding time of cultured clams is prolonged according to the retained water in the pond, higher survival and growth rate are expected. However, the high labor cost for pond construction has limited the farm scale. And during low tide the pond water become stagnant which it is unsuitable for high-density farming (You et al. 2002). In winter, emphasis should be paid to the solidity of the pond walls and to prevent the deposition of sludge. Sediment should be firm enough to prevent water leakage.

Clam culture density varies greatly in different regions and is determined by culture method and conditions. Farming density on mudflats can be estimated with 100–150 kg/ha production for large size clam individuals, that is, when harvested, about 4.5 million individuals will be collected per hectare. Sowing is generally during low tide on cloudy days and clam seeds are evenly distributed into the pond (Fig. 4.9).

4.4.2.3 Aquaculture Management

Scatter: clam seeds are captured every 10 to 15 days with mesh drip bags and then re-distributed to a larger farming area to adjust farm density for growth promotion. During this process, competitive species and predators are removed, and the epiphytic organisms on clam shells are cleaned.

Salinity maintenance: during the rainy season, high precipitation may dilute the seawater. To maintain a certain degree of salinity, if the proportion of seawater fell below 5 psu for more than 3 days, 600 kg/ha of salt should be added to the pond during low tide.

Fig. 4.9 Pond farming in Zhejiang province (From Mao), in the picture, the pond is surrounded by the channels with water, and when clam farming starts, the central pond will be submerged with seawater

Aestivate and overwintering: in South China, the summer sunlight often overheats the retained surface water covering the clam field and leads to high mortality. Therefore, in clam field farming, seawater retention should be avoided. In clam pond farming, on the contrary, water storage should be increased to keep a certain water depth and to avoid a sharp water temperature rise in the pond. Along the North China coast, overwintering is the major concern. Clam farms are moved to the subtidal zone to keep away from low surface temperature. Clam seed overwintering ponds are usually built in the intertidal zone and overwintering migration should be completed before October. Late transportation may increase the seed mortality due to freezing. During the overwintering process, pond water depth is kept between 20–30 cm with no leakage allowed.

Predator capture: clam farms need to be inspected frequently. Predators such as other clams (*Musculus senhousia*), starfish, crabs, fish (*Acanthogobius* spp.), red snails (*Rapana* spp.) and others are cleaned manually. If necessary, mesh cover or other methods can be applied to prevent clams from being preyed upon.

Harvest: The commercial specifications for cultured cockle clam are set as: shell length > 2.5 cm and reach 200 ind./kg. It usually takes 2–3 culture years in South China and 3–4 culture years in North China to satisfy commercial requirements. Afterwards the net profit decreases due to lower growth rate. Generally, clams are harvested in the winter fatness period with good taste. Southern harvest occurs from

December until the following March; in the north it is from November to December. After 3 years of farming, more than 20 tons of clams can be harvested from each hectare.

4.5 Conclusions

In recent years, the species of cultured bivalve in China has been continuously increased from 10 species in the 1950s to more than 70 species now. The bivalve production has been gradually increasing. The bivalve aquaculture production in 2015 was 12.4 Mt., accounting for 66.0% of the total marine aquaculture production.

Bivalve cultured in China have obvious geographical distribution characteristics, among which clams and oysters are all over the country culture species.

Longline culture and bottom culture are the major methods of bivalve farming in China. The main longline cultured bivalves include oysters, scallops, mussels, etc.; bottom cultured ones include clams such as *R. philippinarum* and *T. granosa*.

Artificial breeding techniques of bivalves including oysters, clams and scallops have been extensively applied in China, and has supplied the majority of the seed sources of almost all the main cultured bivalves. Pacific oysters are the representative species of longline cultured bivalves; Manila clam *R. philippinarum* is the representative species of bottom cultured bivalve.

Acknowledgements The authors are grateful to the reviewers for their valuable comments on the manuscript. We also appreciate the support from Haifeng Jiao of Ningbo Academy of Oceanology and Fishery for providing materials of thick shell mussel culture; This work was supported by Nation Natural Science Foundation of China (NSFC)-Shandong Joint Fund for Marine Ecology and Environment Science (U1606404), Central Public-interest Scientific Institution Basal Research Fund, YSFRI, CAFS (20603022017002) and Central Public-interest Scientific Institution Basal Research Fund, CAFS (No. 1620022017041).

References

Bureau of Fisheries, Ministry of Agriculture (2004–2016) China Fishery Statistical Yearbook. Agriculture Press, Beijing

Cigarría J, Fernández J (1998) Manila clam (*Ruditapes Philippinarum*) culture in oyster bags: influence of density on survival, growth and biometric relationships. J Mar Biol Assoc U K 78:551–560

Cigarría AJ, Fernández JM (2000) Management of Manila clam beds: I. Influence of seed size, type of substratum and protection on initial mortality. Aquaculture 182(182):173–182

FAO (2014) Fishery and aquaculture statistics [Global capture production 1950–2013] (FishStatJ). In: FAO Fisheries and Aquaculture Department [online or CD-ROM]. Rome. Updated 2014. http://www.fao.org/fishery/statistics/software/fishstatj/en

Gao YT, Xu GX, Fang JZ (1982) Preliminary investigation of the Pacific oyster natural reproduction. Mar Fish 3:110–114 (in Chinese)

Li HL (2006) Farming technology of *Crassostrea gigas*. J Biol 41(4):50–51 in Chinese
Lian W, Mao YZ (2010) Aquaculture teachniques of *Crassostrea gigas* and common problems. Mod Agric Sci Technol 5:302–303 (in Chinese)
Liu CG (1959) The earliest literature of aquaculture worldwide- scientific value of "Pisciculture" by Fan Li. China Fish 22:45–46 (in Chinese)
Mitchell ME, Godcharles MF, Bullock LH, Murphy MD (1992) Age, growth, and reproduction of jewfish *Epinephelus itajara* in the eastern Gulf of Mexico. Fish Bull 90:243–249
Mojica RM Jr, Nelson WG (1993) Environmental effects of a hard clam (*Mercenaria mercenaria*) aquaculture site in the Indian River Lagoon, Florida. Aquaculture 113:313–329
Spencer BE, Edwards DB, Millican PF (1992) Protecting Manila clam (Tapes philippinarum) beds with plastic netting. Aquaculture 105:251–268
Sui XL, Wang ZS, Ma T, Chen ZQ, Xu Q (1997) The main factors affecting on artificial seedling rearing of Pacific Oyster (*Crassostrea gigas*). J Dalian Fish Univ 12(4):15–20 (in Chinese with English abstract)
Tang QS, Han D, Mao YZ, Zhang WB, Shan XJ (2016) Species composition, non-fed rate and trophic level of Chinese aquaculture. J Fish Sci China 23(4):729–758 (in Chinese with English abstract)
Wang RC, Wang ZP (2008) Science of marine shellfish culture. Press in Ocean University of China, Qingdao
Yan XW (2005) The culture biology and technology and selective breeding in Manila clam, *Ruditapes philippinarum*. Institute of Oceanology, Chinese Academy of Science. Doctor, 206 (in Chinese with English Abstract)
Yan XW (2014) Current situation, problem and prospect of clam breeding industry [C]. Abstract set from symposium exchange of "The ecological security of oceanography and limnology under the global change". Nanjing (in Chinese)
Yang AG, Niu XD, Shen JF, Sun SG (1995) Study on cultchless spat and culture techniques of Pacific Oyster *Crassostrea gigas*. J Fish Sci China 3:29–35 (in Chinese with English Abstract)
You ZJ, Wang YN, Chen J (2002) Growth of *Tegillarca granosa* in the pond culture of Leqing Bay. J Fish China 26(5):440–447 (in Chinese with English abstract)
Zhang YH, Yu ST, Zhu JX (2006) Experiment on the industrial seed rearing of *Ruditapes philippinarum*. Shandong Fish 3:2–7 (in Chinese)
Zhou DT, Wang JR, Gao RC, Qiu WR (1998) A research on rotational culture of the Clam *Ruditapes Philippinarum* and Porphyra in Open Bay. J Shanghai Fish Univ 7(4):306–310 in Chinese with English abstract
Zhu C, Southgate P, Li T (2019) Production of pearls. In: Smaal A et al (eds) Goods and services of marine bivalves. Springer, Dordrecht, pp 73–93

Chapter 5
Production of Pearls

Changbo Zhu, Paul C. Southgate, and Ting Li

Abstract The pearl is known as the queen of jewels, and has been used for adornment and as a symbol of material wealth throughout human history. Pearls are formed by the secretion of nacre from epidermal cells within mollusc mantle tissue. But particular conditions are required for loose natural pearls to form and this occurrence is rare. However, utilization of this process for cultured pearl production now supports industries in more than 30 countries including China, Japan, Australia, Indonesia, French Polynesia, Philippines, Cook Islands, Thailand, Malaysia, India, Sri Lanka, Myanmar and Mexico, of which China has the largest production. Analysis of FAO global statistics shows that in the past decade (from 2005 to 2014), the average annual output of Chinese pearls was 3540 tonnes (t) valued at 15 million USD. This output accounted for over 98% of global cultured pearl output, of which freshwater pearls accounted for 99.5%. Japan has been the world's major marine pearl producer for over a century, and has developed advanced technology in pearl oyster culture and pearl production. In the past decade, the average annual value of marine cultured pearl production in Japan was 127 million USD, accounting for 51.6% of global pearl output value. Average annual production of marine cultured pearls was 23 t in Japan, 18.6 t in China and 12.9 t in French Polynesia. Chinese pearl production is typified by a high-yield, low-value industry structure. Overall, global pearl production fell by 60% while output value fell by 39% over the past

C. Zhu (✉)
Key Laboratory of South China Sea Fishery Resources Exploitation & Utilization, Ministry of Agriculture, Guangzhou, People's Republic of China

South China Sea Fisheries Research Institute, CAFS, Guangzhou, China
e-mail: changbo@scsfri.ac.cn

P. C. Southgate
Faculty of Science, Health, Education and Engineering, University of the Sunshine Coast, Sippy Downs, QLD, Australia
e-mail: psouthgate@usc.edu.au

T. Li
Key Laboratory of South China Sea Fishery Resources Exploitation & Utilization, Ministry of Agriculture, Guangzhou, People's Republic of China
e-mail: liting@scsfri.ac.cn

A. C. Smaal et al. (eds.), *Goods and Services of Marine Bivalves*,
https://doi.org/10.1007/978-3-319-96776-9_5

decade. Cultured pearl production typically includes five stages: oyster selection, nucleus implanting, nurturing, harvesting and pearl processing, of which nucleus implantation is the key step. Compared with other aquaculture sectors, pearl production has a complex process and a relatively long farming cycle which make it economically risky. Pressures to increase production, as well as external pressures such as urbanization, have placed pressures on the pearling industry that require appropriate management practices that support sustainable industry growth.

Abstract in Chinese 珍珠被誉为宝石中的皇后,是由贝类外套膜壳侧上皮细胞的分泌物形成,具有装饰、药用、美容保健三大功用和悠久的历史文化。目前世界上养殖珍珠的国家和地区有中国、日本、澳大利亚、法属波利尼西亚、库克群岛、泰国、马来西亚、印尼,印度、斯里兰卡、缅甸、菲律宾和墨西哥等30多个。其中,中国珍珠产量和养殖规模最大,根据FAO全球统计数据分析,过去十年(2005-2014)中国珍珠平均年产量3540吨(产值约1500万美元),占世界平均年总产量的98%以上,其中99.5%为淡水珍珠。日本是珍珠养殖的传统大国,其珍珠养殖和加工技术世界先进,过去十年来平均总产值达1.27亿美元,约占世界珍珠产业年平均总产值的51.6%。世界海水珍珠平均年产量:日本23吨,中国18.6吨,法属波利尼西亚12.9吨。淡水珍珠主要产自中国。由于各种环境和经济因素,近年来全球海、淡水珍珠产量均逐年下降。总体而言,在过去的十年间世界珍珠总产量下降60%,产值下降了39%。珍珠的生产过程包括:育苗、插核、育珠、收获、加工5个阶段,其中,插核是珍珠养殖的关键技术。相比其他产业,珍珠养殖生产周期长,流程复杂,具有较高的经济风险。受到业内提高产量的压力以及城市化等外部环境压力的双重作用,珍珠产业必然需要采取合适的管理调控措施以支撑产业的可持续发展。

Keywords Pearl culture · Pearl oysters · Pearl mussels · Pteriidae · Unionidae · Margaritiferidae

Keywords in Chinese 珍珠养殖 • 海水珠母贝 • 淡水珠母贝 • 珍珠贝科 • 珠蚌科 • 珍珠螺科

5.1 History of Pearl Production

The pearl is known as the queen of jewels, a symbol of material wealth throughout human history, yet one that is distinctively associated with modern civilization as the only gem produced by mankind. Many ancient civilizations had their own myths and legends about pearls, and showed great appreciation for them (Zhang and Fang 2003; Strack 2006). Before pearls were artificially cultured, they were collected rarely and by chance from oysters gathers from their natural habitat for food, or for the mother-of-pearl lining their shells that was used for decorative purposes (Strack 2008). Historical sources of pearls collected in this way included the Gulf of Mannar, between India and Sri Lanka, the Bay of Bengal, the Egyptian coast

Fig. 5.1 Pendants of pearl Buddhas; blister pearls produced on the inner shell surfaces of freshwater mussels. (Source: Guangdong Ocean University)

(Red Sea) and the Persian Gulf (Saudi Arabian coast) (Matlins 2001; Strack 2008) where the economy was particularly dependent on pearl fishing prior to the twentieth century (Carter 2005). The scene changed dramatically in the early 1900s when natural pearl fisheries became increasingly exhausted and many countries that had a long history of pearling became less significant in a world market that was increasingly dominated by cultured pearls.

China was the first country to culture pearls and people in the Song Dynasty (960–1279 AD) already knew how to grow blister pearls on the inner shell surfaces of freshwater mussels (Xie and Min 2003). In the late thirteenth century (Ming Dynasty) this primitive technique continued to be used to produce pearl Buddhas (Fig. 5.1) that were sold in temple markets (Abbott 1972; Alagarswami 1987). Modern round pearl cultivation owes its founding and status to development of the Mise-Nishikawa-method in Japan in the early 1900s (Taylor and Strack 2008). Commercial production of cultured marine pearls using this method was pioneered by Kokichi Mikimoto (1859–1954). Considered a national hero in Japan, and the 'father' of modern cultured pearl production, Mikimoto opened a new era for pearl cultivation (Alagarswami 1987; Wang et al. 1993) that today supports a global multi-million-dollar industry, producing pearls in more than 30 countries and offering economic opportunities to coastal communities in less developed countries (Southgate 2007). The major cultured pearl producing countries now include China, Japan, Australia, Indonesia, French Polynesia, Cook Islands, Philippines, India, Sri Lanka, Myanmar, Thailand, Malaysia and Mexico (Gervis and Sims 1992; Southgate 2007; Southgate et al. 2008a).

Japan was the world's major cultured marine pearl growing country for over a century, and developed advanced and systematic technology in pearl culture (Fassler 1992). The Japanese cultivated their first spherical marine pearls in 1907 (Wang et al. 1993) using the Akoya pearl oyster (*Pinctada fucata martensii*) and Akoya pearls have been mass-produced since 1945. Annual output reached its peak in 1966, with production of 127 tonnes (t) (Mizumoto 1979; Alagarswami 1987), but since then output has decreased substantially to a level of 20–25 t per year in 2014 (Southgate et al. 2008a; FAO 2016). Although Japan has made several refinements to methods used for oyster husbandry and pearl production, the pearl cultivation techniques used today differ little from those developed a hundred years ago (Taylor and Strack 2008). Because of the monopolistic marketing and strict technology-protection policy in the early years of Japan's cultured pearl production, Japan still remains the dominant force in today's pearl industry (Gervis and Sims 1992; FAO 2016).

China was one of the first countries in the world to harvest and use marine pearls. The earliest record of Chinese pearl collection dates back to 2200 BC, and from the Han Dynasty (206 BC–220 AD) onward, the Chinese collected marine pearls in the South China Sea. However, modern marine pearl culture (spherical pearl cultivation) only began in 1949, after the founding of the People's Republic. It began to flourish after successful artificial breeding of Akoya pearl oysters in 1965 and, by the end of the 1990s, Akoya pearl production was greater than 20 t per annum (Wang et al. 2007; Southgate et al. 2008a).

Other countries such as Indonesia, Australia, French Polynesia and the Cook Islands, which have a relatively brief history of pearl cultivation, now play rather significant roles in the world's pearl market (Strack 2008). Indonesia and Australia are the major producers of white 'South Sea pearls' from the silver or gold-lip pearl oyster *Pinctada maxima* (Southgate et al. 2008a), that are the largest and most valuable of culture pearls. Indonesia currently produces almost 4 t of cultured pearls from *P. maxima* per annum. Smaller producers of cultured pearls from *P. maxima* include Myanmar, Malaysia and Papua New Guinea, with China also having success in developing round pearl culture using this species (Xie and Min 2003; Southgate et al. 2008a). French Polynesia, the Cook Islands and some other Pacific island nations have produced 'black' South Sea pearls from the black-lip pearl oyster, *Pinctada margaritifera*, since the mid-1970s (Southgate et al. 2008a). Pearl culture has become a major export earner for both nations (McElroy 1990; Gervis and Sims 1992), and is second in value only to tourism in French Polynesia. Development of pearl oyster culture offers economic and livelihood opportunities in smaller Pacific nations and research in the western Pacific, in particular, has helped develop commercial pearl culture in smaller nations such as Fiji (Southgate et al. 2008a).

China is by far the major producer of cultured freshwater pearls. Freshwater pearl culture using natural pearl mussels was first demonstrated in Guangdong in 1958, and a significant breakthrough in the artificial breeding technology of Unionidae mussels was later achieved in Zhejiang in late 1970s (Bai et al. 2014). Since then, freshwater pearls have been produced on a large scale with annual output of more than 300 t in 1984 overtaking that of Japan (Hua and Gu 2002; Bai et al. 2014). By the 1990s, China had over 1000 pearl mussel farms and annual output has increased to over 2000 t within 40 years (Yang et al. 2003).

5.2 Mother of Pearl

Almost all species of mollusc are capable of producing pearl-like objects, technically termed "calcareous concretions" (McGladdery 2007). However, those of value and of interest as gemstones are limited to those produced by species capable of secreting nacre or mother-of-pearl (MOP), sometimes referred to as 'mother-of-pearl shell'. Two different groups of MOP shell are widely used for pearl cultivation: (1) marine pearl oysters of the family Pteriidae; and (2) freshwater pearl mussels of the families Unionidae and Margaritiferidae.

5.2.1 Marine Pearl Oyster

Several pearl oyster species from the family Pteriidae have been extensively exploited for pearl production for over a century. From the genus *Pinctada*, important commercial species include the Akoya pearl oyster *Pinctada fucata/martensii*, the gold or silver-lip pearl oysters, *Pinctada maxima*, and the black-lip pearl oyster, *Pinctada margaritifera*. While *Pteria penguin and Pteria sterna* are used for commercial pearl production to a lesser degree (Southgate et al. 2008a; Wada and Temkin 2008). These species support pearl production across a wide area of the Indo-Pacific including the Pacific coast of Mexico. The culture methods used for these species are well established (Southgate 2008; Southgate et al. 2008a) and the methods used for production of their various pearl products are described by Taylor and Strack (2008).

Pinctada fucata/martensii: these oysters are best considered a 'species complex' (Wada and Temkin 2008) and are the most commonly utilized for commercial pearl production. This species complex ranges from the Western Atlantic region (Caribbean region, Gulf of Mexico), Western Pacific Ocean (Korea, Japan, southern China and Australia) to the Indian Ocean, including the Red Sea and Persian Gulf (Gervis and Sims 1992; Wada and Temkin 2008). *Pinctada martensii* (Fig. 5.2) is

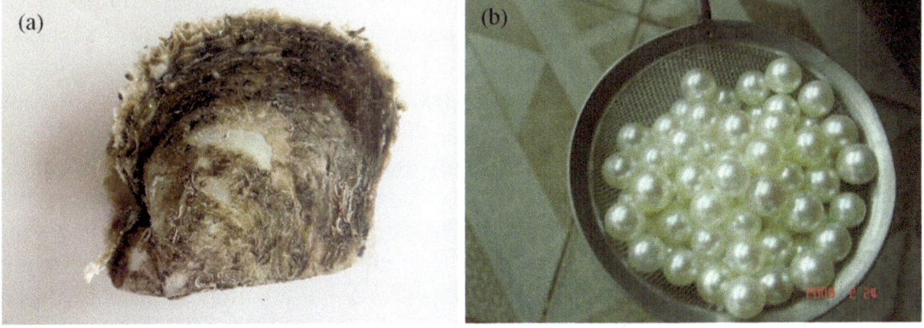

Fig. 5.2 Specimen of *Pinctada martensii* (**a**) and Akoya pearls produced by *P. martensii* (**b**). (Photo: Dahui Yu, South China Sea Fisheries Research Institute)

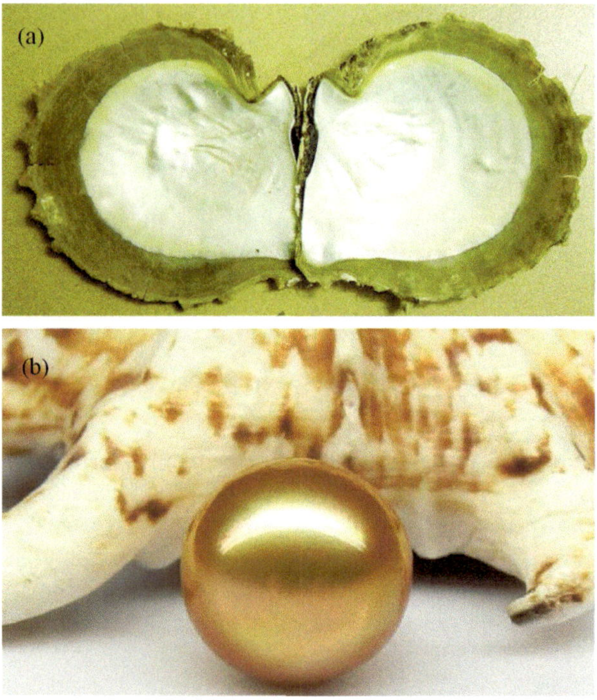

Fig. 5.3 Specimen of *Pinctada maxima* (**a**) and a gold South Sea pearl produced by *P. maxima* (**b**). (Source: Guangdong Ocean University)

found in Japan and China; it is a variety of *Pinctada fucata*, one of the smallest among pearl producing oysters, and is used in both Japan and China for the production of Akoya pearls (Kripa et al. 2007; Southgate et al. 2008a).

Pinctada maxima: the largest pearl oyster species (Fig. 5.3) is used for production of golden and silver South Sea pearls, mainly produced in Indonesia, northern Australia, Philippines, Malaysia and Myanmar (Southgate et al. 2008a). This species produces the largest and most valuable of cultured pearls (Fig. 5.3b).

Pinctada margaritifera: this is the second largest of the pearl oysters that has a broad distribution across the Indo-Pacific, from the eastern Pacific Ocean to the east coast of Africa and the Red Sea (Gervis and Sims 1992; Wada and Temkin 2008). This species is particularly abundant in the atolls of Polynesia where it supports significant production of 'black' or 'Tahitian' pearls in French Polynesia and the Cook Islands. It is also cultured for commercial round pearl production in Fiji where it produces a unique range of colours that is distinct from Polynesian pearls.

Pteria penguin: this species is commonly known as the winged pearl oysters or penguin's wing oyster (Fig. 5.4), is widely distributed in Southeast Asia, particularly China, Japan, Thailand, Indonesia, The Philippines, Malaysia, and Australia. The Japanese name for this species '*mabé gai*' and it is traditionally used for half-pearl or mabé production (Fig. 5.4b). Because of its anatomical structure, this species is difficult to use for round pearl production and reports of successful round pearl production from this species are limited (Liang et al. 2008; Xie et al. 2012).

Fig. 5.4 Specimen of *Pteria penguin* (**a**) and half-pearl or mabé produced by *P. penguin* (**b**). (Source: Guangdong Ocean University)

Fig. 5.5 Cultured round and near-round pearls produced from the Rainbow-lipped pearl oyster, *Pteria sterna*, in Mexico. (Source: Douglas McLaurin, Perlas del Mar de Cortez, Mexico)

Pteria sterna: this species is restricted to the Gulf of California, Mexico (Urban 2000; Mao et al. 2004) and, like *Pteria penguin*, it was initially utilized for half-pearl production (Ruiz-Rubio et al. 2006). However, the high quality of nacre produced by this species prompted research towards commercial production of round and near-round pearls which was successful (Kiefert et al. 2004) and resulted in some of the most colorful cultured pearls (Fig. 5.5).

5.2.2 Freshwater Pearl Mussels

Freshwater pearl mussels can be cultured in freshwater ponds, rivers, or lakes. Over 98% of freshwater pearls are produced in China (FAO 2016), with the remainder produced in Japan, Australia, America, Vietnam, and other countries (Yang et al. 2003). In China, there are more than 100 species of freshwater mussels, but only about ten are used for commercial pearl production. They belong to the families Unionidae and Margaritiferidae, and include *Hyriopsis cumingii* (Triangle sail mussel), *Cristaria plicata*, *Lamprotula leai*, *Lamprotula rochechouarti*, and *Margaritiana dahurica* (Bai et al. 2014). Among these, the most productive is *Hyriopsis cumingii* (Fig. 5.6), followed by *Cristaria plicata*. In China, these two species are readily obtained by pearl farmers, easy to operate, and in relative terms have a higher pearl production rate and produce better quality pearls than other species (Xu et al. 2011). Other than these endemic mussels, several species have been introduced to China for pearl production including *Hyriopsis schlegelii* (native to Japan) which has a strong nacre secretion ability, and *Potamilus alatus* (native to North America) which can produce high quality black freshwater pearls. *Hyriopsis schlegelii and Margaritiana dahurica* are the most commonly used mussels for pearl production in Japan (Alagarswami 1970; Huang 2008).

Fig. 5.6 Specimen of the Triangle sail mussel *Hyriopsis cumingii*. (Source: Hua and Gu 2002)

5.3 Pearl Production

The pearl is unique, since it is the only gem formed inside a living organism. It results from the secretion and deposition of nacre by the epidermal cells of mollusc mantle tissue; the same process that is involved in shell formation. Pearls have the same physical properties and composition of natural shell nacre, with calcium carbonate as the main component and pearl formation had thus been termed biomineralization (Taylor and Strack 2008). Pearl cultivation is based on the natural ability of the mantle tissue of the Pteriidae, Unionidae and Margaritiferidae to secrete nacre, and technical intervention to provide a suitable substrate and environment for nacre secretion. Today's cultured pearls can be divided into three major categories:

1. half-pearls or mabé;
2. beaded 'round' cultured pearls, including most of the marine pearls; and
3. non-beaded freshwater cultured pearls, such as Biwa (Japanese freshwater) pearls and Chinese freshwater pearls.

Half-pearl production involves adhesion of semi-spherical nuclei to the inner shell surface of an oyster. The oyster is then placed back into its culture environment and a period of 10–12 months is generally required for it to adequately cover the nucleus with nacre to form the half-pearl (Ruiz-Rubio et al. 2006; Fig. 5.5b). It is usual for multiple nuclei (usually up to five) to be implanted into one oysters (Ruiz-Rubio et al. 2006; Kishore et al. 2015), and anesthetics are sometime used to relax oysters prior to implantation (Kishore et al. 2015) to minimize oyster stress, and to improve operator access to the inner shell surface for nucleus placement.

5.3.1 *Production Cycle of Pearls*

Production of beaded and non-beaded cultured pearls is more technically demanding than half-pearl production and generally includes five major stages: oyster selection, nucleus implantation, nurturing, harvesting, and pearl processing (Fig. 5.7).

Oyster Selection Selection of suitable host oyster/mussel for pearl production. There are two sources of mollusc stock for pearl production: (1) collection from the wild, such as in Australia and French Polynesia; oysters are collected as adults or as juveniles and grown to a size suitable for pearl production (Southgate 2008); and (2) produce seed/spat (juveniles) through artificial propagation in a dedicated hatchery facility or from 'spat collection' programs. The latter relies on deployment of appropriate substrates to the water column, at an appropriate time, to provide substrates

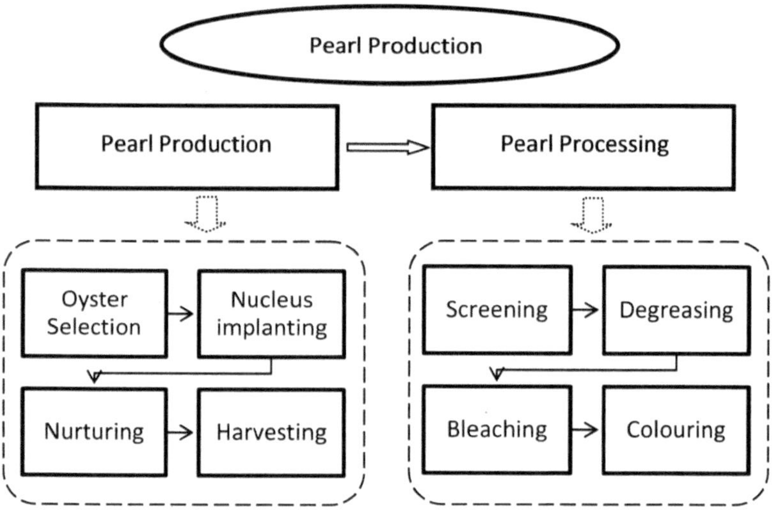

Fig. 5.7 Diagrammatic representation of the stages of pearl production

Fig. 5.8 A pearl oyster farm in South China. (Source: Guangdong Ocean University)

for larval recruitment (Southgate 2008). Juveniles are then grown to a size suitable for pearl production. At least a year and a half is needed for pearl oyster larvae to grow to a size appropriate for pearl production (Yin et al. 2012). Freshwater mussels are much faster growing than pearl oysters and become suitable for pearl production within a year (Huang 2008). The aquatic environment for farmed molluscs must be clean, have a suitable water temperature, and be free of harmful organisms with minimal fouling and predation (Fig. 5.8).

Nucleus Implantation This is the key step in cultured pearl production. In order to grow marine pearls, a tiny piece of mantle tissue (called a graft, or 'saibo' in Japanese), approximately 3 × 3 mm in size, is removed from a suitable donor oyster and implanted with a spherical polished shell-bead or nucleus into the gonad of a recipient or host oyster. For freshwater pearls to grow, a piece of mantle graft alone serves the same purpose, so a nucleus is not a pre-requisite for pearl production. There are still some technical difficulties associated with growing beaded pearls within the visceral mass of freshwater mussels because of their physiological structure (Xie et al. 2015). A period of 'conditioning' or pre-operative treatment is often needed to prepare oysters/mussels for implantation, and appropriate post-operative husbandry reduces stress and helps maximise nucleus/graft tissue retention after implantation (Taylor and Strack 2008; Liang et al. 2016). Survival rate and nucleus retention rate of implanted oysters are strongly correlated with factors such as size and age of oysters, size of nucleus and grafting method (Yukihira and Klumpp 2006; Kripa et al. 2007; Liang et al. 2015).

Nurturing After the nucleus is inserted, implanted oysters/mussels need to be carefully nurtured in a resting zone for at least 2 weeks, a critical period for mortality and nucleus rejection, then returned to the ocean in an area of calm water at a depth of 2–3 m (Wang et al. 1993). Appropriate water temperature is critical for survival of implanted oysters and optimal nacre secretion rate in *P. maxima* occurs at 25–30 °C, when nacre is first secreted onto the nucleus from around 45 days after operation (Liu et al. 2012). In *P. margaritifera*, graft tissue proliferates to create a 'pearl-sac' that completely covers the nucleus within 14 days of grafting, when the epithelial cells responsible for nacre secretion are fully developed; however, first nacre secretion onto the nucleus was not observed until 32 days after grafting (Kishore and Southgate 2016). Nucleated oysters are generally cultured for a further 1–2 years before resulting marine pearls are harvested. A culture period of 1–5 years is usually required for freshwater pearl production depending on culture method and species (Xu et al. 2011; Yin et al. 2012; Lin et al. 2016a).

Harvesting in winter or when water temperature is relatively low, the nacre secretion rate slows, resulting in a more detailed, smooth, and lustrous pearl surface. Thus colder conditions are the best time to harvest pearls (Wang et al. 1993). Akoya and South Sea pearls are grown within the gonad tissue of host oysters (Taylor and Strack 2008). They are grown one pearl at a time which limits the number of pearls at harvest. Oysters that produce high quality South Sea pearls are often implanted with a new, larger bead, then returned to the water for another 2–3 years of growth for the next pearl producing cycle (Taylor and Strack 2008; Lin et al. 2016a). It is possible using this method to produce up to four pearls from a single oyster and Kishore et al. (2015) reported improvements in both pearl size and shape in 'second-graft' pearls produced by *P. margaritifera*. Freshwater pearls are grown in the mantle, where up to 20 grafts may be implanted within each of the two mantle lobes. On this basis, freshwater mussels have a substantially higher pearl yield than marine oysters with usually more than 10 pearls harvested from one mussel (Lin et al. 2016b).

Fig. 5.9 Akoya pearls before (**a**) and after (**b**) bleaching procedure. (Source: Guangdong Ocean University)

Pearl Processing Due to variations in colour and the degree of surface defects, more than 90% of cultured pearls cannot be used directly to produce jewelry or other products (Huang 2008). However, raw pearls may have to be processed to improve their quality to meet the standards of gem-quality merchandise, and pearl enhancement is routinely used for Akoya pearls and freshwater pearls (Strack 2006). Pearl processing techniques may include screening, degreasing, decontamination, bleaching, whitening, colouring etc. (Tang et al. 2016). Pearl appearance and value can be greatly improved by these technical procedures, which enhance colour and surface texture (Fig. 5.9). While fine-quality cultured pearls (marine and freshwater) are selected to make jewelry, small non-beaded cultured pearls, which have little value, may be processed into drugs and cosmetics (Yang et al. 2016).

South Sea pearls are generally not treated in their countries of origin and are promoted as having minimal enhancement consisting of washing and polishing only (Taylor and Strack 2008); however, South Sea pearls are treated by a number of Japanese pearl companies (Strack 2006). As well as colour, pearl luster can be enhanced by mechanical polishing and through the use of solvents and polishing materials such as bees wax.

5.3.2 *Output and Value*

In the past decade, Japan, French Polynesia and China have been the three major marine pearl producing countries, but over 98% of pearls produced worldwide are freshwater pearls from China (Fig. 5.10). Annual output of Chinese pearls averaged 3540 t of which freshwater pearls accounted for 99.5%. Since 2007, China's marine pearl production declined significantly, from 34.5 t in 2006 to 3.7 t in 2014 (Fig. 5.11). Marine pearl output also decreased in Japan from 29 t to 20 t per year over the same period, but increased in French Polynesia from 9 t per year to 15 t per year (Fig. 5.11). The United States, Japan, Switzerland, Germany, Hong Kong,

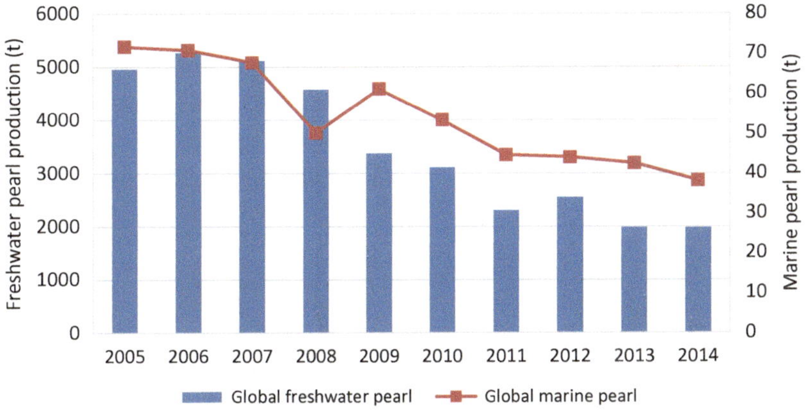

Fig. 5.10 Global pearl production (2005–2014) (All data from FAO 2016), data includes China, Japan and French Polynesia, the three major global pearl producers. Data for other countries, which only accounted for a small part of total pearl output, were not in the database or were incomplete, and were therefore not included. Half-pearl or mother of pearl production was also not included, likewise in Figs. 5.11 and 5.12)

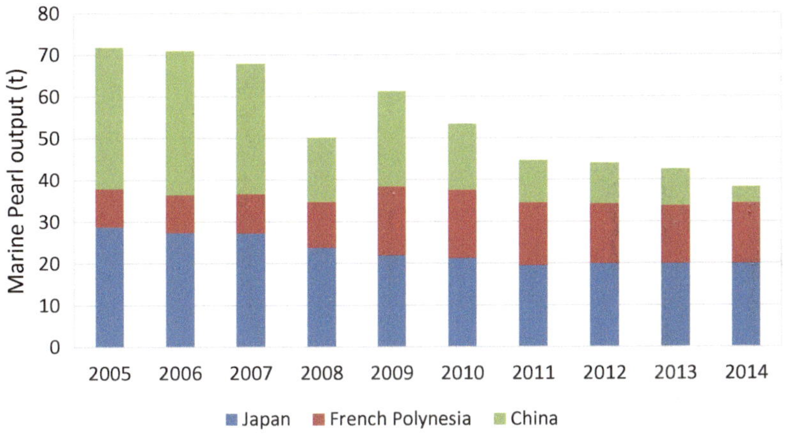

Fig. 5.11 Global production of marine pearls (2005–2014)

France, Britain, Belgium, Italy, Spain, Canada, India and Saudi Arabia are the world's largest pearl importers, ranked by volume (Lin 2004).

Japan was not only one of the largest pearl producers and exporters, but also the largest importer and processing centre of pearls, playing a significant role as a distribution hub in the global pearl industry. In addition, Japan controls the world's leading technology in pearl oyster breeding and pearl production and processing (Fassler 1992) and held about 51.6% of the world's output value in the past decade. The average annual value of pearl production was 127 million USD in Japan, 104 million USD in French Polynesia, and 15 million USD in China (Fig. 5.12).

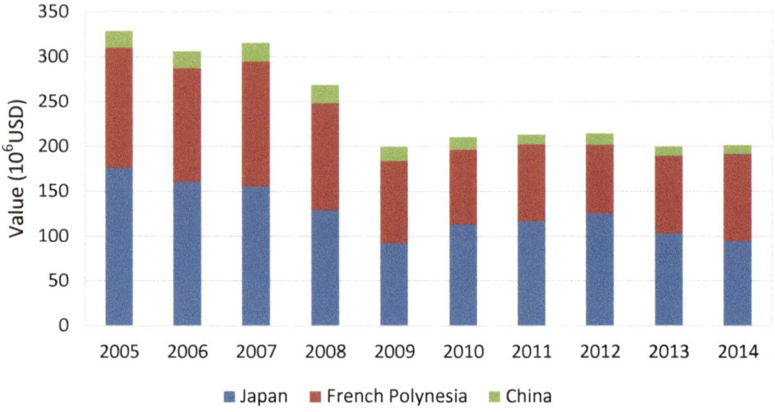

Fig. 5.12 Total output value of global pearl production (2005–2014)

Although China is the largest producer and exporter of pearls, because the market price of freshwater pearls is much lower than that of marine pearls, over 98% of the world's pearl production only corresponds to about 5% of the output value (Fig. 5.12). The extremely unbalanced development between Chinese low-end pearl production and high-quality pearl production has resulted in a high-yield, low-value industrial structure. Overall, global pearl production fell by 60% (Fig. 5.10), while output value fell by 39% over the past decade (Fig. 5.12).

5.4 Goods from Pearls

5.4.1 Types and Value

Based on their method of formation, pearls can be divided into natural (or wild) pearls and cultured pearls. The value of a pearl is determined by a combination of a number of characteristics including size, luster, color, amount of surface flaws, and shape. Pearls are usually categorized into eight basic shapes: round, semi-round, button, drop, pear, oval, baroque, circled and double-boulder (Pan et al. 1994).

Natural pearls form within oysters and mussels when nacre-secreting epithelial cells are transferred into the viscera by 'accidental' means, and their continued secretion of nacre forms a pearl over time (Taylor and Strack 2008). Transfer of epithelial cells may result from the actions of predators or parasites, for example, or from foreign materials that become lodged within oyster tissues. Formed without human intervention, natural pearls take different shapes, and perfectly spherical natural pearls are extremely rare, and highly valuable. Hundreds of pearl oysters or mussels must be gathered and opened to find even one natural pearl; yet for many centuries, this was the only way to obtain pearls, and the reason pearls were so highly regarded in the past. Natural pearls can be distinguished from cultured pearls by X-ray that will reveal no nucleus or curved cavity structures in the centre of the pearl, and a uniform, onion-like structure (Krzemnicki et al. 2010).

Cultured pearls result from a tissue implant and human intervention that utilizes the ability of mollusc tissue to produce and secrete nacre. Marine cultured pearls generally follow the shape of the implanted nucleus resulting in round or near-round pearls. The output of 'round' marine pearls is relatively low, about 54 t per year globally (Fig. 5.11). But marine pearls are much more valuable than freshwater pearls, especially high quality South Sea pearls from *Pinctada maxima* (gold or silver, 10–20 mm in diameter) and Tahitian 'black' pearls (9–20 mm in diameter) (Southgate 2007). They are famous for their unique color and luster, and are the largest, rarest, and most valuable cultured pearls in the pearl market. Freshwater cultured pearls are rarely round, mostly pear-shaped or oval, and the overall quality is poor. Generally, only 1–2 pearls meeting gem-quality standards can be obtained from 100 raw freshwater pearls (Yang et al. 2003). Common natural colors of freshwater pearls are white, pink and purple, and some progress has been made towards cultivating high-quality round freshwater pearls in China (Xie 2010; Xie et al. 2015; Lin et al. 2016b).

5.4.2 Services

Decoration Pearls are the most versatile of gemstones, with three major functions that have been developed over thousands of years. Pearls are used for decoration like all the other gems, and infer a sense of status and material wealth. Pearls with a special luster may become beautiful ornaments and have been used to decorate items such as the crowns of monarchs as symbols of elegance and nobility. Pearls and MOP shells may function as collector's items as one of their services to humans; this aspect is addressed in Duncan and Ghys (2019).

Medical and Biomedical Applications Pearls are used to produce medicine. The history of pearl medicine in China goes back more than 2000 years (Pan et al. 1994). Pearls are a product of the defense mechanism of organic immune systems, and studies of their medicinal value have shown distinct anti-oxidant and anti-inflammatory effects. Extracts from pearls have been used in variety of clinical treatments for ulceration, cataracts, and tumours (Lin 2004; Zheng and Mao 2004).

The process of nacre formation or biomineralization progresses from secretion of a fluid, through film formation and mineralization, to formation of the mature nacre structure composed of sequential layers of aragonite tablets (Fougerouse et al. 2008). Improved understanding of this process, and the unique qualities of nacre, have stimulated considerable interest in the potential biomedical applications of nacre, including its possible use as a substitute for human bone and in bone repair (Southgate et al. 2008b). For example, pearl oyster nacre has been shown to induce mineralization by human osteoblasts (Lopez et al. 1992), to be cyto-compatible with human bone (Cognet et al. 2003), and to stimulate bone repair (Lamghari et al. 1999) and form a dual biomineralized unit (with bone) in sheep (Lopez et al. 2004). Such research has potential for significant biomedical outcomes.

Body Care Pearls are also used in many body care therapies because they are rich in elements that are beneficial to the human body, particularly the skin. Pearls consist of calcium, over 20 different trace elements, more than 15 amino acids, alkaline phosphates, and natural taurine (Huang 2008; Zhang et al. 2014), and meet important health requirements. Pearls may be processed into powdered products for skin whitening, and as calcium supplements (Yang et al. 2016).

Bioremediation Because of their high filtration rates and their ability to accumulate heavy metals, pollutants (including nutrients) and bacteria, pearl oysters and mussels have considerable potential for bioremediation of polluted coastal environment (O'Connor and Gifford 2008). Marine pearl oysters are particularly well suited to this role because their pumping and filtration rates are among the highest reported for bivalve molluscs, and up to ~22 L per hour per oyster (Lucas 2008). They are able to process large quantities of water, removing particulates and being exposed to large quantities of pollutants. They also have a high requirement for nitrogen and phosphorous and an ability to remove large quantities of these nutrients from the water column (Gifford et al. 2005). Pearl oyster based 'bioremediation' also has advantages because oysters used in this way can still produce valuable products (e.g. pearls and MOP), but they do not need to be a product suitable for human consumption (O'Connor and Gifford 2008). The potential for such bioremediation systems was demonstrated by Gifford et al. (2005) who reported that for each tonne of Akoya pearl oysters harvested, 7.4 kg of nitrogen, 0.5 kg of phosphorous and up to 0.7 kg of metals were removed from the water.

Like other bivalves, pearl oysters/mussels have the ability to improve water quality by transforming suspended particulate matter (including microalgae) into faeces and pseudofaeces through biodeposition, which is a very important component of the biogeochemical processes of coastal ecosystems (Ferreira and Bricker 2016). Biodeposition by shellfish is addressed in detail in other chapters of this book.

5.5 Problems and Perspectives

Pearl farming is a very challenging and labour-intensive activity. In general, post-operative survival of nucleated oysters is less than 70% and, of these, 30–40% are likely to reject the implanted nucleus, 20% will produce salable pearls, but only 5% will produce top quality gemstones (Fassler 1992; Norton et al. 1996). For pearl production from *P. margaritifera* for example, it is generally accepted that 5% of the total pearl harvest will generate around 95% of farm profits (Haws 2002). Compared with other aquaculture sectors, pearl production has a more complex procedure and a longer farming cycle which increases economic risk. Urbanization and industrialization in traditional pearl farming areas, and stressed and impoverished coastal environments have led to some serious problems and big challenges for sustainable development of the pearl aquaculture industry. These include reduced oysters supply to the industry, reduced growth rates and survival and reduced pearl yield and quality. Increased reliance on hatchery-produced juveniles brings its own potential problems relating to artificial selection and inbreeding. Striving for increased production has resulted in over-stocking of pearl farms leading to a shortage of nutrients and ecological deterioration at the farm site that increases the probability of epidemic disease outbreaks. Production pressures may also encourage shortening of the pearl production cycle resulting in pearls with a thin nacre covering that do not pass the product inspection standard. Finally, consistent production of high quality pearls relies on the availability and skills of professional pearl grafting technicians (Porter 1991; Fassler 1992). Expansion of pearl farming in some areas has led to a shortage of well-trained technicians and the use of inexperienced technicians or grafting by under-trained farmers. All the above issues have the potential to affect pearl production and quality, as well as farm profitability, and may explain the decline of pearl production in the past decade, especially in China.

If we consider the present situation of China's pearl industry as an example, Liusha Bay was the main production base of Chinese marine pearls (Zhu et al. 2011) and, in 2010, accounted for over 80% of national production. Chinese Akoya pearls, known as 'Nanzhu' in China, referred specifically to pearls produced from *Pinctada martensii* in the Beibu Gulf area. Liusha Bay is located at the junction of Xuwen County and the southwestern of Leizhou City sea area (20°22′–20°31′N, 109°55′E–110°1′E), in the southwestern part of Leizhou Peninsula in Guangdong province. Liusha Bay is a semi-enclosed system with a total area of 69 km^2, an annual average water temperature of 26.4 °C, and its natural geographical condition is particularly suited for *Pinctada* oyster culture. It was historically the production center of Akoya pearls in China, with a farming area of around 20 km^2. However, since 2000, the industry in Liusha has plunged into serious recession. In the late 1990s, a number of alternative aquaculture commodities and cultivation techniques were introduced into Liusha Bay, including cage finfish culture and scallop (*Argopecten irradians concentricus*) culture. Because of the relatively long farming cycle of pearl culture, a complicated situation of multiple aquaculture structures developed in Liusha Bay. In 2007–2008, a series of natural disasters made the situation even worse for pearl production in Liusha and a large number of traditional pearl farmers diverted into cage fish culture, resulting in a dramatic increase in the number of cages in Liusha Bay. As a result, the culture space for *Pinctada martensii* was significantly reduced, and potential food resources for pearl oysters were largely consumed by cultured scallops. Furthermore, the sediment environment worsened, pearl oyster growth slowed and survival decreased (Luo et al. 2014). The Nanzhu cultured pearl industry currently faces a major threat.

The pearl industry of the future will continue to face the dilemma of productivity and reduced profitability, unless radical remedial measures are taken to improve the culture environment and standards. Sustainable development of the pearl culture industry requires management measures that are guided by scientific development relating to breeding and husbandry of pearl oysters and mussels, as well as product processing, and marketing. In addition, appropriate management must also consider social, economic, and environmental factors. Pearls should continue to shine in the modern commodity market, and continue to decorate human civilization.

Acknowledgements The authors acknowledge the support from the Guangdong Pearl Industry Development Project (Z2015013). We are grateful to Dr. Stein Mortensen for his valuable comments on the manuscript. The authors additionally thank Mr. Dahui Yu, Mr. Douglas McLaurin, Mr. Perlas del Mar de Cortez and Guangdong Ocean University for providing photos of pearls and pearl oysters for this review.

References

Abbott RT (1972) Kingdom of the seashell. Hamlyn Publishing Group Ltd, London, 256 pp

Alagarswami K (1970) Pearl culture in Japan and its lessons for India. Mar Biol Assoc India 3:975–993

Alagarswami K (1987) Perspectives in pearl culture. Contributions in Marine Sciences, felicitation volume:37–49

Bai ZY, Wang G, Liu XJ, Li JL (2014) The status and development trend of freshwater pearl seed industry in China. J Shanghai Ocean Univ 6:874–881 (in Chinese)

Carter R (2005) The history and prehistory of pearling in the Persian Gulf. J Econ Soc Hist Orient 48(2):139–209

Cognet JM, Fricain JC, Reau AF, Lavignolle B, Baquey C, Lepeticorps Y (2003) *Pinctada margaritifera* nacre (mother-of-pearl): physico-chemical and biomechanical properties, and *in vitro* cytocompatibility. Revue de Chirurgie Orthopedique et Reparatrice de L'Appareil Moteur 89:346–352

Cranford PJ (2019) Magnitude and extent of water clarification services provided by bivalve suspension feeding. In: Smaal A, Ferreira JG, Grant J, Petersen JK, Strand O (eds) Goods and services of marine bivalves. Springer, Cham, pp 119–141

Duncan P, Ghys A (2019) Shells as collectors items. In: Smaal A, Ferreira JG, Grant J, Petersen JK, Strand O (eds) Goods and services of marine bivalves. Springer, Cham, pp 381–411

FAO (2016) Global aquaculture production online dataset. Food and Agriculture Organization of the United Nations, Rome http://www.fao.org/fishery/statistics/global-aquaculture-production/en

Fassler RC (1992) Farming jewels: the aquaculture of pearls, new opportunities arise as pearl culture spreads beyond Japan. SPC Pearl Oyster Inf Bull 5:29–31

Ferreira JG, Bricker SB (2016) Goods and services of extensive aquaculture: shellfish culture and nutrient trading. Aquac Int 24(3):803–825

Fougerouse A, Rousseau M, Lucas JS (2008) Soft tissue anatomy, shell structure and biomineralization. In: Southgate PC, Lucas JS (eds) The pearl oyster. Elsevier Press, Oxford, pp 77–102

Gervis MH, Sims NA (1992) The biology and culture of pearl oysters (Bivalvia: Pteriidae). ICARM Stud Rev 21:22–41

Gifford S, Dunstan H, O'Connor W, Macfarlane GR (2005) Quantification of *in situ* nutrient and heavy metal remediation by a small pearl oyster farm at Port Stephen, Australia. Mar Pollut Bull 50:417–422

Haws M (2002) The basic methods of pearl farming: a layman's manual. Center for Tropical and Subtropical Aquaculture, Hilo, pp 5–13

Hua D, Gu R (2002) Freshwater pearl culture and production in China. Aquacult Asia 7(1):1–8

Huang Y (2008) Dan Shui Zhen Zhu. Zhejiang Science and Technology, HangZhou, pp 26–28 (in Chinese)

Jansen Henrice M, Strand Ø, van Broekhoven W, Strohmeier T, Verdegem MC, Smaal AC (2019) Feedbacks from filter feeders: review on the role of mussels in cycling and storage of nutrients in oligo- meso- and eutrophic cultivation areas. In: Smaal A, Ferreira JG, Grant J, Petersen JK, Strand O (eds) Goods and services of marine bivalves. Springer, Cham, pp 143–177

Kiefert L, McLaurin-Moreno D, Arizmendi E, Hanni HA, Elen S (2004) Cultured pearls from the Gulf of California, Mexico. Gems Gemmol 40:26–38

Kishore P, Southgate PC (2016) A detailed description of pearl-sac development in the black-lip pearl oyster, *Pinctada margaritifera* (Linnaeus 1758). Aquac Res 47(7):2215–2226

Kishore P, Southgate PC, Seeto J, Hunter J (2015) Factors influencing the quality of half-pearls (mabé) produced by the winged pearl oyster, *Pteria penguin* (Röding, 1758). Aquac Res 46(4):769–776

Kripa V, Mohamed KS, Appukuttan KK, Velayudhan TS (2007) Production of Akoya pearls from the Southwest coast of India. Aquaculture 262(2–4):347–354

Krzemnicki MS, Friess D, Chalus P, Hänni HA, Karampelas S (2010) X-ray computed microtomography: distinguishing natural pearls from beaded and non-beaded cultured pearls. Gems Gemol 46(2):128–134

Lamghari M, Almeida MJ, Berland S, Huet H, Laurent A, Milet C, Lopez E (1999) Stimulation of bone marrow cells and bone formation by nacre: *in vivo* and *in vitro* studies. Bone 25:91S–94S

Liang FL, Deng CM, Fu S, Liu Y (2008) Preliminary study on techniques of round pearl production with *Pteria penguin*. Mar Sci Bull 27(2):91–96 (in Chinese)

Liang FL, Wang QG, Deng YW, Xie SH (2015) Effects of nucleus size, nucleus implantation location and age on round nucleated pearl production of *Pinctada maxima*. J Guangdong Ocean Univ 35(4):51–55 (in Chinese)

Liang FL, Wang QG, Deng YW, Xie SH (2016) Pre-operative condition and nucleus size on round nucleated pearl production of *Pteria penguin*. J Aquac 37(1):30–34 (in Chinese)

Lin Z (2004) Zhen Zhu Sheng Chan Ji Shu. China Agriculture Press, Beijin, 171 pp. (in Chinese)

Lin WC, Xie SH, Du XD, Wang QG, Fu S (2016a) Factors affecting pearl production performance of *pinctada maxima*. Chin Agricult Sci Bull 32:29–33 (in Chinese)

Lin WC, Wang QG, Xie SH, Deng YW, Liang FL (2016b) Study on pearl cultivation techniques of nucleated pearl from the mantle of *Hyriopsis cumingii*. Trans Oceanol Limnol 5:100–103

Liu Y, Zhang CF, Jiao ZY, Deng CM (2012) Culturing round pearl with *Pinctada maxima* (Jameson). Mar Sci 36(4):30–36 (in Chinese)

Lopez E, Vidal B, Berland S, Camprasse S, Camprasse G, Silve C (1992) Demonstration of the capacity of nacre to induce bone formation by human osteoblasts maintained *in vitro*. Tissue Cell 24:667–679

Lopez E, Milet C, Lamghari M, Mouries LP, Borzeix S, Berland S (2004) The dualism of nacre. Bioceramics 16:733–736

Lucas JS (2008) Feeding and metabolism. In: Southgate PC, Lucas JS (eds) The pearl oyster. Elsevier Press, Oxford, pp 103–130

Luo Z, Zhu C, Guo Y, Su L, Li J, Ou Y (2014) Distribution characteristics of C, N and P in Liusha Bay surface sediment and their pollution assessment. South China Fish Sci 10(3):1–8 (in Chinese)

Mao Y, Liang FL, Fu S, Yu XY, Ye FL, Deng CM (2004) Preliminary studies on rainbow-pearl of penguin wing oyster *Pteria penguin*. Chin J Zool 39(1):100–102 (in Chinese)

Matlins AL (2001) The pearl book: the definitive buying guide – how to select, buy, care for and enjoy pearls. Gemstone Press, 3–19 pp

McElroy S (1990) The Japanese pearl market. Infofish Int 90(6):17–23

McGladdery SE (2007) Why the interest in pearl oyster health? In: Bondad-Reantaso MG, McGladdery SE, Berthe FCJ (eds) Pearl oyster health management: a manual. FAO fisheries technical paper. No 503, 3–6 pp

Mizumoto S (1979) Pearl farming in Japan. In: Pillay TVR, Dill WA (eds) Advances in aquaculture. Fishing News Books Ltd, Farnham, pp 381–385

Norton JH, Dashorst M, Lansky TM, Mayer RJ (1996) An evaluation of some relaxants for use with pearl oysters. Aquaculture 144:39–52
O'Connor WA, Gifford SP (2008) Environmental impacts of pearl farming. In: Southgate PC, Lucas JS (eds) The pearl oyster. Elsevier Press, Oxford, pp 497–525
Pan BY, Huang WG, Wen ZF (1994) Zhen Zhu Shi Yong Xin Ji Shu. China Agriculture Science and Technique Press, Beijin, pp 14–37 (in Chinese)
Petersen JK, Holmer M, Termansen M, Hasler B (2019) Nutrient extraction through bivalves. In: Smaal A, Ferreira JG, Grant J, Petersen JK, Strand O (eds) Goods and services of marine bivalves. Springer, Cham, pp 179–208
Porter B (1991) The black-pearl connection. Connoisseur 4:9
Ruiz-Rubio H, Acosta-Salmon H, Olivera A, Southgate PC, Rangel-Dávalos C (2006) The influence of culture method and culture period on quality of half-pearls ('mabe') from the winged pearl oyster *Pteria sterna*, Gould, 1851. Aquaculture 254(1–4):269–274
Southgate PC (2007) Overview of the cultured marine pearl industry. In: Bondad-Reantaso MG, McGladdery SE, Berthe FCJ (eds) Pearl oyster health management: a manual. FAO fisheries technical paper. 503:7–17
Southgate PC (2008) Pearl oyster culture. In: Southgate PC, Lucas JS (eds) The pearl oyster. Elsevier Press, Oxford, pp 231–268
Southgate PC, Strack E, Hart A, Wada KT, Monteforte M, Carino M, Langy S, Lo C, Acosta-Salmon H, Wang A (2008a) Exploitation and culture of major commercial species. In: Southgate PC, Lucas JS (eds) The pearl oyster. Elsevier Press, Oxford, pp 303–355
Southgate PC, Lucas JS, Torrey RD (2008b) Future developments. In: Southgate PC, Lucas JS (eds) The pearl oyster. Elsevier Press, Oxford, pp 555–565
Strack E (2006) Pearls. Ruhle-Diebener-Verlag GmbH & Co. KG, Stuttgart, 707 pp
Strack E (2008) Introduction. In: Southgate PC, Lucas JS (eds) The pearl oyster. Elsevier Press, Oxford, pp 1–35
Tang HN, Yang L, Wan TT, Cai Y, Li SD, Wu H (2016) Progress of pearl processing technology, vol 3. Shandong Chemical Industry, Shandong, pp 43–46 (in Chinese)
Taylor J, Strack E (2008) Pearl production. In: Southgate PC, Lucas JS (eds) The pearl oyster. Elsevier Press, Oxford, pp 273–302
Urban HJ (2000) Culture potential of the pearl oyster *(Pinctada imbricata)* from the Caribbean. I. Gametogenic activity, growth, mortality and production of a natural population. Aquaculture 189(3):361–373
Wada KT, Temkin I (2008) Taxonomy and phylogeny. In: Southgate PC, Lucas JS (eds) The pearl oyster. Elsevier Press, Oxford, pp 37–76
Wang RC, Wang ZP, Zhang JZ (1993) Hai Shui Bei Lei Yang Zhi Xue. Ocean University of China, Qingdao, p 205, 259 pp. (in Chinese)
Wang A, Shi Y, Wang Y, Gu Z (2007) Present status and prospect of Chinese pearl oyster culturing. Aquaculture 2007. World Aquaculture Society, San Antonio, 643 pp
Xie SH (2010) Large-scale cultivation techniques of nucleated freshwater pearl in China. J Guangdong Ocean Univ 30(1):55–58 (in Chinese)
Xie YK, Min ZY (2003) Advances in pearl research in China. J Putian Univ 3:34–38 (in Chinese)
Xie SH, Deng CM, Liang FL, Fu S, Lin WC (2012) Experiment on pearl production of pearl oyster *Pteria penguin*. J Guangdong Ocean Univ 32(6):33–38 (in Chinese)
Xie XZ, Shi ZY, Wang YS (2015) Culture of large nucleated pearl in visceral mass of freshwater pearl mussel *Hyriopsis cumingii* (Lea). Chin J Fish 28(1):45–48 (in Chinese)
Xu QQ, Guo LG, Xie J, Zhao CY (2011) Relationship between quality of pearl cultured in the triangle mussel *Hyriopsis cumingii* of different ages and its immune mechanism. Aquaculture 315:196–200
Yang PH, Li MJ, Wu WX, Xie CH, Wang XY (2003) Industry of freshwater pearl problem, development that face trend and countermeasure in China. J Hunan Univ Arts Sci (Nat Sci Ed) 15(4):33–34 (in Chinese)
Yang AQ, Shen YQ, Zhang LH, Mo JH, Chen ZX, Wang J (2016) Study on the Skin Care Functions of Active Ingredients in Pearl Extractive 1:58–61 (in Chinese)

Yin LP, Deng YW, Du XD, Wang QH (2012) Effects of age on growth, survival and pearl production and traits in pearl oyster *Pinctada martensii*. J Fish Sci China 19(4):715–720 (in Chinese)

Yukihira H, Klumpp DW (2006) The pearl oysters, *Pinctada maxima* and *Pinctada. margaritifera*, respond in different ways to culture in dissimilar environments. Aquaculture 252(2–4):208–224

Zhang GF, Fang AP (2003) Sustainable development of Chinese freshwater pearl industry in 21st century. J Jinhua Polytechnic 3(4):24–29 (in Chinese)

Zhang E, Huang FQ, Wang ZT, Li Q (2014) Characteristics of trace elements in freshwater and seawater cultured pearls. Spectrosc Spectr Anal 34(9):2544–2547 (In Chinese)

Zheng QY, Mao YM (2004) Comparison of component, action and effects between freshwater and seawater pearl. Shanghai J Tradit Chin Med 38(3):54–55 (in Chinese)

Zhu CH, Shen YC, Xie EY, Ye N, Wang Y, Du XD, Wu ZH (2011) Aquaculture carrying capacity of *Pinctada martensii* in Liusha Bay of Zhanjiang. J Trop Oceanogr 3:76–81 (in Chinese)

Chapter 6
Biotechnologies from Marine Bivalves

Paola Venier, Marco Gerdol, Stefania Domeneghetti, Nidhi Sharma, Alberto Pallavicini, and Umberto Rosani

Abstract Bivalve molluscs comprise more than 9000 extant species. A number of them are traditionally farmed worldwide and are fundamental in the functioning of benthic ecosystems. The peculiarities of marine bivalves have inspired versatile biotechnological tools for coastal pollution monitoring and several new biomimetic materials. Moreover, large amounts of sequence data available for some farmed bivalve species can be used to unveil the organism's responses to environmental factors (e.g. global climate change, emergence of new infectious agents and other production problems). In bivalves, data from genomics and transcriptomics increases more quickly than data from other omics, and permit new bioinformatics inferences, real comparative genomics and the study of molecules suitable for biotechnological innovations. Bivalves (and their microorganism communities) produce a variety of bioactive peptides, proteins and metabolites. Among them, the numerous families of antimicrobial peptides identified in the Mediterranean mussel likely contribute to its vigour and could assist with the identification of molecular scaffolds for innovative pharmaceuticals, nutraceuticals and constructs suitable for other applications.

海水双壳贝类相关的生物技术 双壳类软体动物由9, 000多种现存物种组成。其中一些全球分布物种有着比较悠久的养殖历史，并且是底栖生态系统的基础物种。海水双壳贝类的生长及生理学特性为海岸污染监测和创新仿生物材料研发提供了多种多样的生物技术工具。受到海水双壳贝类生物学特性的启发，研究人员研发了一些用于沿海污染监测的通用生物技术工具及数种新仿生材料。此外，大量的养殖双壳贝类的测序数据可以用来揭示生物体对环境因素变化的响应(如全球气候变化，新型传染病和其他养殖问题)。双壳贝类

P. Venier (✉) · S. Domeneghetti · U. Rosani
Department of Biology, University of Padova, Padova, Italy
e-mail: paola.venier@unipd.it; stefania.domeneghetti@unipd.it; umberto.rosani@unipd.it

M. Gerdol · A. Pallavicini
Department of Life Sciences, University of Trieste, Trieste, Italy
e-mail: mgerdol@units.it; pallavic@univ.trieste.it

N. Sharma
Regional Centre for Biotechnology, NCR Biotech Science Cluster, Faridabad, Haryana (NCR Delhi), India

A. C. Smaal et al. (eds.), *Goods and Services of Marine Bivalves*,
https://doi.org/10.1007/978-3-319-96776-9_6

的转录组学和基因组资源比其它组学数据的增长要快得多，从而使针对这一动物类群的新的生物信息学预测、真正的比较基因组学和适用于生物技术创新的分子研究成为可行。双壳贝类(及其微生物群落)会产生多种生物活性肽、蛋白质和代谢物。其中，在地中海贻贝中鉴别出多种可能有助于增强贻贝活力的抗菌肽家族成员，为开发新型药物、营养制品等提供了分子骨架模 板.

Keywords Marine bivalve molluscs · Biotechnology · *Mytilus* · *Crassostrea* · *Ruditapes* · DNA microarray · High-throughput sequencing · Byssus · Biomimetic · Antimicrobial

关键词 软体双壳贝类 • 生物技术，贻贝 • 牡蛎；蛤 • DNA微阵列 • 高通量测序 • 足丝 • 仿生 • 抗菌剂

6.1 Introduction

Technologies based on the peculiarities of marine bivalves not only provide services and products of current use but are expected to grow in the future, owing to the great exploration power of current omics strategies (high-throughput production of different sorts of molecular data aimed at the complete interpretation of biological structures, functions, and dynamics) and to the surprising advances of life sciences, material and nanomaterial sciences and microelectronics engineering. Undeniably, the growing number of bivalve-inspired innovations add value to animal species already identified as fundamental components of marine benthic ecosystems and regarded as a strategic food resource for the future (the European aquaculture production of marine molluscs reached 572,957 tons, nearly 3.5% of the global amount, with an estimated value of 972,987 USD in 2016) (FAO 2018).

6.2 Living Monitors and Source of Versatile Biotechnological Tools

Since the mid '70s, filter-feeding bivalves such as mussels and clams started to be used as pollution sentinels because they integrate in space and time the contaminant mixtures present in the surrounding water and sediments, respectively (Goldberg and Bertine 2000). Complementary to the analysis of toxicants in the soft tissues (Guéguen et al. 2011; Melwani et al. 2014), various pollution biomarkers have been developed and a number of them has been validated (Moore et al. 2006; Banni et al. 2007; Bolognesi and Hayashi 2011) and combined (Pytharopoulou et al. 2008; Okay et al. 2016) to rank coastal sites according to the intensity of toxicant-induced adverse effects.

Table 6.1 Gene expression datasets and DNA microarray platforms available for selected marine bivalves

	Datasets	Microarrays
Crassostrea gigas	833	20[a,b]
Crassostrea virginica	668	3[b]
Mytilus galloprovincialis	480	20[a]
Ruditapes philippinarum	340	10[a]
Mytilus californianus	196	5[a]
Mytilus edulis	163	5[a]
Ruditapes decussatus	141	7
Mytilus trossulus	122	2[a]
Pinctada maxima	89	4
Pinctada fucata	34	3
Mercenaria mercenaria	32	1
Chamelea gallina	32	1
Pinctada martensii	22	2

From Gene Expression Omnibus at Aug 2018 (www.ncbi.nlm.nih.gov)
[a]GPL22172 probes from *Crassostrea angulata, Crassostrea ariakensis, C. gigas, C. virginica, M. californianus, Mytilus chilensis, Mytilus coruscus, M. edulis, M. galloprovincialis, M. trossulus* and *Venerupis (Ruditapes) philippinarum*
[b]GPL3994 probes from *C. gigas* and *C. virginica*

Over time, the increasing availability of nucleotide sequence data inspired the production of DNA microarrays, adaptable biotechnological tools made of spotted DNA/cDNA or *in situ* synthesized oligonucleotides (Table 6.1). Such predefined assemblies of molecular probes allow the multiple and quantitative assessment of gene expression levels, among other purposes.

The hybridization of processed RNA samples on DNA microarray slides could discriminate *Mytilus* mussels and *Ruditapes* clams sampled at different distance from a petro-chemical district in the Venice lagoon area (Venier et al. 2006; Milan et al. 2015), supporting the use of transcriptional profiles in environmental monitoring and suggesting an innovative way to assess quality and the possible illegal origin of traded stocks.

Tissue- stage- and sex-specific transcript profiles obtained by DNA microarrays can assist management actions and sustainability plans in the farming of bivalves. For instance, they have been used to understand the partial sterility of triploid oysters and genes related to growth and reproduction (Dheilly et al. 2014; Guan et al. 2017; Tong et al. 2015) or the oyster response to pathogens and stress factors negatively impacting the production rates (Venier et al. 2011; Anderson et al. 2015; Romero et al. 2015; Pardo et al. 2016). Relevant to the growth of the pearl oyster *Pinctada fucata*, gene expression profiles obtained during larval development highlighted new aspects of shell formation mechanisms (Liu et al. 2015).

Both high-throughput sequencing and a DNA microarray were used to investigate the early mussel response to algal toxins with the aim of developing new monitoring tools for okadaic acid, a heat-stable phosphatase inhibitor causing diarrhetic shellfish poisoning (Suarez-Ulloa et al. 2015). A total of "1,066,985" nucleotide sequences (at 10.08.2018) and "3,478" GEO datasets (at 10.08.2018) are available at NCBI for Bivalvia (10 Aug 2018) and the genomes of nine marine bivalves (oysters: *C. gigas*, *C. virginica*, *P. fucata martensii*; mussels: *Bathymodiolus platifrons*, *M. galloprovincialis*, *Modiolus philippinarum*, *Limnoperna fortunei*; scallops: *Mizuhopecten yessoensis*; clam *Ruditapes philippinarum*) have been completed or drafted (Zhang et al. 2012; Takeuchi et al. 2012; Murgarella et al. 2016; Mun et al. 2017; Sun et al. 2017; Wang et al. 2017a, b; Du et al. 2017).

Different from the DNA microarray analysis, high-throughput sequencing can lead to gene discovery and to the validation of population genetics markers for breeding programmes. The identification of single nucleotide polymorphisms (SNPs, codominant-inherited molecular features very abundant in animal genomes) in bivalves is just a preliminary step, before starting to validate their association with valuable quantitatively inherited traits or with stress-responsive genes, and to proceed with fine linkage mapping and population genetics analyses (Coppe et al. 2012; Ge et al. 2015; Nie et al. 2015; Dong et al. 2016; Fan et al. 2016; Wang et al. 2016a, b; Qi et al. 2017; Gutierrez et al. 2017; Azéma et al. 2017).

Although proteomics, metabolomics and epigenetics studies in marine bivalves are at their onset (Gómez-Chiarri et al. 2015; Digilio et al. 2016; Dineshram et al. 2016; Vincenzetti et al. 2017), in the near future they could reinforce and widen the existing assortment of bivalve services and products. In essence, the comprehensive knowledge of the vital processes in marine bivalves is a fundamental research strategy, consistent with the growth of a sustainable and innovative blue economy for the future. To confirm the continuous attention to marine bivalves and their expanding roles, they have been proposed in Northern Europe as living monitors of multidrug-resistant *Escherichia coli* and other *Enterobacteriaceae* spp. (Grevskott et al. 2017).

In the following section, we present a paradigmatic case which illustrates how the natural properties of bivalve byssus has guided the development of new materials of practical use.

6.3 Byssal Threads and Adhesive Plaques as Archetypes for New Biomimetics

Some freshwater and marine bivalves such as *Dreissena polymorpha*, *Perna viridis* and *Mytilus* spp. anchor themselves to hard substrates by means of silk-like byssus threads, having remarkable mechanical properties, and adhesive plaque proteins, functioning as an underwater superglue.

Descriptions of the general structure and microscopical anatomy of mussel byssus date back to 1711 and 1877, respectively, but only in the early 1950s investigations based on mechanical, chemical and enzymatic assays, histological and histochemical techniques, polarized light and X-ray diffraction, paved the way to bivalve-inspired materials for medical and non-medical applications (Fig. 6.1)

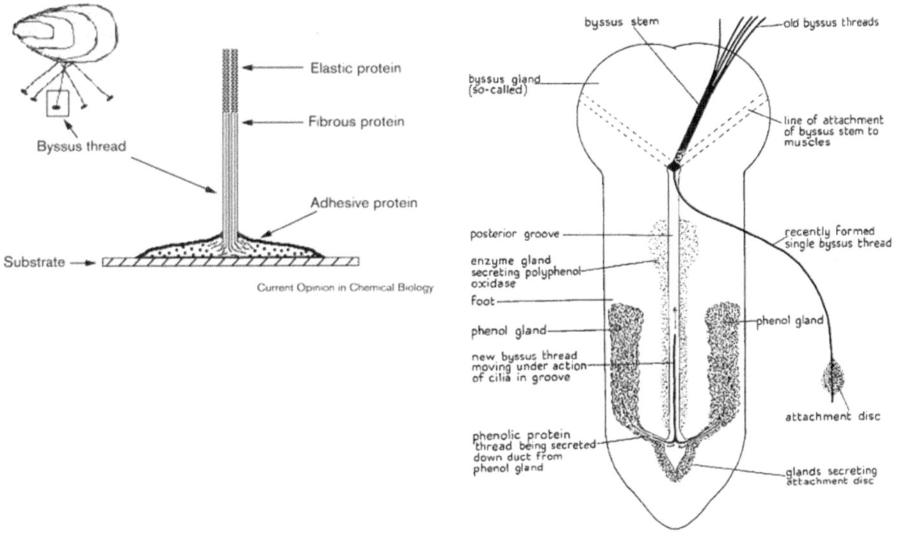

Fig. 6.1 Graphical representations of mussel byssus threads (**left**, as reported in Deming 1999) and anatomy of the byssus production in *Mytilus* (**right**, as reported in Smyth 1954). Gland tissue cells, detectable in precise zones of the mussel foot, emit a thread-like protein secretion along the foot groove whereas cells coating the foot groove secrete the protein components of the terminal adhesive plaque (disk). The byssus thread is released when it occupies the whole groove length

(Brown 1952; Smyth 1954; Deming 1999; Lee et al. 2011; Kord Forooshani and Lee 2017).

The proteinaceous byssus fibers comprise a proximal stem region, a mid-thread region and the terminal adhesive plaque. Mussel byssogenesis occurs in the post-larval stages within minutes by coordinated secretion and extracellular solidification of a composite fluid released by three pedal glands into the distal depression and ventral groove of the foot organ (Silverman and Roberto 2010; Priemel et al. 2017). More than ten types of secreted proteins compose the mussel byssus, including fibrillar collagens, non-collagenous thread matrix proteins and polyphenolic proteins of the thin cuticle surrounding the stretchy fibrous core and the adhesive plaque. As a result of post-translational hydroxylation of tyrosine, L-3,4-dihydroxyphenylalanine (L-DOPA) is a main component of the latter proteins, commonly named mussel foot proteins (Mfp, not to be confused with other proteins with the same acronym) or mussel adhesive proteins.

The unusual resistance of such fibrous and adhesive structure against predators and the mechanical force of waves and currents has considerably stimulated multidisciplinary investigations aimed to develop innovative biomimetic materials (Degtyar et al. 2014; Reinecke et al. 2016; Priemel et al. 2017). In the byssus thread, non-covalent protein–metal interactions stabilize the main constituent proteins and contribute to their tensile strength and self-healing properties. In detail, the thread core is made by bundles of collagenous proteins (preCols) having a central collagen domain with a typical Gly-X-Y triple helical repeat and flanking domains. Among other features, all preCols have N- and C-termini enriched in histidine, the amino acid most likely involved in coordination bonds with transition metal ions such as Zn and Cu. In essence, highly directional and dynamic protein–metal coordination bonds generate cross-linking and hierarchical structuring of byssal protein blocks, with the metal site geometry and activity governed by local charges, helical dipoles and other conformational protein elements. Rupture and rapid restructuring of coordination bonds between histidine residues and Zn^{2+} sustain the self-healing of byssus and, as expected, such self-healing can be inhibited by removing metal ions with ethylenediaminotetraacetic acid or by lowering the pH, a condition known to hamper histidine–metal bonding (Degtyar et al. 2014; Reinecke et al. 2016).

In the byssus plaque of *Mytilus* species, at least six Mfp rich in DOPA and cationic amino acids contribute with specialized roles to the adhesion in wet conditions to hard substrates (Table 6.2). The catechol moiety of L-DOPA permits the formation of hydrogen bonds and the interactions with other aromatic rings and with positively charged ions such as Cu^{2+}, Zn^{2+}, Mn^{2+} and Fe^{3+} among others. At sea water pH (mildly basic), these chemical events result in stable coordination complexes (e.g. DOPA oxidation coupled with the reduction of coordinated Fe^{3+} ions) and cross-linking (e.g. catechols oxidized to quinones can react with various nucleophilic groups and produce intermolecular/interfacial covalent bonds). After secretion, the spontaneous DOPA-Fe cross-linking in the byssus coating acts like a protective varnish as a result of attained hardness and extensibility. The local distribution of different Mfp and the significant presence of positively charged ions in the byssus plaque additionally stabilize its foamy structure and boost cohesive interactions and, hence, enhance the strong (wet) adhesion to hard surfaces (Lee et al. 2011; Reinecke et al. 2016; Kord Forooshani and Lee 2017; Priemel et al. 2017).

Using Mf3 as an example, the multiple alignment of 36 protein sequences available in GenBank highlights fully conserved amino acid residues and variable sequence traits (Fig. 6.2).

In essence, the byssus threads and their terminal plaques have emerged as a model for the development of self-healing polymers and water-resistant adhesive materials (Holten-Andersen et al. 2011; Danner et al. 2012; Guerette et al. 2013; Park et al. 2013; Liu et al. 2014; Fullenkamp et al. 2014; Schmidt et al. 2014; Wu et al. 2014; Nichols 2015; Ryu et al. 2015; Grindy et al. 2015; Miller et al. 2015; Tian et al. 2015; Krogsgaard et al. 2016; Liu et al. 2016; Xu et al. 2016; Zhang et al. 2017b; Waite 2017). In both cases, the coordination of metal ions plays a fundamental role; however, the occurring chemical events and final material properties depend on metals and ligands, their molar ratio, pH and redox reactions. Actually, catechols are regarded as suitable anchoring groups for surface modification, although their metal-binding strength depends on the oxidation status. Other byssogenic bivalves produce somewhat different foot proteins yet capable of strong adhesion, e.g. pvfp-1 from *Perna viridis* contains C(2)-mannosyl-7-hydroxytryptophan, Man7OHTrp, instead of DOPA, and trimerized chains instead of monomeric chains (Hwang et al. 2012). Deep understanding of the complex chemico-physical processes underlying the byssus formation as well as comparative data deriving from the omics technologies (Schultz and Adema 2017) should provide additional hints for a step-by-step

Table 6.2 Some data on the mussel foot proteins (from Kord Foreooshani and Lee 2017)

	Mfp-1	Mfp-2	Mfp-3	Mfp-4	Mfp-5	Mfp-6
Molecular weight (Kda)	108[a]	42–47[a]	5–7[a,b]	90–93[c]	8.9	11[c]
Isoelectric point[c]	10.5	9.5	nd	nd	9	9.5
Secondary structure	Very little	Highly repetitive motifs; 6 mol % Cys	No repeats; 30–35 variants rich in DOPA (>20 to 28 mol %): MFP-3f and Mfp-3s are rich in Gly (25–29 mol %), MFP-3f is highly hydrophililic; MFP-3s is polar but hydrophobic	His-rich decapeptide tandeml y repeated more than 36 times	Just 2 closely related variants; rich in DOPA (30 mol%), cationic ami no acids (27.7 mol %) and phosphoserine (≈4.8 mol%); hydrophilic	Rich in Tyr (20 mol %) mostly not converted in DOPA (3 mol %) and in Cys (11 mol%); the richest in charged aminoacids (23 mol% cationic, 16 mol% anionic)
Proposed role	Protective coating	It is the most abundant protein (≈25 wt %); its disulphide bonds support plaque integrity	It contributes to adhesion at the plaque-surface interphase	Exceptional binding to transition metal ions, functional bridge between thread (PreCol) and plaque proteins	It contributes to adhesion at the plaque-surface interphase	It contributes to adhesion at the plaque-surface interphase; it likely controls the redox chemistry of DOPA in the other plaque proteins

[a]in *Mytilus edulis*
[b]in *Mytilus californianus*
[c]*from* Lee et al. (2011)

development of useful novelties. As long as the new materials mimic natural substances and processes, they should have a great chance to be efficiently produced in environmentally friendly conditions and to be biodegradable. The development of wet adhesive materials using molluscan models could enable the development of new surgical adhesives, artificial joints, contact lenses, dental sealants and hair and skin conditioners (Wu et al. 2014; Nichols 2015; Ryu et al. 2015; Grindy et al. 2015; Miller et al. 2015; Tian et al. 2015). Moreover, byssus-inspired bioadhesive polymers, polymer blends and micro- or nano-structures have been proposed to fabricate new drug delivery or diagnostic systems including the encapsulation of

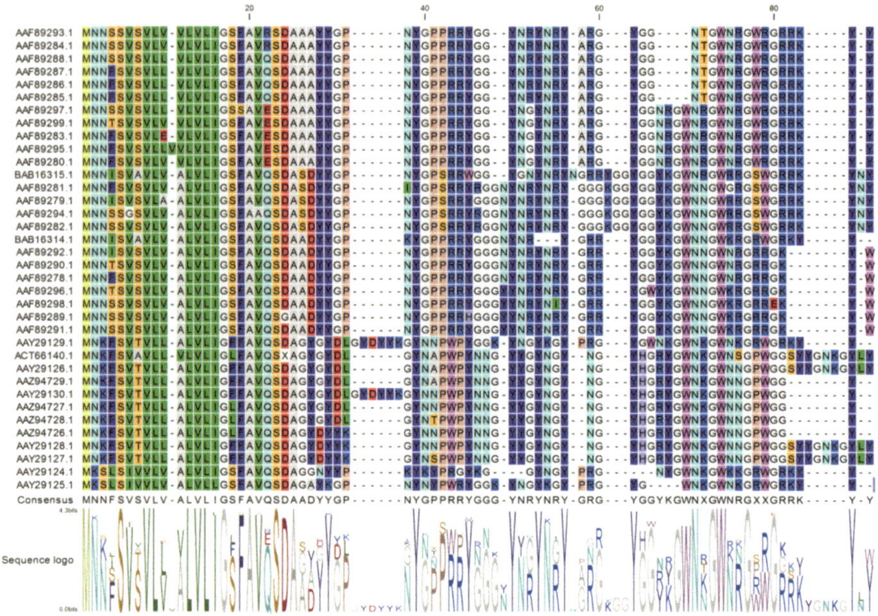

Fig. 6.2 Multiple alignment of amino acid sequences of 36 mussel foot proteins (Mfp 3). GenBank accession number, consensus sequence and sequence logo (i.e. graphical representation of the conservation extent of each protein residue) are reported

therapeutic, prophylactic, diagnostic agents to deliver bioactive components expected to be released upon contact with mucosal tissues of aquatic organisms. One could also imagine the development of biodegradable and nutritionally attractive feed formulations containing biocidal or antibiotic compounds and/or microbes, for the prevention and control of invasive non-indigenous species or for selective nutritional feed ingredients for more efficient growth of farmed species (Ma et al. 2016; Wang et al. 2016a, b, 2017a, b; Li et al. 2017; Luo and Liu 2017; Zhang et al. 2017a). Patents describing byssus-inspired inventions are exemplified in Table 6.3.

Reversing the scope, new lubricant-infused coatings are now suggested as an effective strategy to prevent the mussel adhesion and, hence, to mitigate marine biofouling (Amini et al. 2017).

6.4 Antimicrobials and Other Bioactive Molecules from Marine Bivalves Are Valuable Assets

The search of bioactive molecules of marine origin dates back to the past century but continues to generate pharmaceutics of human use and new compounds (1340 in 2015) (Liu et al. 2009; Mayer et al. 2010; García-Fernández et al. 2016; Kwon et al. 2016; Anjum et al. 2017; Blunt et al. 2017; Kang et al. 2017).

Table 6.3 Examples of patents describing byssus-inspired inventions (from Google patents)

Patent	Registration date	Pubblication date	Candidate Appointee	Title
US5049504	30/05/1990	17/09/1991	Genex Corporation	Bioadhesive coding sequencces
US5202236	25/05/1990	13/04/1993	Enzon Labs Inc.	Method of producing bioadhesive protein
US6987170B1	09/08/2004	17/01/2006	Battelle Energy Alliance, Llc.	Cloning and expression of recombinant adhesive protein Mefp-1 of the blue mussel, Mytilus edulis
WO2005056708A2	09/12/2004	23/06/2005	Spherics, Inc.	Bioadhesive polymers with catechol functionality
WO2007002318A2	23/06/2006	04/01/2007	Spherics, Inc.	Bioadhesive polymers
CA 2864891A1	21/02/2013	29/08/2013	Advanced Bionutrition Corporation and others	Compositions and methods for target delivering a bioactive agent to aquatic organisms
US20160115196A1	28/05/2014	28/04/2016	Ramot At Tel-Aviv University Ltd.	Self-assembled micro-and nanostructures

Marine species including plants, animals and microorganisms (mostly unculturable and unknown) are a rich source of gene-encoded products and metabolites whose molecular moieties mediate biological activities potentially exploitable for new inventions or for the repositioning/reinvention of known bioactive components (pharmaceuticals and nutraceuticals, among others). For instance, inhibitors of proteases and voltage-gated ion channels have been isolated from marine venomous animals such as sea anemones and *Conus* snails and are currently studied for their therapeutical and biotechnological potential (Liu et al. 2009; García-Fernández et al. 2016; Kwon et al. 2016). In the '90s, the cloning of the green florescent protein from the jellyfish *Aequoria victoria* and production of mutants opened the way to use these chromo proteins as probes in cell and tissue imaging (Prasher et al. 1992; Verkhusha and Lukyanov 2004; Chen et al. 2013). Both discoveries have driven significant advancements in the field of life sciences. In the discovery phase, the bioactivity is often claimed following *in vitro* demonstration of antibacterial/ antifungal/ antiviral, anti-proliferative and anti-tumor properties, although the latter must be demonstrated *in vivo* with adequate study design and high costs. It should be noted that different human ethnic groups have traditionally used molluscs and mollusc extracts for their anti-inflammatory, immune-modulatory and wound healing properties. Molluscan species were estimated to be the source of more than 1145 products by 2014. Liprinol® and Biolane Seatone from the green-lipped mussel *Perna canaliculus* exemplify marketed products of current use, the potent analgesic ziconotide from *Conus* snails has been clinically tested and approved by the Food and Drug Administration whereas other compounds are under trial (Ahmad et al. 2018).

Owing to their filtering activity, marine bivalves interact with putative pathogens including bacteria and viruses, and, thus, are expected to possess effective defence mechanisms. Nowadays, bioinformatic approaches accelerate the identification and guide the functional characterization of bioactive molecules from non-model bivalve species. In *Mytilus galloprovincialis*, the Mediterranean mussel, many families of putative cysteine-stabilized antimicrobials have been described. Mytilins, defensins, myticins and mytimycins were reported in the '90s (Hubert et al. 1996; Charlet et al. 1996) whereas big defensins, mytimacins, CRP I and the linear myticalin peptides were more recently discovered (Gerdol et al. 2012; Gerdol et al. 2015; Leoni et al. 2017). Among all of them, myticin C displayed high gene transcript polymorphism, constitutive and microbe-inducible expression, chemokine-like and antiviral activities. Although the action mode of myticin C is still unclear, an engineered construct with superior antiviral activity has been developed (Pallavicini et al. 2008; Novoa et al. 2016). As additional example, Mytichitin CB from *Mytilus coruscus* is a chitotriosidase-like antimicrobial which displays antifungal activity whose recombinant production should permit its full characterization (Qin et al. 2014; Meng et al. 2016).

While no mussel antimicrobial peptide (AMP) has been commercially exploited yet, some pilot studies have been carried out over the years, demonstrating the potential biotechnological applications of engineered peptides. Indeed, synthetic mytilin-derived peptides were capable or reducing mortality in virus-infected shrimp (white-spot syndrome) (Dupuy et al. 2004). Interesting antiviral, antibacterial and antiprotozoan activities also have been demonstrated for engineered defensin and mytilin variants (Dupuy et al. 2004; Liu et al. 2010).

Additional bivalve molecules could be regarded as having therapeutic potential. For instance, the mussel MytiLec-1 is a galactose-binding lectin able to inhibit the growth of both Gram-positive and Gram-negative bacteria (Hasan et al. 2016) and, at the same time, able to bind Burkitt's lymphoma and breast cancer cells expressing globotriose on their surface, significantly inducing apoptosis (Hasan et al. 2015; Liao et al. 2016; Chernikov et al. 2017). These remarkable properties have led to the computational design of an artificial β-trefoil lectin, named Mitsuba, capable of recognizing globotriose-expressing cancer cells, as an initial step for the development of effective MytiLec-1-based cancer treatment or diagnostics tools (Terada et al. 2017).

Other molluscan lectins with biotechnological potential are two C-type lectins from *C. gigas* (CgCLec-4, CgCLec-5), which exhibited anti-microbial (agglutinating) activity against bacteria and fungi (Jia et al. 2016). One extrapallial protein (C1Q-domain containing protein) of the mussel hemolymph serum (MgEP) was also demonstrated to act as an opsonin and to promote interactions between a suspected *Vibrio* pathogen and *Mytilus* hemocytes (Canesi et al. 2016).

In addition to ethanolic extracts, hydrolysates obtained by enzymatic digestion from bivalves and other marine invertebrates, revealed tens of antioxidant peptides which could benefit health or be used to produce novel food products (Chai et al. 2017; Odeleye et al. 2016; Wu and Huang 2017). Almost certainly, there are many more bioactive mollusc/bivalve components yet to be investigated. Regardless of the current state of knowledge of molluscan bioactives, we should never forget the possibility of toxic substances co-occurring in the same biological matrix.

6.5 Conclusions and Perspectives

This paper has presented a historical and conceptual timeline of the products and services provided by marine bivalve molluscs, focusing the attention to biotechnological innovations for a sustainable future. Marine bivalves with their associated microorganisms are central in the marine trophic networks, from the shoreline to the deep ocean. Bivalve species are traditionally fished and farmed worldwide as seafood since ancient times whereas their use as water pollution sentinels was established far more recently. Our time testifies great progresses in life sciences and, accordingly, further research on marine bivalves will likely confirm them as rich source of bioactive compounds and as interesting models for technological innovations (Imhoff et al. 2011; Desriac et al. 2014; Newman 2016). Today, the CRISP/CAS genome editing biotechnology represents a new revolutionary strategy also to engineer and implement bivalve-inspired products (Mojica and Montoliu 2016; Singh et al. 2018). As our knowledge base expands based on a multifaceted blue economy, there is little doubt that discoveries in this field will lead to societal and economic benefit in the near future.

Acknowledgments We thank the Editor of this Springer book for his attention to the author's work. This work was partially funded by the University of Padova-Department of Biology (BIRD168432-2016 to U. Rosani) and by the University of Trieste (Finanziamento di Ateneo per la Ricerca Scientifica 2015 to M. Gerdol). The affiliation institutions had no role in the chapter contents. The authors acknowledge prof. A. Alfaro for the review and her valuable comments.

References

Ahmad TB, Liu L, Kotiw M, Benkendorff K (2018) Review of anti-inflammatory, immune-modulatory and wound healing properties of molluscs. J Ethnopharmacol 210:156–178. https://doi.org/10.1016/j.jep.2017.08.008

Amini S, Kolle S, Petrone L, Ahanotu O, Sunny S, Sutanto CN, Hoon S, Cohen L, Weaver JC, Aizenberg J, Vogel N, Miserez A (2017) Preventing mussel adhesion using lubricant-infused materials. Science 357(6352):668–673. https://doi.org/10.1126/science.aai8977

Anderson K, Taylor DA, Thompson EL, Melwani AR, Nair SV, Raftos DA (2015) Meta-analysis of studies using suppression subtractive hybridization and microarrays to investigate the effects of environmental stress on gene transcription in oysters. PLoS One 10(3):e0118839. https://doi.org/10.1371/journal.pone.0118839

Anjum K, Abbas SQ, Akhter N, Shagufta BI, Shah SAA, Hassan SSU (2017) Emerging biopharmaceuticals from bioactive peptides derived from marine organisms. Chem Biol Drug Des 90(1):12–30. https://doi.org/10.1111/cbdd.12925

Azéma P, Lamy JB, Boudry P, Renault T, Travers MA, Dégremont L (2017) Genetic parameters of resistance to *Vibrio aestuarianus*, and OsHV-1 infections in the Pacific oyster, *Crassostrea gigas*, at three different life stages. Genet Sel Evol 49(1):23. https://doi.org/10.1186/s12711-017-0297-2

Banni M, Dondero F, Jebali J, Guerbej H, Boussetta H, Viarengo A (2007) Assessment of heavy metal contamination using real-time PCR analysis of mussel metallothionein mt10 and mt20 expression: a validation along the Tunisian coast. Biomarkers 12(4):369–383. https://doi.org/10.1080/13547500701217061

Blunt JW, Copp BR, Keyzers RA, Munro MHG, Prinsep MR (2017) Marine natural products. Nat Prod Rep 34(3):235–294. https://doi.org/10.1039/c6np00124f

Bolognesi C, Hayashi M (2011) Micronucleus assay in aquatic animals. Mutagenesis 26(1):205–213. https://doi.org/10.1093/mutage/geq073

Brown CH (1952) Some structural proteins of *Mytilus edulis*. J Cell Sci s3–93:487–502

Canesi L, Grande C, Pezzati E, Balbi T, Vezzulli L, Pruzzo C (2016) Killing of *Vibrio* cholerae and *Escherichia coli* strains carrying D-mannose-sensitive ligands by *Mytilus* hemocytes is promoted by a multifunctional hemolymph serum protein. Microb Ecol 72(4):759–762. https://doi.org/10.1007/s00248-016-0757-1

Chai TT, Law YC, Wong FC, Kim SK (2017) Enzyme-assisted discovery of antioxidant peptides from edible marine invertebrates: a review. Mar Drugs 15(2). https://doi.org/10.3390/md15020042

Charlet M, Chernysh S, Philippe H, Hetru C, Hoffmann JA, Bulet P (1996) Innate immunity. Isolation of several cysteine-rich antimicrobial peptides from the blood of a mollusc *Mytilus edulis*. J Biol Chem 271:21808–21813

Chen SF, Ferré N, Liu YJ (2013) QM/MM study on the light emitters of aequorin chemiluminescence, bioluminescence, and fluorescence: a general understanding of the bioluminescence of several marine organisms. Chemistry 19(26):8466–8472. https://doi.org/10.1002/chem.201300678

Chernikov O, Kuzmich A, Chikalovets I, Molchanova V, Hua KF (2017) Lectin CGL from the sea mussel *Crenomytilus grayanus* induces Burkitt's lymphoma cells death via interaction with surface glycan. Int J Biol Macromol 104(Pt A):508–514. https://doi.org/10.1016/j.ijbiomac.2017.06.074

Coppe A, Bortoluzzi S, Murari G, Marino IA, Zane L, Papetti C (2012) Sequencing and characterization of striped venus transcriptome expand resources for clam fishery genetics. PLoS One 7(9):e44185. https://doi.org/10.1371/journal.pone.0044185

Danner EW, Kan Y, Hammer MU, Israelachvili JN, Waite JH (2012) Adhesion of mussel foot protein Mefp-5 to mica: an underwater superglue. Biochemistry 51(33):6511–6518. https://doi.org/10.1021/bi3002538

Degtyar E, Harrington MJ, Politi Y, Fratzl P (2014) The mechanical role of metal ions in biogenic protein-based materials. Angew Chem Int Ed Engl 53(45):12026–12044. https://doi.org/10.1002/anie.201404272

Deming T (1999) Mussel byssus and biomolecular materials. Curr Opin Chem Biol 3(1):100–105

Desriac F, Le Chevalier P, Brillet B, Leguerinel I, Thuillier B, Paillard C, Fleury Y (2014) Exploring the hologenome concept in marine bivalvia: haemolymph microbiota as a pertinent source of probiotics for aquaculture. FEMS Microbiol Lett 350(1):107–116. https://doi.org/10.1111/1574-6968

Dheilly NM, Jouaux A, Boudry P, Favrel P, Lelong C (2014) Transcriptomic profiling of gametogenesis in triploid Pacific oysters *Crassostrea gigas*: towards an understanding of partial sterility associated with triploidy. PLoS One 9(11):e112094. https://doi.org/10.1371/journal.pone.0112094

Digilio G, Sforzini S, Cassino C, Robotti E, Oliveri C, Marengo E, Musso D, Osella D, Viarengo A (2016) Haemolymph from *Mytilus galloprovincialis*: response to copper and temperature challenges studied by (1)H-NMR metabonomics. Comp Biochem Physiol C Toxicol Pharmacol 183–184:61–71. https://doi.org/10.1016/j.cbpc.2016.02.003

Dineshram R, Chandramouli K, Ko GW, Zhang H, Qian PY, Ravasi T, Thiyagarajan V (2016) Quantitative analysis of oyster larval proteome provides new insights into the effects of multiple climate change stressors. Glob Chang Biol 22(6):2054–2068. https://doi.org/10.1111/gcb.13249

Dong YH, Yao HH, Sun CS, Lv DM, Li MQ, Lin ZH (2016) Development of polymorphic SSR markers in the razor clam (*Sinonovacula constricta*) and cross-species amplification. Genet Mol Res 15(1). https://doi.org/10.4238/gmr.15017285

Du X, Fan G, Jiao Y, Zhang H, Guo X, Huang R, Zheng Z, Bian C, Deng Y, Wang Q, Wang Z, Liang X, Liang H, Shi C, Zhao X, Sun F, Hao R, Bai J, Liu J, Chen W, Liang J, Liu W, Xu Z, Shi Q, Xu X, Zhang G, Liu X (2017) The pearl oyster *Pinctada fucata martensii* genome and multi-omic analyses provide insights into biomineralization. Gigascience 6(8):1–12. https://doi.org/10.1093/gigascience/gix059

Dupuy JW, Bonami JR, Roch P (2004) A synthetic antibacterial peptide from *Mytilus galloprovincialis* reduces mortality due to white spot syndrome virus in palaemonid shrimp. J Fish Dis 27(1):57–64

Fan SG, Wei JF, Guo YH, Huang GJ, Yu DH (2016) Development of coding single nucleotide polymorphic markers in the pearl oyster *Pinctada fucata* based on next-generation sequencing and high-resolution melting analysis. Genet Mol Res 15(4). https://doi.org/10.4238/gmr15049054

FAO (2018) Fisheries and Aquaculture Information and Statistics Branch. http://www.fao.org, accessed on line 08/2018

Fullenkamp DE, Barrett DG, Miller DR, Kurutz JW, Messersmith PB (2014) pH-dependent cross-linking of catechols through oxidation via $Fe3+$ and potential implications for mussel adhesion. RSC Adv 4(48):25127–25134. https://doi.org/10.1039/c4ra03178d

García-Fernández R, Peigneur S, Pons T, Alvarez C, González L, Chávez MA, Tytgat J (2016) The Kunitz-type protein ShPI-1 inhibits serine proteases and voltage-gated potassium channels. Toxins (Basel) 8(4):110. https://doi.org/10.3390/toxins8040110

Ge J, Li Q, Yu H, Kong L (2015) Identification of single-locus PCR-based markers linked to shell background color in the pacific oyster (*Crassostrea gigas*). Mar Biotechnol (NY) 17(5):655–662. https://doi.org/10.1007/s10126-015-9652-x

Gerdol M, De Moro G, Manfrin C, Venier P, Pallavicini A (2012) Big defensins and mytimacins, new AMP families of the Mediterranean mussel *Mytilus galloprovincialis*. Dev Comp Immunol 36(2):390–399. https://doi.org/10.1016/j.dci.2011.08.003

Gerdol M, Puillandre N, De Moro G, Guarnaccia C, Lucafò M, Benincasa M, Zlatev V, Manfrin C, Torboli V, Giulianini PG, Sava G, Venier P, Pallavicini A (2015) Identification and characterization of a novel family of cysteine-rich peptides (MgCRP-I) from *Mytilus galloprovincialis*. Genome Biol Evol 7(8):2203–2219. https://doi.org/10.1093/gbe/evv133

Goldberg ED, Bertine KK (2000) Beyond the mussel watch – new directions for monitoring marine pollution. Sci Total Environ 247(2-3):165–174

Gómez-Chiarri M, Guo X, Tanguy A, He Y, Proestou D (2015) The use of -omic tools in the study of disease processes in marine bivalve mollusks. J Invertebr Pathol 131:137–154. https://doi.org/10.1016/j.jip.2015.05.007

Grevskott DH, Svanevik CS, Sunde M, Wester AL, Lunestad BT (2017) Marine bivalve mollusks as possible indicators of multidrug-resistant *Escherichia coli* and other species of the *Enterobacteriaceae* family. Front Microbiol 8:24. https://doi.org/10.3389/fmicb.2017.00024

Grindy SC, Learsch R, Mozhdehi D, Cheng J, Barrett DG, Guan Z, Messersmith PB, Holten-Andersen N (2015) Control of hierarchical polymer mechanics with bioinspired metal-coordination dynamics. Nat Mater 14(12):1210–1216. https://doi.org/10.1038/nmat4401

Guan Y, He M, Wu H (2017) Differential mantle transcriptomics and characterization of growth-related genes in the diploid and triploid pearl oyster *Pinctada fucata*. Mar Genomics 33:31–38. https://doi.org/10.1016/j.margen.2017.01.001

Guéguen M, Amiard JC, Arnich N, Badot PM, Claisse D, Guérin T, Vernoux JP (2011) Shellfish and residual chemical contaminants: hazards, monitoring, and health risk assessment along French coasts. Rev Environ Contam Toxicol 213:55–111. https://doi.org/10.1007/978-1-4419-9860-6_3

Guerette PA, Hoon S, Seow Y, Raida M, Masic A, Wong FT, Ho VH, Kong KW, Demirel MC, Pena-Francesch A, Amini S, Tay GZ, Ding D, Miserez A (2013) Accelerating the design of biomimetic materials by integrating RNA-seq with proteomics and materials science. Nat Biotechnol 31(10):908–915. https://doi.org/10.1038/nbt.2671

Gutierrez AP, Turner F, Gharbi K, Talbot R, Lowe NR, Peñaloza C, McCullough M, Prodöhl PA, Bean TP, Houston RD (2017) Development of a medium density combined-species SNP array for pacific and european oysters *Crassostrea gigas* and *Ostrea edulis*. G3 (Bethesda) 7(7):2209–2218. https://doi.org/10.1534/g3.117.041780

Hasan I, Sugawara S, Fujii Y, Koide Y, Terada D, Iimura N, Fujiwara T, Takahashi KG, Kojima N, Rajia S, Kawsar SM, Kanaly RA, Uchiyama H, Hosono M, Ogawa Y, Fujita H, Hamako J, Matsui T, Ozeki Y (2015) MytiLec, a mussel R-type lectin, interacts with surface glycan Gb3 on Burkitt's lymphoma cells to trigger apoptosis through multiple pathways. Mar Drugs 13(12):7377–7389. https://doi.org/10.3390/md13127071

Hasan I, Gerdol M, Fujii Y, Rajia S, Koide Y, Yamamoto D, Kawsar SM, Ozeki Y (2016) cDNA and gene structure of MytiLec-1, A bacteriostatic R-Type Lectin from the Mediterranean mussel (*Mytilus galloprovincialis*). Mar Drugs 14(5). https://doi.org/10.3390/md14050092

Holten-Andersen N, Harrington MJ, Birkedal H, Lee BP, Messersmith PB, Lee KY, Waite JH (2011) pH-induced metal-ligand cross-links inspired by mussel yield self-healing polymer networks with near-covalent elastic moduli. Proc Natl Acad Sci USA 108(7):2651–2655. https://doi.org/10.1073/pnas.1015862108

Hubert F, Noel T, Roch P (1996) A member of the arthropod defensin family from edible Mediterranean mussels (*Mytilus galloprovincialis*). Eur J Biochem 240(1):302–306 Erratum in (1996) Eur J Biochem 240:815

Hwang DS, Zeng H, Lu Q, Israelachvili J, Waite JH (2012) Adhesion mechanism in a DOPA-deficient foot protein from green mussels. Soft Matter 8(20):5640–5648. https://doi.org/10.1039/C2SM25173F

Imhoff JF, Labes A, Wiese J (2011) Bio-mining the microbial treasures of the ocean: new natural products. Biotechnol Adv 29(5):468–482. https://doi.org/10.1016/j.biotechadv.2011.03.001

Jia Z, Zhang H, Jiang S, Wang M, Wang L, Song L (2016) Comparative study of two single CRD C-type lectins, CgCLec-4 and CgCLec-5, from pacific oyster *Crassostrea gigas*. Fish Shellfish Immunol 59:220–232. https://doi.org/10.1016/j.fsi.2016.10.030

Kang HK, Kim C, Seo CH, Park Y (2017) The therapeutic applications of antimicrobial peptides (AMPs): a patent review. J Microbiol 55(1):1–12. https://doi.org/10.1007/s12275-017-6452-1

Kord Forooshani P, Lee BP (2017) Recent approaches in designing bioadhesive materials inspired by mussel adhesive protein. J Polym Sci A Polym Chem 55(1):9–33. https://doi.org/10.1002/pola.28368

Krogsgaard M, Nue V, Birkedal H (2016) Mussel-inspired materials: self-healing through coordination chemistry. Chemistry 22(3):844–857. https://doi.org/10.1002/chem.20150338

Kwon S, Bosmans F, Kaas Q, Cheneval O, Conibear AC, Rosengren KJ, Wang CK, Schroeder CI, Craik DJ (2016) Efficient enzymatic cyclization of an inhibitory cystine knot-containing peptide. Biotechnol Bioeng 113(10):2202–2212. https://doi.org/10.1002/bit.25993

Lee BP, Messersmith PB, Israelachvili JN, Waite JH (2011) Mussel-inspired adhesives and coatings. Annu Rev Mater Res 41:99–132. https://doi.org/10.1146/annurev-matsci-062910-100429

Leoni G, De Poli A, Mardirossian M, Gambato S, Florian F, Venier P, Wilson DN, Tossi A, Pallavicini A, Gerdol M (2017) Myticalins: a novel multigenic family of linear, cationic antimicrobial peptides from marine mussels (*Mytilus* spp.). Mar Drugs 15(8). https://doi.org/10.3390/md15080261

Li L, Yan B, Yang J, Huang W, Chen L, Zeng H (2017) Injectable self-healing hydrogel with antimicrobial and antifouling properties. ACS Appl Mater Interfaces 9(11):9221–9225. https://doi.org/10.1021/acsami.6b16192

Liao JH, Chien CT, Wu HY, Huang KF, Wang I, Ho MR, Tu IF, Lee IM, Li W, Shih YL, Wu CY, Lukyanov PA, Hsu ST, Wu SH (2016) A multivalent marine Lectin from *Crenomytilus grayanus* possesses anti-cancer activity through recognizing Globotriose Gb3. J Am Chem Soc 138(14):4787–4795. https://doi.org/10.1021/jacs.6b00111

Liu Z, Xu N, Hu J, Zhao C, Yu Z, Dai Q (2009) Identification of novel I-superfamily conopeptides from several clades of *Conus* species found in the South China Sea. Peptides 30(10):1782–1787. https://doi.org/10.1016/j.peptides.2009.06.036

Liu M, Wu M, Zhou S, Gao P, Lu T, Wang R, Shi G, Liao Z (2010) Designation, solid-phase synthesis and antimicrobial activity of Mytilin derived peptides based on Mytilin-1 from *Mytiluscoruscus*. Sheng Wu Gong Cheng Xue Bao 26(4):550–556. In Chinese

Liu Z, Qu S, Zheng X, Xiong X, Fu R, Tang K, Zhong Z, Weng J (2014) Effect of polydopamine on the biomimetic mineralization of mussel-inspired calcium phosphate cement *in vitro*. Mater Sci Eng C Mater Biol Appl 44:44–51. https://doi.org/10.1016/j.msec.2014.07.063

Liu J, Yang D, Liu S, Li S, Xu G, Zheng G, Xie L, Zhang R (2015) Microarray: a global analysis of biomineralization-related gene expression profiles during larval development in the pearl oyster, *Pinctada fucata*. BMC Genomics 16:325. https://doi.org/10.1186/s12864-015-1524-2

Liu M, Zeng G, Wang K, Wan Q, Tao L, Zhang X, Wei Y (2016) Recent developments in polydopamine: an emerging soft matter for surface modification and biomedical applications. Nanoscale 8(38):16819–16840. https://doi.org/10.1039/c5nr09078d

Luo C, Liu Q (2017) Oxidant-induced high-efficient mussel-inspired modification on PVDF membrane with superhydrophilicity and underwater superoleophobicity characteristics for oil/water separation. ACS Appl Mater Interfaces 9(9):8297–8307. https://doi.org/10.1021/acsami.6b16206

Ma H, Luo J, Sun Z, Xia L, Shi M, Liu M, Chang J, Wu C (2016) 3D printing of biomaterials with mussel-inspired nanostructures for tumor therapy and tissue regeneration. Biomaterials 111:138–148. https://doi.org/10.1016/j.biomaterials.2016.10.005

Mayer AMS, Glaser KB, Cuevas C, Jacobs RS, Kem W, Little RD, McIntosh JM, Newman DJ, Potts BC, Shuster DE (2010) The odyssey of marine pharmaceuticals: a current pipeline perspective. Trends Pharmacol Sci 31(6):255–265. https://doi.org/10.1016/j.tips.2010.02.005

Melwani AR, Gregorio D, Jin Y, Stephenson M, Ichikawa G, Siegel E, Crane D, Lauenstein G, Davis JA (2014) Mussel Watch update: long-term trends in selected contaminants from coastal California, 1977-2010. Mar Pollut Bull 81(2):291–302. https://doi.org/10.1016/j.marpolbul.2013.04.025

Meng DM, Dai HX, Gao XF, Zhao JF, Guo YJ, Ling X, Dong B, Zhang ZQ, Fan ZC (2016) Expression, purification and initial characterization of a novel recombinant antimicrobial peptide Mytichitin-A in *Pichia pastoris*. Protein Expr Purif 127:35–43. https://doi.org/10.1016/j.pep.2016.07.001

Milan M, Pauletto M, Boffo L, Carrer C, Sorrentino F, Ferrari G, Pavan L, Patarnello T, Bargelloni L (2015) Transcriptomic resources for environmental risk assessment: a case study in the Venice lagoon. Environ Pollut 197:90–98. https://doi.org/10.1016/j.envpol.2014.12.005

Miller DR, Das S, Huang KY, Han S, Israelachvili JN, Waite JH (2015) Mussel coating protein-derived complex coacervates mitigate frictional surface damage. ACS BiomaterSci Eng 1(11):1121–1128. https://doi.org/10.1021/acsbiomaterials.5b00252

Mojica FJ, Montoliu L (2016) On the origin of CRISPR-Cas technology: from prokaryotes to mammals. Trends Microbiol 24(10):811–820. https://doi.org/10.1016/j.tim.2016.06.005

Moore MN, Allen JI, McVeigh A, Shaw J (2006) Lysosomal and autophagic reactions as predictive indicators of environmental impact in aquatic animals. Autophagy 2(3):217–220

Mun S, Kim YJ, Markkandan K, Shin W, Oh S, Woo J, Yoo J, An H, Han K (2017) The whole-genome and transcriptome of the manila clam (*Ruditapes philippinarum*). Genome Biol Evol 9(6):1487–1498. https://doi.org/10.1093/gbe/evx096

Murgarella M, Puiu D, Novoa B, Figueras A, Posada D, Canchaya C (2016) A first insight into the genome of the Filter-Feeder Mussel *Mytilus galloprovincialis*. PLoS One 11(3):e0151561. https://doi.org/10.1371/journal.pone.0151561. Correction in (2016) PLoS One 11(7):e0160081.https://doi.org/10.1371/journal.pone.0160081

Newman DJ (2016) Predominately uncultured microbes as sources of bioactive agents. Front Microbiol 7:1832. https://doi.org/10.3389/fmicb.2016.01832

Nichols WT (2015) Designing biomimetic materials from marine organisms. J Nanosci Nanotechnol 15(1):189–191

Nie Q, Yue X, Liu B (2015) Development of *Vibrio* spp. infection resistance related SNP markers using multiplex SNaPshot genotyping method in the clam *Meretrix meretrix*. Fish Shellfish Immunol 43(2):469–476. https://doi.org/10.1016/j.fsi.2015.01.030

Novoa B, Romero A, Álvarez ÁL, Moreira R, Pereiro P, Costa MM, Dios S, Estepa A, Parra F, Figueras A (2016) Antiviral activity of myticin C peptide from Mussel: an ancient defense against Herpesviruses. J Virol 90(17):7692–7702. https://doi.org/10.1128/JVI.00591-16

Odeleye T, Li Y, White WL, Nie S, Chen S, Wang J, Lu J (2016) The antioxidant potential of the New Zealand surf clams. Food Chem 204:141–149. https://doi.org/10.1016/j.foodchem.2016.02.120

Okay OS, Ozmen M, Güngördü A, Yılmaz A, Yakan SD, Karacık B, Tutak B, Schramm KW (2016) Heavy metal pollution in sediments and mussels: assessment by using pollution indices and metallothionein levels. Environ Monit Assess 188(6):352. https://doi.org/10.1007/s10661-016-5346-8

Pallavicini A, Costa Mdel M, Gestal C, Dreos R, Figueras A, Venier P, Novoa B (2008) High sequence variability of myticin transcripts in hemocytes of immune-stimulated mussels suggests ancient host-pathogen interactions. Dev Comp Immunol 32(3):213–226. https://doi.org/10.1016/j.dci.2007.05.008

Pardo BG, Álvarez-Dios JA, Cao A, Ramilo A, Gómez-Tato A, Planas JV, Villalba A, Martínez P (2016) Construction of an *Ostrea edulis* database from genomic and expressed sequence tags (ESTs) obtained from *Bonamia ostreae* infected haemocytes: development of an immune-enriched oligo-microarray. Fish Shellfish Immunol 59:331–344. https://doi.org/10.1016/j.fsi.2016.10.047

Park JY, Yeom J, Kim JS, Lee M, Lee H, Nam YS (2013) Cell-repellant dextran coatings of porous titania using mussel adhesion chemistry. Macromol Biosci 13(11):1511–1519. https://doi.org/10.1002/mabi.201300224

Prasher DC, Eckenrode VK, Ward WW, Prendergast FG, Cormier MJ (1992) Primary structure of the *Aequorea victoria* green-fluorescent protein. Gene 111(2):229–233

Priemel T, Degtyar E, Dean MN, Harrington MJ (2017) Rapid self-assembly of complex biomolecular architectures during mussel byssus biofabrication. Nat Commun 8:14539. https://doi.org/10.1038/ncomms14539

Pytharopoulou S, Sazakli E, Grintzalis K, Georgiou CD, Leotsinidis M, Kalpaxis DL (2008) Translational responses of *Mytilus galloprovincialis* to environmental pollution: integrating the responses to oxidative stress and other biomarker responses into a general stress index. Aquat Toxicol 89(1):18–27. https://doi.org/10.1016/j.aquatox.2008.05.013

Qi H, Song K, Li C, Wang W, Li B, Li L, Zhang G (2017) Construction and evaluation of a high-density SNP array for the Pacific oyster (*Crassostrea gigas*). PLoS One 12(3):e0174007. https://doi.org/10.1371/journal.pone.0174007

Qin CL, Huang W, Zhou SQ, Wang XC, Liu HH, Fan MH, Wang RX, Gao P, Liao Z (2014) Characterization of a novel antimicrobial peptide with chitin-biding domain from *Mytilus coruscus*. Fish Shellfish Immunol 41(2):362–370. https://doi.org/10.1016/j.fsi.2014.09.019

Reinecke A, Bertinetti L, Fratzl P, Harrington MJ (2016) Cooperative behaviour of a sacrificial bond network and elastic framework in providing self-healing capacity in mussel byssal threads. J Struct Biol 196(3):329–339. https://doi.org/10.1016/j.jsb.2016.07.020

Romero A, Forn-Cuní G, Moreira R, Milan M, Bargelloni L, Figueras A, Novoa B (2015) An immune-enriched oligo-microarrayanalysis of gene expression in Manila clam (*Venerupis philippinarum*) haemocytes after a *Perkinsus olseni* challenge. Fish Shellfish Immunol 43(1):275–286. https://doi.org/10.1016/j.fsi.2014.12.029

Ryu JH, Hong S, Lee H (2015) Bio-inspired adhesive catechol-conjugated chitosan for biomedical applications: a mini review. Acta Biomater 27:101–115. https://doi.org/10.1016/j.actbio.2015.08.043

Schmidt S, Reinecke A, Wojcik F, Pussak D, Hartmann L, Harrington MJ (2014) Metal-mediated molecular self-healing in histidine-rich mussel peptides. Biomacromolecules 15(5):1644–1652. https://doi.org/10.1021/bm500017u

Schultz JH, Adema CM (2017) Comparative immunogenomics of molluscs. Dev Comp Immunol 75:3–15. https://doi.org/10.1016/j.dci.2017.03.013

Silverman HG, Roberto FF (2010) Byssus formation in mussel. In: von Byern J, Grunwald I (eds) Biological adhesive systems: from nature to technical and medical application, ch 18. Springer, Wien/New York, pp 273–285, ISBN 978-3-8091-1041-4

Singh RK, Lee JK, Selvaraj C, Singh R, Li J, Kim SY, Kalia VC (2018) Protein engineering approaches in the post-genomic era. Curr Protein Pept Sci 19(1):5–15. https://doi.org/10.2174/1389203718666161117114243

Smyth A (1954) Technique for the histochemical demonstration of polyphenol oxidase and its application to egg-shell formation in helminths and byssus formation in *Mytilus*. J Cell Sci 95(2):139–152

Suarez-Ulloa V, Fernandez-Tajes J, Aguiar-Pulido V, Prego-Faraldo MV, Florez-Barros F, Sexto-Iglesias A, Mendez J, Eirin-Lopez JM (2015) Unbiased high-throughput characterization of mussel transcriptomic responses to sublethal concentrations of the biotoxin okadaic acid. PeerJ 3:e1429. https://doi.org/10.7717/peerj.1429

Sun J, Zhang Y, Xu T, Zhang Y, Mu H, Zhang Y, Lan Y, Fields CJ, Hui JHL, Zhang W, Li R, Nong W, Cheung FKM, Qiu JW, Qian PY (2017) Adaptation to deep-sea chemosynthetic environments as revealed by mussel genomes. Nat Ecol Evol 1(5):121. https://doi.org/10.1038/s41559-017-0121

Takeuchi T, Kawashima T, Koyanagi R, Gyoja F, Tanaka M, Ikuta T, Shoguchi E, Fujiwara M, Shinzato C, Hisata K, Fujie M, Usami T, Nagai K, Maeyama K, Okamoto K, Aoki H, Ishikawa T, Masaoka T, Fujiwara A, Endo K, Endo H, Nagasawa H, Kinoshita S, Asakawa S, Watabe S, Satoh N (2012) Draft genome of the pearl oyster *Pinctada fucata*: a platform for understanding bivalve biology. DNA Res 19(2):117–130. https://doi.org/10.1093/dnares/dss005

Terada D, Voet ARD, Noguchi H, Kamata K, Ohki M, Addy C, Fujii Y, Yamamoto D, Ozeki Y, Tame JRH, Zhang KYJ (2017) Computational design of a symmetrical β-trefoil lectin with cancer cell binding activity. Sci Rep 7(1):5943. https://doi.org/10.1038/s41598-017-06332-7

Tian Y, Shen S, Feng J, Jiang X, Yang W (2015) Mussel-inspired gold hollow superparticles for photothermal therapy. Adv Healthc Mater 4(7):1009–1014. https://doi.org/10.1002/adhm.201400787

Tong Y, Zhang Y, Huang J, Xiao S, Zhang Y, Li J, Chen J, Yu Z (2015) Transcriptomics analysis of *Crassostrea hongkongensis* for the discovery of reproduction-related genes. PLoS One 10(8):e0134280. https://doi.org/10.1371/journal.pone.0134280

Venier P, De Pittà C, Pallavicini A, Marsano F, Varotto L, Romualdi C, Dondero F, Viarengo A, Lanfranchi G (2006) Development of mussel mRNA profiling: can gene expression trends reveal coastal water pollution? Mutat Res 602(1–2):121–134. https://doi.org/10.1016/j.mrfmmm.2006.08.007

Venier P, Varotto L, Rosani U, Millino C, Celegato B, Bernante F, Lanfranchi G, Novoa B, Roch P, Figueras A, Pallavicini A (2011) Insights into the innate immunity of the Mediterranean mussel *Mytilus galloprovincialis*. BMC Genomics 12:69. https://doi.org/10.1186/1471-2164-12-69

Verkhusha VV, Lukyanov KA (2004) The molecular properties and applications of Anthozoa fluorescent proteins and chromoproteins. Nat Biotechnol 22(3):289–296. https://doi.org/10.1038/nbt943

Vincenzetti S, Felici A, Ciarrocchi G, Pucciarelli S, Ricciutelli M, Ariani A, Polzonetti V, Polidori P (2017) Comparative proteomic analysis of two clam species: *Chamelea gallina* and *Tapes philippinarum*. Food Chem 219:223–229. https://doi.org/10.1016/j.foodchem.2016.09.150

Waite JH (2017) Mussel adhesion – essential footwork. J Exp Biol 220(Pt 4):517–530. https://doi.org/10.1242/jeb.134056

Wang J, Li L, Zhang G (2016a) A high-density SNP genetic linkage map and QTL analysis of growth-related traits in a hybrid family of oysters (*Crassostrea gigas* × *Crassostrea angulata*) using genotyping-by-sequencing. G3 (Bethesda) 6(5):1417–1426. https://doi.org/10.1534/g3.116.026971

Wang Y, Chen Z, Luo G, He W, Xu K, Xu R, Lei Q, Tan J, Wu J, Xing M (2016b) In-Situ-generated vasoactive intestinal peptide loaded microspheres in mussel-inspired polycaprolactone nanosheets creating spatiotemporal releasing microenvironment to promote wound healing and angiogenesis. ACS Appl Mater Interfaces 8(11):7411–7421. https://doi.org/10.1021/acsami.5b11332

Wang R, Song X, Xiang T, Liu Q, Su B, Zhao W, Zhao C (2017a) Mussel-inspired chitosan-polyurethane coatings for improving the antifouling and antibacterial properties of polyethersulfone membranes. CarbohydrPolym 168:310–319. https://doi.org/10.1016/j.carbpol.2017.03.092

Wang S, Zhang J, Jiao W, Li J, Xun X, Sun Y, Guo X, Huan P, Dong B, Zhang L, Hu X, Sun X, Wang J, Zhao C, Wang Y, Wang D, Huang X, Wang R, Lv J, Li Y, Zhang Z, Liu B, Lu W, Hui Y, Liang J, Zhou Z, Hou R, Li X, Liu Y, Li H, Ning X, Lin Y, Zhao L, Xing Q, Dou J, Li Y, Mao J, Guo H, Dou H, Li T, Mu C, Jiang W, Fu Q, Fu X, Miao Y, Liu J, Yu Q, Li R, Liao H, Li X, Kong Y, Jiang Z, Chourrout D, Li R, Bao Z (2017b) Scallop genome provides insights into evolution of bilaterian karyotype and development. Nat Ecol Evol 1(5):120. https://doi.org/10.1038/s41559-017-0120

Wu S, Huang X (2017) Preparation and antioxidant activities of oligosaccharides from *Crassostrea gigas*. Food Chem 216:243–246. https://doi.org/10.1016/j.foodchem.2016.08.043

Wu C, Han P, Liu X, Xu M, Tian T, Chang J, Xiao Y (2014) Mussel-inspired bioceramics with self-assembled Ca-P/polydopamine composite nanolayer: preparation, formation mechanism, improved cellular bioactivity and osteogenic differentiation of bone marrow stromal cells. Acta Biomater 10(1):428–438. https://doi.org/10.1016/j.actbio.2013.10.013

Xu M, Zhai D, Xia L, Li H, Chen S, Fang B, Chang J, Wu C (2016) Hierarchical bioceramic scaffolds with 3D-plotted macropores and mussel-inspired surface nanolayers for stimulating osteogenesis. Nanoscale 8(28):13790–13803. https://doi.org/10.1039/c6nr01952h

Zhang G, Fang X, Guo X, Li L, Luo R, Xu F, Yang P, Zhang L, Wang X, Qi H, Xiong Z, Que H, Xie Y, Holland PW, Paps J, Zhu Y, Wu F, Chen Y, Wang J, Peng C, Meng J, Yang L, Liu J, Wen B, Zhang N, Huang Z, Zhu Q, Feng Y, Mount A, Hedgecock D, Xu Z, Liu Y, Domazet-Lošo T, Du Y, Sun X, Zhang S, Liu B, Cheng P, Jiang X, Li J, Fan D, Wang W, Fu W, Wang T, Wang B, Zhang J, Peng Z, Li Y, Li N, Wang J, Chen M, He Y, Tan F, Song X, Zheng Q, Huang R, Yang H, Du X, Chen L, Yang M, Gaffney PM, Wang S, Luo L, She Z, Ming Y, Huang W, Zhang S, Huang B, Zhang Y, Qu T, Ni P, Miao G, Wang J, Wang Q, Steinberg CE, Wang H, Li N, Qian L, Zhang G, Li Y, Yang H, Liu X, Wang J, Yin Y, Wang J (2012) The oyster genome reveals stress adaptation and complexity of shell formation. Nature 490(7418):49–54. https://doi.org/10.1038/nature11413

Zhang C, Li HN, Du Y, Ma MQ, Xu ZK (2017a) $CuSO_4H_2O_2$-triggered polydopamine/poly(sulfobetaine methacrylate) coatings for antifouling membrane surfaces. Langmuir 33(5):1210–1216. https://doi.org/10.1021/acs.langmuir.6b03948

Zhang C, Lv Y, Qiu WZ, He A, Xu ZK (2017b) Polydopamine coatings with nanopores for versatile molecular separation. ACS Appl Mater Interfaces 9(16):14437–14444. https://doi.org/10.1021/acsami.7b03115

Part II
Regulating Services

Chapter 7
Introduction to Regulating Services

Øivind Strand and Joao G. Ferreira

Abstract Bivalves are foundation species with important regulating functions in the ecosystem. This is due to their function as filter feeders, their capacity to extract particles, to regenerate as well as store nutrients and – for the epibenthic bivalves –, their capacity to form hard structures. These services can be applied in many ways as is exemplified in this section. It seems likely that more applicable functions will emerge from the studies reviewed in this section.

Keywords Eutrophication · IMTA · Nutrient cycling · Eco-engineering

The regulating services from bivalves originate from their effects and controlling functions on ecosystem processes and natural cycles.

In natural habitats where bivalves dominate, they may control functions related to

1. physical properties of bottom habitats e.g. reef building
2. geochemical processes in the sediment
3. benthos and its coupling to the pelagic environment

In bivalve aquaculture, regulating services are typically seen when large biomasses are grown for human consumption or in production for energy and feed. But there is also a range of examples where services are shown from more extensive culture initiatives related to enhancement and restoration of bivalve populations, indigenous and invasive. In this chapter, the authors view regulating services from bivalves for a large range of spatial scales, from intensive land-based culture systems to narrow embayments and open sea ecosystems.

Ø. Strand (✉)
Institute of Marine Research, Bergen, Norway
e-mail: oivind.strand@imr.no

J. G. Ferreira
DCEA, FCT, New University of Lisbon, Monte de Caparica, Portugal
e-mail: joao@hoomi.com

A. C. Smaal et al. (eds.), *Goods and Services of Marine Bivalves*,
https://doi.org/10.1007/978-3-319-96776-9_7

As most bivalves are efficient filter feeders on suspended particles, they may exert substantial effects and control of primary production processes, and concentrations of particulate matter. Elevated concentrations of particles and thereby turbidity caused by eutrophication often appears and is visible in densely inhabited coastal areas. The control of such conditions by bivalves is a classic example often promoted as a service regulating and mitigating eutrophication and other undesirable environmental conditions (Petersen et al. 2014). However, it may also be found that consolidation of phytoplankton and particulate organic matter into waste particles (undigested as faeces and uningested as pseudofaeces) redirect part of the undesired particle concentration which we wish to mitigate towards the benthic food web, potentially causing problems. Depending on the type of environment and dispersion pattern of particles, such biodeposition can cause hypoxic or anoxic bottom conditions, which may require further mitigation. These associated biodeposition processes are often ignored. From a wider perspective, biodeposits are sites of mineralization and regeneration of inorganic nutrients. Hence, filtration of suspended particles, biodeposition and biomineralization can provide a positive feedback mechanism in food production, as the regenerated nutrients stimulate primary production that provides feed for the bivalves: feedbacks through filter feeding. The question is how these processes perform under various conditions, in particular with respect to background nutrient concentrations in oligotrophic or eutrophic environments.

Bivalves have been regarded as a prime candidate as extractive species in Integrated Multi-Trophic Aquaculture (IMTA) (Chopin 2013). They may capture waste particles directly from farm discharge and eventually they can extract products coming from another trophic level that converts the waste (bacteria, phytoplankton, zooplankton). Their possible role in providing regulating services when cultured in IMTA vary with type of culture, waste source, environmental conditions, and is certainly also governed by the wider socio-economic differences in aquaculture practice in Asia and the Western World.

Marine and coastal environments play a vital role in regulating the global climate via the carbon cycle, in which organisms precipitating calcium carbonate receive special interest. Bivalves are regarded to be particularly susceptible to ocean acidification, often related to their importance in coastal ecosystems and in aquaculture production. In addition, their role has been emphasised in redirecting suspended particles to the benthic environment and burying organic carbon that potentially can increase carbon sequestration.

The regulating services from bivalves are in many cases based on mitigation of eutrophication and recirculation of anthropogenic waste from land or coastal activities. The distribution of waste sources, hydrodynamic patterns and the physical domain under consideration determine the complexity of the ecosystem and thereby possibilities to assess regulating services. The ecological complexity is often disregarded, and the valuation of the service tends to be based on limited assessments. In developing concepts of regulating services from bivalve shellfish there might be a sequence from general assumptions to model results and extrapolations from experimental studies to full scale environments. The ultimate challenge is to carry out measurements and validation supporting how target processes are affected and mitigated at the full scale.

References

Chopin T (2013) Integrated multi-trophic aquaculture – ancient, adaptable concept focuses on ecological integration. Glob Aquacu Advocate 16:16–17

Petersen JK, Hasler B, Timmermann K, Nielsen P, Tørring DB, Larsen MM, Holmer M (2014) Mussels as a tool for mitigation of nutrients in the marine environment. Marine Pollution Bulletin 82:137–143

Chapter 8
Magnitude and Extent of Water Clarification Services Provided by Bivalve Suspension Feeding

Peter J. Cranford

Abstract Studies in bivalve ecology have emphasized that phytoplankton dynamics in coastal regions may be strongly coupled with bivalve suspension feeding activity to the extent that the bivalves play a major ecological role in controlling phytoplankton biomass. The water clarification capacity of natural and cultured bivalve populations serves as the foundation for what is considered to be a manageable bioengineering tool for mitigating the major symptoms of eutrophication and thereby providing positive ecosystem-scale services. Although often predicted, suspended particle depletion by bivalve aggregations has only recently been measured directly. Field observations of food depletion by bivalve aggregations confirm the large capacity for water clarification. However, progressively increasing the standing stock of bivalves to achieve greater water clarification benefits eventually lead to inefficiencies in bivalve feeding related to increased flow reduction from structure drag, which facilitates an increase in water re-filtration. These physical and biological processes ultimately constrain the maximum water clarification capacity of the population to levels that can be substantially less than previously predicted. Positive ecosystem services from bivalve grazing are likely to occur in many coastal areas experiencing eutrophication (e.g. Lindahl 2011). However, it is important to take an ecosystem-based management approach that also considers the potential for adverse environmental interactions that may be associated with intensive and extensive suspended bivalve operations.

Abstract in Chinese 摘要:双壳贝类的生态学研究表明,沿海地区的浮游植物种群动力学过程与滤食性双壳贝类的摄食活动密切相关,双壳贝类在调控浮游植物生物量方面有着重要的生态作用。野生和养殖的双壳贝类种群的滤食作用能够减轻水域富营养化,从而提供了积极的生态系统水平的服务功能。尽管在之前有过许多理论预测,但是直到最近我们才可以直接测量双壳贝类对水体中悬浮颗粒物的滤食量 。在养殖水域的观测结果表明双壳贝类通过滤食而净化的水体相当可观。然而,双壳贝类的净水能力与养殖量并非线性关系,过量的贝类养殖会导致养殖双壳贝类净水能力下降,其中一个主要原因

P. J. Cranford (✉)
St. Andrews Biological Station, St. Andrews, NB, Canada
e-mail: Peter.Cranford@dfo-mpo.gc.ca

A. C. Smaal et al. (eds.), *Goods and Services of Marine Bivalves*,
https://doi.org/10.1007/978-3-319-96776-9_8

是大规模的养殖设施会对水流产生阻碍作用,从而导致贝类对海水进行重复过滤。这些流体动力学以及生物过程会制约双壳贝类种群的最大净水效率,使之大大低于预期水平。养殖双壳贝类提供的生态系统服务功能对许多富营养化水域有着积极的控制效果(如Lindahl2011)。尽管基于生态系统水平的管理方法可以较好的进行双壳贝类养殖规划,但是我们仍然需要考虑高密度养殖对环境带来的其他方面的负面影响。.

Keywords Bivalve · Suspension feeding · Ecosystem services · Water clarification · Food depletion · Feeding physiology · *Mytilus*

关键词 双壳贝类 • 悬浮颗粒物滤食 • 生态系统服务 • 净水功能 • 食物消耗 • 摄食生理学 • 贻贝

8.1 Introduction

Suspension feeding, by definition, results in some degree of particle reduction in the surrounding water. This small-scale water clarification capacity of individual bivalves serves as the foundation for what is considered to be a manageable bioengineering tool for providing positive ecosystem-scale services. Dense bivalve aggregations, whether natural or introduced, can play a central role in some coastal ecosystems largely as a result of how their suspension feeding activities interact directly and indirectly with energy flow and nutrient cycling (e.g. Dame 1993, 1996; Newell 2004, chapters herein). Bivalve cultivation is widely considered to be an efficient means of nutrient removal in coastal areas affected by eutrophication as a result of the consumption of anthropogenic sources of nitrogen and phosphorous contained within phytoplankton and the sequestration of these excess nutrients into a harvestable bivalve biomass (Lindahl et al. 2005; Petersen et al. 2016). A primary symptom of eutrophication is the presence of high phytoplankton levels. The consumption of excess phytoplankton by bivalves represents a top-down control that increases water clarification and provides numerous ecological services including:

- Reduction in oxygen deficits in water and sediments by phytoplankton respiration during night, and the sedimentation of blooms that impact benthic habitat and communities through organic enrichment and microbial degradation (Newell 2004).
- Increased light penetration to the bottom, which enables recovery and expansion of sensitive sea grass and macroalgae that provide beneficial habitat for fish and invertebrates and promote increased biodiversity (Newell 2004; Petersen et al. 2016).
- Reduction in the occurrence of nuisance/toxic algal blooms (Edebo et al. 2000; Petersen 2004).
- Moderation of the infectious presence of some microbial pathogens (Pietrak et al. 2012).

- Sequestration of biotoxins and particle-reactive contaminants through bioaccumulation in bivalves and land-based detoxification (Lindahl et al. 2005)

Nutrient extraction by bivalves represents a cumulative effect from bivalve feeding activity over the full growth cycle, and can be readily quantified from harvest data. Water clarification (i.e. particle depletion and food reduction), however, represents a transient effect on the suspended particle field in a given area that is forced by a constantly changing balance between the rates of particle removal and resupply. Large-scale food depletion by the feeding activity of bivalves has largely been predicted based on mathematical modelling in the context of estimating the production carrying capacity of bivalve aquaculture areas (e.g. Grant et al. 2008). However, this approach poses many challenges as a result of the need to accurately characterize numerous model forcing functions (Ferreira et al. 2008). These functions include spatial and temporal parameters describing the physical, morphological and physiological characteristics of the bivalve population, as well as ecological and hydrodynamic parameters describing the model domain and exchanges with the outside region. The prediction of water clarification rates (i.e. clearance rate; L h^{-1}) of individual bivalves is also complicated by non-linear feeding responses to multiple internal and external forcing functions (e.g. temperature, salinity, food availability and composition, energy demands, and body size) that vary over different spatial and temporal scales (Cranford et al. 2011).

Until recently, particle depletion by bivalve aggregations was "rarely studied and seldom demonstrated" (Petersen et al. 2008). High temporal and spatial variability in many of the physiological and hydrodynamic processes that control particle depletion provide a significant challenge to directly measuring particle depletion at most spatial scales. This is particularly evident at scales that would clearly indicate an ecosystem service. However, direct evidence of the magnitude and extent of suspended particulate matter depletion by bivalve populations has increased in recent years, resulting in new insights on related ecological services. Given the available literature, this review focused largely on observations of food depletion by mussel (*Mytilus* spp.) aquaculture activities. Suspended mussel aquaculture practices are similar to those envisaged for eutrophication remediation, except that stocking densities for the latter may be higher (Nielsen et al. 2016). Unlike aquaculture, a bioremediation farm is believed to be less constrained by the negative feedback of excessive food depletion on the productivity, meat quality and appearance of the bivalves (Petersen et al. 2016).

8.2 Particles Captured by Suspension Feeding Bivalve Molluscs

Basic knowledge on the capacity of different species to capture different sources of marine particles is of obvious importance in determining a species ability to remove excessive quantities of those particles. Particle retention efficiency measurements, which are also reported as capture efficiency, provide information on the percentage of available suspended particles (numbers or concentration) that are captured as water is pumped through the mantle cavity and processed by the feeding apparatus. Retention efficiency (*RE*) depends on particle size and the relationship has generally been described as a rapid decrease below a threshold size between 3 and 7 µm. All particles larger than this threshold are effectively retained (e.g. Vahl 1972; Møhlenberg and Riisgård 1978; Riisgård 1988). *RE* measurements depend on a number of methodological factors (e.g. body size, pumping rate, flow-through feeding chamber geometry, static chamber sampling interval, etc.) and measured *RE* values are standardized by scaling the particle size showing maximal retention to 100% and normalising values obtained for the remaining particles to the maximal retention. Standardized *RE* values for several bivalve species are shown in Fig. 8.1. Particle retention in *M. edulis* has been reported to rapidly decrease below approximately 4 µm diameter (Fig. 8.1a) and this relationship with particle size has become firmly established in the literature. However, several field studies have reported distinctly different *RE* size distributions. Strohmeier et al. (2012) showed that blue mussels (*M. edulis*) in a Norwegian fjord exhibited a seasonally variable particle retention size spectra with a more gradual increase in *RE* from small to large particles and maximum *RE* occurring for particle sizes between 7 and 35 µm. These authors suggested that these seasonal changes in *RE* coincided with changes in the ambient particle size distribution. Cranford et al. (2016) assessed possible sources of error in *RE* measurements using natural seston and artificial diets and confirmed that the *RE* spectra of mussels is not always maximal for all particle sizes larger than 4 µm (Fig. 8.1b). Although it is still unclear if variable particle retention is

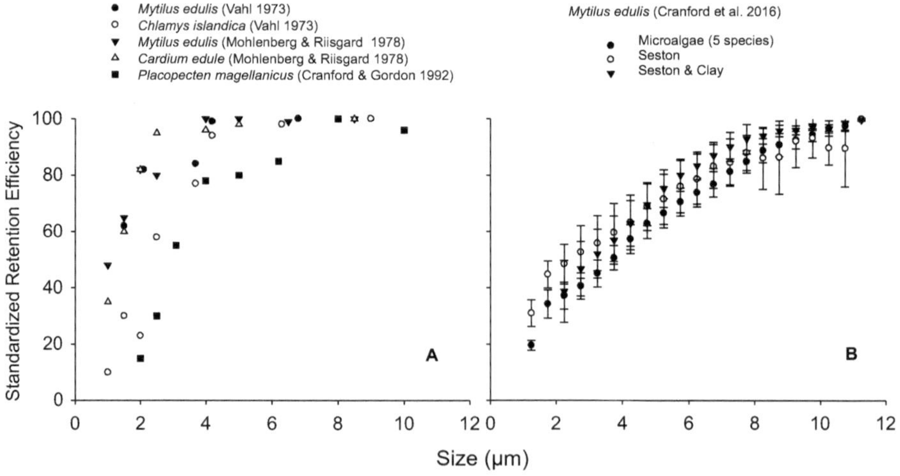

Fig. 8.1 (**a**) Relationship between standardized retention efficiency (%) and particle size for the indicated suspension feeding bivalve species (redrawn from references provided). (**b**) Relationships for *Mytilus edulis* from an oligotrophic Norwegian fjord fed three diets (mean ± SD; redrawn from data provided in Cranford et al. 2016)

physiologically controlled or results from some exogenous factor(s) influencing the feeding mechanism, the above evidence shows that the *RE* size spectra of bivalves does not consistently follow the traditional response shown in Fig. 8.1a (Cranford and Gordon 1992; Lucas et al. 1987; Stenton-Dozey and Brown 1992; Zhang et al. 2010; Strohmeier et al. 2012; Cranford et al. 2016).

The observations that the particle retention characteristics of some bivalve species can exhibit spatial and temporal variability has important implications for not just determining what particles are being effectively cleared from suspension by bivalves, but also for understanding how fast these particles are cleared from suspension. Cranford et al. (2016) showed that mussel clearance rates were underestimated by 25% if it was assumed that all particles larger than 4 μm particles were effectively retained. Similarly, seasonal clearance rates of mussels were shown to be underestimated by an average of 26% (0–48%) as a result of the same erroneous assumption (Strohmeier et al. 2012). The accurate quantification of feeding rates and efficiencies represents a critical first step towards understanding the biofiltration capacity of bivalve populations and related ecosystem services.

8.3 The Bivalve Feeding Zone

Bivalves only directly affect particle concentrations within their feeding zone. This zone is defined here as the distance to which they can capture food and directly alter the suspended particle field through their inhalent and exhalent flows. The feeding zone has been described using a variety of modelling, laboratory and field techniques. Fluorescence-based flow visualization has been used to provide a general characterization of the dynamic characteristics of the siphon jet flows in model bivalves (e.g. Monismith et al. 1990; O'Riordan et al. 1993). Dye studies in a flume with the cockle *Clinocardium nuttallii* indicated that flows from the incurrent siphon entrained fluid up to 4 cm laterally and 2 cm vertically (Ertman and Jumars 1988). Concentration boundary layers can form around dense assemblages of bivalves as a result of particle capture. Muschenheim and Newell (1992) measured suspended particulate matter concentrations in the bottom 50 cm of the water column above a mussel bed and detected strong vertical concentration gradients that indicated an effective feeding zone of 7 cm distance from the seabed. Measurements of vertical gradients in particle concentrations above other mussel beds have detected a depletion gradient up to 11 and 20 cm from the seabed (Saurel et al. 2013 and Petersen et al. 2013, respectively). This depletion-gradient approach has also been applied to suspended mussel ropes through the use of the 'siphon mimic' technique of Petersen et al. (2008). These experiments showed depletion of phytoplankton around mussel (*M. edulis*) ropes in a long-line farm in a highly eutrophic Danish fjord (Nielsen et al. 2016) and in a mussel (*M. galloprovincialis*) raft in Spain (Petersen et al. 2008). Both studies indicated an effective feeding zone of 10 to 20 cm, however, the average chlorophyll *a* depletion of 27–44% in the feeding zone in the Danish mussel farm was considerably lower than the 63–74% reduction measured at the Spanish raft.

The magnitude of particle depletion within the feeding zone reflects a balance between the rates of particle capture by the bivalves and resupply from external waters (Fréchette et al. 1989; Muschenheim and Newell 1992; Saurel et al. 2013). The rate of particle capture by a specific bivalve species depends on a wide range of endogenous and exogenous factors (reviewed by Cranford et al. 2011). Similarly, the rate that food particles are supplied to the feeding zone is a function of multiple factors that control ecosystem productivity and multi-scale hydrological processes that include sedimentation, resuspension, advection, diffusion and mixing (Fig. 8.2). A poorly studied process that is particularly important at determining the magnitude of food depletion at the scale of the feeding zone is water re-filtration by individual and neighboring siphonate bivalves. Re-filtration is generally overlooked when scaling individual feeding rates up to entire populations despite observations that the feeding zones of multiple individuals often overlap. Bivalve species with short siphons, such as mussels, will experience the highest amount of water re-filtration (Monismith et al. 1990; O'Riordan et al. 1993). The impact of this small-scale re-filtration behavior on predicting the ecosystem services of bivalves will be discussed in the following sections.

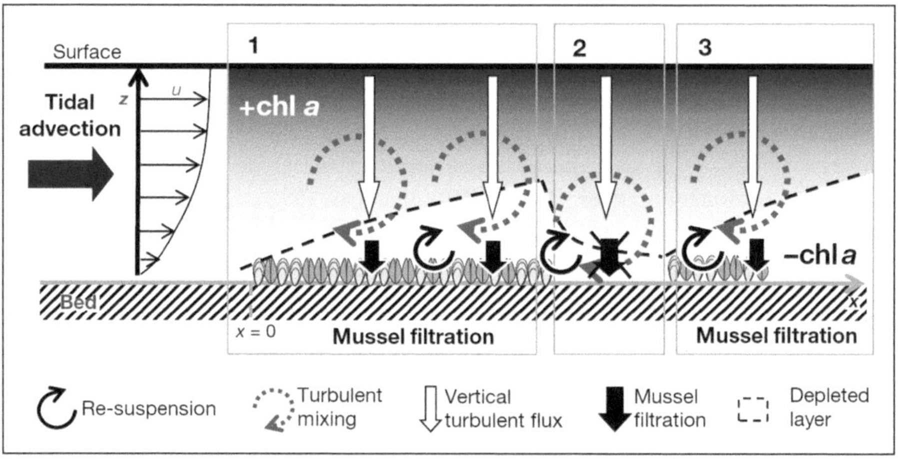

Fig. 8.2 Illustration of the particle concentration boundary layer over a mussel bed with a unidirectional flow showing some of the processes controlling the magnitude of particle depletion. The example includes a bare patch (Zone 2) in the middle of two mussel beds (Zones 1 and 3). *x* is mussel bed length, *z* is water depth and *u* is current velocity (reprinted from Saurel et al. 2013)

8.4 Local-Scale Particle Depletion

Given the substantial effect that dense bivalve aggregations can have on particle concentrations within their feeding zone, they may also be expected to influence the particle field at the local scale (defined here as a single bivalve aggregation or farm) through transport and mixing processes. At this scale, the potential for benthic populations to increase water column clarity and control excess phytoplankton is constrained by their spatial separation from much of the water column, vertical mixing processes and the available space for bivalves (Petersen 2004). Suspended culture operations, however, can provide bivalves with direct access to particles throughout the water column and can greatly increase bivalve biomass and biofiltration capacity per unit area relative to benthic populations. Particle depletion at the farm scale will occur if the time-scale for the stocked biomass to clear the farm volume (clearance time) is faster than the time-scale of hydrodynamic flushing (water residence time). Given the typical size of individual farms, it can be assumed that the time-scale for phytoplankton growth (turnover time) is of little relevance to the replenishment of particles inside the farm compared with hydrodynamic flushing.

Concentrated arrays of mussels suspended from long-lines and rafts have been cited to cause significant levels of seston depletion for several decades (e.g. Rosenberg and Loo 1983; Rodhouse et al. 1985; Navarro et al., 1991; Fréchette et al. 1991). However, the available empirical data has yielded variable results. For example, Ogilvie et al. (2000) and Grange and Cole (1997) found significant differences between phytoplankton abundance inside and outside mussel long-line farms in Beatrix Bay, New Zealand while Murdoch and Oliver (1995) could not find any differences. Similarly, Schröder et al. (2014) detected reductions in chlorophyll *a* levels downstream of a mussel farm, relative to upstream concentrations, but reported the opposite result for Secchi depth measurements at the same locations. Such discrepancies in the literature may be explained by differences in sampling approaches. Measuring food depletion on scales greater than the bivalve feeding zone can be problematic as temporal and spatial variability in natural food supplies in coastal regions have a tendency to mask the depletion signal (Bacher et al. 2003; Nielsen et al. 2016). Water sampling and particle sensing at discrete sampling stations may not adequately document spatial variability, and additional temporal variability can confound the interpretation of spatial data collected over the sampling period. However, in areas with low natural seston variability and a relatively high effect size (degree of particle depletion), this sampling approach has provided insights on the water clarification capacity of bivalve populations. For example, Strohmeier et al. (2005) collected chlorophyll *a* fluorescence profiles at discrete sampling stations along the length of a long-line mussel farm and showed that more than 30% of the incoming phytoplankton biomass was depleted within the first 30-m of the farm (Fig. 8.3). Petersen et al. (2008) also measured fluorescence profiles and collected water samples for chlorophyll *a* analysis within and around a Spanish mussel raft. These authors reported 10 to 45% reduction in phytoplankton biomass in the middle of the raft compared with reference stations.

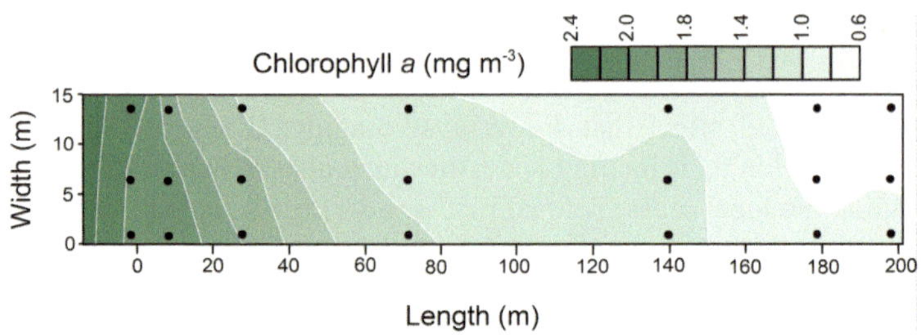

Fig. 8.3 Contour plot showing the decline in mean chlorophyll *a* concentrations at 4 m depth inside a mussel long-line farm as the water flows along the length of the farm from left to right. The symbol • indicates the location of measurements. Redrawn from Strohmeier et al. (2005)

A second approach used to document local-scale particle depletion is to place electronic particle sensors (e.g. chlorophyll *a* fluorometers and turbidity sensors) inside and outside farms for an extended period to provide information on both natural and bivalve-induced temporal variations in the suspended particle field. The disadvantage of this approach is that a large number of sensor moorings are required to provide sufficient horizontal and vertical coverage to adequately identify all sources of variations in particle concentrations and to determine the mean effect of the entire bivalve population. The size of most long-line mussel farms makes this approach impractical given that slight differences in the location of instruments can lead to contradictory results (Ogilvie et al. 2000). However, mussel rafts are considerably more compact than long-line farms and the moored sensor approach has proven effective at revealing the temporal dynamics and magnitude of seston and phytoplankton depletion inside the rafts. Cranford et al. (2014) used an array of moored *in situ* sensors (current speed, chlorophyll *a* fluorescence, optical scattering and laser diffraction particle counters) to study food depletion dynamics at two mussel rafts (*M. galloprovincialis*) in the Ría de Betanzos, Spain. Phytoplankton (chlorophyll *a*) and total suspended particulate matter depletion was similar at both rafts and averaged 40% and 17%, respectively. Food depletion was highest for particles in the 4–45 μm size range, which included much of the available phytoplankton, whereas the total particulate matter was dominated by larger particles that were shown to be less effectively cleared by the mussels.

A third sampling approach that has proven to be capable of quantifying both the magnitude and spatial extent of water clarification at the farm-scale is based on conducting rapid high-resolution synoptic surveys of chlorophyll *a* and/or turbidity using *in situ*, airborne or satellite sensors (Gibbs 2007; Grant et al. 2007; Cranford et al. 2008; Gernex et al. 2017). This survey approach has been particularly useful for studying the water clarification capacity of long-line mussel farms as it provides spatial data over a large area in a short period. The survey data can be statistically interpolated into a map that shows the particle field inside the farm and in adjacent waters. This mapping approach essentially combines control-impact (farm vs. reference) and gradient experimental sampling designs for detecting a non-point source effect. Grant et al. (2007) and Gernex et al. (2017) employed hyperspectral remote sensing to show significant changes in phytoplankton concentrations within a mussel lease in eastern Canada (Tracadie Bay) and in an oyster farm in France (Bourgneuf Bay), respectively. Data from specific spectral bands were used to generate an algorithm for chlorophyll *a* concentrations in coastal waters and these data suggest a decline in phytoplankton biomass within the farm. A more direct approach to mapping water clarification by mussels is described by Gibbs (2007) and Cranford et al. (2008) who employed vessel mounted and towed particle sensors, respectively, to rapidly survey particle concentrations within and around suspended mussel farms. Data collected in a bivalve growing area in the Marlborough Sounds with a continuous sampling pumped fluorometer showed patches of water depleted in phytoplankton (Gibbs 2007). The undulating towed-vehicle approach has the added advantage of providing particle concentration data throughout the full depth range of the mussel culture (Cranford et al. 2008; Nielsen et al. 2016). This 3-D sampling approach with a light, towed sensor vehicle has been employed at several long-line mussel culture sites in Norway (Cranford et al. 2008), The Netherlands (Cranford, unpublished data) and Denmark (Nielsen et al. 2016). Examples of phytoplankton and total particulate matter concentrations around mussel farms, obtained with this approach, are shown in Fig. 8.4.

The depth-averaged spatial distribution of suspended particle in a mussel farm shows substantial variations over time as a result of short-term fluctuations in current speed and direction and seasonal changes in the standing stock of mussels (e.g. Fig. 8.4a and b). The highest degree of food depletion often appears in the center of the farm (e.g. Fig. 8.4a and c), possibly as a result of increased water retention inside the farm or the enhanced mixing of water on the down-current side of the farm. While the concentration maps provide intuitive visual evidence of the water clarification capacity of long-line mussel farms, accurate calculations of total particle removal by the farmed mussel population requires adhering to certain methodological assumptions. Particle depletion calculations require comparison of concentrations inside and outside the farmed area, which assumes that the total surveyed area represents a single uniform water mass. This can be confirmed from contour maps of water density, which are calculated from CTD data that are also collected using the tow vehicle. For those surveys that show a plume of depleted water exiting the farmed area (e.g. Fig. 8.4b), the plume can be excluded from the reference data set. Based on this approach, Nielsen et al. (2016) provided the first direct estimates of the food depletion percentage averaged over the total farm volume. This mussel farm was specifically operated to capture and extract excess phytoplankton and nutrients from a eutrophic fjord. Particle concentrations inside the farmed volume were between 13 and 31% lower than in the reference area around the farm. The similar spatial distribution of both chlorophyll *a* and turbidity depletion (e.g. Fig. 8.4b) and the similar degrees of phytoplankton and total particle depletion provided a high degree of confidence in these farm-scale food depletion estimates. Although this farm has been shown to remove large quantities of excess nutrients (11–17 t N y^{-1} and 0.6–0.9 t P y^{-1}; Petersen et al. 2016), the mean reduction in chlorophyll *a* concentration did not change the water quality status of the site as defined by EU and Danish Nature Agency thresholds. The farm and reference sites were both classified as "moderate" during the sampling periods (6.0 to

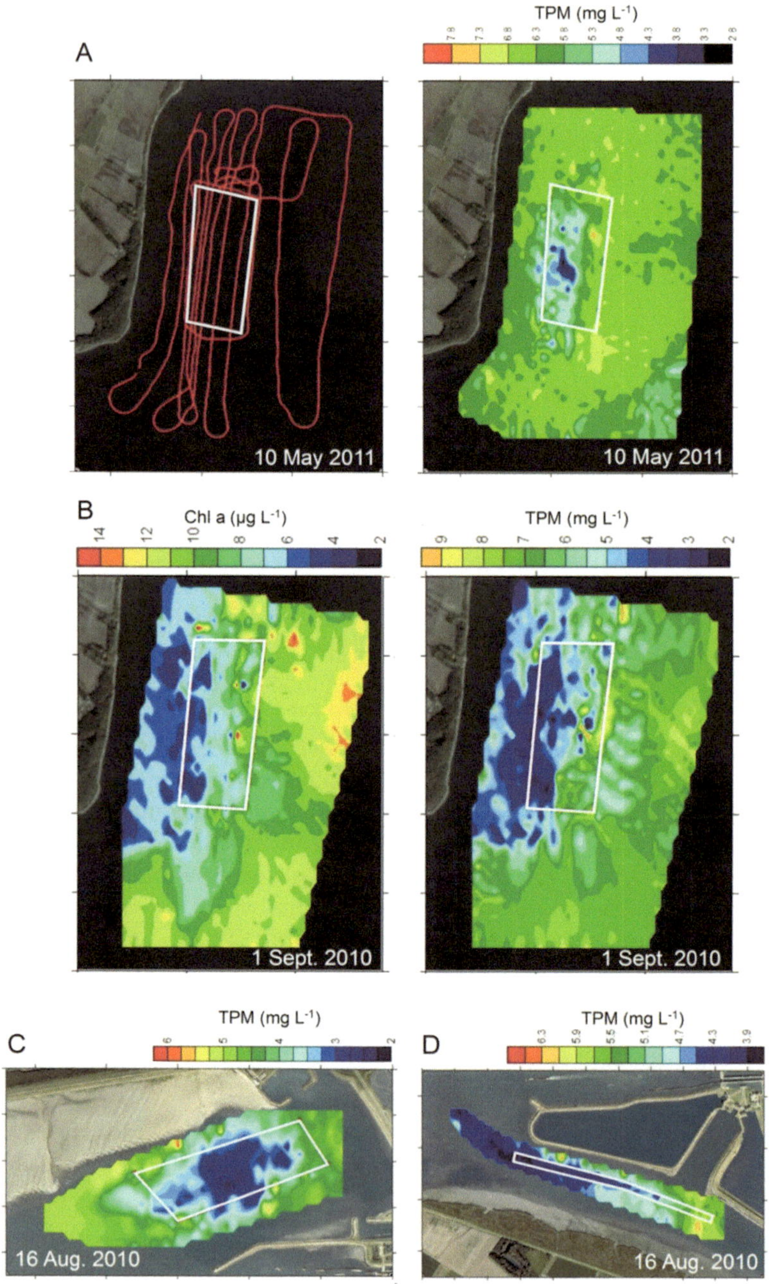

Fig. 8.4 Total particulate matter (TPM) and chlorophyll *a* (Chl *a*) concentrations averaged over the depth of suspended mussel farming areas (white polygon) in the Skive Fjord, Denmark (Plots **a** and **b**; reprinted from Nielsen et al. 2016) and in the Oosterchelde, The Netherlands (Plots **c** and **d**; Cranford, unpublished data). Plot **a** shows both the sensor tow track (left, red line) used to construct the concentration map (right)

9.0 μg L^{-1} thresholds). For the farm to reduce the status from "bad" to "moderate", chlorophyll *a* levels would have to be consistently reduced by 33%. This percentage reduction increases to 50% for changing the status from "poor" to "bad" at this site (9.0–18.0 μg L^{-1} thresholds).

8.5 Ecosystem-Scale Particle Depletion

Many coastal regions contain dense aggregations of bivalve filter-feeders that, under certain conditions, have the potential to affect suspended particulate concentrations at the ecosystem scale. Early conclusions on the capacity of natural populations to exert bay-scale grazing control on the phytoplankton and other suspended particulate matter largely originated from relatively simple predictions of the time it takes for a known population to clear the bay volume of particles (Cloern 1982; Officer et al. 1982; Nichols 1985; Hily 1991; Smaal and Prins 1993; Dame 1996; Dame and Prins 1998; Prins et al. 1998; Newell 2004). Observations of a feedback effect on bivalve growth, presumable resulting from bivalve populations over-exploiting their food supply, provided additional early evidence of the capacity of farmed bivalves to exert large-scale control on suspended particle dynamics. Carrying capacity modelling, which is often based on estimating the food depletion potential of bivalve farms, has subsequently flourished and supported general conclusions on the top-down grazing control of phytoplankton stocks and primary production (Grant and Filgueira 2011). The direct measurement of embayment-scale particle depletion by intensive bivalve farming has only recently become possible and is described in the following case studies.

One of the most intensive bivalve growing regions in Canada is Tracadie Bay, where approximately 45% of the bay volume is leased for long-line mussel culture (Fig. 8.5). Numerous ecosystem and carrying capacity modelling exercises for this shallow coastal lagoon have indicated a large controlling effect of the mussels on the phytoplankton (e.g. Cranford et al. 2007; Grant et al. 2008). Direct evidence of the substantial phytoplankton clearance capacity of the suspended mussel culture activities in this embayment was obtained by measuring the decline in phytoplankton in the embayment over a period of time when there were no important external sources of phytoplankton entering the system. The horizontal and vertical distribution of chlorophyll *a* in this embayment was rapidly surveyed at high resolution with a towed undulating sensor vehicle. Given the complex orientation of mussel lines in Tracadie Bay, it was not possible to completely survey the entire region in the same way as described above for individual farms. However, data obtained by resampling the same 4 km long tow track revealed a 29% reduction in phytoplankton concentrations in the bay (averaged across depth and distance) in a 3.75 h period during ebb tide (Fig. 8.5).

A second case study on the coastal-scale influence of mussel culture operations on suspended particulate matter comes from the floating raft (bateas) culture of *M. galloprovincialis* in the Galician region of Spain. This region supplies almost half of

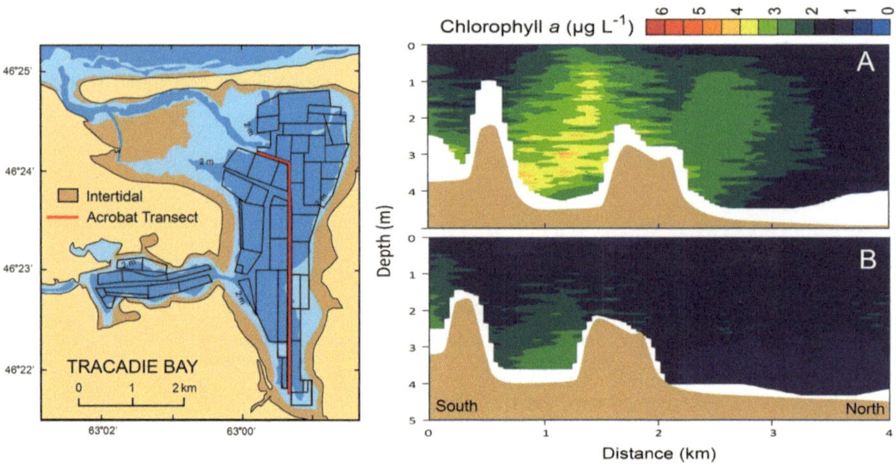

Fig. 8.5 Left: Map of Tracadie Bay showing areas leased for mussel culture and the 4 km tow vehicle transect (red line). Right: Chlorophyll *a* concentrations in the water column on June 18, 2003 along the transect at (**a**) high tide and (**b**) 3.75 h later on ebb tide (P. Cranford, unpublished data)

the mussels produced in all the European Union and the annual production in excess of 200,000 tonnes is second only to bivalve production in China (Cranford et al. 2014). The high productivity of mussel culture in the Galician coastal region can be attributed to natural hydrographic conditions that promote nutrient fertilization and enhanced primary production in the rías. Studies on food depletion from mussel rafts in the Ría de Ares-Betanzos, Spain were conducted as part of a Canada/Spain research collaboration. This is a relatively undeveloped aquaculture area compared with other estuaries in the region and represents approximately 3% of mussel production in Galicia with a total of 147 rafts that produce approximately 10,000 tonnes of mussels annually (Labarta et al. 2004). This case study focused on a cluster of 86 rafts in a coastal embayment near Lorbé (Fig. 8.6). The towed sensor vehicle was used to survey water column properties (CTD, chlorophyll *a* and total particulate matter) to a depth of 12 m in the area around the 86 rafts and in a reference area located outside the raft cluster (enclosed by dashed line in Fig. 8.6). Water column physical properties were similar in both the raft and reference areas during both surveys, indicating the presence of a single water mass throughout the survey domain.

Separate surveys conducted during ebb and flood tidal stages on 14 October, 2010 revealed lower chlorophyll *a* and total particulate matter concentrations in the coastal embayment containing the mussel rafts than in the outside reference area (Figs. 8.6 and 8.7). This was evident in both the surface layer (1–2 m depth; Fig. 8.6) and in deeper water (4–10 m depth; Fig. 8.7). An increasing landward depletion of particulate matter was generally indicated in both surveys as the inflowing water passed successive mussel rafts. The distribution of total particulate matter showed

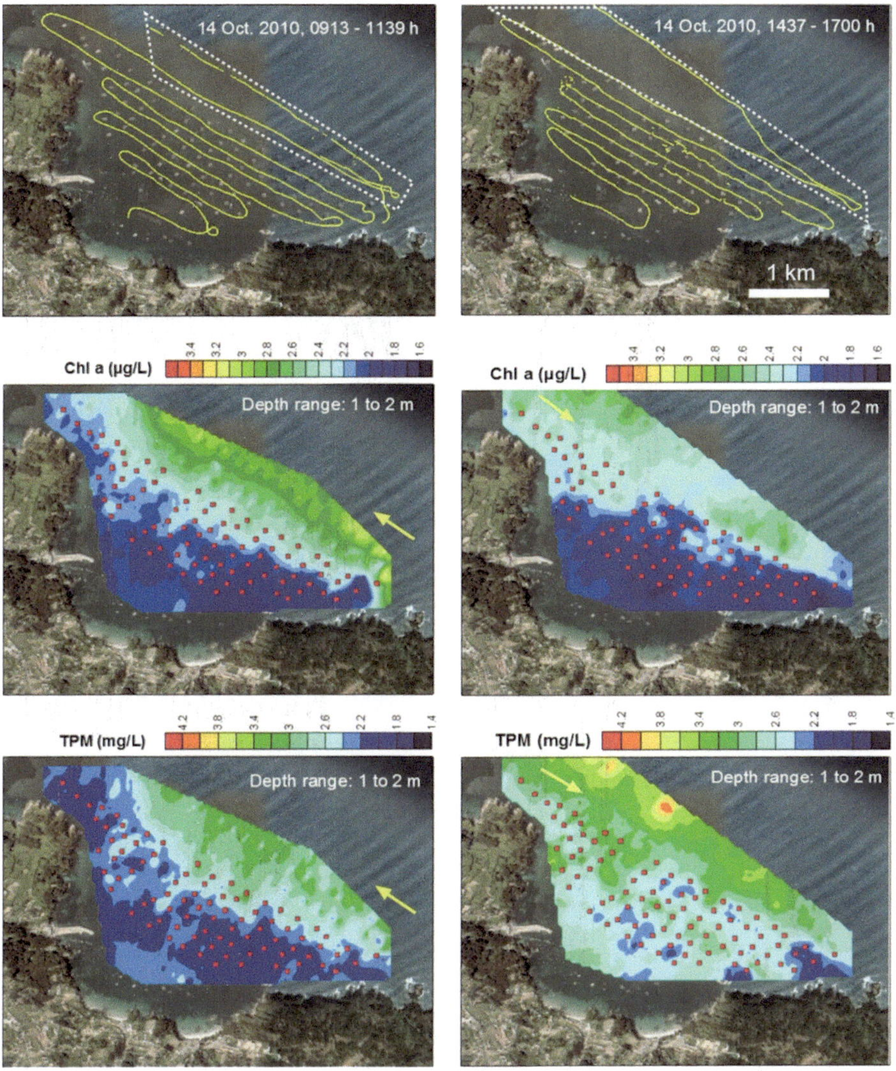

Fig. 8.6 Chlorophyll *a* (Chl *a*) and total particulate matter (TPM) distributions in surface water (1–2 m depth) around mussel rafts (red squares) in the Ría de Ares-Betanzos, Spain. The town of Lorbé can be seen in the lower left. The Acrobat tow tracks are shown in the upper plots along with the reference data area (dashed white polygon). The left and right plots represent ebb and flood tide conditions on 14 October 2010, respectively. The yellow arrows indicate the approximate tidal direction

greater spatial variability in the inner part of the embayment than chlorophyll *a*, perhaps as a result of land runoff, and/or the resuspension of sediments by wave action near the coastline. Consequently, the spatial distribution of chlorophyll *a* provided the most accurate indication of the water clarification capacity of the

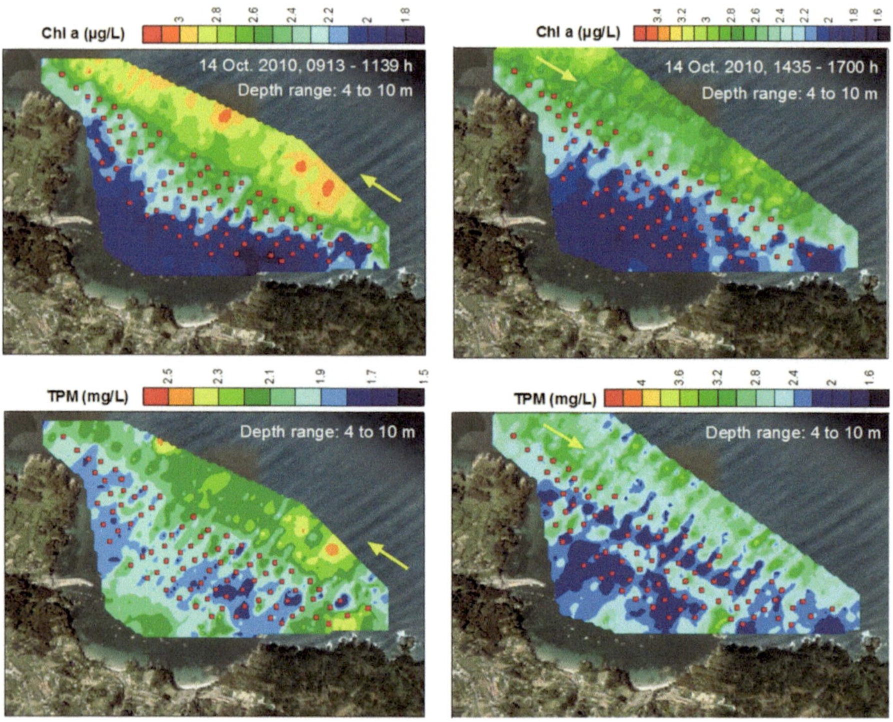

Fig. 8.7 Same as Fig. 8.6 except that the particle concentration data are from the 4 to 10 m depth range and the tow tracks (same as Fig. 8.6) are not shown

Table 8.1 Average (± 1 SD) chlorophyll *a* (Chl *a*) and total particulate matter (TPM) concentrations in the reference and remaining (mussel rafts) sections of the sensor tow path shown in Fig. 8.7. The average percentage reduction of each particle type was calculated for different depth zones and tidal phases

	Reference area		Raft area		Percentage reduction	
Data set	Chl *a* (µg L^{-1})	TPM (mg L^{-1})	Chl *a* (µg L^{-1})	TPM (mg L^{-1})	Chl *a*	TPM
Ebb tide						
1–2 m	2.69 (0.21)	2.72 (0.36)	2.10 (0.18)	2.27 (0.19)	22.0	16.4
4–10 m	2.67 (0.26)	2.05 (0.19)	2.29 (0.31)	1.93 (0.18)	14.0	5.7
Flood tide						
1–2 m	2.38 (0.17)	3.11 (0.34)	2.00 (0.37)	2.66 (0.37)	16.0	14.3
4–10 m	2.70 (0.30)	2.56 (0.45)	2.12 (0.23)	2.23 (0.38)	21.5	12.8

mussel rafts. Table 8.1 provides summary data on average concentration values in the reference and raft regions and the overall percentage reduction of particles by the cultured mussels. Although some areas of this embayment exhibited ca. 40% lower chlorophyll *a* concentrations than measured in the reference area (Figs. 8.6 and 8.7), average chlorophyll *a* levels in the raft region were between 14 and 22%

lower than measured in the outside area (Table 8.1). The high flux of particles into the raft area, facilitated by both high phytoplankton concentrations in the Ría de Ares-Betanzos and rapid tidal flushing, prevents higher levels of food depletion and facilitates the high growth rates and productivity of mussels in this, and other, Galician rías.

8.6 Self-Limitation of Water Clarification Capacity

The increasing availability of empirically derived parameters controlling the particle depletion capacity of bivalve farms is helping to increase certainty in model predictions. Measurements of particle depletion inside mussel rafts (*M. galloprovincialis*) in the Ría de Betanzos showed that phytoplankton depletion varied substantially over a tidal cycle, largely as a result of tidal variations in current speed (Cranford et al. 2014). These authors developed a raft-scale food depletion model that provided average values that were within 5 and 11% of the measured phytoplankton depletion for the two rafts. However, this level of accuracy was only achieved after accounting for the effect of raft-induced flow reduction on water re-filtration by the cultured mussels. The capacity of raft structures, including the mussels themselves, to slow the incoming current speed to the extent that the mussels begin to significantly re-filter water at certain tidal stages was shown to represent an important self-limitation on the maximum degree of food depletion. The raft model consistently overestimated the observed food depletion without considering this constraint. Bivalve clearance rate measurements are conducted in a manner that intentionally excludes any possibility of water re-filtration. However, re-filtration does occur in nature and results in a reduction in particle capture rate (filtration rate) even though clearance rate remains constant. The model correction required to account for water re-filtration will depend on local stocking densities, but the raft study indicated a linear decrease in the 'realized' clearance rate as current velocities fall below 2 cm s^{-1} (Cranford et al. 2014). Inclusion of a re-filtration function in the raft model resulted in particle depletion within the raft area being restricted to below a maximum of 56%.

Model development was also conducted in combination with direct measurements of food depletion at the commercial-scale nutrient extractive mussel farm in the highly eutrophic Skive Fjord, Denmark (Nielsen et al. 2016). As noted above, particle concentrations inside this 18.8 ha farm were between 13 and 31% lower than in the reference areas around the farm (Figs. 8.4a and b). The food depletion model was developed for this farm to estimate the optimal mussel density required to maximize removal of excess phytoplankton. Rather than using the conventional approach of parameterizing the model with experimental mussel clearance rates from the literature, realized clearance rates were calculated from the measured food depletion, current velocity and mussel density. These realized clearance rates were 40–74% lower than typical experimental values. Nielsen et al. (2016) suggested that this was the result of a high degree of water re-filtration during passage through this farm. Flow conditions in the Skive Fjord are primarily driven by fresh water runoff and the current speeds inside the farm averaged just 2.7 cm s^{-1} during the two periods when farm-scale food depletion was measured. At these low flows, the high stocking density of mussels (300–600 mussels m^{-3}) will re-filter a large fraction of the water entering their feeding zone. Water re-filtration by mussels results in particle capture rates being considerably lower than expected based on the rate of water processing by the feeding apparatus (i.e. experimental clearance rate). This reduction in particle filtration rate, combined with the relatively low rate of transport of particles to the feeding zone (i.e. low current speed) likely contributed to the relatively low food depletion percentage at this site (averaged 22%; Nielsen et al. 2016) compared with the Spanish rafts (averaged 40%; Cranford et al. 2014). The raft sites were stocked at a density of approximately 300 mussels m^{-3}. Although somewhat counter intuitive, the higher depletion at the raft site can be explained by the higher current speeds replenishing water in the feeding zone and thereby enabling the full capacity of the mussel to capture particles.

8.7 Ecosystem-Based Assessment of Biofiltration Services

Positive ecological benefits from the water clarification capacity of dense bivalve aggregations, whether predicted or observed, should not be assumed without first undertaking a detailed assessment of a wide range of potential ecosystem responses. Evidence of negative effects from the biofiltration activities of bivalves have largely been consigned to a few intensively cultured areas in shallow, poorly flushed waters; where the biodeposition of undigested organic matter in bivalve faeces can affect sediment geochemistry and benthic infaunal communities (e.g. Cranford et al. 2009; Hargrave et al. 2008; Burkholder and Shumway 2011). However, to ensure that human activities are carried out in a sustainable manner, an ecosystem-based and knowledge-based approach is essential for environmental management decision making. Consequently, some additional pathways of effect are considered in the following discussion to promote an objective assessment of the net ecological services provided by intensive bivalve grazing activities.

The reduced turbidity and increased light penetration to the bottom from bivalve grazing is expected to have a positive effect by extending the depth range of benthic macrophytes and microphytobenthos (Newell 2004; Newell and Koch 2004). Deslous-Paoli et al. (1998) reviewed studies on macrophyte distributions during the development of extensive oyster and mussel culture operations in the Thau Lagoon, France. They concluded that *Zostera* spp. extended its distribution from shallow regions to areas up to 5 m deep. However, De Casabianca et al. (1997) noted a shift from the dominance of *Zostera* in this same lagoon to communities composed of opportunistic algae (*Ulva* and *Gracilaria* spp.). Macrophytes also tend to be absent under extensive aquaculture structures as a result of shading (Deslous-Paoli et al. 1998). Although increased light penetration may be expected to permit an increase in microphytobenthos growth, this process may promote nitrogen retention in the system as opposed to the desired removal of nitrogen through denitrification by the sediment microbial community (Newell 2004). Similarly, the mussels can direct substantially more nitrogen to sediments in their biodeposits than is removed in the harvest, resulting in increased nitrogen retention and recycling in the coastal zone (Cranford et al. 2007). Consequently, the desired ecosystem service of bivalve stocks to promote excess nitrogen removal through water clarification may not be achieved under some conditions.

The numerous studies in the Oosterschelde estuary (The Netherlands), as described in Smaal et al. (2013), collectively provide a case study on some additional large-scale ecological interactions associated with the biofiltration activities of cultured mussels. Nutrient concentrations in this estuary are generally low but primary production is nutrient limited only for short periods because of the regulating role of bivalves. Bivalve grazing and their effects on nutrient regeneration initially stimulated primary production and phytoplankton turnover. However, the total filtration capacity of bivalves stocks in the estuary increased by 30% between 1995 and 2009, and a point was reached where grazing pressure controlled primary production (Smaal et al. 2013). The resulting switch from bottom-up to top-down (grazing) control of the phytoplankton was cited as the most likely cause for the observed 49% decline in primary production in the basin and possibly contributed to a 38% decrease in the annual growth of wild commercial cockles.

Ecosystem-scale phytoplankton depletion by bivalves may be expected to be accompanied by a shift in phytoplankton composition towards small algal cells. Bivalves do not effectively capture small nanoplankton and all types of picoplankton (photoautotrophic and heterotrophic). Picophytoplankton may be expected to thrive under high bivalve grazing pressure because the pelagic nano-protists that feed on their populations are also a trophic resource for bivalves (Dupuy et al. 1999; Maar et al. 2007; Nielsen and Maar 2007; Trottet et al. 2008). The rapid nutrient uptake and growth rates of small cells (Stockner 1988) can also be enhanced by bivalve-mediated effects on light penetration and nutrient regeneration. Enclosure experiments in which *M. edulis* were sufficiently abundant to deplete nano- and microphytoplankton, showed that the picoplankton became dominant (Riemann et al. 1988; Olsson et al. 1992; Granéli et al. 1993). Size-selective bivalve grazing has resulted in high abundance of picophytoplankton in the Thau Lagoon, France (Courties et al. 1994; Vaquer et al. 1996; Dupuy et al., 2000; Souchu et al. 2001), in land-locked Norwegian oyster ponds (Klaveness 1990), and in several other estuaries in Canada, Itay, and The Netherlands (Cranford et al. 2008; Caroppo 2000; Smaal et al. 2013). A shift towards small algal cells over a scale of 20 km was observed to accompany the significant depletion of phytoplankton in water passing a large natural mussel bed in the turbulent Öresund strait (Norén et al. 1999). These results largely confirm the hypothesis that intense bivalve grazing gives small phytoplankton a competitive advantage such that they dominate under conditions where bivalves exert significant control over the larger phytoplankton. The ecosystem consequences of such an alteration to the base of the food chain are largely unknown.

A weight of evidence exists to conclude that extensive bivalve aquaculture activities in several coastal areas may alter ecosystem structure and function as a result of their grazing influence on low-trophic level resources; including the phytoplankton, pelagic ciliates and flagellates, zooplankton, and detritus. A general conclusion from ecosystem modelling is that bivalve aquaculture routes energy flow towards benthic food webs instead of the pelagic (e.g. Dame 1996; Cranford et al. 2007; Filgueira and Grant 2009). Ecosystem-scale control of the phytoplankton and other pelagic trophic resources by suspended bivalves would represent a significant trophic interaction that may upset critical ecological equilibria, possibly resulting in cascading food web changes. Gibbs (2007) noted that several possible consequences can result from the ecosystem effects of intensive bivalve grazing; (1) reducing/replacing the role of zooplankton, (2) shifting benthic communities from filter- to deposit-feeders, and (3) redirecting energy flow and nutrient cycling in the microbial loop. Bivalve culture competes with the ecological role of mesozooplankton and may, under certain conditions, replace that role (Jiang and Gibbs 2005; Gibbs 2007). The transfer of energy up to other trophic levels through the consumption of bivalves by predators is considerably weaker than the transfer of energy through zooplankton. The magnitude of ecological services associated with bivalve grazing are always site-specific with the net result depending on multiple ecosystem processes. Positive effects of bivalve grazing are likely to occur in areas where bivalve populations serve as a manageable biofilter that improves water quality and lessens coastal eutrophication (e.g. Lindahl 2011). However, it is important to be mindful of the potential for less desirable ecosystem interactions that may be associated with intensive and extensive suspended bivalve aquaculture activities.

8.8 Conclusions

Studies in bivalve ecology have emphasized that phytoplankton dynamics in coastal regions may be strongly coupled with bivalve filter-feeding activity to the extent that the bivalve community plays a major ecological role in controlling phytoplankton biomass and trophic structure. Suspended culture operations provide the greatest potential for maximizing the water clarification services of bivalve filter-feeders owing to the direct access of dense populations to particulate matter throughout the water column. Direct measurements of the water clarification capacity of mussel farms have revealed up to 80% particle depletion inside some sections of the farm. When averaged across the total farm volume, some the most intensive suspended mussel aquaculture operations currently in production have been observed to reduce suspended particle concentrations by 13–31%. The spatial extent and magnitude of this control on the phytoplankton is always site- and time-specific as a result of factors controlling food consumption (e.g. intensity of culture, food availability and composition, and temperature) and food resupply processes such as tidal flushing and primary production. Coastal ecosystems are often highly productive and dynamic and exhibit a large facility to replenish bivalve-mediated food depletion through hydrological processes. Bivalve feeding can also exert a positive feedback on primary production (Dame 1996). Progressively increasing the standing stock of bivalves to achieve greater water clarification benefits will eventually lead to inefficiencies in bivalve feeding related to increased flow reduction from structure drag, which facilitates an increase in water re-filtration. These physical and biological processes ultimately constrain the maximum water clarification capacity of the population. Nutrient extraction in the bivalve harvest represents one of the most promising measures for controlling the consequences of anthropogenic nutrient supply to coastal waters (Petersen et al. 2016). Attempts to also maximize the ecological services from water clarification should consider possible interactions with nutrient extraction. This aligns with the concept of production carrying capacity in which aquaculture farm production (i.e. nutrient extraction) is maximized by preventing excessive food depletion (water clarification).

Acknowledgements The contributions of the many individuals that facilitated and participated in the synoptic surveys of food depletion at mussel aquaculture farms summarized in this review are gratefully recognized. Surveys conducted in Norway were orchestrated by Øivind Strand and Tore Strohmeier through funding by the Research Council of Norway. The work in Denmark were conducted as part of the Mussels-Mitigation and Feed for Husbandry (MuMiHus) project led by Jens Petersen and funded by the Danish Council for Strategic Research. Phytoplankton depletion studies in Spain were conducted under a Canadian-Spanish joint research project (Ecological Sustainability of Suspended Mussel Aquaculture-ESSMA) with assistance from Maria José Fernández-Reiriz and Uxio Labarta and funding from Fisheries and Oceans Canada. Studies in The Netherlands were assisted by Pauline Kamermans and Karin Troost as part of a project funded by the Dutch Ministry of Economic Affairs. Dr. Ingvar Olsen and dr Fred Jean are acknowledged for their review on the manuscript.

References

Bacher C, Grant J, Hawkins AJS, Fang J, Zhu M, Besnard M (2003) Modelling the effect of food depletion on scallop growth in Sungo Bay (China). Aquat Living Resour 16:10–24

Burkholder JM, Shumway SE (2011) Bivalve shellfish aquaculture and eutrophication. In: Shumway SE (ed) Shellfish aquaculture and the environment. Wiley-Blackwell, Hoboken

Caroppo C (2000) The contribution of picophytoplankton to community structure in a Mediterranean brackish environment. J Plankton Res 22:387–397

Cloern JE (1982) Does the benthos control phytoplankton biomass in South San Francisco Bay? Mar Ecol Prog Ser 9:191–202

Courties C, Vaquer A, Lautier J, Troussellier M, Chrétiennot-Dinet MJ, Neveux J, Machado C, Claustre H (1994) Smallest eukaryotic organism. Nature 370:255

Cranford PJ, Gordon DC (1992) The influence of dilute clay suspensions on sea scallop (*Placopecten magellanicus*) feeding-activity and tissue-growth. Neth J Sea Res 30:107–120

Cranford PJ, Strain PM, Dowd M, Hargrave BT, Grant J, Archambault MC (2007) Influence of mussel aquaculture on nitrogen dynamics in a nutrient enriched coastal embayment. Mar Ecol Prog Ser 347:61–78

Cranford PJ, Li W, Strand Ø, Strohmeier T (2008) Phytoplankton depletion by mussel aquaculture: high resolution mapping, ecosystem modeling and potential indicators of ecological carrying capacity. ICES CM 2008/H:12:1–5

Cranford PJ, Hargrave BT, Doucette L (2009) Benthic organic enrichment from suspended mussel (*Mytilus edulis*) culture in Prince Edward Island, Canada. Aquaculture 292:189–196

Cranford PJ, Ward JE, Shumway SE (2011) Bivalve filter feeding: variability and limits of the aquaculture biofilter. In: Shumway SE (ed) Shellfish aquaculture and the environment. Wiley-Blackwell, Hoboken
Cranford PJ, Duarte P, Robinson SMC, Fernández-Reiriz FJ, Labarta U (2014) Suspended particulate matter depletion and flow modification inside mussel (*Mytilus galloprovincialis*) culture rafts in the Ría de Betanzos. Spain J Exp Mar Biol Ecol 452:70–81
Cranford PJ, Strohmeier T, Filgueira R, Strand Ø (2016) Potential methodological influences on the determination of particle retention efficiency by suspension feeders: *Mytilus edulis* and *Ciona intestinalis*. Aquat Biol 25:61–73
Dame RF (1993) Bivalve filter feeders in estuarine and coastal ecosystem processes. Springer, New York
Dame RF (1996) Ecology of marine bivalves. An ecosystem approach. CRC Press, Boca Raton
Dame RF, Prins TC (1998) Bivalve carrying capacity in coastal ecosystems. Aquat Ecol 31:409–421
De Casabianca T, Laugier T, Collart D (1997) Impact of shellfish farming eutrophication on benthic macrophyte communities in the Thau lagoon, France. Aquacult Int 5:301–314
Deslous-Paoli JM, Souchu P, Mazouni N, Juge C, Dagault F (1998) Relations milieu-ressources: impact de la conchyliculture sur un environnement launaire mediterraneen (Thau). Oceanol Acta 21:831–843
Dupuy C, Le Gall S, Hartmann HJ, Bréret M (1999) Retention of ciliates and flagellates by the oyster *Crassostrea gigas* in French Atlantic coastal ponds: protists as a trophic link between bacterioplankton and benthic suspension-feeders. Mar Ecol Prog Ser 177:165–175
Dupuy C, Vaquer A, Lam HT, Rougier C, Mazouni N, Lautier J, Collos Y, Le Gall S (2000) Feeding rate of the oyster *Crassostrea gigas* in a natural planktonic community of the Mediterranean Thau Lagoon. Mar Ecol Prog Ser 205:171–184
Edebo L, Haamer J, Lindahl O, Loo LO, Piriz L (2000) Recycling of macronutrients from sea to land using mussel cultivation. Int J Environ Pollut 13:190–207
Ertman SC, Jumars PA (1988) Effects of bivalve siphonal currents on the settlement of inert particles and larvae. J Mar Res 46:797–813
Ferreira JG, Hawkins AJS, Monteiro P, Moore H, Service M, Pascoe PL, Ramos L, Sequeira A (2008) Integrated assessment of ecosystem-scale carrying capacity in shellfish growing areas. Aquaculture 275:138–151
Filgueira R, Grant J (2009) A box model for ecosystem-level management of mussel carrying capacity in a coastal bay. Ecosystems 12:1222–1233
Fréchette M, Butman CA, Geyer WR (1989) The importance of boundary-layer flows in supplying phytoplankton to the benthic suspension feeder, *Mytilus edulis* L. Limnol Oceanogr 34:19–36
Fréchette M, Booth DA, Myrand B, Bermard H (1991) Variability and transport of organic seston near a mussel aquaculture site. ICES Mar Sci Symp 192:24–32
Gernex P, Doxaran D, Barillé L (2017) Shellfish aquaculture from space: potential of Sentinel2 to monitor tide-driven changes in turbidity, chlorophyll concentration and oyster physilogical response at the scale of an oyster farm. Front Mar Sci 4:1–15
Gibbs MT (2007) Sustainability performance indicators for suspended bivalve aquaculture activities. Ecol Indic 7:94–107
Granéli E, Olsson P, Carlsson P, Graneli W, Nylander C (1993) Weak 'top-down' control of dinoflagellate growth in the coastal Skagerrak. J Plankton Res 15:213–237
Grange K, Cole R (1997) Mussel farming impacts. Aquaculture Update 19:1–3
Grant J, Filgueira R (2011) The application of dynamic modeling to prediction of production carrying capacity in shellfish farming. In: Shumway SE (ed) Shellfish aquaculture and the environment. Wiley-Blackwell, Hoboken
Grant J, Bugden G, Horne E, Archambault M-C, Carreau M (2007) Remote sensing of particle depletion by coastal suspension-feeders. Can J Fish Aquat Sci 64:387–390
Grant J, Bacher C, Cranford PJ, Guyondet T, Carreau M (2008) A spatially explicit ecosystem model of seston depletion in dense mussel culture. J Mar Systems 73:155–168
Hargrave BT, Doucette LI, Cranford PJ, Milligan TG (2008) Influence of mussel aquaculture on sediment organic enrichment in a nutrient-rich coastal embayment. Mar Ecol Prog Ser 365:137–149

Hily C (1991) Is the activity of benthic suspension feeders a factor controlling water quality in the Bay of Brest? Mar Ecol Prog Ser 69:179–188
Jiang W, Gibbs MT (2005) Predicting the carrying capacity of bivalve shellfish culture using a steady, linear food web model. Aquaculture 244:171–185
Klaveness D (1990) Size structure and potential food value of the plankton community to *Ostrea edulis* L. in a traditional Norwegian Østerspoil. Aquaculture 86:231–247
Labarta U, Fernández Reiriz MJ, Pérez Camacho A, Pérez Corbacho E (2004) Bateeiros, Mar, mejillón. Una perspectiva bioeconómica (Serie Estudios Sectoriales). Fundación Caixa Galicia. A Coruña, España
Lindahl O (2011) Mussel farming as a tool for re-eutrophication of coastal waters: experiences from Sweden. In: Shumway SE (ed) Shellfish aquaculture and the environment. Wiley-Blackwell, Hoboken
Lindahl O, Hart R, Hernroth B, Kollberg S, Loo LO, Olrog L, Rehnstam-Holm A, Svensson J, Svensson S, Syversen U (2005) Improving marine water quality by mussel farming: a profitable solution for Swedish society. Ambio 34:131–138
Lucas MI, Newell RC, Shumway SE, Seiderer LJ, Bally R (1987) Particle clearance and yield in relation to bacterioplankton and suspended particulate availability in estuarine and open coast populations of the mussel *Mytilus edulis*. Mar Ecol Prog Ser 36:215–224
Maar M, Nielsen TG, Bolding K, Burchard H, Visser AW (2007) Grazing effects of blue mussel *Mytilus edulis* on the pelagic food web under different turbulence conditions. Mar Ecol Prog Ser 339:199–213
Møhlenberg F, Riisgård HU (1978) Efficiency of particle retention in 13 species of suspension feeding bivalves. Ophelia 17:239–246
Monismith SG, Koseff JR, Thompson JK, O'Riordan KA, Nepf HM (1990) A study of model bivalve siphonal currents. Limnol Oceanogr 35:680–696
Murdoch R, Oliver M (1995) Study of chlorophyll concentrations within and around mussel farms: Beatrix Bay, Pelorus Sound. 1995r6-WN, National Institute of Water and Atmospheric Research, Wellington
Muschenheim D, Newell C (1992) Utilization of seston flux over a mussel bed. Mar Ecol Prog Ser 85:131–136
Navarro E, Iglesias J, Perez-Camacho A, Labarta U, Beiras R (1991) The physiological energetics of mussels (*Mytilus galloprovincialis* L) from different cultivation rafts in the Ria de Arosa (Galicia, NW Spain). Aquaculture 94:197–212
Newell RIE (2004) Ecosystem influences of natural and cultivated populations of suspension-feeding bivalve molluscs: a review. J Shellfish Res 23:51–61
Newell RIE, Koch EW (2004) Modeling seagrass density and distribution in response to changes in turbidity stemming from bivalve filtration and seagrass sediment stabilization. Estuaries 27:793–806
Nichol FH (1985) Increased benthic grazing: an alternative explanation for low phytoplankton biomass in northern San Francisco Bay during the 1976-1977 drought. Est Coast Shelf Sci 21:379–388
Nielsen TG, Maar M (2007) Effects of a blue mussel *Mytilus edulis* bed on vertical distribution and composition of the pelagic food web. Mar Ecol Prog Ser 339:185–198
Nielsen P, Cranford PJ, Maar M, Petersen JK (2016) Magnitude, spatial scale and optimization of ecosystem services from a nutrient extraction mussel farm in the eutrophic Skive Fjord, Denmark. Aqua Env Int 8:311–329
Norèn F, Haamer J, Lindahl O (1999) Changes in the plankton community passing a *Mytilus edulis* mussel bed. Mar Ecol Prog Ser 191:87–194
O'Riordan CA, Monismith SG, Koseff JR (1993) A study of concentration boundary-layer formation over a bed of model bivalves. Limnol Oceanogr 38:1712–1729
Officer CB, Smayda TJ, Mann R (1982) Benthic filter feeding: a natural eutrophication control. Mar Eco Prog Ser 9:203–210
Ogilvie SC, Ross AH, Schiel DR (2000) Phytoplankton biomass associated with mussel farms in Beatrix Bay, New Zealand. Aquaculture 181:71–80

Olsson P, Graneli E, Carlsson P, Abreu P (1992) Structure of a post spring phytoplankton community by manipulation of trophic interactions. J Exp Mar Biol Ecol 158:249–266

Petersen JK (2004) Grazing on pelagic primary producers – the role of benthic suspension feeders in estuaries. In: Nielsen SL, Banta G, Pedersen MF (eds) Estuarine nutrient cycling: the influence of primary producers. Kluwer Academic, Dordrecht, pp 129–152

Petersen JK, Nielsen TG, van Duren L, Maar M (2008) Depletion of plankton in a raft culture of *Mytilus galloprovincialis* in Ria de Vigo, NW Spain. I. Phytoplankton. Aquat Biol 4:113–125

Petersen JK, Maar M, Ysebaert T, Herman P (2013) Near-bed gradients in particles and nutrients above a mussel bed in the Limfjorden: influence of physical mixing and mussel filtration. Mar Ecol Prog Ser 490:137–146

Petersen JK, Saurel C, Nielsen P, Timmermann K (2016) The use of shellfish for eutrophication control. Aquac Int 24:857–878

Pietrak MR, Molly SD, Bouchard DA, Singer JT, Bricknell I (2012) Potential role of *Mytilus edulis* in modulating the infectious pressure of Vibrio *anguillarum* 02β on an integrated multi-trophic aquaculture farm. Aquaculture 326–329:36–39

Prins TC, Smaal AC, Dame RF (1998) A review of the feedbacks between bivalve grazing and ecosystem processes. Aquat Ecol 31:349–359

Riemann B, Nielsen TG, Horsted SJ, Bjernsen PK, Pock-Steen J (1988) Regulation of phytoplankton biomass in estuarine enclosures. Mar Ecol Prog Ser 48:205–215

Riisgård HU (1988) Efficiency of particle retention and filtration-rate in 6 species of northeast American bivalves. Mar Ecol Prog Ser 45:217–223

Rodhouse PG, Roden CM, Hensey MP, Ryan TH (1985) Production of mussels, *Mytilus edulis,* in suspended culture and estimates of carbon and nitrogen flow: Killary Harbour, Ireland. J Mar Biol Ass UK 65:55–68

Rosenberg R, Loo L-O (1983) Energy flow in a *Mytilus edulis* culture in western Sweden. Aquaculture 53:151–161

Saurel C, Petersen JK, Wiles P, Kaiser MJ (2013) Turbulent mixing limits mussel feeding: direct estimates of feeding rate and vertical diffusivity. Mar Ecol Prog Ser 485:105–121

Schröder T, Stank J, Schernewski G, Krost P (2014) The impact of a mussel farm on water transparency in the Kiel Fjord. Ocean Coast Manag 101:42–52

Smaal AC, Prins TC (1993) The uptake of organic matter and the release of inorganic nutrients by suspension feeding bivalve beds. In: Dame RF (ed) Bivalve filter feeders in estuarine and coastal ecosystem processes. Springer, Heidelberg, pp 273–298

Smaal AC, Schellekens T, van Stralen MR, Kromkamp JC (2013) Decrease of the carrying capacity of the Oosterschelde estuary (SW Delta, NL) for bivalve filter feeders due to overgrazing? Aquaculture 404–405:28–34

Souchu P, Vaquer A, Collos Y, Landrein S, Deslous-Paoli J-M, Bibent B (2001) Influence of shellfish farming activities on the biogeochemical composition of the water column in Thau lagoon. Mar Ecol Prog Ser 218:141–152

Stenton-Dozey JME, Brown AC (1992) Clearance and retention efficiency of natural suspended particles by the rock-pool bivalve *Venerupis corrugatus* in relation to tidal availability. Mar Ecol Prog Ser 82:175–186

Stockner JG (1988) Phototrophic picoplankton: and overview from marine and freshwater ecosystems. Limnol Oceanogr 33:765–775

Strohmeier T, Aure J, Duinker A, Castberg T, Svardal A, Strand Ø (2005) Flow reduction, seston depletion, meat content and distribution of diarrhetic shellfish toxins in a long-line blue mussel (*Mytilus edulis*) farm. J Shellfish Res 24:15–23

Strohmeier T, Strand Ø, Alunno-Bruscia M, Duinker A, Cranford PJ (2012) Variability in particle retention efficiency by the mussel *Mytilus edulis*. J Exp Mar Biol Ecol 412:96–102

Trottet A, Roy S, Tamigneaux E, Lovejoy C, Tremblay R (2008) Impact of suspended mussels (*Mytilus edulis* L.) on plankton communities in a Magdalen islands lagoon (Québec, Canada): A mesocosm approach. J Exp Mar Biol Ecol 365:103–115

Vahl O (1972) Efficiency of particle retention in *Mytilus edulis* L. Ophelia 10:17–25

Vaquer A, Troussellier M, Courties C, Bibent B (1996) Standing stock and dynamics of picophytoplankton in the Thau Lagoon (Northwest Mediterranean Coast). Limnol Oceanogr 41:1821–1828

Zhang JH, Fang JG, Liang XM (2010) Variations in retenti on efficiency of bivalves to different concentrations and organic content of suspended particles. Chin J Oceanol Limnol 28:10–17

Chapter 9
Feedbacks from Filter Feeders: Review on the Role of Mussels in Cycling and Storage of Nutrients in Oligo- Meso- and Eutrophic Cultivation Areas

Henrice Maria Jansen, Øivind Strand, Wouter van Broekhoven, Tore Strohmeier, Marc C. Verdegem, and Aad C. Smaal

Abstract Cultured and wild bivalve stocks provide ecosystem services through regulation of nutrient dynamics; both by regeneration of nutrients that become available again for phytoplankton production (positive feedback), and by extraction of nutrients through filtration and storage in tissue (negative feedback). Consequently, bivalves may fulfil a role in water quality management. The magnitude of regulating services by filter feeding bivalves varies between coastal ecosystems. This review uses the blue mussel as a model species and evaluates how cultured mussel stocks regulate nutrient dynamics in oligo- meso- and eutrophic ecosystems. We thereby examine (*i*) the eco-physiological response of mussels, and (*ii*) the positive and negative feedback mechanisms between mussel stocks and the surrounding ecosystem. Mussel culture in nutrient-poor areas (deep Norwegian fjords) are compared with cultures in other coastal systems with medium- to rich nutrient conditions. It was found that despite differences in eco-physiological rates under nutrient-poor

H. M. Jansen (✉)
Institute of Marine Research (IMR), Bergen, Norway

Wageningen UR – Wageningen Marine Research (WMR), Yerseke, The Netherlands
e-mail: henrice.jansen@wur.nl

Ø. Strand · T. Strohmeier
Institute of Marine Research (IMR), Bergen, Norway
e-mail: oivind.strand@imr.no; tore.strohmeier@imr.no

W. van Broekhoven · A. C. Smaal
Wageningen UR – Wageningen Marine Research (WMR), Yerseke, The Netherlands

Department of Aquaculture and Fisheries, Wageningen University,
Wageningen, The Netherlands
e-mail: wouter2.vanbroekhoven@wur.nl; aad.smaal@wur.nl

M. C. Verdegem
Department of Aquaculture and Fisheries, Wageningen University,
Wageningen, The Netherlands
e-mail: marc.verdegem@wur.nl

A. C. Smaal et al. (eds.), *Goods and Services of Marine Bivalves*,
https://doi.org/10.1007/978-3-319-96776-9_9

conditions (higher clearance, lower egestion, similar excretion and tissue storage rates), the proportion of nutrients regenerated was similar between (deep) nutrient-poor and (shallow) nutrient-rich areas. Of the filtered nutrients, 40–50% is regenerated and thus made available again for phytoplankton growth, and 10–50% of the filtered nutrients is stored in tissue and could be removed from the system by harvest. A priori, we inferred that as a consequence of low background nutrient levels, mussels would potentially have a larger effect on ecosystem functioning in nutrient-poor systems and/or seasons. However, this review showed that due to the physical characteristics (volume, water residence time) and low mussel densities in nutrient-poor Norwegian fjord systems, the effects were lower for these sites, while estimates were more profound in shallow nutrient-rich areas with more intensive aquaculture activities, especially in terms of the negative feedback mechanisms (filtration intensity).

Abstract in Chinese 养殖及野生的双壳类动物通过调节环境营养物质动力学过程来提供生态系统服务:其中包括向环境释放营养物质促进浮游植物生长(正反馈)以及通过滤食将环境中的营养物质转化为软组织进行储存(负反馈)。因此,双壳贝类可以作为水质调控的工具物种发挥作用。 双壳贝类滤食所产生的调节作用与效果因所处不同的近岸生态系统而异。本文以紫贻贝为参考物种,阐述了养殖的贻贝种群如何调控不同营养水平的生态系统营养动力过程。内容包括:贻贝的生态生理响应;不同种群数量的贻贝与周围生态系统之间的正负反馈机制。我们对贫营养地区(挪威深海峡湾)的养殖贻贝与其他沿海中等营养水平和富营养状况下的养殖贻贝进行了比较。结果表明,尽管在营养不良条件下,贻贝的摄食生态生理效率存在差异(更高的滤食率,较低的排粪率,相似的排泄和组织储存效率),但是在营养贫乏水域(水深较深)和营养充足水域(水深较浅),贻贝向环境释放的营养物质的比例大致相同。在被滤食的营养物质中,大约40-50%再生并被浮游植物生长利用,大约10-50%的滤食营养物质被储存在组织中,通过收获从生态系统中移出。种种迹象表明,贻贝可能会对营养贫乏的生态系统功能有较大的影响。但需要指出的是,尽管挪威峡湾内的营养较匮乏,但由于其水文特征(水体体积,水滞留时间等)和较低的贻贝养殖密度,贻贝养殖对峡湾的生态环境影响较低,而在浅海营养丰富的水域,由于养殖规模和密度的增加,贻贝强大的滤水能力对生态系统的影响更 大。.

Keywords Nitrogen · mytilus · Eco-physiology · Ecosystem interactions · Sink and source

关键词 氮 • 贻贝 • 生理生态学 • 生态系统相互作用 • 汇与源

9.1 Introduction

Suspension-feeding bivalves have the potential to influence ecosystem functioning due to their eco-physiological responses and role in nutrient cycling (Dame 1996; Newell 2004). Filtration by bivalves may depress phytoplankton biomass, while at the same time nutrient regeneration by bivalves may stimulate phytoplankton production (Asmus and Asmus 1991; Prins et al. 1995; Shumway 2011). These processes are regarded as the positive and negative feedback mechanisms of bivalves onto phytoplankton populations (Dame 1996). The capacity to influence ecosystem functioning is particularly evident in areas with concentrated bivalve communities (Smaal and Prins 1993; Dame and Prins 1998), such as in aquaculture settings. Mussels dominate bivalve production in many regions (see Wijsman et al. 2019), hence this paper uses the blue mussel *Mytilus* spp. as model species to discuss the role of bivalve cultivation in nutrient cycling. Whether the feedback processes contribute to a desirable regulation of the system (service) or results in an undesirable effect (impact) depends on the environmental characteristics of a site and the scale of culture activities (Newell 2004). Most mussel cultivation sites are situated in nutrient-rich coastal areas that are influenced by river run-off, thereby taking advantage of high primary production rates to achieve rapid growth (Saxby 2002; Smaal 2002), yet commercial mussel cultivation does exist in oligotrophic ecosystems (Strohmeier et al. 2008; Brigolin et al. 2009). Such differences in ecosystem characteristics indicate that the same process in some systems can be regarded as a regulating ecosystem service while in other systems it is rather a negative ecosystem impact (see Fig. 9.1). Under excessive nutrient availability, filtration of phytoplankton (negative feedback) may help to prevent or overcome eutrophication problems (particularly when coupled with harvesting of the biomass), wherefore this has been recognized as an ecosystem service of mussel aquaculture (Lindahl et al. 2005; Ferreira et al. 2014; Petersen et al. 2014). At the same time, in oligotrophic (nutrient-poor) systems mussel filtration can impose an ecosystem impact when it leads to depletion of phytoplankton and carrying capacity is exceeded. In these nutrient-poor systems, regeneration of nutrients is considered an ecosystem service as it may boost primary production, and result in higher mussel yields.

This paper aims to evaluate the regulating functions of mussel aquaculture through the two major pathways (filtration, nutrient regeneration) as a function of ecosystem trophic status (from nutrient-poor, to nutrient-rich). A relatively large set of literature is available presenting eco-physiological rates measured in nutrient-rich conditions (a.o. Bayne and Scullard 1977; Hawkins and Bayne 1985; Dame et al. 1991; Smaal and Vonck 1997; Filgueira et al. 2010), but because little information

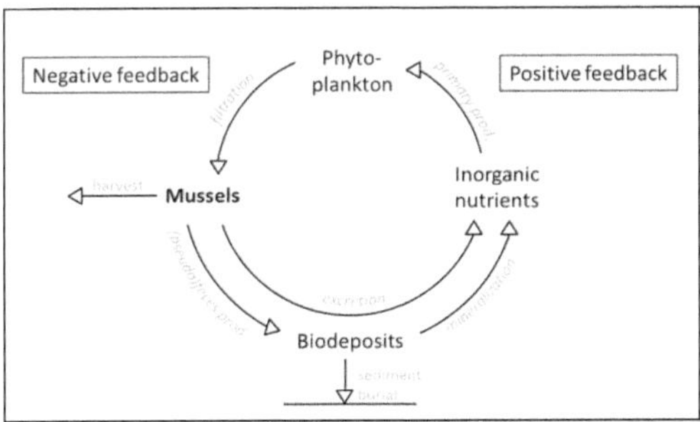

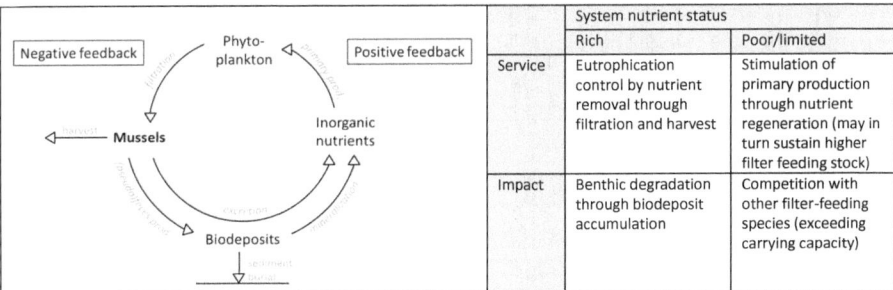

	System nutrient status	
	Rich	Poor/limited
Service	Eutrophication control by nutrient removal through filtration and harvest	Stimulation of primary production through nutrient regeneration (may in turn sustain higher filter feeding stock)
Impact	Benthic degradation through biodeposit accumulation	Competition with other filter-feeding species (exceeding carrying capacity)

Fig. 9.1 Feedback loop of filter feeder activity on filter feeder growth linked to potential ecosystem services and ecosystem impacts for nutrient-rich and nutrient-poor systems

was available for nutrient-poor conditions, most information in this paper was drawn from oligotrophic Norwegian fjords (Strohmeier et al. 2009; Jansen et al. 2012a, b). The *first section* provides a review of eco-physiological rates and discusses whether and how the functioning of mussels differs between eutrophic and oligotrophic conditions. Specific emphasis is thereby given to differences between measurements on individuals compared to entire communities. Physiological processes are generally studied at the level of the organism (Dame 1996; Gosling 2015), but extrapolating "average" individual rates to yield population estimates neglects community specific effects such as refiltration or metabolic activity of associated fauna and microbial decomposition of organic material on mussel cultures (Richard et al. 2006; Jansen et al. 2011). The *second section* of this review evaluates interactions between mussel cultivation and the surrounding ecosystem with particular reference to ecosystem services and impacts. To this end, the positive and negative feedback mechanisms of mussel culture on phytoplankton are compared between areas spanning a gradient from nutrient-poor to nutrient-rich. *At last*, perspectives on the role of mussel cultivation on nutrient cycling are provided.

9.2 Mussels as Intermediaries in Nutrient Cycling (Eco-Physiology)

The major eco-physiological pathways in which mussels interact with coastal nutrient cycling are; *(i)* filtration of seston (particulate nutrients) from the water column, *(ii)* nutrient storage in mussel tissue (assimilation), and growth, *(iii)* excretion of inorganic metabolic waste products, and *(iv)* production and mineralization of biodeposits (reviews by Prins et al. 1998; Newell 2004). The mussel *Mytilus edulis* is one of the most studied bivalves in terms of its eco-physiological responses (Bayne 1998; Shumway 2011; Gosling 2015). These studies have shown that mussels tolerate a wide range of environmental conditions, facilitated by a remarkable plasticity of their physiological responses. This physiological plasticity can vary between populations, among individuals of the same population, and due to seasonal changes and variation in the natural environment (Hawkins and Bayne 1992; Shumway 2011). In the following section eco-physiological rates are reviewed for mussels as a function of trophic status of the culture environment, thereby specifically addressing differences between individual and community scale measurements.

9.2.1 Filtration

Bivalve feeding has been extensively studied at the level of individual animals (see review by Cranford et al. 2011). Strohmeier et al. (2009, 2015) showed that mussels can display high feeding rates and high net absorption efficiencies under oligotrophic and low seston conditions despite contradicting feeding paradigms for mussels; Table 9.1 and the review by Cranford et al. (2011) show that clearance rates

Table 9.1 Clearance rates in mussel cultivation areas

Area	Country	Species	Food source	Clearance rates [l g^{-1} h^{-1}]		Ref
Measurements on individuals						
Åfjord	NO	*M. edulis*	Natural seawater	5.4	(3.2–8.4)	1
Austevoll	NO	*M. edulis*	Natural seawater	6.4	(3.0–9.6)	2
Oosterschelde	NL	*M. edulis*	Natural seawater		(1.4–2.8)	3
Oosterschelde	NL	*M. edulis*	Natural + *P tricornutum*	1.5	(0.3–3.5)	4
Oosterschelde	NL	*M. edulis*	Natural seawater	2.6	(1.3–3.5)	5
Oosterschelde	NL	*M. edulis*	Natural + *S costatum*		(5.0–8.5)	6
Lynher estuary	UK	*M. edulis*	Natural seawater		(1.0–2.5)	7
Aiguillon	FR	*M. edulis*	Natural + *S costatum*		(9.6–11.0)	6
Ria de Arousa	ESP	*M. galloprovincialis*	Mix sediment & *I galbana*	5.0–5.8		8
New Foundland	CA	*M. edulis*	Natural seawater		(1.5–2.0)	9
Nova Scotia	CA	*M. edulis*	Natural seawater		(1.0–8.0)	10
New Foundland	CA	*M. edulis*	Natural seawater		(0.2–3.5)	10
Great Entry Lagoon	CA	*M. edulis*	Algae mix		(3.0–4.5)	11
Amherst Basin	CA	*M. edulis*	Algae mix		(2.5–4.0)	11
Beatrix Bay	NZ	*P. canaliculus*	Natural seawater		(0.8–3.9)	12
Measurements on communities (benthic mussel beds)						
Sylt	DEN	*M. edulis*	Natural seawater	1.1		13
Waddensea	NL	*M. edulis*	Natural seawater	1.5	(0.7–1.9)	14
Oosterschelde	NL	*M. edulis*	Natural seawater	2.2	(1.1–4.8)	5
Marennes-Oleron	FR	*M. edulis*	Natural seawater	1.8	(1.0–2.9)	15
Measurements on communities (suspended ropes)						
Åfjord	NO	*M. edulis*	Natural seawater	1.5	(1.0–2.1)	1
Oosterschelde	NL	*M. edulis* spat	Natural seawater		(2.4–30.7)	16
Waddensea	NL	*M. edulis* spat	Natural seawater	0.8		17
Havre-aux-Maisons	CA	*M. edulis*	Natural seawater		(1.7–6.3)	18

Data were standardized to L g^{-1} tissue DW h^{-1}. Weight conversion factors reported by Ricciardi and Bourget (1998) were applied. Values are presented as mean (minimum – maximum), and empty cells indicate that rates were not determined. Country codes (also for following tables): *NO* Norway, *SW* Sweden, *DEN* Denmark, *GER* Germany, *NL* The Netherlands, *NIR* Northern Ireland, *UK* United Kingdom, *FR* France, *ESP* Spain, *IT* Italy, *CA* Canada, *USA* United States, *AU* Australia, *NZ* New Zealand, *JP* Japan

1 (Jansen 2012); 2 (Strohmeier et al. 2009); 3 (Smaal and Vonck 1997); 4 (Smaal et al. 1997); 5 (Prins et al. 1996); 6 (Petersen et al. 2004); 7 (Bayne and Widdows 1978); 8 (Filgueira et al. 2008); 9 (Thompson 1984); 10 (MacDonald and Ward 2009); 11 (Tremblay et al. 1998); 12 (James et al. 2001); 13 (Asmus et al. 1990); 14 (Prins et al. 1994); 15 (Smaal and Zurburg 1997); 16 (van Broekhoven et al. 2014); 17 (Jacobs et al. 2015); 18 (Trottet et al. 2008a)

reported for individual mussels under oligotrophic conditions in Norway were among the highest reported for this species. Jansen (2012) confirmed high feeding rates for individual animals under oligotrophic conditions, but also demonstrated that community-scale rates under field conditions were 2 to 3 times lower (Table 9.1). Prins et al. (1996) showed that community estimates for benthic mussel beds in eutrophic cultivation areas were also lower than measurements on individuals, and Jacobs et al. (2015) concluded that low feeding rates measured on suspended spat collector communities were the result of refiltration within the culture community. Others have also hypothesized that lower community-scale clearance rates could be related to crowding affecting water exchange and/or refiltration (Frechette et al. 1992; Cranford et al. 2011). While the accuracy of various methods for determination of clearance rates for individuals have been the subject of debate during the last decade (Riisgard 2001; Petersen 2004; Petersen et al. 2004; Riisgard 2004; Cranford et al. 2011), there is good evidence for differences in feeding rates between individuals and communities that merit further study.

9.2.2 Nutrient Storage in Mussel Tissue

Surprisingly few studies report on the nutrient composition of mussel tissue, but the concentrations reported seem to correspond between the different cultivation areas (Table 9.2). These estimates do no account for nutrient storage in byssus or shell (Hawkins and Bayne 1985). Seasonal changes in nutrient composition are primarily driven by endogenous processes, and seasonal nutrient composition as well as

Table 9.2 Nutrient composition in mussel tissue in mussel cultivation areas

Area	Country	Species	Carbon [mg g^{-1}]	Nitrogen [mg g^{-1}]	Phosphorus [mg g^{-1}]	Ref.
Austevoll	NO	*M. edulis*	438 *(402–469)*	106 *(94–123)*	7 *(5–11)*	1
Whitsand Bay	UK	*M. edulis*	440 *(400–470)*	80 *(55–110)*		2
Oosterschelde	NL	*M. edulis*	448 *(113–623)*	102 *(68–126)*	7 *(5–12)*	3
Oosterschelde	NL	*M. edulis* spat		97 (92–104)	7.5 (6.6–8.4)	4
Ria de Arosa	ESP	*M. galloprovincialis*	448			5
Western Australia	AU	*M. edulis*	333	101	4	6
Mahurangi Harb.	NZ	*A. zelandica*	396	71		7

Data were standardized to mg element g^{-1} tissue DW. Weight conversion factors by Ricciardi and Bourget (1998) were applied. Values are presented as mean (*minimum – maximum*), and empty cells indicate that concentrations were not determined. Country codes given in Table 9.1
1 (Jansen et al. 2012a); 2 (Hawkins et al. 1985); 3 (Smaal and Vonck 1997); 4 Van Broekhoven (unpublished data); 5(Tenore et al. 1982); 6 (Vink and Atkinson 1985); 7 (Gibbs et al. 2005)

metabolic requirements associated with the reproductive cycle are similar for mussels under both nutrient-poor (Jansen et al. 2012a) and nutrient-rich conditions (Kuenzler 1961; Hawkins et al. 1985; Smaal and Vonck 1997).

9.2.3 Excretion of Inorganic Nutrients

Respiration and nutrient excretion rates of individual mussels measured under nutrient-poor conditions (Table 9.3) are within the range reported for nutrient-rich areas (Table 9.3, see also Burkholder and Shumway 2011), albeit toward the lower end. The slightly lower rates are likely related to the relatively cold and oligotrophic Norwegian fjords, as respiration and excretion rates of mussels are influenced by fluctuations in temperature (Widdows and Bayne 1971; Leblanc et al. 2003) and food supply (Bayne et al. 1993; Lutz-Collins et al. 2009; Jansen et al. 2012a). Eco-physiological models are often used to integrate responses of individual mussels with fluctuations in environmental conditions (Beadman et al. 2002; Dowd 2005). Jansen (2012) applied and validated a model normally used to simulate mussel responses in nutrient-rich areas (Filgueira and Grant 2009), and found that the model accurately predicted excretion rates under nutrient-poor conditions. This demonstrates that metabolic responses in mussels are comparable between cultivation areas of different trophic status, as the model is based on generic equations.

Mussel cultures are complex community structures, which besides the mussels include bacteria, epifauna, epiflora, and trapped biodeposits, which also contribute to nutrient exchange rates (Richard et al. 2006, 2007). The contribution of decomposing biodeposits (see also next section) to community nutrient release rates is particularly evident in the case of bottom cultures, where nearly all egested material is trapped in the community matrix. Indeed, the relatively high release rates for nutrients from bottom cultures are primarily attributed to decomposition of biodeposits (Asmus et al. 1990; Prins and Smaal 1994). Nutrient recycling from the organic matter trapped in suspended cultures is relatively low (Jansen 2012), which seems reasonable as the majority of biodeposits sink to the seafloor resulting in lower biodeposits on suspended mussel culture compared to benthic mussel cultures. Van Broekhoven et al. (2014) concludes that the combined activity of biodeposit decomposition and fauna on mussel spat collectors are either very small or scaled proportionally with mussel biomass or activity, whilst respiration and nutrient release rates are likely dominated by mussel spat activity. Richard et al. (2006, 2007), on the other hand, relate the high nitrate and nitrite fluxes of suspended mussel cultures in Canada to decomposition of organic material trapped in the community matrices.

Abundance and species composition of fauna associated with mussel cultures varies between seasons and farming locations, adding both temporal and spatial components to mussel farming dynamics (Cayer et al. 1999; Khalaman 2001; Richard et al. 2006; Lutz-Collins et al. 2009; Jansen et al. 2011). Jansen (2012) finds that during periods of high fouling abundance, ascidian (*Ciona intestinalis*)

Table 9.3 Respiration and inorganic nutrient release rates of different species of mussels and culture types in mussel cultivation areas

Area	Country	Species	Temperature [°C]	Respiration [μmol g^{-1} h^{-1}]	TAN excretion [μmol g^{-1} h^{-1}]	PO_4 excretion [μmol g^{-1} h^{-1}]	Si excretion [μmol g^{-1} h^{-1}]	Ref
Measurements on individuals								
Austevoll	NO	*M. edulis*	3–19	14.2 (5.7–27.8)	0.7 (0.3–1.8)	0.07 (<0–0.24)	–	1
Austevoll	NO	*M. edulis*	5–20	25.9 (12.6–48.1)	(1.8–2.6)			2
Åfjord	NO	*M. edulis*	12	48.9				3
Waddensea	NL	*M. edulis*	June and Sept		(0.8–5.0)	(0.02–0.17)		4
Waddensea	NL	*M. edulis*	3–24	(10.0–70.0)				5
Oosterschelde	NL	*M. edulis*	5–18	21.3 (10.3–36.0)	1.0 (0.2–3.1)	0.07 (0–0.13)	–	6
Oosterschelde	NL	*M. edulis*	1–20	26.3 (15.6–53.1)	1.1 (0.9–1.6)			7
South	UK	*M. edulis*	8–20	(22.3–71.5)	(0.1–2.9)			8
Whitsand Bay	UK	*M. edulis*			0.9 (0.3–2.1)			9
Whitsand Bay	UK	*M. edulis*	9–15	9.8 (3.1–17.2)	0.7 (0.1–1.2)			10
Whitsand Bay	UK	*M. edulis*	9-13	4.6 (4.2–8.3)	0.4 (0.2–0.5)			11
Lynher river	UK	*M. edulis*	11–21		(0.3–2.7)			12
Lynher river	UK	*M. edulis*	8–15		(0.4–1.3)			13
Lynher estuary	UK	*M. edulis*	5-25	(18.8–34.8)	(0.6–2.8)			14
Swansey Bay	UK	*M. edulis*	.		(1.6–2.1)			15
Heacam Bay	UK	*M. edulis*	15	(17.9–44.7)	(0.1–0.6)			16
Ria de Arosa	ESP	*M. galloprovincialis*	July		(0.1–0.2)			17
Ria de Arosa	ESP	*M. galloprovincialis*	14–15		(0.4–0.6)			18
New Foundland	CA	*M. edulis*	0–15	(8.9–35.7)	(0.1–0.9)			19
Great Entry Lagoon	CA	*M. edulis*	20	(44.7–160.8)	(0.7–7.9)			20
Amherst Basin	CA	*M. edulis*	20	(35.7–80.4)	(0.7–2.5)			20
Nova Scotia	CA	*M. edulis*	0–15	8.0 (3.3–12.1)	1.4 (0.5–2.5)			21

Beatrix Bay	NZ	*P. canaliculus*	11–17	31.8 (22.3–38.7)	(1.6–4.4)			
Western Australia	AU	*M. edulis*	15–20	6.7		0.02		22
Measurements on communities (benthic mussel beds)								
Baltic		*M. edulis*			(0.1–3.5)	(0.01–0.50)		23
Sylt	DEN	*M. edulis*	13–19		1.2 *(0.02–5.0)*			24
South	DEN	*M. edulis*	1–18	(0–12.5)	(0.1–3.2)	(0.10–0.53)		25
Waddensea	GER	*M. edulis*			1.2 (0–5.0)	0.10 (0–0.60)	0.6 (<0–1.4)	26
Waddensea	NL	*M. edulis*	June and Sept		(1.7–14.4)	(0.08–0.50)		4
Waddensea	NL	*M. edulis*	June-Sept		4.4		2.5	27
Oosterschelde	NL	*M. edulis*			5.6	1.70	2.3	27
Oosterschelde	NL	*M. edulis*			(0.9-15.8)	(0.03–0.68)	(<0–3.0)	28
Marennes-Oleron	FR	*M. edulis*	M-O-J-O		(0–7.3)			29
Narragansett Bay	USA	*M. edulis*	15		3.1			30
Measurements on communities (suspended ropes)								
Austevoll	NO	*M. edulis*	3–19	16.2 (3.4–28.7)	0.8 (0.2–1.8)	0.06 (0.00–0.15)	~0	3, 31
Åfjord	NO	*M. edulis*	12	17.1	1.1	0.11		3
Oosterschelde	NL	*M. edulis* spat	18–21	(72–381)	(5–70)	(0.12–6.4)	(0.1–5.3)	32
Sacca di Goro	IT	*M. edulis*	8–27	(25.1–26.9)	(3.2–7.6)			33
GreatEntry Lagoon	CA	*M. edulis*	16–19	(53.0–92.4)	(1.7–11.6)	(0.22–0.34)	(0.0–0.7)	34

Rates were standardized to μmol g^{-1} tissue DW h^{-1}. Where needed weight conversion factors by Ricciardi and Bourget (1998) were used. Values are presented as mean (minimum – maximum), and empty cells indicate that rates were not determined. Country codes are given in Table 9.1

1 (Jansen et al. 2012a); 2 (Strohmeier 2009); 3(Jansen 2012); 4 (Prins and Smaal 1994); 5 (Devooys 1976); 6 (Smaal and Vonck 1997); 7 (Smaal et al. 1997); 8 (Bayne and Widdows 1978); 9 (Kreeger et al. 1995); 10 (Hawkins et al. 1985); 11 (Hawkins and Bayne 1985); 12 (Bayne and Scullard 1977); 13 (Livingstone et al. 1979); 14 (Widdows 1978); 15 (Bayne et al. 1979); 16 (Gabbott and Bayne 1973); 17 (Lum and Hammen 1964); 18 (Labarta et al. 1997); 19 (Thompson 1984); 20 (Tremblay et al. 1998); 21 (Hatcher et al. 1994); 22 (Vink and Atkinson 1985); 23 (Kautsky and Wallentinus 1980); 24 (Asmus et al. 1990); 25 (Schluter and Josefsen 1994); 26 (Asmus et al. 1990); 27 (Dame et al. 1991); 28 (Prins and Smaal 1990); 29 (Smaal and Zurburg 1997); 30 (Nixon et al. 1976); 31 (Jansen et al. 2011); 32 (van Broekhoven et al. 2014); 33 (Nizzoli et al. 2006); 34 (Richard et al. 2006)

metabolism contributes up to 18% of total nitrogen released from suspended mussel culture communities. The contribution of the associated fauna to nutrient cycling cannot, therefore, be ignored. This is also acknowledged by Tang et al. (2011) who estimate that tissue carbon content of fouling ascidians is approximately 6.4% of the carbon production in scallops in Sungo Bay (China). A full understanding and prediction of nutrient regeneration by mussel culture communities requires more information on faunal growth, abundance, and metabolic dynamics within and across cultivation areas.

9.2.4 Biodeposit Release and Mineralisation

Biodeposit production represents a significant pathway in bivalve nutrient cycling (Kuenzler 1961; Prins and Smaal 1994; Cranford et al. 2007). Biodeposition rates under oligotrophic conditions, as measured in the laboratory for individual mussels, are in range with, but not at the maximum rates reported for other areas, whereas the organic matter content (OM) is relatively high (Table 9.4). The latter is likely related to high OM in the food source (~60–70%; Strohmeier et al. 2009, 2015) and the fact that pseudofaeces production is mostly absent under oligotrophic conditions. Seasonal fluctuations in biodeposition rates seem related to changes in food quantity and quality, rather than to temperature (Jansen et al. 2012b). This is consistent with Strohmeier et al. (2009), who suggest that the feeding response to low food concentrations (i.e.oligotrophic conditions) is likely the determining factor for total ingestion, rather than temperature.

Although measurements of mussel biodeposits are essential to understand and quantify their contribution to regeneration of nutrients, little has been published on biodeposit quality and their decay rates (reviewed by McKindsey et al. 2011) and more recently reported by Jansen et al. (2012b) and van Broekhoven et al. (2015). Nutrient concentrations in biodeposit depend on the concentration and type of diet the mussels feed on (Miller et al. 2002; Giles and Pilditch 2006) and therefore varies between seasons (Jansen et al. 2012b) and systems (Table 9.4). It has been suggested that mineralization rates of biodeposits are related to the presence of resident gut bacteria that can be voided from the mussel's digestive system along with the faecal pellets (Harris 1993). However, mineralization rates of fresh biodeposits increase considerably after an initial lag phase of one or two days (Fabiano et al. 1994; Carlsson et al. 2010; van Broekhoven et al. 2015), suggesting that a period of microbial growth may also be due to additional colonization by external microbes during the lag phase (Canfield et al. 2005). Since mineralization rates depend on the presence of microbes on either the benthic or the suspended mussel culture (Giles and Pilditch 2006; Carlsson et al. 2010; Jansen et al. 2012b), decomposition will be more rapid than in the water phase (van Broekhoven et al. 2015). The proportion of carbon and nitrogen decomposed as a function of available (labile) organic nutrients in biodeposits is relatively similar between oligotrophic (Jansen et al. 2012b) and eutrophic environments (Giles and Pilditch 2006; Carlsson et al. 2010; van

Table 9.4 Biodeposition and biodeposit composition in mussel cultivation areas

Area	Country	Species	Biodeposition [mg g^{-1} tissue d^{-1}]	OM [%]	Carbon [mg g^{-1} biodep]	Nitrogen [mg g^{-1} biodep]	Phosphorus [mg g^{-1} biodep]	Silicon [mg g^{-1} biodep]	Ref
Austevoll	NO	*M. edulis*	32 (11–72)	36 (22–48)	135 (62–194)	15 (7–23)	1.3 (0.8–1.7)		1
Askö, Baltic	SW	*M. edulis*	31 (7–104)	19 (8–45)	129 (50–200)	15 (8–21)	1.9 (1.0–3.0)		2
Oosterschelde	NL	*M. edulis*	Feces Pseudofeces	20 26	52 55	4.8 5.4	1.4 1.4	42 31	3
Bedford Basin	CA	*M. edulis*	(0–20)	(30–70)					4
Mahone Bay	CA	*M. edulis*	(0–80)	(10–70)					4
GreatEntry Lagoon	CA	*M. edulis*	54 (18–114)	22 (20–25)					5
Logy Bay (NF)	CA	*M. modiolus*	5 (1–8)	17 (13–23)	69 (47–103)	8 (5–12)		205 (100–335)	6
Queele Estuary	CH	*M. chilensis*		21	60	4			7
Firth of Thames	NZ	*P. canaliculu*		10	25	3			8
Mutsu Bay	JP	*M. edulis*	(6–116)						9

Data were standardized to mg DW biodeposit g^{-1} tissue DW d^{-1} (biodeposition rates), percentage (organic matter content), and mg element g^{-1} biodeposit DW (organic nutrient content). Where needed weight conversion factors by Ricciardi and Bourget 1998were used. Values are presented as mean (minimum – maximum), and empty cells indicate that rates were not determined. Country codes are given in Table 9.1

1 (Jansen et al. 2012a, b); 2 (Kautsky and Evans 1987); 3(van Broekhoven et al. 2015); 4 (Cranford and Hill 1999); 5 (Callier et al. 2006); 6 (Navarro and Thompson 1997; during springbloom conditions); 7 (Jaramillo et al. 1992); 8 (Giles and Pilditch 2006); 9 (Tsuchiya 1980).

Broekhoven et al. 2015) (Table 9.5). However, under oligotrophic conditions, the amount of nutrient released per gram biodeposit will be higher due to the higher concentrations of nutrients in the mussel biodeposits (Table 9.4). Phosphorus mineralization patterns are inconclusive among studies, likely as a result of the potential for phosphate to bind to sediment and other organic material (Sundby et al. 1992). Profound seasonal differences (up to a factor 80) are observed for silicon release rates by Jansen et al. (2012b), and is assumed to be high when mussel food contains a large fraction of diatoms (Navarro and Thompson 1997). Proportional silicon mineralization rates are 1.4 times higher for feces than pseudofeces, while proportional nitrogen and phosphate mineralization rates were similar for feces and pseudofeces (van Broekhoven et al. 2015). Hypothesised causes are breakdown of the organic matrix by digestive bacterial activity (Bidle and Azam 1999) selection during the feeding process for less recalcitrant diatom frustules, and fragmentation of diatom frustules during the digestive process (as speculated by Dame et al. 1991). Since the proportion of pseudofeces rises with increasing food concentration above a certain level (Foster-Smith 1975; Tsuchiya 1980), the role of mussels in terms of Si regeneration may be proportionally greater at lower food levels (assuming a similar food composition).

9.3 Ecosystem Effects of Nutrient Cycling by Mussels

The previous section demonstrated that mussels contribute to nutrient cycling by translocation, transformation and remineralization of nutrients. These processes related to the mussel's physiology interact with nutrient cycling in coastal ecosystems through various feedback systems influencing primary production (see reviews by Prins et al. 1998; Newell 2004). Consequently, intensive cultivation of mussels will affect the ecosystem; for example, by altering the carrying capacity (Smaal and Heral 1998; Grant and Filgueira 2011). The feeding activity of mussel communities may influence the abundance of phytoplankton and thereby inhibit primary production ('top-down' pathway or negative feedback). Furthermore, Cranford et al. (2009) reported a shift towards a phytoplankton population dominated by picophytoplankton in bays with high densities of mussel cultivation and related this to high grazing activity of the cultured stocks. Meanwhile, mussel excretion and mineralisation of biodeposits result in the regeneration of nutrients, which may stimulate primary production ('bottom-up' pathway or positive feedback). Not all ingested nutrients are regenerated in a short cycle; a part is retained by the mussel community or in a non-decomposed fraction of biodeposits, and a part may be permanently removed from the system, e.g. when mussels are harvested. Mussel communities can therefore act as a 'source' and as a 'sink' for nutrients within the ecosystem. The specific pathways contributing to sinks/sources depend on physical features (e.g. depth) of the area and the culture type applied (Table 9.6). Given that phytoplankton use nutrients in specific proportions (Redfield ratio; Redfield et al. 1963), the 'bottom-up' stimulation by bivalve nutrient regeneration is influenced by both nutrient availability and stoichiometry of regenerated nutrients. It has been argued

Table 9.5 Biodeposit remineralization rates in mussel cultivation areas

Area	Country	Species	Type	Temp (°C)	Unit	CO_2 release	TAN release	PO_4 release	$Si(OH)_4$ release	Ref.
Austevoll	NO	*M. edulis*	feces	5,10,15	mmol g^{-1} d^{-1}	3.3 (2.0–4.3)	0.17 (0.12–0.21)	0.06 (0.01–0.08)	3.9 (0.1–11.5)	1
					%	24% (15–31)	17% (10–20)			
Oosterschelde	NL	*M. edulis*	feces pseudofeces feces pseudofeces	20	μmol g^{-1} d^{-1} μmol g^{-1} d^{-1}% %	2.5 2.7	13.1% 12.4%	8.7% 7.9%		2
Great Entry Lagoon	CA	*M. edulis*	biodeposit	Jun-Aug	mmol g^{-1} d^{-1}	(max 4.5)	(max 0.3)	(max 0.02)	(max 1.0)	3
Roskilde & Limfjord	DEN	*M. edulis*	biodeposit	8-10	%	(25–38%)				4
Firth of Thames	NZ	*P. canaliculus*	biodeposit	20	%	40%	18%			5

Data were standardized either to release rate per g biodeposit DW per day or to fraction of initial nutrient content in the biodeposits (e.g. % = TAN/PON*100) for feces or pseudofeces ('biodeposit' indicates that it was unknown whether feces or a mix of (pseudo)feces was incubated). Values are presented as mean (minimum – maximum), and empty cells indicate that rates were not determined. Country codes are given in Table 9.1

1 (Jansen et al. 2012b); 2(van Broekhoven et al. 2015); 3 (recalculated from Callier et al. 2009); 4 (Carlsson et al. 2010); 5 (Giles and Pilditch 2006)

Table 9.6 Nutrient source and sink processes by water depth system and mussel culture type

Depth system	Culture type	Regeneration (source)	Retention (sink)	Removal (sink)
Shallow	Bottom	*Benthic* – CO_2 (DIC) & NH_4 & PO_4 excretion mussels & fauna – CO_2 (DIC), NH_4, PO_4 & Si biodeposit mineralization – NO_2/NO_3 nitrification of NH_4	*Benthic* – PO_4 binding to sediment – POC, PON, POP, PSi burial of biodeposits	*Benthic* – N_2 from nitrification/ denitrification of NH_4 – PON, PON, POP harvest mussel tissue
Shallow	Suspended	*Pelagic* CO_2 (DIC) & NH_4 & PO_4 excretion mussels & fauna		*Pelagic* – PON, PON, POP harvest mussel tissue
		Benthic – CO_2 (DIC), NH_4, PO_4 & Si biodeposit mineralization – NO_2/NO_3 nitrification from NH_4	*Benthic* – PO_4 binding to sediment – POC, PON, POP, POSi burial of biodeposits	*Benthic* – N_2 nitrification/ denitrification from NH_4
Deep	Suspended	*Pelagic* – CO_2 (DIC) & NH_4 & PO_4 excretion mussels & fauna		*Pelagic* – PON, PON, POP harvest mussel tissue
			Benthic (deep fjord basin) – POC, PON, POP, POSi burial of biodeposits – CO_2 (DIC), NH_4, PO_4 & Si biodeposit mineralization	

that both feedback control mechanism on phytoplankton can stabilize ecosystems (Herman and Scholten 1990) with 'top-down' and 'bottom-up' pathways occurring simultaneously. This section evaluates the pathways and magnitude of the feedback mechanisms in different mussel cultivation areas, and assesses if trophic status of the ecosystem is an important driver for defining ecosystem services and ecosystem impacts.

9.3.1 *Physical and Environmental Characteristics of Mussel Cultivation Areas*

The extent to which bivalves influence the ecosystem is largely defined by physical and environmental conditions (Newell 2004), which vary considerably among bivalve cultivation areas (Table 9.7). The majority of mussel cultivation areas are

Table 9.7 Physical characteristics of mussel cultivation areas

Area	Country	Type	Water depth [m]	Volume system [10^6 m^3]	Residence time [d]	Ref
Lysefjord – total	NO	Fjord	(460 max)	9100	7 year	1
Lysefjord – above sill	NO	Fjord	14	880	11	1
Åfjord – total	NO	Fjord	50 (120 max)	807	150	2
Åfjord – above sill	NO	Fjord	20	250	5	2
Limfjorden	DEN	Estuary with multiple basins	5	7100	225	3
Sylt	DEN		2	7	0.5	4
Oosterschelde	NL	Estuary	9	2740	40 (10–150)	5, 6
Wadden Sea	NL	Bay	3	4020	10 (5–15)	6
Carlingford Louch	IR	Estuary	(35 max)	460	14–26	7
Louch Foyle	IR	Bay	(19 max)	752	4–30	7
Bay of Brest	FR	Bay	10	1480	17	8
Thau Lagoon	FR	Lagoon	4	300	90–120	9
Marennes-Oleron	FR		5	675	7	10
Ria de Arosa	ESP	Bay, upwelling,	19	4335	23	11
N. Adriatic Sea	IT	Open Sea	22	–	–	12
Tracadie Bay	CA	Bay	2.5 (6 max)	41	4–10	13
Great Entry Lagoon	CA	Two-lagoon system	6	117	20–30	14
Saldanha Bay	SA	Two-bay system, upwelling	10 (30 max)	596	6–10	15
Firth of Thames	NZ	Estuary	(50 max)	16,500	12	16

Country codes are given in Table 9.1
1 (Aure et al. 2001); 2 (Aure pers. comm.); 3 (Wiles et al. 2006, Maar et al. 2010); 4 (in Smaal and Prins 1993); 5 (Smaal et al. 2001); 6 (Dame et al. 1991); 7 (Ferreira et al. 2007); 8 (in Smaal and Prins 1993); 9 (Thouzeau et al. 2007); 10 (in Smaal and Prins 1993); 11(Ferreira et al. 2007); 11 (AlvarezSalgado et al. 1996a, Figueiras et al. 2002); 12 (Brigolin 2007); 13 (Filgueira and Grant 2009); 14 (eastern basin; pers. comm. T. Guyondet); 15 (Shannon and Stander 1977, Monteiro et al. 1998); 16 (Zeldis 2005)

shallow mesotidal bays or estuaries. Due to the variation in physical conditions of the shallow bays and estuaries, water residence times vary from 1 day to several months. Oligotrophic fjord systems are exceptional when compared to "coastal plain estuaries" due to the large depths (100–1000 m). Many Norwegian fjords have a sill at the mouth of the fjord which limits renewal of the deepwater basin, resulting in relatively long residence times in terms of months and years for the whole system, whereas residence times are much shorter in terms of days and weeks for the upper and intermediate layers.

Annual primary production rates vary between 73 and 1245 g C m^{-2} y^{-1} for the different mussel cultivation areas, with rates reported for Norwegian fjord systems in the lower region (Table 9.8). Background nutrient levels in most areas are influenced by anthropogenic nutrient sources, with the exception of most Norwegian fjord systems (Aksnes et al. 1989). Wassmann (2005) shows that estuaries and coastal ecosystems are now the most nutrient-enriched ecosystems in the world, which he attributes primarily to land-based nutrient sources. Limfjorden (Denmark), for example, receives approximately 20,000 ton N y^{-1} from land-based sources, and the increased nitrogen input during the most recent decades resulted in high phytoplankton biomass levels, sustaining high densities of mussels up to levels causing hypoxia-induced mortality (Christiansen et al. 2006). The highest primary production rates are reported for Ria-de-Arousa and Saldahna Bay, which are coastal bays that benefit from upwelling of deep nutrient-rich water. The coastal upwelling along the South African coastline (Benguela current system) supplies a flux of approximately 1819 ton NO_3-N y^{-1} into Saldanha Bay (Monteiro et al. 1998). Areas that benefit from coastal upwelling are among the most productive and successful mussel farming areas (Figueiras et al. 2002; Saxby 2002).

The pathways for 'nutrient regeneration' differ between shallow and deep systems as a consequence of depth, stratification, mixing of the water column, and on the resulting presence or absence of benthic-pelagic coupling (see also Table 9.6). Benthic nutrient regeneration can play an important role in shallow coastal ecosystems with well-mixed water columns, as it may provide up to 80% of the nutrients required for primary production (Jensen et al. 1990; Zeldis 2005; Giles 2006). In contrast, benthic regeneration does not contribute to the nutrient pools in the euphotic zone of Norwegian fjords when the water column is stratified (Aure et al. 1996; Asplin et al. 1999). Euphotic zones of fjord systems are nutrient-limited for extended periods of the year (Paasche and Erga 1988; Sætre 2007), resulting in low Chl *a* concentrations (Erga 1989; Aure et al. 2007).

9.3.2 Nutrient Sinks and Sources

Physiological processes such as inorganic nutrient excretion, biodeposition (and subsequent remineralisation processes), and growth of tissue material (see also previous section) interact with physical features of the area and the culture type applied (Table 9.6) to drive the fraction of ingested nutrients that becomes regenerated, and thus becomes available as a source of nutrients to the ecosystem. Figure 9.2 (left panels) provides an overview of the relative importance of the physiological processes involved in nutrient cycling by mussel cultures. The processes have been expressed as fractions, with the sum of the three processes giving 100%. It is thereby assumed that the sum of the three processes equals ingestion (in accordance with

Table 9.8 Biochemical characteristics of mussel cultivation areas

Area	Country	Trophic classification	Time	PP [g C m^{-2} y^{-1}]	SPM [mg l^{-1}]	Chl *a* [µg l^{-1}]	POC [mg l^{-1}]	PON [mg l^{-1}]	DIN [µM]	PO_4 [µM]	Si [µM]	Ref
Austevoll	NO	Oligotrophic	Annual		0.4 (0.2–1.1)	1 (0–8)	0.2 (0.1–0.6)	0.02 (0.01–0.05)	2.1 (0–7)	0.2 (0–0.6)	2.4 (0–8.5)	1
Lysefjord	NO	Oligotrophic	Annual	100–140	(0.8-4)	1–1.5 (0.9-6.5)	(0.15–.05)	(0.01–0.02)	(0–2)	(0.02–0.03)	(1–5)	2
Åfjord	NO	Mesotrophic	Seasonal			1.2 (0.1–4.0)	0.3 (0.1–1.5)	0.04 (0.01–0.2)	2.4 (0.5–7.3)	0.3 (0.1–0.5)	3.2 (0.9–5.5)	3
Sylt	DEN		July–Aug	73	30 (17–202)		1.0 (0.3–7.5)	0.15 (0.05–0.6)				4
Limfjorden	DEN	Eutrophic	Annual	284 (0–1460)	5.6	>10 (0–50)			37 (0–100)	0.5 (0–9)		5
Oosterschelde	NL	Eutrophic	Annual	200 (115–456)	6 (5–6)	5	1	0.1	30	1.5 (1-6)	15 (1-40)	6
W. Waddenzee	NL	Eutrophic	Annual Springbl	200	36 (6–120)	(3–13)			17 (0-64) 67	(0.5-3) 0.8	16.3	7 8
Carlingford Lough	UK	Eutrophic	Annual		7.6	2.3			8.1	0.6		9
Lough Foyle	UK	Eutrophic	Annual		15.5	3.2			35.1	1.1		9
Thau Lagoon	FR	Eutrophic	Annual	400	(0.5–5)	(0–20)	0.3 (0.1–0.7)	0.04 (0.01–0.13)	1.8 (1–12)			10
Ria de Arousa	ESP	Eutrophic	Seasonal	99 (0–1351)	1.1 (0.5–2.6)	4.6 (0.1–34)	0.3	(0.01–0.07)	(2–12)			11
N. Adriatic Sea	IT	–	Annual			(0.4–16)	0.2	0.03 (0.02–0.1)				12
Tracadie Bay	CA	Eutrophic	Seasonal	318 (18–1204)	3.3	2.9 (1–12)		0.1	5 (1–14)	0.3 (0.1–0.6)	2 (0.2–6)	13

(continued)

Table 9.8 (continued)

Great Entry Lagoon	CA	Oligotrophic	Jun–Oct	(50–220)	(4–27)	1.8 (0.8–3.1)	0.4		0.3	0.3 (0.1–0.5)	1.1 (0.4–2.5)	14
New Foundland	CA		Seasonal		4.3 (2.2–6.5)	(0.2–5)						15
Saldanha Bay	SA	Eutrophic	Seasonal	1240 (581–5875)	3.6 (Feb)	8.6 (0.4–5.9)			(0–35)			16
Firth of Thames	NZ	Mesotrophic	Annual	168 (69–384)	(3–10)				1.5	0.3		17

Trophic classification according to Nixon (1995). Primary production (PP), Suspended Particulate Material (SPM), Organic Material (OM), Chlorophyll *a* (Chl *a*), Particulate Organic Carbon and Nitrogen (POC, PON), Dissolved Inorganic Nitrogen (DIN), Phosphate (PO_4) and Silicate (Si). Values are presented as mean (minimum – maximum). Country codes are given in Table 9.1

1 (Strohmeier 2009, Jansen et al. 2012a); 2 (Aure et al. 2001, 2007, Strohmeier unpubl data; values provided for upper 10 m); 4 (Asmus et al. 1990, Smaal and Prins 1993); 5 (Olesen 1996, Wiles et al. 2006, Maar et al. 2010); 6 (Smaal and Vonck 1997, Smaal et al. 2001, Wetsteyn et al. 2003); 7 (Dame et al. 1991, Philippart et al. 2007; waterbase NL); 8 (Philippart et al. 2007); 9 (Ferreira et al. 2007); 10 (Souchu et al. 2001, Plus et al. 2006); 11 (Smaal and Prins 1993, AlvarezSalgado et al. 1996b, Figueiras et al. 2002, 2010); 12 (Brigolin 2007); 13 (Bates and Strain 2006, Cranford et al. 2007, Cranford unpubl data, Harris unpubl data); 14 (Tremblay et al. 1998, Callier et al. 2006, Trottet et al. 2007); 15 (Thompson 1984, Navarro and Thompson 1997); 16 (Monteiro et al. 1998, Pitcher and Calder 1998, Probyn unpubl data); 17 (Zeldis 2005)

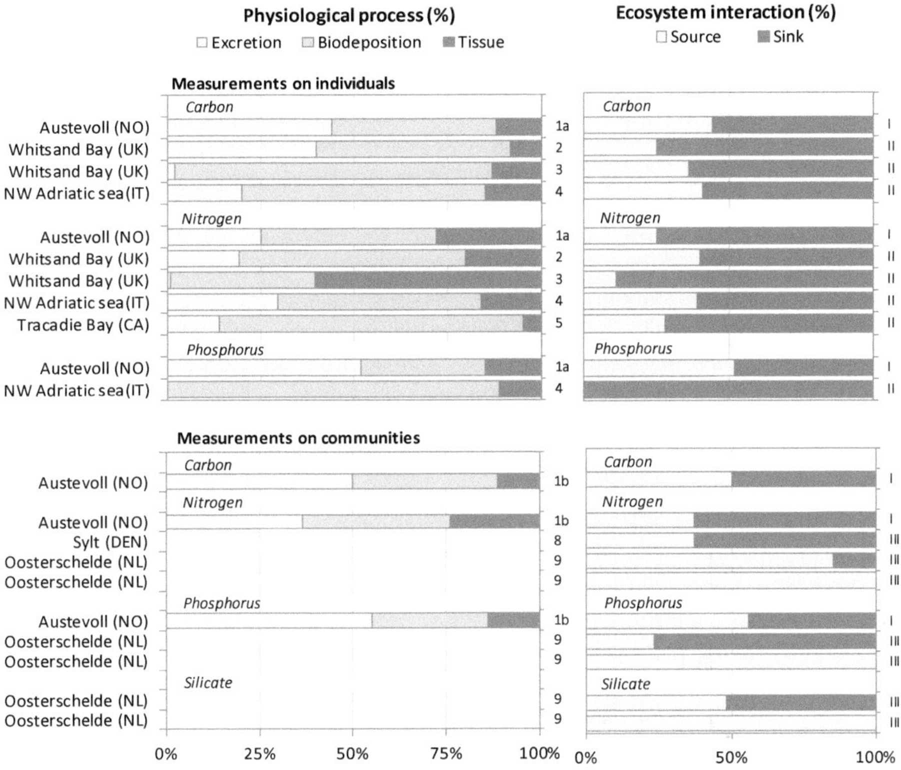

Fig. 9.2 Relative importance of physiological processes (left panels) and ecosystem interactions (right panels) for mussels (*Mytilus* spp.) across cultivation areas (left panels) for individual and community scale measurements. Data originates from budget analysis studies of which reference numbers are indicated on the secondary vertical axis in the left panels (see Tables 1–5 for full references). Ecosystem interactions refer to the fraction of ingested nutrients which is either recycled and available for phytoplankton growth (source), or is permanently lost from the system (sink). The calculation of source and sink fractions takes account of the physical characteristics of the system under consideration (depth, benthic-pelagic coupling) and consequently the fate of remineralized biodeposits. The type of calculation applied to each system is indicated on the secondary vertical axis of the right panels, according to

[I]Source = Excretion; Sink = Biodeposition + Tissue growth.

[II]Source = Excretion + remineralization; Sink = Tissue growth + (Biodeposition – remineralization) (assuming mineralization rates of 32% for C, 17% for N, and 0% for P; see Table 9.5)

[III]Based on *in situ* measurements of uptake and release rates in benthic tunnels.

Kreeger et al. 1995). Under oligotrophic conditions, less than 50% of the captured nutrients are expelled with biodeposits, which is lower than the other areas where more than 50% and, in certain cases, up to 80% of the ingested nutrients are expelled with biodeposits (Fig. 9.2). The right hand panels of Fig. 9.2 present the fractions of ingested nutrients either recycled as a source of nutrients, or retained or removed as sinks of nutrients (sum is 100%). Whether remineralisation of biodeposits acts as a source of nutrients available for phytoplankton growth depends on the system

(Table 9.6). Excretion of inorganic nutrients always acts as a source, while nutrient removal when mussels are harvested is always considered a sink. Biodeposition can result in both nutrient sources and sinks, depending on interactions with benthic processes: nutrients are either returned to the water column, buried in the sediment, or released in gaseous form (N_2). In deep fjords, biodeposits sink to the seafloor and as a consequence of limited benthic-pelagic coupling it is assumed that remineralized nutrients will not be available for phytoplankton growth. The estimates presented in Table 9.2 do not account for loss of mussels from the culture structures (Frechette 2012), nor for nutrient storage in byssus or shell (Hawkins and Bayne 1985); so that harvest values will be either slightly over or underestimated.

Firstly, measurements are considered for individual mussels (Fig. 9.2, upper panels). It is estimated that in deep fjord systems, approximately half of the ingested carbon and phosphorus, and 25% of nitrogen is regenerated (Fig. 9.2). Lower regeneration values for nitrogen are related to the capture and storage of nitrogen in tissue material (Jansen et al. 2012a). Mineralization of biodeposits does not significantly contribute to the source of recycled nutrients in deep fjord systems, because the majority of nutrients sink to the seafloor and regenerated nutrients are not returned to the euphotic zone of fjord systems within short time intervals due to stratification of the water column. For on-bottom and suspended cultivation of mussels in shallow areas, benthic biodeposit decomposition has been shown to significantly contribute to total nutrient regeneration (Asmus et al. 1990; Baudinet et al. 1990; Hatcher et al. 1994; Prins and Smaal 1994; Giles et al. 2006; Richard et al. 2007). Combining nutrients released by biodeposit remineralisation with those released by direct excretion results in relatively similar ‘source’ values for carbon and nitrogen regeneration in oligotrophic fjords and shallow eutrophic areas. All regenerated carbon is assumed to contribute to the source of recycled nutrients. This assumption is reasonable for Norwegian fjord systems which are generally considered to be weak absorbers of atmospheric CO_2, whereas in some eutrophic estuaries CO_2 might be released to the atmosphere since these systems often have oversaturated pCO_2 levels (Frankignoulle et al. 1998). In these estuaries, release of CO_2 by eco-physiological processes represents a sink process, and values presented in Fig. 9.2 might underestimate the carbon sink for these cases (see also Filgueira et al. 2019).

Secondly, measurements are considered for mussel communities (Fig. 9.2, lower panels). Nutrient regeneration rates for suspended cultures are defined in a similar manner as for individuals (see subscript Fig. 9.2). Regeneration by benthic communities is defined as the difference between uptake of organic material and release of inorganic nutrients, and has been determined using benthic tunnel measurements in the Oosterschelde (Netherlands) and Sylt (Denmark). A high degree of variability between measurements has been observed with occasionally higher release rates than uptake rates (source >100%), likely induced by mineralization of biodeposits or dead mussels trapped within the culture structures. An extensive seasonal study on nutrient cycling by oyster *Crassostrea virginica* reefs in the North Inlet estuary (South Carolina; Dame et al. 1989), using similar benthic tunnel measurements, indicate that 66% of nitrogen and 8% of phosphorus taken up by the reef is regenerated as ammonia and phosphate, respectively. Studies performed on benthic cultures

(Dame et al. 1989; Asmus et al. 1990; Prins and Smaal 1994) also pointed out that sediment processes may bind, and thus retain, phosphate, and that denitrification processes may lead to a loss of gaseous nitrogen from the system by the formation of N_2. The effects of bivalve cultures on denitrification rates have not been fully characterised (Newell 2004) and previous studies of sediments underlying suspended mussel cultures have been inconsistent, showing either increase (Kaspar et al. 1985; Giles et al. 2006) or decrease (Christensen et al. 2003).

9.3.3 Stoichiometry of Regenerated Nutrients

The previous section pointed out that mussel communities can act as a source of regenerated nutrients. The nutrients are regenerated in different proportions (stoichiometry), which may differ to varying degrees from the stoichiometry of the inorganic nutrient pool in the ambient water (Prins et al. 1998; Jansen et al. 2011). On a large scale, the average stoichiometric composition of phytoplankton is described by Redfield's ratio (Redfield ratio 106C:16Si:16 N:1P; Redfield et al. 1963). However, the stoichiometric composition of individual phytoplankton species, and therefore their nutrient requirements, may deviate from this ratio (Falkowski 2000). Changes in stoichiometry of available inorganic nutrients may affect phytoplankton growth (Goldman et al. 1979), and in this way could potentially induce a shift in the composition of phytoplankton species.

Figure 9.3 presents dissolved inorganic N:P ratios in the water at various mussel cultivation areas, and for the purposes of this review we assume that ratios below Redfield's ratio (N:P = 16) are indicative of more nitrogen-limited systems, whereas ratios above this ratio are indicative of more phosphorus-limited systems. Most of the mussel cultivation areas show a N:P ratio < 16, which is consistent with the common observation of nitrogen limitation in marine environments (Nixon et al. 1996). The assumption that phosphorus is generally sufficiently available in coastal waters (Nixon et al. 1996), does not seem to hold for all of the coastal waters used for shellfish cultivation; the Wadden Sea during spring bloom, Lough Foyle, and the Northern Adriatic Sea have been reported to be phosphorus-limited (Ferreira et al. 2007; Philippart et al. 2007; Brigolin et al. 2009).

N:P ratios of regenerated nutrients determined for individual mussels and for mussel communities are presented in Fig. 9.3 by broken and by solid lines, respectively. There are no cases where the N:P ratio of the net release by individual mussels or by mussel communities exceeds the Redfield's ratio, indicating that mussel activity is not likely to increase the ratio of N:P in the water. In most cases the N:P ratios of the regenerated nutrients (lines) differ from the ambient water (bars). The N:P ratios of nutrients released by suspended mussel communities (Austevoll, Great Entry Lagoon) are higher than ratios of nutrients released by benthic communities (Oosterschelde, Sylt; Fig. 9.3). In one case, the Oosterschelde estuary in the Netherlands, measurements have been made for both suspended mussel communities and mussel beds. The suspended community releases N and P in a ratio of

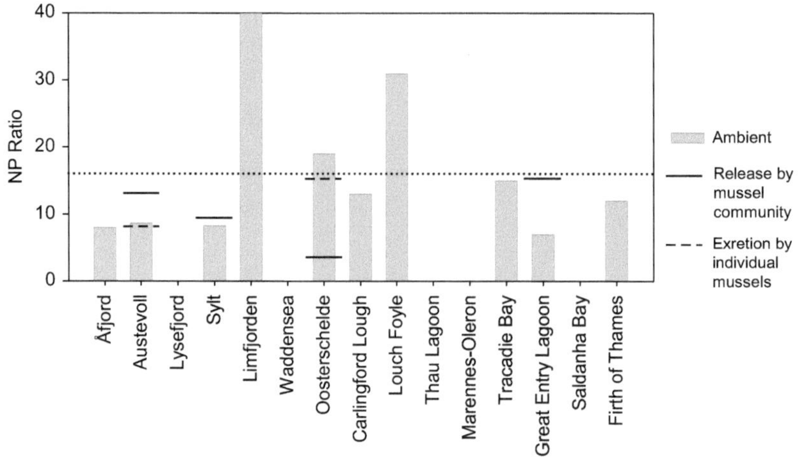

Fig. 9.3 Annual N:P [DIN:DIP] stoichiometry in the water at various mussel cultivation areas (bars), with release rates measured for individual mussels (broken lines) and mussel communities (solid lines). Horizontal dotted line indicates the Redfield ratio (N:P=16). References are given in Table 9.3

approximately 7, whilst the ratio of N and P released from mussel beds is lower (Fig. 9.3). Removal of nitrogen through denitrification processes has been suggested as a cause for the low N:P ratio measured in mussel beds (Asmus et al. 1990; Prins and Smaal 1994).

Measurements of phosphate dynamics over sediments underneath mussel farms have shown release in some cases (Baudinet et al. 1990; Souchu et al. 2001; Richard et al. 2007), and an apparent balance or an uptake in others (Hatcher et al. 1994; Mazouni et al. 1996; Giles and Pilditch 2006). Asmus et al. (1995) attributed differences in phosphorus fluxes to site-specific environmental characteristics. A balance or an uptake of phosphate can be related to the buffering capacity of sediments, caused by absorption of phosphate by iron hydroxides or calcite occurring in the oxidized surface layer of marine sediments (Sundby et al. 1992). This suggests that phosphate dynamics vary according to the location where decomposition takes place. Benthic mineralized phosphate may become trapped in the sediment, while pelagic mineralized phosphate is likely to become available in the water column.

Silicon does not play a role in physiology of mussels (Prins and Smaal 1994; Jansen et al. 2012a), and, therefore, all ingested silicon is expected to be egested with biodeposits. Decomposing biodeposits show high release rates of silicate (Jansen et al. 2012b; van Broekhoven et al. 2015, see also Table 9.5). In contrast to nitrogen and phosphorus, silicon mineralisation from biodeposits is thought to be driven primarily by chemical dissolution rather than microbial processing (van Broekhoven et al. 2015). In deep stratified systems, biodeposits (including all of the captured silicon, but not all of the carbon, nitrogen, and phosphorus) are transported to the bottom of the basin and regenerated nutrients, including silicon, do not become regenerated in the euphotic zone. This may potentially suppress the development of siliceous phytoplankton diatoms and favour development of non-siliceous phytoplankton such as flagellates and dinoflagellates (Turner et al. 1998). In shallow estuaries, biodeposit remineralization contributes to the pool of regenerated silicate (Asmus et al. 1990; Prins and Smaal 1994), which reduces the potential of silicate limitation in those areas (Prins et al. 1995).

9.3.4 Significance at Ecosystem Scale

The previous sections have discussed the *potential* effects of mussel communities on nutrient cycling in coastal ecosystems, irrespective of mussel abundance or dimensions of the system. In order to be able to evaluate system-wide interactions, estimates for the bivalve standing stock are an essential parameter (Table 9.9); although the majority of these values are associated with a large uncertainty. Combining standing stock estimates with dimensions of the systems (Table 9.7) provides area and volume-based biomass density estimates (Table 9.9). The Wadden Sea (NL) and several systems in France are important mussel cultivation areas in terms of total harvest quantities. However, these systems are also characterized by co-culture or co-existence of several bivalve species (e.g. *Crassostrea gigas* or *Ensis* sp.). As the current review focusses on mussels, systems where mussels comprise a minor proportion of total bivalve biomass were excluded from the analysis of mussel-ecosystem interactions. Mussel biomass density is highest in the eutrophic estuaries in Tracadie Bay (Canada) and the small coastal inlet Sylt (Germany), whereas biomass density in oligotrophic fjord systems is among the lowest reported.

Interactions are firstly evaluated by the total food uptake relative to the total food available (Fig. 9.4a, Smaal and Prins 1993; Dame and Prins 1998), which can also be described as an indicator for the 'top-down' influence on phytoplankton or 'negative feedback mechanism'. In the Norwegian fjords (Åfjord and Lysefjord) clearance times (CT) are longer than water residence times (RT) and primary production times (PPT) despite oligotrophic conditions, indicating that mussel cultures do not dominate food dynamics in these fjord systems. This is different from many other systems where clearance times are shorter than residence times (CT/RT <1). This confirms studies by Smaal and Prins (1993), Dame and Prins (1998) who report that clearance times are shorter than the residence times for most mussel cultivation areas. However, for most areas primary production is faster than mussel feeding (CT/PPT>1) indicating that the food source is renewed faster than it is filtered. Limfjorden has the longest residence times (almost one year), and a high mussel biomass which together result in high food uptake relative to residence times (CT/RT$\ll1$) indicating that the system is potentially regulated by mussel filtration. However, high nutrient loading in this system results in high primary production rates (Maar et al. 2010) which subsequently indicates that mussels do not overgraze phytoplankton populations (CT/PPT$\gg1$).

Secondly, mussel-ecosystem interactions were evaluated by nitrogen (DIN) turnover time (Dame 1996) relative to the residence time (Fig. 9.3b). This indicator can

Table 9.9 Bivalve density in mussel cultivation areas

Area	Country	Species	Culture type	Harvest (WW) [ton y^{-1}]	Standing stock (DW) [ton]	[g m^{-2}]	[g m^{-3}]	Ref
Lysefjord	NO	*M. edulis*	Rope		94	2.1	0.1	1
Åfjord	NO	*M. edulis*	Rope	1200	109*	7.8	0.4	2
Limfjorden	DEN	*M. edulis*	Bottom	90,000	2509*	1.6	0.4	3
		C. gigas		580	6*	0.0	0.0	
Sylt	DEN	*M. edulis*	Bottom		189		26.3	4
Oosterschelde	NL	*M. edulis*	Bottom	25,000	6061	17.3	2.2	5
		C. giga	Bottom		2424	2.4	0.3	
		cockles	Bottom		848	6.9	0.9	
Wadden Sea	NL	*M. edulis*	Bottom + rope		5018	3.6	1.3	6
		M. arenaria	Natural		8419	6.0	2.1	
		Ensis	Natural		12,880	9.1	3.2	
		Other bivalves	Natural		5799	4.1	1.4	
Carlingford Lough	UK	*M. edulis*	Bottom + rope	2500	209*	4.3	0.5	7
		C. gigas	Trestles	320	27*	0.6	0.06	
Belfast Lough	UK	*M. edulis*	Bottom	15,318	1281*		1.7	8
		C. gigas	Trestles	50	4*		0.006	
Bay of Brest	FR	Various			13,275	90	8.9	9
Thau Lagoon	FR	*C gigas + M. edulis*		13,500				10
Marennes-Oleron	FR	*M. edulis*			242		0.4	11
		C. gigas			2424		3.6	
		Other bivalves			788		1.2	
Ria de Arosa	SP	*M. galloprovincialis*	Raft	172,500	4809*	19.6	1.1	12
Tracadie Bay	CA	*M. edulis*	Rope	1943	261	15.9	6.4	13
Great Entry Lagoon	CA	*M. edulis*	Rope	180	15*	0.5	0.1	14
Firth of Thames	NZ	*P. canaliculus*	Rope	9000	251*	0.2	0.02	15

Density is expressed in terms of harvest rate (ton WW y^{-1}), and in standing stock for the whole system (ton DW), per unit area (g DW m^{-2}) and per unit volume (g DW m^{-3}). For the Norwegian fjords, only the water volume above the sill was used in the calculations. Asterisk (*) indicates that standing stock was reconstructed based on harvest, length of the production cycle and WW/DW conversion factors by Ricciardi and Bourget (1998). Country codes are given in Table 9.1

1 (Strohmeier et al. 2005; pers. comm Strohmeier); 2 (pers. comm. M. Hoem and A. Koteng); 3 (Dolmer and Geitner 2004); 4 (in Smaal and Prins 1993); 5 (Smaal et al. 2001); 6 (Philippart et al. 2007, Schellekens et al. 2014); 7 & 8 (Ferreira et al. 2007); 9 (in Smaal and Prins 1993); 10 (Thouzeau et al. 2007); 11 (Smaal and Zurburg 1997); 12 (Figueiras et al. 2002); 13 (Cranford et al. 2007); 14 (Trottet et al. 2008b); 15 (Zeldis 2005)

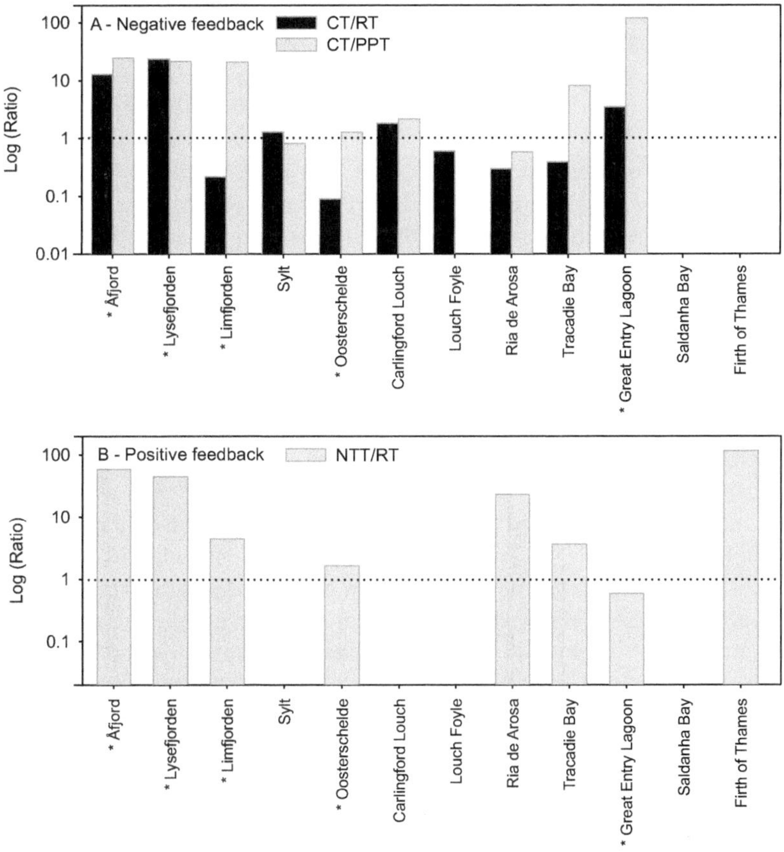

Fig. 9.4 Mussel-ecosystem interactions expressed by indicators for negative and positive regulation of primary production, calculated according to Dame and Prins (1998), Smaal and Prins (1993) and Dame (1996) based on the following parameters

Residence time (RT) = Time to exchange water body
Clearance time (CT) = Time to filter the water body
= (system volume) / (CR × mussel biomass)
Primary production time (PPT) = Time to renew phytoplankton (Bp/P)
= ($POC_{phytopl.}$ × volume system) / (Primary production × Area system) *with the assumption: 40* $mgPOC_{phytopl}\ mgChla^{-1}$
Nitrogen turnover time (NTT) = Time to renew DIN
= (DIN × system volume) / (DIN Release × Mussel biomass)

The extent to which mussel populations have a regulating function in the ecosystem is evaluated by the ratios between the parameters:

CT/RT >1 : no/minor regulation CT/RT <1 : phytoplankton potentially regulated by mussel filtration
CT/PPT >1 : no/minor regulation CT/PPT <1 : phytoplankton is overgrazed
NTT/RT >1 : no/minor regulation NTT/RT<1 : mussels potentially driving nutrient cycling

References are given in Tables 1-9. Asterisk (*) indicates that community-scale rates were applied.

describe the potential extent of 'bottom-up' stimulation of phytoplankton production or the 'positive feedback mechanism'. The total DIN pool in the ambient water was lowest in Åfjord, Lysefjord and the Firth of Thames, so that a quantity of regenerated nitrogen from mussel cultures could make a proportionally greater contribution to its availability. However, mussel density in these areas is also low (<0.4 g DW m^{-3}). As a result, nitrogen turnover times remain long relative to water residence times (NTT/RT>40), indicating a limited effect of mussels on nutrient cycling. Low DIN concentrations are reported for Great Entry Lagoon resulting in a high NTT value, suggesting a relatively high effect on the DIN pool (NTT/RT<1). However, this outcome may be skewed because ambient values are based on the period June-October, thus excluding the higher winter values. Besides Great Entry Lagoon, the relative effect of regeneration processes (NTT/RT) is most pronounced in the Oosterschelde estuary and Tracadie Bay, indicating that mussels may influence nutrient cycling although NTT/RT values did not fall below 1. These are shallow estuaries/bays with high mussel cultivation activity, as indicated by the high relative mussel density (2-6 g DW m^{-3}, Table 9.9).

This analysis of positive and negative feedback mechanisms of mussels acting on phytoplankton growth (Fig. 9.4) addresses some consequences of mussel populations for ecosystem functioning, but it is based on a static approach. However, marine systems are complex, and suspended organic matter and inorganic nutrient concentrations are subject to physical, biochemical and eco-physiological processes and fluctuate over both temporal and spatial scales. It should be noted that the literature presented here represents integrated annual values, whereas in fact most of the parameters fluctuate over temporal scales. Prins and Smaal (1994) address the importance of seasonality in terms of the contribution of mussels to nutrient regeneration in the Oosterschelde, demonstrating that mussel beds could account for almost half of the total DIN regeneration of the system, but only during summer when nutrients are limiting. Similarly, Jansen et al. (2011) demonstrate that at the scale of one mussel farm in a Norwegian fjord, the contribution of mussels to the inorganic nutrient pool is insignificant during winter conditions but substantial during summer. This is a result of the combination of low nutrient concentrations (nutrient limitation) in the ambient water, high metabolic activity of the mussel population, and high biomass and metabolic activity of fouling organisms.

9.4 Perspective on the Regulating Services of Mussels in Nutrient-Poor and Nutrient-Rich Cultivation Areas

The extent to which bivalve suspension feeders fulfil a regulative role varies between coastal ecosystems (Dame and Prins 1998). Trophic status (nutrient-poor to nutrient-rich) of a system influences the regulating potential for mussels in two ways: (1) the eco-physiological response may vary as a function of ambient nutrient (and thus food) concentrations, and (2) nutrient regeneration has a proportionally greater effect when ambient concentrations are low.

9.4.1 Physiological Response

The high feeding rates observed in oligotrophic areas suggest that the physiological response of mussels under low nutrient conditions may differ from areas with higher nutrient concentrations. As model results indicated that metabolic responses are comparable between cultivation areas, this suggests that the slightly lower rates observed for oligotrophic areas are simply a result of low food concentrations rather than a specific response related to the trophic status of the system. Also, nutrient composition of the mussel tissue is similar in oligo- and eutrophic areas, and appears to be endogenously regulated and driven primarily by reproductive processes. Mussels are able to efficiently use the low-concentration but high-quality food sources in oligotrophic systems, resulting in low biodeposit production (in absolute and in relative terms). In eutrophic areas, up to 95% of the filtered nutrients can be expelled with biodeposits in certain cases, which is partly due to pseudofaeces production, while in oligotrophic areas less than 50% of all ingested nutrients is expelled with faecal material.

9.4.2 System Feedbacks

Differences in eco-physiological rates under oligotrophic as compared to eutrophic conditions (higher clearance, lower egestion, approximately similar excretion, and similar storage in tissue) may lead to distinct mussel-ecosystem interactions. Proportionally more nutrients are excreted as metabolic waste products under oligotrophic conditions (e.g. NH_4), potentially resulting in a higher positive feedback and thus enhanced primary production. In deep fjord systems, the pool of nutrients available for phytoplankton growth is only supplied by directly excreted inorganic metabolic waste products, while in shallow areas remineralization of biodeposits may also contribute to the pool. Ecosystem interactions are here defined as the fraction of ingested nutrients either recycled and again available for primary production (source) or permanently removed from the system (sink). The current review showed that through these mechanisms the ecosystem interactions are comparable between deep oligotrophic and shallow eutrophic systems. This indicates that the theoretical role of mussels in nutrient cycling and positive feedback processes is relatively similar across mussel cultivation areas. Furthermore, stoichiometry of regenerated nutrients (C>N>P) is generally different from that observed in the ambient water and from the Redfield ratio. This indicates that mussel cultures have the potential to influence phytoplankton community composition by causing shifts in the proportional availability of C, N, P, and Si. The oligotrophic fjord systems are examples where silicate limitation, potentially induced by mussel activity, may suppress diatoms while favouring (dino)flagellate development, while in shallow estuaries this phenomenon is expected to be of less importance due to the contribution of regenerated silicate through biodeposit decomposition.

Evaluation of the regulating potential of mussel cultures at the ecosystem level is based on indicators for negative (CT/RT and CT/PPT) and positive (NTT/RT) feedback processes on primary production. These indicators for mussel-ecosystem interaction demonstrate that despite low background nutrient levels, mussel aquaculture in Norwegian fjord systems at present has limited effects owing to low mussel densities and physical characteristics of the fjords (large volume, short residence times of the upper water layer). Estimates for mussel-ecosystem interactions are more profound in shallow nutrient-rich areas with high mussel biomass, especially in terms of the negative feedback mechanisms through filtration of phytoplankton. The significance of the positive feedback mechanism (nutrient regeneration) has a strong seasonal component as many mussel cultivation systems are nitrogen-limited during summer periods when mussel activity is high. These comparisons between cultivation areas suggest that physical characteristics of the site in combination with mussel density better define the feedback to the ecosystem, and hence the regulating potential of mussel cultures, rather than trophic state.

Acknowledgements The authors are grateful to dr Sandy Shumway and dr John Icely for their valuable comments on the manuscript.

References

Aksnes DL, Aure J, Kaartvedt S, Richard J (1989) On the significance of advection for the carrying capacity of a fjord. Mar Ecol Prog Ser 50:263–274

AlvarezSalgado XA, Roson G, Perez FF, Figueiras FG, Pazos Y (1996a) Nitrogen cycling in an estuarine upwelling system, the Ria de Arousa (NW Spain). 1. Short-time-scale patterns of hydrodynamic and biogeochemical circulation. Mar Ecol Prog Ser 135:259–273

AlvarezSalgado XA, Roson G, Perez FF, Figueiras FG, Rios AF (1996b) Nitrogen cycling in an estuarine upwelling system, the Ria de Arousa (NW Spain). 2. Spatial differences in the short-time-scale evolution of fluxes and net budgets. Mar Ecol Prog Ser 135:275–288

Asmus RM, Asmus H (1991) Mussel beds – limiting or promoting phytoplankton. J Exp Mar Biol Ecol 148:215–232

Asmus H, Asmus RM, Reise K (1990) Exchange processes in an intertidal mussel bed: a Sylt-flume study in the Wadden Sea. Berichte der Biologischen Anstalt Helgoland 6:1–79

Asmus H, Asmus RM, Zubillaga GF (1995) Do mussel beds intensify the phosphorus exchange between sediment and tidal waters? Ophelia 41:37–55

Asplin L, Salvanes AGV, Kristoffersen JB (1999) Nonlocal wind driven fjord-coast advection and its potential effect on plankton and fish recruitment. Fish Oceanogr 8:255–263

Aure J (pers. comm.) Based on model simulations *Ancylus FjordEnv 3.3.* http://www.ancylus.net

Aure J, Molvaer J, Stigebrandt A (1996) Observations of inshore water exchange forced by a fluctuating offshore density field. Mar Pollut Bull 33:112–119

Aure J, Strand Ø, Skaar A (2001) Future opportunities for aquaculture in Lysefjorden (in Norwegian). Fisken og Havet (report), Book 9. Institute of Marine Research

Aure J, Strand O, Erga SR, Strohmeier T (2007) Primary production enhancement by artificial upwelling in a western Norwegian fjord. Mar Ecol Prog Ser 352:39–52

Bates SS, Strain PM (2006) Nutrients and phytoplankton in Prince Edward Island inlets during late summer to fall: 2001–2003. Can Tech Rep Fish Aquat Sci

Baudinet D, Alliot E, Berland B, Grenz C, Plantecuny MR, Plante R, Salenpicard C (1990) Incidence of Mussel Culture on Biogeochemical Fluxes at the Sediment-Water Interface. Hydrobiologia 207:187–196

Bayne BL (1998) The physiology of suspension feeding by bivalve molluscs: an introduction to the Plymouth "TROPHEE" workshop. J Exp Mar Biol Ecol 219:1–19

Bayne BL, Scullard C (1977) Rates of nitrogen-excretion by species of *Mytilus* (bivalvia - mollusca). J Mar Biol Assoc UK 57:355–369

Bayne BL, Widdows J (1978) Physiological ecology of 2 populations of *Mytilus edulis* L. Oecologia 37:137–162

Bayne BL, Moore MN, Widdow J, Livingstone DR, Salkeld PN (1979) Measurements of the responses of individuals to environmental stress and pollution: studies with bivalve molluscs. Phylosophical Trans R Soc Lond B Biol Sci 286:563–581

Bayne BL, Iglesias JIP, Hawkins AJS, Navarro E, Heral M, Deslouspaoli JM (1993) Feeding-behavior of the mussel, *Mytilus edulis*, responses to variations in quantity and organic content of the seston. J Mar Biol Assoc UK 73:813–829

Beadman HA, Willows RI, Kaiser MJ (2002) Potential applications of mussel modelling. Helgol Mar Res 56:76–85

Bidle KD, Azam F (1999) Accelerated dissolution of diatom silica by marine bacterial assemblages. Nature 397:508–511

Brigolin D (2007) Development of integrated numerical models for the sustainable management of marine aquaculture. PhD thesis. A.A. 2003/2004 – A.A. 2006/2007, Universita Ca' Foscari Venezia, Italy

Brigolin D, Dal Maschio G, Rampazzo F, Giani M, Pastres R (2009) An individual-based population dynamic model for estimating biomass yield and nutrient fluxes through an off-shore mussel (*Mytilus galloprovincialis*) farm. Estuar Coast Shelf Sci 82:365–376

Callier MD, Weise AM, McKindsey CW, Desrosiers G (2006) Sedimentation rates in a suspended mussel farm (Great-Entry Lagoon, Canada): biodeposit production and dispersion. Mar Ecol Prog Ser 322:129–141

Callier MD, Richard M, McKindsey CW, Archambault P, Desrosiers G (2009) Responses of benthic macrofauna and biogeochemical fluxes to various levels of mussel biodeposition: An in situ "benthocosm" experiment. Mar Pollut Bull 58:1544–1553

Canfield DE, Kristensen E, Thamdrup B (2005) Aquatic geomicrobiology, vol 48. Elsevier Academic Press, London

Carlsson MS, Glud RN, Petersen JK (2010) Degradation of mussel (*Mytilus edulis*) fecal pellets released from hanging long-lines upon sinking and after settling at the sediment. Can J Fish Aquat Sci 67:1376–1387

Cayer D, MacNeil M, Bagnall AG (1999) Tunicate fouling in Nova Scotia aquaculture: a new development. J Shellfish Res 18:327

Christensen PB, Glud RN, Dalsgaard T, Gillespie P (2003) Impacts of longline mussel farming on oxygen and nitrogen dynamics and biological communities of coastal sediments. Aquaculture 218:567–588

Christiansen T, Christensen TJ, Markrager S, Petersen JK, Mouritsen LT (2006) Limfjorden in 100 year (in Danish). Faglig rapport fra DMU, nr 578

Cranford PJ, Hill PS (1999) Seasonal variation in food utilization by the suspension-feeding bivalve molluscs *Mytilus edulis* and *Placopecten magellanicus*. Mar Ecol Prog Ser 190:223–239

Cranford PJ, Strain PM, Dowd M, Hargrave BT, Grant J, Archambault MC (2007) Influence of mussel aquaculture on nitrogen dynamics in a nutrient enriched coastal embayment. Mar Ecol Prog Ser 347:61–78

Cranford P, Hargrave B, Li W (2009) No mussel is an island. ICES Insight 46:44–49

Cranford P, Ward JE, Shumway SE (2011) Bivalve filter feeding: variability and limits of the aquaculture biofilter. In: Shumway SE (ed) Shellfish aquaculture and the environment. Wiley-Blackwell, London

Dame R (1996) Ecology of marine bivalves: an ecosystem approach. CRC Press, Boca Raton

Dame R, Prins T (1998) Bivalve carrying capacity in coastal ecosystems. Aquat Ecol 31:409–421

Dame RF, Spurrier JD, Wolaver TG (1989) Carbon, nitrogen and phosphorus processing by an oyster reef. Mar Ecol Prog Ser 54:249–256
Dame R, Dankers N, Prins T, Jongsma H, Smaal A (1991) The influence of mussel beds on nutrients in the Western Wadden Sea and Eastern Scheldt Estuaries. Estuaries 14:130–138
Devooys CGN (1976) Influence of temperature and time of year on oxygen-uptake of sea mussel *Mytilus edulis*. Mar Biol 36:25–30
Dolmer P, Geitner K (2004) Integrated coastal zone management of cultures and fishery of mussels in Limfjorden, Denmark. ICES CM 2004/V:07. Theme Session V: Towards sustainable aquaculture
Dowd M (2005) A bio-physical coastal ecosystem model for assessing environmental effects of marine bivalve aquaculture. Ecol Model 183:323–346
Erga SR (1989) The importance of external physical controls on vertical distribution of phytoplankton and primary production in fjords of western Norway. Dr Scint thesis. University of Bergen
Fabiano M, Danovaro R, Olivari E, Misic C (1994) Decomposition of fecal matter and somatic tissue of Mytilus-Galloprovincialis – changes in Organic-matter composition and microbial succession. Mar Biol 119:375–384
Falkowski P (2000) Rationalizing nutrient ratios in unicellular algae. J Phycol 36:3–6
Ferreira JG, Hawkins AJS, Monteiro P, Service M, Moore H, Edwards A, Gowen R, Lourenco P, Mellor A, Nunes JP, Pascoe PL, Ramos L, Sequeira A, Simas T, Strong J (2007) SMILE – Sustainable Mariculture in northern Irish Louch Ecosystems
Ferreira JG, Saurel C, Silva J, Nunes JP, Vazquez F (2014) Modelling of interactions between inshore and offshore aquaculture. Aquaculture 426:154–164
Figueiras FG, Labarta U, Reiriz MJF (2002) Coastal upwelling, primary production and mussel growth in the Rias Baixas of Galicia. Hydrobiologia 484:121–131
Filgueira R, Grant J (2009) A box model for ecosystem-level management of mussel culture carrying capacity in a Coastal Bay. Ecosystems 12:1222–1233
Filgueira R, Labarta U, Fernandez-Reiriz MJ (2008) Effect of condition index on allometric relationships of clearance rate in *Mytilus galloprovincialis* Lamarck, 1819. Revista De Biologia Marina Y Oceanografia 43:391–398
Filgueira R, Fernandez-Reiriz MJ, Labarta U (2010) Clearance rate of the mussel *Mytilus galloprovincialis*. II. Response to uncorrelated seston variables (quantity, quality, and chlorophyll content). Cienc Mar 36:15–28
Filgueira R, Strohmeier T, Strand Ø (2019) Regulating services of bivalve molluscs in the context of the carbon cycle and implications for ecosystem valuation. In Smaal et al (eds) Goods and services of marine bivalves. Springer, Cham, pp 231–251
Foster-Smith RL (1975) The effect of concentration of suspension on the filtration rates and pseudofaecal production for *Mytilus edulis* L., *Cerastoderma edule* L. and *Venerupis pullastra* (Montagu). J Exp Mar Biol Ecol 17:1–22
Frankignoulle M, Abril G, Borges A, Bourge I, Canon C, DeLille B, Libert E, Theate JM (1998) Carbon dioxide emission from European estuaries. Science 282:434–436
Frechette M (2012) Self-thinning, biodeposit production, and organic matter input to the bottom in mussel suspension culture. J Sea Res 67:10–20
Frechette M, Aitken AE, Page L (1992) Interdependence of food and space limitation of a benthic suspension feeder – consequences for self-thinning relationships. Mar Ecol Prog Ser 83:55–62
Gabbott PA, Bayne BL (1973) Biochemical effects of temperature and nutritive stress on *Mytilus edulis* L. J Mar Biol Assoc UK 53:269–286
Gibbs M, Funnell G, Pickmere S, Norkko A, Hewitt J (2005) Benthic nutrient fluxes along an estuarine gradient: influence of the pinnid bivalve Atrina zelandica in summer. Mar Ecol Prog Ser 288:151–164
Giles H (2006) Dispersal and remineralisation of biodeposits: ecosystem impacts of mussel aquaculture. PhD, The University of Waikato, Hamilton, New Zealand
Giles H, Pilditch CA (2006) Effects of mussel (*Perna canaliculus*) biodeposit decomposition on benthic respiration and nutrient fluxes. Mar Biol 150:261–271

Giles H, Pilditch CA, Bell DG (2006) Sedimentation from mussel (*Perna canaliculus*) culture in the Firth of Thames, New Zealand: Impacts on sediment oxygen and nutrient fluxes. Aquaculture 261:125–140

Goldman JC, McCarthy JJ, Peavey DG (1979) Growth-rate influence on the chemical composition of phytoplankton in oceanic waters. Nature 279:210–215

Gosling EM (2015) Bivalve Molluscs; biology, ecology and culture. Blackwell Publishing, Cornwall

Grant J, Filgueira R (2011) The application of dynamic modelling to predicition of production carrying capacity in shellfish farming. In: Shumway SE (ed) Shellfish aquaculture and the environment. Wiley-Blackwell, London

Harris JM (1993) The presence, nature, and role of gut microflora in aquatic invertebrates – a synthesis. Microb Ecol 25:195–231

Hatcher A, Grant J, Schofield B (1994) Effects of suspended mussel culture (*Mytilus* Spp) on Sedimentation, benthic respiration and sediment nutrient dynamics in a Coastal Bay. Mar Ecol Prog Ser 115:219–235

Hawkins AJS, Bayne BL (1985) Seasonal-variation in the relative utilization of carbon and nitrogen by the mussel *Mytilus-edulis* – budgets, conversion efficiencies and maintenance requirements. Mar Ecol Prog Ser 25:181–188

Hawkins AJS, Bayne BL (1992) Physiological interrelations, and the regulation of production. In: Gosling E (ed) The mussel *Mytilus*: ecology, physiology, genetics and culture. Elsevier, Amsterdam, pp 171–212

Hawkins AJS, Salkeld PN, Bayne BL, Gnaiger E, Lowe DM (1985) Feeding and resource allocation in the mussel *Mytilus edulis* – evidence for time-averaged optimization. Mar Ecol Prog Ser 20:273–287

Herman PMJ, Scholten H (1990) Can suspension-feeders stabilise estuarine ecosystems? In: Barnes M, Gibson RN (eds) Trophic relationships in the marine environment. Aberdeen University Press, Aberdeen

Jacobs P, Troost K, Riegman R, van der Meer J (2015) Length- and weight-dependent clearance rates of juvenile mussels (Mytilus edulis) on various planktonic prey items. Helgol Mar Res 69:101–112

James MR, Weatherhead MA, Ross AH (2001) Size-specific clearance, excretion, and respiration rates, and phytoplankton selectivity for the mussel *Perna canaliculus* at low levels of natural food. N Z J Mar Freshw Res 35:73–86

Jansen HM (2012) Bivalve nutrient cycling – nutrient turnover by suspended mussel communities in oligotrophic fjords. PhD, Wageningen University

Jansen H, Strand Ø, Strohmeier T, Krogness C, Verdegem M, Smaal A (2011) Seasonal variability in nutrient regeneration by mussel *Mytilus edulis* rope culture in oligotrophic systems. Mar Ecol Prog Ser 431:137–149

Jansen HM, Strand O, Verdegem M, Smaal A (2012a) Accumulation, release and turnover of nutrients (C-N-P-Si) by the blue mussel (*Mytilus edulis*) under oligotrophic conditions. J Exp Mar Biol Ecol 416–417:185–195

Jansen HM, Verdegem MCJ, Strand Ø, Smaal AC (2012b) Seasonal variation in mineralization rates (C-N-P-Si) of mussel *Mytilus edulis* biodeposits. Mar Biol 159:1567–1580

Jaramillo E, Bertran C, Bravo A (1992) Mussel biodeposition in an Estuary in Southern Chile. Mar Ecol Prog Ser 82:85–94

Jensen MH, Lomstein E, Soerensen J (1990) Benthic NH4+ and NO3- flux following sedimentation of a spring phytoplankton bloom in Aarhus Bight, Denmark. Mar Ecol Prog Ser 61:87–96

Kaspar HF, Gillespie PA, Boyer IC, Mackenzie AL (1985) Effects of mussel aquaculture on the nitrogen-cycle and benthic communities in Kenepuru sound, Marlborough sounds, New-Zealand. Mar Biol 85:127–136

Kautsky N, Evans S (1987) Role of biodeposition by *Mytilus Edulis* in the circulation of matter and nutrients in a baltic coastal ecosystem. Mar Ecol Prog Ser 38:201–212

Kautsky N, Wallentinus I (1980) Nutrient release from Baltic *Mytilus*-red algal community and its role in benthic and pelagic productivity. Ophelia 1:17–30

Khalaman VV (2001) Fouling communities of mussel aquaculture installations in the White Sea. Russ J Mar Biol 27:227–237

Kreeger DA, Hawkins AJS, Bayne BL, Lowe DM (1995) Seasonal variation in the relative utilization of dietary-protein for energy and biosynthesis by the mussel *Mytilus edulis*. Mar Ecol Prog Ser 126:177–184

Kuenzler EJ (1961) Phosphorus budget of a mussel population. Limnol Oceanogr 6:400–415

Labarta U, FernandezReiriz MJ, Babarro JMF (1997) Differences in physiological energetics between intertidal and raft cultivated mussels Mytilus galloprovincialis. Mar Ecol Prog Ser 152:167–173

Leblanc AR, Landry T, Miron G (2003) Fouling organisms of the blue mussel *Mytilus edulis*: their effect on nutrient uptake and release. J Shellfish Res 22:633–638

Lindahl O, Hart R, Hernroth B, Kollberg S, Loo LO, Olrog L, Rehnstam-Holm AS, Svensson J, Svensson S, Syversen U (2005) Improving marine water quality by mussel farming: a profitable solution for Swedish society. Ambio 34:131–138

Livingstone DR, Widdows J, Fieth P (1979) Aspects of nitrogen-metabolism of the common mussel *Mytilus edulis* – adaptation to abrupt and fluctuating changes in salinity. Mar Biol 53:41–55

Lum SC, Hammen CS (1964) Ammonia excretion of Lingula. Comp Biochem Physiol 12:185–190

Lutz-Collins V, Quijon P, Davidson J (2009) Blue mussel fouling communities: polychaete composition in relation to mussel stocking density and seasonality of mussel deployment and sampling. Aquac Res 40:1789–1792

Maar M, Timmermann K, Petersen JK, Gustafsson KE, Storm LM (2010) A model study of the regulation of blue mussels by nutrient loadings and water column stability in a shallow estuary, the Limfjorden. J Sea Res 64:322–333

MacDonald BA, Ward JE (2009) Feeding activity of scallops and mussels measured simultaneously in the field: Repeated measures sampling and implications for modelling. J Exp Mar Biol Ecol 371:42–50

Mazouni N, Gaertner JC, DeslousPaoli JM, Landrein S, d'Oedenberg MG (1996) Nutrient and oxygen exchanges at the water-sediment interface in a shellfish farming lagoon (Thau, France). J Exp Mar Biol Ecol 205:91–113

McKindsey CW, Archambault P, Callier MD, Olivier F (2011) Influence of suspended and off-bottom mussel culture on the sea bottom and benthic habitats: a review. Can J Zool 89:622–646

Miller DC, Norkko A, Pilditch CA (2002) Influence of diet on dispersal of horse mussel Atrina zelandica biodeposits. Mar Ecol Prog Ser 242:153–167

Monteiro PMS, Spolander B, Brundrit GB (1998) Shellfish mariculture in the Benguela system: estimates of nitrogen-driven new production in Saldanha Bay using two physical models. J Shellfish Res 17:3–13

Navarro JM, Thompson RJ (1997) Biodeposition by the horse mussel *Modiolus modiolus* (Dillwyn) during the spring diatom bloom. J Exp Mar Biol Ecol 209:1–13

Newell RIE (2004) Ecosystem influences of natural and cultivated populations of suspension-feeding bivalve molluscs: a review. J Shellfish Res 23:51–61

Nixon SW (1995) Coastal marine eutrophication – a definition, social causes, and future concerns. Ophelia 41:199–219

Nixon SW, Oviatt CA, Garber J, Lee V (1976) Diel metabolism and nutrient dynamycs in a salt-marsh embayment. Ecology 57:740–750

Nixon SW, Ammerman JW, Atkinson LP, Berounsky VM, Billen G, Boicourt WC, Boynton WR, Church TM, Ditoro DM, Elmgren R, Garber JH, Giblin AE, Jahnke RA, Owens NJP, Pilson MEQ, Seitzinger SP (1996) The fate of nitrogen and phosphorus at the land sea margin of the North Atlantic Ocean. Biogeochemistry 35:141–180

Nizzoli D, Bartoli M, Viaroli P (2006) Nitrogen and phosphorous budgets during a farming cycle of the Manila clam *Ruditapes philippinarum*: an in situ experiment. Aquaculture 261:98–108

Olesen B (1996) Regulation of light attenuation and eelgrass *Zostera marina* depth distribution in a Danish embayment. Mar Ecol Prog Ser 134:187–194

Paasche E, Erga SR (1988) Phosphorus and nitrogen limitation of phytoplankton in the inner Oslo fjord (Norway). Sarsia 73:229–243

Petersen JK (2004) Methods for measurement of bivalve clearance rate – hope for common understanding. Mar Ecol Prog Ser 276:309–310

Petersen JK, Bougrier S, Smaal AC, Garen P, Robert S, Larsen JEN, Brummelhuis E (2004) Intercalibration of mussel *Mytilus edulis* clearance rate measurements. Mar Ecol Prog Ser 267:187–194

Petersen JK, Hasler B, Timmermann K, Nielsen P, Torring DB, Larsen MM, Holmer M (2014) Mussels as a tool for mitigation of nutrients in the marine environment. Mar Pollut Bull 82:137–143

Philippart CJM, Beukema JJ, Cadee GC, Dekker R, Goedhart PW, vanIperen JM, Leopold MF, Herman PMJ (2007) Impacts of nutrient reduction on coastal communities. Ecosystems 10:95–118

Pitcher GC, Calder D (1998) Shellfish mariculture in the Benguela system: phytoplankton and the availability of food for commercial mussel farms in Saldanha Bay, South Africa. J Shellfish Res 17:15–24

Plus M, La Jeunesse I, Bouraoui F, Zaldivar JM, Chapelle A, Lazure P (2006) Modelling water discharges and nitrogen inputs into a Mediterranean lagoon – Impact on the primary production. Ecol Model 193:69–89

Prins TC, Smaal AC (1990) Benthic-pelagic coupling: the release of inorganic nutrients by an intertidal bed of *Mytilus edulis*. In: Barnes M, Gibson RN (eds) Trophic relationships in marine environments. Aberdeen University Press, Aberdeen, pp 89–103

Prins TC, Smaal AC (1994) The role of the blue mussel *Mytilus edulis* in the cycling of nutrients in the Oosterschelde Estuary (the Netherlands). Hydrobiologia 283:413–429

Prins TC, Dankers N, Smaal AC (1994) Seasonal varitaion in the filtration-rates of a semi-natural mussel bed in relation to seston composition. J Exp Mar Biol Ecol 176:69–86

Prins TC, Escaravage V, Smaal AC, Peeters JCH (1995) Nutrient cycling and phytoplankton dynamics in relation to mussel grazing in a mesocosm experiment. Ophelia 41:289–315

Prins TC, Smaal AC, Pouwer AJ, Dankers N (1996) Filtration and resuspension of particulate matter and phytoplankton on an intertidal mussel bed in the Oosterschelde estuary (SW Netherlands). Mar Ecol Prog Ser 142:121–134

Prins T, Smaal A, Dame R (1998) A review of feedbacks between bivalve grazing and ecosystem processes. Aquat Ecol 31:349–359

Redfield AC, Ketchum BH, Richards FA (1963) The influence of organisms on the composition of seawater, vol 2. Wiley-Interscience, New York

Ricciardi A, Bourget E (1998) Weight-to-weight conversion factors for marine benthic macroinvertebrates. Mar Ecol Prog Ser 163:245–251

Richard M, Archambault P, Thouzeau G, Desrosiers G (2006) Influence of suspended mussel lines on the biogeochemical fluxes in adjacent water in the Iles-de-la-Madeleine (Quebec, Canada). Can J Fish Aquat Sci 63:1198–1213

Richard M, Archambault P, Thouzeau G, McKindsey CW, Desrosiers G (2007) Influence of suspended scallop cages and mussel lines on pelagic and benthic biogeochemical fluxes in Havre-aux-Maisons Lagoon, Iles-de-la-Madeleine (Quebec, Canada). Can J Fish Aquat Sci 64:1491–1505

Riisgard HU (2001) On measurement of filtration rates in bivalves – the stony road to reliable data: review and interpretation. Mar Ecol Prog Ser 211:275–291

Riisgard HU (2004) Intercalibration of methods for measurement of bivalve filtration rates – a turning point. Mar Ecol Prog Ser 276:307–308

Sætre R (2007) The Norwegian coastal current-oceanography and climate. Tapir Academic press, Trondheim, p 159

Saxby SA (2002) A review of food availability, sea water characteristics and bivalve growth performance at coastal culture sites in temperate and warm temperate regions of the world. Government of Western Australia – Department of Fisheries

Schellekens T, van Stralen M, Kesteloo-Hendrikse J, Smaal A (2014) Analyse historische data Oosterschelde en Waddenzee. IMARES Rapport C189/13 (in Dutch)

Schluter L, Josefsen SB (1994) Annual variation in condition, respiration and remineralization of *Mytilus edulis* L in the Sound, Denmark. Helgolander Meeresuntersuchungen 48:419–430

Shannon LV, Stander GH (1977) Physical and chemical characteristics of water in Saldanha Bay and Langebaan Lagoon. Trans R Soc S Afr 42:441–459
Shumway SE (ed) (2011) Shellfish aquaculture and the environment. Wiley-Blackwell, London
Smaal AC (2002) European mussel cultivation along the Atlantic coast: production status, problems and perspectives. Hydrobiologia 484:89–98
Smaal A, Heral M (1998) Modelling bivalve carrying capacity. Aquat Ecol 31:1386–2588
Smaal AC, Prins TC (1993) The uptake of organic matter and the release of inorganic nutrients by bivalve suspension feeders, vol G33. Springer, Berlin/Heidelberg
Smaal AC, Vonck A (1997) Seasonal variation in C, N and P budgets and tissue composition of the mussel *Mytilus edulis*. Mar Ecol Prog Ser 153:167–179
Smaal AC, Zurburg W (1997) The uptake and release of suspended and dissolved material by oysters and mussels in Marennes-Oleron Bay. Aqua Liv Res 10:23–30
Smaal AC, Vonck A, Bakker M (1997) Seasonal variation in physiological energetics of *Mytilus edulis* and *Cerastoderma edule* of different size classes. J Mar Biol Assoc UK 77:817–838
Smaal A, van Stralen M, Schuiling E (2001) The interaction between shellfish culture and ecosystem processes. Can J Fish Aquat Sci 58:991–1002
Souchu P, Vaquer A, Collos Y, Landrein S, Deslous-Paoli JM, Bibent B (2001) Influence of shellfish farming activities on the biogeochemical composition of the water column in Thau lagoon. Mar Ecol Prog Ser 218:141–152
Strohmeier T (2009) Feeding behaviour and bioenergetic balance of the great scallop (*Pecten maximus*) and the blue mussel (*Mytilus edulis*) in a low seston environment and relevance to suspended shellfish aquaculture. PhD. University of Bergen, Norway
Strohmeier T, Aure J, Duinker A, Castberg T, Svardal A, Strand O (2005) Flow reduction, seston depletion, meat content and distribution of diarrhetic shellfish toxins in a long-line blue mussel (Mytilus edulis) farm. J Shellfish Res 24:15–23
Strohmeier T, Duinker A, Strand O, Aure J (2008) Temporal and spatial variation in food availability and meat ratio in a longline mussel farm (*Mytilus edulis*). Aquaculture 276:83–90
Strohmeier T, Strand O, Cranford P (2009) Clearance rates of the great scallop (*Pecten maximus*) and blue mussel (*Mytilus edulis*) at low natural seston concentrations. Mar Biol 156:1781–1795
Strohmeier T, Strand O, Alunno-Bruscia M, Duinker A, Rosland R, Aure J, Erga S, Naustevoll L, Jansen H, Cranford PJ (2015) Response of Mytilus edulis to enhanced phytoplankton availability by controlled upwelling in an oligotrophic fjord. Mar Ecol Prog Ser 518:139–152
Sundby B, Gobeil C, Silverberg N, Mucci A (1992) The phosphorus cycle in coastal marine sediments. Limnol Oceanogr 37:1129–1145
Tang QS, Zhang JH, Fang JG (2011) Shellfish and seaweed mariculture increase atmospheric CO2 absorption by coastal ecosystems. Mar Ecol Prog Ser 424:97–104
Tenore KR, Boyer LF, Cal RM, Corral J, Garciafernandez C, Gonzalez N, Gonzalezgurriaran E, Hanson RB, Iglesias J, Krom M, Lopezjamar E, McClain J, Pamatmat MM, Perez A, Rhoads DC, Desantiago G, Tietjen J, Westrich J, Windom HL (1982) Coastal upwelling in the Rias Bajas, Nw Spain – Contrasting the benthic regimes of the Rias De Arosa and De Muros. J Mar Res 40:701–772
Thompson RJ (1984) The reproductive cycle and physiological ecology of the mussel *Mytilus edulis* in a subarctic, non-estuarine environment. Mar Biol 79:277–288
Thouzeau G, Grall J, Clavier J, Chauvaud L, Jean F, Leynaert A, ni Longphuirt S, Amice E, Amouroux D (2007) Spatial and temporal variability of benthic biogeochemical fluxes associated with macrophytic and macrofaunal distributions in the Thau lagoon (France). Estuar Coast Shelf Sci 72:432-446
Tremblay R, Myrand B, Sevigny JM, Blier P, Guderley H (1998) Bioenergetic and genetic parameters in relation to susceptibility of blue mussels, *Mytilus edulis* (L.) to summer mortality. J Exp Mar Biol Ecol 221:27–58
Trottet A, Roy S, Tamigneaux E, Lovejoy C (2007) Importance of heterotrophic planktonic communities in a mussel culture environment: the Grande Entree øagoon, Magdalen Islands (Quebec, Canada). Mar Biol 151:377–392

Trottet A, Roy S, Tamigneaux E, Lovejoy C, Tremblay R (2008a) Impact of suspended mussels (*Mytilus edulis* L.) on plankton communities in a Magdalen Islands lagoon (Quebec, Canada): a mesocosm approach. J Exp Mar Biol Ecol 365:103–115

Trottet A, Roy S, Tamigneaux E, Lovejoy C, Tremblay R (2008b) Influence of suspended mussel farming on planktonic communities in Grande-Entree Lagoon, Magdalen Islands (Quebec, Canada). Aquaculture 276:91–102

Tsuchiya M (1980) Biodeposit production by the mussel *Mytilus edulis* L on rocky shores. J Exp Mar Biol Ecol 47:203–222

Turner RE, Qureshi N, Rabalais NN, Dortch Q, Justic D, Shaw RF, Cope J (1998) Fluctuating silicate : nitrate ratios and coastal plankton food webs. Proc Natl Acad Sci USA 95:13048–13051

van Broekhoven W, Troost K, Jansen H, Smaal A (2014) Nutrient regeneration by mussel Mytilus edulis spat assemblages in a macrotidal system. J Sea Res 88:36–46

van Broekhoven W, Jansen H, Verdegem M, Struyf E, Troost K, Lindeboom H, Smaal A (2015) Nutrient regeneration from feces and pseudofeces of mussel Mytilus edulis spat. Mar Ecol Prog Ser 534:107–120

Vink S, Atkinson MJ (1985) High dissolved C-P excretion ratios for large benthic marine-invertebrates. Mar Ecol Prog Ser 21:191–195

Wassmann P (2005) Cultural eutrophication: perspectives and prospects. In: Wassmann P, Olli K (eds) Drainage basin inputs and eutrophication: an intergarted approach. University of Tromsø Tromsø, Norway

Wetsteyn LPMJ, Duin RNM, Kromkamp JC, Latuhihin MJ, Peene J, Pouwer A, Prins TC (2003) Investigation on carrying capacity of the Oosterschelde estuary (in Dutch). Report RIKZ/2003049. Institute for the coastal zone and sea (RIKZ)

Widdows J (1978) Combined effects of body size, food concentration and season on physiology of *Mytilus edulis*. J Mar Biol Assoc UK 58:109–124

Widdows J, Bayne BL (1971) Temperature acclimation of *Mytilus edulis* with reference to its energy budget. J Mar Biol Assoc UK 51:827–843

Wijsman J, Troost K, Fang J, Roncarati A (2019) Global production of marine bivalves. Trends and challenges. In Smaal et al (eds) Goods and services of marine bivalves. Springer, Cham, pp 7–26

Wiles PJ, van Duren LA, Hase C, Larsen J, Simpson JH (2006) Stratification and mixing in the Limfjorden in relation to mussel culture. J Mar Syst 60:129–143

Zeldis J (2005) Magnitudes of natural and mussel farm-derived fluxes of carbon and nitrogen in the Firth of Thames. NIWA client report: CHC2005-048

Chapter 10
Nutrient Extraction Through Bivalves

Jens Kjerulf Petersen, Marianne Holmer, Mette Termansen, and Berit Hasler

Abstract Ecosystem services provided by marine bivalves in relation to nutrient extraction from the coastal environment have gained increased attention to mitigate adverse effects of excess nutrient loading from human activities, such as agriculture and sewage discharge. These activities damage coastal ecosystems and require action from local, regional, and national environmental management. Marine bivalves filter particles like phytoplankton, thereby transforming particulate organic matter into bivalve tissue or larger faecal pellets that are transferred to the benthos. Nutrient extraction from the coastal environment takes place through two different pathways: (i) harvest/removal of the bivalves – thereby returning nutrients back to land; or (ii) through increased denitrification in proximity to dense bivalve aggregations, leading to loss of nitrogen to the atmosphere. Active use of marine bivalves for nutrient extraction may include a number of secondary effects on the ecosystem, such as filtration of particulate material. This leads to partial transformation of particulate-bound nutrients into dissolved nutrients via bivalve excretion or enhanced mineralization of faecal material. In this chapter, concepts in relation to nutrient extraction by bivalves are presented and discussed in relation to nutrient cycling and additional effects of enhancing bivalve communities. In addition, meth-

J. K. Petersen (✉)
Danish Shellfish Centre, Institute of Aquatic Resources, Danish Technical University, DK, Nykoebing Mors, Denmark
e-mail: jekjp@aqua.dtu.dk

M. Holmer
Department of Biology, University of Southern Denmark, Odense, Denmark
e-mail: holmer@biology.sdu.dk

M. Termansen
Section for Environment and Natural Resources, Department of Food and Resource Economics, University of Copenhagen, Copenhagen, Denmark
e-mail: mt@ifro.ku.dk

B. Hasler
Environmental Social Science, Department of Environmental Science, Aarhus University, Roskilde, Denmark
e-mail: bh@envs.au.dk

A. C. Smaal et al. (eds.), *Goods and Services of Marine Bivalves*,
https://doi.org/10.1007/978-3-319-96776-9_10

ods to valorise nutrient extraction by bivalves are evaluated. Examples of calculations of the value of nutrient extraction by bivalves are presented.

Abstract in Chinese 摘要:农业和污水排放等人类活动造成的水体富营养化现象已经严重威胁到近海生态系统健康状况,亟需采取有效的环境管理策略进行应对,在这样的背景下,海水双壳类在移除营养物质方面的生态服务功能(受到了越来越多的关注。海水双壳贝类可以通过滤食行为将颗粒有机物转化成软体组织或较易沉淀的较大的粪便颗粒。近岸海域中营养物质的移除包括两种途径:i)收获/移除养殖的双壳贝类,由此将营养盐返回到陆地;或者ii)高密度养殖的双壳贝类能够增强氮的反硝化作用,使水体中的氮释放到大气中。但是,需要指出的是,双壳贝类在发挥净水作用的同时也会产生一系列相关的次生效应,如贝类通过摄食、排泄、粪便的矿化作用等将营养物质由颗粒态转化为溶解态。在本章,我们将介绍并讨论双壳贝类移除营养物质的过程和机制,包括营养盐的循环过程和大量贝类聚集时产生的附加效应。另外,我们评估了双壳贝类净水能力的价值和效率,并给出了贝类净化海域环境的价值计算实例。.

Keywords Mitigation · Bivalve farming · Denitrification · Nutrient cycling · Economic valuation · Abatement policies

关键词 缓解作用 • 双壳贝类养殖 • 脱氮作用 • 营养物质循环 • 经济评估 • 减排政策

10.1 Introduction

Excess loading of nutrients is one the largest concerns for the marine environment worldwide (Cloern 2001; Duarte et al. 2009). In the coastal zone, nitrogen loading from human activities within contributing watersheds and atmospheric deposition have prompted regulators, managers, and the political system to set standards for water quality and enforce legislation to prevent further deterioration of the marine environment. On a larger scale, examples include the legislative actions enforced by the European Union, e.g. the Nitrate Directive (Anon. 1991), the Water Framework Directive (Anon. 2000) and the Marine Strategy Framework Directive (Anon. 2008) that all aim to reduce nutrient loading – in particular nitrogen – to the marine environment as a means to improve water quality. Traditional measures utilized to reduce nutrient loading to the marine environment are land based. These are directed either towards point sources like sewage treatment plants, or diffuse emissions mainly from cultivated land. Abatement measures for diffuse sources comprise a long list; including restrictions in fertilization, restriction in the periods where fertilization is allowed, requirements for catch crops and winter green fields, wetland restoration and wetland reconstruction, afforestation, and fallowing of intensively cultivated fields. With increasing marginal costs for implementing traditional land-based abatement measures (Hasler et al. 2012), it is appealing to look for alternatives, such as mitigation measures in the recipient water bodies. Strategies less costly than traditional abatement measures are attractive in coastal zones where population densities are low. Finally, internal loading from sediments in areas that have been affected by decades of excess nutrient loading is a problem for water quality that can only be dealt with by marine mitigation measures.

In this context, nutrient extraction services provided by bivalves become interesting. Through their filtering of water, bivalves remove a proportion of the phytoplankton that in large concentrations otherwise is part of the negative effects of excess nutrient loading. By clearing the water column of particles, bivalves contribute to reductions in turbidity and concentrations of particulate organic nutrients, like nitrogen and phosphorous (see e.g. Dame 2012 and references therein). The filtered material is either not ingested and ejected as pseudo faeces or is ingested and digested, then transformed into bivalve tissue or faecal material that settles in proximity of the bivalves. Nutrients in the ingested material that is transformed into bivalve tissue are immobilized, hence temporarily not accessible for primary production. If the bivalves are removed from the water column, e.g. through harvest, the nutrients are permanently made inaccessible. The material ejected as faeces or pseudo faeces can enter nutrient cycles that may result in either permanent burial in the sediment or removal through chemical processes; i.e. denitrification. Both processes will result in a nutrient extraction service provided by the bivalves that potentially can be used as a mitigation tool by managers seeking means of remediating effects of excess nutrient loading to coastal ecosystems. This can be realized as

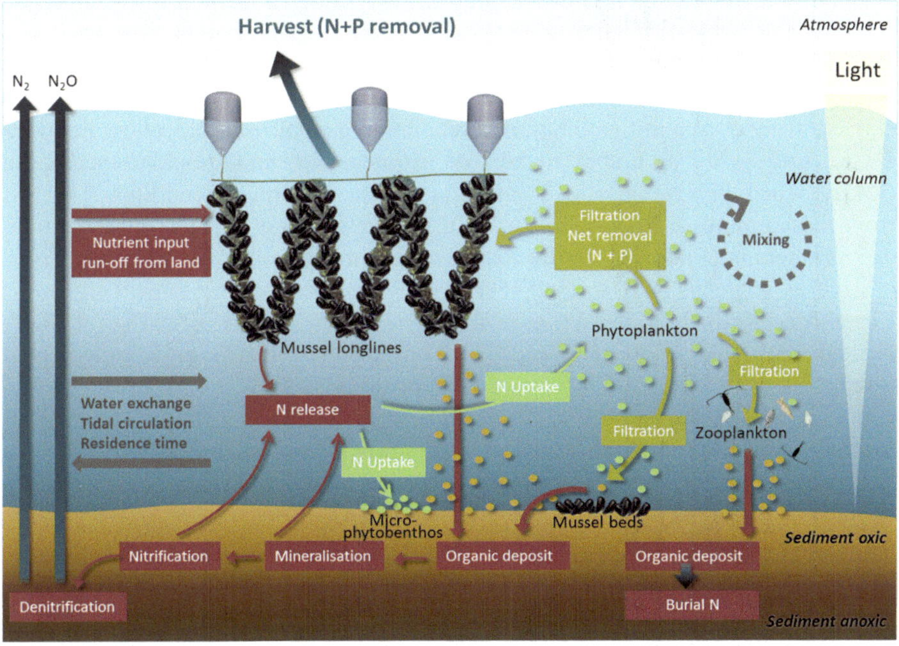

Fig. 10.1 Nutrient extraction services provided by bivalves. Blue mussels are used as examples but other bivalves like oysters can also provide these nutrient extraction services

either bivalve aquaculture or by promoting or restoring natural bivalve populations (Fig. 10.1); e.g. oyster reefs or mussel beds.

10.2 Nutrient Extraction Through Bivalve Aquaculture

Nutrient extraction through mussel farming or other forms of bivalve aquaculture is based on two simple principles: (1) By providing substrate for mussel or oyster larvae to settle on or by other means of actively increasing recruitment, (e.g. deploying seed from hatcheries) resulting in new bivalve biomass being produced; and (2) mass balance, i.e. the nutrients stored in bivalve tissue are extracted from the water body where bivalves are produced at harvest. Irrespective of the efficiency of the bivalves in transforming particle-bound nutrients into tissue, including the loss of nutrients as dissolved nutrients from the bivalves as excretion (see e.g. Cranford et al. 2007), there will be a net nutrient extraction at the water body scale when the bivalves are harvested (e.g. Holmer et al. 2015; Guyondet et al. 2015).

Bivalve aquaculture for consumption is performed in many different ways, from artisanal seeding of infauna clams to offshore mussel farming on specialized long-line systems. Common for all types is that the main aim is to produce an optimal product suited for human consumption. There is thus less focus on maximum biomass removal or nutrient extraction. However, as long as the culturing activity has resulted in a new production of bivalves and the bivalves are harvested, it can be assumed that nutrients have been removed from the system, or more precisely, have been recycled back to land.

A special application of bivalve aquaculture is bivalve farming aimed at nutrient extraction either as a mitigation tool in relation to general eutrophication, i.e. from diffuse sources, or as a specialized tool in integrated multi trophic aquaculture (IMTA) where bivalve farming is intended to remove particle bound nutrients lost from fish farming (e.g. Chopin 2013). In bivalve farming for nutrient extraction, excess amounts of nutrients in the coastal waters are considered as a resource to be utilized as recycled back to land (Hart 2003; Petersen 2004; Møhlenberg 2007; Lindahl and Kollberg 2009; Weber et al. 2010; Rose et al. 2014, Bricker et al. 2014, Kellogg et al. 2014, Petersen et al. 2016). Nutrients in the marine environment – but originating from land – are taken up by the bivalves as particles, preferably phytoplankton, and returned back to land after harvest. Bivalve farming for nutrient extraction will thus immobilize nutrients lost to the aquatic environment, store them in bivalve tissues, and to a lesser extent in shell (and byssus), and bring them back to land when harvested. Back on land, the bivalves can be used for various purposes, e.g. food, feed, or otherwise; and thus provide additional goods and services. This concept has been termed mitigation, bioremediation, bio-extraction, bio harvesting, agro-aqua recycling, or compensation aquaculture (Petersen et al. 2014); and the whole process is based on a mass balance principle in the recipient water body. Furthermore, there is not necessarily a direct link between the nutrient source and the nutrient extraction process. In IMTA, bivalve farming is physically directed toward marine point sources of nutrients, like aquaculture of fed cultivated animals i.e. fish and shrimps (Chopin 2013; Troell et al. 2009). It should be noted, however, that bivalves only take up nutrients as particulate matter and will thus only be able to directly use the nutrients emitted from a fish farm to a negligible extent. Additionally, due to hydrodynamic constraints, cultured bivalves in IMTA farms will have difficulties filtering major parts of the particulate waste material pool released from fish farms (Cranford et al. 2013; Petersen et al. 2016). Hence, the mitigation of nutrient release from a fish farm in IMTA also works on the mass balance principle, and not as a measure to remove nitrogen and phosphorous molecules released from the fish farm (Cranford et al. 2013; Petersen et al. 2016).

Farming bivalves with the main aim of extracting nutrients from the aquatic environment is different in nature from most commercial bivalve farming, which is mainly performed for human consumption (Farber et al. 2006) or a recent farming practice aimed at providing seed for on-bottom culture of blue mussels (Capelle 2017). Commercial bivalve aquaculture aims for uniform size, high quality and good appearance; where the product is very dependent on the market. Mitigation bivalves are produced to remove as much nutrients as possible at the lowest costs in order to be an efficient tool from a management point of view. The resulting product may not necessarily, or entirely, be suited for human consumption due to its size, heterogeneity, and appearance (Petersen et al. 2016). This has some implications for farming practice, as it for several reasons (e.g. cost-effectiveness), may be preferable to harvest young and small bivalves rather than wait for commercial size:

1. Total bivalve biomass, rather than individual size and quality, matters for harvesting time. Bivalves grow fast in the early stages; relative biomass gain on the production unit will be greater in early stages after recruitment compared to later stages when bivalves are approaching commercial size. Sometime after settling (or deployment), space may become a limiting factor, and density (number of individuals per area settling material) will decrease (e.g. Lahance-Bernard et al. 2010). Biomass may still increase as mussels can grow on top of each other and self-thinning will reduce density without affecting biomass. Ultimately, lack of space or competition for food may become limiting for further biomass increase. In commercial mussel farming, this will result in the farmer either thinning (on net structures) or socking (in long-line units), or losing a part of the crop due to self-thinning as mussels become detached from the settling material (Lahance-Bernard et al. 2010). When the mussels become approximately 1 year old, new spat may start settling on the culture unit; thereby further increasing competition for space. It is thus important in mitigation farming to harvest at the time of maximum biomass.
2. An additional factor may influence harvested biomass: Biofouling. There are many reports on fouling of aquaculture units (see e.g. Locke et al. 2009 and references herein) and the consequences for bivalve aquaculture production (see e.g. Daigle and Herbinger 2009). If bivalve production is affected negatively by fouling, it will affect biomass development and hence the mitigation effect. On the other hand, if the production strategy is designed to promote early harvest, levels of biofouling will be reduced in comparison to present commercial farming practices, due to shorter immersion times of the farm structure. Biofouling may even increase nutrient capture and the subsequent removal when harvested.

3. Total biomass per se is not, however, the only guiding parameter for optimization of farming for nutrient extraction. Nutrient content of the biomass is also important, and different parts of the bivalves contain different amounts of nutrients; for example, the shells of blue mussels have lower nutrient concentrations than blue mussel tissue (Petersen et al. 2014). Tissue weight will fluctuate over the year, resulting in varying total concentrations of nutrients, as tissue content of mussels will depend on size, growth state, and gonadal cycle of the mussels (e.g. Dare and Edwards 1975; Rodhouse et al. 1984). In general, relative tissue content is highest in small, fast-growing bivalves (see e.g. Smaal and Vonck 1997). In addition, blue mussel byssus may add substantially to the total nutrient extraction of blue mussel farming. In a recent experiment in Skive Fjord, Denmark, 12–19% of the nitrogen removal through harvest was from blue mussel byssus (Petersen et al. 2014). In relation to biofouling, nutrient content of the fouling organism will also matter.
4. When bivalve farming is implemented primarily for nutrient extraction, resource allocation for handling the aquaculture unit becomes important in relation to yield in the form of nutrient build-up in the farmed bivalves. Resources include labour, materials like buoys (for keeping the long-lines floating) or on-bottom structures, and boat hours. Beyond a certain point, further investments in labour and/or equipment will not result in increased biomass of bivalves; while preceding that point, the investment will not match the net gain in biomass. Factors determining how long bivalves are to be maintained in the aquaculture unit, rather than being harvested include: relative increase in biomass, tissue content of the bivalves, and environmental conditions like potential ice coverage or increased frequency of storms requiring additional efforts to maintain the biomass.

There is little experimental validation on full scale of bivalve farming with the purpose of nutrient extraction. To our knowledge, the only scientific validation experiment on full production scale of mitigation aquaculture using blue mussels has so far been performed in Skive Fjord, the Limfjorden Denmark (Petersen et al. 2014). Skive Fjord is a shallow estuary with a mean depth of 4.7 m, in the inner part of the Limfjorden (Maar et al. 2010). In the Limfjorden, there are almost no tidal currents and the water column varies between stratified and mixed conditions on a time scale of days to weeks, controlled by differential advection, fresh water input, heating and mixing (Maar et al. 2010; Wiles et al. 2006; Møhlenberg 1999). Skive Fjord is highly nutrient-enriched, characterized by high chlorophyll *a* concentrations and high primary production throughout the year and seasonal hypoxia occurring in late summer (Møhlenberg 1999; Maar et al. 2010; Holmer et al. 2015). Production of blue mussels (*Mytilus edulis*) in this trial took place on approx. 90 km of settling material deployed in May 2010 on 90 long-lines in an approximately 19 ha aquaculture unit. During the production period – from deployment of settling material in May 2010 to test harvest in October 2010 and March 2011 and final harvest in May 2011 – there was no intermediate handling (e.g. socking or thinning) of the settled mussels. The only handling of the aquaculture farm during the course of the production period was ordinary maintenance, in particular adding support buoys (buoying up) as mussels grew. By May 2011, approximately 1100 t of fresh mussels could be harvested corresponding to 16 t of N and 0.7 t of P. The efficiency of the aquaculture unit corresponds to a removal of 0.6–0.9 t N ha^{-1} $year^{-1}$ and 0.03–0.05 t P ha^{-1} $year^{-1}$ (Petersen et al. 2014). This is more area-efficient in nutrient removal compared to most land based abatement measures, such as establishing riparian wetlands or buffer strips, which is estimated to remove 0.1 and 0.04 t N ha^{-1} $year^{-1}$, respectively (Petersen et al. 2014). Despite observed depletion of phytoplankton both on the micro scale (close to the mussels) and on the farm scale (Nielsen et al. 2016), there was no evidence of food limitation in the farm (Fig. 10.2). Measurements of spatial variations in mussel biomass throughout the year showed no significant differences in mussel biomass between farm sections as well as between edges, and the centre of the mussel farm. Thus, reduced growth of mussels positioned downstream was not observed in the mussel farm in Skive Fjord, as observed/modelled in other mussel cultivation units (Heasman et al. 1998; Fuentes et al. 2000; Strohmeier et al. 2005; Aure et al. 2007; Petersen et al. 2008a, b; Strohmeier et al. 2008; Rosland et al. 2011). A food depletion model indicated that total mussel filtration rates could be increased by 80–120% without exceeding threshold for the necessary food supply to maintain growth (Nielsen et al. 2016); exhibiting options for further improvement of area efficiency of the mussel production/nutrient extraction (e.g. by approximately doubling the standing stock of mussels within the farm area – if practically possible).

As the Baltic Sea can be considered highly nutrient enriched and suffering from the negative effects of excess nutrient loading to the marine environment, it is logical that mitigation measures in the recipient water body have been considered in the Baltic Sea (Stadmark and Conley 2011). In the Western Baltic Sea, trials using mussel production for nutrient extraction have been carried out in the municipality of Lysekil, Sweden. In the period 2005–11, the municipality was allowed to purchase ecosystem services in the form of nitrogen removal through blue mussel farming from a mussel farmer producing for human consumption (Lindahl 2011). There is no scientific documentation of production volumes and efficiency of the mitigation measure during the trial period, and the trial was terminated before the trial period had expired due to the mussel farmer's financial problems (Kollberg, pers. comm.).

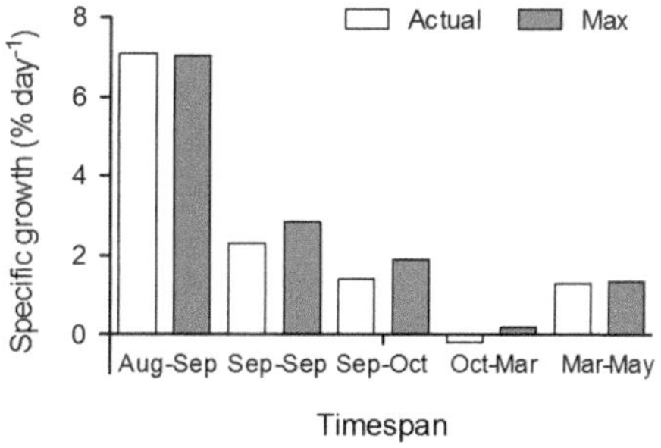

Fig. 10.2 Specific measured mussel growth rates (% day^{-1}) (white bars) and the corresponding potential maximal growth derived from DEB modelling (grey bars) calculated for different timespans between biomass sampling dates (from Nielsen et al. 2016)

The municipality achieved acceptance, in relation to the European Community sewage directive, to exchange nitrogen removal in a sewage treatment plant with nutrient removal through mussel production (Lindahl 2011). However, experiences from the trial indicated that when nutrient extraction is tightly connected to mussel production, primarily aiming at other purposes than nutrient extraction, and the payment of ecosystem services amounts to a minor part of the production costs, there is a high risk of non-compliance with the set goals for nutrient extraction. In the Central and Eastern Baltic, nutrient extraction through mussel production is challenged by low salinity, making production of blue mussels suboptimal (e.g. Maar et al. 2015). According to Lindahl (2012), there have been a number of small trials with blue mussel production from the Great Belt in the west (see also Riisgård et al. 2014) to the Åland archipelago in the east. The trials demonstrated that blue mussels settle and can be grown to sizes leading to substantial biomass accumulation, but growth rates are very low. In the BalticSea2020 project on mussel farming as an environmental measure in the Baltic Sea (http://balticsea2020.org/english/), it was shown that in the Åland archipelago, up to 14 kg of mussel ha^{-1} could be harvested after 2–3 years; however, some trials resulted in less than 10% of this biomass, with mussels of a maximum length of 25 mm. As mussels could be grown on nets up to 4 m in height it was estimated that there is a potential to produce up to 100–150 t blue mussel ha^{-1} over a 2–3-year period corresponding to 1.2–1.8 t N ha^{-1} removal (http://balticsea2020.org/english/). These numbers should probably be considered with some care as they are extrapolated from a rather small test sample.

In the Baltic proper, an alternative option that has been proposed is to farm zebra mussels, *Dreissena polymorpha,* for nutrient extraction. Zebra mussels are widely distributed in the area, from freshwater to brackish and low saline areas, where they can be present in relatively high abundances (Zaiko et al. 2011). The effects of filtration of zebra mussels in freshwater ecosystems are well documented (see e.g. Fahnenstiel et al. 1995; Idrisi et al. 2001; Smith et al. 1998; Caraco et al. 2006; Weber et al. 2010, Pires et al. 2010). From a more theoretical perspective, it has been suggested to use farming of zebra mussels for nutrient extraction (e.g. Stybel et al. 2009; Schernewski et al. 2012). Experiments with farming zebra mussels have been launched in the Oder/Szczecin Lagoon on the border between Germany and Poland and the Curonian Lagoon, Lithuania, but so far with limited data on efficiency (Lindahl 2012). When using zebra mussels for mitigation purposes, and thereby actively taking steps that can result in further proliferation of the species, precautions should be taken that the species is invasive and can cause severe changes to ecosystems. As an invasive species, a large body of scientific literature has documented the changes that zebra mussels can cause in recipient ecosystems. The financial costs preserved by using zebra mussels for mitigation purposes in relation to eutrophication effects may be cancelled-out by increased control of the undesired effects, in systems where they are invasive. The same principle would apply for using the Pacific oyster (*Crassostrea gigas*) as a mitigation crop in areas, where it is not native. Pacific oysters are today a commercial crop in many countries and one of the largest global aquaculture crops, making a direct comparison with zebra mussels difficult. However, a number of countries still prohibit aquaculture of Pacific oysters; and the damage Pacific oysters can cause in coastal ecosystems are well documented (e.g. Herbert et al. 2016). In areas where they are not endemic, the spread of the Pacific oyster should thus in principle not be enhanced as a means to harvest goods and services from bivalve aquaculture, for the same reasons that apply to zebra mussels.

A special case of nutrient extraction is the potential use of blue mussel spat collectors in Dutch on-bottom culture. In Dutch on-bottom blue mussel production, the mussel seed fishery will (as a consequence of a national compromise between NGOs, industry and the government) gradually be replaced by spat collection on floating spat collectors (Capelle 2017). Spat collection resembles production for nutrient extraction as the primary purpose is to maximize viable mussel spat (i.e. biomass), rather than selectivity for size and quality of the mussels. Thus, to a large degree, Dutch on-bottom culture can be considered as new production, especially in the Oosterschelde where natural spat fall is limited (Capelle 2017). This method generally yields biomass production of 1.5–2.5 kg harvested mussel per kg seeded (Capelle 2017); as such, there is in principle a net nutrient extraction also after a relay period on the bottom. However, in the first approximately 2 months after relay of the seeded mussels, there is a loss of 60–69% of the seeded mussels coming from spat collectors (Capelle et al. 2016). Some of the loss will result in increased nutrient regeneration; therefore, the extraction effect of this aquaculture practice is debatable. If relative biomass production approaches 1, on-bottom culture becomes less relevant in a nutrient extraction perspective, and its primary ecosystem service will be provisioning.

10.3 Nutrient Extraction Through Altered Nutrient Cycling

The basic principle of nutrient extraction provided by bivalves through altered nutrient cycling is that aggregations of bivalves (e.g. in bivalve beds or in/below aquaculture units) augmenting the capture of organic material. This mechanism leads to altered biogeochemical processes and subsequently loss of nitrogen through enhanced denitrification (Rose et al. 2014). This type of nutrient extraction can further be pursued as goods and services; provided by artificially established or re-established bivalve beds, e.g. oyster reefs (Kellogg et al. 2014) or by bivalve aquaculture (Humphries et al. 2016). Enhanced denitrification is of further interest if the cultured bivalves, as recently demonstrated for *Crassostrea virginica* and *Crassostrea gigas*, can contribute to denitrification themselves (Caffrey et al. 2016).

Chemical Reactions in the Sediment

Nitrification	$NH_{4+} + 2O_2 \rightarrow NO_{3-} + H_2O + 2H$
Denitrification	$NO_{3-} \rightarrow NO_{2-} \rightarrow NO \rightarrow N_2O \rightarrow N_2$
ANAMMOX	$NH_{4+} + NO_{2-} \rightarrow N_2 + H_2O$
Dissimilatory nitrate reducation to ammonium	$NO_{3-} \rightarrow NO_{2-} \rightarrow NH_{4+}$

Denitrification is a suboxic process. In shallow and often turbulent coastal ecosystems, denitrification is confined to a narrow zone in the surface sediments, typically from a few millimetres to a few centimetres below the sediment surface. Denitrification requires nitrate as an electron acceptor, which is either produced through nitrification, referred to as coupled nitrification-denitrification, or supplied by diffusion from the water column into the sediments. Nitrification only occurs under oxic conditions in the sediments, and rates of nitrification show large seasonal variation controlled by water temperature, ammonium availability, and oxygen concentrations in the sediments. Under eutrophic conditions with high sediment oxygen consumption in the summer, nitrification may be inhibited; thereby diminishing coupled nitrification-denitrification and leading to low rates of denitrification during the summer. Rates of denitrification also exhibit large seasonal variation, but less predictable, as high nitrate concentrations in the water column in the spring may stimulate rates, independent of water temperature and nitrification rates. Both nitrification and denitrification rates typically increase concurrently with transfer of organic matter to the sediments and linked remineralisation and ammonium availability; but only to a certain extent, when nitrification is inhibited due to low oxygen availability in the sediments. In this case, dissimilatory nitrate reduction to ammonium (DNRA) becomes a dominant process, eventually resulting in an elevated transfer of ammonium from the sediments to the water column. Nitrogen cycling in organically enriched sediments (e.g. from sedimentation of biodeposits from bivalve aggregates) may thus be very different from surrounding non-impacted sediments. Organically enriched sediments can be either larger sinks of nitrogen through enhanced denitrification, removing nitrogen from the marine area through N_2 production, or larger sources of nitrogen to the water column by enhanced DNRA and NH_4^+ production. Permanent nitrogen removal from the marine environment, as in denitrification, also occurs in the anammox process, where N_2 is formed in the sediments by bacteria using NH_4^+ and NO_2^-. Anammox is, however, most important for nitrogen removal in oligotrophic systems, where it may contribute up to 80% of the nitrogen removal compared to <20% in organic enriched systems (Dalsgaard et al. 2005); and has been found to play only a minor role in N_2 removal in bivalve sediments (Minjead et al. 2009; Higgins et al. 2013).

Due to the increased availability of organic matter and ammonium in or below bivalve aggregations, there is a potential for stimulated nitrification and denitrification. Studies with measurements of rates of nitrification and denitrification demonstrate variable response to bivalve aggregations, as rates can be reduced or enhanced depending on bivalve species, sediment conditions, and environmental factors (Table 10.1). Giles and Pilditch (2006) examined the effects of mussel (*Perna canaliculus*) aquaculture on sediment oxygen uptake and nutrient fluxes, and found extensive seasonal variation with higher rates of nitrogen release from the sediments at the farm in spring and autumn, but lower during the summer compared to a reference. They suggested that lower nitrogen efflux during the summer was due to enhanced denitrification, reducing the efflux of dissolved inorganic nitrogen compounds, but overall the farm sediments contributed to greater nutrient regeneration compared to the reference site. In this case, denitrification was probably

Table 10.1 Effect on denitrification rates (μmol N_2-N m^{-2} d^{-1}) associated with bivalves from natural reefs and aquaculture

Source	Bivalve	Bottom type	Denitrification enhancement μmol N_2-N m^{-2} d^{-1}
Natural reef			
Sisson et al. 2011	Oyster	Reef	124
Piehler and Smyth 2011	Oyster	Sediments	13.5-95.8
Kellogg et al. 2013	Oyster	Reef	199.2-1486.4
Smyth et al. 2013	Oyster	Sediments	39.9-188.9
Smyth et al. 2015	Oyster	Restored reef, Sediments	180
Aquaculture			
Holyoke 2008	Oyster	-	-37.7-6.3
Higgins et al. 2013	Oyster	-	-58.9-82.8
Testa et al. 2015	Oyster	Sediments	-0.8
Mortazavi et al. 2015	Oyster	Sediments	Ca. 25-100
Lindemann et al. 2016	Oyster	Sediments	357-2143
Smyth et al. 2016	Oyster	Reef	Ca. 100
Murphy et al. 2016a	Hard clam	Sediments	-240-480
Carlsson et al. 2012	Mussel	Sediments	-29.0-41.7
Nizzoli et al. 2006	Clam	Sediments	20-80
Nizzoli et al. 2006	Mussel	Sediments	-10-50

Rates were calculated as the difference between the bivalve site and the rate at the control site (without bivalves) with positive values indicating enhancement and negative ones (in red) indicating reduction in net denitrification rate. Rates are either average of light and dark incubations or only dark incubations and range is between minimum and maximum enhancement in the given study (e.g. seasonal variation)

enhanced by the higher organic matter input to the sediments, but without compromising the nitrification rates. So, higher denitrification rates delivered intensified nitrogen regeneration in the sediments. Similarly, studies of hard clams (*Mercenaria mercenaria*) showed increased fluxes of ammonium and phosphate compared to uncultivated sediments; denitrification rates were also enhanced, but only for parts of the growth season (Table 10.1, Murphy et al. 2016b). In the same study, DNRA was stimulated throughout the growth season and appeared to be the favoured nitrogen cycling process over denitrification, enhancing nitrogen flux to the water column. Welsh and Castadelli (2004) reported enhanced nitrification and coupled nitrification-denitrification from several different bivalves, and suggested that animal-associated nitrogen cycling contributed significantly to nitrogen regeneration in these systems. Furthermore, studies of the manila clam (*Ruditapes philippinarum*) showed that the clams contributed 64–133% of the total rates of sediment oxygen uptake, nitrogen regeneration, nitrification, and denitrification. This indicates that clam biomass/density play a crucial role in nitrogen cycling in bivalve farming areas (Welsh et al. 2015). Enhanced rates were due to metabolic activity of clams and bacterial activity hosted on the clams. The clam sediments were significant sources of both N_2 and N_2O gasses through enhanced nitrification and denitrification. Yet, as N_2O is a greenhouse gas, this contribution is important to consider in environmental impact assessments of bivalve culturing.

Special attention has been devoted to natural or re-established oyster reefs on the North American east coast (see e.g. Kellogg et al. 2014; Smyth et al. 2015). In a feature paper, Kellogg et al. (2013) estimated annual denitrification rates in restored oyster reefs in the Choptank River, Chesapeake Bay, USA, resulting in removal of approximately 0.5 t N ha^{-1} $year^{-1}$ more than at control plots (Kellogg et al. 2013). This corresponds to removal rates from mussel farming as described above. Besides the uncertainty of measuring the effect of oyster reefs (see e.g. Hoellein and Zarnoch 2014; Smyth et al. 2015; Lindemann et al. 2016), there are some caveats in extrapolating this number to larger areas. The rates measured by Kellogg et al. (2013) are in the high end of results when compared to other studies (see Kellogg et al. 2014), and may not be entirely representative for all coastal areas. Further, denitrification rates are variable between reefs/areas; there are large differences between seasons making integration over entire years problematic, and there may also be differences depending on methods (Kellogg et al. 2014; Humphries et al. 2016). The differences between reefs may be explained by their position above or below the euphotic zone (Newell et al. 2005), the actual physical structure of the reef, as well as bioturbation and feeding activities of associated fauna in and around the reefs (e.g. Nizzoli et al. 2006; Smyth et al. 2016). It has generally been concluded that denitrification rates are enhanced in natural or restored oyster reefs compared to rates in oyster aquaculture, probably due to inhibition of nitrification in the more anoxic aquaculture sediments (Higgins et al. 2013; Kellogg et al. 2014; Smyth et al. 2016). However, recent studies have demonstrated comparable denitrification rates in both restored reefs and oyster aquaculture (Humphries et al. 2016). By adding the apparent discrete effect of oysters to denitrification rates (Caffrey et al. 2016), and to a much lesser extent solely empty shells, bivalve mediated denitrification can be considered as an important nutrient extraction service.

10.4 Additional Mitigation Benefits

Bivalve aggregations or bivalve aquaculture may not only facilitate nutrient extraction either through harvest of bivalves, or as enhanced nitrogen loss to the atmosphere, but may also mitigate effects of excess nutrient loading by filtering the water column and thus removing phytoplankton. This is an important aspect of the ecosystem services provided by bivalves as phytoplankton concentrations directly or indirectly serve as ecosystem health indicator, and high concentrations are seen as an indication of adverse effects. For example, in the EU Water Framework Directive, concentration of chlorophyll *a* is an intercalibrated indicator in the Baltic ecoregion, and high concentrations are also indirectly influencing the depth limit of eelgrass, which is another indicator.

The effects of bivalve suspension feeding on water column phytoplankton concentrations were first described for South San Francisco Bay (Cloern 1982). Using a simple model describing change in phytoplankton concentration, where dispersive transport and zooplankton grazing balance growth rate, calculated concentrations of phytoplankton were much higher than actually observed concentrations in the bay (ibid). By estimating the filtration capacity of benthic suspension-feeders, primarily clams, it was shown that these had the capacity to clear the water column more than once per day. It was further described that invasion of a non-indigenous clam in the northern part of the bay resulted in persistently low levels of phytoplankton (Alpine and Cloern 1992). Since then, a number of studies have demonstrated the impact of grazing exerted by benthic bivalves on the overlaying water column, and on the basin scale (e.g. Hily 1991; Møhlenberg 1995; Ackerman et al. 2001; Petersen et al. 2013). An illustrative example is from Ringkøbing Fjord, Denmark (Petersen et al. 2008a, b), where a small change in sluice practice allowed slightly more saline water from the North Sea to enter the estuary, causing a small increase in salinity. This allowed for massive recruitment of the clam *Mya arenaria*. With the invasion of clams, benthic grazing became the key feature of the biological structure, causing a sudden regime shift from a bottom-up controlled turbid state, into a top-down controlled clear water state. Mean annual concentration in chlorophyll *a* dropped concurrently and significantly from 52.3 $\mu g\ l^{-1}$ in the period 1989–94 to 8.7 $\mu g\ l^{-1}$ in 1997–2004 concomitant with the increase in benthic grazing capacity. In the years around the change in sluice management, the change in mean annual concentration of chlorophyll *a* was especially evident, with a decrease from 64.6 $\mu g\ l^{-1}$ in 1995 to 21.0 $\mu g\ l^{-1}$ in 1996 and 7.6 $\mu g\ l^{-1}$ in 1997. Phytoplankton composition and zooplankton abundance were also affected by the change following the invasion of clams.

The impact of clearance by large populations of bivalves on water column concentrations of phytoplankton/particulate matter at the basin scale will depend not only on the size of the populations and their clearance capacity, but also on water residence time in the basin. Bivalve top-down control of phytoplankton biomass can be achieved when clearance time, i.e. time needed for the bivalve standing stock to clear the entire water column, is shorter than residence time, or primary production time, defined as rate of renewal of the phytoplankton biomass (Dame and Prins 1998). Experimentally it has been demonstrated that under well-mixed conditions, a mussel standing stock with a potential clearance time of 20–35% d^{-1} of the entire water volume is enough to control phytoplankton biomass under conditions where primary production is not limited by nutrient concentrations (e.g. Cloern 1982; Prins et al. 1995; Prins et al. 1998; Wang and Wang 2011). Similar conclusions can be drawn from modelling exercises (e.g. Herman and Scholten 1990), indicating that increasing nutrient loading under conditions with high suspension-feeding pressure will only marginally change phytoplankton concentrations.

The clearance effects of suspended cultures of bivalves on water column concentrations of phytoplankton have also been documented in the scientific literature (see e.g. Heasman et al. 1998; Cranford et al. 2008; Grant et al. 2008; Petersen et al. 2008a, b; Cranford et al. 2014; Nielsen et al. 2016). In Skive Fjord, where to date the most extensive experiment with farming of mussels for mitigation purposes took place, depletion of phytoplankton could be observed on all scales; from nearby the mussel lines on the micro scale to farm scale (Nielsen et al. 2016). Farm-scale depletion was detected and visualized based on intensive 3D spatial surveys of the distribution of Secchi depth, chlorophyll *a* and total suspended particulate matter concentrations, both inside and outside the farmed area. Depletion of phytoplankton concentrations within the farm was measured, with average depletion levels of 13–31%; while some areas exhibited >50% depletion. The depletion effects were most pronounced within the farm. Additional model studies showed that summer chlorophyll *a* concentrations were reduced by 30%, and Secchi depth (Fig. 10.3) was improved by 16% relative to a reference situation without the mussel farm (Nielsen et al. 2016). The environmental effects of mussel clearance were however not only evident on the farm scale, but also on the basin scale. The area affected by mussel clearance reached to the shoreline, thereby potentially increasing areas suitable for submerged vegetation (Petersen et al. 2016). Adding more mitigation farms to the model would increase the effect on chlorophyll *a* concentration and light attenuation in the Skive Fjord estuary (Timmermann pers. comm.). Thus, given sufficient capacity, farming of bivalves can act as a control mechanism for the effects of nutrient enrichment, like increased phytoplankton biomass, as could the natural population of clams in Ringkøbing Fjord. As such, a strategy for mitigating effects of excess nutrient enrichment can be to establish extractive bivalve aquacultures, especially in relation to internal loading and diffusive sources. In relation to production carrying capacity of mitigation aquacultures – i.e. where mussel productivity is limited by a shortage of phytoplankton – it is not a major concern in mitigation aquacultures in contrast to commercial mussel production. On the contrary, it can be considered as the objective of mitigation mussel farming, as the aim of this type of aquaculture is to remove nutrients and improve water transparency. If carrying capacity on the basin scale becomes an issue, and the production volume decreases

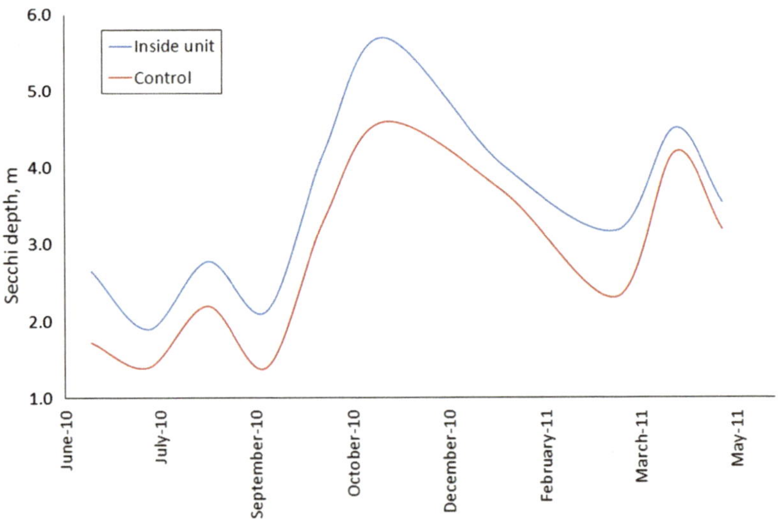

Fig. 10.3 Secchi depth inside production unit in Skive Fjord compared to a control station during a production period from May 2010 to May 2011

and/or environmental parameters like water transparency improves, the purposes of the mitigation farming have been realized, and mitigation aquaculture can be discontinued. In any case, heavily nutrient-enriched systems of interest in this context will require extensive mitigation aquaculture in order to approach limitation of production carrying capacity (Petersen et al. 2016).

10.5 Nutrient Extraction and Nutrient Cycling

The basis of understanding nutrient cycling and potential nutrient extraction in relation to bivalve aggregations – either as aquaculture units or as dense beds/populations – is the trapping of suspended particles in the water column through suspension-feeding, the partitioning of the trapped particles into bivalve tissue (and shell), and as waste products, either dissolved through excretion or as particulate matter as faecal pellets or pseudo faeces. Depending on water transport rate, water depth, and potential resuspension, faecal material will be concentrated in or nearby bivalve aggregations (Chamberlain et al. 2001; McKindsey et al. 2011). Excretion from bivalves is a relatively fast process (compared to regeneration of particle waste products), where particle bound nutrients captured by the bivalves are transformed to dissolved nutrients (mainly ammonia), and thus easily accessible to primary production. Excretion from bivalves is in the order of 10% of the ingested material and has been shown to account for up to 82% of the nitrogen regeneration in mussel farming (Holmer et al. 2015). The turnover of nutrients from excretion is increased compared to release from biodeposits, where decomposition has to take place before nutrients become available in dissolved forms, leading to a slower turnover compared to direct excretion.

Solid waste products from bivalve aggregates will settle on the bottom below or next to the bivalve aggregations. Nutrient regeneration in the surrounding sediments are typically enhanced by the higher quantity (e.g. Hatcher et al. 1994; Grant et al. 1995) and higher quality (e.g. Carlsson et al. 2009) of organic material produced as a result of bivalve digestion, compared to locations lacking larger bivalve populations. Rates of nutrient regeneration reflect the activity of the bivalves, with large seasonal variations controlled by, for example, food availability, water temperatures, and environmental conditions. The regeneration of nitrogen is particularly critical in coastal ecosystems as nitrogen loading from land is high, and are affected by eutrophication with nitrogen as the most important limiting nutrient (Conley et al. 2009; Carstensen et al. 2013; Murphy et al. 2016a). Potential enhanced nitrogen regeneration is thus important to take into account when evaluating the history of coastal areas with major losses of bivalves (e.g. loss of oyster reefs, blue mussel beds; Caffrey et al. 2016), when planning restoration projects (oyster reefs, blue mussel beds; Kellogg et al. 2014, Smyth et al. 2016), or when applying bivalves for mitigation purposes (Stadmark and Conley 2011; Petersen et al. 2012; Petersen et al. 2014). On the other hand, bivalves have a high content of nitrogen in their tissues and shells due to their high protein content and when harvested, significant amounts of nitrogen are permanently removed from the marine environment (Kellogg et al. 2013; Holmer et al. 2015; Petersen et al. 2014). Furthermore, increased denitrification may contribute to a net removal of nitrogen from the ecosystem, particularly if the rates are stimulated during critical periods of the growth season for primary production, such as during summer where nitrogen is the limiting nutrient of phytoplankton production. The general conclusions of the published literature indicate that the regeneration of nitrogen is higher in bivalve aggregates and the nearby surroundings compared to reference sites, and the aggregations/sediments should be considered as net contributors to nitrogen in the water column, when summing up over a production/growth season. There are, however, several important considerations to be taken into account. First, most studies with in situ measurements of denitrification show that rates are enhanced in aggregations/sediments (e.g. Carlsson et al. 2012; Kellogg et al. 2013; Welsh et al. 2015), and the net removal of nitrogen from the marine environment is thus more than just the harvest of bivalves (Table 10.1). The area-specific denitrification rates are typically enhanced, with 25–260% compared to reference conditions (Carlsson et al. 2012). Kellogg et al. (2014) estimated that annual denitrification rates were enhanced from 2.7 to 55.6 g N m^{-2} $year^{-1}$ in oyster reefs. These rates can be quite significant in comparison to nitrogen loading from land – e.g. 1.4–60.1 g N m^{-2} $year^{-1}$ on the East coast of US (Carmichael et al. 2012) and 0.5–100 g N m^{-2} $year^{-1}$ in Danish estuaries (Timmermann pers. comm). Second, bivalve aggregations are concentrators of organic matter (phytoplankton and seston) in the ecosystem, due to their filtering of large volumes of water followed by sedimentation of organic matter in much smaller area, thereby concentrating organic matter enrichment of the ecosystem to a limited area. Biodeposits are heavy and sink to the sediment on the scale of minutes from long-line cultures, and rapidly settle after resuspension events (e.g. Giles and Pilditch 2006; Carlsson et al. 2012). High sedimentation rates thus confine organic enrichment to the immediate vicinity of the aggregations. In contrast, due to increased capture of particles in the bivalve culture, sedimentation will be reduced outside the unit, i.e. on basin scale. The increased particle capture in the bivalve culture will also lead to increased water transparency, promote light penetration and hence reduce nutrient regeneration further afield from bivalve aggregations. This may be beneficial towards the internal loading of marine areas, which is reduced if thresholds of nitrification are not exceeded, allowing coupled nitrification-denitrification to proceed, and remove nitrogen from the area through N_2 production (Carlsson et al. 2012) and similarly for the redox-sensitive release of phosphorus (Holmer et al. 2003). Reduction in sedimentation outside bivalve aggregations can be difficult to detect and cannot be deduced from differences in control vs. affected sites in areas where bivalve aggregations are already present; it should be measured before initiating bivalve production in an area. To our knowledge, only a few studies have addressed the potential effects of bivalve aggregations on concentrating sedimentation in hot spots, and comparing these effects with overall basin scale sedimentation and nutrient regeneration outside the bivalve aggregations. Murphy et al. (2016a) suggested that the net import of particles to support hard clam production contributed to increased nitrogen regeneration in the study area. From a mass balance point of view, sedimentation outside the aggregations must, however, be reduced in comparison to a situation without bivalve aggregations. The local effect will depend on a number of factors including water retention time in the specific basin, organic content of the sediments, nutrient input to the basin, water depth, and stratification. One important possible effect of reduced sedimentation, particularly under eutrophic conditions, is minimizing the risk of oxygen depletion events. Oxygen depletion, where the benthic fauna and flora die-off, generally results in high internal nutrient loading. The release of inorganic nitrogen and phosphorous from the sediments to the water column increases the risk of stimulating blooms of phytoplankton or opportunistic macro algae, and initiating a negative feedback loop maintaining high internal loading. By reducing the internal loading at the basin scale, water quality improves, resulting in higher water transparency and growth of benthic vegetation in deeper waters. Such a scenario can be considered a positive feedback on the ecosystem, as benthic vegetation slows the regeneration of nutrients to the water column, particularly during summer months with high productivity in the vegetation. This eventually leads to longer periods of nutrient limitation of phytoplankton, and thus higher ecological quality of the specific area.

Mass balance calculations of the effects of bivalve aggregations on the basin scale are available in the literature (e.g. Cranford et al. 2007; Brigolin et al. 2009; Holmer et al. 2015; Guyondet et al. 2015). These calculations take into account both nutrient removal through harvest and nutrient regeneration in the bivalve structures, as well as the surrounding sediments, but without accounting for reduced sedimentation outside the bivalve aggregations. These studies indicate that natural bivalve reefs and bivalve aquaculture contribute to a net nitrogen removal at the basin scale through harvest and denitrification, despite increased nitrogen regeneration in the water column and sediments (Holmer et al. 2015; Guyondet et al. 2015). The net nitrogen removal capacity, however, varies between studies from negligible (e.g. Cranford et al. 2007) to important (e.g. Guyondet et al. 2015; Holmer et al. 2015). All studies consider the decrease in phytoplankton concentration as the most important effect of bivalve aggregations on ecosystem processes at the basin scale. These studies also highlight the effects of increased sedimentation and stimulated nutrient regeneration in bivalve aggregations, for example, leading to a higher flux of nitrogen to the water column and to the sediments (e.g. Murphy et al. 2016a). Guyondet et al. (2015) observed that the intensive mussel farming in St Peter's Bay in Eastern Canada maintained phytoplankton biomass at levels corresponding to the 1980s, when aquaculture had not yet developed and nitrogen loading was half of the present level. Basin scale sedimentation in St Peter's Bay was reduced by 14%, and it was concluded that cultivated mussels play an important role in remediating the negative impacts of land-derived nutrient loading in this area, as the mussel farming in St Peter's Bay could counteract a doubling in nitrogen loading. Similarly, mussel farming in the eutrophic Limfjorden, Denmark improved water quality and increased light penetration, promoting the light conditions for benthic vegetation in the area (Petersen et al. 2016). In this study, the uptake of nitrogen in the sediments was stimulated, possibly due to high rates of denitrification, and thereby removing a larger fraction of nitrogen from the fjord compared to the absence of mussel farming. The farm thus contributed to water quality improvements by removal of organic bound nitrogen in phytoplankton, as well as stimulating removal of inorganic nitrogen in the sediments (Holmer et al. 2015). Such recent studies suggest that mussel farming under eutrophic conditions has broad potential for mitigation of excess load of nutrients in marine areas; and increase in mussel farming may reduce effects of eutrophication.

Understanding the overall effects of natural beds of bivalves and/or aquaculture of bivalves on nitrogen cycling in the local environment can be complicated, as multiple factors affect the cycling of nitrogen in the environment. Removal of nitrogen through harvest is relatively easy to measure and extrapolate from single longlines/bivalve aggregates to farms/reefs and farming areas, whereas the effects on water quality and nutrient regeneration can be more difficult to document. The net depositional flux of organic matter is a central parameter driving nutrient regeneration in the sediments, but it is difficult to measure in shallow waters due to methodological constraints and dynamic processes, such as resuspension and advection affecting sedimentation on short and long-term time scales. Modelling is therefore becoming ever more important in management of coastal waters (e.g. Cranford et al. 2007; Guyondet et al. 2015). By using a model where sediment trap deployments were combined with a sediment flux model in an area with oysters (*Crassostrea virginica*), Testa et al. (2015) demonstrated that resuspension and transport effectively removed oyster biodeposits from the studied farms, resulting in limited local environmental impact as there was no long-term sediment accumulation near the oysters, creating hot spots for nutrient recycling. Guyondet et al. (2015) applied a coupled hydrodynamic-biogeochemical model in an area with mussel aquaculture and found that mussel harvest extracts nitrogen resources equivalent to 42% of river inputs and 46.5% of phytoplankton primary production. Based on the limited number of studies at the basin scale, and case studies of individual reefs and individual farms, it is apparent that natural bivalve beds/reefs and/or culturing units of bivalves act as net sinks of nitrogen at the basin scale, due to removal of nitrogen by harvest of bivalve biomass as well as enhanced denitrification. Nonetheless, more mass balance and modelling studies are needed to account for the large spatial and seasonal variation in rates of nitrogen cycling and processes affecting nitrogen cycling measured so far.

10.6 The Economic Value of Bivalve Nutrient Extraction

The economic value of a natural resource is reflected through the flow of services to people, derived from the resource, and can be thought of as the interest on a natural resource asset. We are now accustomed to call this interest *ecosystem services*. While the economic interpretation is simple and intuitive, there are a number of challenges in identifying the economic values, and important caveats related to the existing valuation methods. There are primarily two types of ecosystem services derived from bivalves: provisioning and regulating services. Provisioning services

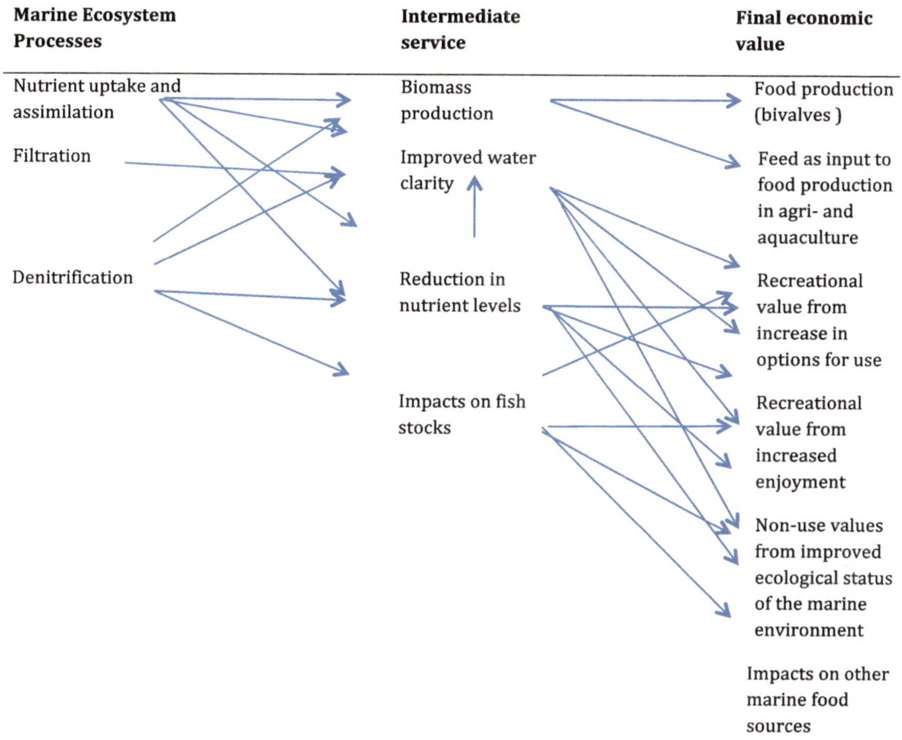

Fig. 10.4 Linking processes to services to economic values. The arrows are illustrative and not a complete mapping of the interconnection between the different aspects

are the production value of the bivalves themselves for human consumption and potentially for other purposes like a feed ingredient for fish, pigs, and poultry. These are private goods, in an economic sense, and markets reflect the economic value of the production. It is important to notice that these services are not entirely provided by marine ecosystems, as labour and capital inputs are needed to convert the ecosystem processes to the final economic good. This means that the economic value of the marine space for bivalve production needs to take into account the costs of the inputs in production.

The other important marine ecosystem service that bivalve production provides is the regulation of water quality associated with bivalve filtration of the water column, and the associated nutrient extraction and denitrification. The filtration effects and the extraction of nutrients are not inherently an economic good, but an intermediate service that contributes to improved water quality and the associated increase in human uses and enjoyment of the marine environment. Any economic valuation of bivalve services should reflect the value of the final goods related to production and water quality improvements (Fig. 10.4). Figure 10.4 illustrates that there is a multitude of processes, services and values involved. While they are all interconnected, there is not a 1:1 relationship between the processes, the services, and the values. This implies that when economists seek to value a particular economic good, they select methods to capture the values outlined in the third column of Fig. 10.2. Due to the lack of 1:1 relationships, each study will not capture every economic value aspects of marine ecosystem processes.

Valuation approaches that have frequently been used for marine ecosystem services are the stated-preference methods. Stated-preference methods are environmental valuation methods based on surveys. These surveys collect data on people's stated rankings of, or choices between, different hypothetical changes in the state of the environment and payments for the change in the associated environmental quality. Such methods have been used to measure the value of clear water by coastal recreationists and other users, but also the more intangible benefits of clear water on biodiversity that do not necessarily depend on recreational use. Clear water also has aesthetic value, which might influence the value from recreation. However, the use of the sea for bivalve production might also be associated with disutility, as the area will not be available for other purposes such as recreation and fisheries. This disutility has not yet, to the author's knowledge, been studied and quantified. A number of studies have attempted to estimate the use and non-use monetary values by estimating the willingness to pay for water clarity improvements. One example is Söderqvist (1996, updated in Söderqvist and Hasselström 2008), who made such an attempt for the Baltic Sea in the mid 1990'ies. Another is the more recent study of Ahtiainen et al. (2014) who aimed to value achieving good ecological status in 2050 also in the Baltic Sea. Most stated-preference valuation studies have focused on measuring the improvement in clarity of the water or achievement of good ecological status. Attributing these economic values to ecosystem processes, and ultimately production of bivalves, is challenging. Söderqvist's (1996) study estimates the economic value of a 50% reduction of the nutrient load to the Baltic Sea, which at that time was the load reduction target to achieve good water quality. The study was updated in 2008 (Gren 2008), and is in fact one of the few studies that estimate the value of good water quality in terms of the value per kg N reduced. This value can in turn be used as the value of 1 kg nitrogen assimilated and removed by bivalves, using the measurements of the effect on nitrogen assimilation and denitrification by bivalves. Using the contingent valuation method, Söderqvist (1996) estimated the willingness to pay to be 12–24 € kg^{-1} N (reported in Gren 2008).

The Söderqvist (1996) study may no longer reflect the present use and non-use value of clear water. Furthermore, the values may vary between locations due to the differences in the number of people exposed and variation in their values and socio-economic characteristics. However, this is only one of the reasons why the economic values of nitrogen reduction are not constant across space. In addition, it is questionable if the biophysical relationships are applicable for all locations, as the nitrogen reduction required to obtain clear water and good ecological status varies between locations. As an example, the nitrogen reduction required to obtain good ecological status in different parts of the Limfjorden (including Skive Fjord) in Denmark varies by a factor 3. Furthermore, as the relationship between the response in water quality and nitrogen reduction might not be linear, this adds further complexity to the valuation task. Overall, these observations imply that the value of reducing nitrogen in different marine ecosystems will likely vary to a substantial degree. This is in sharp contrast to studies valuing reductions in CO_2 emissions. For carbon, it is valid to derive unit values of reductions independent of the location where the emissions are reduced. However, the valuation task related to the two types of emission also has similarities as people could still have different willingness to pay for the ecosystem service. In the context of CO_2 emissions, this would reflect the difference in peoples' willingness to give up current consumption to reduce the risk of climate change in the future.

An alternative and frequently used approach, when it is difficult to measure the willingness to pay per kg emission reduced, is cost-based methods. One of these approaches is the substitute cost method, which is based on measurements of the alternative costs of achieving the ecosystem service by using other means, such as reductions in agricultural nutrient loads. This method is appropriate under the assumption that the cost-estimate of achieving an improvement in ecosystem service provision reflects the marginal costs of an optimal investment decision. If this is the case, the cost estimate reflects the sum of the marginal individual willingness to pay for the service. If it is reasonable to assume that the current ecological status is lower than the societal optimal level, cost based estimates can be used as a conservative value estimate. Studies, such as Ahtiainen et al. (2014) support this assumption, as the willingness to pay for clear water and reduced eutrophication up to the level of good ecological status was higher than the costs of obtaining the required nutrient loads in all countries around the Baltic Sea.

Objectives to achieve good ecological status in a coastal area represent a societal level commitment to invest in improved water quality. If the gap between current ecological status and good ecological status (as in the EU Water Framework Directive) can be expressed as targets for nitrogen load reduction, marginal costs of achieving load reductions through land based measures (e.g. agriculture) can be used as estimates of the value of nitrogen extraction using bivalves. The implication of this is that the value of bivalve nutrient extraction is a function of the nutrient load targets and not a constant value. The higher the required nutrient load targets are to achieve good ecological status, the higher the value of bivalve generated nutrient extraction and improvement of water clarity will be. Such marginal value functions have been estimated for the Baltic Sea (e.g. Hasler et al. 2014). They indicate a marginal value of 24 € kg^{-1} N to obtain load reduction to the level required by HELCOM's international agreement on nutrient load reductions. This estimate is an average estimate for all countries around the Baltic Sea. Studies at a much more detailed level in the Limfjorden indicate lower marginal costs. Hasler et al. (2015) estimated a cost of 12 EUR € kg^{-1} N. To estimate the value of nitrogen extraction and removal by mussels, the production costs should be subtracted. The costs of producing mussels for nitrogen mitigation in Skive fjord in Limfjorden is estimated to be in the range 14–20 € kg^{-1} N (Petersen et al. 2014); i.e. the value of nitrogen removal is negative. It is important to note, however, that in this example it is assumed that a market for the produced mussels does not exist; neither for feed or human consumption. Break-even is reached at a sales price of approximately 0,13 € kg^{-1} mussel. Furthermore, the filtration effect can be included. In Petersen et al.

(2014), the mussel clearance effect on Secchi depth has been calculated; including this effect reduces the costs per kg nitrogen to 2 € kg^{-1} N. However, as pointed out by Petersen et al. (2014), these estimates should not be included in cost-effectiveness analysis of nitrogen removal, as these indirect effects do not remove nutrients from the ecosystem.

Gren et al. (2009) also estimated marginal costs functions to assess the value of nutrient extraction by mussel farms at the Baltic Sea drainage basin scale. When Gren et al. (2009) subtract the costs of producing the mussels from the value, the results suggest that the value of bivalve nitrogen extraction is still positive; varying between 1.7 and 24.7 € kg^{-1} N. The range of values depends on the assumption about production costs, whether a market exists for the harvested bivalves, and whether the market is for human consumption or feed. For all scenarios, they assume that 1 kg of live mussels contains between 8.5 and 12 g N, 0.6–0.8 g P and about 40–50 g C with reference to Lutz (1980) and Haamer (1996). In the scenario where no market for the products exists, the value is estimated to be within a range of 0.02–0.11 € kg^{-1} live mussel, reflecting a value of nitrogen extraction between 1.7 and 2.4 € kg^{-1} N. When markets exist, the value range is between 0.12 and 0.21 € kg^{-1} live mussel, reflecting values of nitrogen between 9.2 and 12.9 € kg^{-1} N. Gren et al. (2009) also distinguishes between high and low production costs for the mussels: In the scenario where mussels are sold on a market, the range is estimated to be between 10.0 and 14.1 € kg^{-1} N for high production costs, and between 17.5 and 24.7 € kg^{-1} N for low production costs. They further estimate the value for different parts of the Baltic Sea leading to even larger variations in the value between regions, attributed to differences in the nutrient load reduction targets that vary between sea regions and the differences in the levels of the production costs between the sites. This illustrates that variability in recipient sensitivity to nutrient load levels determines the value of bivalves as extraction aquaculture. Finally, spatial heterogeneity in production costs between locations, due to differences in growth conditions and differences in labour costs also play an important role in determining the economic value of bivalve extraction. This is important, as labour costs constitute the largest part of the production costs (Petersen et al. 2014; Gren et al. 2009).

Grabowski et al. (2012) estimate the value of ecosystem services provided by another type of bivalves; oyster beds. As part of their study, they report the value of denitrification in the oyster beds, and estimate the denitrification in the oyster beds from literature. The value is estimated using substitute costs, based on the average trading price per kilogram. The data from the trading programme have also been used by Piehler and Smyth (2011) giving a value of 13€ kg^{-1} in the Nutrient Offset Credit Program. These studies therefore also used the cost-based approach, but only for the denitrification contribution of oysters, i.e. smaller proportion of the potential for nitrogen extraction and removal by bivalves. They calculate the value of a number of ecosystem services delivered by bivalves, including the intermediate service, nitrogen removal, and conclude that this is worth between \$1385 and 6716 ha^{-1} $year^{-1}$. A major constraint in calculations of the economic value of nutrient extraction through denitrification is that it is difficult to assess a precise amount of nitrogen removed through the denitrification process. Pollack et al. (2013) have also used a

Table 10.2 Summary of examples of studies valuing effects of bivalves

Study	Region	Economic good	Method applied	Value estimated € kg^{-1} N
Söderqvist (1996), Söderqvist and Hasselström (2008)	Baltic Sea	Water clarity	Contingent valuation	12–14
Gren et al. (2009)	Baltic Sea	Achieve good ecological status	Substitute costs using agricultural mitigation costs	1.7–24.7
Piehler and Smyth (2011)	North Carolina	Not specified	Emissions trading	13.0
Grabowski et al. (2012)	SE United States	Not specified	Substitute costs using nitrogen emission trading markets	26,08 ($1385–6716 ha^{-1})
Pollack et al. (2013)	Mission-Aransas estuary, Texas	Not specified	Replacement costs	6.99
Petersen et al. (2014), Hasler et al. (2015)	Skive Fjord Denmark	Achieve good ecological status	Substitute costs using agricultural mitigation costs	−8-20

cost based approach to estimate the alternative cost of removing nitrogen from the Mission-Aransas Estuary in Texas using biological nitrogen removal processes in a wastewater treatment plant. Unlike extraction through bivalve aquaculture, where the extracted volume can be measured easily, nutrient extraction through denitrification requires measurements and modelling that can be subjected to debate about the actual amount of nitrogen removed.

The overview of existing studies (see also Table 10.2) illustrates that primary studies aiming at valorisation of nutrient extraction services by bivalves are rare. If the ecosystem services provided by bivalves are to be used as an active mitigation measure combatting excess nutrient loading to coastal waters, it is crucial that methods for valorisation and exact accounting of the overall services provided include the effects of potential enhanced nutrient retention and nutrient recycling in the ecosystem. Only then can these services be assessed in an unbiased, reliable, and cost-effective way.

10.7 Outlook – The Role of Bivalves in Abatement Policies

The value of nutrient regulation by bivalves can be utilized in nutrient reduction policies. This type of mitigation measure requires that bivalve producers be compensated for the direct costs involved in provision of this service, or paid the societal value of improving water quality through nutrient removal. There are different potential institutional set-ups for such compensations, dependent on the actors involved in payments for the services. The arrangement could transpire through negotiations between the public body responsible for meeting nutrient load reductions and the bivalve/mussel producers; or directly through trade between emitters of nutrients from land (farmers, waste water treatment plants) and bivalve producers (often facilitated by a public body). Trade between emitters and bivalve producers involve a purchase of offsets from bivalve producers, allowing emitters to reduce their abatement efforts accordingly. Permits are traded when the price of the offset offered by bivalve producers are lower than the marginal costs of reducing nutrient loads through other measures. The incentives for trade therefore depend on the need for nutrient load reductions in the specific water body, and the costs of alternative measures. Trade mechanisms are rarely used in nutrient regulation in Europe and only a few examples exist worldwide (e.g. Piehler and Smyth 2011; Grabowski et al. 2012; Shortle 2013; Duhon et al. 2015; Ferreira and Bricker 2016). However, there is an increasing focus on this instrument to promote more cost-effective solutions in nutrient regulation. The use of offsets for nutrient abatement by bivalves has been tested in Sweden (Lindahl and Kollberg 2009), and is currently being tested in ongoing experimentation Sweden and Finland (Nutritrade 2017, http://nutritrade-baltic.eu/wp-content/uploads/sites/34/2017/06/EUSBSR-Annual-Forum_NutriTrade-pilot-mussel.pdf). There are ample experiences of market based mechanisms in other environmental policy areas (e.g. biodiversity conservation) in both the US and Europe (Pöll et al. 2016). These experiences should be used to explore the risks and potentials of this type of regulation in the aquatic environment.

Acknowledgements This manuscript was supported by Innovation Fund Denmark project 6150-00008B MuMiPro "Mussel farming – mitigation and protein source for organic husbandry". This manuscript was also a result of the BONUS OPTIMUS project supported by BONUS (Art 185), funded jointly by EU and Innovation Fund Denmark.

References

Ackerman JD, Loewen MR, Hamblin PF (2001) Benthic-pelagic coupling over a zebra mussel reef in western Lake Erie. Limnol Oceanogr 46:892–904

Ahtiainen H, Artell J, Czajkowski M, Hasler B et al (2014) Benefits of meeting nutrient reduction targets for the Baltic Sea – a contingent valuation study in the nine coastal states. J Environ Econ Pol 3(3):278–305

Alpine AE, Cloern JE (1992) Trophic interactions and direct physical effects control phytoplankton biomass and production in an estuary. Limnol Oceanogr 37:946–955

Anonymous (1991) Council Directive 91/676/EEC of 12 December 1991 concerning the protection of waters against pollution caused by nitrates from agricultural sources. Off J L375

Anonymous (2000) Directive 200/60/EEC of the European Parliament and of the Council of 23 October 2000 establishing a framework for Community action in the field of water policy. Off J Eur Commun L327/1

Anonymous (2008) Directive 2008/56/EC of the European Parliament and of the Council of 17 June 2008 establishing a framework for community action in the field of marine environmental policy (Marine Strategy Framework Directive). Off J Eur Commun L164/19

Aure J, Strohmeier T, Strand Ø (2007) Modelling current speed and carrying capacity in long-line blue mussel (*Mytilus edulis*) farms. Aquac Res 38:304–312

Bricker SB, Rice KC, Bricker OP III (2014) From headwaters to coast: influence of human activities on water quality of the Potomac River estuary. Aquat Geochem 20(1–2):291–323

Brigolin D, Maschio GD, Rampazzo F, Giani M, Pastres R (2009) An individual-based population dynamic model for estimating biomass yield and nutrient fluxes through an off-shore mussel (*Mytilus galloprovincialis*) farm. Estuar Coast Shelf Sci 82:365–376

Caffrey JM, Hollibaugh JT, Mortazavi B (2016) Living oysters and their shells as sites of nitrification and denitrification. Mar Pollut Bull 112:86–90. https://doi.org/10.1016/j.marpolbul.2016.08.038

Capelle JJ (2017) Production efficiency of mussel bottom culture. PhD Dissertation, University of Wageningen

Capelle JJ, Scheiberlic G, Weijsman JWM, Smaal A (2016) The role of shore crabs and mussel density in mussel losses at a commercial intertidal mussel plot after seeding. Aquacult Internat 24:1459–1147

Caraco NF, Cole JJ, Strayer DL (2006) Top-down control from the bottom: regulation of eutrophication in a large river by benthic grazing. Limnol Oceanogr 51:664–670

Carlsson MS, Holmer M, Petersen JK (2009) Seasonal and spatial variation of benthic impacts of mussel long-line farming in a eutrophicated Danish fjord, Limfjorden. J Shellfish Res 28(4):791–801

Carlsson MS, Glud RN, Petersen JK (2010) Degradation of mussel (*Mytilus edulis*) fecal pellets released from hanging long-lines upon sinking and after settling at the sediment. Can J Fish Aquat Sci 67:1376–1387

Carlsson MS, Engström P, Lindahl O, Ljungqvist L, Petersen JK, Svanberg L, Holmer M (2012) Effects of mussel farms on the benthic nitrogen cycle on the Swedish west coast. Aquacult Environ Interact 2:177–191

Carmichael RH, Walton W, Clark H, Ramcharan C (2012) Bivalve-enhanced nitrogen removal from coastal estuaries. Can J Fish Aquat Sci 69:1131–1149

Carstensen J, Krause-Jensen D, Markager S et al (2013) Water clarity and eelgrass responses to nitrogen reductions in the eutrophic Skive Fjord, Denmark. Hydrobiologia 704:293–309

Chamberlain J, Fernandes TF, Read P et al (2001) Impacts of biodeposits from suspended mussel (*Mytilus edulis* L.) culture on the surrounding surficial sediments. ICES J Mar Sci 58:411–416

Chopin T (2013) Integrated multi-trophic aquaculture. In: Christou P, Savin R, Costa-Pierce BA, Misztal I, Whitelaw CBA (eds) Sustainable food production. Springer, New York, pp 184–205

Cloern JE (1982) Does the benthos control phytoplankton in South San Francisco Bay? Mar Ecol Progr Ser 9:191–202

Cloern JE (2001) Our evolving conceptual model of the coastal eutrophication problem. Mar Ecol Prog Ser 210:223–253

Conley DJ, Paerl HW, Howarth RW et al (2009) Controlling eutrophication: nitrogen and phosphorous. Science 323(5917):1014–1015

Cranford PJ, Strain PM, Dowd M et al (2007) Influence of mussel aquaculture on nitrogen dynamics in a nutrient enriched coastal embayment. Mar Ecol Prog Ser 347:61–78

Cranford PJ, William L, Strand Ø, Strohmeier T (2008) Phytoplankton depletion by mussel aquaculture: high resolution mapping, ecosystem modeling and potential indicators of ecological carrying capacity. ICES CM2008/H:12

Cranford PJ, Reid GK, Robinson SMC (2013) Open water integrated multi-trophic aquaculture: constraints on the effectiveness of mussels as an organic extractive component. Mar Ecol Prog Ser 4:163–173

Cranford PJ, Duarte P, Robinson SMC et al (2014) Suspended particulate matter depletion and flow modification inside mussel (*Mytilus galloprovincialis*) culture rafts in the Riá de Betanzos, Spain. J Exp Mar Biol Ecol 452:70–81

Daigle R, Herbinger CM (2009) Ecological interactions between the vase tunicate (*Ciona intestinalis*) and the farmed blue mussel (*Mytilus edulis*) in Nova Scotia, Canada. Aquat Invasions 4:177–187

Dalsgaard T, Thamdrup B, Canfield DE (2005) Anaerobic ammonium oxidation (anammox) in the marine environment. Res Microbiol 156:457–464

Dame RF (2012) Ecology of marine bivalves: an ecosystem approach. CRC Press, London

Dame RF, Prins TC (1998) Bivalve carrying capacity in coastal ecosystems. Aquat Ecol 31:409–421

Dare PJ, Edwards BD (1975) Seasonal changes in flesh weight and biochemical composition of mussels (*M. edulis* L.) in Conway estuary, North Wales. J Exp Mar Biol Ecol 18:89–97

Duarte CM, Conley DJ, Carstensen J, Sanchez-Camacho M (2009) Return to neverland: shifting baselines affect eutrophication restoration targets. Estuar Coasts 32:29–36

Duhon M, Mcdonald H, Kerr S (2015) Nitrogen trading in Lake Taupo: an analysis and evaluation of an innovative water management policy. Motu working paper 15-07. Motu Economic and Public Policy Research

Fahnenstiel GL, Lang GA, Nalepa TF, Jahnengen TH (1995) Effects of zebra mussel (*Dreissena polymorpha*) colonization on water quality parameters in Saginaw Bay, Lake Huron. J Great Lakes Res 21(4):435–448

Farber S, Costanza R, Childers DL et al (2006) Linking ecology and economics for ecosystem management. Bioscience 56:121–133

Ferreira JG, Bricker SB (2016) Goods and services of extensive aquaculture: shellfish culture and nutrient trading. Aquacu Int 24:803–825

Fuentes J, Gregorio V, Giráldez R, Molares J (2000) Within-raft variability of the growth rate of mussels, *Mytilus galloprovincialis*, cultivated in the Ría de Arousa (NW Spain). Aquaculture 189:39–52

Giles H, Pilditch CA (2006) Effects of mussel (*Perna canaliculus*) biodeposit decomposition on benthic respiration and nutrient fluxes. Mar Biol 150:261–271

Grabowski JH, Brumbaugh RD, Conrad RF et al (2012) Economic valuation of ecosystem services provided by oyster reefs. Bioscience 62:900–909

Grant J, Hatcher A, Scott DB et al (1995) A multidisciplinary approach to evaluating impacts of shellfish aquaculture on benthic communities. Estuaries 18(1A):124–144

Grant J, Bacher C, Cranford PJ et al (2008) A spatially explicit ecosystem model of seston depletion in dense mussel culture. J Mar Sys 73:155–168

Gren I-M (2008). Costs and benefits from nutrient reductions to the Baltic Sea. The Swedish Environmental Protection Agency, Stockholm, Sweden. http://www.naturvardsverket.se/Documents/publikationer/978–91–620–5877–7.pdf

Gren I-M, Lindahl O, Lindqvist M (2009) Values of mussel farming for combating eutrophication: an application to the Baltic Sea. Ecol Engineer 35:935–945

Guyondet T, Comeau LA, Bacher C et al (2015) Climate change influences carrying capacity in a coastal embayment dedicated to shellfish aquaculture. Estuar Coasts 38:1593–1618

Haamer J (1996) Improving water quality in a eutrophied fjord system with mussel farming. Ambio 25:356–362

Hart R (2003) Dynamic pollution control – time lags and optimal restoration of marine ecosystems. Ecol Econom 47:9–93

Hasler B, Smart JCR, Fonnesbech-Wulff A et al (2012) Regional cost-effectiveness in transboundary water quality management for the Baltic Sea. http://www.worldwaterweek.org/documents/.../IWREC/BeritHasler.pdf

Hasler B, Smart JCR, Fonnesbech-Wulff A et al (2014) Hydro-economic modelling of cost-effective transboundary water quality management in the Baltic Sea. Water Res Econom 5:1–23

Hasler B, Hansen LB, Andersen HE, Konrad M (2015) Modellering af omkostningseffektive reduktioner af kvælstoftilførslerne til Limfjorden: Dokumentation af model og resultater. Aarhus University, DCE - Danish Centre for Environment and Energy (in Danish)

Hatcher A, Grant J, Schofield B (1994) Effects of suspended mussel culture (*Mytilus* spp.) on sedimentation, benthic respiration and sediment nutrient dynamics in a coastal bay. Mar Ecol Prog Ser 115(3):219–235

Heasman KG, Pitcher GC, Mcquaid CD, Hecht T (1998) Shellfish mariculture in the Benguela system: raft culture of *Mytilus galloprovincialis* and the effect of rope spacing on food extraction, growth rate, production, and condition of mussels. J Shellfish Res 17:33–39
Herbert RJH, Humphreys JH, Davies CJ et al (2016) Ecological impacts of non-native Pacific oysters (*Crassostrea gigas*) and management measures for protected areas in Europe. Biodivers Conserv 25(14):2835–2865
Herman PMJ, Scholten H (1990) Can suspension-feeders stabilise estuarine ecosystems. In: Barnes MA, Gibson RN (eds) Trophic relations in the marine environment. Aberdeen University Press, Aberdeen, pp 104–116
Higgins CB, Tobias C, Piehler M et al (2013) Effect of aquacultured oyster biodeposition on sediment N_2 production in Chesapeake Bay. Mar Ecol Prog Ser 473:7–27
Hily C (1991) Is the activity of benthic suspension feeders a factor controlling water quality in the Bay of Brest. Mar Ecol Prog Ser 69:179–188
Hoellein TJ, Zarnoch CB (2014) Effect of eastern oysters (*Crassostrea virginica*) on sediment carbon and nitrogen dynamics in an urban estuary. Ecol Appl 24(2):271–286
Holmer M, Ahrensberg N, Jørgensen NP (2003) Impacts of mussel dredging on sediment dynamics in an eutrophic Danish fjord. Chem Ecol 19:343–362
Holmer M, Thorsen SW, Carlsson MS, Petersen JK (2015) Pelagic and benthic nutrient regeneration processes in mussel cultures (*Mytilus edulis*) in a eutrophic coastal area (Skive Fjord, Denmark). Estuar Coasts 38(5):1629–1641
Holyoke RR (2008) Biodeposition and biogeochemical processes in shallow mesohaline sediments of Chesapeake Bay. Ph.D. Dissertation, University of Maryland, College Park
Humphries AT, Ayvazian SG, Carey JC et al (2016) Directly measured denitrification reveals oyster aquaculture and restored oyster reefs remove nitrogen at comparable high rates. Front Mar Sci 3:74. https://doi.org/10.3389/fmars.2016.00074
Idrisi N, Mills EL, Rudstam LG, Stewart DJ (2001) Impact of zebra mussels (*Dreissena polymorpha*) on the pelagic lower trophic levels of Oneida Lake, New York. Can J Fish Aquat Sci 58:1430–1441
Kellogg ML, Cornwell JC, Owens MS, Paynter KT (2013) Denitrification and nutrient assimilation on a restored oyster reef. Mar Ecol Prog Ser 480:1–19
Kellogg ML, Smyth AR, Luckenbach MW et al (2014) Use of oysters to mitigate eutrophication in coastal waters. Estuar Coast Shelf Sci 151:156–168
Lahance-Bernard M, Daigle G, Himmelman JH, Frechette M (2010) Biomass–density relationships and self-thinning of blue mussels (*Mytilus* spp.) reared on self-regulated longlines. Aquaculture 308:34–43
Lindahl O (2011) Mussel farming as a tool for re-eutrophication of coastal waters: experiences from Sweden. In: Shumway S (ed) Shellfish aquaculture and the environment. Wiley, London, pp 217–237
Lindahl O (ed) (2012) Mussel cultivation. In: Submariner compendium: an assessment of innovative and sustainable uses of Baltic marine resources. Maritime Institute in Gdansk. ISBN: 978-83-62438-14-3. https://www.submariner-network.eu/images/downloads/submariner_compendium_web.pdf
Lindahl O, Kollberg S (2009) Can the EU agri-environmental aid program be extended into the coastal zone to combat eutrophication? Hydrobiologia 629:59–64
Lindemann S, Zarnoch CB, Castignetti D, Hoellein TJ (2016) Effect of eastern oysters (*Crassostrea virginica*) and seasonality on nitrite reductase gene abundance (nirS, nirK, nrfA) in an urban estuary. Estuar Coasts 39:218–232
Locke A, Hanson JM, Macnair NG, Smith AH (2009) Rapid response to non-indigenous species. 2. Case studies of invasive tunicates in Prince Edward Island. Aquat Invasions 4(1):249–258
Lutz RA (ed) (1980) Mussel culture and harvest: a north American perspective. Elsevier Scientific Publishing, Amsterdam
Maar M, Timmermann K, Petersen JK et al (2010) A model study of the regulation of blue mussels by nutrient loadings and water column stability in a shallow estuary, the Limfjorden. J Sea Res 64:322–333

Maar M, Saurel C, Landes A et al (2015) Growth potential of blue mussels (*M. edulis*) exposed to different salinities evaluated by a dynamic energy budget model. J Mar Sys 148:48–55
McKindsey CW, Archambault P, Callier MD, Olivier F (2011) Influence of suspended and off-bottom mussel culture on the sea bottom and benthic habitats: a review. Can J Zool 89:622–646
Minjead L, Michotey VD, Garcia N, Bonin PC (2009) Seasonal variation in di-nitrogen fluxes and associated processes (denitrification, anammox and nitrogen fixation) in sediment subject to shellfish farming influences. Aquat Sci 71:425–435
Møhlenberg F (1995) Regulating mechanisms of phytoplankton growth and biomass in a shallow estuary. Ophelia 42(1):239–256
Møhlenberg F (1999) Effect of meteorology and nutrient load on oxygen depletion in a Danish micro-tidal estuary. Aquat Ecol 33:55–64
Møhlenberg SJ (2007) Blue mussel cultivation for nitrogen removal in fjords. Assessment of an alternative measure to comply with the water framework directive using Odense Fjord as a case study. Msc Dissertation, Copenhagen University, Denmark
Mortazavi B, Ortmann AC, Wang L et al (2015) Evaluating the impact of oyster (*Crassostrea virginica*) gardening on sediment nitrogen cycling in a subtropical estuary. Bull Mar Sci 91(3):323–341
Murphy AE, Anderson IC, Smyth AR, Luckenbach MW (2016a) Microbial nitrogen processing in hard clam (*Mercenaria mercenaria*) aquaculture sediments: the relative importance of denitrification and dissimilatory nitrate reduction to ammonium (DNRA). Limnol Oceanogr 61:1589–1604
Murphy AE, Emery KA, Anderson IC et al (2016b) Quantifying the effects of commercial clam aquaculture on C and N cycling: an integrated ecosystem approach. Estuar Coasts 39:1746–1761
Newell RIE, Fisher TR, Holyoke RR, Cornwell JC (2005) Influence on eastern oysters on nitrogen andphosphorous regenration in Chesapeake Bay, USA. In: Dame R, Olenin S (eds) The comparative role of suspension feeders in ecosystems. NATO science series IV. Earth and environmental sciences, vol 47. Springer, Amsterdam, pp 93–120
Nielsen P, Cranford PJ, Maar M, Petersen JK (2016) Magnitude, spatial scale and optimization of ecosystem services from a nutrient extraction mussel farm in the eutrophic Skive Fjord, Denmark. Aquacult Environ Interact 8:311–329
Nizzoli D, Welsh DT, Fano EA, Viaroli P (2006) Impact of clam and mussel farming on benthic metabolism and nitrogen cycling, with emphasis on nitrate reduction pathways. Mar Ecol Prog Ser 315:151–165
Petersen JK (2004) Grazing on pelagic primary producers - the role of benthic suspension feeders in estuaries. In: Nielsen SL, Banta G, Pedersen MF (eds) Estuarine nutrient cycling: the influence of primary producers. Kluwer Academic, Dordrecht, pp 129–152
Petersen JK, Hansen JW, Laursen MB et al (2008a) Regime shift in a coastal marine ecosystem. Ecol Appl 18(2):497–510
Petersen JK, Nielsen TG, van Duren L, Maar M (2008b) Depletion of plankton in a raft culture of *Mytilus galloprovincialis* in Riá de Vigo, NW Spain. I. Phytoplankton. Aquat Biol 4:113–125
Petersen JK, Timmermann K, Carlsson MS et al (2012) Mussel farming can be used as mitigation tool – a reply. Mar Pollut Bull 64:452–454
Petersen JK, Maar M, Ysebart T, Herman PMJ (2013) Near-bed gradients in particles and nutrients above a mussel bed in the Limfjorden: influence of physical mixing and mussel activity. Mar Ecol Prog Ser 490:137–146
Petersen JK, Hasler B, Timmermann K et al (2014) Mussels as a tool for mitigation in the marine environment. Mar Pollut Bull 82:137–143
Petersen JK, Saurel C, Nielsen P, Timmermann K (2016) The use of shellfish for eutrophication control. Aquacult Internat 24(3):857–878
Piehler MF, Smyth AR (2011) Habitat-specific distinctions in estuarine denitrification affect both ecosystem function and services. Ecosphere 2:12

Pires LMD, Ibeling BW, van Donk E (2010) Zebra mussels as a potential tool in the restoration of eutrophic shallow lakes dominated by toxic cyanobacteria. In: van der Velde G, Rajagopal S, Bij de Vaate A (eds) The zebra mussel in Europe. Backhuys Publishers, Leiden, pp 331–342
Pöll CE, Willner W, Wrbka T (2016) Challenging the practice of biodiversity offsets: ecological restoration success evaluation of a large-scale railway project. Landsc Ecol Eng 12:85–97. https://doi.org/10.1007/s11355-015-0282-2
Pollack JB, Yoskowitz D, Hae-Cheol K, Montagna PA (2013) Role and value of nitrogen regulation provided by oysters (*Crassostrea virginica*) in the mission-Aransas estuary, Texas, USA. PLoS One 8(6):e65314. https://doi.org/10.1371/journal.pone.0065314
Prins TC, Escaravage V, Smaal AC, Peeters JCH (1995) Nutrient cycling and phytoplankton dynamics in relation to mussel grazing in a mesocosm experiment. Ophelia 41:289–315
Prins TC, Smaal AC, Dame RF (1998) A review of the feedbacks between bivalve grazing and ecosystem processes. Aquat Ecol 31:349–359
Riisgård HU, Lundgreen K, Larsen PS (2014) Potential for production of 'mini-mussels' in Great Belt (Denmark) evaluated on basis of actual and modelled growth of young mussels *Mytilus edulis*. Aquacult Internat 22:859–885
Rodhouse PG, Roden CM, Hensey MP, Ryan TH (1984) Resource allocation in Mytilus edulis on the shore and in suspended culture. Mar Biol 84(1):27–34
Rose JM, Bricker SB, Tedesco MA, Wikfors GH (2014) A role for shellfish aquaculture in coastal nitrogen management. Environ Sci Technol 48(5):2519–2525
Rosland R, Bacher C, Strand Ø et al (2011) Modelling growth variability in longline mussel farms as a function of stocking density and farm design. J Sea Res 66:318–330
Schernewski G, Stybel N, Neumann T (2012) Zebra mussel farming in the Szczecin (Oder) Lagoon: water-quality objectives and cost-effectiveness. Ecol Soc 17:4 https://doi.org/10.5751/ES-04644-170204
Shortle J (2013) Economics and environmental markets: lessons from water-quality trading. Agricult Res Econ Rev 42(1):57–74
Sisson GM, Kellogg ML, Luckenbach MW et al (2011) Assessment of oyster reefs in Lynnhaven River as a Chesapeake Bay TDML best management practice. Technical report. Virginia Institute of Marine Science, Gloucester Point, VA
Smaal AC, Vonck APMA (1997) Seasonal variation in C, N and P budgets and tissue composition of the mussel *Mytilus edulis*. Mar Ecol Prog Ser 153:167–179
Smith TE, Stevenson RJ, Caraco NF, Cole JJ (1998) Changes in phytoplankton community structure during the zebra mussel (*Dreissena polymorpha*) invasion of the Hudson River (New York). J Plankton Res 20(8):1567–1579
Smyth AR, Thompson SP, Siporin KN et al (2013) Assessing nitrogen dynamics throughout the estuarine landscape. Estuar Coasts 36:44–55
Smyth AR, Piehler MF, Grabowski JH (2015) Habitat context influences nitrogen removal by restored oyster reefs. J Appl Ecol 52(3):716–725
Smyth AR, Geraldi NR, Thompson SP, Piehler MF (2016) Biological activity exceeds biogenic structure in influencing sediment nitrogen cycling in experimental oyster reefs. Mar Ecol Prog Ser 560:173–183
Söderqvist T (1996). Contingent valuation of a less eutrophicated Baltic Sea. Beijer discussion paper series no. 88, Beijer International Institute of Eco logical Economics, The Royal Swedish Academy of Sciences, Stockholm
Söderqvist T, Hasselström L (2008). The economic value of ecosystem services provided by the Baltic Sea. Swedish Environmental Protection Agency, Stockholm, Sweden
Stadmark J, Conley DJ (2011) Mussel farming as a nutrient reduction measure in the Baltic Sea: consideration of nutrient biogeochemical cycles. Mar Pollut Bull 62:1385–1388
Strohmeier T, Aure J, Duinker A et al (2005) Flow reduction, seston depletion, meat content and distribution of diarrhetic shellfish toxins in a long-line blue mussel (*Mytilus edulis*) farm. J Shellfish Res 24:15–23

Strohmeier T, Duinker A, Strand Ø, Aure J (2008) Temporal and spatial variation in food availability and meat ratio in a longline mussel farm (*Mytilus edulis*). Aquaculture 276:83–90
Stybel N, Fenske C, Schernewski G (2009) Mussel cultivation to improve water quality in the Szczecin Lagoon. J Coast Res SI56:1459–1463
Testa JM, Brady DC, Cornwell JC et al (2015) Modeling the impact of floating oyster (*Crassostrea virginica*) aquaculture on sediment-water nutrient and oxygen fluxes. Aquacult Environ Interact 7:205–222
Troell M, Joyce A, Chopin T et al (2009) Ecological engineering in aquaculture – potential for integrated multi-trophic aquaculture (IMTA) in marine offshore systems. Aquaculture 297:1–9
Wang B, Wang Z (2011) Long-term variations in chlorophyll a and primary productivity in Jiaozhou Bay, China. J Mar Biol 2011:1–7
Weber A, Smit MGD, Collombon MT (2010) Eutrophication and algal blooms: zebra mussels as a weapon. In: van der Velde G, Rajagopal S, Bij de Vaate A (eds) The zebra mussel in Europe. Backhuys Publishers, Leiden, pp 343–347
Welsh DT, Castadelli G (2004) Bacterial nitrification activity directly associated with isolated benthic marine animals. Mar Biol 144:1029–1037
Welsh DT, Nizzoli D, Fano EA, Viaroli P (2015) Direct contribution of clams (*Ruditapes philippinarum*) to benthic fluxes, nitrification, denitrification and nitrous oxide emission in a farmed sediment. Estuar Coast Shelf Sci 154:84–93
Wiles PJ, van Duren LA, Häse C, Larsen J, Simpson JH (2006) Stratification and mixing in the Limfjorden in relation to mussel culture. J Mar Sys 60:129–143
Zaiko A, Lehtiniemi M, Narscius A, Olenin S (2011) Assessment of bioinvasion impacts on regional scale: a comparative approach. Biol Invasions 13:1739–1765

Chapter 11
Perspectives on Bivalves Providing Regulating Services in Integrated Multi-Trophic Aquaculture

Øivind Strand, Henrice M. Jansen, Zengjie Jiang, and Shawn M. C. Robinson

Abstract The concept of integrating species into one culture system originates from Asia and the Middle East. Development of integrated aquaculture involving marine bivalves is relatively new, going back to the late 1980s in China and 1990s in the Western world. In this chapter, we present four cases of integrated multi-trophic aquaculture (IMTA) where bivalves are involved in providing regulating services: i) shrimp culture in ponds, ii) cascading pond systems, iii) open-water caged finfish culture and iv) bay-scale culture systems. The bay-scale integrated culture system in Sanggou Bay in China represents commercial IMTA where a range of different regulating services are provided by the bivalves. Bivalves use degraded fragments derived from cultured kelp and organic waste products from fish farming, and play an important role in the ecosystem processes of the bay. The provision of regulating services in shrimp and cascading ponds is evident as the system configurations allow for biogeochemical processing of waste to maximize extraction by the bivalves. The current configurations used in open-water finfish cage culture suggest that adaptation of concepts allowing for control of effluent water, producing longer contact times and increased biogeochemical processing of the waste products, will dominate future IMTA development. If global bivalve culture production is sustained, we will likely see more regulating services from

Ø. Strand (✉)
Institute of Marine Research, Bergen, Norway
e-mail: oivinds@imr.no

H. M. Jansen
Wageningen UR, Wageningen Marine Research, Yerseke, The Netherlands
e-mail: henrice.jansen@wur.nl

Z. Jiang
Yellow Sea Fisheries Research Institute, Chinese Academy of Fishery Sciences, Qingdao, China
e-mail: jiangzj@ysfri.ac.cn

S. M. C. Robinson
Department of Fisheries and Oceans Biological Station, Fisheries and Oceans Canada, St. Andrews Biological Station, St. Andrews, NB, Canada
e-mail: shawn.robinson@dfo-mpo.gc.ca

A. C. Smaal et al. (eds.), *Goods and Services of Marine Bivalves*,
https://doi.org/10.1007/978-3-319-96776-9_11

bivalves in IMTA systems, as new opportunities may arise for developing novel IMTA configurations and concepts.

Abstract in Chinese 摘要:将不同类型的生物组合到一个养殖系统的理念起源于亚洲和中东。包含滤食性贝类的海水综合养殖方式最早可追溯到20世纪80年代的中国和90年代的西方国家。本章列举了包含滤食性贝类的四种典型多营养层次综合养殖模式(Integrated Multi-trophic Aquaculture, IMTA),包括:i)池塘虾类养殖;ii)级联式池塘养殖系统,iii)开放海域鱼类网箱养殖,iv)海湾养殖。中国的桑沟湾是成功实现IMTA产业化的典型海湾,滤食性贝类通过同化养殖海带产生的碎屑和鱼类养殖过程中产生的有机废物,担负着调节海湾生态系统状态的重要功能。在虾类和串联式池塘养殖系统中,滤食性贝类提供的调节服务功能也非常明显,这主要得益于养殖系统的合理化设计,充分利用了生物地球化学过程来实现滤食性贝类对废物利用效率的最大化。目前基于开放海域鱼类网箱养殖IMTA的经验表明,未来IMTA的发展将趋向于养殖水体富营养化的控制,延长营养物质在各营养层级生物间的接触时间和养殖废物的生物地球化学过程等。如果全球双壳贝类的养殖产量保持持续增长态势,更多新型的IMTA模式将会陆续出现,这也为我们发掘贝类在IMTA系统中更多的调节服务功能提供了新机遇。

Keywords IMTA · Waste recirculation · Extraction efficiency · Biogeochemical processing · Sequential culture

关键词 多营养层次综合养殖 • 养殖废物循环利用 • 利用效率 • 生物地球化学过程 • 可持续养殖

11.1 Introduction

The concept of integrating different species in aquaculture has its roots in ancient traditions in China and other parts of Asia and the Middle East, going as far back as the origin of aquaculture. In 2200–2100 B.C., the document *You Hou Bin* detailed the integration of fish with aquatic plants and vegetable production in China and images on tombs in Egypt showed evidence of historical culture, growing tilapia in conjunction with agricultural activities in 1550 to 1070 B.C. (Bardach et al. 1972; Chopin 2013). In the "Complete Book on Agriculture" by Guangqi Xu, published in 1639, it was said that "the optimized ratio for stocking silver carp and grass carp was 600:200, and only the grass carp was fed with grass" (Zhu and Dong 2013). Experiential and practical knowledge of the farmers have been the basis for the polyculture inventions and traditions, conceived to provide regulating services like mitigating waste materials entering the farming environment, controlling phytoplankton blooms and recirculating nutrient resources. Today a wide variety of polyculture is practiced in many Asian countries (Troell et al. 2009; Soto 2009), mostly dominated China. The classic polyculture model, which essentially includes the co-culturing of species at the same trophic level and/or belonging to different trophic levels, has widely been applied in freshwater aquaculture all over China, at a production level of about 30 million tonnes in 2015 (Wartenberg et al. 2017). Development of integrated culture involving marine bivalves is relatively new in China, going back to late 1980s (Fang et al. 2016). It is, however, based on the philosophy, principles and strong knowledge base from the ancient traditions and can best be exemplified by the bivalve – macroalgae – fish cage combination used in Sanggou Bay in the Shandong province and in Zhelin Bay in the Guangdong province (Zhou et al. 2006; Fang et al. 2016).

The development of modern aquaculture in the Western world differs from the Asian model as commercial systems have typically been characterized by increasing intensification of monoculture production. Co-culture and ecosystem-integrative concepts have been researched on an experimental level and promoted as an alternative mitigation strategy to improve sustainability and potentially increase profitability (Ridler et al. 2007), but have rarely developed to the commercial level. Early work in North America on land-based polyculture was done by the team of John Ryther who pioneered the concept of treating nutrients from sewage from urban areas (Boston, Massachussetts) using biological filters, including six species of bivalves (Ryther et al. 1972; Goldman et al. 1974; Ryther et al. 1975; Mann and Ryther 1977; Ryther 1981). This work continued in Israel where researchers began to look at intensive multi-species aquaculture in desert climates with an emphasis on water conservation and nutrient control (Krom et al. 1985; Krom and Neori 1989; Neori et al. 1989; Israel et al. 1995; Shpigel and Neori 1996; Shpigel et al. 1996). From this previous work, the term "integrated multi-trophic aquaculture" (IMTA) eventually emerged (Chopin et al. 2001) and is now a widely accepted label for the practice. Currently, it is slowly being implemented in commercial farming operations in the western world, while in Asia, it has become common to use the IMTA term on systems originally called polyculture. The IMTA concept involved the arrangement of species belonging to different trophic levels where the integrated culture was facilitating conversion of various wastes produced into animal and seaweed biomass (different trophic positions), creating additional aquaculture revenue for the farmer and removing some of the excess nutrients from the environment (Chopin et al. 2001; Troell et al. 2003; Soto 2009). As most bivalves are efficient at filtering particles suspended in the water column and some species are possible to culture in high densities, mussels were initially proposed as an early candidate for IMTA to regulate the fine organic particulate waste (faeces or excess feed) from finfish culture thereby mitigating the farming impact on the environment. Several pilot-scale farms were set up in various parts of the world to test this concept (Fig. 11.1). This approach was also supported by studies demonstrating that bivalves may exert substantial influence on primary production processes and concentrations of particulate matter (Dame 1996; Prins et al. 1998). Waste products that elevate natural concentrations of particles or nutrients stimulating plankton production will theoretically also contribute to higher food availability for bivalve production. Consequently, bivalves have been proposed as a candidate for mitigating and recycling waste in aquaculture, thereby providing a regulating service.

Fig. 11.1 Farming mussels in association with salmon farms in the Bay of Fundy, (**a**) aerial shot of a salmon farm in the Bay of Fundy with 4 rafts of mussels on the down-stream end, (**b**) a mussel raft showing the arrangement of mussel lines hanging within an empty fish cage collar and the suspension system using floats, (**c**) close-up of the mussel lines showing the mussel socks hanging down from the top line that is supported by the buoys, (**d**) close-up of one of the mussel socks hanging from the top line on one of the rafts. (Photo credit: S.M.C Robinson, DFO)

With experience gained from testing the IMTA concept in varying environments, the approaches and understanding of IMTA principles are continually evolving and have broadened (Chopin 2013; Jansen et al. 2015; Fang et al. 2016). The use of bivalves in IMTA development might be characterized as being in its infancy, as bivalves are being studied for their ability to directly capture of organic particulates from farm sites and also in the larger scale of relative nutrient extraction at the bay level without specific requirements on proximity to a farm and nutritional connectivity (Chopin 2013). Other potential benefits provided by bivalves in IMTA are improved perception of sustainable production by the public (Yip et al. 2017), extraction of pathogens and salmon lice from finfish aquaculture (Molloy et al. 2011, 2014; Bartsch et al. 2013; Webb et al. 2013), their role in the carbon cycle with consequences for CO_2 sequestering and climate change (Jiang et al. 2015; see also Filgueira et al. 2019) and new socio-economic approaches to motivate industry to adopt IMTA (Shi et al. 2013; Hughes and Black 2016).

In this chapter, we present four cases of IMTA where bivalves are involved in providing regulating services. The cases represent a range of culture configurations varying in scale, ecosystems and control of water transport. The cases are pond culture, cascading pond systems promoting micro-algae production, open-water caged finfish culture and finally a bay-scale culture system. Investigating the perspectives of these cases and their characteristics, we assess the scales in which bivalves in IMTA may provide regulating services.

11.1.1 Pond–Scale Systems: Shrimp–Bivalve IMTA

Soto (2009) reviewed the research, implementation and prospects of integrated marine and brackish-water aquaculture in tropical regions, including the co-culture of shrimp and fish with filter feeders (mussel, oyster) and seaweed. A review of integrated shrimp-oyster farming (Table 11.1) suggests that there is a significant potential for oysters to remove particulate material from shrimp culture effluents,

Table 11.1 Bioww-mitigation potential of bivalves in combination with shrimp

		Bacteria	Suspended solids	Chl *a*	Total N	Total P	References
Oysters *Saccostrea commercialis*	Lab scale Flow through After sedimentation	65	81	61	33	44	Jones et al. (2002)
Oysters *Saccostrea commercialis*	Lab test recirculation After sedimentation	88	84	96			Jones et al. (2002)
Oysters *Crassostrea gigas*	Lab scale After sedimentation		71	100			Ramos et al. (2009)
Oysters *Crassostrea rhizophorae*	Lab scale After sedimentation		41	51			Ramos et al. (2009)
Oysters *Crassostrea virginica*	Shrimp pond coupled to raceways		41		−9	41	Kinne (2001)
Oysters *Saccostrea commercialis*	Lab scale After sedimentation	70	88	92	33	37	Jones et al. (2001)
Oysters *Crassostrea rhizophorae*	Experimental in sedimentation tank		~75	~75			de Azevedo et al. (2015)

Numbers are given in percentage reduction by the bivalve unit.

demonstrating the regulating services of bivalves in these integrated cultivation systems. It should be noted though that most of the studies defined removal rates after sedimentation of particulate matter and these rates can therefore not be directly related to total waste production. Growth and physiological state of oysters were generally good, but under conditions with high particle loading, growth was inhibited (Jones et al. 2001, 2002). Similarly, nutritional stress was observed for mussels solely being fed with solid fish wastes (Both et al. 2011). It therefore seems beneficial to allow the shrimp effluent to settle before bivalve biofiltration in order to improve growth and reduce stress (Jones et al. 2002). In Mexico, the black clam (*Chione fluctifraga*) was found to be feasible in co-culture IMTA pond systems with the white shrimp (*Litopenaeus vannamei*) through improving the water quality and increasing the production rate of the shrimp, although this was still at an experimental scale (Martinez-Cordova and Martinez-Porchas 2006; Martinez-Cordova et al. 2011, 2013).

Despite the fact that the potential for using filter feeders in re-circulated shrimp systems has been shown in several small and experimental settings (*e.g.* in Thailand, China, Vietnam, Malaysia, Mexico, Australia; see Soto 2009), Soto (2009) concluded that virtually no commercial practices can be found.

11.1.2 Cascading-Pond Systems: Linking Fish and Bivalves Through Phytoplankton Production

Stimulation of phytoplankton production in culture operations provides yet another food resource in integrated bivalve systems (Delia et al. 1977; Goldman and Ryther 1976; Ryther et al. 1972, 1975; Milhazes-Cunha and Otero 2017), and is commonly applied in semi-closed cultivation systems such as ponds. Phytoplankton assimilates the inorganic waste streams originating from fish and shrimp culture, while in turn, serves as a valuable food source for bivalves. The combination of phytoplankton and animal (*i.e.* shrimp, carp, tilapia and other planktivorous fish) production in ponds has been practiced for millennia in China (Neori et al. 2004) and recent trials in the Haiyang city of Shandong province where the effluent water from tanks holding fish flows into cascading-pond ponds with scallop culture show promising results. Phytoplankton blooms in these systems are often uncontrolled and are generally characterized by the lack of nutritionally-desirable microalgae species (Goldman and Ryther 1976; Benemann 1992). When bivalves are integrated with fish cultivation, often a settling basin or a foam fractionator is situated between the fish and the bivalve cultivation units to allow settlement of particulate wastes (Shpigel and Blaylock 1991; Hussenot et al. 1998; Lefebvre et al. 2000; Jones et al. 2001). Although bivalves can feed on both fin fish organic wastes and phytoplankton, a diet solely based on the former is not desirable. Both et al. (2011) indicated that mussels may become nutritionally stressed when only fed with organic wastes from cod farming and Handå et al. (2012a) showed that growth was lower for mussels fed with salmon faeces compared to those given microalgae or salmon feed. This is not particularly surprising since several studies have shown that bivalves need nutrients such as essential fatty acids (Caers et al. 1998, 2003; Milke et al. 2004; Nevejan et al. 2003) that are often retained by organisms and not readily available in faecal pellets (Reid et al. 2013). Apart from removing particulate wastes, the settling ponds also promote a more stable and diverse phytoplankton production compared to the production in the fish ponds (Shpigel and Blaylock 1991; Lefebvre et al. 2000). Recent developments in pond aquaculture include the increase in culture robustness of phytoplankton by controlling and monitoring a known mixture of phytoplankton species, thus avoiding culture crashes (Milhazes-Cunha and Otero 2017). These types of systems typically consist of a series of cascading ponds, where the effluent water from the fish pond (or tank) flows into a pond where phytoplankton production takes places and finally this water is directed towards the bivalve ponds. Separate ponds for phytoplankton production allow better control and, by introducing phytoplankton reactors and (small) inoculation ponds, a population dominated by microalgae species with high nutritional value can be realized (Hussenot et al. 1998). This could also imply that specific nutrients need to be supplemented to the fish waste water to realize the optimal nutrient balance for the desired phytoplankton species (*e.g.* silicate for diatom growth) (Lefebvre et al. 1996; Hussenot et al. 1998). Separation of phytoplankton and bivalve ponds is necessary to give phytoplankton the opportunity to grow and multiply before filtration by the bivalves.

Milhazes-Cunha and Otero (2017) reviewed the biomitigation potential of integrated fish-phytoplankton-bivalve systems indicating that nutrient removal efficiencies are generally high (>90%) for recirculating aquaculture systems. This is higher compared to cascading-pond systems which have lower removal efficiencies (67% ammonia, 47% phosphate) (Hussenot et al. 1998). Shpigel et al. (1993) demonstrated that for a pond system gilthead seabream (*Sparus aurata*) and the Japanese

Fig. 11.2 Left: Four cascading-pond systems each consisting of three interlinked ponds in The Netherlands. The system is designed for cultivation of the common sole (*Solea solea*), the king ragworm (*Alitta virens*) (pond 1) and the Manila clam (*Venerupis philippinarum*) (pond 3) by reusing fish waste streams to stimulate phytoplankton production (pond 2). Right: Manila clams cultured in the cascading-pond system

oyster (*Crassostrea gigas*), including a sedimentation tank, 11% of the total waste nitrogen (TN) was removed, but it was unknown how much of the inorganic waste stream this constituted.

Growth of bivalves is generally good in fish-phytoplankton-bivalve integrated systems (Shpigel and Blaylock 1991; Jara-Jara et al. 1997; Shpigel et al. 1993) and no microbiological contamination of rearing waters or bivalves has been observed (Courtois et al. 2003). The combination of phytoplankton and bivalves can thus remove substantial fractions of the (inorganic) waste streams from fish aquaculture while at the same time resulting in a valuable crop (Fig. 11.2). However, bivalves also produce metabolic waste products in the form of inorganic (NH_4) and organic (faeces) nutrients. Like fish faeces, part of the faecal material will be broken down by bacteria and other microorganisms and contribute to the total pool of inorganic nutrients. In estuaries, approximately half of the particulate nitrogen bivalves feed on is regenerated in inorganic forms (Jansen 2012). It is unknown how much of the particulate nutrients are being regenerated in pond systems. To remove the remaining inorganic nutrients, several studies have therefore integrated a seaweed or periphyton compartment following the bivalve ponds (Shpigel et al. 1993; Levy et al. 2017).

11.1.3 Open–Water Caged Finfish Aquaculture: Salmon–Bivalve IMTA

Open-water cage culture represents the dominant global production method of fed marine finfish where the environment inside the cage is largely dependent on the exchange rate of various water quality variables (Oppedal et al. 2011). This exchange is essential to avoid depletion of oxygen, vital for respiratory needs of the fish, and to ensure waste product discharge from the net pens. Faeces and uneaten feed constitutes the majority of the particulate load while excreted ammonia dominates the dissolved waste fraction. The composition of these nutrients is dependent on the feed and species in culture (Wang et al. 2013). About 60% of the nitrogen in the feed supplied to the farmed salmonids in Norwegian aquaculture is released as waste, 15% particulate and 40–45% dissolved (Wang et al. 2012). This discharge of effluent waste has prompted concerns on environmental impacts which has led to the development of monitoring and regulating systems to manage the industry (Folke et al. 1994; Holmer 2010) and initiatives to develop mitigation approaches like IMTA. In this case, bivalves were proposed to act as a regulating service by extracting these particulates from the waste streams emanating from the cages.

Studies of bivalve performance in suspended culture downstream from open-water finfish net pens, to extract waste particles of feed and fish faeces, have been carried out in a range of environments and cage arrangements (Fig. 11.1) ranging up to 50 m in diameter and 25-m deep, comprising a volume of 36,000 m^3 (Handå et al. 2012b) and smaller volumes of about 50 m^3 (Jiang et al. 2012). Some studies fed the cultured fish with trash fish (Gao et al. 2006; Jiang et al. 2012) while the larger sized companies used modern commercial feeds with total amounts of 5216 tonnes for farms with eight cages (50 m in diameter) over a study period of 13 months (Handå et al. 2012b).

The studies of bivalves cultivated in open water IMTA systems have shown varying results with respect to benefits in bivalve growth, ranging from positive (Gao et al. 2006; Sara et al. 2009; Handå et al. 2012b; Lander et al. 2012; Jiang et al. 2012) to no effect (Taylor et al. 1992; Parsons et al. 2002; Navarette-Mier et al. 2010; Cheshuk et al. 2003). Enhanced growth of bivalves seems to only occur at distances very close to the cages and decreases quickly at distances much less than the spatial dimension of the fish-cage arrangements (Sara et al. 2009; Handå et al. 2012b; Lander et al. 2012; Jiang et al. 2012). The recent use of tracer techniques (stable isotopes, fatty acid profiling and DNA), in attempts to assess the assimilation of waste products by extracting bivalves, has generally indicated that contribution of aquaculture-derived nutrients to bivalve nutrition is relatively small (Handå et al. 2012b; Woodcock et al. 2017).

The dispersion patterns of the particulate waste leaving the cages and its availability to bivalves intended for extraction in IMTA have recently been examined in several studies (Reid et al. 2009; Cranford et al. 2013; Brager et al. 2015; Jansen et al. 2016a; Brager et al. 2016; Filgueira et al. 2017). In general, the larger and heavier particles sink faster while the finer material remains suspended for longer periods of time and therefore travels over longer distances from the cages (Bannister et al. 2016). An extensive study of temporal variability in waste concentrations in the water column at open-water fish farms in eastern Canada and Norway indicated that temporal variations in suspended particulate material (SPM) around the farms were largely driven by natural processes and that the addition of fish wastes had a negligible effect on background SPM concentrations (Brager et al. 2016). The authors concluded that there is little rationale for introducing bivalves in IMTA to mitigate the horizontal flux of small particulate fish wastes, confirming earlier modelling studies (Troell and Norberg 1998). The rapid dilution of nutrients away from fish cages has been documented by some of the work looking at therapeutant dispersion with a high dilution rate happening in minutes (Page et al. 2014). Cranford et al. (2013) identified constraints on the capacity of mussels (*Mytilus edulis*) to capture and absorb organic fish waste under open-water IMTA scenarios. They demonstrated how waste particle capture by mussels is severely limited by the time available to intercept solid wastes contained in the horizontal flux of the particles. Increasing the waste extraction efficiency by using higher mussel biomass may ultimately be constrained by current velocity, available IMTA farm space, negative feedback effects on fish culture from flow reduction caused by mussel culture, and depletion of their particulate food supply to a level that will limit production. Cranford et al. (2013) also argued that the proportion of organic fish faeces relative to ambient seston concentration and seston organic content affects the ability of mussels to absorb more IMTA-generated waste than they egest as mussel faeces. Consequently, the biomitigation potential of mussels will be greatest where seston abundance is low and the organic content of IMTA waste is high. This was also pointed out by Filgueira et al. (2017) who simulated pumping rate (*e.g.* ingestion) of mussels in a finfish-bivalves IMTA configuration with different background seston concentrations. From their modelling study exploring different spatial arrangements of an IMTA case, they concluded that waste mitigation would be best achieved by placing extractive species such as deposit feeders on the seabed directly beneath the cages rather than using suspension filter feeders to extract the horizontal flux of waste, although one study found that scallops (*Placopecten magellanicus*) would grow and survive well directly under fish cages (Robinson et al. 2011). Handå et al. (2012a) found a more pronounced incorporation of nutrients in the tissues and better growth in shell length of mussels from salmon feed compared to salmon faeces, which suggests that mussels will utilize fish feed more efficiently than faecal particles when cultured in IMTA. Assuming that bivalves efficiently encounter waste particles, Reid et al. (2013) suggested that estimating the dietary quality of the waste particles provides useful information for assessing the mitigation potential of filter feeders and inferring a nutrient reduction potential. They assessed that the percentage of fish culture solids in an extractive species' diet that must be exceeded for mussel culture to reduce the net IMTA site organic load is 14.5% for salmon faeces and high-quality seston, 19.6% for salmon faeces and low quality seston, 11.5% for salmon feed fines and high-quality seston, and 15.6% for salmon feed fines and low-quality seston.

11.1.4 Bay-Scale Interactions: Fish-Bivalve-Seaweed Cultivation in Sanggou Bay, China

China's leading case for a truly commercial, engineered IMTA system is Sanggou Bay (Wartenberg et al. 2017), located on the eastern coast of the Shandong peninsula facing towards the Yellow Sea. The bay is famous for its mariculture and development of polyculture and IMTA concepts for over 30 years (Fang et al. 2016). Sanggou Bay is now one of the most important and dense farming areas in China and is a model globally. The bivalve culture in the bay is evidently integrated with the other main group cultured, the macroalgae.

Table 11.2 Summary of aquaculture in Sanggou Bay, China

Cultured species	Stocking period	Harvesting period	Culture period	Production (tonne year^{-1})
Crassostrea gigas	May	March	1–2 year	~60,000
Chlamys farreri	May	March	1–2 year	~15,000
Saccharina japonica	November	May	7 month	~84,500 dry weight
Gracilaria lemaneiformis	June	October	5 month	~25,000
Paralichthys olivaceus	May	October	6 month	~24,000

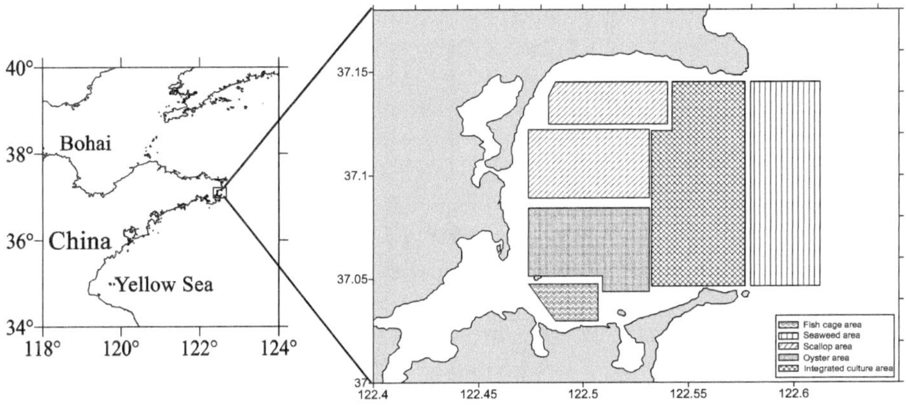

Fig. 11.3 Layout map of aquaculture practices in Sanggou Bay, Shandong Province, China

The bay is 140 km^2, with an average depth of 7 m and a maximum depth of 20 m at the entrance of the bay. It receives freshwater from one large and a few smaller rivers with the main input occurring during summer. The sediment is dominated by mud and sand. The main farmed species are kelp (*Saccharina japonica*), red algae (*Gracilaria lemaneiformis*), Farreri's scallop (*Chlamys farreri*), and Pacific oyster (*Crassostrea gigas*) (Table 11.2), which are all cultured from longline systems. Fish culture in cages is now dominated by Japanese flounder (*Paralichthys olivaceus*), although the Japanese pufferfish (*Fugu rubripes*) has previously been farmed. Kelp monoculture occurs mainly near the mouth and outside of the bay (Fig. 11.3), bivalves are mainly raised near the head of the bay and the middle part is characterized by a co-culture of kelp and bivalves. Fish cages are situated south west in the bay, and bivalves and seaweed are cultivated on long lines around the fish cages. The bivalves are mainly cultured in nets hung from longlines and kelp is tied to ropes and grows vertically in the water column.

Mahmood et al. (2016) used a stable isotopic technique to study pathways of organic matter (OM) in Sanggou Bay in order to better understand the role of fish-bivalve-seaweed IMTA practices related to assimilation and accumulation of OM in the cultured species during the summer and winter seasons. They indicated that 90% of carbon and 60% of nitrogen in the diet of bivalves originated from fish faeces and uneaten particles from trash fish during the summer. Alternative sources of OM in the winter season, during low temperatures, may be from detritus lost in large-scale cultivation of kelp. The bivalves cultured in Sanggou Bay are important in reducing OM, but it is suggested that they may also be able to increase production and survival rate of other species in the IMTA system by maintaining high water quality, thereby improving the economic benefit of the entire system (Mahmood et al. 2016). A study in the adjacent Ailian Bay showed that the assimilation efficiency of the Pacific oyster for fish-aquaculture-derived organic matter was 54% (10% waste feed and 44% fish faeces) (Jiang et al. 2012). Given that 50% of the total solid nutrient loads from fish cages are assumed to be within the suitable size range that can be efficiently retained by the gills, the oysters will theoretically be able to recover 27% of the total particulate organic matter released from fish cages if the waste source is directed towards the location where bivalves are cultured. Bivalves functioning as recyclers of organic matter could contribute to environmentally-sustainable aquaculture and could increase the profitability of fish cultivation.

The detritus lost during the kelp growth cycle is regarded as an important food resource for the filter-feeding bivalves (Xu et al. 2016). Using the stable isotope technique, it has been demonstrated that the diet of filter feeders inhabiting natural kelp forest habitats and adjacent environments was largely based on kelp detritus (Fredriksen 2003; Schaal et al. 2009; Miller and Page 2012). Xu et al. (2016) evaluated the trophic importance of kelp (*S. japonica*) fragments to the co-cultured scallop *C. farreri* in Sanggou Bay and showed with stable isotope techniques that the diet of scallops consisted of 14–43% of kelp-derived organic carbon. Additionally, substantial amounts of dissolved organic carbon (DOC) are released to the surrounding water by kelp (Mahmood et al. 2017). DOC can directly be taken up by bivalves, in addition to particulate organic matter (Roditi et al. 2000), and Mahmood et al. (2017) indicated that the bivalves farmed in Sanggou Bay act both as a source and a sink of DOC, with the highest removal rate of 60% occurring in the bivalve culture area. There are a number of additional positive interactions between bivalves and seaweeds. Bivalve respiration (see Filgueira et al. 2019) generates CO_2 and also releases other metabolic waste products such as ammonia, all of which can serve as an input for growth of seaweeds. Jiang et al. (2014) reported that a scallop (*C. farreri*) population in Sanggou Bay sequestered 78.1 ± 5.8 g C·m^{-2}·year^{-1} deposited in the shell, while the CO_2 fluxes due to calcification and respiration resulted in 54.0 ± 4.0 g C·m^{-2}·year^{-1} and 71.7 ± 6.5 g C·m^{-2}·year^{-1}, respectively. In this context, the CO_2 released from the bivalves can provide part of the dissolved inorganic carbon (DIC) requirement of the seaweed. The macroalgae harvest from the bay is an important component providing powerful support for revealing the role of Sanggou Bay in the carbon cycle (Jiang et al. 2015). In terms of the bay scale, Sanggou Bay acted as a net DIC sink with an annual mean uptake estimated at 139,000 tonnes (Jiang et al. 2015).

11.2 Discussion

The four cases presented in this chapter show a variety of IMTA configurations, environments and socio-economic settings where bivalves are positioned to exploit aquaculture waste products, and thereby potentially provide regulating services. An assessment of how the bivalves provide regulating services will rest on the definitions applied to IMTA, which can range from the direct capture of the particulate waste on the farm, to removal of an equivalent amount of the effluent-related nutrients in the far-field by harvesting the bivalves. The latter scenario can occur regardless of distance and connectivity to the actual waste nutrients, where it can also support sustained ecosystem functioning, depending on the scales of extraction involved. Also, aspects related to traditions and philosophy of integrating aquaculture (like in Asia) and the state of integrated aquaculture development will influence how regulating services are perceived. The wide ranging and sometimes ambiguous nature of the IMTA definition and questions on how much extraction, in our case by bivalves, is enough to qualify for the definition, have frequently been raised (Chopin 2013; Reid et al. 2013; Jansen et al. 2015). Ultimately, the benchmark for comparison will likely be made to monoculture systems growing comparable amounts of biomass of the same species, such as the pioneering work done in Sanggou Bay (Shi et al. 2013). Considering that IMTA may mitigate undesirable impacts, a reference state of environmental condition may be needed, depending on the socio-economic setting and regulatory requirements. The environmental hazard or impact to be mitigated by the bivalves will therefore, in most cases, need to be identified to justify the development of IMTA principles.

Adapting principles of IMTA to local environments and regulatory frameworks seems to be crucial for the successful development of integrated aquaculture systems. The success of IMTA in Sanggou Bay (Fang et al. 2016) is based on a complex set of factors such as the existing high variety of species cultured, inherent philosophy among farmers of combining species in culture, ability to rapidly adapt to environmental changes, a pliant regulatory framework and a socio-economic system promoting multi-species culture. The Sanggou Bay case represents full-scale commercial IMTA where a range of goods and services from bivalves can be achieved. Although there is a need for understanding the role of bivalves in the ecosystem when assessing regulating services in this coastal bay, other factors (e.g. socio-economic issues) seem to be the main driver for the development. In this case, recirculation and recycling of waste nutrients is as important as any direct extraction of aquaculture waste providing regulating services from IMTA in Sanggou Bay.

One comparison that can be made among the case studies, relates to the efficiency of using bivalves to capture waste particles directly from the farm discharge before they are assimilated or bio geochemically cycled, compared to extraction of products coming from another trophic level that converts the waste (bacteria, phytoplankton, zooplankton). Direct capture has typically been anticipated for the open-water cage finfish aquaculture case, while the fish waste products stimulating phytoplankton production that is then extracted by bivalves is achieved in the cascading-pond system. The efficiencies in removal of waste experienced in these two cases are strikingly different, mainly caused by the ability to direct water flow in the cascading-pond system determining particle dispersion and thereby ability to maintain the availability of the converted particles for extraction by the bivalves. The pattern of particles horizontally dispersed from open-water cage finfish aquaculture explains the marginal estimates of waste removal (Troell and Norberg 1998; Cranford et al. 2013; Brager et al. 2015, 2016; Filgueira et al. 2017). In contrast, the cascading-pond systems with integrated fish-phytoplankton-bivalves show generally high removal efficiencies (Milhazes-Cunha and Otero 2017) supporting the concept of sequential control of the effluent water to maintain the nutrient quantity and quality through the biogeochemical cycle and thereby maximize extraction of the waste by the bivalves through greater contact times. The provision of regulating service in the cascading-pond system is evident and supports the earlier studies of Ryther (1981).

There is a consensus that extractive species in open-water cage finfish aquaculture should be placed underneath the cages where most of the organic waste flux goes, rather than trying to extract the horizontal flux which is marginal in terms of total particulate waste amounts (Cubillo et al. 2016; Brager et al. 2016; Filgueira et al. 2017). The gradient of increased waste flux towards the vertical plane from the cages is also affected by the size distribution of the waste particles that are smallest in the horizontal plane and largest in the vertical plane from the cages, thereby influencing the ability of bivalves to extract the waste (Bannister et al. 2016). Of course, an option always exists to resize the larger waste particles into smaller ones through the manipulation of the binders in the diets resulting in looser (smaller) or more compact (larger) faecal pellets (Appleford and Anderson 1997; Brinker 2007; Brinker and Friedrich 2012; Brinker et al. 2005; Dias et al. 1998; Rodehutscord et al. 2000). Size of the waste particles also determines how fast assimilation by the bivalves and bacterial degradation occur which, together with the dispersion patterns from the cages, will influence the ability of bivalves to directly capture the waste. The challenges of using bivalves to effectively capture and feed on highly-dispersive waste particles from open-water finfish cages seems overwhelming for current practices (Troell et al. 2009; Cranford et al. 2013; Filgueira et al. 2017). This conclusion assumes, however, that future technology will be based on the status quo open-water net cages with high water exchange. But it is possible that new concepts and designs may arise for open-water cages where particles may exit the cages in a more controlled manner. This would likely increase the potential efficiency of assimilation of farm waste by bivalves. Today, due to various environmental challenges with using open-water cage culture systems (e.g. diseases, parasites, organic loading), efforts are now being encouraged to focus on developing new technology, including closed containment systems at sea mainly to reduce disease and parasite interaction with the environment (Lekang et al. 2016). These enclosed systems will require handling and treatment of the waste nutrients, so knowledge on various IMTA concepts converting waste nutrients into feed for bivalves will have more potential, similar to the cascading-pond system. The technology development on sea-based closed containment is expected to diversify future finfish production systems with possibly a higher proportion of the production including options for controlling the effluent waste water. Such systems allow for the development of IMTA concepts with a much higher potential for sequential control of the effluent water and higher removal efficiency of waste than in current open-cage systems.

The role of bivalves in the bay-scale integrated aquaculture production system in Sanggou Bay is evident as a provider of regulating services. These services include: (1) bivalves using degraded fragments derived from the cultured kelp, (2) bivalves directly using organic waste products from fish farming, (3) bivalve harvest removes nutrients supporting sustained functioning of the ecosystem. Bivalve farming ultimately also provides regulating services on extracting nutrients derived from the populated surrounding land area of the bay. Mahmood et al. (2016) estimated that

72% of particulate organic matter in the bay during the summer season originated from land and their results indicated that ~80% of the particulate organic matter, including faecal material and riverine material, is extracted by cultured oysters and scallops. The interaction between the bivalves and the microbial food web was elucidated in experimental mesocosm and flow-through system studies indicating how farmed scallops (*C. farreri*), through phosphorous egestion and size selection of particles, affected the different microbial components (Lu et al. 2015; Jiang et al. 2017). This impact on the "protozoan trophic link" may enable a positive feedback by energy transfer from the microbial loop to the scallops. Protists (nanoflagellates and ciliates) were the dominant source of carbon retained by the scallops (49%). Dissolved organic carbon released from phytoplankton and seaweeds can also serve as energy sources for micro-heterotrophic organisms available as food for the scallops. Of recent and increasing interest is also the role of bivalve respiration and calcification processes to the carbon cycle in this bay, and its importance in how low-trophic aquaculture (bivalves and seaweed) at a coastal scale affects carbon sequestering and climate change (Jiang et al. 2014). These studies demonstrate how bivalves in Sanggou Bay may provide regulating services at the same time as providing provisioning services through their role in processes of carbon cycling related to environmental and climate-change issues.

The regulating services provided by bivalves in Sanggou Bay are assessed, based on investigations and IMTA culture practice over more than three decades (Fang et al. 2016). The ancient history of integrated culture and the inherent approach in China to combine species to maximize yield are essential factors in explaining their success in developing IMTA. Considering the long history of national need for increased food production as the main driver for the dramatic expansion of aquaculture in the coastal zone (Liu and Su 2017), IMTA concepts have been a key component to mitigate the often severe challenges related to environmental impacts and related socio-economic issues. In a recent review, Wartenberg et al. (2017) listed the most adverse impacts of suspended mariculture in China and how these could be mitigated through the application of IMTA systems. The main impacts identified were chemical, ecological, physical and socio-economic. Out of eighteen measures recommended for improving suspended mariculture, IMTA was most frequently considered to have capabilities for bioremediation and increased farm production. The challenges facing the expansion of commercial IMTA included lack of new technology, limited skills development, limited production of low trophic-level species, biogeographic and temporal barriers and negative system feedbacks. They concluded that implementing commercial IMTA is a promising measure for reducing the impacts of suspended mariculture because it presents a range of secondary benefits that can improve the overall sustainability of aquaculture in the coastal zone. Fang et al. (2016) and Wartenberg et al. (2017) clearly demonstrate the existing and future potential for provisions of regulating services by bivalves in IMTA.

The position of China as the dominant global aquaculture producer is expected to continue into the foreseeable future (FAO 2016; Wartenberg et al. 2017), based on its need for internal food production. Global aquaculture production is dominated by low-trophic resources, with bivalves among the most important contributors.

If the bivalve culture position is sustained and the development of IMTA in China is realized, as projected by Wartenberg et al. (2017), we will likely see more regulating services from bivalves in IMTA and new opportunities for developing novel IMTA configurations and concepts with bivalves playing a central role providing such services. The current knowledge of open-water finfish cage culture and the low efficiency of direct capture of waste suggest that adaptation of production systems allowing for sequential control of effluent water, thereby maintaining higher contact times of bivalves with the nutrients and biogeochemical processing of the waste products, will dominate future IMTA.

Bivalves are a dominant aquaculture group worldwide and because they efficiently consume food that is relatively low in the food chain, they may play a key role in the anticipated contribution from aquaculture to the increasing global demand for human food in the coming century (Wijsman et al. 2019). There will be a range of challenges to be solved for this development, among them technology, spatial issues, disease control, government policies and regulations, eutrophication and resource recirculation. Innovative approaches to integrate bivalve aquaculture with other marine sectors (Buck et al. 2017; Jansen et al. 2016b) to optimize the ecological efficiency of the increasing production will be essential to ensure sustainable expansion and obtaining the regulating services from future bivalve aquaculture.

Acknowledgements The authors are grateful to two reviewers for constructive comments on the manuscript.

References

Appleford P, Anderson TA (1997) Apparent digestibility of tuna oil for common carp, *Cyprinus carpio* – effect of inclusion level and adaptation time. Aquaculture 148:143–151

Bannister RJ, Johnsen IA, Hansen PK, Kutti T, Asplin L (2016) Near-and far-field dispersal modelling of organic waste from Atlantic salmon aquaculture in fjord systems. ICES J Mar Sci 73:2408–2419

Bardach JE, Ryther JH, Mclarney WO (1972) Aquaculture – the farming and husbandry of freshwater and marine organisms. Wiley, New York, 351p

Bartsch A, Robinson SMC, Liutkus M, Ang KP, Webb J, Pearce CM (2013) Filtration of sea louse, *Lepeophtheirus salmonis*, copepodids by the blue mussel, *Mytilus edulis*, and the Atlantic Sea scallop, *Placopecten magellanicus*, under different flow, light and copepodid-density regimes. J Fish Dis 36:361–370

Benemann JR (1992) Microalgae aquaculture feeds. J Appl Phycol 4(3):233–245

Both A, Parrish CC, Penney RW, Thompson RJ (2011) Lipid composition of *Mytilus edulis* reared on organic waste from a *Gadus morhua* aquaculture facility. Aquat Living Resour 24(3):295–301

Brager LM, Cranford PJ, Grant J, Robinson SMC (2015) Spatial distribution of suspended particulate wastes at open-water Atlantic salmon and sablefish aquaculture farms in Canada. Aquacult Environ Interact 6:135–149

Brager LM, Cranford PJ, Jansen HM, Strand Ø (2016) Temporal variations in suspended particulate waste concentrations at open water fish farms in Canada and Norway. Aquac Environ Interact 8:437–452

Brinker A (2007) Guar gum in rainbow trout (*Oncorhynchus mykiss*) feed: the influence of quality and dose on stabilisation of faecal solids. Aquaculture 267:315–327

Brinker A, Friedrich C (2012) Fish meal replacement by plant protein substitution and guar gum addition in trout feed. Part II: effects on faeces stability and rheology. Biorheology 49:27–48

Brinker A, Koppe W, Rosch R (2005) Optimised effluent treatment by stabilised trout faeces. Aquaculture 249:125–144

Buck BH, Nevejan N, Wille M, Chambers MD, Chopin T (2017) Offshore and multi-use aquaculture with extractive species: seaweeds and bivalves. In: Buck BH Langan R (eds) Aquaculture perspective of multi-use sites in the open ocean. https://doi.org/10.1007/978-3-319-51159-7_2

Caers M, Coutteau P, Lombeida P, Sorgeloos P (1998) The effect of lipid supplementation on growth and fatty acid composition of *Tapes philippinarum* spat. Aquaculture 162:287–299

Caers M, Coutteau P, Sorgeloos P, Gajardo G (2003) Impact of algal diets and emulsions on the fatty acid composition and content of selected tissues of adult broodstock of the Chilean scallop *Argopecten pupuratus* (Lamarck, 1819). Aquaculture 217:437–452

Cheshuk BW, Pursera GJ, Quintana R (2003) Integrated open-water mussel (*Mytilus planulatus*) and Atlantic salmon (*Salmo salar*) culture in Tasmania, Australia. Aquaculture 218:357–378

Chopin T (2013) Integrated multi-trophic aquaculture – ancient, adaptable concept focuses on ecological integration. Global Aquaculture Advocate pp 16–17

Chopin T, Buschmann AH, Halling C, Troell M, Kautsky N, Neori A, Kraemer GP, Zertuche-Gonzalez JA, Yarish C, Neefus C (2001) Integrating seaweeds into marine aquaculture systems: a key towards sustainability. J Phycol 37:975–986

Courtois O, Piquet JC, Roesberg D, Hussenot J (2003) Microbiological survey of an integrated aquaculture system involving marine fish-microalgae-bivalve mollusc. In: Chopin T, Reinertsen H (eds) Aquaculture Europe 2003: Beyond monoculture, Trondheim, Norway. European Aquaculture Society. EAS Special Publication, 33:158–159

Cranford PJ, Reid GK, Robinson SMC (2013) Open water integrated multi-trophic aquaculture: constraints on the effectiveness of mussels as an organic extractive component. Aquacult Environ Interact 4:163–173

Cubillo AM, Ferreira JG, Robinson SMC, Pearce CM, Corner CA, Johansen J (2016) Role of deposit feeders in integrated multi-trophic aquaculture — a model analysis. Aquaculture 453:54–66

Dame R (1996) Ecology of marine bivalves: an ecosystem approach. CRC Press, Boca Raton

de Azevedo RV, Tonini WCT, Martins Dos Santos MJ, Braga LGT (2015) Biofiltration, growth and body composition of oyster *Crassostrea rhizophoraein* effluents from shrimp *Litopenaeus vannamei*. Rev Ciênc Agron 46(1):193–203

Delia CF, Ryther JH, Losordo TM (1977) Productivity and nitrogen balance in large-scale phytoplankton cultures. Water Res 11:1031–1040

Dias J, Huelvan C, Dinis MT, Metailler R (1998) Influence of dietary bulk agents (silica, cellulose and a natural zeolite) on protein digestibility, growth, feed intake and feed transit time in European seabass (*Dicentrarchus labrax*) juveniles. Aquat Living Resour 11:219–226

Fang J, Zhang J, Xiao T, Huang D, Liu S (2016) Integrated multi-trophic aquaculture (IMTA) in Sanggou Bay, China. Aquacult Environ Interact 8:201–205

FAO (2016) The state of world fisheries and aquaculture 2016. Contributing to food security and nutrition for all. Food and Agriculture Organization of the United Nations, Rome

Filgueira R, Guyondet T, Reid GK, Grant J, Cranford PJ (2017) Vertical particle fluxes dominate integrated multi-trophic aquaculture (IMTA) sites: implications for shellfish–finfish synergy. Aquacu Environ Interact 9:127–143

Filgueira R, Strohmeier T, Strand Ø (2019) Regulating services of bivalve molluscs in the context of the carbon cycle and implications for ecosystem valuation. In Smaal et al (eds) Goods and services of marine bivalves. Springer, Cham, pp 231–251

Folke C, Kautsky N, Troell M (1994) The costs of eutrophication from salmon farming: implications for policy. J Environ Manag 40:173–182

Fredriksen S (2003) Food web studies in a Norwegian kelp forest based on stable isotope ($\delta^{13}C$ and $\delta^{15}N$) analysis. Mar Ecol Prog Ser 260:71–81

Gao QF, Shin PKS, Lin GH, Chen SP, Cheung SG (2006) Stable isotope and fatty acid evidence for uptake of organic waste by green-lipped mussels *Perna viridis* in a polyculture fish farm system. Mar Ecol Prog Ser 317:273–283

Goldman JC, Ryther JH (1976) Temperature influenced species competition in mass cultures of marine phytoplankton. Biotechnol Bioeng 18:1125–1144

Goldman JC, Tenore KR, Ryther JH, Corwin N (1974) Inorganic nitrogen removal in a combined tertiary treatment marine aquaculture system. 1. Removal efficiencies. Water Res 8:45–54

Handå A, Ranheim A, Olsen AJ, Altin D, Reitan KI, Olsen Y, Reinertsen H (2012a) Incorporation of salmon fish feed and faeces components in mussels (*Mytilus edulis*): implications for integrated multi-trophic aquaculture in cool-temperate North Atlantic waters. Aquaculture 370-371:40–53

Handå A, Min H, Wang X, Broch OJ, Reitan KI, Reinertsen H, Olsen Y (2012b) Incorporation of fish feed and growth of blue mussels in close proximity to salmon aquaculture: implications for integrated multi-trophic aquaculture in Norwegian coastal waters. Aquaculture 356-357:328–341

Holmer M (2010) Environmental issues of fish farming in offshore waters: perspectives, concerns and research needs. Aquac Environ Interact 1:57–70

Hughes AD, Black KD (2016) Going beyond the search for solutions: understanding trade-offs in European integrated multi-trophic aquaculture development. Aquac Environ Interact 8:191–199

Hussenot J, Lefebvre S, Brossard N (1998) Open-air treatment of wastewater from land-based marine fish farms in extensive and intensive systems: current technology and future perspectives. Aquat Living Resour 11:297–304

Israel AA, Friedlander M, Neori A (1995) Biomass yield, photosynthesis and morphological expression of *Ulva lactuca*. Bot Mar 38:297–302

Jansen HM (2012) Bivalve nutrient cycling – translocation, transformation and regeneration of nutrients by suspended mussel communities in oligotrophic fjords. PhD thesis, Wageningen University, The Netherlands

Jansen H, Handå A, Husa V, Broch OJ, Hansen PK, Strand Ø (2015) What is the way forward for IMTA development in Norwegian Aquaculture? Aquaculture Europe 2015 Conference "Aquaculture, Nature and Society", October 20–23, 2015, Rotterdam, The Netherlands

Jansen HM, Reid GK, Bannister RJ, Husa V, Robinson SMC, Cooper JA, Quinton C, Strand Ø (2016a) Discrete water quality sampling at open-water aquaculture sites: limitations and strategies. Aquac Environ Interact 8:463–480

Jansen HM, Van Den Burg S, Bolman B, Jak RG, Kamermans P, Poelman M, Stuiver M (2016b) The feasibility of offshore aquaculture and its potential for multi-use in the North Sea. Aquac Int 24(3):735–756

Jara-Jara R, Pazos AJ, Abad M, Garcia-Martin LO, Sanchez JL (1997) Growth of clam seed (*Ruditapes decussatus*) reared in the wastewater effluent from a fish farm in Galicia (N.W. Spain). Aquaculture 158:247–262

Jiang ZJ, Wang GH, Fang JG, Mao YZ (2012) Growth and food sources of Pacific oyster *Crassostrea gigas* integrated culture with sea bass *Lateolabrax japonicus* in Ailian Bay, China. Aquac Int 21:45–52

Jiang ZJ, Fang JG, Han TT, Li JQ, Mao YZ, Du MR (2014) The role of *Gracilaria lemaneiformis* in eliminating the dissolved inorganic carbon released from calcification and respiration process of *Chlamys farreri*. J Appl Phycol 26(1):545–550

Jiang ZJ, Li JQ, Qiao XD, Wang GH, Bian DP, Jiang X, Liu Y, Huang DJ, Wang W, Fang JG (2015) The budget of dissolved inorganic carbon in the shellfish and seaweed integrated mariculture area of Sanggou Bay, Shandong, China. Aquaculture 446:167–174

Jiang ZJ, Du MR, Fang JH, Gao YP, Li JQ, Zhao L, Fang JG (2017) Size fraction of phytoplankton and the contribution of natural plankton to the carbon source of Zhikong scallop *Chlamys*

Farreri in mariculture ecosystem of the Sanggou Bay. Acta Oceanologica Sinica. 36(10):97–105 https://doi.org/10.1007/s13131-017-0970-x
Jones AB, Dennison WC, Preston NP (2001) Integrated treatment of shrimp effluent by sedimentation, oyster filtration and macroalgal absorption: a laboratory scale study. Aquaculture 193(1–2):155–178
Jones AB, Preston NP, Dennison WC (2002) The efficiency and condition of oysters and macroalgae used as biological filters of shrimp pond effluent. Aquac Res 33:1–19
Kinne PN, Tzachi M. Samocha, Ed R. Jones & Craig L. Browdy (2001) Characterization of intensive shrimp pond effluent and preliminary studies on biofiltration, N Am J Aquac 63:25–33
Krom MD, Neori A (1989) A total nutrient budget for an experimental intensive fishpond with circularly moving seawater. Aquaculture 83:345–358
Krom MD, Porter C, Gordin H (1985) Description of the water quality conditions in a semi-intensively cultured marine fish pond in Eilat, Israel. Aquaculture 49:141–157
Lander TR, Robinson SMC, Macdonald BA, Martin JD (2012) Enhanced growth rates and condition index of blue mussels (*Mytilus edulis*) held at integrated multitrophic aquaculture sites in the bay of Fundy. J Shellfish Res 31:997–1007
Lefebvre S, Hussenot J, Brossard N (1996) Water treatment of land-based fish farm effluents by outdoor culture of marine diatoms. J Appl Phycol 8:193–200
Lefebvre S, Barille L, Clerc M (2000) Pacific oyster (Crassostrea gigas) feeding responses to a fish-farm effluent. Aquaculture 187:185–198
Lekang OI, Salas-Bringas C, Bostock JC (2016) Challenges and emerging technical solutions in on-growing salmon farming. Aquac Int 24:757–766
Levy A, Milstein A, Neori A, Harpaz S, Shpigel M, Guttman L (2017) Marine periphyton biofilters in mariculture effluents: nutrient uptake and biomass development. Aquaculture 473:513–520
Liu H, Su J (2017) Vulnerability of China's nearshore ecosystems under intensive mariculture development. Environ Sci Pollut Res 24:8957–8966
Lu JC, Huang L, Xiao T, Jiang Z, Zhang W (2015) The effects of Zhikong scallop (*Chlamys farreri*) on the microbial food web in a phosphorus-deficient mariculture system in Sanggou Bay, China. Aquaculture 448:341–349
Mahmood T, Fang J, Jiang Z, Zhang J (2016) Carbon, nitrogen flow and trophic relationship among the cultured species in an integrated multitrophic aquaculture (IMTA) bay. Aquac Environ Interact 8:207–219
Mahmood T, Fang J, Jiang Z, Ying W, Zhang J (2017) Seasonal distribution, sources and sink of dissolved organic carbon in integrated aquaculture system in coastal waters. Aquac Int 25:1–15
Mann R, Ryther JH (1977) Growth of six species of bivalve mollusks in a waste recycling aquaculture system. Aquaculture 11:231–245
Martinez-Cordova LR, Martinez-Porchas M (2006) Polyculture of Pacific white shrimp, *Litopenaeus vannamei*, giant oyster, *Crassostrea gigas* and black clam, *Chione fluctifraga* in ponds in Sonora, Mexico. Aquaculture 258:321–326
Martinez-Cordova LR, Lopez-Elias JA, Martinez-Porchas M, Bernal-Jaspeado T, Miranda-Baeza A (2011) Studies on the bioremediation capacity of the adult black clam, *Chione fluctifraga*, of shrimp culture effluents. Revista De Biologia Marina Y Oceanografia 46:105–113
Martinez-Cordova LR, Enriquez-Ocana LF, Lopez-Rascon F, Lopez-Elias JA, Martinez-Porchas M (2013) Overwintering the black clam *Chione fluctifraga* in a tidal shrimp pond and in an estuary, using suspended and bottom systems. Aquaculture 396:102–105
Milhazes-Cunha H, Otero A (2017) Valorisation of aquaculture effluents with microalgae: the integrated multi-trophic aquaculture concept. Algal Res 24:416–424
Milke LM, Bricelj VM, Parrish CC (2004) Effects of microalgal diets and fatty acid composition on the growth performance of postlarval and juvenile bay scallops *Argopecten irradians*. J Shellfish Res 23:303
Miller RJ, Page HM (2012) Kelp as a trophic resource for marine suspension feeders: a review of isotope-based evidence. Mar Biol 159:1391–1402

Molloy SD, Pietrak MR, Bouchard DA, Bricknell I (2011) Ingestion of *Lepeophtheirus salmonis* by the blue mussel *Mytilus edulis*. Aquaculture 311:61–64
Molloy SD, Pietrak MR, Bouchard DA, Bricknell I (2014) The interaction of infectious salmon anaemia virus (ISAV) with the blue mussel, *Mytilus edulis*. Aquac Res 45:509–518
Navarette-Mier F, Sanz-Lázaro C, Marín A (2010) Does bivalve mollusc polyculture reduce marine fin fish farming environmental impact? Aquaculture 306:101–107
Neori A, Krom MD, Cohen I, Gordin H (1989) Water quality conditions and particulate chlorophyll-a of new intensive seawater fishponds in Eilat, Israel – daily and diel variations. Aquaculture 80:63–78
Neori A, Chopin T, Troell M, Buschmann A, Kraemer G, Halling C, Shpigel M, Yarish C (2004) Integrated aquaculture: rationale, evolution and state of the art emphasizing seaweed biofiltration in modern mariculture. Aquaculture 231:361–391
Nevejan N, Saez I, Gajardo G, Sorgeloos P (2003) Supplementation of EPA and DHA emulsions to a *Dunaliella tertiolecta* diet: effect on growth and lipid composition of scallop larvae, *Argopecten purpuratus* (Lamarck, 1819). Aquaculture 217:613–632
Oppedal F, Dempster T, Stien LH (2011) Environmental drivers of Atlantic salmon behaviour in sea-cages. Rev Aquac 311:1–18
Page F, Chang BD, Beattie M, Losier R, Mccurdy P, Bakker J, Haughn K, Thorpe B, Fife J, Scouten S, Bartlett G, Ernst B (2014) Transport and dispersal of sea lice bath therapeutants from salmon farm net-pens and well-boats operated in Southwest New Brunswick: a mid-project perspective and perspective for discussion. DFO Can Sci Advis Sec Res Doc. 2014/102. v 63 p
Parsons GJ, Shumway SE, Kuenstner S, Gryska A (2002) Polyculture of sea scallops (*Placopecten magellanicus*) suspended from salmon cages. Aquac Int 10:65–77
Prins T, Smaal A, Dame R (1998) A review of feedbacks between bivalve grazing and ecosystem processes. Aquat Ecol 31:349–359
Ramos R, Vinatea L, Seiffert W, Beltrame E, Santos Silva J, Ribeiro Da Costa RH (2009) Treatment of shrimp effluent by sedimentation and oyster filtration using *Crassostrea gigas* and *C. rhizophorae*. Braz Arch Biol Technol 52:775–783
Reid GK, Liutkus M, Robinson SMC, Chopin TR, Blair T, Lander T, Mullen J, Page F, Moccia RD (2009) A review of the biophysical properties of salmonid faeces: implications for aquaculture waste dispersal models and integrated multi-trophic aquaculture. Aquac Res 40:257–273
Reid GK, Robinson SMC, Chopin T, Macdonald BA (2013) Dietary proportion of fish culture solids required by shellfish to reduce the net organic load in open-water integrated multi-trophic aquaculture: a scoping exercise with co-cultured Atlantic salmon (*Salmo salar*) and blue mussel (*Mytilus edulis*). J Shellfish Res 32:509–517
Ridler N, Wowchuk M, Robinson B, Barrington K, Chopin T, Robinson S, Page F, Reid G, Szemerda M, Sewuster J, Boyne-Travis S (2007) Integrated multi – trophic aquaculture (IMTA): a potential strategic choice for farmers. Aquac Econ Manag 11:99–110
Robinson SMC, Martin JD, Cooper JA, Lander TR, Reid GK, Powell F, Griffin R (2011) The role of 3-D habitat in the establishment of integrated multi-trophic aquaculture (IMTA) systems. Aquacul. Assoc. Canada Bull 109:23–29
Rodehutscord M, Gregus Z, Pfeffer E (2000) Effect of phosphorus intake on faecal and non-faecal phosphorus excretion in rainbow trout (*Oncorhynchus mykiss*) and the consequences for comparative phosphorus availability studies. Aquaculture 188:383–398
Roditi HA, Fisher NS, Sañudo-Wilhelmy SA (2000) Uptake of dissolved organic carbon and trace elements by zebra mussels. Nature 407:78–80
Ryther JH (1981) Mariculture, ocean ranching, and other culture-based fisheries. Bioscience 31:223–230
Ryther JH, Tenore KR, Dunstan WM, Huguenin JE (1972) Controlled eutrophication – increasing food production from sea by recycling human wastes. Bioscience 22:144–152
Ryther JH, Goldman JC, Gifford CE, Huguenin JE, Wing AS, Clarner JP, Williams LD, Lapointe BE (1975) Physical models of integrated waste recycling, marine polyculture systems. Aquaculture 5:163–177

Sara G, Zenone A, Tomasello A (2009) Growth of *Mytilus galloprovincialis* (mollusca, bivalvia) close to fish farms: a case of integrated multi-trophic aquaculture within the Tyrrhenian Sea. Hydrobiologia 636:129–136

Schaal G, Riera P, Leroux C (2009) Trophic significance of the kelp *Laminaria digitata* (Lamour.) for the associated food web: a between-sites comparison. Estuar Coast Shelf Sci 85:565–572

Shi HH, Zheng W, Zhang XL, Zhu MY, Ding DW (2013) Ecological-economic assessment of monoculture and integrated multi-trophic aquaculture in Sanggou Bay of China. Aquaculture 410:172–178

Shpigel M, Blaylock RA (1991) The use of the Pacific oyster *Crassostrea gigas*, as a biological filter for marine fish aquaculture pond. Aquaculture 92:187–197

Shpigel M, Neori A (1996) The integrated culture of seaweed, abalone, fish and clams in modular intensive land-based systems. 1. Proportions of size and projected revenues. Aquac Eng 15:313–326

Shpigel M, Lee J, Soohoo B, Fridman R, Gordin H (1993) The use of effluent water from fishponds as a food source for the pacific oyster *Crassostrea gigas* Tunberg. Aquac Fish Manag 24:529–543

Shpigel M, Neori A, Marshall A (1996) The suitability of several introduced species of abalone (Gastropoda:Haliotidae) for land-based culture with pond-grown seaweed in Israel. Isr J Aquacult-Bamidgeh 48:192–200

Soto D (2009) Integrated mariculture: a global review, FAO fish Aquacult tech pap 529. FAO, Rome

Taylor BE, Jamieson G, Carefoot TH (1992) Mussel culture in British Columbia: the influence of salmon farms on growth of Mytilus edulis. Aquaculture 108:51–66

Troell M, Norberg J (1998) Modelling output and retention of suspended solids in an integrated salmon–mussel culture. Ecol Model 110:65–77

Troell M, Halling C, Neori A, Buschmann AH, Chopin T, Yarish C, Kautsky N (2003) Integrated mariculture: asking the right questions. Aquaculture 226:69–90

Troell M, Joyce A, Chopin T, Neori A, Buschmann AH, Fang JG (2009) Ecological engineering in aquaculture— potential for integrated multi-trophic aquaculture (IMTA) in marine offshore systems. Aquaculture 297:1–9

Wang X, Olsen LM, Reitan KI, Olsen Y (2012) Discharge of nutrient wastes from salmon farms: environmental effects, and potential for integrated multi-trophic aquaculture. Aquac Environ Interact 2:267–283

Wang X, Andresen K, Handå A, Jensen B, Reitan KI, Olsen Y (2013) Chemical composition of feed, fish and faeces as input to mass balance estimation of biogeneic waste discharge from an Atlantic salmon farm with an evaluation of IMTA feasibility. Aquac Environ Interact 4:147–162

Wartenberg R, Feng L, Wu JJ, Mak YL, Chan LL, Telfer TC, Lam PKS (2017) The impacts of suspended mariculture on coastal zones in China and the scope for Integrated Multi-Trophic Aquaculture, ecosystem health and sustainability. https://doi.org/10.1080/20964129.2017.1340268

Webb JL, Vandenbor J, Pirie B, Robinson SMC, Cross SF, Jones SRM, Pearce CM (2013) Effects of temperature, diet, and bivalve size on the ingestion of sea lice (*Lepeophtheirus salmonis*) larvae by various filter-feeding shellfish. Aquaculture 406:9–17

Wijsman J, Troost K, Fang J, Roncarati A (2019) Global production of marine bivalves. Trends and challenges. In Smaal et al (eds) Goods and services of marine bivalves. Springer, Cham, pp 7–26

Woodcock SH, Troedsson C, Strohmeier T, Balseiro P, Skaar KS, Strand Ø (2017) Combining biochemical methods to trace organic effluent from fish farms. Aquacu Environ Interact 9:429–443

Xu Q, Gao F, Yang H (2016) Importance of kelp-derived organic carbon to the scallop *Chlamys farreri* in an integrated multi-trophic aquaculture system. Chin J Oceanol Limnol 34:322–329

Yip W, Knowler D, Haider WG, Trenholm R (2017) Valuing the willingness-to-pay for sustainable seafood: integrated multitrophic versus closed containment aquaculture. Can J Agric Econ 65:93–117

Zhou Y, Yang H, Hu H, Liu Y, Mao Y, Zhou H, Xu X, Zhang F (2006) Bioredimation potential of the macroalga *Gracilaria lemaneiformis* (Rhodophyta) integrated into fed fish culture in coastal waters of North China. Aquaculture 252:264–276

Zhu C, Dong S (2013) Aquaculture site selection and carrying capacity management in the People's Republic of China. In: Ross RG, Telfer TC, Falconer L, Soto D, Aguilar-Manjarrez J (eds) Site selection and carrying capacities for inland and coastal aquaculture, pp 219–230. FAO/Institute of Aquaculture, University of Stirling, Expert Workshop, 6–8 December 2010. Stirling, the United Kingdom of Great Britain and Northern Ireland. FAO Fisheries and Aquaculture Proceedings No. 21. Rome FAO 282 pp

Chapter 12
Regulating Services of Bivalve Molluscs in the Context of the Carbon Cycle and Implications for Ecosystem Valuation

R. Filgueira, T. Strohmeier, and Ø. Strand

Abstract The role of marine bivalves in the CO_2 cycle has been commonly evaluated as the balance between respiration, shell calcium carbonate sequestration, and CO_2 release during biogenic calcification; however, this individual-based approach neglects important ecosystem interactions that occur at the population level, e.g. the interaction with phytoplankton populations and benthic-pelagic coupling, which in turn can significantly alter the CO_2 cycle. Therefore, an ecosystem approach that accounts for the trophic interactions of bivalves, including the role of dissolved and particulate organic and inorganic carbon cycling, is needed to provide a rigorous assessment of the role of bivalves as a potential sink of CO_2. Conversely, the discussion about this potential role needs to be framed in the context of non-harvested vs. harvested populations, given that harvesting represents a net extraction of matter from the ocean. Accordingly, this chapter describes the main processes that affect CO_2 cycling and discuss the role of non-harvested and harvested bivalves in the context of sequestering carbon. A budget for deep-fjord waters is presented as a case study.

Abstract in Chinese 摘要:海水双壳贝类在二氧化碳循环中的作用通常根据基于呼吸作用、钙化作用和钙化期间二氧化碳释放进行评价。 然而,这种基于个体的评估方法并没有考虑种群水平的贝类与生态系统的相互作用。例如,贝类与浮游植物种群和底栖生物的相互作用,这种相互作用可以明显改变CO_2循环过程。因此,需要建立一套综合考虑溶解有机碳、溶解无机碳、颗粒有机碳、颗粒无机碳等碳存在形态的生态系统方法来评估双壳贝类潜在的碳汇作用。然而,关于这种潜在作用的讨论需要在区分自然种群和养殖种群的情况下进行,因为养殖种群的最终收获其实是从海水中进行相关营养成分的净提取。因此,本章介绍了影响CO_2循环的主要过程,并讨论了自然和养殖的双壳贝类在碳移除过程中的作用。在挪威峡湾内的一个双壳贝类养殖区的碳收支会作为一个案例研究进行展示。

R. Filgueira (✉)
Marine Affairs Program, Dalhousie University, Halifax, NS, Canada

Institute of Marine Research, Bergen, Norway

T. Strohmeier · Ø. Strand
Institute of Marine Research, Bergen, Norway
e-mail: tore.strohmeier@imr.no; oivind.strand@imr.no

A. C. Smaal et al. (eds.), *Goods and Services of Marine Bivalves*,
https://doi.org/10.1007/978-3-319-96776-9_12

Keywords Bivalve · CO_2 · Carbon cycling · Carbon trading system

关键词 双壳贝类 • 二氧化碳 • 碳循环 • 碳交易系统

12.1 Introduction

Bivalves are soft-bodied organisms protected by an external shell consisting of two hinged valves. The ratio shell:tissue in terms of weight is different across species and is habitat dependent within species (e.g. Newell and Hidu 1982; Rodhouse et al. 1984; MacDonald and Thompson 1985; Penney et al. 2008). For example, mussels cultured in suspended structures tend to have lighter shells than those in natural populations, which could be related to the feeding conditions in aquaculture facilities promoting faster growth and thinner shells (Aldrich and Crowley 1986), but also to the reduced predation pressure (Lowen et al. 2013). The shells of cultured bivalves can generally be considered residues although they are sometimes used as by-products in construction and agriculture (e.g. Rodríguez Álvaro et al. 2014; Varhen et al. 2017). Taking into account the global annual production of cultured bivalves is ~14 $x10^6$ tons, including clams, cockles, oysters, mussels and scallops (www.fao.org reporting 2015 data) and assuming an average contribution of shell to total body weight of 50% (general ballpark figure given that this varies greatly between species), shell represents a residue (potential by-product) of ~7×10^6 tons, of which 95% is calcium carbonate.

The shell is an exoskeleton that offers protection against predators and adverse environmental conditions. Adductor muscles are attached to the shell providing the animal with the capability to close their valves, isolating the internal tissues from the environment, although the effectiveness varies across species. In the case of scallops, the rapid contraction of the adductor muscle forces the valves to quickly squeeze the intervalvar fluid, which creates a water jet that propels the scallop, providing them with swimming capabilities (Guderley and Tremblay 2016). The different shell shapes across bivalve species allowed this class of molluscs to colonize a variety of habitats (Stanley 1970). Marine bivalves are widely distributed from tropical to boreal waters, and can be found inhabiting a variety of substrates, ranging from rocky to soft bottoms, infaunal and epifaunal. Most marine bivalves are suspension-feeders and can reach high densities in the wild, e.g. oyster reefs or mussel beds. At high density, they are ecosystem engineers (*sensu* Jones et al. 1994). Bivalves can modify the physical environment, for example by preventing erosion (Jones et al. 1994). They can also modify the available resources for other species, by controlling phytoplankton populations and/or altering nutrient cycling (Mann and Powell 2007; Filgueira et al. 2015). Consequently, the effects of bivalves on biogeochemical cycles goes beyond the individual scale. Accordingly, an ecosystem scale approach in which these feedbacks are included becomes imperative when studying the implications of marine bivalves in biogeochemical cycles.

The role of bivalves as ecosystem engineers and the need for an ecosystem approach become even more relevant when bivalves are cultured at high densities. Although the same ecosystem process can be conceptually applied to wild and cultured populations, the higher densities in aquaculture sites can significantly alter the magnitude of biogeochemical fluxes. For example, although cultured bivalves can exert a bottom-up nutrient control in stimulating primary production (Cranford et al. 2007; Jansen 2012), this positive effect is density dependent, with a resulting high bivalve biomass causing a reduction in primary production (Burkholder and Shumway 2011; Smaal et al. 2013). Given their ideal growing conditions, growth rates of cultured populations are usually higher than for wild populations; however, the most critical aspect of cultured bivalves is that their biomass is extracted from the ocean, a relevant consideration when comparing the role of wild versus cultured populations in biogeochemical cycles. The shells of wild bivalves will eventually dissolve in seawater, but those of cultured bivalves may end up on land. Note that some wild populations may also end up on land when they are commercially exploited (e.g. mussel or scallop dredging). Therefore, in this chapter bivalves will be considered according to two main categories: non-harvested (wild populations that are not harvested) and harvested (cultured and wild populations that are harvested). Separation of non-harvested and harvested populations is critical when evaluating the role of bivalves from each group in the CO_2 cycle and, in general, when valuing ecosystem services.

The goal of this chapter is to describe the role of bivalves in the CO_2 cycle with special emphasis on the specific role of their shells and the implications for ecosystem services valuation. To achieve this, the chapter has been structured accordingly:

- The role of calcifying organisms in the CO_2 budget.- which describes the chemistry of shell formation.
- The influence of organic carbon on CO_2 fluxes.- which describes the main processes involving organic carbon that are relevant to the CO_2 cycle.
- Ecosystem services of non-harvested and harvested populations.- which describes the implications of harvesting bivalves as a food source in the context of a holistic valuation of ecosystem services.
- Case study – Norwegian cultured mussels.- in which the rationale described in previous sections is applied to the case of Norwegian cultured mussels.
- Conclusions.- which summarizes the most relevant findings of the chapter.

12.2 The Role of Calcifying Organisms in the CO_2 Budget

Calcifying organisms are directly involved in two processes that release CO_2. First, CO_2 is released via the catabolism of ingested organic matter:

$$CH_2O + O_2 \rightarrow CO_2 + H_2O \tag{12.1}$$

and, second, it is released via calcium carbonate ($CaCO_3$) formation by biogenic calcification:

$$Ca^{2+} + 2HCO_3^- \leftrightarrow CaCO_3 + CO_2 + H_2O \quad (12.2)$$

This release of CO_2 also induces shifts in the carbonate system:

$$CO_2 + H_2O \leftrightarrow H_2CO_3 \leftrightarrow H^+ + HCO_3^- \leftrightarrow 2H^+ + CO_3^{2-} \quad (12.3)$$

These processes depend on environmental conditions such as pH, alkalinity, salinity, and temperature (Millero 1995; Lerman and Mackenzie 2005; Dickson 2010; Mackenzie and Andersson 2013).

The balance between the CO_2 released in respiration and biogenic calcification and the net C sequestered as calcium carbonate have been used to evaluate the role of bivalves in the CO_2 cycle. The available studies in which these processes have been quantified for bivalves is reviewed in Table 12.1. The units from the different studies have been converted to g C m^{-2} y^{-1} for comparative purposes (conversion factors: 12 g C in 100 g $CaCO_3$; 12 g C in 1 mol CO_2). With the exception of the estimations from Hily et al. (2013), all other studies suggest that sequestration minus biocalcification and respiration is negative (Table 12.1), which suggests that bivalves are net generators of CO_2. Hily et al. (2013) suggested that under specific environmental conditions *Crassostrea gigas* and *Mytilus edulis* can sequester carbon effectively after accounting for biocalcification and respiration. The disagreement between Hily et al. (2013) and the other studies (Table 12.1) seems to be related to the respiration flux in Hily et al. (2013) which is especially obvious when comparing the ratio between sequestration and respiration. The respiration values in Hily et al. (2013) are extremely low compared to the other studies (Table 12.1) when considering the carbon that is sequestered in the shell. This is even more striking given the fact that most of these studies, including Hily et al. (2013), use the same empirical equation proposed by Schwinghamer et al. (1986) to estimate respiration. Nevertheless, aside from Hily et al. (2013), the level at which bivalves release CO_2 is species dependent, with a net carbon release ranging from 0.35 to 2.45 gC $m^{-2}year^{-1}$ per 1 gC $m^{-2}year^{-1}$ (Table 12.1). The results of several studies (see Table 12.1) demonstrate that bivalves are CO_2 generators when the balance strictly focuses on this inorganic form of carbon at the individual level.

Solely from the individual perspective, it makes sense that a filter feeder is a net generator of CO_2. The deposition of calcium carbonate generates a small net sequestration explicitly resulting from individual biocalcification given that the precipitation of 1 mol of $CaCO_3$ releases approximately 0.6 mol of CO_2 (Ware et al. 1992). But this net sequestration (1.0–0.6 = 0.4 mol of CO_2 per mol of $CaCO_3$) is not enough to compensate the CO_2 that is released due to the catabolism of organic matter. Nevertheless, scaling these numbers up from the individual to the ecosystem level is not a trivial task. In a controversial paper, Tang et al. (2011) proposed that bivalve (and seaweed) aquaculture could increase atmospheric CO_2 absorption within coastal ecosystems. These authors did not account for the release of CO_2 via

Table 12.1 Carbon fluxes in different bivalve species: sequestration (carbon content in the shell), biocalcification (carbon released during biogenic calcification), respiration (carbon released through respiration of organic matter), balance (sequestration minus biocalcification and respiration), ratio balance/sequestration, and bibliographic references

Species (Habitat)	Sequestration gC $m^{-2}year^{-1}$	Biocalcification gC $m^{-2}year^{-1}$	Respiration gC $m^{-2}year^{-1}$	Balance gC $m^{-2}year^{-1}$	Balance/ Sequestration	References
Potamocorbula amurensis	23.9[a]	18.0	37.0	−31.1	−1.30	Chauvaud et al. (2003)
Mytilus edulis (sheltered)	3.8	2.3[a]	1.9	−0.4	−0.09	Hily et al. (2013)
Mytilus edulis (semiexposed)	129.2	77.4 [a]	44.3	7.6	0.06	Hily et al. (2013)
Mytilus edulis (exposed)	45.0	27.0 [a]	19.6	−1.6	−0.03	Hily et al. (2013)
Crassotrea gigas (sheltered)	286.8	172.0 [a]	11.9	103.0	0.36	Hily et al. (2013)
Chlamys farreri	78.1	54.0	71.7	−47.6	−0.61	Jiang et al. (2014)
Crassostrea gigas	15.5[a]	11.1	32.7	−28.3	−1.83	Lejart et al. (2012)
Ruditapes philippinarum	98.2	66.7	272.4	−241.0	−2.45	Mistri and Munari (2012)
Arculata senhousia	46.0	11.7	50.4	−16.1	−0.35	Mistri and Munari (2013)
Mytilus galloprovincialis	1639.2	1041.6	2253.6	−1656.0	−1.01	Munari et al. (2013)

[a]shell dissolution is included in this term

respiration in their budget (see Mistri and Munari 2013; Munari et al. 2013) but they argued for the inclusion of some relevant ecosystem effects when scaling up from the individual to the ecosystem level. For example, Tang et al. (2011) suggested that in a strongly autotrophic system, CO_2 released by carbonate precipitation may be used by photosynthetic organisms, resulting in a lower transfer of CO_2 from water to the atmosphere. They also suggested that removing shells from the oceans presents a long-term carbon sink. The slow dissolution of shells in the oceans, e.g. ~29 years for a 4-year old oyster excluding abrasion effects from waves and dissolution after burial (Suykens et al. 2011), provides a buffering capacity of respiratory acids to the environment (Waldbusser et al. 2013). Consequently, this removal can cause a loss of alkalinity regeneration and buffering of metabolic acids, which could affect ecosystem functioning (Waldbusser et al. 2013). These effects on water chemistry highlight that a simple multiplicative extrapolation from the individual to the ecosystem level oversimplifies the role of bivalves in the ecosystem. As stated by Lejart et al. (2012), the contribution of *C. gigas* to total carbon fluxes should be estimated for the entire community and not just for oysters. In addition, as stated by Waldbusser et al. (2013), the final destination of the shells can be relevant for ecosystem functioning and consequently has a feedback on the bivalves themselves. Clearly an integrated approach is required in which the ecosystem as well as anthropogenic aspects are simultaneously considered.

12.3 The Influence of Organic Carbon on CO_2 Fluxes

The strong coupling between inorganic and organic carbon cycles is fundamental for scaling up from individual to population fluxes. This is even more critical in aquaculture sites, where bivalve populations are artificially maintained at generally high densities. The ecosystem role, and implications on the CO_2 cycle, of dense bivalve populations can be very complex due to cascading effects, e.g. indirect effects on fish species via zooplankton consumption (Gibbs 2007; Kluger et al. 2017). Only the direct bivalve ecophysiological processes will be discussed in this chapter. The five main, direct ecophysiological processes of bivalves within the carbon cycle are: (1) respiration, which implies a net release of CO_2 (discussed above); (2) biocalcification, which involves a net sequestration of carbon (discussed above); (3) food ingestion; (4) rejection of uningested food; and (5) egestion of unabsorbed food. In addition, an indirect link with the carbon cycle is carried out by excreted nutrients (Fig. 12.1). Although ingestion, rejection, egestion, and excretion are not directly involved in the inorganic carbon cycle, they are key processes for phytoplankton dynamics, which in turn play a key role in the CO_2 cycle.

Bivalve ingestion may cause a direct top-down control on zooplankton (Maar et al. 2008) and phytoplankton populations (Dame 1996; Dame and Prins 1998; Newell 2004; Petersen et al. 2008; Huang et al. 2008). The net effects are strongly dependent on bivalve biomass and its relation to local environmental conditions, mainly water residence time and phytoplankton production rates (Dame and Prins

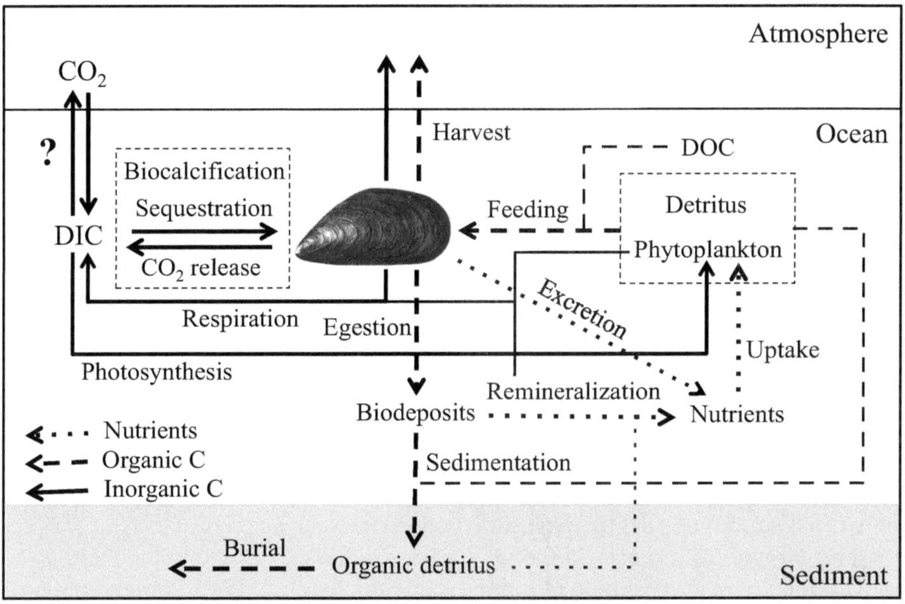

Fig. 12.1 Ecosystem approach to carbon cycling (continuous and dashed lines for organic and inorganic carbon, respectively) and feedbacks of mussel aquaculture on the pool of inorganic nutrients (dotted line). (Adapted for C from Cranford et al. 2012)

1998), which represent the renewal of planktonic resources driven by allochthones and autochthonous processes, respectively. If filtration capacity dominates over renewal, planktonic communities could be negatively affected (e.g. Heral 1993; Prins et al. 1998; Maar et al. 2007, 2010). This effect on planktonic biomass could have a direct effect on CO_2 dynamics although secondary local drivers could also exert a significant influence on the net fluxes. For example, in nutrient-limited systems, the reduced phytoplankton population could accelerate its turnover rate by using the additional available nutrients, which in turn could result in the same levels of CO_2 fixation as for a larger population (Newell 2004). Contrarily, in light-limited systems, the increase in filtration pressure usually causes a decrease in phytoplankton biomass and primary production (Cloern et al. 2007; Smayda 2008). This effect can be relaxed if filtration activity is sufficient to increase water clarity and light penetration (Cerco and Noel 2007; Schröder et al. 2014), which could stimulate phytoplankton growth and consequently CO_2 fixation. In addition to the changes in biomass, the structure of phytoplankton communities could also be affected due to the increasing retention efficiency from small to large particles (Jacobs et al. 2015; Cranford et al. 2016). This differential retention efficiency may benefit the relative abundance of the smallest planktonic species (e.g. Vaquer et al. 1996; Smaal et al. 2013; Froján et al. 2014); however, this is a site-specific effect, as demonstrated by Sonier et al. (2016), who could not find any changes in the ratio picoplankton:nanoplankton in a densely cultured site in Atlantic Canada. In any case, the potential alteration of a phytoplankton community could have an effect on CO_2 fluxes.

During the feeding process, phytoplankton and particulate organic matter are consolidated into pseudofaeces (rejected uningested material) and faeces (egested unabsorbed material). These biodeposits sink to the seafloor and their fates are highly dependent upon local environmental conditions (Carlsson et al. 2009, 2010; Jansen 2012). The hydrodynamic regime is relevant not only for determining the horizontal advection of the biodeposits (Pearson and Black 2001; Grant et al. 2005), but also for their potential disaggregation (Driscoll 1970). The remineralization of the biodeposits begins in the water column and consequently the amount of organic matter that reaches the bottom is dependent on water depth. This vertical flux is critical for pelagic-benthic coupling and consequently for CO_2 dynamics. For example, in shallow systems, biodeposits accumulated on the seafloor are exposed to very dynamic conditions in which resuspension and mixing can play important roles in determining remineralization rates or organic matter (Findlay and Watling 1997). In contrast, in deep fjord-type systems, sedimentation of biodeposits could transfer carbon to deep waters, potentially reaching the sediment (Sepúlveda et al. 2005), which can be considered as a carbon sequestering compartment. In addition to hydrodynamics and depth, other local conditions such as grain size, temperature, dissolved oxygen, presence/absence of seagrass, infauna, etc. determine the assimilative capacity of the benthos (Kusuki 1981; Souchu et al. 2001; Mitchell 2006). These local processes, in conjunction with bay-scale aspects such as terrestrial organic inputs and stoichiometry of nutrient inputs, define bay-scale dynamics and ultimately ocean-atmosphere CO_2 fluxes (Laruelle et al. 2010; Bauer et al. 2013).

The remineralization of biodeposits on the seafloor enhances the fluxes of nutrients under highly dense bivalve populations (e.g. Carlsson et al. 2009; Alonso-Pérez et al. 2010). In addition, bivalve ammonia excretion constitutes another source of nitrogen that can be directly used by phytoplankton (e.g. Smaal and Prins 1993; Sara 2007). Nitrogen is probably the most limiting nutrient in coastal marine ecosystems in the temperate zone (Howarth and Marino 2006). Therefore, in nutrient-limited systems, bivalve ammonia excretion can enhance primary production (Smaal 1991; Prins et al. 1995, 1998; Pietros and Rice 2003). This bottom-up control on phytoplankton populations has been demonstrated for aquaculture sites emplaced in nutrient-limited systems such as in Grande-Entrée Lagoon (Canada, Trottet et al. 2008) or Narragansett Bay (USA, Oviatt et al. 2002). Bottom-up control effectively accelerates phytoplankton turnover and primary production rates, which directly increase the net CO_2 fixation via photosynthesis, thereby accelerating carbon assimilation into the biosphere.

12.4 Ecosystem Services of Non–Harvested and Harvested Populations

The chemical and ecological aspects discussed above can be directly applied to both non-harvested and harvested populations; however, the final destination of the bivalve is a critical aspect that needs to be considered when valuing ecosystem services. For example, as stated above, the final destination of the shells can be relevant for water chemistry and consequently for ecosystem functioning (Waldbusser et al. 2013). In the case of non-harvested populations, the shells remain in the ocean but the final destinations of harvested bivalves are diverse, from waste to building materials (e.g. Rodríguez Álvaro et al. 2014; Varhen et al. 2017), agricultural usage or the production of lime (calcium oxide CaO), which could be used to remove phosphates from rural watersheds (Abeynaike et al. 2011). This difference is fundamental for the shell, but it is even more critical when the meat of the bivalve is part of the equation. In the case of non-harvested bivalves, the tissue will become part of the food web via predation and decomposition after death. Food provision via the meat of harvested bivalves is the primary goal of culturing bivalves. Although these differences are meaningless when discussing the role of bivalves as a whole in the CO_2 cycle, they become very important when valuing ecosystem services. Therefore, in the case of non-harvested bivalves that are not harvested to provide food, the analysis of their role on the CO_2 cycle should only include the chemical and ecological aspects discussed above. In the case of harvested bivalves, however, a clear distinction between the tissue, which is the main product of this economic activity, and the shell, which usually is considered waste, can be made when valuing their ecosystem services.

In the most extreme scenario, it can be argued that the shell has no marketable value and should be considered waste. In that situation, the carbon sequestered in the shell could be used to valorize the waste and create a by-product for carbon sequestration. Consequently, in that scenario all the CO_2 released from biocalcification and respiration should be accounted towards the CO_2 budget of the product, the meat. This would result in valuing the waste (shell) as a by-product that constitutes a net sink of carbon independent of the CO_2 released during the biocalcification and respiration. An alternative, and probably more logical, accountability would be to split the CO_2 fluxes towards shell and meat as a function of the biological processes involved in their formation. This implies splitting all the ecosystem fluxes and respiration among shell and meat as a function of their energetic demand. Splitting the energetic demand of a bivalve between shell and meat is not straightforward. It is commonly accepted that most of the energy is allocated towards maintenance, tissue growth and reproduction rather than shell growth. Nevertheless, the exact fraction of total energy that is invested in shell growth is unknown in part because any estimation is highly dependent on local environmental conditions. For example, habitat (Fig. 12.2, Rodhouse et al. 1984), feeding conditions (Aldrich and Crowley 1986), hydrodynamics (Steffani and Branch 2003) and predation pressure (Lowen et al. 2013) can all affect the energy allocation towards shell.

The lack of specific studies on energy allocation and the effects of local conditions on growth investment becomes a serious limitation when trying to split carbon fluxes between shell and tissue. The available data are limited to the estimations by Hawkins and Bayne (1992) who suggested that *Mytilus edulis* could spend more than 20% of the energy that is available for growth on shell formation. This matches with the calculations of Duarte et al. (2010), who indirectly estimated that *Mytilus galloprovincialis* could invest 20–28% of the energy that is available for growth in

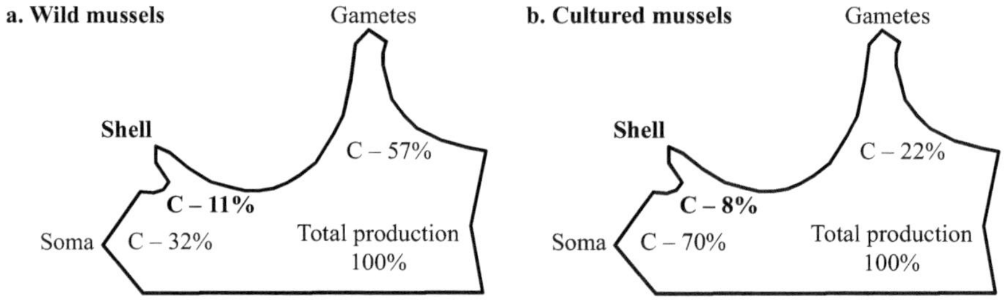

Fig. 12.2 Allocation of carbon in wild and cultured *Mytilus edulis*. (From Rodhouse et al. 1984)

shell formation. It is important to highlight that these estimations establish the energy that is available for growth as a bottom line for the calculations, in other words, the available energy after paying maintenance, digestion/absorption and growth costs (Scope For Growth, Winberg 1960). The shell does not require any maintenance costs, with the exception of repairing mechanical damage, and consequently allocating 20–28% (based on Duarte et al. 2010) of the total CO_2 fluxes towards shell would overestimate the energetic requirements of shell growth. Accordingly, for the following estimated calculations, 10% has been assumed as the percentage of the total energetic demands that is allocated towards shell (with the remaining 90% allocated towards maintenance and tissue growth).

As explained above, all the processes in the full ecosystem approach towards the quantification of CO_2 fluxes should be split between tissue and shell according to this 10/90% estimation. For these preliminary calculations and for simplicity, the following calculations have included only biocalcification and respiration in the CO_2 budget, following the approach presented in Table 12.1. Accordingly, the respiration values provided in Table 12.1 have been re-calculated in Table 12.2 by considering only 10% of the total respiration, which would represent the CO_2 flux that corresponds to the shell energetic requirements. The datasets from Hily et al. (2013) have been removed from this table due to the uncertainties highlighted above. Splitting respiration provides a general budget for shell CO_2 fluxes (Table 12.2) rather than for the whole individual (Table 12.1). According to the Table 12.2 calculations and in the context of harvested bivalves, the shells, which are waste of an industrial process, could be considered net sinks of CO_2 and consequently valorized as by-products. It should be re-emphasized that this reasoning is based on the assumption that humans culture bivalves with the aim of producing food and not sequestering CO_2 and consequently, from the perspective of ecosystem services the CO_2 generated through respiration should be split between meat and shell.

The next logical question is: is this sequestered carbon relevant from a global perspective? As stated above, cultured bivalves produce ~7×10^6 tons of shell per year. Taking into account that 95% is calcium carbonate, and 12% of that is carbon, shells contain 8×10^5 tons of carbon per year. Assuming that shell growth demands 10% of total energy and the net sequestration of carbon in the shell is ~21% (aver-

Table 12.2 Carbon fluxes for shells in different bivalve species: sequestration (carbon content in the shell), biocalcification (carbon released during biogenic calcification), respiration (10% of carbon released through respiration of organic matter), balance (sequestration minus biocalcification and respiration), ratio balance/sequestration, and bibliographic references

Species	Sequestration g C $m^{-2}year^{-1}$	Biocalcification g C $m^{-2}year^{-1}$	Respiration g C $m^{-2}year^{-1}$	Balance g C $m^{-2}year^{-1}$	Balance/ Sequestration	References
Potamocorbula amurensis	23.9	18.0	3.7	2.2	0.09	Chauvaud et al. (2003)
Chlamys farreri	78.1	54.0	7.2	16.9	0.22	Jiang et al. (2014)
Crassostrea gigas	15.5	11.1	3.3	1.1	0.07	Lejart et al. (2012)
Ruditapes philippinarum	98.2	66.7	27.2	4.2	0.04	Mistri and Munari (2012)
Arculata senhousia	46.0	11.7	5.0	29.3	0.64	Mistri and Munari (2013)
Mytilus galloprovincialis	1639.2	1041.6	225.4	372.2	0.23	Munari et al. (2013)

age value of Balance/Sequestration column in Table 12.2), 1.71 × 10^5 tons of carbon, or 6.3 × 10^5 tons of CO_2 equivalent, per year are effectively sequestered in shells of cultured bivalves. In economic terms, the impact highly depends upon the carbon initiative that values a ton of CO_2, which can range from US$131 in the Swedish carbon tax, to US$1 in Mexico (World Bank et al. 2016). Assuming an average value of US$24 per ton of CO_2 (average for Denmark, France, United Kingdom, British Columbia and Ireland; World Bank et al. 2016) the global value of the carbon effectively sequestered in shells of cultured bivalves is ~15.7 million US$ per year. This amount represents less than 0.01% of the total bivalve aquaculture value.

12.5 Case-Study: Norwegian Cultured Mussels

Marine carbon burial is the main natural mechanism of long-term organic carbon sequestration (Berner 1982; Hedges et al. 1997). Fjords are deep, glacially carved estuaries situated at high latitudes. Smith et al. (2015) estimated that 18 × 10^6 tons carbon is buried in fjord sediment each year. This is equivalent to 11% of the annual marine carbon burial globally, and makes the fjord organic carbon burial rate 100 times more efficient than the global ocean average, per unit area. As stated above, local conditions are critical for the implications of cultured bivalves on the CO_2 cycle. The estimation of the CO_2 budget of mussel (*Mytilus edulis*) farming in a Norwegian fjord has been selected as a case-study to guide the application of the rationale described in this study to a cultured system. It is important to emphasize that due to the effects of the cultured species and local conditions, the following calculations cannot be extrapolated to other bivalves or locations.

The CO_2 budget is based on the life history of a 2-year old mussel, the typical lifespan of mussels in suspended culture in Norway. The mussel is harvested before reproduction, and obtains a dry shell weight (DSW) of 4.8 g and a dry tissue weight of 1.0 g. It is assumed that 85% of farmed mussels will be harvested and 15% will fall off their ropes during strong winds and wave action, farm operation, density control/thinning, harvest and predation (Strohmeier et al. 2008). As a consequence of low food quantity and high food quality, mussels have not been reported to reject uningested food in Norwegian fjords (Strohmeier et al. 2015). Following the rationale described above, the CO_2 fluxes were split between the shell and the tissue according to their presumed energetic demand.

12.5.1 *Respiration*

Throughout their life history, mussels consume oxygen and release CO_2 as a result of the catabolism of organic matter. The oxygen required for the mussel growth has been estimated by allometric scaling (Bayne and Widdows 1978; Thompson 1984;

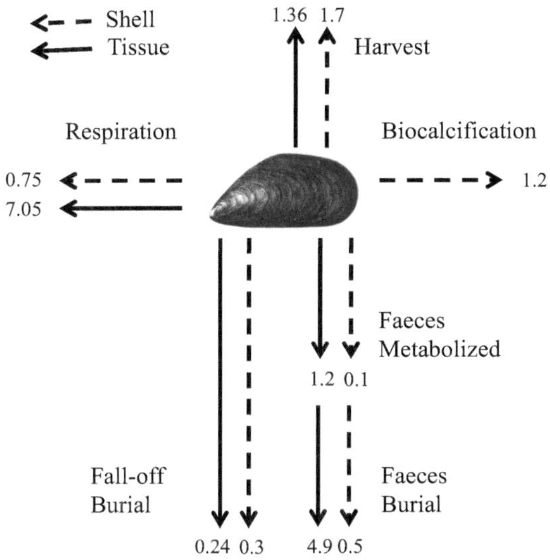

Fig. 12.3 CO_2 fluxes (total g CO_2 per mussel) in cultured Norwegian mussel

Smaal et al. 1997) and a Dynamic Energy Budget (DEB) model parameterized for Norwegian mussels (Rosland et al. 2009), using a seasonal time series of mussel growth and ecophysiology that included respiration data (Strohmeier 2009; Strohmeier et al. 2015). The results indicated a cumulative oxygen consumption of 4.5 to 8.8 g. Assuming a respiratory quotient towards herbivory (0.85, Galtsoff 1964), this results in a cumulative release of 0.12 to 0.23 mol or 5.3 to 10.3 g CO_2. Splitting the CO_2 fluxes between shell and tissue according to the 10/90% outlined above, the allocated catabolism of the shell represents 0.5 to 1 g CO_2 (mean 0.75, Fig. 12.3) and the catabolism of the tissue from 4.8 to 9.3 g CO_2 (mean 7.05, Fig. 12.3).

12.5.2 The Shell

The deposit of $CaCO_3$ in a 4.8 g mussel shell sequesters 0.55 g carbon or 2.0 g CO_2. The flux of CO_2 due to shell formation to land (harvest) and seabed (fall off) is 1.7 and 0.3 g CO_2, respectively (Fig. 12.3). The amount of CO_2 released during biocalcification for the same individual is 1.2 g. Therefore, the net sequestration in the shell is 0.8 g CO_2. Including the associated cost of respiration (10%) to the net sequestration of CO_2 in the shell results in a balance (sequestration minus biocalcification and respiration) in the range − 0.2 to 0.3 g CO_2, which accounts for all the relevant fluxes at the individual level needed to define the CO_2 budget (e.g. Table 12.1). Under the assumption that 85% of the mussels are harvested and 15% fall off, the mean balance indicates a net flux of 0.04 g CO_2 to land (harvest) and 0.01 g CO_2 to the seabed (fall off).

12.5.3 The Tissue

The carbon content of the tissue shows seasonal variation, with a mean value of 0.44 g C per gram of dry weight (range 0.40–0.47, Jansen 2012). The mean carbon content of a 1 g of mussel's tissue in terms of dry weight is thereby 0.44 g, corresponding to 1.61 g CO_2. The flux of CO_2 to land (harvest) and seabed (fall off) is thus 1.36 and 0.24 g CO_2, respectively (Fig. 12.3). Inclusion of the associated cost of catabolism (90%) results in a net balance (sequestration minus catabolism) in the range from −7.7 to −3.2 g CO_2. The mean balance indicates a net flux of −4.9 g CO_2 to land (harvest) and − 0.5 g CO_2 to seabed (fall off).

12.5.4 Egestion of Unabsorbed Food

The cumulative mass and carbon content of fecal pellets has been estimated based on Jansen et al. (2012) and the DEB model (Rosland et al. 2009). The results indicate that a mussel egests 12.9–13.7 g faeces over the 2 year period in terms of dry weight. The fecal pellets comprise a C fraction of 13.5% (Jansen 2012). The cumulative egestion is thereby 1.7–1.9 g C or 6.4–7.0 g CO_2. Faeces contain fresh biological material, and may be used as a food source by other organisms until they are buried in the sediment. Faeces were assumed to enter the pelagic environment after being "trapped" on the mussel collectors for a brief period of time, then they sink towards the seabed (Jansen et al. 2012). The sinking velocity of fecal pellets, obtained for mussels grazing on natural seston, has been reported at 3.9 mm s^{-1} or 337 m d^{-1} (Carlsson et al. 2010). Overall, the residence time for faeces in the pelagic environment was set to two days, representing the average depth of a Norwegian fjord of about 300 m.

Faeces contain a labile faction that can be fully catabolized in oxygenated water on a timescale ranging from 5 to 15 days, depending on the season (Jansen 2012). Here a constant decay and 10 days to fully catabolize the fecal matter is assumed. Taking into account the average sinking time of 2 days, 80% of the faeces will reach the seabed (mean 1.4 g C or 5.4 g CO_2), while 20% will be metabolized in the pelagic environment (0.4 g C or 1.3 g CO_2). Splitting these fluxes according to the associated energy demand of shell (10%) and tissue (90%), 0.1 and 1.2 g CO_2 of the fecal matter will be respired in the pelagic environment, and 0.5 and 4.9 g CO_2 will reach the sediment for shell and tissue, respectively (Fig. 12.3). In deep anoxic fjords a high carbon burial rate can be expected, and in this budget it is assumed that the carbon that reaches the seabed is not metabolized further.

12.5.5 General Budget in the Context of Ecosystem Services

The balance for the shell and tissue was estimated separately as: +burial of fall off mussel +burial of faeces +harvest of mussel –biocalcification –respiration –faeces respired in the water column. The balance for shell is +0.45 g CO_2 (+0.3 + 0.5 + 1.7–1.2 –0.75 –0.1) suggesting that mussel shell of cultured mussels in a 2 year cycle in a Norwegian fjord can be considered a net sink of CO_2, assuming that the harvested shells are disposed of in a way that can be considered sequestered material, e.g. concrete. The balance for tissue is −3.11 g CO_2 (+0.24 + 4.9 + 0–0 –7.05 –1.2). Note that for tissue, CO_2 towards the term 'harvest of mussel' has not been included in the budget. This flow of CO_2 is assumed to be consumed and respired in the short term and consequently not sequestered in the long term. Accounting shell and tissue together, this budget confirms that mussels are, as expected, net sources of CO_2.

This budget includes the traditional fluxes of respiration and biocalcification (e.g. Table 12.1), but also an additional direct ecosystem flux, the egestion of unabsorbed food. Given that in Norwegian waters the rejection of uningested food is negligible, the impact of bivalve ingestion and ammonia excretion would be the only two additional processes to assess for a holistic ecosystem approach to the CO_2 budget. The CO_2 fluxes have been split according to the biological process involved in the formation of the shell and tissue, based on their anticipated energetic demands (see text above). In valuing the ecosystem service of mussel farming in the carbon cycle a distinction has been made between the shell (waste) and the tissue (food). Following this rationale, the goods and services of mussel farming in deep fjords includes the valorization of the shells as a net sink of CO_2.

The more holistic ecological approach reveals a previously unaccounted for, yet significant indirect carbon sequestration by deposition of mussel faeces in sediment. Given the assumption that all mussel faeces are buried in deep fjords, the sediment may sequester more than 60% of the total CO_2 respired. In environments comprising high food quantities, mussels can produce a significant amount of pseudofaeces (rejected uningested material) in addition to fecal matter and thereby increase the organic flux to the seabed (Galimany et al. 2013). If this particulate matter, faeces and pseudofaeces, sinks into an environment where it is not further catabolized, then a net CO_2 sequestering from mussel farming is plausible. This may serve as an example to encourage an ecosystem approach towards the quantification of bivalve CO_2 fluxes.

A typical Norwegian mussel farm produces a volume from 50 to 150 tons each year equating to a mean farm production of about 6.25 x10^6 mussels. Valuing the ecosystem service of mussel farming in the carbon cycle, the shells sequester 2.8 tons of CO_2, or 146 US$ per year (assuming US$53 per ton of CO_2 in Norway according to World Bank et al. (2016)).

12.6 Conclusions

As expected and as proved in the literature (Table 12.1), given their nature as primary consumers, bivalves release CO_2. The sequestration of CO_2 in the shell is not enough to compensate the release generated during the respiration of organic matter. Note that the use of the term "production" was avoided within the manuscript in the context of bivalves "producing" CO_2. This has been done intentionally to avoid negative connotations associated with being a CO_2 generator. As discussed, all primary consumers release CO_2 that was captured by primary producers. Accordingly, a better term could be "recycling" CO_2 rather than a term that suggests the production of new CO_2. In any case, in the context of ecosystem services, there are two fundamental aspects that should be also taken into account when estimating a CO_2 budget: the consideration of ecosystem processes when scaling individual fluxes to the population level (e.g. Lejart et al. 2012), and the final destination of the bivalves (e.g. Waldbusser et al. 2013), that is, bivalves harvested for food production or non-harvested bivalves. Ecosystem processes involving bivalves are relevant and alter the CO_2 cycle via filtration and/or nutrient cycling (Lejart et al. 2012). Consequently, they should be considered when the CO_2 budget is calculated for bivalve populations. When valuing ecosystem services, it has been recognized that humans harvest bivalves to provide food and consequently shells should be considered waste. Accordingly, a different CO_2 budget should be calculated for product (tissue) and waste (shell).

Under these considerations, bivalve shells can be considered net sinks of CO_2 and consequently provide additional ecosystem services besides the food provided by the tissue. A full life cycle analysis should be performed to account for the emissions required to properly dispose of the shells. The 0.45 g CO_2 sequestered by the shell of each cultured mussel in Norway is hardly significant taking into account that a regular car produces more than 100 g CO_2 per km. For example, since 2015 European Union law requires that new cars do not emit more than an average of 130 g CO_2 per km, with a target of 95 g CO_2 per km by 2021 (European Commission, Climate Action). Even when these numbers are extrapolated to the global scale, a conservative extrapolation of the individual bivalve budget to the global production would result in a sequestration of 6.3×10^5 tons of CO_2 per year, ~15.7×10^6 US$/year. In different units, this is equivalent to the annual emissions of 242,307 cars driving an average of 20,000 km each. Although this is far from solving a global problem, everything counts. In addition, it is important to re-emphasize that this comes at no cost or effort given that bivalves are cultured to produce food.

Acknowledgments We would like to thank Sandra Shumway and Gary Wikfors for their constructive and valuable reviews of the chapter. The work was partly based on results from two projects funded by the Research Council of Norway: Carrying Capacity in Norwegian Aquaculture (CANO) (project no. 173537) and Growth performance and detoxification of mussels cultured in a fjord enhanced by forced upwelling of nutrient rich deeper water (GATE) (project No. 196560). Further support was provided by the Natural Sciences and Engineering Research Council of Canada (NSERC Discovery Grant to RF).

References

Abeynaike A, Wang L, Jones MI, Patterson DA (2011) Pyrolysed powdered mussel shells for eutrophication control: effect of particle size and powder concentration on the mechanism and extent of phosphate removal. Asia Pac J Chem Eng 6:231–286

Aldrich JC, Crowley M (1986) Condition and variability in *Mytilus edulis* (L.) from different habitats in Ireland. Aquaculture 52:273–286

Alonso-Pérez F, Ysebaert T, Castro CG (2010) Effects of suspended mussel culture on benthic-pelagic coupling in a coastal upwelling system (Rı'a de Vigo, NW Iberian Peninsula). J Exp Mar Biol Ecol 382:96–107

Bauer JE, Cai WJ, Raymon PA, Bianchi TS, Hopkinson CS, Regnier PAG (2013) The changing carbon cycle of the coastal ocean. Nature 504:61–70

Bayne BL, Widdows J (1978) Physiological ecology of two populations of *Mytilus edulis* L. Oecologia 37(2):137–162

Berner RA (1982) Burial of organic carbon and pyrite sulfur in the modern ocean: its geochemical and environmental significance. Am J Sci 282:451–473

Burkholder JM, Shumway SE (2011) Bivalve shellfish aquaculture and eutrophication. In: Shumway SE (ed) Shellfish aquaculture and the environment. Wiley, New York, pp 155–215

Carlsson MS, Holmer M, Petersen JK (2009) Seasonal and spatial variations of benthic impacts of mussel longline farming in a eutrophic Danish fjord, Limfjorden. J Shellfish Res 28:791–801

Carlsson MS, Glud RN, Petersen JK (2010) Degradation of mussel (*Mytilus edulis*) fecal pellets released from hanging long-lines upon sinking and after settling at the sediment. Can J Fish Aquat Sci 67(9):1376–1387

Cerco CF, Noel MR (2007) Can oyster restoration reverse cultural eutrophication in Chesapeake Bay? Estuar Coasts 30:331–343

Chauvaud L, Thompson JK, Cloern JE, Thouzeau G (2003) Clams as CO_2 generators: the *Potamocorbula amurensis* example in San Francisco Bay. Limnol Oceanogr 48:2086–2092

Cloern JE, Jassby AD, Thompson JK, Hieb KA (2007) A cold phase of the East Pacific triggers new phytoplankton blooms in San Francisco Bay. P Natl Acad Sci USA 104:18561–18565

Cranford PJ, Strain PM, Dowd M, Hargrave BT, Grant J, Archambault MC (2007) Influence of mussel aquaculture on nitrogen dynamics in a nutrient enriched coastal embayment. Mar Ecol Prog Ser 347:61–78

Cranford PJ, Kamermans P, Krause G, Mazurié J, Buck BH, Dolmer P, Fraser D, Nieuwenhove KV, O'Beirn FX, Sanchez-Mata A, Thorarisdóttir GG, Strand Ø (2012) An ecosystem-based approach and management framework for the integrated evaluation of bivalve aquaculture impacts. Aquac Environ Interact 2:193–213

Cranford PJ, Strohmeier T, Filgueira R, Strand Ø (2016) Potential methodological influences on the determination of particle retention efficiency by suspension feeders: *Mytilus edulis* and *Ciona intestinalis*. Aquat Biol 25:61–73

Dame RF (1996) Ecology of marine bivalves: an ecosystem approach. CRC Press, Boca Raton

Dame RF, Prins TC (1998) Bivalve carrying capacity in coastal ecosystems. Aquat Ecol 31:409–421

Dickson AG (2010) The carbon dioxide system in seawater: equilibrium chemistry and measurements. In: Riebesell U, Fabry VJ, Hansson L, Gattuso JP (eds) Guide to best practices for ocean acidification research and data reporting. Publications Office of the European Union, Luxembourg, pp 17–52

Driscoll EG (1970) Selective bivalve shell destruction in marine environments, a field study. J Sediment Petrol 40(3):898–905

Duarte P, Fernández-Reiriz MJ, Filgueira R, Labarta U (2010) Modelling mussel growth in ecosystems with low suspended matter loads. J Sea Res 64:273–286

Filgueira R, Comeau LA, Guyondet T, Mckindsey CW, Byron CJ (2015) Modelling carrying capacity of bivalve aquaculture: a review of definitions and methods. In: Meyers R (ed) Encyclopedia of sustainability science and technology. Springer, New York

Findlay RH, Watling L (1997) Prediction of benthic impact for salmon net-pens based on the balance of benthic oxygen supply and demand. Mar Ecol Prog Ser 155:147–157

Froján M, Arbones B, Zúñiga D, Castro CG, Figueiras FG (2014) Microbial plankton community in the Ría de Vigo (NW Iberian upwelling system): impact of the culture of *Mytilus galloprovincialis*. Mar Ecol Prog Ser 498:43–54

Galimany E, Rose JM, Dixon MS, Wikfors GH (2013) Quantifying feeding behaviour of ribbed mussels (*Geukensia demissa*) in two urban sites (Long Island Sound, USA) with different seston characteristics. Estuar Coasts 36:1265–1273

Galtsoff PS (1964) Transport of water by the gills and respiration. In: Galtsoff PS (ed) The American oyster, *Crassostrea virginica* Gmelin, Fishery bulletin 64. U.S. G.P.O, Washington, DC

Gibbs MT (2007) Sustainability performance indicators for suspended bivalve aquaculture activities. Ecol Indic 7:94–107

Grant J, Cranford PJ, Hargrave B, Carreau M, Schofield B, Armsworthy S, Burdett-Coutts V, Ibarra D (2005) A model of aquaculture biodeposition for multiple estuaries and field validation at blue mussel (*Mytilus edulis*) culture sites in eastern Canada. Can J Fish Aquat Sci 62:1271–1285

Guderley HE, Tremblay I (2016) Swimming in scallops. In: Shumway SE, Parsons GJ (eds) Scallops: biology, ecology, aquaculture, and fisheries. Elsevier, San Diego, pp 535–566

Hedges JI, Keil RG, Benner R (1997) What happens to terrestrial organic matter in the ocean? Org Geochem 27:195–212

Hawkins AJS, Bayne B (1992) Physiological interrelations, and the regulation of production. In: Gosling E (ed) The mussel mytilus: ecology, physiology, genetics and culture (developments in aquaculture and. Fisheries science 25). Elsevier, Amsterdam, pp 171–222

Heral M (1993) Why carrying capacity models are useful tools for management of bivalve molluscs culture. In: Dame RF (ed) Bivalve filter feeders in estuarine and coastal ecosystem processes. Springer, Berlin, pp 455–477

Hily C, Grall J, Chavaud L, Lejart M, Clavier J (2013) CO_2 generation by calcified invertebrates along rocky shores of Brittany, France. Mar Freshw Res 64:91–101

Howarth RW, Marino R (2006) Nitrogen as the limiting nutrient for eutrophication in coastal marine ecosystems: evolving views over three decades. Limnol Oceanogr 51:364–376

Huang CH, Lin HJ, Huang TC, Su HM, Hung JJ (2008) Responses of phytoplankton and periphyton to system-scale removal of oyster-culture racks from a eutrophic tropical lagoon. Mar Ecol Prog Ser 358:1–12

Jacobs P, Riegman R, van der Meer J (2015) Impact of the blue mussel *Mytilus edulis* on the microbial food web in the western Wadden Sea. Mar Ecol Prog Ser 527:119–131

Jansen HM (2012) Bivalve nutrient cycling. Wageningen University. PhD dissertation

Jansen HM, Strand Ø, Verdegem M, Smaal A (2012) Accumulation, release and turnover of nutrients (C-N-P-Si) by the blue mussel *Mytilus edulis* under oligotrophic conditions. J Exp Mar Biol Ecol 416:185–195

Jiang ZJ, Fang JG, Han TT, Mao YZ, Li JQ, Du MR (2014) The role of *Gracilaria lemaneiformis* in eliminating the dissolved inorganic carbon released from calcification and respiration process of *Chlamys farreri*. J Appl Phycol 26:545–550

Jones CG, Lawton JH, Shachak M (1994) Organisms as ecosystem engineers. Oikos 69:373–386

Kluger L, Filgueira R, Wolff M (2017) Integrating the concept of resilience into the ecosystem-based approach for bivalve aquaculture management. Ecosystems 20:1364–1382

Kusuki Y (1981) Fundamental studies on the deterioration of oyster growing grounds. Bull Hiroshima Fish Exp Stn 11:1–93

Laruelle GG, Dürr HH, Slomp CP, Borges AV (2010) Evaluation of sinks and sources of CO_2 in the global coastal ocean using a spatially-explicit typology of estuaries and continental shelves. Geophys Res Lett 37:L15607

Lejart M, Clavier J, Chauvaud L, Hily C (2012) Respiration and calcification of *Crassostrea gigas*: contribution of an intertidal invasive species to coastal ecosystem CO_2 fluxes. Estuar Coasts 35:622–632

Lerman A, Mackenzie FT (2005) CO_2 air-sea exchange due to calcium carbonate and organic matter storage, and its implications for the global carbon cycle. Aquat Geochem 11:345–390

Lowen JB, Innes DJ, Thomson RJ (2013) Predator-induced defenses differ between sympatric *Mytilus edulis* and *M. trossulus*. Mar Ecol Prog Ser 475:135–143

Maar M, Nielsen TG, Bolding K, Burchard H, Visser AW (2007) Grazing effects of blue mussel *Mytilus edulis* on the pelagic food web under different turbulence conditions. Mar Ecol Prog Ser 339:199–213

Maar M, Nielsen TG, Petersen JK (2008) Depletion of plankton in a raft culture of *Mytilus galloprovincialis* in Ría de Vigo, NW Spain. II. Zooplankton. Aquat Biol 4:127–141

Maar M, Timmermann K, Petersen JK, Gustafsson KE, Storm LM (2010) A model study of the regulation of the blue mussels by nutrient loadings and water column stability in a shallow estuary, the Limfjorden. J Sea Res 64:322–333

Macdonald BA, Thompson RJ (1985) Influence of temperature and food availability on the ecological energetics of the giant scallop *Placopecten magellanicus*. 1. Growth-rates of shell and somatic tissue. Mar Ecol Prog Ser 25:279–294

Mackenzie FT, Andersson AJ (2013) The marine carbon system and ocean acidification during Phanerozoic time. Geochem Perspect 2:1–22

Mann R, Powell EN (2007) Why oyster restoration goals in the Chesapeake Bay are not and probably cannot be achieved. J Shellfish Res 26:905–917

Millero FJ (1995) Thermodynamics of the carbon dioxide system in the oceans. Geochim Cosmochim Acta 59:661–677

Mistri M, Munari C (2012) Clam farming generates CO_2: a study case in the Marinetta lagoon (Italy). Mar Pollut Bull 64:2261–2264

Mistri M, Munari C (2013) The invasive bag mussel *Arcuatula senhousia* is a CO_2 generator in near-shore ecosystems. J Exp Mar Biol Ecol 440:164–168

Mitchell IM (2006) In situ biodeposition rates of Pacific oysters (*Crassostrea gigas*) on a marine farm in southern Tasmania (Australia). Aquaculture 257:194–203

Munari C, Rossetti E, Mistri M (2013) Shell formation in cultivated bivalves cannot be part of carbon trading systems: a study case with *Mytilus galloprovincialis*. Mar Environ Res 92:264–267

Newell RIE (2004) Ecosystem influences of natural and cultivated populations of suspension-feeding bivalve molluscs: a review. J Shellfish Res 23:51–61

Newell CR, Hidu H (1982) The effects of sediment type on growth-rate and shell allometry in the soft shelled clam *Mya arenaria* L. J Exp Mar Biol Ecol 65:285–295

Oviatt C, Keller A, Reed L (2002) Annual primary production in Narragansett Bay with no bay-wide winter–spring phytoplankton bloom. Estuar Coast Shelf Sci 54:1013–1026

Pearson TH, Black KD (2001) The environmental impact of marine fish cage culture. In: Black KD (ed) Environmental impacts of aquaculture. Academic, Sheffield, pp 1–31

Penney RW, Hart MJ, Templeman ND (2008) Genotype-dependent variability in somatic tissue and shell weights and its effect on meat yield in mixed species *Mytilus edulis* L., *M. trossulus* (Gould), and their hybrids cultured mussel populations. J Shellfish Res 27:827–834

Petersen JK, Nielsen TG, van Duren L, Maar M (2008) Depletion of plankton in a raft culture of *Mytilus galloprovincialis* in Ría de Vigo, NW Spain. I. Phytoplankton. Aquat Biol 4:113–125

Pietros JM, Rice MA (2003) The impacts of aquacultured oysters, *Crassostrea virginica* (Gmelin, 1791) on water column nitrogen and sedimentation: results of as mesocosm study. Aquaculture 220:407–422

Prins TC, Escaravage V, Smaal AC, Peters JCH (1995) Nutrient cycling and phytoplankton dynamics in relation to mussel grazing in a mesocosm experiment. Ophelia 41:289–315

Prins TC, Smaal AC, Dame RF (1998) A review of the feedbacks between bivalve grazing and ecosystem processes. Aquat Ecol 31:349–359

Rodhouse PG, Roden CM, Hensey MP, Ryan TH (1984) Resource allocation in *Mytilus edulis* on the shore and in suspended culture. Mar Biol 84:27–34

Rodríguez Álvaro R, Seara Paz G, Pérez Ordóñez JL (2014) Morteros para revestimiento con árido procedente de concha de mejillón. Escola Universitaria de Arquitectura Técnica de A Coruña. Universadide de A Coruña

Rosland R, Strand Ø, Alunno-Bruscia M, Bacher C, Strohmeier T (2009) Applying Dynamic Energy Budget (DEB) theory to simulate growth and bio-energetics of blue mussels under low seston conditions. J Sea Res 62:49–61

Sara G (2007) A meta-analysis on the ecological effects of aquaculture on the water column: dissolved nutrients. Mar Environ Res 63:390–408

Schröder T, Stank J, Schernewski G, Krost P (2014) The impact of a mussel farm on water transparency in the Kiel Fjord. Ocean Coast Manag 101:42–52

Schwinghamer P, Hargrave B, Peer D, Hawkins CM (1986) Partitioning of production and respiration among size groups of organisms in an intertidal benthic community. Mar Ecol Prog Ser 31:131–142

Sepúlveda J, Pantoja S, Hughen K, Lange C, Gonzalez F, Muñoz P, Rebolledo L, Castro R, Contreras S, Ávila A, Rossel P, Lorca G, Salamanca M, Silva N (2005) Fluctuations in export productivity over the last century from sediments of a southern Chilean fjord (44°S). Estuar Coast Shelf Sci 65:587–600

Smaal AC (1991) The ecology and cultivation of mussels: new advances. Aquaculture 94:245–261

Smaal AC, Prins TC (1993) The uptake of organic matter and the release of inorganic nutrients by bivalve suspension feeder beds. In: Dame RF (ed) Bivalve filter feeders in estuarine and coastal ecosystem processes. Springer, Berlin, pp 271–298

Smaal AC, Vonck A, Bakker M (1997) Seasonal variation in physiological energetics of *Mytilus edulis* and *Cerastoderma edule* of different size classes. J Mar Biol Assoc UK 77:817–838

Smaal AC, Schellekens T, van Stralen MR, Kromkamp JC (2013) Decrease of the carrying capacity of the Oosterschelde estuary (SW Delta, NL) for bivalve filter feeders due to overgrazing? Aquaculture 404-405:28–34

Smayda TJ (2008) Complexity in the eutrophicatio-harmful algal bloom relationship, with comment on the importance of grazing. Harmful Algae 8:140–151

Smith RW, Bianchi RS, Allison M, Savage C, Galy V (2015) High rates of organic carbon burial in fjord sediments globally. Nat Geosci 8:450–453

Sonier R, Filgueira R, Guyondet T, Tremblay R, Olivier F, Meziane T, Starr M, Leblanc AR, Comeau LA (2016) Picoplankton contribution to *Mytilus edulis* growth in an intensive culture environment. Mar Biol 163:72

Souchu P, Vaquer A, Collos Y, Landrein S, Deslous-Paoli JM, Bibent B (2001) Influence of shellfish farming activities on the biogeochemical composition of the water column in Thau Lagoon. Mar Ecol Prog Ser 218:141–152

Stanley SM (1970) Relation of shell form to life habits of the Bivalvia (Mollusca). The Geological Society of America, Inc. Memoir 125

Steffani CN, Branch GM (2003) Growth rate, condition, and shell shape of *Mytilus galloprovincialis*: responses to wave exposure. Mar Ecol Prog Ser 246:197–209

Strohmeier T (2009) Feeding behavior and bioenergetic balance of the great scallop (*Pecten maximus*) and the blue mussel (*Mytilus edulis*) in a low seston environment and relevance to suspended shellfish aquaculture. The University of Bergen. PhD dissertation

Strohmeier T, Duinker A, Strand Ø, Aure J (2008) Temporal and spatial variation in food availability and meat ratio in a longline mussel farm (*Mytilus edulis*). Aquaculture 276:83–90

Strohmeier T, Strand Ø, Alunno-Bruscia M, Duinker A, Rosland R, Aure J, Erga SR, Naustvoll LJ, Jansen HM, Cranford PJ (2015) Response of *Mytilus edulis* to enhanced phytoplankton availability by controlled upwelling in an oligotrophic fjord. Mar Ecol Prog Ser 518:139–152

Suykens K, Schmidt S, Delille B, Harlay J, Chou L, De Bodt C, Fagel N, Borges AV (2011) Benthic remineralization in the northwest European continental margin (northern Bay of Biscay). Cont Shelf Res 31:644–658

Tang Q, Zhang J, Fang J (2011) Shellfish and seaweed mariculture increase atmospheric CO_2 absorption by coastal ecosystems. Mar Ecol Prog Ser 424:97–104

Thompson RJ (1984) The reproductive cycle and physiological ecology of the mussel *Mytilus edulis* in a subarctic, non-estuarine environment. Mar Biol 79:277–288

Trottet A, Roy S, Tamigneaux E, Lovejoy C, Tremblay R (2008) Influence of suspended mussel farming on planktonic communities in Grande-Entrée Lagoon, Magdalen Islands (Québec, Canada). Aquaculture 276:91–102

Vaquer A, Troussellier M, Courties C, Bibent B (1996) Standing stock and dynamics of picophytoplankton in the Thau Lagoon (Northwest Mediterranean coast). Limnol Oceanogr 41:1821–1828

Varhen C, Carrillo S, Ruiz G (2017) Experimental investigation of Peruvian scallop used as fine aggregate in concrete. Constr Build Mater 136:533–540

Waldbusser GG, Powell EN, Mann R (2013) Ecosystem effects of shell aggregations and cycling in coastal waters: an example of Chesapeake Bay oyster reefs. Ecology 94:895–903

Ware JR, Smith SV, Reaka-Kudla MK (1992) Coral reefs: sources or sinks of atmospheric CO_2? Coral Reefs 11:127–130

Winberg GG (1960) Rate of metabolism and food requirements of fishes. Transl Ser Fish Res Board Can 194:1–202

World Bank, Ecofys and Vivid Economics (2016) State and trends of carbon pricing 2016 (October). World Bank, Washington, DC

Chapter 13
Habitat Modification and Coastal Protection by Ecosystem-Engineering Reef-Building Bivalves

Tom Ysebaert, Brenda Walles, Judy Haner, and Boze Hancock

Abstract Reef-building bivalves like oysters and mussels are conspicuous ecosystem engineers in intertidal and subtidal coastal environments. By forming complex, three-dimensional structures on top of the sediment surface, epibenthic bivalve reefs exert strong bio-physical interactions, thereby influencing local hydro- and morphodynamics as well as surrounding habitats and associated species. The spatial impact of the ecosystem engineering effects of reef-building bivalves is much larger than the size of the reef. By influencing hydrodynamics oysters and mussels modify the sedimentary environment far beyond the boundaries of the reef, affecting morphological and ecological processes up to several hundreds of meters.

Being key-stone species in many coastal environments, reef-building bivalves are increasingly recognized for their role in delivering important ecosystem services that serve human wellbeing. Here we focus on two services, namely the regulating service coastal protection (coastal erosion prevention, shoreline stabilization) and the supporting habitat for species service (enhancement of biodiversity and diversification of the landscape). Due to their wave dampening effects, reef-building bivalve reefs are increasingly used for shoreline protection and erosion control along eroding coastlines, as an alternative to artificial shoreline hardening. The

T. Ysebaert (✉)
Wagenungen UR, Wageningen Marine Research, Wageningen University & Research, AB, Yerseke, The Netherlands

NIOZ Yerseke, Royal Netherlands Institute for Sea Research and Utrecht University, AC, Yerseke, The Netherlands
e-mail: tom.ysebaert@wur.nl

B. Walles
Wageningen UR, Wageningen Marine Research, Yerseke, The Netherlands
e-mail: brenda.walles@wur.nl

J. Haner
The Nature Conservancy, Coastal Programs Office, Mobile, AL, USA

B. Hancock
The Nature Conservancy, URI Graduate School of Oceanography, Narragansett, RI, USA
e-mail: bhancock@TNC.ORG

A. C. Smaal et al. (eds.), *Goods and Services of Marine Bivalves*,
https://doi.org/10.1007/978-3-319-96776-9_13

facilitative interactions at long-distances of bivalve reefs provide biodiversity benefits and more specifically facilitate or protect other valuable habitats such as intertidal flats, sea grasses, saltmarshes and mangroves.

Two case studies are used to demonstrate how bivalve reefs can be restored or constructed for shoreline protection and erosion control, thereby focusing on oyster reefs: (1) Oyster reefs for shoreline protection in coastal Alabama, USA, and (2) Oyster reefs as protection against tidal flat erosion, Oosterschelde, The Netherlands.

It is argued that bivalve reefs should be promoted as nature-based solutions that provide biodiversity benefits and coastal protection and help in climate change mitigation and adaptation. In order to successfully restore these habitats practitioners should consider a general framework in which habitat requirements, environmental setting and long-distance interdependence between habitats are mutually considered.

Abstract in Chinese 牡蛎和贻贝这类可以形成贝礁的双壳贝类是潮间带和潮下带海岸环境中出色的生态系统工程师。通过在沉积物表面上形成复杂的三维结构，浅海双壳贝礁发挥着强大的生物-物理作用，并影响当地的水文和形态动力学以及周围的生境和相关物种。造礁双壳类生态系统工程的空间影响远大于生物贝礁的本身尺度。牡蛎和贻贝可以通过影响贝礁体周边的水动力状态从而改变离贝礁本体较远区域的沉积环境，他们对底质形态和生态过程的影响范围可达礁体周边数百米。

作为诸多沿海环境中的关键物种，造礁双壳贝类担负着非常重要的生态系统服务功能，因此越来越受到人们的重视，。本文中我们着重于双壳贝类的两个生态服务，即海岸带保护的调节服务（预防海岸侵蚀，稳固海岸线）和栖息地维护服务（增强生物多样性和景观多样化）。作为人造海岸线硬化的一种替代方案，双壳贝礁越来越多地被用于易受波浪冲刷侵蚀岸线地带的防护。大规模的双壳贝礁有助于增加海岸带的生物多样性，且更好地保护了潮间带滩涂，海草，盐沼和红树林等宝贵的栖息地。

本文利用两个案例研究说明如何恢复或建造双壳贝礁以保护海岸线和控制侵蚀，以牡蛎礁为例：(1)在美国沿海阿拉巴马州用于海岸线保护的牡蛎礁；(2)在荷兰的东斯海尔德水道应对潮滩侵蚀的牡蛎礁。

一些观点认为应该推广双壳贝礁，他们可以作为提供生物多样性和沿海保护的自然方案，并可以帮助减缓和适应气候变化。为了成功恢复这些栖息地，相关人员应该考虑一个总体框架，在这个框架中，栖息地要求，环境背景和较远栖息地之间相互依赖性应当进行综合考量。

Keywords Ecosystem engineers · Oyster reefs · Ecosystem services · Coastal protection · Facilitation of habitats

关键词 生态工程 • 牡蛎礁 • 生态服务 • 海岸线保护 • 栖息地恢复

13.1 Bivalves as Ecosystem Engineers

Ecosystem engineers are organisms that have morphological and/or behavioral traits that enable them to modify, maintain and/or create habitats (Jones et al. 1994). They induce physical state changes in abiotic and biotic materials, thereby regulating the availability of resources to other organisms. This alters the availability of ecological niches to other species, facilitating certain species and inhibiting others (Bruno et al. 2003; Bouma et al. 2009). Jones et al. (1994, 1997) distinguished two types of ecosystem engineers. Autogenic ecosystem engineers change environments through their physical structures (e.g. vegetation, coral reefs), in other words they are part of the engineered habitat, whereas allogenic ecosystem engineers transform living or non-living materials from one physical state to another (e.g. beavers, bioturbating worms).

Bivalves are conspicuous ecosystem engineers that often have both autogenic and allogenic characteristics. In general, in soft-sediment environments, bivalves can be divided into epibenthic and endobenthic organisms depending on whether they spend most of their lifetime above or below the sediment, respectively. Many endobenthic bivalves such as cockles and clams modify the sedimentary habitat through their behavior and can be considered allogenic ecosystem engineers (Bouma et al. 2009). They affect a number of resources mainly through bioturbation and bioirrigation (Reise 2002; Ciutat et al. 2007; Montserrat et al. 2009), but in high densities they also can increase sediment stability (Donadi et al. 2013). The most prominent epibenthic ecosystem engineers inhabiting bare coastal sediments are reef building bivalves. The best-known examples of reef-building bivalves are intertidal and shallow subtidal mussels (*Mytilidae*) and oysters (*Ostreidae*). These epibenthic ecosystem engineers modify the sedimentary habitat mainly through their physical structures, and thus are true autogenic ecosystem engineers (Gutiérrez et al. 2003, Bouma et al. 2009). At the same time they are active filter feeders, removing large quantities of suspended material from the water column, influencing water clarity and quality, and producing faecal and pseudofaecal biodeposits that accumulate in the reef and its surroundings (Newell 2004; Kellogg et al. 2013). So, these epibenthic ecosystem engineers have both autogenic and allogenic properties.

Coastal engineers characterise structures such as coral and oyster reefs as low crested or submerged breakwaters and have modelled and validated the effect of these structures on wave height and current velocities (Van der Meer et al. 2005). However, the literature describing the impact of natural reefs on disrupting wave energy is much better developed for coral reefs than bivalve reefs (Sheppard et al. 2005; Rogers et al. 2013; Ferrario et al. 2014). This has included how reef parameters and morphology influence the physics of wave attenuation and flooding (Monismith 2007; Lowe et al. 2010; Quataert et al. 2015).

In this paper, we focus on the ecosystem engineering effects of intertidal and shallow subtidal reef-building bivalves in soft-sediment environments, thereby focusing on the bio-physical interactions these organisms have with their environment and two ecosystem services these interactions provide, (1) coastal protection (i.e. erosion control and shoreline stabilization) and (2) habitat for species by facilitation of other marine habitats and species (i.e. increasing biodiversity at large landscape scale). Using two case studies, one in the US and one in the Netherlands, we subsequently show how constructed oyster reefs contribute to shoreline and salt marsh protection or erosion control.

The shift in species composition at the spatial scale of the engineered habitat, this is within the footprint of the reef structure, is discussed in two other papers within this book, namely the effect on associated benthic macroinvertebrates (Craeymeersch and Jansen 2019) and the effect on finfish and crustacean production (Hancock and zu Ermgassen 2019).

13.2 Characteristics of Epibenthic, Reef–Building Bivalves

Epibenthic, reef-forming bivalves create spatially and topographically complex habitats and can be found in a wide range of spatial scales, from small clumps to large patches to extensive beds and reefs that cover thousands of square meters and extend kilometres in length (Gutiérrez et al. 2003, 2011) (Fig. 13.1). Bivalves attach

Fig. 13.1 Examples of epibenthic bivalve reefs: (top left) extensive mussel beds (*Mytilus edulis*) in the Wadden Sea, The Netherlands (K. Troost – WMR, (top right) extensive oyster reefs (*Crassostrea gigas*) in the Oosterschelde, Netherlands (T. Ysebaert – WMR), (bottom left) Fringing reef, Georgia, USA (M. Spalding – TNC), (bottom right) oyster reef, Georges Bay, Tasmania, Australia (C. Gillies – TNC)

themselves to the substrate or to each other via byssal threads (e.g. mussels like *Mytilus edulis*, *Musculista senhousia*) or via calcification, this is cementing themselves to the substrate or to each other (oysters like *Crassostrea gigas*, *Crassostrea virginica*, *Ostrea edulis*). This makes individual mussels more mobile compared to oysters, on the other hand reefs of mussels will be less persistent compared to oysters, whose structure will persist for a long time, even after dying off of the reef.

The often distinct spatial patterns observed in reef-building bivalves are caused by feedback loops and self-organization processes at different spatial scales and lead to complex habitat forms (Bertness and Grosholz 1985; van de Koppel et al. 2005, 2008). Mussels like the Blue mussel (*Mytilus edulis*) typically attach themselves to a hard substrate or to each other by forming byssal threads. To find protection or food, mussels can move by releasing the byssal threads and using its foot to move to a new location. Liu et al. (2014) demonstrated that in mussel beds, self-organization generates spatial patterns at two characteristic spatial scales: small-scale, net-shaped patterns due to individual movements (i.e. behavioural aggregation) of individuals, and large-scale, regular banded patterns due to the interplay between-mussel facilitation and long-distance competition for algae. The interplay between self-organizing processes at individual and ecosystem level is therefore a key determinant of the functioning and resilience of mussel beds (Liu et al. 2014). A decrease of water depth over blue mussel *Mytilus edulis* mounds results in an increase in water flow, enhancing food transport to the mussels on top of mounds. This locally lowers algal depletion resulting in higher net growth on top of the mounds (positive feedback loop) (Liu et al. 2012).

Oysters typically build reefs through the gregarious settlement of multiple generations onto existing oyster substrate. The shells provide a solid substrate on which new oyster larvae can attach, increasing larval survival by providing shelter from predators and preventing burial in sediment (Mann and Powell 2007). Each generation of larvae settle on top of the preceding generation so that the reef grows vertically and becomes highly complex, with many structural irregularities and infoldings (Dame 2005; Rodriguez et al. 2014). Once settled, oysters are less mobile compared to mussels. But also oyster reefs appear in distinct reef morphologies and spatial patterns, thought to arise from feedback mechanisms between oysters and local hydrodynamics (Lenihan 1999; Dame 2005; Colden et al. 2016). Colden et al. (2016) describe three different reef types or morphologies for the Eastern oyster (*Crassostrea virginica*) based on historical work by Grave (1905): string reefs, fringe reefs and patch reefs.

13.3 Interaction with the Local Environment

Reef-building bivalves add hard substrate to an otherwise soft, more unstable, and often relatively flat bottom in sedimentary environments. By forming complex, three-dimensional structures on top of the sediment surface, epibenthic bivalve reefs influence local hydrodynamics and sediment dynamics (Lenihan 1999; Gutiérrez et al. 2003). Epibenthic bivalves enhance near-bed turbulence and vertical mixing in the lower water column and slow down current velocities through increased roughness of the bed (Widdows et al. 2002). Epibenthic bivalve reefs, especially oyster reefs, are often mentioned to act as breakwaters that can attenuate waves (Grabowski and Peterson 2007; Scyphers et al. 2011), but empirical evidence is scarce. Flume studies by Borsje et al. (2011) and Manis et al. (2015) showed that oysters effectively reduce wave energy compared to bare sediment. Lunt et al. (2017) measured in the field wave height and current speed at the windward and leeward side of oyster (*Crassostrea virginica*) reefs in St Charles and Aransas Bays (Texas, USA), with average wave height (average: leeward 0.05 m, windward 0.22 m) and current speed (average: leeward 3.75 cm s^{-1}, windward 6.68 cm s^{-1}) highest on the windward side of oyster reefs.

Epibenthic bivalve reefs exert a strong effect on the benthic-pelagic coupling through their suspension feeding behavior, feeding on seston in the water column and transferring undigested organic and inorganic material in their faeces and pseudofaeces to the sediment surface (Dame 1996; Newell 2004). When abundant, bivalves can exert top-down control on the phytoplankton, thereby reducing turbidity and increasing water transparency (see Cranford 2019).

13.4 Ecosystem Engineers Offer Essential Ecosystem Services Including Coastal Protection and Habitat for Species

Coastal ecosystems are increasingly recognized as essential elements in ecosystem restoration and coastal adaptation (Cheong et al. 2013; Temmerman et al. 2013). The capacity of marshes and shellfish reefs to maintain their own habitat via biophysical feedbacks, and their ability to grow with sea-level rise is seen as an important advantage over traditional man-made hard engineering structures (Rodriguez et al. 2014; Kirwan et al. 2016; Walles et al. 2015b). Ecosystem engineers, in particular, are increasingly recognized for their role in delivering important ecosystem services that serve human wellbeing. In case of epibenthic bivalve reefs, ecosystem services include a myriad of provisioning, regulating, habitat and cultural services (Coen and Luckenbach 2000, Coen et al. 2007, Beck et al. 2009, Grabowski et al. 2012, and papers within this book). Here we focus on two services, namely the regulating service *coastal protection* (coastal erosion prevention, shoreline stabilization) and the supporting/habitat service *habitat for species* (enhancement of biodiversity and diversification of the landscape). The latter specifically deals with the extending effects outside the boundary of the reef structure itself. Figure 13.2 visualizes these ecosystem services in a cross-section along a shoreline.

In their natural setting, bivalve reefs attenuate wave energy and stabilize sediments, protecting adjacent habitats such as intertidal flats, sea grasses and salt

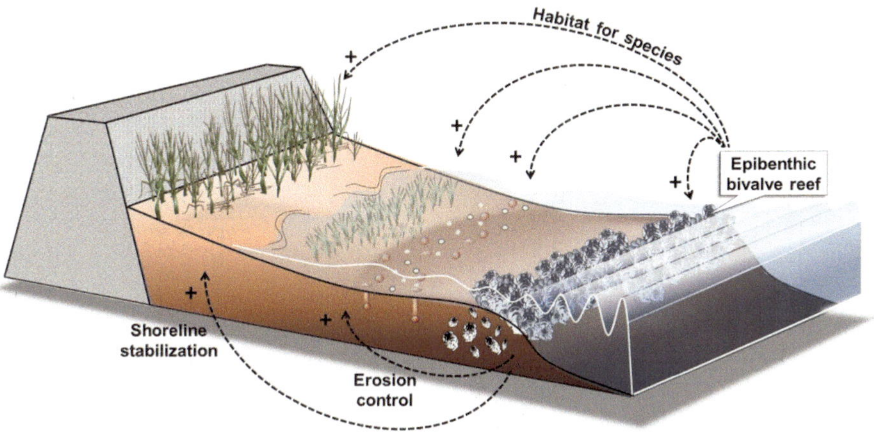

Fig. 13.2 Visualisation of the ecosystem services delivered by epibenthic bivalve reefs. Reefs provide coastal protection through erosion control and shoreline stabilization, and modify the physical landscape by ecosystem engineering, thereby providing habitat for species by facilitative interactions with other habitats such as tidal flat benthic communities, sea grasses and marshes

marshes against erosion. As coastal erosion is an increasing problem worldwide, due to human interventions, worsened by climate induced sea level rise and more frequent storms, shellfish reefs are therefore an attractive living shoreline approach that can be used for shoreline protection and erosion control along eroding coastlines, offering an alternative to artificial shoreline hardening (Meyer et al. 1997; Piazza et al. 2005; Scyphers et al. 2011; La Peyre et al. 2015; Walles et al. 2016a) (Fig. 13.3). For example, 30% of the oyster restoration projects in the US involve coastal protection as one of the targets (http://projects.tnc.org/coastal/). Besides relatively small restoration projects, increasingly large projects are in development. In Alabama there is a long range goal to construct 100 miles of oyster reef for shoreline protection and conservation of coastal marshes and seagrasses. This work has developed into large shoreline protection projects using oyster reef in all the US Gulf of Mexico states, valued at over $178 Million US (See case study 1). In the Netherlands, 1.3 km of oyster reefs were constructed in the Eastern Scheldt for erosion control of intertidal flats (Walles et al. 2016a, see case study 2).

Another important ecosystem service delivered by bivalve reefs is the provision of habitat(s) for species, leading to enhancement of biodiversity and diversification of the landscape. Indeed, the spatial impact of the ecosystem engineering effects of reefs is much larger than the size of the reef footprint. By influencing hydrodynamics oysters and mussels modify the sedimentary environment and physical landscape far beyond the boundaries of the reef, affecting morphological and ecological processes up to several hundreds of meters. Biodeposition of organic material in the form of faeces and pseudofaeces changes the sediment composition and enhance primary production of microphytobenthos in the surroundings (Newell 2004; Donadi et al. 2013; Engel et al. 2017). Walles et al. (2015a) demonstrated alteration of tidal flat morphology by Pacific oyster *Crassostrea gigas* reefs up to tens or

Fig. 13.3 Examples of reefs constructed for shoreline protection and erosion control. (**a**) Reefs constructed for shoreline protection at Swift Tract Alabama, USA (M-K Brown, TNC); (**b**) Swift Tract reef demonstrating wave attenuation (M-B Charles, TNC); (**c**) Reef grown on concrete domes for shoreline protection, Tampa Bay, Florida, USA (B. Hancock, TNC); (**d**) Concrete domes with 3 years of oyster growth, Tampa Bay, FL, USA (B. Hancock, TNC); (**e**) Reef consisting of gabions filled with oyster shells constructed for erosion control of an intertidal mudflat of Viane, Oosterschelde, The Netherlands (T. Ysebaert, WMR); (**f**) Reef at Viane after 3 years of oyster growth, Oosterschelde, The Netherlands (B. Walles, WMR)

hundreds of meters in the Oosterschelde estuary (The Netherlands). Nieuwhof et al. (2017) demonstrated increased water storage capacity on intertidal flats as a consequence of the enhanced engineering by shellfish reefs. In the Wadden Sea zonation of biological communities was observed as a consequence of the long-distance, cross-habitat interactions with mixed blue mussel *Mytilus edulis* and Pacific oyster *Crassostrea gigas* reefs (van de Koppel et al. 2015). These interactions were shown to be scale-dependent: at short distances cockle (*Cerastoderma edule*) abundance was suppressed, whereas at larger distances increased spatfall and better survival of adult cockles was observed due to reduced sediment erosion (Donadi et al. 2013). This in turn influenced higher trophic levels, such as shorebirds that feed upon cockles (Donadi et al. 2013, van der Zee et al. 2012). Other studies showed cross-habitat interactions between epibenthic, suspension feeding bivalves and seagrass habitats (Peterson and Heck 2001; Newell and Koch 2004) and marshes (Meyer et al. 1997; Piazza et al. 2005). Therefore, at the landscape level, ecosystem engineering typically enhances environmental heterogeneity, thereby increasing niche opportunities and eventually the diversity of communities and habitats such as sea grasses, marshes and mangroves (van de Koppel et al. 2015). The facilitation of these vegetated habitats in turn also deliver essential ecosystem services (Barbier et al. 2011) and coastal plant communities are increasingly recognized for their capacity for climate change adaptation and mitigation (Duarte et al. 2013). Marshes and mangroves overall have higher potential as natural defences for coastal protection, but restoration of bivalve reefs in front of these vegetated habitats can strengthen their growth and survival. Therefore, epibenthic bivalve reefs are as such not suitable for use as primary flood defences, due to their relatively low position in the intertidal zone, but indirectly, through their long-distance effects and facilitative interactions on other habitats, will add to the protection and flood defence of our coasts and improve the resilience of ecosystem-based coastal defence practices (Temmerman et al. 2013).

13.5 Study Case 1: Oyster Reefs for Shoreline and Salt Marsh Protection in Coastal Alabama, USA

In the northern Gulf of Mexico (USA), several coastal habitats have suffered declines of more than 50% over the last century, including seagrass, wetlands, and oyster reefs. Locally, Alabama has lost some 70% of its seagrass habitat, 85% of its oyster habitat, and thousands of acres of wetlands. Loss of nearshore habitats has been caused by decreased water quality, dredge-and-fill activities, construction of seawalls, jetties and groins among other causes, but in the case of oyster habitat mainly from overfishing (Beck et al. 2011). Increased erosion is due to increased ship wakes from the channel to the Alabama Port in Mobile as well as seasonal storms, sea level rise and climate change, and has been exacerbated by the loss of habitat such as oyster reef that can reduce wave energy before it reaches the coast. The declines, along with increasing appreciation of the value of the services these habitats provide, have prompted increased efforts to restore these habitats and ecosystem services. A goal of many habitat restoration projects has been shoreline protection (e.g. living shorelines, constructing oyster reef to act as breakwaters and beneficial use of dredge material) (Kroeger and Guannel 2014). Coastal Alabama hosts some advanced and well documented ecological studies that demonstrate the potential for restoring degraded habitats and enhancing many ecosystem services for community resilience, wave attenuation, shoreline protection, and fish production (Coen and Luckenbach 2000; Piazza et al. 2005; Grabowski and Peterson 2007; Gregalis et al. 2009; Powers et al. 2009; Scyphers et al. 2011; Grabowski et al. 2012; La Peyre et al. 2014; Scyphers et al. 2015; zu Ermgassen et al. 2016).

In 2009 the Obama administration provided funds for economic stimulus to counter the recession triggered by the real estate collapse in 2008 (https://www.treasury.gov/initiatives/recovery/Pages/recovery-act.aspx). Some of these funds were directed toward habitat restoration and used to move what had generally been 'proof of concept' scale work to an ecosystem scale. In Mobile Bay, AL, a coalition of partners led by The Nature Conservancy used this opportunity to implement large scale oyster habitat restoration directed primarily toward shoreline protection (Kroeger and Guannel 2014). This approach has been continued through the successful engagement of municipalities and communities across the Alabama coast, the development and implementation of watershed management plans (WMP) and additional shoreline protection projects. Socioeconomic studies of coastal Alabama residents have demonstrated an awareness of the decline of coastal habitats, a broad appreciation that they provide important ecosystem services, and substantial support for enhancing coastal resilience (Scyphers et al. 2014a, 2014b).

A substantial investment (~$50 M US dollar) has been made in more than 13 completed and 5 progressing restoration projects in coastal Alabama to the end of 2017 (Table 13.1). Data from 13 projects completed by multiple partners is currently being synthesised to determine which techniques have worked best for oyster recruitment, fisheries production and shoreline protection, including an economic valuation of the outcomes. At the time of writing, project monitoring has highlighted a need for some specific enhancements on the initial habitat restoration projects to improve functionality, as well as to add additional benefits beyond the original objectives. For these existing restoration projects, adaptive management strategies informed by long-term monitoring will be implemented to enhance these current investments and further improve future projects.

For example, in early 2011 following the Deepwater Horizon oil spill, at a site known as Helen Wood Park, 12 intertidal reefs were constructed to protect the

Table 13.1 Total RESTORE Act Investments in Oyster Reefs by State for the US Gulf of Mexico

State	Number of projects	Total $$ (approved/allocated/spent)	Total miles	Total acres
Florida	5	$17,330,718	9.95	4
Alabama	9	$53,829,415	10.13	0
Mississippi	4	$85,820,460	18.8	100
Louisiana[a]				
Texas	10	$16,183,000	8.21	50
TOTAL	28	$178,163,593	47.09	154

https://www.restorethegulf.gov/sites/default/files/FPL_forDec9Vote_Errata_04-07-2016.pdf; http://www.nfwf.org/gulf/pages/gulf-projects.aspx; http://www.alabamacoastalrestoration.org/NRDA/Prior-Announcements (Phase I, III and IV))

[a]Louisiana has substantial investment but data are not readily available

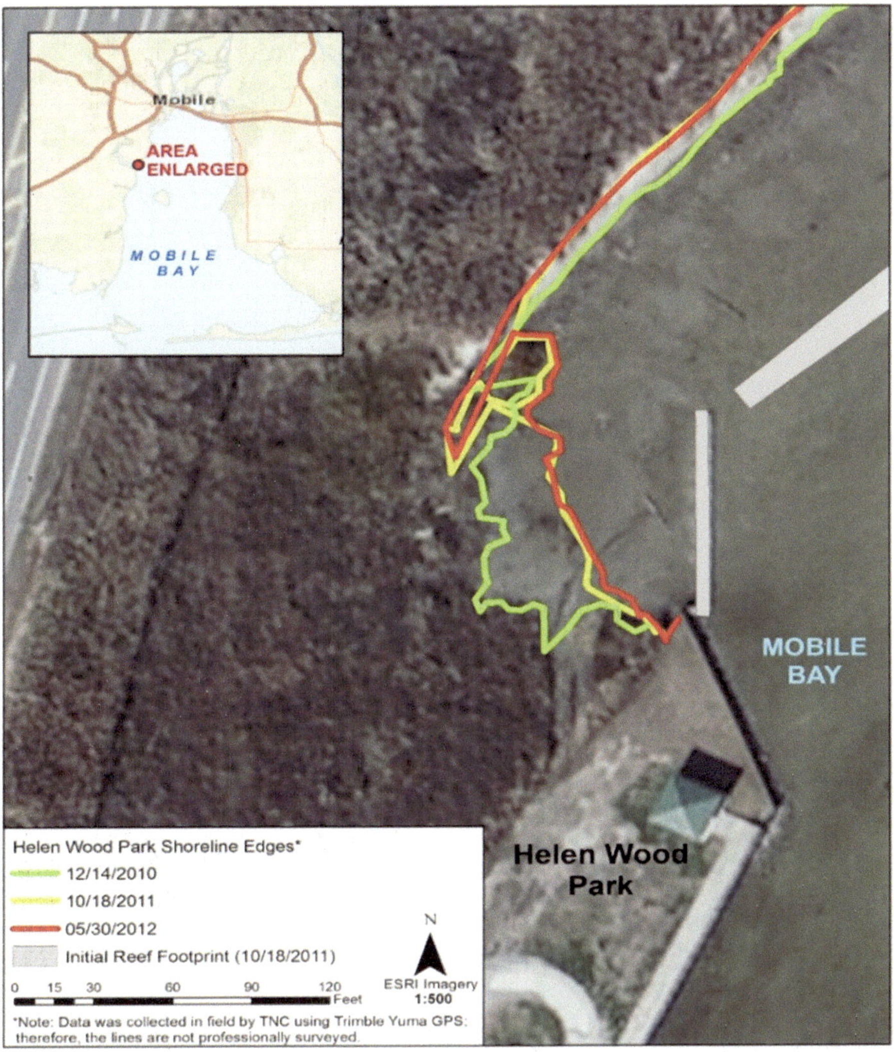

Fig. 13.4 Shoreline change at Helen Wood Park's south marsh December 2010, October 2011 and May 2012. The constructed oyster reefs are shown as grey rectangles

392 m shoreline and provide habitat. The site has been impacted by adjacent development (seawall) and has experienced significant erosion. The reefs measured ~25 m long and 2–3 m wide with 7.5 m gaps. A combination of natural and engineered materials was used, with 4 reefs constructed using concrete Reefballs, 7 reefs were constructed using bagged oyster shell, and 1 reef being partly made of Reefballs and partly bagged oyster shells. Shoreline position, reef footprints and recruitment of sessile organisms was monitored annually. Approximately one year post-construction, the marsh at the south end had expanded by some 7% and accumulation of sediment leeward of the structures had occurred (Fig. 13.4).

Fig. 13.5 Reef footprints at Helen Wood Park with purple lines showing as built outline in spring 2011. The green line (2014) shows changes after 3 years. The reefs measured ~25 m long and 2–3 m wide with 7.5 m gaps. The two reefs on the left are bagged oyster shells and the reef on the right made of Reefballs. The second reef on the right is partly made of Reefballs (right side) and partly bagged oyster shells (left side)

As monitoring continued, by year 3, the bagged shell reefs had noticeably broken down. Their footprint expanded as the shell spread, and the vertical structure had decreased (Fig. 13.5). While the use of bagged oyster shell was preferred for conspecific recruitment, as well as aesthetics, this substrate broke down after 3–4 years, providing reduced wave attenuation and shoreline protection (Unpublished data).

The long-term monitoring conducted at the Helen Wood Park site was used to inform subsequent projects to increase project success and longevity. For example, the 940 m Swift Tract project was completed in the summer of 2012. The construction technique utilized at the Swift Tract consisted of Hesco barriers, galvanized steel modular baskets that were installed and then filled with gabion stone. A 0.15 m layer of oyster shell was placed on top of the gabion stone within the cages. Pockets on the front and rear sides of the Hesco barriers ~0.15 m deep were filled with oyster shell. Five individual segments were constructed, each ~125 m long and 5.5 m wide, with 12 m gaps between. This hybrid technique used cages and rock to retain vertical structure and oyster shell for oyster and mussel settlement.

This shoreline at Swift Tract had been experiencing erosion, so one of the project objectives was to reduce wave energy impacting the shore. Historically (1957–

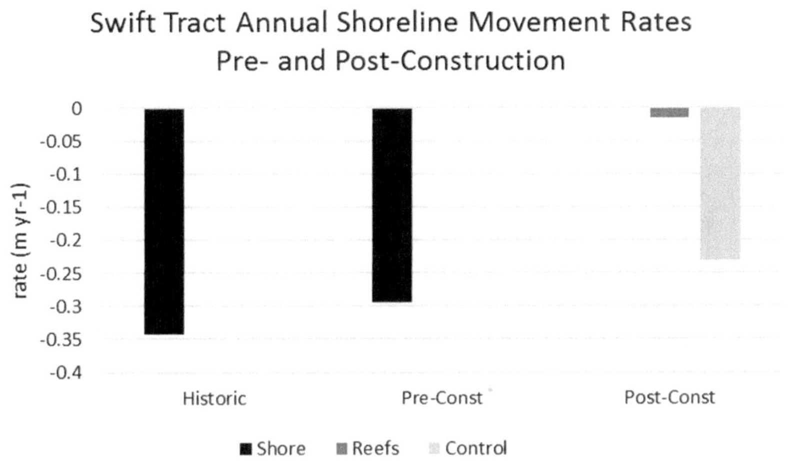

Fig. 13.6 Annual rate of shoreline loss at Swift Tract during two time periods prior to construction of reef structures, with the reduced rate of post-construction shoreline loss associated with the reefs compared to adjacent control sites

1981), just under 0.35 m of shoreline was lost each year. In the time immediately prior to reef construction (1981–2011), the rate of loss was approximately 0.3 m/yr. With 4 years of data following reef construction, the annual shoreline loss has dropped to ~0.02 m/yr. (Fig. 13.6). The materials used have not lost their vertical structure during the first 5 years of monitoring. It is possible that given a longer monitoring period (>4 years) we may find that the shoreline has stabilized.

While progress has been made toward reducing shoreline erosion in Alabama, a coordinated and comprehensive effort to restore shorelines is necessary to help maximize these benefits for both the communities and the natural habitats. Like traditional bulkheads, nature-based or hybrid shoreline restoration options are not one-size-fits-all. By developing a ready-to-implement plan for a range of individual sites, communities and residents will have a resource available to guide restoration activities that help make their communities more resilient, while also improving habitat and water quality. Broader implementation of sound restoration techniques at the place-based level can:

- improve coastal and community resilience by incorporating multiple individual efforts to maintain and improve municipal Community Rating System (CRS) rankings for coastal insurance incentives,
- enhance coastal habitat, such as marsh, to uptake nonpoint source nutrients and pathogens and improve water quality before runoff enters coastal tributaries,
- improve estuarine habitat for fish and shellfish, coastal birds and other wildlife.

These projects, with the benefit of the lessons learned and adaptive management, can provide long term benefits for the broader community by enhancing fisheries, providing recreational opportunities, improving water quality, retaining property and land values, and increasing community resilience. Collectively, these benefits contribute to a stronger economy and better quality of life.

13.6 Study Case 2: Oyster Reefs as Protection Against Tidal Flat Erosion, Oosterschelde, The Netherlands

In the Netherlands three intertidal oyster reefs were constructed in 2010 in the Oosterschelde to investigate their contribution to coastal protection, acting as a natural buffer for erosion control. The Oosterschelde is a coastal bay (tidal range at the study site 3 m, salinity 30 ppt) that suffers from eroding intertidal flats, due to infrastructural works (i.e. construction of a storm-surge barrier) in the 1980s (Nienhuis and Smaal 1994; Walles et al. 2016a). Tidal flats are eroding on average 1 cm. $yr.^{-1}$ in height, and it is predicted that more than half of the 11,000 hectares will be lost by the end of this century. This has consequences for both nature values as well as coastal protection, as the tidal flats are foraging grounds for an internationally important number of waders like Oystercatchers (*Haematopus ostralegus*), and at the same time protect the dikes against wave erosion.

The constructed reefs, made of 25 cm high gabions filled with Pacific oyster shells, provided substrate for settlement of new oyster recruits. To attenuate waves, the 200 m long reefs were positioned perpendicular to the dominant wave direction at three different elevations: 23%, 35% and 50% emersion time (Walles et al. 2016a) (Fig. 13.3). Over a 5-year period, the development of these reefs and their effect on tidal flat morphology (erosion/sedimentation) were monitored.

Reef development was related to vertical position in the intertidal zone. Recruitment rate, shell growth, and the condition of the oysters (*Crassostrea gigas*) were correlated with tidal emersion (Walles et al. 2016b). The reef positioned at 50% tidal emersion lacked sufficient settlement of new oyster recruits to maintain the reef structure. As such, the constructed reef shows a continuous deterioration. The reefs positioned at 35% and 23% tidal emersion had sufficient recruitment to maintain their structures. Oyster grew at both reefs, however a loss in weight and a low condition of the oysters on the reef at 35% tidal emersion indicated stress at this elevation (Fig. 13.7). Rodriguez et al. 2014 showed that reefs have an upper limit with respect to tidal range, a so-called growth ceiling above which reefs cannot grow as stress from limited inundation is too high. For natural reefs in the Oosterschelde the growth ceiling occurs above 60% tidal emersion (Walles et al. 2016b). Based on growth rates observed on the lowest reef, a vertical accretion of the reef base in the order of 7.0–16.9 mm $year^{-1}$ is expected (Walles et al. 2016a). This is an underestimation of the real reef growth as vertical accretion of the taphonomically active zone is much higher. Over the course of 5 years this reef increased on average 10 cm (unpublished data).

Measurements of waves windward and leeward of the reefs showed that wave attenuation is depending on the size of the incoming waves and the water depth over

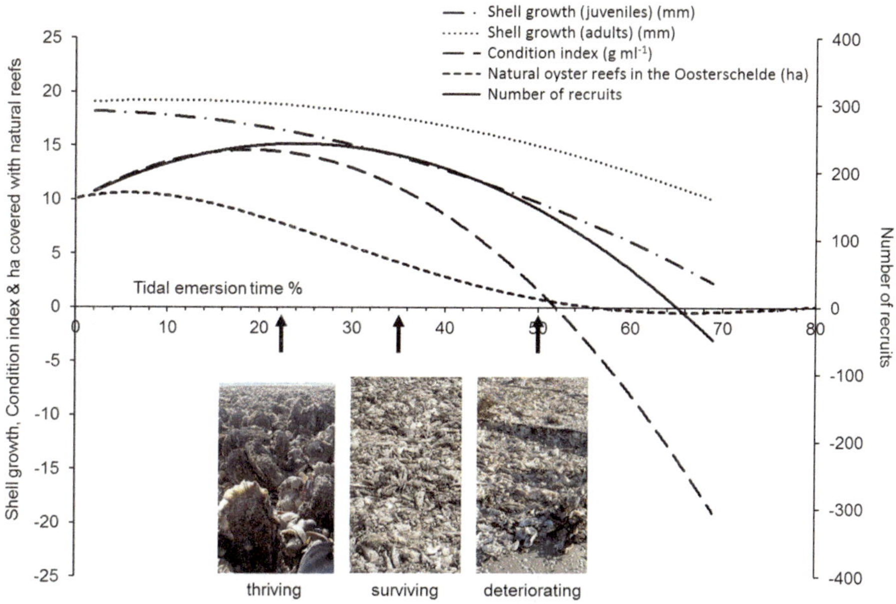

Fig. 13.7 The occurrence of Pacific Oyster reefs and the response curves of recruitment, shell growth condition index of oysters, and reef area, along a tidal emersion gradient in the Oosterschelde (adapted from Walles et al. 2016b). Pictures of the reefs positioned along the tidal emersion gradient show the different development stages of the reefs, with a thriving population at 23% tidal emersion and a deteriorating reef at 50% tidal emersion

the reef structure. Close to submergence, reefs potentially attenuate waves up to 95% of the incoming waves. For larger water depths, the attenuation depends on the wave height and wave length. We found that wave height leeward of the reef relates to water depth on top of the reef with a factor of 0.3. This is comparable to rubble mount breakwaters (Van der Meer et al. 2005). For example, for a water depth of 1 m on top of the reefs, waves smaller than 30 cm will not be influenced by the reef anymore.

Elevation measurements showed that the reef positioned at 23% tidal emersion, effectively reduced erosion leeward of the reef, as predicted (Walles et al. 2015a). Up to 90 m leeward of the reef there was a reduction of 51 ± 29% in the erosion measured. The reef however influences the morphology over a much larger area, up to 360 m leeward of the reef.

13.7 Management Applications and Considerations

There is growing interest in the restoration and conservation of epibenthic bivalve reefs for coastal protection and shoreline stabilization, also because of their potential to adapt to climate change impacts and sea level rise. This is evidenced by the growing number of projects and studies done worldwide, as well as the increasing funding for nature-based defences or living shoreline initiatives. Despite increasing interest in restoration, the success of bivalve reef restoration is variable (Coen et al. 2007; Mann and Powell 2007). Yet, there are few syntheses of information on what kind of projects meet the goals or are cost effective (Grabowski et al. 2012; Bayraktarov et al. 2016). Data from monitoring the success or failure of restoration projects are not always consistent or readily available, or projects lack sound monitoring efforts at all (Baggett et al. 2015). In order to properly develop guidelines and methodologies for project design, monitoring whether a restoration project designed for coastal protection (or another ecosystem service) has achieved its stated objectives is necessary and should be incorporated in each project setup. A general framework that considers both project success and cost-benefits of restoring epibenthic bivalve reefs for coastal protection or facilitation of other habitats is needed for each project. Therefore, it is necessary to consider carefully how to design and implement restoration projects given the stated goals and targeted ecosystem services. Understanding the interaction between reef-forming ecosystem engineers and the surrounding environment is of utmost importance and several aspects need to be considered when restoring or creating epibenthic bivalve reefs for coastal protection or facilitation of other habitats. *First*, using epibenthic bivalve reefs for coastal protection or facilitation of other habitats requires knowledge about the habitat requirements of the targeted bivalve species, or 'site suitability' for the selected species (La Peyre et al. 2015). Recruitment, growth and survival of epibenthic bivalves depend on many factors, including salinity, temperature, dissolved oxygen, hydrodynamics, tidal emersion and wave exposure (Baggett et al. 2015; Walles et al. 2016c; Theuerkauf et al. 2017). In addition, biological factors, such as availability of larvae, predation pressure and diseases will influence the long-term survival and sustainability of the constructed reef. This all determines whether a certain site is suitable for the construction of a bivalve reef.

Second, to get optimal effect on shoreline stabilization or facilitation of other habitats, reefs should be constructed at the right position in the intertidal or shallow-subtidal zone. Wave dampening by shellfish reefs depends on water depth (tidal range), wave height/length, and reef height/width. The latter implies that design parameters are also crucial aspects for successful restoration for shoreline protection. Based on the slope of the area the long-term distance of influence by the reef can be estimated. One should realise that there are limitations on the extent that oyster reefs offer shoreline protection or erosion control. When positioned properly, and with the right conditions, reefs can dampen waves considerably, interacting over long distances with the seabed, reducing erosion and facilitating neighbouring habitats and ecosystems such as tidal flats, seagrasses and salt marshes. In other cases, for instance when the bed slope is very steep or the tidal range large, the effect will be strongly reduced. Additionally, the role of reefs in stabilizing the substrate underneath their footprint is a currently under-appreciated service that helps in armouring the coast against erosion.

Thirdly, when one knows where a bivalve reef can potentially grow and act as coastal defence or habitat facilitator, the restoration itself also requires attention. Restoration of epibenthic bivalve reefs can be done in several ways. For oysters, typically some kind of substrate is offered on which oyster larvae can settle. Oyster larvae preferentially settle on conspecifics, so adding loose oyster shells is often used in low-dynamic, sheltered environments, whereas shells packed in bushels or gabions are needed in more dynamic situations. Often there is a shortage of oyster shells, and other substrates are being used such as reef balls, oyster castles, etc. (zu Ermgassen et al. 2016). For restoration of mussel beds, it has been proven that taking into account the behavioural self-organisation of mussels improved restoration success considerably (De Paoli et al. 2017).

13.8 Conclusions

In conclusion, epibenthic bivalve reefs are conspicuous ecosystem engineers that modify the soft-sediment environment in which they live to a great extent. Both their physical structure and suspension feeding activity strongly influence the neighbouring sedimentary environment up to 100 s of meters outside their own footprint. Wave attenuation by the reefs reduces erosion, facilitating other species and protecting habitats such as seagrasses and salt marshes. These ecosystem engineering effects of epibenthic bivalve reefs are increasingly recognized and bivalve reefs are promoted as nature-based solutions that provide biodiversity benefits, support many ecosystem services including coastal protection, and contribute to climate change mitigation and adaptation. Much has been learned about how bivalve reefs function and deliver these services, which has accelerated growth and interest in bivalve reef restoration projects ((zu Ermgassen et al. 2016). Objectives for restoration of these reefs should be framed on the basis of the desired ecosystem services provision. This requires a general framework in which habitat requirements of the considered species, environmental setting and long-distance interdependence between habitats are mutually considered (zu Ermgassen et al. 2016).

The coast cannot be restored to its historic condition, but innovation and coordination of techniques and projects can enhance ecosystem function and quality of life and make coastal communities more resilient in the decades to come.

Acknowledgements This work was supported by the Netherlands Organisation for Scientific Research (NWO) via the project "EMERGO – Ecomorphological functioning and management of tidal flats" (850.13.020). The authors are grateful to dr M. Baptist and Dr. B Borsje for reviewing this paper.

References

Baggett LP, Powers SP, Brumbaugh RD, Coen LD, Deangelis BM, Greene JK, Hancock BT, Morlock SM, Allen BL, Breitburg DL, Bushek D (2015) Guidelines for evaluating performance of oyster habitat restoration. Restor Ecol 23:737–745

Barbier EB, Hacker SD, Kennedy C, Koch EW, Stier AC, Silliman BR (2011) The value of estuarine and coastal ecosystem services. Ecol Monogr 81:169–193

Bayraktarov E, Saunders MI, Abdullah S, Mills M, Beher J, Possingham HP, Mumby PJ, Lovelock CE (2016) The cost and feasibility of marine coastal restoration. Ecol Appl 26:1055–1074. https://doi.org/10.1890/15-1077

Beck MW, Brumbaugh RD, Airoldi L, Carranza A, Coen LD, Crawford C, Defeo O, Edgar GJ, Hancock B, Kay MC, Lenihan HS, Luckenback MW, Toropova CL, Zhang G (2009) Shellfish reefs at risk: a global analysis of problems and solutions. The Nature Conservancy, Arlington

Beck MW, Brumbaugh RD, Airoldi L, Carranza A, Coen LD, Crawford C, Defeo O, Edgar GJ, Hancock B, Kay MC, Lenihan HS, Luckenbach MW, Toropova WL, Zhang G, Guo X (2011) Oyster reefs at risk and recommendations for conservation, restoration, and management. Bioscience 61:107–116

Bertness MD, Grosholz E (1985) Population dynamics of the ribbed mussel, *Geukensia demissa*: the costs and benefits of an aggregated distribution. Oecologia 67:192–204

Borsje BW, van Wesenbeeck BK, Dekker F, Paalvast P, Bouma TJ, van Katwijk MM, de Vries MB (2011) How ecological engineering can serve in coastal protection. Ecol Eng 37:113–122

Bouma TJ, Olenin S, Reise K, Ysebaert T (2009) Ecosystem engineering and biodiversity in coastal sediments: posing hypotheses. Helgol Mar Res 63:95–106

Bruno JF, Stachowicz JJ, Bertness MD (2003) Inclusion of facilitation into ecological theory. Trends Ecol Evol 18:119–125

Cheong S-M, Silliman B, Wong PP, van Wesenbeeck B, Kim C-K, Guannel G (2013) Coastal adaptation with ecological engineering. Nat Clim Chang 3:787–791

Ciutat A, Widdows J, Pope ND (2007) Effect of *Cerastoderma edule* density on near-bed hydrodynamics and stability of cohesive muddy sediments. J Exp Mar Biol Ecol 346:114–126

Coen LD, Luckenbach MW (2000) Developing success criteria and goals for evaluating oyster reef restoration: ecological function or resource exploitation? Ecol Eng 15:323–343

Coen LD, Brumbaugh RD, Bushek D, Grizzle R, Luckenbach MW, Posey MH, Powers SP, Tolley SG (2007) Ecosystem services related to oyster restoration. Mar Ecol-Prog Ser 341:303–307

Colden AM, Fall KA, Cartwright GM, Friedrichs CT (2016) Sediment suspension and deposition across restored oyster reefs of varying orientation to flow: implications for restoration. Estuar Coasts 39:1435–1448

Craeymeersch J, Jansen H (2019) Bivalve shellfish assemblages as hotspots for biodiversity. In: Smaal et al (eds) Goods and services of marine bivalves. Springer, Cham, pp 275–294

Cranford P (2019) Magnitude and extent of water clarification services provided by bivalve suspension feeding. In: Smaal et al (eds) Goods and services of marine bivalves. Springer, Cham, pp 119–141

Dame RF (1996) Ecology of marine bivalves: an ecosystem approach. CRC Press, Boca Raton 254 pp

Dame RF (2005) Oyster reefs as complex ecological systems. In: Dame RF, Olenin S (eds) The comparative roles of suspension-feeders in ecosystems, NATO science series IV: earth and environmental series, vol 47. Springer, Dordrecht

De Paoli H, Van der Heide T, Van den Berg A, Silliman BR, Herman PMJ, Van de Koppel J (2017) Behavioral self-organization underlies the resilience of a coastal ecosystem. Proc Natl Acad Sci U S A. https://doi.org/10.1073/pnas.1619203114

Donadi S, van der Heide T, van der Zee EM, Eklöf JS, de Koppel JV, Weerman EJ, Piersma T, Olff H, Eriksson BK (2013) Cross-habitat interactions among bivalve species control community structure on intertidal flats. Ecology 94:489–498

Duarte CM, Losada IJ, Hendriks IE, Mazarrasa I, Marba N (2013) The role of coastal plant communities for climate change mitigation and adaptation. Nat Clim Chang 3:961–968

Engel FG et al (2017) Mussel beds are biological power stations on intertidal flats. Estuar Coast Shelf Sci 191:21–27

Ferrario F, Beck MW, Storlazzi CD, Micheli F, Shepard CC, Airoldi L (2014) The effectiveness of coral reefs for coastal hazard risk reduction and adaptation. Nat Commun 5:3794. https://doi.org/10.1038/ncomms4794

Grabowski JH, Peterson CH (2007) Restoring oyster reefs to recover ecosystem services. In: Cuddington K, Byers JE, Wilson WG, Hastings A (eds) Theoretical ecology series. Academic, Burlington, pp 281–298

Grabowski JH, Brumbaugh RD, Conrad RF, Keeler AG, Opaluch JJ, Peterson CH, Piehler MF, Powers SP, Smyth AR (2012) Economic valuation of ecosystem services provided by oyster reefs. Bioscience 62:900–909

Grave C (1905) Investigation for the promotion of the oyster industry of North Carolina. In: U.S. fish commission report for 1903, U.S. Commission for Fish and Fisheries, pp 247–341

Gregalis KC, Johnson MW, Powers SP (2009) Restored oyster reef location and design affect responses of resident and transient Fish, Crab, and Shellfish species in Mobile Bay, Alabama. Trans Am Fish Soc 138(2):314–327. https://doi.org/10.1577/T08-041.1

Gutiérrez JL, Jones CG, Strayer DL, Iribarne OO (2003) Mollusks as ecosystem engineers: the role of Shell production in aquatic habitats. Oikos 101:79–90

Gutiérrez JL, Jones CG, Byers JE, Arkema KK, Berkenbusch K, Committo JA, Duarte CM, Hacker SD, Hendriks IE, Hogarth PJ, Lambrinos JG, Palomo MG, Wild C (2011) Physical ecosystem engineers and the functioning of estuaries and coasts. In: Wolanski E, McLusky D (eds) Treatise on estuarine and coastal science. Elsevier, Amsterdam, pp 53–81

Hancock B, zu Ermgassen P (2019) Enhanced production of finfish and large crustaceans by bivalve reefs. In: Smaal et al (eds) Goods and services of marine bivalves. Springer, Cham, pp 295–312

Jones CG, Lawton JH, Shachak M (1994) Organisms as ecosystem engineers. Oikos 69:373–386

Jones CG, Lawton JH, Shachak M (1997) Positive and negative effects of organisms as physical ecosystem engineers. Ecology 78:1946–1957

Kellogg ML, Cornwell JC, Owens MS, Paynter KT (2013) Denitrification and nutrient assimilation on a restored oyster reef. Mar Ecol Prog Ser 480:1–19. https://doi.org/10.3354/meps10331

Kirwan ML, Temmerman S, Skeehan EE, Guntenspergen GR, Fagherazzi S (2016) Overestimation of marsh vulnerability to sea level rise. Nat Clim Chang 6:253–260

Kroeger T, Guannel G (2014) Fishery enhancement and coastal protection services provided by two restored Gulf of Mexico oyster reefs. In: Ninan K (ed) Valuing ecosystem services-methodological issues and case studies. Edward Elgar, Cheltenham, pp 334–357 464 pp

La Peyre MK, Humphries AT, Casas SM, La Peyre JF (2014) Temporal variation in development of ecosystem services from environmental management oyster reef restoration. Ecol Eng 63:34–44. https://doi.org/10.1016/j.ecoleng.2013.12

La Peyre MK, Serra K, Joyner TA, Humphries A (2015) Assessing shoreline exposure and oyster habitat suitability maximizes potential success for sustainable shoreline protection using restored oyster reefs. PeerJ 3:e1317

Lenihan HS (1999) Physical-biological coupling on oyster reefs: how habitat structure influences individual performance. Ecol Monogr 69:251–275

Liu Q-X, Weerman EJ, Herman PMJ, Olff H, van de Koppel J (2012) Alternative mechanisms alter the emergent properties of self-organization in mussel beds. Proc R Soc B Biol Sci 279:2744–2753

Liu QX, Herman PMJ, Mooij WM, Huisman J, Scheffer M, Olff H, Van de Koppel J (2014) Pattern formation at multiple spatial scales drives the resilience of mussel bed ecosystems. Nat Commun 5:5234. https://doi.org/10.1038/ncomms6234

Lowe RJ, Hart C, Pattiaratchi CB (2010) Morphological constraints to wave-driven circulation in coastal reef-lagoon systems: a numerical study. J Geophys Res 115:C09021. https://doi.org/10.1029/2009JC005753

Lunt J, Reustle J, Smee DL (2017) Wave energy and flow reduce the abundance and size of benthic species on oyster reefs. Mar Ecol Prog Ser 569:25–36
Manis JE, Garvis SK, Jachec SM, Walters LJ (2015) Wave attenuation experiments over living shorelines over time: a wave tank study to assess recreational boating pressures. J Coast Conserv 19:1–11
Mann R, Powell EN (2007) Why oyster restoration goals in the Chesapeake Bay are not and probably cannot be achieved. J Shellfish Res 26:905–917
Meyer DL, Townsend EC, Thayer GW (1997) Stabilization and erosion control value of oyster cultch for intertidal marsh. Restor Ecol 5:93–99
Monismith SG (2007) Hydrodynamics of coral reefs. Annu Rev Fluid Mech 39:37–55. https://doi.org/10.1146/annurev.fluid.38.050304.092125
Montserrat F, Van Colen C, Provoost P, Milla M, Ponti M, Van den Meersche K, Ysebaert T, Herman PMJ (2009) Sediment segregation by biodiffusing bivalves. Estuar Coast Shelf Sci 83:379–391
Newell RIE (2004) Ecosystem influences of natural and cultivated populations of suspension-feeding bivalve molluscs: a review. J Shellfish Res 23:51–61
Newell RIE, Koch EW (2004) Modeling seagrass density and distribution in response to changes in turbidity stemming from bivalve filtration and seagrass sediment stabilization. Estuaries 27:793–806
Nienhuis PH, Smaal AC (1994) Oosterschelde estuary (the Netherlands): a case-study of a changing ecosystem. Kluwer Academic Publishers, Dordrecht
Nieuwhof S, van Belzen J, Oteman B, van de Koppel J, Herman PMJ, van der Wal D (2017) Shellfish reefs increase water storage capacity on intertidal flats over extensive spatial scales. Ecosystems. https://doi.org/10.1007/s10021-017-0153-9
Peterson BJ, Heck KL (2001) Positive interactions between suspension-feeding bivalves and seagrass: a facultative mutualism. Mar Ecol Prog Ser 213:143–155
Piazza BP, Banks PD, La Peyre MK (2005) The potential for created oyster shell reefs as a sustainable shoreline protection strategy in Louisiana. Restor Ecol 13:499–506
Powers S, Peterson C, Grabowski J, Lenihan H (2009) Success of constructed oyster reefs in no-harvest sanctuaries: implications for restoration. Mar Ecol Prog Ser 389:159–170 http://www.jstor.org/stable/24873610
Quataert E, Storlazzi C, van Rooijen A, Cheriton O, van Dongeren A (2015) The influence of coral reefs and climate change on wave-driven flooding of tropical coastlines. Geophys Res Lett 42:2015GL064861. https://doi.org/10.1002/2015GL064861
Reise K (2002) Sediment mediated species interactions in coastal waters. J Sea Res 48:127–141
Rodriguez AB, Fodrie FJ, Ridge JT, Lindquist NL, Theuerkauf EJ, Coleman SE, Grabowski JH, Brodeur MC, Gittman RK, Keller DA, Kenworthy MD (2014) Oyster reefs can outpace sea-level rise. Nat Clim Chang 4:493–497
Rogers JS, Monismith SG, Feddersen F, Storlazzi CD (2013) Hydrodynamics of spur and groove formations on a coral reef. J Geophys Res Ocean 118:3059–3073. https://doi.org/10.1002/jgrc.20225
Scyphers SB, Powers SP, Heck KL, Byron D (2011) Oyster reefs as natural breakwaters mitigate shoreline loss and facilitate fisheries. PLoS One 6:e22396
Scyphers SB, Picou JS, Brumbaugh RD, Powers SP (2014a) Integrating societal perspectives and values for improved stewardship of a coastal ecosystem engineer. Ecol Soc 19(3):38. https://doi.org/10.5751/ES-06835-190338
Scyphers SB, Powers SP, Heck KL Jr (2014b) Ecological value of submerged breakwaters for habitat enhancement on a residential scale. Environ Manag. https://doi.org/10.1007/s00267-014-0394-8
Scyphers SB, Powers SP, Heck KL (2015) Ecological value of submerged breakwaters for habitat enhancement on a residential scale. Environ Manag 55:383–391. https://doi.org/10.1007/s00267-014-0394-8
Sheppard C, Dixon DJ, Gourlay M, Sheppard A, Payet R (2005) Coral mortality increases wave energy reaching shores protected by reef flats: examples from the Seychelles. Estuar Coast Shelf Sci 64:223–234. https://doi.org/10.1016/j.ecss.2005.02.016

Temmerman S, Meire P, Bouma TJ, Herman PMJ, Ysebaert T, De Vriend HJ (2013) Ecosystem-based coastal defence in the face of global change. Nature 504:79–83

Theuerkauf SJ, Eggleston DB, Puckett BJ, Theuerkauf KW (2017) Wave exposure structures oyster distribution on natural intertidal reefs, but not on hardened shorelines. Estuar Coasts 40:376–386

van de Koppel J, Rietkerk M, Dankers N, Herman PMJ (2005) Scale dependent feedback and regular spatial patterns in young mussel beds. Am Nat 165:E66–E77

van de Koppel J, Gascoigne JC, Theraulaz G, Rietkerk M, Mooij WM, Herman PMJ (2008) Experimental evidence for spatial self-organization and its emergent effects in mussel bed ecosystems. Science 322:739–742

van de Koppel J, van der Heide T, Altieri AH, Eriksson BK, Bouma TJ, Olff H, Silliman BR (2015) Long-distance interactions regulate the structure and resilience of coastal ecosystems. Annu Rev Mar Sci 7:139–158

Van der Meer JW, Briganti R, Zanuttigh B, Wang B (2005) Wave transmission and reflection at low-crested structures: design formula, oblique wave attack and spectral change. Coast Eng 52:915–929

Walles B, Salvador De Paiva J, van Prooijen B, Ysebaert T, Smaal A (2015a) The ecosystem engineer *Crassostrea gigas* affects tidal flat morphology beyond the boundary of their reef structures. Estuar Coasts 38:941–950

Walles B, Mann R, Ysebaert T, Troost K, Herman PMJ, Smaal A (2015b) Demography of the ecosystem engineer *Crassostrea gigas*, related to vertical reef accretion and reef persistence. Estuar Coast Shelf Sci 154:224–233

Walles B, Troost K, van den Ende D, Nieuwhof S, Smaal AC, Ysebaert T (2016a) From artificial structures to self-sustaining oyster reefs. J Sea Res 108:1–9

Walles B, Smaal AC, Herman PMJ, Ysebaert T (2016b) Niche dimension differs among life-history stages of Pacific oysters in intertidal environments. Mar Ecol Prog Ser 562:113–122

Walles B, Fodrie FJ, Nieuwhof S, Jewell OJW, Herman PMJ, Ysebaert T (2016c) Guidelines for evaluating performance of oyster habitat restoration should include tidal emersion: reply to Baggett et al. Restor Ecol 24(1):4–7

Widdows J, Lucas JS, Brinsley MD, Salkeld PN, Staff FJ (2002) Investigation of the effects of current velocity on mussel feeding and mussel bed stability using an annular flume. Helgol Mar Res 56:3–12

Zee E, Heide T, Donadi S, Eklöf J, Eriksson B, Olff H, Veer H, Piersma T (2012) Spatially extended habitat modification by intertidal reef-building bivalves has implications for consumer-resource interactions. Ecosystems 15:664–673

zu Ermgassen P, Hancock B, DeAngelis B, Greene J, Schuster E, Spalding M, Brumbaugh R (2016) Setting objectives for oyster habitat restoration using ecosystem services: a manager's guide. The Nature Conservancy, Arlington. 76pp. Available at http://oceanwealth.org/tools/oyster-calculator/

Chapter 14
Bivalve Assemblages as Hotspots for Biodiversity

J. A. Craeymeersch and H. M. Jansen

Abstract Many bivalve species occur in aggregations, and locally cover large parts of the seafloor. Above a certain density they provide a distinct, three-dimensional structure and the aggregations are called bivalve beds or reefs. These persistent aggregations form a biogenic habitat for many other species. Bivalve beds, therefore, often have, in comparison with the surrounding areas, a high biodiversity value and can be seen as hotspots for biodiversity. Bivalve have a wide global distribution, on rocky and sedimentary coasts. Different processes and mechanisms influence the presence of associated benthic fauna. This paper reviewed the main drivers that influence the biodiversity, such as the bivalve species involved, the density, the size and the age of the bed, the depth or height in the tidal zone and the substratum type.

Bivalve beds not only occur naturally in many subtidal and intertidal areas around the world, but mussels and oysters are also extensively cultured. Addition of physical cultivation structures in the water column or on the bottom allows for development of substantial and diverse communities that have a structure similar to that of natural beds. Dynamics of culture populations may however differ from natural bivalve reefs as a result of culture site and/or maintenance and operation like harvesting of the bivalve cultures. We used the outcome of the review on the drivers for wild assemblages to evaluate trade-offs between bivalve aquaculture and biodiversity conservation. Studies comparing natural and cultured assemblages proved to allow for a better understanding of the effect of the culture strategies and, consequently, to forward sustainable bivalve cultures. This is illustrated by a case study in the Dutch Wadden Sea.

Abstract in Chinese 多数双壳类是群聚的，在这些贝类出现的区域，通常会覆盖海底的大部分地区。当种群数量超过一定密度时，双壳贝类会形成一种独

J. A. Craeymeersch (✉)
Wageningen UR, Wageningen Marine Research, Yerseke, The Netherlands
e-mail: johan.craeymeersch@wur.nl

H. M. Jansen
Institute of Marine Research (IMR), Bergen, Norway

Wageningen UR – Wageningen Marine Research (WMR), Yerseke, The Netherlands
e-mail: henrice.jansen@wur.nl

A. C. Smaal et al. (eds.), *Goods and Services of Marine Bivalves*,
https://doi.org/10.1007/978-3-319-96776-9_14

特的三维结构聚合体，这种聚合体被称为双壳贝床或贝礁，这些聚合体为多其他物种提供栖息地。因此，与周边地区相比，双壳贝床具有很高的生物多样性价值，可以被看作是生物多样性的热点区域。双壳贝类是全球性物种，在岩石底质和沉积海岸地区均有广泛分布。不同双壳贝床的形成过程和机制会影响相关的底栖动物群落结构。本文综述了影响双壳贝床/礁生物多样性的主要驱动因素，包括贝类的种类，密度，贝床/礁的大小和年龄，在潮间带所处的深度和高度以及底质类型等。

双壳贝床不仅自然分布于世界各地的潮间带和潮下带，而且还广泛用于养殖，例如贻贝和牡蛎等。在水体内或海底投放的人工养殖结构可以形成与天然贝床结构类似的多种多样的生物群体。但是由于养殖过程中的养殖区域选择、日常维护、收获等人为干扰因素，养殖的双壳贝类的种群动力学过程与自然贝礁并不相同。我们利用野生贝礁作为驱动因素对双壳贝类养殖与生物多样性保护之间的权衡进行了评估。自然和养殖情况下环境状况的对比研究有助于更好地摸清养殖活动的环境效应，研究结果对于贝类养殖业的可持续发展具有非常重要的指导意义。本文将通过荷兰瓦登海的一个研究实例进行说明。

Keywords Bivalves · Biodiversity · Natural beds · Reefs · Aquaculture · Wadden Sea

关键词 双壳贝类 • 生物多样性 • 自然贝床 • 贝礁 • 水产养殖 • 瓦登海

14.1 Introduction

14.1.1 Background

Ecosystem services have become a key focus in resource management, conservation planning and environmental decision analysis. Biodiversity itself is valued by humans in many ways for the key ecosystem services it provides, and thus is important to include in any assessment that seeks to identify and quantify the value of ecosystems to humans (http:fws-case-12nmsu.edu/CASE/ES). Although in some cases weak no or even negative correlations were found between biodiversity and ecosystem services (Manhaes et al. 2016), evidence is growing that biodiversity supports ecosystem services delivery. Worm et al. (2006), for instance, found positive relationships between diversity and ecosystem functions and services. High-diversity systems consistently provide more services with less variability and, thus, species diversity has a buffering impact on the resistance and recovery of ecosystem services. Moreover, the authors did not find evidence for redundancy at high levels of biodiversity: the improvement of services (such as fished taxa richness and productivity in catch) was continuous on a log-linear scale. These results fit into the predictions of competition theory that greater diversity leads to greater ecosystem stability and lower species stability, among others due to the so-called portfolio effect (Tilman et al. 2006). Thus, hotspots of biodiversity – i.e. areas with a relatively high biodiversity value (Johnson 2013) – are likely to provide many ecosystem services.

Several bivalve species occur in aggregations, and locally cover large parts of the surface. Above a certain density they provide a distinct, three-dimensional structure and the aggregations are called bivalve beds (Cohen et al. 2007). Mytilid mussels form aggregations by attaching byssal threads to the substratum and conspecifics (Buschbaum et al. 2009; Ysebaert et al. 2009; Lancaster et al. 2014). Oysters are another important group of aggregating species living attached to hard substrata, including living and old shells and conspecifics. The larvae get attached to the substratum by a kind of 'cement' produced by a gland in the food (Walne 1974). These persistent aggregations form – in contrast to aggregations of more mobile bivalve species such as sea scallops (Brocken and Kenchingon 1999) - a biogenic habitat for many other species. Bivalve beds, therefore, often have a relatively high biodiversity value compared to surrounding areas and can be seen as hotspot for biodiversity (Bruno et al. 2003; Johnson 2013). Indeed, several authors report that mussel beds on rocky shores and sedimentary coasts harbour more diverse communities than surrounding rock or tidal flats (see e.g. Buschbaum et al. 2009 and the references therein). The magnitude differs depending on a number of biological, ecological or bio-geographical aspects.

Bivalves have a wide global distribution, on rocky and sedimentary coasts, and not only abundance of wild populations is significant, but also of cultured stocks. Though aquaculture is frequently judged for its ecological impacts, it is increasingly recognized that cultured bivalve stocks can also provide a variety of ecosystem services. From a biodiversity perspective, fisheries of natural bivalve stocks can negatively impact biodiversity, while at the same time biodiversity can be high at culture plots or suspended longlines, suggesting that ecosystems may also benefit from aquaculture activities.

14.1.2 Scope and Aim of Review

Biodiversity, and the associated ecosystem services, are not only provided by natural bivalve assemblages but also by aquaculture communities. To assess the role of bivalve aquaculture in biodiversity conservation it is essential to understand the drivers that determine settlement and succession of associated species on bivalve beds. On the one hand are drivers linked to natural processes in each cultivation area (i.e. geographical location, water temperature, depth etc), on the other hand cultivation activities (i.e. seed collection, relay/resocking, harvest, predation control) may also interfere with biodiversity succession. Studies on biodiversity development on aquaculture structures often have a limited temporal resolution, we therefore evaluate the natural biodiversity drivers for wild assemblages and use this to evaluate trade-offs between bivalve aquaculture and biodiversity conservation. The final section of this chapter presents a case study from the Dutch Wadden Sea where the effects of both mussel seed fisheries and bottom cultivation on biodiversity reduction and/or stimulation were evaluated on a bay wide scale.

14.2 Drivers for Biodiversity in Natural Bivalve Assemblages

Persistent bivalve beds are highly structured compared to the surrounding areas, physically change the environment and, thus, create unique habitats (Buschbaum et al. 2009; Ysebaert et al. 2009). Different processes and mechanisms influence the presence of associated benthic fauna. Relative importance of each mechanism will determine the (combined) outcome of the ecosystem engineering effect of the mussels (Ysebaert et al. 2009).

In sedimentary environments, epibenthic bivalve beds provide a major hard substratum on the sediment surface (Buschbaum et al. 2009). Biogenic habitat also offers shelter and predator refuge for mobile epibenthos, which might be predators of the mussels themselves (e.g. crabs and starfish), and are thus also attracted by the mussels as prey (Beadman et al. 2004; Ysebaert et al. 2009). In the intertidal zone, the complex structure provides refuge from tidal stress, and the habitat created is much cooler and more humid than elsewhere during low tides (Cole 2010; Arribas et al. 2013; Jungerstam et al. 2014).

Biodeposition caused by the bivalves will locally change the sediment composition, due to an enrichment of the sediment. As a result, several studies observed a decline in polychaetes, or shift from a community dominated by polychaetes to one dominated by oligochaetes (Commito and Boncavage 1989; Dittmann 1990; Ragnarsson and Raffaelli 1999; Ysebaert et al. 2009) within the bivalve beds. The increased supply of mussel deposits and organic matter may also be an additional food source within bivalve patches: the associated fauna depends on mussel deposition for 24 to 31% of its energy demand (Norling and Kautsky 2007). As a result, biodeposition may have an additional positive effect on diversity (Buschbaum et al. 2009). Thus, bivalve beds provide ecosystem services to the benthic community beyond the physical habitat provided by shells alone (Spooner and Vaughn 2006).

Activity by bivalves themselves might influence the settlement of other species. Dense suspension-feeding bivalves reduce the probability of successful larval settlement by any larvae, including their own. Several authors hypothesize that infaunal species that form cocoons, brood, fragment asexually, or disperse at large postlarval stages may be relatively more abundant in mussel beds than species with planktonic larval dispersal, although this enhancement might also simply be related to the higher spatial complexity of the bivalve bed (Dittmann 1990; Dolmer 2002; Thiel and Ullrich 2002; Ysebaert et al. 2009).

Thus, modification of the physical environment by habitat-forming species may have cascading effects on the associated fauna, in most cases increasing species diversity (Cole and McQuaid 2010; Arribas et al. 2013). On sediment dominated tidal flats as in the Wadden Sea, *Mytilus edulis* beds are seen as 'islands of biodiversity' (Bushbaum and Nehls 2003 in Markert et al. 2010). Higher biodiversity has been reported with increasing structural complexity even within the same species (Tsuchiya and Nishihira 1986; Markert et al. 2010).Some invasive engineers, however, may decrease the complexity of habitats by replacing more heterogeneous native species or assemblages, resulting in decreased species diversity as shown for plantations (Crooks 2002).

The effect on biodiversity might differ between species, the density and the size of the bed or the age of the assemblage. There might also be differences between intertidal and subtidal beds, the position on the (soft) sediment (epibenthic vs endobenthic), and regional or local conditions. In the next paragraphs we will review these aspects.

14.2.1 Bivalve Species

A few studies compared beds of different species in the same region. Markert et al. (2010) compared established *Crassostrea*-reefs and native *Mytilus*-beds. The authors report higher diversity values in oyster beds, and these findings are discussed in terms of differences in ecosystem engineering by *C. gigas* versus *M. edulis*. The *Crassostrea*-reef might have influenced the frequency of epibenthic organisms by providing a more complex habitat matrix with an extended hard substrate surface. The geometry of *Crassotrea* shells offers various cryptic microhabitats most suitable for colonization by several vermicular organisms. In contrast to mainly horizontal surfaces which occur in *Mytilus*-patches, vertically oriented *Crassostrea* shells show complex patterns of current flow. Several species, such as suspension-feeding organisms like *Polydora*, may benefit from these conditions, thus resulting in a higher diversity of the epibiont community. Compared to a *Mytilus*-bed, the superficial structure of a *Crassostrea*-reef increases bottom roughness and water turbulences. Thus, more biodeposits could have been exported from *Crassostrea*-patches than from *Mytilus*-patches. *M. edulis* are more frequently affected by burial. The mussels themselves are able to move back to the surface but attached organisms may suffer. In *Crassostrea*-patches, there is a permanently sediment-free upper shell surface, which may have contributed to the richer epibenthos in *Crassostrea*-patches.

Arribas et al. (2013) compared beds of 2 coexisting mytilids on intertidal rocks. *Brachidontes rodriguezii* and *Perumytilus* (*Brachidontes*) *purpuratus*, along the northern Argentinian coast. Although these species are very similar in their biological and ecological function, the fauna associated with their matrices are very different. Some species were found associated with only 1 species of mussel, e.g. the bivalve *Lasaea adansoni* with *Perumytilus purpuratus*, or *Mytilus edulis* with *Brachidontes rodriguezii*.

Jungerstam et al. (2014) on the other hand did not find evidence of a strong mussel species effect on associated communities in rocky shore mussel assemblages in South Africa.

When comparing biodiversity of bivalve beds of different regions, the degree of diversity may depend strongly on the regional spectrum of species and the ability of these species to adapt to the engineered conditions within mussel beds. Soft bottom mussel beds may constitute physically similar habitats through the world but the responses of other benthic species may not be the same, and thus the arising mussel bed communities arise by site-specific rules (Buschbaum et al. 2009). In a study comparing mytilid beds in the North Sea (*M. edulis*), the southern Chilean coast (*Perumytilus purpuratus*, *M. chilensis*), the Yellow Sea (*Musculista senhousia*), and the coast of southern Australia (*Xenostrobus incostans*), these authors did find higher diversity than surrounding areas for mussel beds in the North Sea and at the Chilean coast. For mussel beds in the other regions the number of associated species were only slightly higher (Australia) and even somewhat lower (Yellow Sea) than in adjacent sediments. Comparisons might, however, not only be hampered by regional species' pools, but by differences in e.g. bivalve density, path sizes, age of the bivalve bed, tidal height and substrate type.

14.2.2 Bivalve Density and Patch Size

Some studies related differences in diversity to the bivalve density at the time of the sampling (Commito 1987; Dittmann 1990; Murray et al. 2007), but more recent studies did not find increased diversity at higher bivalve densities, or expressed their doubt. Faunal assemblages associated with ribbed mussel beds along the South American coast varied independently of the density of mussels (Sepúlveda et al. 2016). Asmus (1987) found no correlation between the density of blue mussels and the species number of associated epifauna. The mussel density encountered within a bed at the time of sampling requires careful consideration in view of the fact that the mussel bed will change dynamically due to mussel growth and mortality. As a result, the infaunal assemblages encountered at the time of sampling may reflect not only the mussel density at that time but also the initial mussel stocking density. The latter may have a long-term influence through the biodeposition that has occurred prior to invertebrate sampling (Beadman et al. 2004). In conclusion, any positive (and negative) correlations are thus likely to depend on local physical conditions and larval dispersions, and no general assertions can be made (Murray et al. 2007).

Cole (2010) experimentally compared engineered and unmodified habitat, and different configurations of engineered patches of the marine intertidal mussel *Trichomya hirsuta*. Regularly spaced solitary mussels had more edge and consequently more species, unique species (mostly macroalgal species but also several molluscs, arthropods and polychaetes) and densities of generalists. The findings suggest that the configuration of patches of a habitat is a crucial factor affecting mussel bed biodiversity, and fragmentation of habitat into regularly spaced patches may have a positive influence on biodiversity due to the positive response of other species to habitat edges. The experimental design, however, poorly reflects natural complex structure as described above. Factors affecting the structure of the habitat (bed thickness, age distribution, cover, …) probably have larger effect than patch size itself. Not surprisingly, thus, in literature different relationships are found between patch size, even within a single study area, although always either positive or not significant (see e.g. Tsuchiya and Nishihira 1986; Norling and Kautsky 2008; Koivisto et al. 2011; Jungerstam et al. 2014; Sepúlveda et al. 2016).

14.2.3 Age and Size Structure of the Bivalve Assemblage

With the aging of the mussel assemblage, mussels require more space for attachment and some individuals in the periphery of the patch are pushed out while some inside the patch are shifted. This results in a multi-layered bivalve bed. It also results in more space and larger amounts of sediments and shell fragments (Tsuchiya and Nishihira 1986). As mentioned above, this mostly results in an increase in species diversity. However, if recruitment fails, the patch might become mono-layered and poorer in species richness (Tsuchiya and Nishihira 1986).

In the study of Tsuchiya en Nishihira (1986) patches of different age also differed in size: older patches consisted of larger mussels. The size of mussels is, however, not necessarily related to their age (Buschbaum and Saier 2001). To separate the effects of size and age, O'Connor en Crowe (2007) manipulated the age of mussel patches of *Mytilus edulis* and the size of mussels within them to test experimentally the effect of size on the associated assemblages. At one of the two locations, the size of the mussels did affect the abundance of some species, but did not affect species richness. Cole en McQuaid (2010) found the same results in beds of *Mytilus galloprovincialis* and *Perna perna*. Sepúlveda et al. (2016), on the other hand, did not find any significant association of both richness and abundance of the associated fauna with the size or density of ribbed mussels *Aulacomya atra*.

14.2.4 Substrate Type and Stability

Benthic species show distinct distribution patterns in relation to the type of substratum (see e.g.Wood 1987; Künitzer et al. 1992; Reiss et al. 2010) and substrate stability (Arribas et al. 2013). On hard substrates, bivalve beds are obviously epibenthic. Soft bottom mussel beds may be endobenthic, with a diversity of transitions between endobenthic and epibenthic mode. In endobenthic beds, most individuals are positioned below the sediment surface. Thus, diversity may depend on the epi- versus endobenthic traits (Buschbaum et al. 2009). Low substrate stability consequently results in unstable habitat for the associated fauna, directly as a consequence of increased susceptibility of the bivalve bed to dislodgment, or indirectly as a consequence of differences in the amount of sediment trapped. On rocky substrates along the northern Argentinian coast, a relationship was found between rock hardness, the amount of sediment trapped and the biological assemblage. Species composition was different and total abundance was lower at the shore with the lowest rock hardness and the smallest amount of sediment trapped. Diversity, however, was not significantly different (Arribas et al. 2013). Ysebaert et al. (2009) hypothesized that a decrease of number of endobenthic species due to an increased organic flux to the sediment to be stronger in a low-flow environment than in a high-flow environment where most biodeposition was expected to be swept away with currents. However, this hypothesis was rejected by their results. Apparently (pseudo) fecal material is also deposited nearby the mussel bed (M. *edulis*) under a strong current regime. The physical structure of the dense bed leads to protected conditions. Moreover, the strong hydrodynamic forces lead to much higher suspended matter concentrations in the water and thus increase the biodeposition rates of mussels as compared to the quiet clearer waters in calm conditions.

In literature both a rich associated assemblage of species of bivalve beds are reported as well as similar or reduced diversity in comparison to the surrounding sedimentary environments. This suggests that mussel beds in sedimentary environments may not invariably be hot diversity spots.. Buschbaum et al. (2009) found, for instance, enhanced species richness and diversity in epibenthic *Mytilus edulis* beds, and lower in *Musculista senhousia*. Other studies showed the opposite: higher species richness inside *M. senhousia* beds, decreased diversity in *M. edulis* beds (Crooks 1998; Commito et al. 2005). Apparently the response is not dependent on the species, but the effects on the associated species are site specific (Buschbaum et al. 2009).

14.2.5 Tidal Versus Subtidal

Bivalves in the intertidal zone experience different abiotic conditions then their subtidal conspecifics. Mussels in the intertidal zone experience extremes in temperature, from baking in the sun in summer to freezing in winter. They are subject to freshwater exposure during rainstorms, to risk of being dislodged by waves or battered by logs during storms, and to periodic interruption of feeding, gas exchange, and excretion through tidal cycling (www.asnailsodyssey.com). And, of course, there is a difference in tidal emergence. This might have influence on bivalve bed characteristics, such as growth of the animals, the density or the three-dimensional structure (AIN 2001; Saier et al. 2002). Moreover, species composition changes too along the tidal gradient and, consequently, this may result in differences in the associated species.

The literature mentioned above compared either mussel- and non-mussel covered areas or different bivalve beds, but comparisons were only made within intertidal or subidal areas. We are only aware of one study comparing intertidal with subtidal. Saier et al. (2002) compared studies of intertidal and adjacent shallow subtidal mussel beds in the northern Wadden Sea. They concluded that intertidal and subtidal sites were ecologically different with respect to the mussel bed structure as well as associated organisms. The studies revealed higher densities in intertidal beds of smaller mussels. Subtidal mussels were less fouled. Several sessile species are found only on either intertidal or subtidal beds. Finally, higher diversity and species richness on non-attached epifauna (mobile invertebrates living within the mussel bed matrix) was found subtidally.

14.2.6 Other Factors

When comparing studies on biodiversity aspects of bivalve beds, one should keep in mind that the degree of diversity may strongly depend on the biogeographical/ regional species pool. Even if the mussel beds constitute physically similar habitats the responses of other benthic organisms may not be the same, hence, the associated mussel bed communities arise by site-specific rules (Buschbaum et al. 2009). Sepúlveda et al. (2016), for instance, showed that the mussel-associated fauna along the northern Argentinean coast differed between the northern (Peru and Northern Chile) and southern area (Southern Chile). The differences reflect the well-known Peruvian and Magellanic provinces and show that the associated fauna is highly sensitive to biogeographic signals, despite the fact that the fauna make use of similar bioengineered habitat throughout their geographic ranges..

In conclusion, there seem to be some generic drivers, but one should realize that the influence of the mentioned drivers depend strongly on the local hydrodynamic, topographic and biogeographic conditions.

Fig. 14.1 Mussel aquaculture on bottom plots (left) and suspended ropes (right) demonstrating that cultures include a rich community of flora, epifauna and mobile fauna (crabs, fish) species.©J. Capelle (left) and T. Strohmeier (right)

14.3 Biodiversity Trade–Offs in Cultured Bivalve Assemblages

Bivalve beds naturally occur in many subtidal and intertidal areas around the world, but mussels and oysters are also extensively cultured (Wijsman et al. 2019). Mussel aquaculture is done by means of bottom cultures (by seeding intertidal or subtidal beds), but also by suspended cultures (using rafts or longlines), and cultures on bouchots (Smaal 2002; Beadman et al. 2004) (Fig. 14.1). Addition of physical cultivation structures in the water column or on the bottom allows for development of substantial and diverse communities that have a structure similar to that of natural reefs (Callier et al. 2017): the biogenic structure offers habitat for numerous species, infaunal and epibenthic, hard substrate species, as well as shelter and predator refuge for mobile epibenthos. Apart from the reef building function in the water column, suspended cultures may also create rich fauna communities in the benthic environment through fall-off and enrichment by biodeposition (McKindsey et al. 2011; Callier et al. 2017) similar to enrichment effects observed in direct proximity of natural bivalve reefs (Dittmann 1999, van der Zee et al. 2012, Ysebaert et al. 2019, and references mentioned above). Moreover, some of the epibenthic species, including the cultured bivalves, are attracted by the bivalves as prey. Dynamics of culture populations may however differ from natural bivalve reefs as a result of culture site and/or maintenance and operation of bivalve farms (Callier et al. 2017). Consequently, differences may be expected in the processes that are dominant for driving biodiversity development, and thus in the faunal communities of natural and cultured beds. We can expect this to be the same in cultured bivalve assemblages (Table 14.1).

14.3.1 *Mussel Fisheries on Wild Beds*

Mussel seedbeds, where spatfall, the settling and attachment of young) to the substrate has occurred, are often exploited by dredging the young seed mussels and moving them to areas where growing conditions are more favourable. Surprisingly few studies are available describing the impacts of this type of bivalve dredging on

Table 14.1 Expected influence of cultured bivalve assemblages on biodiversity

Aquaculture strategy	Expectation based on natural beds
Choice of cultured species	Biodiversity varies with species
Density of species	Multi-layered, more complex beds have higher diversity
Predator control	Disturbance limits complexicity and diversity
Relay and harvesting	Disturbance limits complexicity and diversity
Age when harvested	Young, less complex, beds have lower diversity
Site selection	Local conditions are most important driver

changes in the abundance of associated species. A meta-analysis on the effects of different types of trawling and dredging by Collie et al. (2000) concluded that data on impacts and recovery of epifaunal structure-forming benthic communities are indeed lacking.

Dolmer (2002) demonstrated that dredging on commercial-sized mussels negatively influence fauna communities on the short-term, especially for polychaetes. Furthermore, after 4 months the effects of dredging on epibenthos were still evident and included a reduction in density of a number of taxa (sponges, echinoderms, anthozoans, molluscs, crustaceans, and ascidians). A large scale study in the Dutch Wadden Sea suggested that impacts can in certain seasons last up to 1.5 years, while for other seasons it was hard to define any impact at all (see case study below). Collie et al. (2000) concluded that recovery from dredging may take several years.

By collecting spat to stock cultivation sites new habitats are created that will support some of the animals present on the seedbeds (Smaal and Lucas 2000; Murray et al. 2007).

14.3.2 Benthic Cultivation Plots

As for natural beds, it is expected that by increasing mussel biomass through cultivation, the species richness, abundance, and biomass of the associated macrofauna would also increase compared to the surrounding, bare areas. An increase in the number of epibenthic species indeed was found for mussel culture plots in the Oosterschelde (the Netherlands) but not in Limfjorden (Denmark) (Ysebaert et al. 2009). In the latter system, however, epifaunal species are rare, and although no increase in diversity was observed within the mussel plot, species richness was higher at sites with mussels compared to sites with none or almost no mussels. Trianni (1996) observed higher diversity for cultivation sites of on-bottom oyster culture relative to surrounding muddy bottom areas due to increase abundance of epifauna associated with the oyster shells.

Mussel farmers will attempt to lay mussels at a density and tidal height that will realize highest growth and the greatest financial return upon harvest. At high mussel densities multi-layering is likely to occur, which will increase mussel bed complexity (see e.g. Beadman et al. 2004; Smith and Shackley 2004; Ysebaert et al. 2009), providing habitat for a large number of associated species. If so, the seeding practice might influence the number of species found on a commercial bivalve bed. This conclusion is also made by Murray et al. (2007) based on the findings of Tsuchiya en Nishihira (1986) in natural beds that mussel patch size is positively correlated with species richness. However local differences do exists as on intertidal mudflats in north Wales, UK, highest number of species was found at beds with low mussel cover (Beadman et al. 2004). These beds have habitats suitable for both the typical mudflat fauna and the typical mussel bed fauna by the extra microhabitats provided within the isolated clumps of mussels. Species richness declined with increased area of mussels, hence the more positive benefits of increased habitat complexity are apparently out-weighted by negative factors, such as a highly anoxic environment and competition for food and space. Murray et al. (2007) investigated mussel and fauna biomass on rafts, and intertidal and subtidal bottom cultures in Maine (USA). They observed both significant positive and negative, as well as no correlations between mussel biomass and associated faunal biomass This indicates that, as for natural beds, the biodiversity associated to bivalve cultures is likely to a large extent driven by local conditions. Ysebaert et al. (2009) also suggests that the impact on the benthic community due to biodeposition is influenced by local topographic and hydrodynamic conditions.

Regular relay and harvesting of cultured beds may, on the other hand, prevent the age and size of the mussel patches increasing above a certain point. It may also make the bed structure less layered and complex. Thus, harvesting is expected to limit the diversity of the associated fauna (Tsuchiya and Nishihira 1986; Smith and Shackley 2004). Though under certain circumstances, a lower cover might result in a higher diversity (Beadman et al. 2004), as mentioned before. Little is known about the temporal dynamics of succession in associated species on commercial bottom plots and how culture practices influence these processes. Questions still remain to be answered include e.g. how much fauna is transported together with seed to the bottom plots, how quickly will a plot be colonised by opportunists and the more resident fauna species, and what are the effects of relaying and predator control?

In bottom cultures predation by starfish and crabs can be significant (Capelle et al. 2016a) and measures are taken to remove starfish with special adapted trawls that basically 'mob' the bottom plots removing part of the starfish population. Yet the efficiency of this method is debated.

14.3.3 Suspended Cultures

The ecological effect of the habitat created by bivalve farming is well-recognized for bottom cultures. Forrest et al. (2009) suggests that there is also evidence of a comparable role for suspended culture structures, intertidal trestles or other intertidal structures used for bivalve cultivation.

In suspended cultures, the physical infrastructures themselves (buoys, ropes and anchors) already provide substrate for many organisms (Murray et al. 2007; Ysebaert et al. 2009), as well as the bivalve populations itself. The settlement of different ascidian, polychaete and crustacean genera reported by Jansen et al. (2011) on suspended mussel ropes, reflected a significant increase in taxonomic richness throughout an annual cycle. This agrees with Taylor et al. (1997), Richard et al. (2006) and Lutz-Collins et al. (2009), which showed that number and composition of fauna associated to bivalve cultures are dependent on culture duration. Intra-annual variation in associated faunal abundance is also observed in suspended oyster culture (Mazouni et al. 2001). Temperature is thought to be an important driver for abundance of associated species, especially for filter-feeding species that attach to the suspended cultures (Khalaman 2001). The average number of fauna genera associated with mussels ropes ranged between 7–10 genera (Richard et al. 2006; Jansen et al. 2011). The proportion of the fouling biomass relative to mussel biomass ranges from 0 to 10% (LeBlanc et al. 2007; Jansen et al. 2011). The presence of deposit-feeding polychaetes, such as *Capitella* and *Neoamphitrite*, indicate that mussel ropes contain large amounts of organic material, thereby indicating that suspended ropes serve as a sediment compartment in the water column (Jansen et al. 2011 and references therein).

The mussel stocking densities used (i.e., the number of mussels per unit length within a sock) influence the growth rate of the mussels and, simultaneously, determine the amount of surface area available for the epifaunal organisms to colonize (Tsuchiya and Nishihira 1986; Thompson and MacNair 2004). Lutz-Collins et al. (2009) studied the effect of mussel density on colonization by polychaetes in Prince Edward Island (Canada). Polynoid worms of the genera *Harmothoe* and *Lepidonotus* were by far the most abundant taxa colonizing mussel socks. Although there were sharp density variations associated with stocking density, these differences were inconsistent and no trends across stocking densities were observed: an increase in stocking density did not seem to be causally related to an increase (or decrease) in the total specific epifaunal densities. Because of this apparent lack of influence, stocking density was considered to be irrelevant. Date along the growing season was in this study the most obvious factor influencing the overall epifaunal composition.

In many cultivation areas mussel lines are at least once being resocked during the culture cycle to grade the mussels and thin densities to prevent drop-off. During this process most of the associated biota will be removed, however, no information on the effects of those management activities are reported in literature.

It should be noted though that fostering biodiversity in suspended cultures might seem somewhat paradoxical from an commercial perspective, as fouling species are in many cases the bane of the aquaculture industry (Durr and Watson 2010; Fitridge et al. 2012). Fouling species might interfere, compete for food sources with the bivalves (e.g. tunicates) or they might predate on the bivalves itself (e.g. start fish or crabs). Particularly in case of suspended cultures methods have been developed to remove the fouling organisms. Generally, control of biofouling in aquaculture is achieved through the avoidance of natural recruitment, physical removal and the use of antifoulants (Fitridge et al. 2012). Methods to remove ascidians, a fouling species that can become dominant on mussel ropes and competes for space and food resources, include freshwater and acid treatments (Carman et al. 2016).

Mussels from rafts or longlines not only have effects on the fauna associated with the cultivation structure but also on the fauna of the sedimentary environment below them. Drop-off from mussels mostly enhances species such as star fish, sea cucumber and crabs (Romero et al. 1982; McKindsey et al. 2011). McKindsey et al. (2011) reviewed the extensive literature on the effects of biodeposition on infauna communities and suggest that, for the most part, community responses follow the Pearson en Rosenberg (1978) model of organic enrichment. As the level of organic input increases, typical soft sediment communities dominated by large filter-feeders are replaced by smaller, more deposit-feeding organisms, starting with small polychaetes (e.g., *Capitella* spp.), shifting to nematodes, and finally ending up with anoxic conditions and mats of the bacteria *Beggiatoa* spp. Though the latter is not frequently observed under mussel farms. Biomass and species richness may increase with limited organic loading whereas abundance may increase with moderate loading as smaller, opportunistic, species come to dominate.

14.4 Case Study: How Doe Benthic Mussel Culture Activities Affect Subtidal Biodiversity in the Western Wadden Sea

Mussel culture in the Dutch Wadden Sea is dominated by bottom cultivation, and mussel seed is traditionally collected from wild subtidal mussel beds, though a shift towards suspended systems for spat collection is implemented. Seed fisheries and management of bottom plots may each have specific effects on reduction or enhancement of biodiversity of associated species. To assess these effects, an integrated approach was applied which provided an ecosystem wide evaluation on the effects of mussel culture activities on biodiversity of infauna and epifauna communities in sublittoral areas of the Dutch Wadden Sea.

14.4.1 Fisheries Impacts on Biodiversity

Mussel seed generally settles in the south-western part of the Wadden Sea, yet the total area of mussel seed beds varies strongly from year to year. Approximately 50% of the beds are characterised as instable, indicating that they will not survive the winter as a result of storms and/or predation. Fisheries takes place two times a year; in autumn the classified instable beds are open to fisheries, and in spring stable beds may be fished, given that there is enough total mussel biomass for birds to feed on (Capelle 2017).

The effect of mussel fishery on mussels, epifauna and infauna species was investigated over a period up to 6 years comparing adjacent plots with and without mussel seed fishery (Craeymeersch et al. 2013; Glorius et al. 2013; Van Stralen et al. 2013) (Fig. 14.2). Only 4 of the 21 areas were fished more than once during the research period. Short to medium-term (weeks to months) effects of fishery activities were observed in terms of total density and in species composition of fauna populations. As most species were positively correlated to the presence of mussels (associated species), changes in species communities were assumed to be correlated to the removal of mussels. Reduction in abundance of anemones was linked to removal or damage by fishing nets. Long term (> 1.5 year) effects were not observed, thereby assuming that mussels that remained on the fished plots provided enough structure for development of associated fauna populations. Observed effects, i.e. different development in open and closed plots, were more profound for plots fished during spring than during autumn because closed (no fishery) plots also changed considerably in terms of mussel and thus associated fauna biomass during winter storms, making it difficult to detect any fishery related impacts.

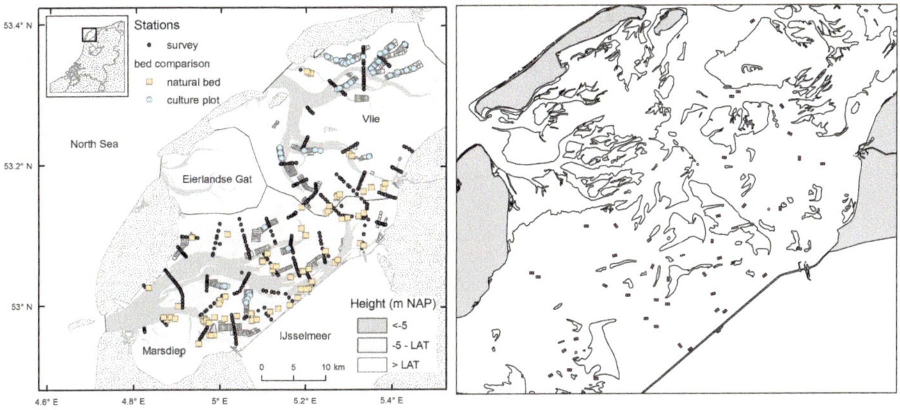

Fig. 14.2 Maps of the western Dutch Wadden Sea. Left:sampling stations of benthic survey in 2008 and from a study comparing natural mussel beds and mussel culture plots (Drent and Dekker 2013b). Right: locations used in studies on the impact of mussel seed fisheries (Craeymeersch et al. 2013)

The study also confirmed large heterogeneity of associated fauna composition within a mussel bed, and large year-to-year variation in species composition, independently of any human impact. It was therefore concluded that overall, any fishery effects seemed to be less important in determining species composition than external factors controlling mortality and recruitment.

14.4.2 Biodiversity on Culture Plots

Moving seed mussels to bottom culture plots enhances the total mussel biomass in the Wadden Sea by 27% compared to a situation where no fisheries exists (Wijsman et al. 2014; Capelle et al. 2016b). This is a consequence of mussel seed fisheries on instable mussel beds, and subsequent transport to bottom plots where the mussels have higher survival rates and where reduction of predators (starfish, crabs) is achieved through effective management strategies.

It is well known that bivalve populations serve as a suitable habitat for a number of species, resulting in high biodiversity within bivalve aggregations (see Sect. 14.2). A field study was performed to test if this also holds for biodiversity on culture plots (Drent and Dekker 2013a). Approximately three times higher total biomass of associated fauna was observed for wild beds (i.e. beds originating from natural spatfall) compared to culture plots, mainly caused by high biomass of endobenthic species on wild beds (Fig. 14.3). However, the total number of species recorded was significantly higher for culture plots (102 for plots, versus 84 for beds), indicating that culture plots do serve as an unique habitat for biodiversity development. A complicating factor for direct comparison is however the spatial distribution; culture plots are mostly located in the north, while wild beds survive best in the South-West of the Wadden Sea, indicating that not only the origin (wild

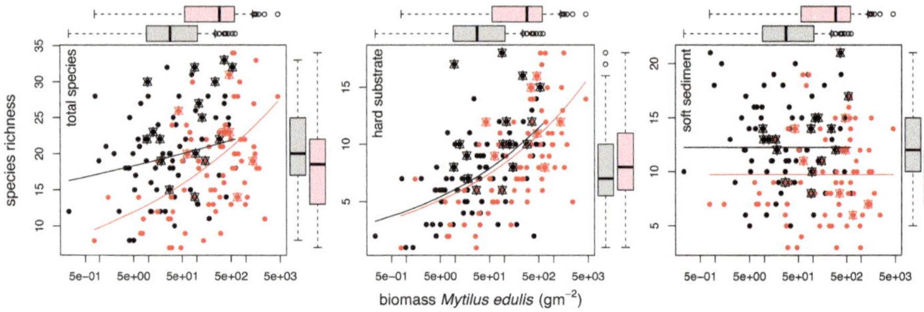

Fig. 14.3 Relationship between total species richness (left), species richness of hard substrate species (middle) and species richness of soft sediment species (right), in a 0.06 core sample and biomass of Mytlius edulis in the same sample. Black dots are for cores outside mussel culture plots and red dots for cores inside mussel culture plots. Lines are fitted GLM model results for inside and outside mussel culture plot observations. Boxes in the margins show the distribution of the observations. Stations with Crassostrea gigas are indicated with stars (Drent and Dekker 2013a)

bed vs culture plot) but also local environmental conditions, like salinity, might drive biodiversity development within the mussel aggregations. Culture plots are situated in higher salinity zones near tidal inlets connecting the Wadden Sea with the North Sea, wild beds in lower salinity zones landwarts of the tidal basins. Comparison of culture plots and wild beds located in proximity of each other show higher species richness for wild beds, indicating that environmental conditions may indeed affect biodiversity development. The overall conclusion of this study was that wild beds and culture plots differ in fauna communities (species and densities), but both form a unique habitat for a diverse population of benthic fauna.

14.4.3 Integrated Assessment

These studies show that the effects of different mussel culture activities vary; seed fisheries on wild beds may have a direct negative impact on biodiversity (short term), but at the same time leads to an increased survival and thus higher total biomass of mussels on the culture plots. High mussel biomass in turn leads to high biodiversity, also on culture plots. Quantitative comparison between those processes is difficult, due to large temporal and spatial differences of the activities. Nevertheless, those results provide valuable guidance for further development of management strategies for nature conservation and sustainable bivalve culture in the sublittoral areas of the Dutch Wadden Sea.

14.5 Concluding Remarks

The influence of different drivers such as the kind of bivalve species, bivalve density, substrate type, tidal zone, etc. … on associated species appear generic while local hydrodynamic, topographic and biogeographic conditions strongly define which driver is dominant for each specific area. This holds for both natural as cultured bivalve assemblages. Even within a small waterbody, such as the western Wadden Sea, differences in biodiversity are found that can partly explained by natural drivers, such as salinity or depth, and another part is the result of local spatial variability. Nevertheless, studies comparing natural and cultured assemblages allow for a better understanding on the effect of the culture strategies and, consequently, to forward sustainable bivalve cultures.

Acknowledgements The authors are grateful to the referees dr J. Drent and dr T. Ysebaert for their valuable comments on the manuscript.

References

AIN (2001) Comparing subtidal and intertidal growth in off bottom oyster culture. FARD AIN 07.2001. 2pp

Arribas LP, Bagur M, Klein E, Penchaszadeh PE, Palomo MG (2013) Geographic distribution of two mussel species and associated assemblages along the northern Argentinean coast. Aquat Biol 18:91–103

Asmus H (1987) Secondary production of an intertidal mussel bed community related to its storage and turnover compartments. Mar Ecol Prog Ser 39:251–266

Beadman HA, Kaiser MJ, Galanidi M, Shucksmith R, Willows RI (2004) Changes in species richness with stocking density of marine bivalves. J Appl Ecol 41:464–475

Brocken F, Kenchingon E (1999) A comparison of scallop (Placopecten magellanicus) population and community characteristics between fished and unfished areas in Lunenburg County. N.S., Canada. Can Tech Rep Fish Aquat Sci 2258:vi ″ 93 p

Bruno JF, Stachowicz JJ, Bertness MD (2003) Inclusion of facilitation into ecological theory. Trends Ecol Evol 18:119–125

Buschbaum C, Saier B (2001) Growth of the mussel Mytilus edulis L. in the Wadden Sea affected by tidal emergence and barnacle epibionts. J Sea Res 45:27–36

Buschbaum C, Dittmann S, Hong JS, Hwang IS, Strasser M, Thiel M, Valdivia N, Yoon SP, Reise K (2009) Mytilid mussels: global habitat engineers in coastal sediments. Helgol Mar Res 63:47–58

Callier M, Byron C, Bengtson D, Cranford P, Cross S, Focken U, Jansen H, Kamermans P, Kiessling A, Landry T (2017) Attraction and repulsion of mobile wild organisms to finfish and shellfish aquaculture: a review. Rev Aquac 0:1–26

Capelle JJ (2017) Production efficiency of mussel bottom culture. University of Wageningen

Capelle JJ, Scheiberlich G, Wijsman JWM, Smaal AC (2016a) The role of shore crabs and mussel density in mussel losses at a commercial intertidal mussel plot after seeding. Aquac Int 24:1459–1472

Capelle JJ, Wijsman JWM, van Stralen MR, Herman PMJ, Smaal AC (2016b) Effect of seeding density on biomass production in mussel bottom culture. J Sea Res 110:8–15

Carman M, Lindel L, Green-Beach E, Starczak V (2016) Treatments to eradicate invasive tunicate fouling from blue mussel seed and aquaculture socks. Manag Biol Invasions 7:101–110

Cohen A, Cosentino-Manning N, Schaeffer K (2007) Shellfish beds. Report on the subtidal habitats and associated biological taxa in San Francisco Bay. pp 50

Cole VJ (2010) Alteration of the configuration of bioengineers affects associated taxa. Mar Ecol Prog Ser 416:127–136

Cole VJ, McQuaid CD (2010) Bioengineers and their associated fauna respond differently to the effects of biogeography and upwelling. Ecology 91:3549–3562

Collie JS, Hall SJ, Kaiser MJ, Poiner IR (2000) A quantitative analysis of fishing impacts on shelf-sea benthos. J Anim Ecol 69:785–798

Commito JA (1987) Adult-larval interactions – predictions, mussels and cocoons. Estuar Coast Shelf S 25:599–606

Commito JA, Boncavage EM (1989) Suspension-feeders and coexisting infauna – an enhancement counterexample. J Exp Mar Biol Ecol 125:33–42

Commito JA, Celano EA, Celico HJ, Como S, Johnson CP (2005) Mussels matter: postlarval dispersal dynamics altered by a spatially complex ecosystem engineer. J Exp Mar Biol Ecol 316:133–147

Craeymeersch J, Jansen J, Smaal A, van Stralen M, Meesters E, Fey F (2013) Impact of mussel seed fishery on subtidal macrozoobenthos in the western Wadden Sea. IMARES report number PR 7 C003/13. 123 pp

Crooks JA (1998) Habitat alteration and community-level effects of an exotic mussel, Musculista senhousia. Mar Ecol Prog Ser 162:137–152

Crooks JA (2002) Characterizing ecosystem-level consequences of biological invasions: the role of ecosystem engineers. Oikos 97:153–166

Dittmann S (1990) Mussel beds – Amensalism or amelioration for intertidal Fauna. Helgoländer Meeresun 44:335–352

Dittmann S (1999) The Wadden Sea ecosystem – stability properties and mechanisms. Springer, Heidelberg

Dolmer P (2002) Mussel dredging: Impact on epifauna in Limfjorden, Denmark. J Shellfish Res 21:529–537

Drent J, Dekker R (2013a) How different are sublitoral *Mytilus edulis* communities of natural mussel beds and mussel culture plots in the western Dutch Wadden Sea. NIOZ Report 2013-6. 94 pp.

Drent J, Dekker R (2013b) Macrofauna associated with mussels, *Mytilus edulis,* in the subtidal of the western Dutch Wadden Sea. NIOZ Report 2013-7. 77 pp.

Durr S, Watson D (2010) Biofouling and antifouling in aquaculture. In: Durr S, Thomason J (eds) Biofouling. Wiley-Blackwell, Oxford

Fitridge I, Dempster T, Guenther J, de Nys R (2012) The impact and control of biofouling in marine aquaculture: a review. Biofouling 28:649–669

Forrest BM, Keeley NB, Hopkins GA, Webb SC, Clement DM (2009) Bivalve aquaculture in estuaries: review and synthesis of oyster cultivation effects. Aquaculture 298:1–15

Glorius S, Rippen AD, van Stralen MR, Jansen J (2013) Deelrapport bodemschaaf en zuigkordata Effecten van mosselzaadvisserij op het bodemleven van de Waddenzee IMARES rapport PR 8 C162/12

Jansen HM, Strand O, Strohmeier T, Krogness C, Verdegem M, Smaal A (2011) Seasonal variability in nutrient regeneration by mussel Mytilus edulis rope culture in oligotrophic systems. Mar Ecol Prog Ser 431:137–149

Johnson M (2013) Biodiversity hotspots. Available from http://www.coastalwiki.org/wiki/Biodiversity_hotspots. Accessed on 6 Apr 2017

Jungerstam J, Erlandsson J, McQuaid CD, Porri F, Westerbom M, Kraufvelin P (2014) Is habitat amount important for biodiversity in rocky shore systems? A study of south African mussel assemblages. Mar Biol 161:1507–1519

Khalaman VV (2001) Fouling communities of mussel aquaculture installations in the White Sea. Russ J Mar Biol 27:227–237

Koivisto M, Westerbom M, Arnkil A (2011) Quality or quantity: small-scale patch structure affects patterns of biodiversity in a sublittoral blue mussel community. Aquat Biol 12:261–270

Künitzer A, Basford D, Craeymeersch JA, Dewarumez JM, Dörjes J, Duineveld GCA, Eleftheriou A, Heip C, Herman P, Kingston P, Niermann U, Rachor E, Rumohr H, de Wilde PAJ (1992) The benthic infauna of the North Sea: species distribution and assemblages. ICES J Mar Sci 49:127–143

Lancaster JE, McCallum S, Lowe A.C., Taylor E A. C, Pomfret J (2014) Development of detailed ecological guidance to support the application of the Scottish MPA selection guidelines in Scotland's seas. Scottish Natural Heritage Commissioned Report No.491. Horse Mussel Beds – supplementary document

LeBlanc N, Davidson J, Tremblay R, McNiven M, Landry T (2007) The effect of anti-fouling treatments for the clubbed tunicate on the blue mussel, Mytilus edulis. Aquaculture 264: 205–213
Lutz-Collins V, Quijon P, Davidson J (2009) Blue mussel fouling communities: polychaete composition in relation to mussel stocking density and seasonality of mussel deployment and sampling. Aquac Res 40:1789–1792
Manhaes AP, Mazzochini GG, Oliveira AT, Ganade G, Carvalho AR (2016) Spatial associations of ecosystem services and biodiversity as a baseline for systematic conservation planning. Divers Distrib 22:932–943
Markert A, Wehrmann A, Kroncke I (2010) Recently established Crassostrea-reefs versus native Mytilus-beds: differences in ecosystem engineering affects the macrofaunal communities (Wadden Sea of Lower Saxony, southern German Bight). Biol Invasions 12:15–32
Mazouni N, Gaertner JC, Deslous-Paoli JM (2001) Composition of biofouling communities on suspended oyster cultures: an in situ study of their interactions with the water column. Mar Ecol Prog Ser 214:93–102
McKindsey CW, Archambault P, Callier MD, Olivier F (2011) Influence of suspended and off-bottom mussel culture on the sea bottom and benthic habitats: a review. Can J Zool 89:622–646
Murray LG, Newell CR, Seed R (2007) Changes in the biodiversity of mussel assemblages induced by two methods of cultivation. J Shellfish Res 26:153–162
Norling P, Kautsky N (2007) Structural and functional effects of Mytilus edulis on diversity of associated species and ecosystem functioning. Mar Ecol Prog Ser 351:163–175
Norling P, Kautsky N (2008) Patches of the mussel Mytilus sp are islands of high biodiversity in subtidal sediment habitats in the Baltic Sea. Aquat Biol 4:75–87
O'Connor NE, Crowe TP (2007) Biodiversity among mussels: separating the influence of sizes of mussels from the ages of patches. J Mar Biol Assoc UK 87:551–557
Pearson TH, Rosenberg R (1978) Macrobenthic succession in relation to organic enrichment and pollution of the marine environment. Oceanogr Mar Biol Annu Rev 16:229–311
Ragnarsson SA, Raffaelli D (1999) Effects of the mussel Mytilus edulis L. on the invertebrate fauna of sediments. J Exp Mar Biol Ecol 241:31–43
Reiss H, Degraer S, Duineveld GCA, Kröncke I, Craeymeersch J, Rachor E, Aldridge J, Cochrane S, Eggleton JD, Hillewaert H, Lavaleye MSS, Moll A, Pholmann T, Robertson M, Smith R, vanden Berghe E, Van Hoey G, Rees HL (2010) Spatial patterns of infauna, epifauna and demersal fish communities and underlying processes in the North Sea. ICES J Mar Sci 67:278–293
Richard M, Archambault P, Thouzeau G, Desrosiers G (2006) Influence of suspended mussel lines on the biogeochemical fluxes in adjacent water in the Iles-de-la-Madeleine (Quebec, Canada). Can J Fish Aquat Sci 63:1198–1213
Romero P, Gonzalezgurriaran E, Penas E (1982) Influence of mussel rafts on spatial and seasonal abundance of Crabs in the Ria De Arousa, Northwest Spain. Mar Biol 72:201–210
Saier B, Buschbaum C, Reise K (2002) Subtidal mussel beds in the Wadden Sea: threatened oases of biodiversity. Wadden Sea Newsl 1:12–14
Sepúlveda R, Camus P, Moreno C (2016) Diversity of faunal assemblages associated with ribbed mussel beds along the South American coast: relative roles of biogeography and bioengineering. Mar Ecol 37:943–956
Smaal AC (2002) European mussel cultivation along the Atlantic coast: production status, problems and perspectives. Hydrobiologia 484:89–98
Smaal AC, Lucas L (2000) Regulation and monitoring of marine aquaculture in The Netherlands. J Appl Ichthyol 16:187–191
Smith J, Shackley SE (2004) Effects of a commercial mussel Mytilus edulis lay on a sublittoral, soft sediment benthic community. Mar Ecol Prog Ser 282:185–191
Spooner DE, Vaughn CC (2006) Context-dependent effects of freshwater mussels on stream benthic communities. Freshw Biol 51:1016–1024
Taylor JJ, Southgate PC, Rose RA (1997) Fouling animals and their effect on the growth of silver-lip pearl oysters, Pinctada maxima (Jameson) in suspended culture. Aquaculture 153:31–40

Thiel M, Ullrich N (2002) Hard rock versus soft bottom: the fauna associated with intertidal mussel beds on hard bottoms along the coast of Chile, and considerations on the functional role of mussel beds. Helgol Mar Res 56:21–30

Thompson R, MacNair N (2004) An overview of the clubbed tunicate (Styela clava) in Prince Edward Island. PEI Department of Agriculture, fisheries, aquaculture and forestry: fisheries and aquaculture division. Technical report #234. 29pp

Tilman D, Reich PB, Knops JMH (2006) Biodiversity and ecosystem stability in a decade-long grassland experiment. Nature 441:629–632

Trianni M (1996) The influence of commercial oyster culture activities on the benthic infauna of Arcata Bay. Master of Science, The Faculty of Humboldt State University,

Tsuchiya M, Nishihira M (1986) Islands of Mytilus-Edulis as a habitat for small intertidal animals – effect of Mytilus age structure on the species composition of the associated Fauna and Community organization. Mar Ecol Prog Ser 31:171–178

van der Zee EM, van der Heide T, Donadi S, Eklof JS, Eriksson BK, Olff H, van der Veer HW, Piersma T (2012) Spatially extended habitat modification by intertidal reef-building bivalves has implications for consumer-resource interactions. Ecosystems 15:664–673

Van Stralen M, Craeymeersch J, Drent J, Glorius S, Jansen JMJ, Smaal A (2013) Het mosselbestand op de PRODUS-vakken en de effecten van de visserij daarop. Marinx-rapport 2013:54, 68 pp–PR6

Walne PR (1974) Culture of bivalve molluscs; 50 years of experience at Conwy. Fishing News Books Ltd, Farnham

Wijsman J, Schellekens T, van Stralen M, Capelle J, Smaal A (2014) Rendement van mosselkweek in de westelijke Waddenzee IMARES Wageningen UR Rapport C047/14. 79 pp.

Wijsman JWM, Troost K, Fang J, Roncarati A (2019) Global production of marine bivalves. Trends and challenges. In: Smaal A et al (eds) Good and services of marine bivalves. Springer, Cham, pp 7–26

Wood EM (1987) Subtidal ecology. Edward Arnold (Publishers) Ltd, London, 125 pp

Worm B, Barbier EB, Beaumont N, Duffy JE, Folke C, Halpern BS, Jackson JBC, Lotze HK, Micheli F, Palumbi SR, Sala E, Selkoe KA, Stachowicz JJ, Watson R (2006) Impacts of biodiversity loss on ocean ecosystem services. Science 314:787–790

Ysebaert T, Hart M, Herman PMJ (2009) Impacts of bottom and suspended cultures of mussels Mytilus spp. on the surrounding sedimentary environment and macrobenthic biodiversity. Helgol Mar Res 63:59–74

Ysebaert T, Walles B, Haner J, Hancock B (2019) Habitat modification and coastal protection by ecosystem-engineering reef-building bivalves. In: Smaal A et al (eds) Good and services of marine bivalves. Springer, Cham, pp 253–273

Chapter 15
Enhanced Production of Finfish and Large Crustaceans by Bivalve Reefs

Boze Hancock and Philine zu Ermgassen

Abstract Several bivalve families include species that occur in sufficient densities to modify the environment and create structured biogenic habitat. These habitats have also suffered among the highest losses of any marine habitat globally. In the case of bivalve reefs, the physical structure provided by the shells, supplied with biodeposits produced from filter feeding, supports a high density of macroinvertebrate prey, as well as providing shelter for many juvenile fish. This combination leads to enhanced fish production when compared to the unstructured sediment; the habitat type which typically replaces bivalve reefs when they are destroyed. Measuring the densities of juvenile fish and crustaceans on oyster reefs, and at unstructured control sites provides a measure of the net increase in juvenile fish and large crustaceans supported by oyster habitat. Applying growth and mortality schedules from fishery stock assessment literature allows an estimate of the increased lifetime production of juveniles by oyster reef habitats. Species may also benefit from oyster reefs at later life history stages, but these potential benefits have not been included in the current estimates of production. Services such as increased fish production have been used to highlight the range of stakeholders, in addition to the oyster fishers, that benefit from oyster habitat. The broader constituent base for bivalve habitats includes groups such as recreational anglers and commercial fishers as well as the industries that support them. Engaging with these stakeholders through quantifying the benefits of bivalve habitats to fisheries has proven an invaluable asset in promoting bivalve habitat restoration globally, as well as in drawing more funding into restoration efforts. Furthermore, quantifying fish production introduces the potential to include habitats such as those produced by bivalves in Ecosystem-based Fisheries Management.

B. Hancock (✉)
The Nature Conservancy, URI Graduate School of Oceanography, Narragansett, RI, USA
e-mail: bhancock@tnc.org

P. zu Ermgassen
Changing Oceans Group, School of Geosciences, University of Edinburgh, Grant Institute, Edinburgh, UK
e-mail: Philine.zuermgassen@cantab.net

A. C. Smaal et al. (eds.), *Goods and Services of Marine Bivalves*,
https://doi.org/10.1007/978-3-319-96776-9_15

Abstract in Chinese 摘要:当某些双壳贝类家族的个体密度达到一定程度时,它们可以改变环境并形成结构化的生物栖息地。这类栖息地的消亡也属于全球性海洋栖息地损失的范畴。双壳贝礁,通过贝壳形成物理结构,以摄食活动产生的生物沉积物作为营养物质来源,为众多的大型无脊椎动物提供,并为许多幼鱼提供栖息场所。与非结构化底质相比,这种底质环境会促进鱼类产量提高; 而当这种贝壳礁被破坏时,栖息地的类型往往也会改变。通过对比测量牡蛎礁和对照地点的幼鱼和甲壳类动物的密度,可以衡量牡蛎礁型栖息地对于幼鱼和大型甲壳类动物净增长的促进作用。应用渔业资源评估文献中的生长和死亡率时间表,我们可以估算牡蛎礁类栖息地对延长幼鱼生命周期的作用。许多物种在生命周期的末期可能会从牡蛎礁中受益。除了收获牡蛎外,一些业主也从牡蛎礁栖息地周围的鱼类产量增加而获益。 除了养殖户以及养殖企业,垂钓爱好者也是从双壳贝礁生态多样性的受益者。通过量化双壳贝礁对养殖企业的积极作用,并与企业进行合作是推动全球双壳贝类栖息地恢复工作的重要渠道和方式。此外,将鱼类养殖业中的各种因素进行量化可以更好地阐述牡蛎礁在生态系统水平渔业管理应用中的潜力。

Keywords Fish production · Bivalve habitat · Oyster reef · Mussel bed · Ecosystem services · Ecosystem-based fisheries management

关键词 鱼类生产,双壳栖息地,牡蛎礁,贻贝床生态系统服务,基于生态系统的渔业管理。

15.1 Bivalves As Ecosystem Engineers Supporting Fish Production

Ecosystem engineers are organisms that modulate the availability of resources to other species, by causing physical state changes in biotic or abiotic materials (Jones et al. 1994). In the case of bivalve reefs,[1] they create and maintain habitat primarily through the deposition of generations of shell (Gutiérrez et al. 2003, Walles et al. 2015), supplemented by a constant supply of biodeposits from filter feeding (Kellogg et al. 2013). The structure created by a matrix of shell provides shelter for many species and the biodeposits, in the form of faeces and pseudofaeces, supply concentrated nutrients to the benthic deposit feeders. It is the combination of shelter and protection from predation, combined with the biodeposits fueling a greater abundance of prey, that has long been considered as driving force for enhanced fish production by bivalve habitat (Humphries et al. 2011, Kesler 2015).

[1]A definition of 'reef', and 'biogenic structure' are given in Appendix 1 of the Natura2000 Marine documents at http://ec.europa.eu/environment/nature/natura2000/marine/docs/appendix_1_habitat.pdf.

Bivalve reefs are alternatively referred to as shellfish reefs in many publications.

A range of bivalves fall into the category of ecosystem engineers. The primary groups that generate habitat are the oysters (Ostreidae) along with many species of mussels (Mytilidae). Other groups form aggregations dense enough to be considered as biogenic structure, such as the pearl oysters (Pteriidae), leaf oysters (Isognomonidae), and fan clams or penn shells (Pinnidae) (Gillies et al. 2015). However, relatively few current examples of high density reefs or beds (where the structured habitat has little vertical elevation) exist for these groups. There is also a lack of information on the historic extent of habitat formed by these groups which makes it difficult to determine their historic importance in forming biogenic structure.

Oysters and mussels are recognized worldwide as generating dense beds or reefs that may develop to a depth of many meters (Büttger et al. 2008, Todorova et al. 2009). As such they are the estuarine and higher latitude analogs of coral reefs, often consisting of substantial calcium carbonate structures with an outer veneer of living bivalves (Stenzel 1971, Walles et al. 2015). Bivalve habitats also support a generally diverse and dense array of associated organisms (Harding and Mann 2001). Although there is strong anecdotal evidence that mussel habitat is important for fish production, there are currently no quantitative measures of this impact. Restoration of mussel bed habitat is currently being undertaken on an experimental scale in Port Phillip Bay Australia (*Mytilus edulis*), and in the Haruki Gulf of New Zealand (*Perna canicularis*), along with the penn shell (*Atrina zelandica*), with the aim of improving water quality and for the production of fish, in particular pink snapper (*Pagrus auratus*), a popular recreational species (www. http://reviveour-gulf.org.nz/). Similarly, the fishing community of El Manglito near La Paz, Mexico have traditionally lived from fishing in the Ensenada and La Paz Bay. This community has linked the functional extinction of the pen shell habitat in the Ensenada de La Paz with the collapse of the finfish stocks in this once productive bay, and are planning the restoration of the pen shell habitat as a recovery strategy for the finfish fishery (B. Hancock, personal communication).

Oyster reefs are the best understood of the biogenic bivalve habitats. They have suffered the greatest losses of any marine habitat that has been examined (Beck et al. 2011) with more than 90% loss in many areas (Beck et al. 2011; zu Ermgassen et al. 2012; Alleway and Connell 2015; Gillies et al. 2015). As in the case with the examples from mussel habitat in New Zealand and the Pinnidae shell habitat in New Zealand and Ensenada de La Paz, Mexico, oyster reefs have long been assumed to be important habitats for fish production (Fig. 15.1). The motivation to quantify fish production has been driven by this understanding of the links between healthy bivalve habitats and fish production, and the potential value of this ecosystem service in supporting the restoration of oyster habitat.

Fig. 15.1 Examples of bivalve reefs that form structured habitat. (**a**) *Pinna bicolor* in Streaky Bay, South Australia (Peter Hunt). (**b**) Pearl Oysters *Pinctada albina* in upper Spencer Gulf South Australia (Heidi Alleway) (**c**) Restored Green Lipped Mussel Bed *Perna canaliculus* in the Hauraki Gulf New Zealand (Richard Robinson) (**d**) Restored mussel bed *Mytilus edulis* in Port Phillip Bay, Victoria, Australia (Simon Branigan) (**e**) Oyster reef *Ostrea angasi* in Georges Bay, Tasmania, Australia (Chris Gillies) and (**f**) Restored oyster reef *Crassostrea virginica* in Virginia, USA (Bo Lusk)

15.2 History of Quantifying Fish Production from Oyster Habitat

Oysters have long been fished by coastal communities (Rick and Erlandson 2009). Since the first century AD the Romans imported oysters from as far afield as southern England (Philpots 1890). The rise of mechanised fishing led to the global collapse of oyster stocks (Beck et al. 2011). Consequently, oyster restoration efforts have long been part of the wild oyster fishery in many parts of the world (Saville-Kent 1894; Ogburn et al. 2007). It is, however, only recently that work in the United States paved the way for oyster habitat restoration for the multiple services provided by this habitat, in addition to the oyster fishery (e.g. enhanced water quality and shoreline protection; Petersen et al. 2019; Ysebaert et al. 2019). This is a conservation action that has gained a high level of support as the ecosystem service benefits that oyster habitats provide have become better understood.

One compelling service is the increased production of finfish and large crustaceans. Much of the early thinking behind quantifying the fish production from oyster habitat came from Federal Government mandates in the United States. One initial driver of oyster habitat restoration was a government-legislated requirement to restore the public resources injured by discrete environmental incidents such as chemical or oil spills, the release of pollutants from an identifiable catastrophic event, or from physical damage to the habitat such as dredging for port expansion or land reclamation (e.g. NOAA 1977). The initial legislation was described in the Comprehensive Environmental Response, Cleanup and Liability Act (CERCLA 1980) and the Oil Pollution Act (OPA 1990). For each incident addressed under such legislation the damage first needed to be quantified, prior to designing restoration to make the community 'whole'. These laws helped develop both the practice of oyster habitat restoration and the initial methods to measure fish production from oyster habitat as an additional service to be accounted for. The acts dictate that restoration is undertaken to compensate the public for losses or injuries to natural resources under public ownership and held in trust by government managers and that restoration includes the services that those natural resources would have provided. This legislation continues to influence the quantification of ecosystem services from multiple habitats, including bivalve habitat, and is being expanded in the US section of the Gulf of Mexico through the Restore Act (2012), legislating the response to the Deep Water Horizon oil spill (available at https://www.treasury.gov/services/restore-act/Pages/home.aspx).

A parallel driver of oyster habitat restoration stems from the conservation communities interest in restoring this previously abundant and ecologically significant habitat. While the increasing number of comprehensive studies documenting and quantifying the loss of oyster habitat are essential for setting a realistic order of magnitude for the amount of habitat that might be restored (Beck et al. 2011; zu Ermgassen et al. 2012, 2013; Alleway and Connell 2015; Gillies et al. 2015), simply understanding the loss does not, by itself, generate the incentive to fund and support restoration at the required scale. Given the expense and time required for successful restoration, it has become increasingly necessary to quantify the services provided by the restored oyster habitat in order to place a monetary value on those services and demonstrate the tangible gain society receives from their restoration. Quantifying the expected tonnage of fish and large crustaceans produced per unit area of restored oyster reef has been a powerful way to demonstrate the value of oyster habitat to society and the return on the investment in oyster habitat restoration (C. Gillies, TNC, personal communication).

An additional motivation to quantify the fish production from marine habitats was the concept of Essential Fish Habitat (EFH) which was introduced to fishery management in many jurisdictions from the mid 1980s (e.g. Minns et al. 2011) and in the US in 1996 through the Magnusson-Stephenson Fishery Conservation and Management Act. The Act linked fish production to habitat, attempting to expand the focus of fishery management to include consideration of the capacity of the ecosystem to produce fish, rather than focusing purely on limiting extraction from stocks of the target species (Pikitch et al. 2004). The legislation also had the effect of focusing attention on how to measure the impact of habitat on fish production (Peterson et al. 2000). The concept of certain habitats being limited but important to one or more life history stages and, therefore, the overall success of a fish population, is among the most fundamental questions in fisheries ecology, and the foundation of the concept of EFH. When applied to the early life history stages, generally referred to as juvenile fish habitat, it recognizes that the early life history stages are typically those with the highest mortality rates and where small changes in survival can have large impacts on the number of individuals surviving to older cohorts.

The changes in habitat dependence of different age classes of many fish and large crustaceans complicates the measurement of the relative values of fish production by habitat for these species. A complete accounting of fish production would require assessing the contribution of habitat to fish production by individual age classes of fish. One alternative is to focus on one year class, simplifying the investigation to a level that can be measured and applied (e.g. Levin and Stunz 2005). The 0+ year class is the most abundant and also usually subject to the highest rates of mortality. Consequently, nursery habitats that impact the survival of this cohort will have the greatest influence on the lifetime production.

The work of Peterson et al. (2003) has been influential in the development of methods for modelling the fish production from oyster habitat. zu Emgassen et al. (2016a) have conducted one of the first meta-analyses of the degree to which oyster habitat enhances the density of young-of-year fish and macro-invertebrates, and the consequent increase in production of those species over their lifetime. This has made oyster reefs a model system for quantifying the magnitude and regional variability in augmented fish production from nursery habitats (Fig. 15.2).

Fig. 15.2 Fish and invertebrates utilizing restored (**a**) Mussel bed *Perna canaliculus* in Hauraki Gulf, New Zealand (Shane Kelly) and (**b**) *Crassostrea virginica* oyster reef Rhode Island USA (Matt Griffin)

15.3 Current Status of Quantifying Fish Production Enhancement by Oyster Habitat

Quantitative data on the degree to which fish and macroinvertebrates are enhanced by bivalve habitats are rare outside of the United States. While there has been recent progress in understanding the role of *Modiolus modiolus* in Europe as an important habitat for the commercially important whelk *Buccinum undatum* (Kent et al. 2016, 2017), for most bivalve habitats outside of the U.S. evidence is limited to historical documentation of species counts (e.g. Moebius 1883; Riesen and Reise 1982). In order to quantify the enhancement of fish and invertebrate production by bivalve habitats, it is necessary to measure the abundance of the target age classes within the habitat, relative to where that habitat is absent. As such, repeated and paired density data from the contrasting habitats are essential in supporting such quantification.

By collating available paired on and off oyster reef fish and invertebrate data from 31 studies in the United States, zu Ermgassen et al. (2016a) identified species for which the juveniles were consistently found at higher abundances on oyster habitats as opposed to unstructured mud and sand habitat. These habitats often replace oyster reefs when lost, and were therefore considered the most suitable control habitat for comparison. As in Peterson et al. (2003), the authors found that the presence of oyster reef enhanced species at both the juvenile and later life history stages. They also found marked differences between biogeographical regions with regards to the species of fish and invertebrates enhanced by oyster reefs, with 12 species in the Mid and South Atlantic and to 19 species in the Gulf of Mexico enhanced as juveniles, and two and five species respectively at later life history stages.

By applying established growth and mortality estimates to the enhanced densities of juveniles found on as opposed to off oyster reefs, zu Ermgassen et al. (2016a) estimated the year on year production of each of the consistently enhanced species (Fig. 15.3). This represents the increased production resulting from the presence of oyster reef habitat as opposed to unstructured benthic habitats. Enhancement can

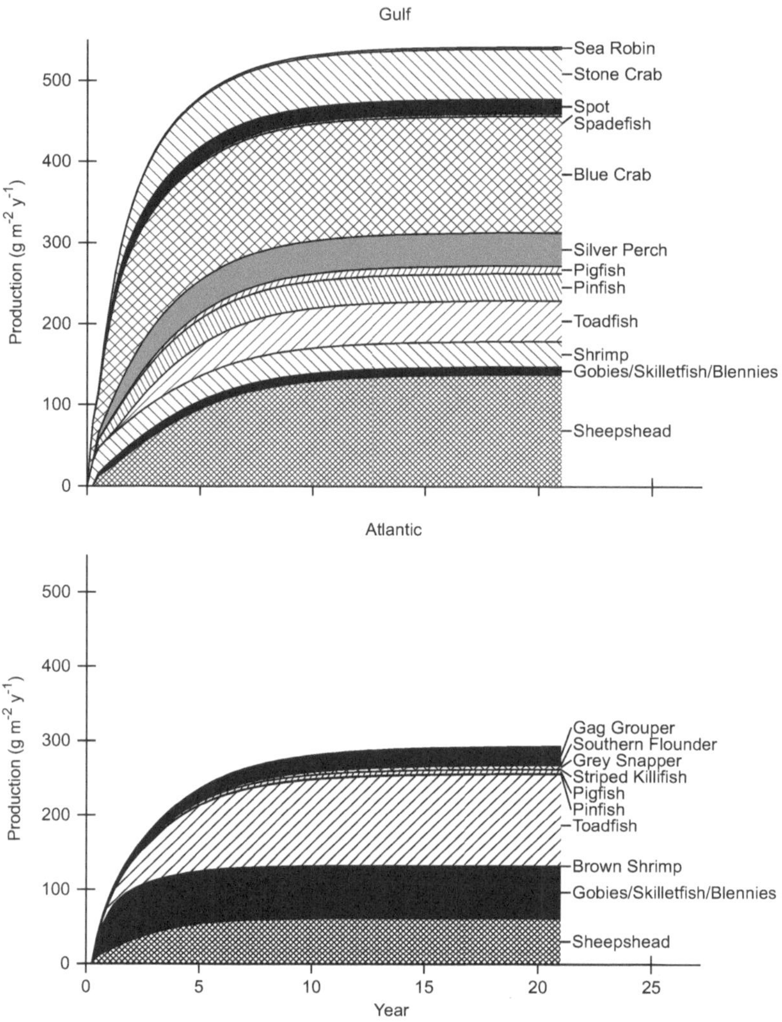

Fig. 15.3 Graphs of the annual enhancement in production per species resulting from the presence of oyster reef as opposed to unstructured benthic habitat for the Gulf of Mexico and the US Atlantic south of Cape Cod. Enhancement can be considered as annual post-restoration enhancement up to the asymptote, which represents the annual enhancement seen at existing natural reefs. The recreationally and commercially important species include stone crab, spot, spadefish, blue crab, silver perch, pigfish, shrimp, sheepshead, gag grouper, southern flounder and grey snapper, with sea robin and toad fish being targeted by a smaller but increasing number of fishers (zu Ermgassen et al. 2016a, updated in zu Ermgassen et al. 2018)

represent either production post-restoration; in which case the production value increases over time as successive generations and years of growth which can be attributed to the oyster habitats accumulate (Fig. 15.3), or existing oyster habitats (the value illustrated at the end of the graph can be attributed year on year). The error distribution was also calibrated to the variance obtained from the original publications. They found that each ha of oyster reef provided on average an additional 2.83 ± 0.57 t/year in the South and Mid-Atlantic and 5.28 ± 1.28 t/year in the Gulf of Mexico (Fig. 15.3). These estimates currently stand as the most developed estimates of the fisheries ecosystem service potential of oyster reef habitats. Of these species enhanced as juveniles, many of them are of direct fisheries value, with others representing forage fish and prey for the higher trophic fisheries species (Fig. 15.3). Only the contribution of the species directly benefiting from oyster reef as a juvenile nursery ground are fully quantified in this approach. The contribution of enhanced growth in later life history stages was not tackled, due to a concern regarding double counting of benefits and the ongoing attraction versus production debate surrounding the role of structured habitats for later life history stages (e.g. Pierson and Eggleston 2014). The full ecosystem service value from the enhancement by oyster reefs on *fisheries* is therefore challenging to quantify, as the total value will be a sum of the quantified juvenile enhancement of fisheries species, the contribution of the enhanced abundance and biomass of prey species consumed once they leave the reef, and the contribution of the enhanced prey items available on the reef for species that associate with the reef at later life history stages. Peterson et al. (2003) attempted to assess this value, but it is generally agreed that more detailed habitat specific life history information would greatly improve the existing estimates. Despite the current model providing a conservative estimate of the augmented fish production from the increased abundance of only one year class it is, none the less, substantial and does provide a measure of fish production on a regional scale that is relevant to policy and management. The model also provides valuable estimates of the uncertainty surrounding the model predictions (Fig. 15.4).

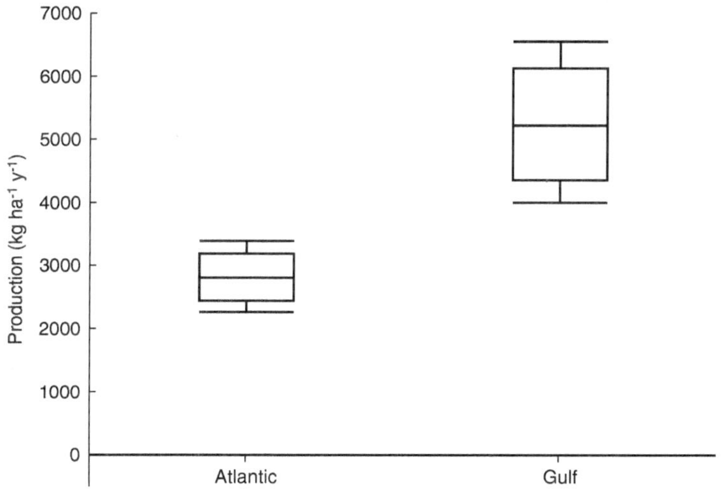

Fig. 15.4 Mean, upper and lower quartile and minimum and maximum estimated enhancement of gross production of fish and large crustacean resulting from the presence of oyster reef as opposed to unstructured benthic habitat in both the Gulf of Mexico and the Mid and South Atlantic region of the USA at t_{max} of the longest-lived species (zu Ermgassen et al. 2016a)

Table 15.1 Species found by zu Ermgassen et al. (2016a) to derive growth enhancement from *C. virginica* oyster reef in the Gulf of Mexico and in the South and Mid-Atlantic (zu Ermgassen 2016a)

	Species	Common name	Proportion of individuals caught on oyster reefs (%)
Gulf of Mexico	*Menticirrhus americanus*	Southern kingfish	52
	Paralichthys lethostigma	Southern flounder	82
	Pogonias cromis	Black drum	75
	Rhinoptera bonasus	Cownose ray	82
	Sciaenops ocellatus	Red drum	63
Atlantic coast	*Centropristis striata*	Black Sea bass	63
	Morone saxatilis	Striped bass	93

The potential importance of oyster reef habitats to species which utilize the habitat at later life history stages can be inferred by calculating the relative amount of time spent within each habitat. While fish may be caught in areas through which they seek to migrate quickly, rather than where they are spending the majority of their time, there is ample evidence of many large species choosing to spend more time over oyster reefs as they are rich foraging ground (e.g. Harding and Mann 2003; George 2007). zu Ermgassen et al. (2016a) therefore calculated the proportion of time spent over oyster reef, or the proportion of adult individuals caught over oyster reef as opposed to neighboring unstructured habitats in order to assess which species oyster reef may be important for at later life history stages. They found that five species from the Gulf of Mexico and two species from the Atlantic coast of the US that spent a large proportion of their time on oyster reefs: between 52% (Southern kingfish, *Menticirrhus americanus*) and 93% (Striped bass, *Morone saxatilis*) (Table 15.1). These species are believed to derive some degree of growth enhancement from the oyster habitat, from feeding disproportionately frequently on oyster reefs.

15.4 Assumptions and Limitations of the Current Approach

The approach developed initially by Peterson et al. (2003) and further developed by zu Ermgassen et al. (2016a) has provided a novel opportunity to gauge the lifetime benefits of nursery habitats to fish and invertebrate populations. The benefits that can be attributed to nursery habitats from reduced juvenile mortality are substantial and otherwise extremely challenging to capture. The approach is, however, dependent on the application of established fish growth and mortality models used in fishery stock assessment and as such should be caveated by the same underlying assumptions. Estimates of fish and invertebrate growth were derived by applying the von Bertalanffy growth curve to juveniles. In order to do so, various life history traits need to be known. These traits (length at infinite age, the constant K, and the length at time equals zero) are themselves estimated and may therefore be subject to some error. Estimated natural mortality is also required for the model. Mortality is one of the greatest sources of uncertainty in fisheries models (Rosenberg and Restrepo 1994), especially at smaller size classes as mortality derived from the field is often reliant on fisheries size classes. In order to reduce, as far as possible, the uncertainty in the mortality estimates applied, zu Ermgassen et al. (2016a) used the size dependent mortality equation developed by Lorenzen (2000), so as to better represent the higher mortality suffered by the small size classes of fish and invertebrates represented in the model.

It is universally known that any model is only as good as the data it uses. While oyster reefs in the U.S. are the best studied in the world, zu Ermgassen et al. (2016a) point out that some of the differences in fish and invertebrate enhancement between regions are still likely due to a lack of data and differing sampling efforts or techniques between regions. As such, the inclusion of more data in the model can only serve to improve the resulting estimates of the benefits of oyster reefs as nursery habitats. The data handling approach used required that species were represented in at least two different estuaries in order for that species to be included in the assessment. It is therefore possible that some rarer species, or species which are not as effectively captured using density-specific capture techniques (e.g. drop traps, seines), are currently missing from the existing estimates. A larger number of studies seeking to quantify the enhancement of the fish and invertebrate community by oyster reefs can only serve to improve the current model.

One important assumption of the model that is highlighted in zu Ermgassen et al. (2016a) is that the bivalve habitat must be limiting in the site of interest. The model provides an estimate of the per unit area enhancement of the fish and invertebrate community by oyster reef habitats. The authors argue that in the current landscape of extreme loss of oyster and other bivalve habitats globally it is likely that, for species whose juveniles are enhanced by oyster reef presence, habitat is in fact limiting. As such the addition of habitat should result in a greater number of individuals surviving to larger size classes. The authors, however, concede that should substantial areas of oyster reef be restored, other factors may well start to limit the production and the assumed linear relationship between habitat area and juvenile enhancement would cease to exist. The point at which this would happen is likely to be highly species dependent and the position of any such threshold, or even how to derive it, remains unknown. As such, it is important to bear this assumption in mind when planning large scale restoration or recovery of oyster habitats, so as not to oversell the potential of oyster reef in supporting fisheries as an ecosystem service in these later stages of oyster habitat recovery.

A further consideration is the effect of habitat redundancy, or the interaction between structured habitats in close proximity, on the nursery function of oyster reefs. Oyster reefs close to alternative structured habitats, such as seagrasses and saltmarshes may not result in the same, or indeed any, observable enhancement of juvenile fish (Grabowski et al. 2005; Geraldi et al. 2009), most likely because the abundance of an equivalent structured habitat can provide similar food and shelter to oyster reefs. This is, however, certainly not always the case, with some studies finding that oyster reefs in seagrass and saltmarsh landscapes enhanced invertebrate (Grabowski et al. 2005) and fish (Stunz et al. 2010) communities. An assessment of the interaction between the different types of essential fish habitat, when they occur in close proximity, will be important for fine tuning the overall estimates of fish production from structured habitat on an estuary scale.

15.5 Making the Results Available

Among the primary motivators for quantifying the production of fish and large crustaceans from bivalve habitat is the need to quantify the services lost (and therefore the services to be restored), following an environmental disaster like an oil spill, and also to help regulators and funders within the conservation community visualise the benefit from restoration. The ability to easily visualise the return on a restoration investment in terms of the fish produced, by species, helps the many stakeholder groups formally involved in habitat restoration, set meaningful goals for restoration on a bay or estuary scale. While the quantity of fish produced from a given restoration scenario can be calculated from the data provided (zu Ermgassen et al. 2016a) or estimated from the graphs in Fig. 15.3, providing a tool that allows this benefit to be immediately calculated for any proposed restoration could benefit those conversations. To facilitate this outcome a coalition of partners has produced a manual describing the multiple services derived from oyster habitat (zu Ermgassen 2016b). Of these services, the filtration (Cranford 2019) and fish production services have been quantified sufficiently accurately to be included in a 'Web Calculator' for the USA (Fig. 15.5, the calculator is available at http://oceanwealth.org/tools/oyster-calculator/). At present estimates of enhanced fish production are available for the US Gulf of Mexico and the US South and mid-Atlantic coasts as described above, and filtration results are available for the US east, gulf and Pacific coasts. The area included in the calculator will be expanded as the relevant filtration algorithms and fish production data become available. As additional data will likely allow fine tuning of the results the calculator and associated manual (zu Ermgassen 2016b) may be updated and the web site should be accessed for the most recent results. Additional services such as denitrification rates (Ferreira and Bricker 2019) will also be added when their quantification is sufficiently understood to be represented at a regional scale.

The calculator is designed to allow users to enter data such as existing oyster density and mean size, expected oyster density and mean size for the restored habitat, and adjust the target % of the estuary volume to be filtered by oysters within the residence time of the estuary (see also Smaal and van Duren 2019). Existing data such as estuary volume, the residence time of water within the estuary, mean summer water temperature, and the historic percentage of estuary filtration achieved by the biomass of oysters present at the earliest available census (generally around 1900), and even recent existing oyster size and density values, are provided where available. The site calculates the area of oyster habitat that would need to be restored

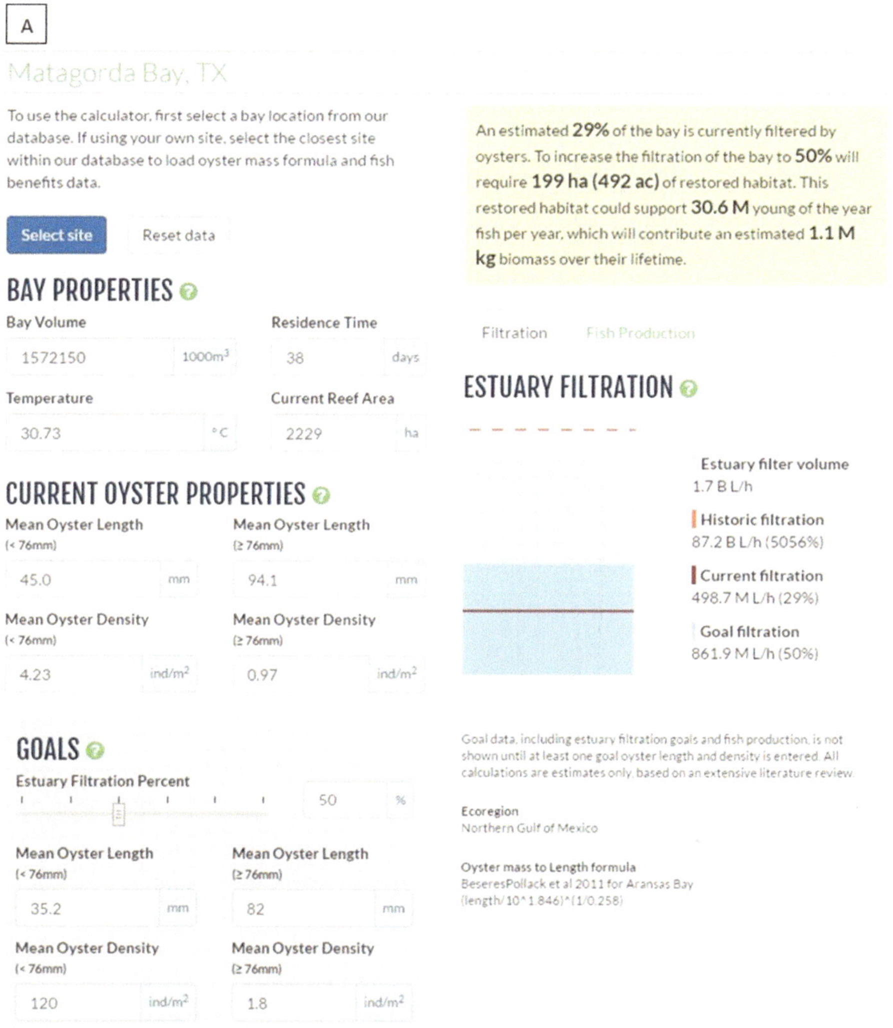

Fig. 15.5 Screen capture of the 'Oyster Calculator' Illustrating pre-loaded data for Matagorda Bay, Texas, with hypothetical data in the 'Goals' section and the 'Estuary Filtration Percent' set to 50%. The output tab for 'Filtration' in X5, (**a**) would normally toggle in the same position as the 'Fish Production' tab in X5, (**b**) with only one output tab visible at a time. Only a portion of the fish production data are shown in X5, (**b**)

to achieve the specified level of filtration and the number and weight of fish, by species, that would be produced from that area of restored oyster habitat. Making these results available in real time during planning conversations is intended to facilitate setting objectives for oyster habitat restoration based on the filtration and fish production services returned at a system scale.

The model results provided in the calculator represent the mean production across the whole of a region. It is therefore critical that local knowledge be used to

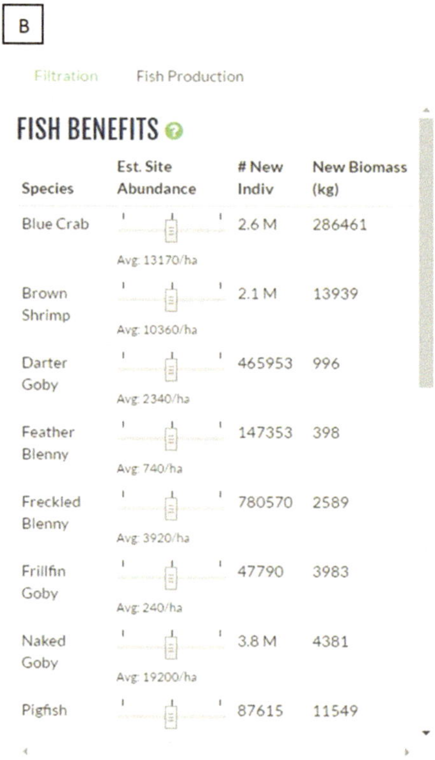

Fig. 15.5 (continued)

adapt the results to more accurately reflect the estimated production at a given site of interest. There are likely to be local factors that affect the availability of one or more species within the suite of species identified as being enhanced by oyster habitat (Humphries and La Peyre 2015). The ability to account for these local variations in species abundance has therefore been built into the calculator. There are options to set the production for any species to zero if the species is deemed absent from the site or rare. There is the option to use the average production from the meta-analysis, or even the upper confidence interval if there is evidence that a species is particularly abundant in the estuary being considered.

15.6 Management Applications

The major threat responsible for the reduction of bivalve habitat globally has been overharvest. In fact, the estimated 85% reduction in oyster habitat over the last approximately 100 years (Beck et al. 2011) is itself an underestimate, as most of the historical surveys undertaken around the end of the 1800s or early 1900s, and used as a baseline measure of the historical extent of oyster habitat, were conducted because of concerns that overfishing had already depleted the oyster stocks (zu

Ermgassen et al. 2012, 2016b). Much of the subsequent depletion of these habitats has occurred because managers of the fisheries have been responding to the inputs and concerns of only one stakeholder group interested in the habitat; the bivalve fishers. Managers, or the politicians they advise, have therefore been focused solely on the landings of bivalves rather than considering those landings as just one of a number of legitimate services to be considered when managing the bivalve resource. Recognizing that the recreational and commercial fishers, that benefit from the fish and crustaceans produced by the oyster habitat, are also stakeholders with a legitimate interest in the bivalve habitat, along with members of the associated fish processing and recreational support industries, has the potential to change the view of managers responsible for that resource. The same can be said for the constituents connected to the other services provided by bivalve habitats described in this book, such as water quality from filtration and denitrification or increased coastal resilience from shoreline protection (Brumbaugh et al. 2010; Ferreira and Bricker 2019). Similarly, managing oyster resources for harvest alone has tended to focus the emphasis on replenishment activities or put-and-take management intended to increase the supply of oysters for harvest or the amount of shell substrate available for the recruitment of juvenile oysters and their subsequent harvest. This does not consider that oyster biomass and the condition of the oyster reef may be important factors influencing the provision of additional services as well as the long-term sustainability of the restored reef (Grabowski et al. 2012). Demonstrating the value of the finfish and crustaceans produced from oyster habitat is a powerful tool for supporting the protection of at least a portion of the remaining bivalve habitat and investing in the restoration of additional habitat.

Having estimates of the fish production from oyster habitat well documented and available, if only for a small region given the global distribution of biogenic bivalve habitats, provides the ability to influence fisheries management in two important ways. It provides the logic for fundamentally changing the paradigm for managing the fishery, based on consideration of the multiple stakeholder groups impacted by changes in the level of services provided by bivalve habitats, in addition to harvest. It also introduces the option of including bivalve habitat in the management considerations for the finfish and large crustacean species supported by those habitats, in a truly Ecosystem Based Fishery Management (EBFM) scenario.

Ecosystem based fisheries management has been a goal for many fisheries managers for many years and has been adopted to various extents by most fisheries management agencies worldwide (e.g. Fletcher et al. 2010). Most management agencies also recognize that there is still a long way to go, in order to approach comprehensive EBFM (Berkes 2012). Developing habitat specific fish production measures for oyster habitat has generated interest in developing similar measures for multiple essential fish habitats. Analogous measures are currently available from seagrass in southern Australia (Blandon and zu Ermgassen 2014) and shrimp from seagrass in Queensland, Australia (Watson et al. 1993). The development of models to estimate the fish and large crustacean production are currently underway for salt marsh and seagrass habitats from the US and from mangrove habitat globally (Hancock and zu Ermgassen, personal communication). The US studies of fish production from salt marsh and seagrass habitats include provision for engaging the fisheries management community in considering how measures of fish production from habitat can be included in fisheries management decisions (NOAA, National Marine Fisheries Service, personal communication), a potentially productive direction for the development of EBFM.

Acknowledgements The authors are grateful to Dr. B. Walles and an anonymous reviewer for their constructive comments on the manuscript.

References

Alleway HK, Connell SD (2015) Loss of an ecological baseline through the eradication of oyster reefs from coastal ecosystems and human memory. Conserv Biol 29(3):795–804. https://doi.org/10.1111/cobi.12452

Beck MW, Brumbaugh RD, Airoldi L, Caranza A, Coen LD, Crawford C, Defeo O, Edgar GJ, Hancock B, Kay M, Lenihan H, Luckenbach MW, Toropova CL, Zhang G (2011) Oyster reefs at risk and recommendations for conservation, restoration and management. Bioscience 61(2):107–116

Berkes E (2012) Implementing ecosystem-based management: evolution or revolution? Fish Fish 13:465–476

Blandon A, zu Ermgassen PSE (2014) Quantitative estimate of commercial fish enhancement by seagrass habitat in southern Australia. Estuar Coast Shelf Sci 141:1–8

Brumbaugh RD, Beck MW, Hancock B, Wrona-Meadows A, Spalding M, zu Ermgassen P (2010) Changing a management paradigm and rescuing a globally imperiled habitat. National Wetlands Newsletter, November–December, pp 16–20

Büttger H, Asmus H, Asmus R, Buschbaum C, Dittmann S, Nehls Helgol G (2008) Community dynamics of intertidal soft-bottom mussel beds over two decades. Mark Res 62:23–36. https://doi.org/10.1007/s10152-007-0099-y

CERCLA (1980) Comprehensive environmental response, cleanup and liability act, Available at https://www.gpo.gov/fdsys/pkg/USCODE-2011-title42/html/USCODE-2011-title42-chap103.htm

Cranford P (2019) Magnitude and extent of water clarification services provided by bivalve suspension feeding. In: Smaal A, Ferreira JG, Grant J, Petersen JK, Strand O (eds) Goods and services of marine bivalves. Springer, Cham, pp 119–141

Ferreira J, Bricker S (2019) Assessment of nutrient trading services from bivalve farming. In: Smaal A, Ferreira JG, Grant J, Petersen JK, Strand O (eds) Goods and services of marine bivalves. Springer, Cham, pp 551–584

Fletcher WJ, Shaw J, Metcalf SJ, Gaughan DJ (2010) An ecosystem based fisheries management framework: the efficient, regional-level planning tool for management agencies. Mar Policy 34(6):1226–1238

George GJ (2007) Acoustic tagging of black drum on Louisiana oyster reefs: movements, site fidelity, and habitat use. Master of Science, Louisiana State University and Agricultural and Mechanical College, 71 pp

Geraldi NR, Powers SP, Heck KL, Cebrian J (2009) Can habitat restoration be redundant? Response of mobile fishes and crustaceans to oyster reef restoration in marsh tidal creeks. Mar Ecol Prog Ser 389:171–180

Gillies CL, Creighton C, McLeod IM (eds) (2015) Bivalve reef habitats: a synopsis to underpin the repair and conservation of Australia's environment, social and economically important bays and estuaries. Centre for Tropical Water and Aquatic Ecosystem Research (TropWATER) Publication, James Cook University, Townsville, 90 pp

Grabowski JH, Hughes AR, Kimbro DL, Dolan MA (2005) How habitat setting influences restored oyster reef communities. Ecology 86(7):1926–1935

Grabowski JH, Brumbaugh RD, Conrad RF, Keeler AG, Opaluch JJ, Peterson CH, Piehler MF, Powers SP, Smyth AR (2012) Economic valuation of ecosystem services provided by oyster reefs. Bioscience 62:900–909

Gutiérrez JL, Jones CG, Strayer DL, Iribarne OO (2003) Mollusks as ecosystem engineers: the role of shell production in aquatic habitats. Oikos 101(1):79–90

Harding JM, Mann R (2001) Oyster reefs as fish habitat: opportunistic use of restored reefs by transient fishes. J Shellfish Res 20(3):951–959

Harding JM, Mann R (2003) Influence of habitat on diet and distribution of striped bass (*Morone saxatilis*) in a temperate estuary. Bull Mar Sci 72:841–851

Humphries AT, La Peyre MK (2015) Oyster reef restoration supports increased nekton biomass and potential commercial fishery value. PeerJ 3:e1111. https://doi.org/10.7717/peerj.1111

Humphries AT, La Peyre MK, Decossas GA (2011) The effect of structural complexity, prey density, and "predator-free space" on prey survivorship at created oyster reef Mesocosms. PLoS One 6:e28339

Jones CG, Lawton JH, Shachak M (1994) Organisms as ecosystem engineers. Oikos 69:373–386

Kellogg ML, Cornwell JC, Owens MS, Paynter KT (2013) Denitrification and nutrient assimilation on a restored oyster reef. Mar Ecol Prog Ser 480:1–19. https://doi.org/10.3354/meps10331

Kent FEA, Gray MJ, Last KS, Sanderson WG (2016) Horse mussel reef ecosystem services: evidence for a whelk nursery habitat supporting a shellfishery. Int J Biodivers Sci Ecosyst Serv Manag 12:172–180. https://doi.org/10.1080/21513732.2016.1188330

Kent FEA, Mair JM, Newton J, Lindenbaum C, Porter JS, Sanderson WG (2017) Commercially important species associated with horse mussel (Modiolus modiolus) biogenic reefs: a priority habitat for nature conservation and fisheries benefits. Mar Pollut Bull 118:71–78. https://doi.org/10.1016/j.marpolbul.2017.02.051

Kesler KE (2015) Influence of the biotic and structural components of Crassostrea virginca on the oyster reef community. Doctor of Philosophy, University of Maryland, 162 pp

Levin PS, Stunz GW (2005) Habitat triage for exploited fishes: can we identify essential "essential fish habitat?". Estuar Coast Shelf Sci 64(1):70–78. https://doi.org/10.1016/j.ecss.2005.02.007

Lorenzen K (2000) Allometry of natural mortality as a basis for assessing optimal release size in fish-stocking programmes. Can J Fish Aquat Sci 57:2374–2381

Minns CK, Randall RG, Smokorowski KE, Clarke KD, Vélez-Espino A, Gregory RS, Courtenay S, LeBlanc P (2011) Direct and indirect estimates of the productive capacity of fish habitat under Canada's Policy for the Management of Fish Habitat: where have we been, where are we now, and where are we going? Can J Fish Aquat Sci 68(12):2204–2227. https://doi.org/10.1139/f2011-130

Moebius K (1883) The oyster and oyster-culture. Report of commissioner of fish and fisheries, pp 683–747

NOAA (1977) Habitat equivalency analysis: an overview, policy and technical paper series, no. 95–1, damage assessment and restoration program. National Oceanographic and Atmospheric Administration, Silver Spring, MD, USA

Ogburn DM, White I, McPhee DM (2007) The disappearance of oyster reefs from eastern Australian estuaries – impact of colonial settlement or mudworm invasion? Coast Manag 35:271–287

OPA (1990) Oil Pollution Act available at https://www.congress.gov/bill/101st-congress/house-bill/1465

Petersen J, Holmer M, Termansen M, Hasler B (2019) Nutrient extraction through bivalves. In: Smaal A, Ferreira JG, Grant J, Petersen JK, Strand O (eds) Goods and services of marine bivalves. Springer, Cham, pp 143–177

Peterson CH, Summerson HC, Thomson E, Lenihan HS, Grabowski J, Manning L, Micheli F, Johnson G (2000) Synthesis of linkages between benthic and fish communities as a key to protecting essential fish habitat. Bull Mar Sci 66(3):759–774

Peterson CH, Grabowski JH, Powers SP (2003) Estimated enhancement of fish production resulting from restoring oyster habitat: quantitative valuation. Mar Ecol Prog Ser 264:249–264

Philpots JR (1890) Oysters and all about them. John Richardson & Co, London/Leicester

Pierson KJ, Eggleston DB (2014) Response of estuarine fish to large-scale oyster reef restoration. Trans Am Fish Soc 143:273–288

Pikitch EK, Santora C, Babcock EA, Bakun A, Bonfil R, Conover DO, Dayton P, Doukakis P, Fluharty D, Heneman B, Houde ED, Link J, Livingston PA, Mangel M, McAllister MK, Pope J, Sainsbury KJ (2004) Ecosystem-based fishery management. Science 305:346–347
Rick TC, Erlandson JM (2009) Coastal exploitation. Science 952:325–326
Riesen W, Reise K (1982) Macrobenthos of the subtidal Wadden Sea: revisited after 55 years. Helgoländer Meeresun 35:409–423
Rosenberg AA, Restrepo VR (1994) Uncertainty and risk evaluation in stock assessment advice for U.S. Marine Fisheries. Can J Fish Aquat Sci 51:2715–2720
Saville-Kent W (1894) Fish and fisheries of Western Australia, from the registrar generals western Australian year-book 1893–4. Commissioner of Fisheries to the West Australian Government
Smaal A, van Duren L (2019) Bivalve aquaculture carrying capacity: concepts and assessment tools. In: Smaal A, Ferreira JG, Grant J, Petersen JK, Strand O (eds) Goods and services of marine bivalves. Springer, Cham, pp 451–483
Stenzel HB (1971) Oysters. In: Moore RC (ed) Treatise on invertebrate paleontology, part N. University of Kansas Press, Kansas, pp 953–1224
Stunz GW, Minello TJ, Rozas LP (2010) Relative value of oyster reef as habitat for estuarine nekton in Galveston Bay, Texas. Mar Ecol Prog Ser 406:147–159
Todorova V, Micu D, Klisurov L (2009) Unique oyster reefs discovered in the Bulgarian Black Sea. Comptes rendus de l'Acad'emie bulgare des Sciences 62(7):871–874
Walles B, Mann R, Ysebaert T, Troost K, Herman PMJ, Smaal AC (2015) Demography of the ecosystem engineer *Crassostrea gigas*, related to vertical reef accretion and reef persistence. Estuar Coast Shelf Sci 154:224–233
Watson RA, Coles RG, Lee Long W (1993) Simulation estimates of annual yield and landed value for commercial penaeid prawns from a tropical seagrass habitat, northern Queensland, Australia. Aust J Mar Freshwat Res 44:211–219
Ysebaert T, Hancock B, Walles B (2019) Habitat modification and coastal protection by ecosystem-engineering reef-building bivalves. In: Smaal A, Ferreira JG, Grant J, Petersen JK, Strand O (eds) Goods and services of marine bivalves. Springer, Cham, pp 253–273
zu Ermgassen PSE, Spalding MD, Blake BD, Coen LD, Dumbauld B, Geiger S, Grabowski JH, Grizzle R, Luckenbach M, McGraw K, Rodney W, Ruesink JL, Powers SP, Brumbaugh R (2012) Historical ecology with real numbers: past and present extent and biomass of an imperiled estuarine ecosystem. Proc R Soc B 279:3393–3400. https://doi.org/10.1098/rspb.2012.0313
zu Ermgassen P, Spalding M, Allison A (2013) The native oyster: Britain's forgotten treasure. Br Wildl 2013:317–324
zu Ermgassen PSE, Grabowski JH, Gair JR, Powers SP (2016a) Quantifying fish and mobile invertebrate production from a threatened nursery habitat. J Appl Ecol 53:596–606. https://doi.org/10.1111/1365-2664.12576
zu Ermgassen P, Hancock B, DeAngelis B, Greene J, Schuster E, Spalding M, Brumbaugh R (2016b) Setting objectives for oyster habitat restoration using ecosystem services: a manager's guide. The Nature Conservancy, Arlington, 76pp
zu Ermgassen PSE, Grabowski JH, Gair JR, Powers SP (2018) Corrigendum. J Appl Ecol. https://doi.org/10.1111/1365-2664.13143

Part III
Cultural Services

Chapter 16
Introduction to Cultural Services

Aad C. Smaal and Øivind Strand

Abstract Cultural services of marine bivalves are of high value as they provide well-being in many different ways. These services are more difficult to quantify but provide a lot of qualities. Shell collectioning, shells as archives, community efforts for bivalve restoration and gardening are some cases of cultural services. Marine bivalves have been recognized as a carrier of a variety of cultures since pre-historic times.

Keywords Shells · Shell collection · Gardening

Cultural services are defined in the Millennium Assessment as the nonmaterial benefits people obtain from ecosystems, such as cultural diversity, spiritual and religious values, knowledge systems, educational values, inspiration, aesthetic values, social relations, sense of place, cultural heritage values, recreation and ecotourism (Millennium Assessment 2005). For marine bivalves many examples exist of cultural services. Shells are well-known collector items. Collecting seashore shells is worldwide spread leisure activity, and an organised profession as well, in the framework of the scientific discipline of malacology. This links to marine bivalves as a source of knowledge. Shells as fossil records have information for evolutionary studies, and their mineral content can reflect past climatological events as long-term archives. Shells are widely used for decoration and in art. Educational programs and community involvement exist in bivalve restoration programs. Sea gardening of marine bivalves is an upcoming leisure activity. Hence cultural services link directly to social structures that provide the framework for the appreciation of these

A. C. Smaal (✉)
Wageningen UR – Wageningen Marine Research (WMR), Yerseke, The Netherlands

Department of Aquaculture and Fisheries, Wageningen University, Wageningen, The Netherlands
e-mail: aad.smaal@wur.nl

Ø. Strand
Institute of Marine Research, Bergen, Norway
e-mail: oivind.strand@imr.no

A. C. Smaal et al. (eds.), *Goods and Services of Marine Bivalves*,
https://doi.org/10.1007/978-3-319-96776-9_16

services. Nonmaterial services may be more difficult to quantify than the other services, yet the benefits for people go far beyond the material benefits, as it concerns the core of human life that is able to reflect on all different services of – in this case – the marine bivalves (Daniel et al. 2012). In this section some examples of cultural services are reviewed, from community activities in different forms to scientific application of shell archaeology.

References

Daniel TC, Muhar A, Arnberger A, Aznar O, Boyd JW, Chan KMA, Costanza R, Elmqvist T, Flint CG, Gobster PH, Grêt-Regamey A, Lave R, Muhar S, Penker M, Ribe RG, Schauppenlehner T, Sikor T, Soloviy I, Spierenburg M, Taczanowska K, Tam J, von der Dunk A (2012) Contributions of cultural services to the ecosystem services agenda. PNAS 109(23):8812–8819

Millennium Ecosystem Assessment (2005) Ecosystems and human Well-being: synthesis. Island Press, Washington, DC

Chapter 17
Socio-economic Aspects of Marine Bivalve Production

Gesche Krause, Bela H. Buck, and Annette Breckwoldt

Abstract This paper provides an overview of a number of socio-economic aspects related to bivalve aquaculture focussing on cultural services these activities provide to the culturing communities. Some direct socio-economic benefits of aquaculture in general exist through its supply of highly nutritious foods and other commercially valuable products. Additionally, it provides a variety of jobs and creates a set of income options. Yet, the question arises how to capture these in a coherent manner - what data is available and applicable to assess sustainable aquaculture in an inclusive way?

Starting with some general information on marine bivalve aquaculture development and the local contexts of the producing (usually coastal) communities, the paper discusses what it takes to generate meaningful information needed for decision-making and governance of the sector. To date, such decisions about marine aquaculture development are still (too) often based on incomplete and short-termed information, particularly in relation to socio-economic dimensions. Consequently, inadequate accounts of how trade-offs are associated with different development options are made. Aquaculture expansion may come at the expense of increased and possibly unsustainable pressure on ecosystem goods and services, ultimately jeopardizing people's food security, health and livelihoods. Its development may therefore generate negative impacts on other industries and people's livelihoods, e.g. fisheries, agriculture, shipping, and tourism. Additionally, in some cases, benefits derived from aquaculture systems are moving away from the local communities directly affected by aquaculture to stakeholders operating at a global market level. These considerations are discussed in this paper. Central focus is placed here on the question of how a more direct way of cultural inclusion of the local (mostly coastal) communities directly involved and dependent on marine bivalve aquaculture could occur.

Exemplified by case-studies, the paper will look at the culturing communities themselves, their everyday challenges, socio-economic controversies and benefits

G. Krause · B. H. Buck · A. Breckwoldt (✉)
Alfred Wegener Institut Helmholtz Center for Polar and Marine Research, Bremerhaven, Germany
e-mail: Gesche.Krause@awi.de; bela.h.buck@awi.de; annette.breckwoldt@awi.de

A. C. Smaal et al. (eds.), *Goods and Services of Marine Bivalves*,
https://doi.org/10.1007/978-3-319-96776-9_17

but also conflicts related to e.g. management and certification schemes. Our focus hereby is exclusively on *cultured bivalves*, not on the many and complex systems around the world where wild bivalves are harvested. Marine bivalves can represent important opportunities for economic activity and social cohesion in coastal rural areas, providing many jobs in those areas that are often otherwise economically depressed. Provided for a good governance set-up, the culturing community thereby contributes to the wellbeing of all its members – which in turn is defined as the willingness of members of a society to cooperate with each other in order to survive and prosper.

Due to its ocean-bound nature, marine bivalve aquaculture could also provide an occupational alternative for displaced fishermen. Its development can preserve the character and ambience of seaside fishing communities, utilize the local acquired knowledge and skills of the coastal folk, and allow the local denizens to remain economically and culturally tied to the marine environment. The consideration on the socio-economics of culturing communities should, however, neither stop at the local level, nor at the border of each country. On a national level, main considerations must stress small-scale units which, due to their size, pose fewer management problems and function with more flexibility. These projects must have a privileged status on domestic markets particularly in developing countries. From then onwards, they hold the potential, via well-developed and sustainable markets and trade pathways, also to extrapolate internationally.

Abstract in Chinese 本文概述了与双壳贝类水产养殖有关的一些社会经济活动，重点是这些活动为进行养殖的社区所提供的文化服务。 一般来说，水产养殖直接的社会经济效益是通过提供高营养食品和其他有商业价值的产品来实现的。 此外，它提供了不同工作岗位，为人们创造了一系列的创收选择。然而，问题在于如何以一种连贯的方式捕捉这些数据——哪些数据是可用的，并且适用于以一种包容的方式评估可持续的水产养殖？

根据一些海洋双壳贝类养殖发展的基本资料信息和养殖（通常是沿海）区域的实际情况出发，本文讨论了如何为有关部门的决策和监管提供有用的信息。迄今为止，关于海水养殖发展的一些决策仍然是基于不完整和短时间内的信息，特别是涉及到社会经济方面。因此，如何权衡决定不同的养殖发展选项时需要足够的支持信息。水产养殖的无序扩张不但对生态系统可持续发展及其产品和服务方面造成压力，威胁到人们的粮食安全，健康和生计，还可能对其他行业和民生例如渔业，农业，航运和旅游业造成负面影响。此外，在某些情况下，水产养殖带来的利好正在从直接从事养殖的当地养殖户惠及到全球市场上的利益相关者。本文讨论了这些考虑因素，重点在于如何寻找一种更加包容的方式将海水双壳贝类养殖与沿海社区发展有机融合在一起。

以个案研究为例，本文将着眼于从事贝类养殖的社区它们每天所面临的问题，社会经济争议和利益以及涉及到管理和认证等方面的冲突。 我们的关注点仅限于养殖的双壳贝类，而不涉及野生双壳贝类的采捕。海水双壳贝类养殖可以提升沿海和农村地区经济活力，为一些经济萧条的地区提供诸多就业机会。 同时，也有助于促使社区建立一个互利合作，共同致富，发展繁荣的良好发展管理体系。

由于海洋区域的特性差异，海洋双壳类养殖也可为流离失所的渔民提供就业岗位。 它的发展可以保护海滨渔业社区的特点和文化氛围，有效利用当地民众掌握的知识和技能，使当地居民无论是在经济上，还是在文化上都与海洋保护密切关系。.

但是对从事双壳贝类养殖社区社会经济方面的考虑，既不能局限于养殖区当地，也不能停留在每个国家的边界层面。在国家层面上，主要考虑的是小型的养殖企业，因为小型的养殖企业规模较小，管理问题较少，运作灵活。 这些项目必须在国内市场，特别是在发展中国家中享有特殊地位。在这种局面下，这些养殖企业有潜力通过良好的可持续的市场和贸易途径拓展国际市场。

Keywords Marine bivalves · Aquaculture · Socio-economic dimensions · Social indicators · Cultural services · Decision support

关键词 海水双壳贝类 • 水产养殖 • 社会经济规模 • 社会指标 • 文化服务 • 决策支持

17.1 Background

> Scotland will have a sustainable, diverse, competitive and economically viable aquaculture industry, of which its people can be justifiably proud. It will deliver high quality, healthy food to consumers at home and abroad, and social and economic benefits to communities, particularly in rural and remote areas. It will operate responsibly, working within the carrying capacity of the environment, both locally and nationally, and throughout its supply chain (Shared Vision, Scottish Government on Scottish Aquaculture[1])

This vision of the Scottish Government towards aquaculture generally sums up nicely the great potential of sustainable culturing; it reflects the definition of sustainable aquaculture, which will be used in this paper. However, to date, aquaculture has not yet fully realized its potential as a source of food, nutrition, and income generation, among other (e.g. technological) reasons often due to the unavailability of the metrics or tools for understanding and assessing the social and economic impacts. This is in stark contrast to the fact that the interest and investment into marine aquaculture to provide humankind's increasing demand for (sea)food is spreading and growing rapidly globally (Anderson 2002; FAO 2016; SAPEA 2017). Thereby, a 'people-policy' gap remains for many aquaculture endeavours (Krause et al. 2015), i.e., the gap in available knowledge and available policies taking up this knowledge in an integrated way, a gap in knowledge exchange between the aquaculture industry, policy makers trying to support aquaculture development and people who depend on aquaculture for a job and/or food source.

Among various institutions around the globe, this gap has seen increasing attention over the past years, leading to more research attention at the human-nature interface. In the following, we consider marine bivalve culture as part of a social-ecological system (SES) in which humans are considered an intrinsic part of the

[1] Strategic Framework for Scottish Aquaculture (www.scotland.gov.uk/library5/environment/sfsa-00.asp).

natural system. SES can be defined as bio-geo-physical territories with their associated social agents and institutions (following Glaser et al. 2012), where the individual parts of the system mutually interact, shape and reshape the resource itself, its goods and benefits, and its governance. While this paper does not intend to apply the SES concept to all these individual parts in the context of bivalve aquaculture – considering aquaculture as part of a SES involves envisioning a paradigm shift from the persisting strong focus on biological, technical, and economic considerations of aquaculture. This shift was the main driver behind developing this paper. As a case in point, the International Council for the Exploration of the Sea (ICES) has increased their activities at this interface for the north Atlantic, leading to the establishment of the Strategic Initiative on the Human Dimension in Integrated Ecosystem Assessments, whose task is to develop strategies to support the integration of social and economic sciences into ICES work. More specifically for aquaculture, the Working Group on Social and Economic Dimensions of Aquaculture (WGSEDA) addresses this gap, with the question of how to balance the negative and positive socio-economic consequences of aquaculture development. Motivation behind its founding was the observation that while in many instances the introduction of aquaculture was technically a success, socio-economic and cultural factors of the technology were not well-adopted by local communities and municipalities. Oftentimes, some of these activities are more visible, such as farm construction, and some to a lesser extent, such as the manufacturing of processing equipment, or hatcheries (Krause et al. 2015).

The question arises how to identify and capture the direct socio-economic benefits of aquaculture in general through its supply of highly nutritious foods and other commercially valuable products whilst providing a variety of jobs and creating a network of income options. What data is available and applicable to assess sustainable aquaculture in an inclusive way? More often than not, available socio-economical relevant data is not regarded as being of relevance to aquaculture, and/or is not being collected at the appropriate scale or level to generate meaningful information needed for decision-making and governance of the sector. Consequently, inadequate accounts are made of how trade-offs relate to different development options. Hence, aquaculture expansion may come at the expense of increased and possibly unsustainable pressure on ecosystem goods and services, ultimately jeopardizing people's food security (and health) and livelihoods (e.g., in events of bivalve diseases, parasite infestation). Its development may therefore generate negative impacts on other industries and respectively related livelihoods, e.g., fisheries, agriculture, and tourism. Additionally, benefits derived from aquaculture systems are in some cases shifting from the local communities directly affected by aquaculture, to stakeholders operating at a global market level.

These considerations form the point-of-departure of this paper, which will focus in a more direct way on the local (mostly coastal) communities directly involved and dependent on marine bivalve aquaculture. In many countries in e.g., Asia, North America, and the Mediterranean, bivalve culture has been the oldest sector of the aquaculture industry. For example, the commercial culture of the Pacific oyster in British Columbia began soon after it was first introduced from the Far East in 1912, and in the Gulf of Taranto in Italy, it was an important large commercial commodity since the Middle Ages (Cataudella and Spagnolo 2013). As such, these examples provide an excellent basis for the focus on cultural services of marine bivalve aquaculture in this book section in general, and the socio-economics of the respective coastal communities in particular. Our focus will be exclusively on *cultured bivalve*, not on the many and complex systems around the world where wild bivalve is harvested.

17.2 What Defines Bivalve Culturing Communities?

For this paper, we define a 'culturing or producing community' as a coastal community anywhere in the world, where most of its local residents are directly (e.g., farm operator) or indirectly (e.g., manufacturer of clam mesh bags or further processing of the harvested crop) dependent on marine bivalve aquaculture, and who receive goods and services from this culture. These can be a Norwegian coastal community directly running "semi-intensive" marine bivalve farms, or a local community in Panama of which most members work in a foreign-owned farms, or an extensive relayed mussel on-bottom farming for a local community in China. The individual ownership settings, responsibilities, time, and finances invested vary depending on the position in the entire process (if existing) – who is doing the culturing, running the businesses, taking care of health standards, doing the marketing, etc.

The overarching question here is the (variable) degree of dependency on marine bivalve aquaculture. Dependency also develops and 'materialises' differently as to whether the producer is purely oriented towards economic gain, or whether his/her motivation is rather a combination of tradition and socio-cultural factors. The level of dependency therefore varies for these communities and for their members, and so does whatever is at stake for them, whether it is the main source of income, or the clean coastal waters for paying tourists to enjoy. Often, the higher the dependency, the higher is the potential of being vulnerable to shocks, and the higher is the responsibility people are willing to take on.

Thus, depending on the contextual setting of bivalve cultivation, the goods and services to these communities vary across scale and time – hence, the identification and value of the cultural service of this activity also surface differently, which pinpoints to the dilemma of capturing the cultural services of bivalve cultivation across different global settings. One point of entry to tackle this is to look into more detail on the socio-economic typology of bivalve cultivation.

17.3 Cultural Services

The broad range of important aspects mentioned in the previous sections has already provided some vital insights into the wide range of cultural services the coastal communities involved in bivalve aquaculture are receiving at present, and could be receiving in the future (Daniel et al. 2012). The Millennium Ecosystem Assessment (2005) defined cultural services as "the non-material benefits people obtain from ecosystems through spiritual enrichment, cognitive development, reflection, recreation, and aesthetic experience, including, e.g., knowledge systems, social relations, and aesthetic values". Understood in this context, cultural services of bivalve aquaculture move way beyond the cultural aspects of shells as money for traditional ceremonies (Duncan and Ghys 2019). Cultural services have "value in their own right, and they can play an important role in motivating public support for the protection of ecosystems" (Daniel et al. 2012, p. 8817). In the context of sustainable bivalve aquaculture, one rather tangible cultural service is that it enables coastal rural communities to stay in their familiar environment and not having to move away to urban areas for employment. Thus, bivalve culture may act as important keystone activity for local meaning-making, shaping the cultural identities of a place and ownership. This again can be linked to some less visible cultural services such as job satisfaction, freedom, way of life, lifelong learning, providing a sense of home, relation to nature, spiritual value of 'being out there', the knowledge of doing something *with* and *for* the marine environment, and for sustaining a healthy food production and a healthy coastal ecosystem. Other examples for cultural services extend even to the visitors of the region and the tourism industry, in that bivalve aquaculture can offer opportunities for tourists to experience aquaculture as an occupation one may not come across very often, and in addition profit from the produced healthy food. This may even lead to promoting local food culture that again shapes cultural identities and place-based meaning making (SAPEA 2017), as well as to external benefits if the bivalve products are used outside the SES where they were produced.

17.4 Socio-economic Controversies: Benefits, Dependencies, Complementarities

Despite some clear and much needed socio-economic benefits from marine bivalve aquaculture, it also competes for economic, social, physical, and ecological resources, can limit the perceived beauty of a seascape, and can result in environmental degradation (Bacher et al. 2014).

The economic effects of marine bivalve culture on the culturing communities can be immense (e.g., in terms of investments, market influence, risks and hazards, benefits) and its repercussions to society vast. Municipalities may oppose establishment of marine bivalve aquaculture unless the benefits to the municipalities are made clear, transparent, and actually stay in the communities.

Bivalve culture has been a vital part of global coastal communities' livelihoods for centuries, and has contributed a vital, sometimes the main, part to local incomes. This can be substantial and means a substantial vulnerability to crises driven by environmental changes (Guillotreau et al. 2018a). In the following, insights from various recent case studies help to shed some light on the broad complexities of the socio-economic effects of bivalve aquaculture, as well as how environmental changes/disasters to such socio-economic important systems can affect the dependent communities (incl. producers, wholesalers and consumers; e.g., Héral and Deslous-Paoli 1991). One important example is the Bonamia outbreak in European oyster cultures, which also led to the introduction of the Pacific oyster (Buestel et al. 2009; Bromley et al. 2016).

During the last decades, harvested and cultured bivalves around the world have been repeatedly struck by mass mortality events/episodes, leading to socio-economic vulnerabilities and the development of adaptation strategies by the culturing communities (Guillotreau et al. 2018a). The various mitigation or adaptation responses of farmers and farming communities have rarely been assessed for their social and ecological aspects (Bundy et al. 2016; Guillotreau et al. 2018b). In France in summer 2008, for example, the consequences of young oyster mass mortality events related to environmental change affected an entire bivalve culturing profession and its related SES. Short-term effective responses came mainly from the industry itself: against all scientific recommendations, the oyster farmers decided to over-invest in hatcheries and spat (oyster seed) collection to compensate for the high mortality rates striking the cultured stocks. This resulted in significant market price increases after the decline of output levels, resulting in better profitability levels for the surviving firms (Guillotreau et al. 2018b).

In Matsushima Bay, north-eastern Japan, oyster farming also constitutes a major activity (Seki 2018). The Great East Japan earthquake and tsunami in March 2011 destroyed fishing boats and oyster farms, as well as sewage treatment facilities. This resulted in coastal pollution and ultimately the spread of Norovirus, resulting in widespread food poisoning caused by consumption of contaminated oysters. Again, the oyster industry was the driver of first responses by adopting a virus inactivation (heat) treatment of shucked oysters (Seki 2018). In this case, however, this treatment substantially modified the oysters, vastly reducing the price at which they could be sold, hence resulting in an income decline for the over 160 farmers in the area. However, alternative and more innovative methods to stop pathogenic pollution are difficult to find, especially with dysfunctional sewage treating facilities.

In another case from the US Pacific Northwest (Washington and Oregon) in summer 2007, substantial production failures of Pacific oyster (*Crassostrea gigas*) larvae in the three main bivalve hatcheries jeopardized an industry worth US $270 million and 3200 jobs in Washington State alone (Cooley et al. 2018). Scientists and industry cooperated and identified ocean acidification to be the major cause, exacerbated by nutrient runoff and sluggish exchange with ocean water in the region. As an immediate response, the hatcheries began ocean acidification monitoring and building hatcheries outside of the region. A state-level panel consisting of scientists, industry representatives, elected officials, and natural resource managers reviewed knowledge around ocean acidification and recommended appropriate responses, e.g. to advance research and monitoring in cooperation with Washington State University. A Marine Resources Advisory Committee was established to contribute to multi-stakeholder consultative policymaking for the aquaculture-dependent industry, and to prioritize ocean acidification in regional-level management efforts. These state-level responses originated primarily from informal governance networks, led by charismatic industry representatives who used their social capital also to exert influence on science and policymaking.

The final example from British Columbia's aquaculture industry (Canada), which dates back to 1912, focuses not on the challenges of adaptation to shocks but is included here to show-case the strong dependency and socio-economic impact bivalve aquaculture systems can have (Vancouver Island University, Centre for Shellfish Research[2]). In this region, the most commonly farmed species since 1912 are the Pacific oysters (*Crassostrea gigas*) and Manila clams (*Venerupis philippinarum*). In 2003, the aquacultural production was equivalent to that of the wild industry. Currently, there are 460 licensed shellfish tenures occupying 2114 ha in British Columbia (BC Agriculture & Lands). Much of the economic benefit and impact associated with the industry remains in the coastal communities and local economies of Vancouver Island and British Columbia's mainland. The great majority of bivalve operators are still small companies, many of which are family-owned, thus providing permanent, year-round employment in areas where jobs are scarce and the percentage of displaced workers is high. On a percentage basis, the bivalve aquaculture industry spends more on wages than other sectors such as conventional agriculture and fishing. Currently, 700–1000 direct jobs can be attributed to this industry, with workers under the age of 30 holding approximately 50% of those jobs. In addition to direct jobs, there are >500 jobs associated with industries that supply and service bivalve aquaculture. This 'spin off' employment is itself also located in rural coastal communities, rather than in the larger urban centres.

17.5 Discussion

The examples outlined above clearly show some of the socio-economic challenges and potential risks affecting the various levels and scales of marine bivalve aquaculture. The strong involvement of local communities and the related institutional arrangements, for example related to property rights, are shaped and reshaped as part of the interactions between users/stakeholders and 'their' marine resources. To highlight some of the more visible challenges, the following two sections will discuss relevant aspects of such 'shaping processes' taking place in the everyday activities of culturing communities (17.5.1) and the involved multidisciplinary research and governance environment (17.5.2).

[2] https://www2.viu.ca/csr

17.5.1 Critical Processes

Aquaculture activities can be at the centre of a number of diverse **conflicts** – as seen in many contemporary public discourses as well as in the examples above. Marine bivalve aquaculture can indeed be both a source as well as a victim of factors leading to conflict.

Pollution, as exemplified above, can be one main source of conflict for the bivalve culturing communities. Especially, in regard to its direct linkage to health risks and biosafety aspects as the more obvious risks, but also financial, legal, and insurance risks for every stakeholder involved in the culturing of bivalves, and the marketing and distribution of the bivalve products. As such, they also indicate some of the obstacles towards the implementation of sustainable aquaculture, but also pathways to some of the potential solutions.

Bivalve cultivation can also be faced with increased social conflicts between the stakeholders involved - farmers, nature conservationists, recreation/tourism, fisheries, shipping (commercial/private) and people aesthetically impacted by installations. These conflicts (e.g., with fisheries, wind farms) are often based on competition for space (and hence substractability, where one user's use directly affects the potential of resource use by another). Most EU countries employ a complex aquaculture planning consultation process to minimize the environmental impact of their culturing developments, and to ensure that the deposit and cultivation of aquaculture animals does not conflict with rights of others (e.g., moorings/boats, farm effluent, sites of scientific interest, tourism). These processes follow, for example the Best Management Practice of the UN Food and Agriculture Organisation, as well as the Best Environmental Practice (BET) and Best Available Technique (BAT) guidelines (FAO Code of Conduct for Responsible Fisheries[3]). In addition, industry codes of practice are designed to encourage sustainability with minimum impact (e.g., see the *Association of Scottish Shellfish Growers Code of Best Practice for shellfish aquaculture*[4]).

A new emerging and promising avenue towards sustainability and the acknowledgement of the positive effects of bivalve cultivation are the **certification schemes** of cultured bivalves. Some of the aspects commonly considered for bivalve certification include land and water use, water pollution, benthic effects, effects on biodiversity, use of antibiotics and other chemicals, and relationships with workers and local communities (Boyd et al. 2005). Further discussions include water use conflicts (see above), public health risks associated with bivalve consumption, and the introduction of non-native species and related genetic alterations, e.g. of oysters (Boyd et al. 2005; Cranford et al. 2012). To date, certification schemes face the challenges of most aquaculture products on what to certify and how to certify aquaculture production itself. Despite these difficulties, it is something many culturists are keen to achieve, and are, more often than not, actively driving this process forward.

[3] http://www.fao.org/docrep/005/v9878e/v9878e00.htm

[4] www.assg.co.uk

This is especially valid, as they already have to work with very high environmental and human health standards (at least in Western societies).

While organic certification is a promising tool towards sustainability, organizations currently use different standards for organic certification, which have to be evaluated. Certification schemes relevant in some way to aquaculture have been reviewed by Corsin et al. (2007) and the World Wildlife Fund (WWF). The latter also co-founded the Aquaculture Stewardship Council in 2010, which targets 'responsible aquaculture' and includes social standards.[5] Other organisations active in this field include the UN Food and Agriculture Organisation, Friend of the Sea,[6] Naturland (the only organic certification scheme at present[7]), Global G.A.P. (mainly focussing on farm assurance with only some sustainability components[8]), and the Aquaculture Certification Council, which, in addition to environmental and safety standards, also includes social standards (recently changed to Best Aquaculture Practices Certification[9]). The Marine Stewardship Council[10] is only interested in ecologically sustainable fisheries, with only capture-based bivalve aquaculture operations certified. The majority of certification schemes have in common that they consider (or at least strive for) ecological sustainability as key requirement for securing long-term socio-economic benefits (and hence cultural services). Some include social standards (closely linked to cultural services), most do not (yet). There remains much scope of improvement of certifications schemes for bivalve aquaculture to cover all the aspects related to the culturing activities, including considerations of the social, economic, and environmental impact of bivalve culture and management (Cranford et al. 2012). This also requires a critical reflection on other (often unintended) consequences of certification schemes, for example, bivalves that used to be available for local consumption, trade or cultural services may be exported to a global market through preferred (high retailer demand) markets for certified aquaculture products. The phenomenon of marine bivalve aquaculture products being shipped around the world may have repercussions on the availability of and access to these bivalves by local communities (Brenner et al. 2014; Muehlbauer et al. 2014). Particularly, where small-scale aquaculture producers are in the employ of companies or depending on middlemen, the economic benefits from harvesting a certified resource may not reach them in full extend.

Sustainability issues related to responsible consumption of cultured bivalves received increasing attention over the course of the last decades. In response to consumer requests for organic products, organic certification of cultured bivalves is therefore gaining speed. The afore-mentioned aspect to local production for global markets, however, does question sustainability and has to be integrated in the evolution of aquaculture certification schemes. In addition, bivalve cultivation activities

[5] www.asc-aqua.org

[6] www.friendofthesea.org

[7] https://naturland.de

[8] www.globalgap.org

[9] https://bapcertification.org

[10] www.msc.org

can have adverse effects on the ecosystem, such as bottom disturbance when dredging for seed, enhanced deposition of organic material in local areas, and reduction of the carrying capacity for other filter feeders, as well as effetcs on the entire ecosystem (e.g., Byron et al. 2011; Filgueira et al. 2015; Smaal and van Duren 2019). Together with potential changes in biodiversity as induced by the introduction of culture facilities and the bivalves themselves, these aspects could again impact the outcome of certification processes.

Directly relevant to the certification schemes for marine bivalve aquaculture, there is a discussion on the reliability of tools and methods for genetic confirmation of **species identification** of cultured bivalves, as the correct species' names are not always available. A review of current methods and recommendations for species identification are needed. Correct names are important for commercial purposes and certification, as well as for disease control and management, thus inherently affecting the culturing activities of the local communities. For example, the introductions of closely related species that can produce infertile hybrids should be avoided.

The introduction and translocation of live bivalve from hatcheries and field sites around the world, can involve the introduction of **non-indigenous species, diseases, parasites, and harmful algae**. Potential implications to wild and cultured stocks include impacts on recruitment, reduced fitness, increased competition, and predation, as well as change in genetic composition, diversity, and polymorphism. Information is gathered and needed on guidelines for, and records of, the transfer of cultured bivalve in ICES countries (ICES 2009). Potential implications and effects of the introduction and transfer of alien species need to be considered to help minimize impacts and guide farmers, aquaculture-dependent communities and policy makers in support of the development of policy decisions on cultured bivalve transfers.

Finally, there are also some positive **practical considerations** that need to be highlighted briefly. Despite these daunting conflicts and challenges, bivalves nevertheless make an excellent candidate for an organic product, as it does not need additional foreign-source input of feed other that naturally occurring phytoplankton. Furthermore, their protein content makes them an interesting option from the point of providing and maintaining food security for the growing world population (SAPEA 2017). In addition, during their life in the coastal zone, they also have a role in ecosystem services such as reducing nutrients in the water column and acting as a carbon sink (see other papers in this book). An oyster farm of about 1 ha can compensate for the nitrogenous waste of 40–50 coastal inhabitants (Shumway et al. 2003; Petersen et al. 2019). In this way, bivalve feeding can also avoid harmful algal blooms. These health and safety aspects are clearly not to be underestimated in their importance to the communities. Bivalve aquacultures operate under public health standards, e.g., waters that are certified under national sanitation programmes, and are thus regularly and strictly monitored. For example, in the USA, the *National Shellfish Sanitation Program* (NSSP) standards exceed those required for swimming; and failure has immediate consequences such as the closing of waters to harvest. The presence of bivalve aquaculture therefore often results in increased awareness and monitoring of environmental marine conditions. No untreated sewage can be tolerated, a different marine stewardship develops, and other harmful inputs into the local waters are regularly monitored alongside. This correlation has often provided political impetus for improvement of sewage and wastewater treatment, not rarely placing bivalve farmers first in the line of defence towards enacting laws on water quality, implementing technological advances, etc.. Many bivalve producing companies have or are developing ‘environmental codes of practice’, including, for example, best management practices, to ensure that as the industry develops, it maintains a responsible environmental record (e.g., in Scotland), which again can facilitate the development of certification schemes. Cultured bivalves therefore do not only represent a valuable food product, their cultivation can also enhance alternative livelihoods in rural areas, provide social welfare, as well as ecological, economic, social and cultural services (e.g., improving social capital related to certification standards).

17.5.2 Working with Socio-economic Indicators?

To support the marine bivalve aquaculture activities of the involved communities, our perceived **role as scientists** in the development of a management approach for bivalve aquaculture impacts is to provide science-based advice (e.g., Kluger et al. 2017) on:

- Effective performance-based approaches and indicators for characterizing ecosystem status and impacts of a highly diverse bivalve aquaculture industry;
- Identifying the potential consequences to coastal marine ecosystems from changes in ecosystem status and impacts, and identifying related thresholds of potential public concern;
- Identifying effective measures for preventing or mitigating any impacts from bivalve aquaculture;
- Reviewing and assessing available management frameworks that facilitate ecological sustainability by considering their capacity to incorporate an ecosystem perspective, societal values, and the economic viability for industry (ICES 2009).

Socio-economic science considerations are paramount in setting critical decision criteria, e.g., what constitutes an unacceptable impact? Deliberations on many components of a pragmatic bivalve aquaculture management framework require the discussion of costs to the diverse industry involved (e.g., for monitoring) and “potential” public concerns (e.g., impact mitigation measures). To help define what level of impacts are ‘acceptable’, ecology and socioeconomics can help in clarifying the values and expectations of different groups, and contribute to the economic evaluation of environmental services. Furthermore, environmental conservation and protection, and other legislations pertaining to the utilization of coastal areas, are clearly important considerations for the selection of indicators, and particularly for the setting of regulatory triggers/thresholds.

An integrated, ecosystem-based bivalve aquaculture management approach requires endorsing socio-economic activities, potential societal consequences, as well as the environmental dimensions of sustainability (Cranford et al. 2012). Indicators of socio-economic issues not only need to measure the operating performance of commercial bivalve cultures, which at its simplest could be summarized using financial ratios, but also the wider impacts of aquaculture on society at large. Indeed, it is precisely these impacts which can be expected to invoke an institutional response intended to alter the way in which aquaculture is regulated and managed.[11] Among the many different indicators proposed in the literature, some are of direct relevance for bivalve culture operations. They are related to four different overarching social dimensions, namely (1) the social acceptability of the bivalve culture, (2) the supply availability to the market, (3) the livelihood security for the local communities, and (4) the economic efficiency of bivalve culture operations. Possible indicators related to these four social dimensions are outlined below (cf. ICES 2009).

1. **Social acceptability** of the bivalve culture operations can be assessed with two indicators:
   - Public attitude towards aquaculture (bivalve culture). This is evaluated by means of regular enquiries, using statistical treatments (Whitmarsh and Wattage 2006).
   - Assessment of emerging and existing conflicts. Bivalve culture may be the origin of visual intrusion, which may affect tourism, or it may compete for space with other coastal activities in a spatially constrained environment. These can be evaluated by means of observations, regular interviews with local stakeholders, and institutional bodies.
2. An indicator on the **supply availability** to the market corresponds to the consumption of bivalve products per capita (in those cases where bivalves are consumed locally) and to their entailing costs for the consumer:
   - The consumption of bivalves is usually computed at national levels, indicating the quantity of food per capita and per year.
   - The consumers' price is based on the trends in wholesale prices. Large national markets publish trade journals from which these data can be obtained.
3. **Livelihood security** for the local communities corresponds to the well-being of the bivalve producer on the local level. Indicators that address this issue pertain to:
   - Income per capita. The importance of aquaculture in supporting local livelihoods is most directly measured by per capita income in this sector. A proxy measure may be derived based on the ratio gross value added (GVA) to employment.

[11] www.ecasa.org.uk

- Employment rate. Total employment is a measure of the scale (or 'importance') of the aquaculture industry in absolute terms. It is an indicator of the number of people that depend on aquaculture directly and indirectly for their livelihood. It has a political as well as an economic significance.

  However, these two indicators do not consider long-term aspects of income provision, inherent to the implications of the term 'livelihood security'. A bivalve culture classified as 'secure' through this may still be vulnerable towards external (e.g., environmental or market-driven) disturbances.

4. One of the most important group of indicators relate to the direct **economic efficiency** of a particular bivalve aquaculture operation. These can be gauged as follows:

   - Productivity ratio. Productivity is a measure of output per unit of input. For instance, trends in labour productivity are an important indicator of technical progress in aquaculture, and productivity differences between farms may indicate which farms are most vulnerable to falling prices and profits.
   - Protection costs. Costs may be incurred in dealing with the environmental impact of aquaculture, and are likely to consist of two elements: (i) Compliance costs incurred bivalve cultures (e.g., arising from the obligation of producers to undertake Environmental Impact Assessments (EIAs)), and (ii) regulation, surveillance and enforcement costs by the respective institutions. Environmental protection costs are the counterpart of environmental damage costs. Thus, an inverse relationship between these can be expected.
   - Profit. Profitability is a basic indicator of financial viability. In the absence of published data, the profitability of a bivalve operation can be addressed and calculated from its different elements (i.e. input costs, pricing of products, etc.)

Finally, a further dimension of socio-economic indicators could be the existence and performance of **financial and social security institutions** for culturists/producers (e.g., FAO 1985), including:

- Specialized banking organisations (e.g., The Fund for Regulation and Organisation of the Market for sea and marine culture products in Spain), to improve the collective infrastructure of the sector;
- Socio-educational programmes to enable the participation and representation of culturists/producers;
- Acknowledgement of culturists' "brotherhoods" and shared reimbursement, which in the Mediterranean date back to the thirteenth century and can have the power to negotiate and ensure authority, e.g., for participation in government decisions (sometimes evolved into professional chambers with extensive management powers).

17.6 Conclusions and Outlook

The information in the section above summarizes some of the complexities surrounding marine bivalve aquaculture. This snapshot already very clearly highlights the contextual nature of the cultural and socio-economic benefits and implications for communities living with and from these cultured products. Marine bivalves can thus represent important opportunities for economic activity and for supporting social cohesion in coastal rural areas, providing potential jobs to areas that may be economically isolated otherwise. If working well, the culturing community hence contributes to the well-being of all its members, by their willingness to cooperate with each other in order to survive and prosper.

Due to its ocean-bound nature, marine bivalve aquaculture could also provide an occupational alternative for migrant or job-seeking fishers (i.e., fishers who did not lose their job due to aquaculture). This statement clearly has to be made with caution in view of evidence that fishers are not farmers and may find it difficult if not impossible to adapt and adopt to commercial bivalve culture. Nonetheless, bivalve aquaculture development can preserve the character and ambience of seaside fishing communities, utilize the local acquired knowledge and skills of the coastal residents, and allow the local denizens to remain economically and culturally tied to the marine environment.

The consideration of the socio-economic aspects of culturing communities should, however not stop at the local level, nor at the border of each country. These small-scale projects must have a privileged status on domestic markets particularly in developing countries. Nevertheless, they can also be extrapolated internationally via well-developed and sustainable markets and trade pathways.

Finally, a further dimension, which is important but goes beyond the scope and objective of this paper, is the growing potential and spread of offshore bivalve aquaculture (e.g., in concepts such as Open Ocean Aquaculture or Open Sea Shellfish Farming). This brings a very different perspective to the discourse and reality of the culturing and producing communities, with implications on their responsibilities and contribution in local and regional marine spatial (and potentially even protected area) planning efforts. This development 'far offshore and away from sensitive ecosystems' has the potential to both reduce and exacerbate user conflicts, for example in terms of employment, ownership (of both equipment and production and planning processes), or technological choices, particularly in developing countries (often more directed towards producing luxury products destined for European countries). This leads, however - from a cultural, economic or ecosystem service point of stance – to the normative questions of how we can evaluate effects of new established marine management strategies such as a marine spatial planning act. What are indicators of the status of social perception of bivalve culture that can help in avoiding conflicts? How do social values and norms as well as administrative organizations in different countries/regions affect trends in the intensity, methodology, structure, and type of aquaculture? Moreover, who decides, what type of social value will be traced in the planning and management process? These are clearly important aspects to consider in the management of shellfish resources and bivalve cultivation in particular, in such a way that they will generate cultural (and other) services in the longer term.

One of the closer objectives will be to identify specific cross-cutting and integrative methods (to also include local historical and long-term data, for example) to support the evaluation of the direct and indirect socio-economic consequences of aquaculture operations at all levels, from the local to the global. In this way, already existing socio-economic data and lessons will not be lost but their applicability used and further developed, to identify current data gaps and more narratives of successful sustainable marine bivalve aquaculture projects.

Acknowledgements The authors are grateful to Dr. L. Kluger and dr ir N. Steins for reviewing the manuscript.

References

Anderson JL (2002) Aquaculture and the future: why fisheries economists should care. Mar Resour Econ 17(2):133–151

Bacher K, Gordoa A, Mikkelsen E (2014) Stakeholders' perceptions of marine fish farming in Catalonia (Spain): a Q-methodology approach. Aquaculture 424–425(0):78–85

Boyd CE, McNevin AA, Clay J, Johnson HM (2005) Certification issues for some common aquaculture species. Rev Fish Sci 13(4):231–279

Brenner M, Fraser D, Van Nieuwenhove K, O'Beirn F, Buck BH, Mazurié J, Thorarinsdottir G, Dolmer P, Sanchez-Mata A, Strand O, Flimlin G, Miossec L, Kamermans P (2014) Bivalve aquaculture transfers in Atlantic Europe. Part B: environmental impacts of transfer activities. Ocean Coast Manag 89(Supplement C):139–146

Bromley C, McGonigle C, Ashton EC, Roberts D (2016) Bad moves: pros and cons of moving oysters – A case study of global translocations of Ostrea edulis Linnaeus, 1758 (Mollusca: Bivalvia). Ocean Coast Manag 122(Supplement C):103–115

Buestel D, Ropert M, Prou J, Goulletquer P (2009) History, status, and future of oyster culture in France. J Shellfish Res 28(4):813–820

Bundy A, Chuenpagdee R, Cooley SR, Defeo O, Glaeser B, Guillotreau P, Isaacs M, Mitsutaku M, Perry RI (2016) A decision support tool for global change in marine systems. The IMBER-ADApT framework. Fish Fish 17:1183–1193

Byron C, Bengtson D, Costa-Pierce B, Calanni J (2011) Integrating science into management: ecological carrying capacity of bivalve shellfish aquaculture. Mar Policy 35(3):363–370

Cataudella S, Spagnolo M (2013) The state of Italian marine fisheries and aquaculture. The Italian Ministry of Agriculture, Food and Forestry Policies, Rome

Cooley S, Cheney JE, Kelly RP, Allison EH (2018) Chapter 2: ocean acidification and Pacific oyster larval failures in the Pacific Northwest United States. In: Guillotreau P, Bundy A, Perry I (eds) Societal and governing responses to global change in marine systems. In: routledge studies in environment, culture, and society. Taylor & Francis, New York, pp 40–53

Corsin F, Funge-Smith S, Clausen J (2007) A qualitative assessment of standards and certification schemes applicable to aquaculture in the Asia-Pacific region. In: RAP Publication (FAO), no. 2007/25/. Asia-Pacific Fishery Commission; FAO, Bangkok (Thailand). Regional Office for Asia and the Pacific, 2007, 98 pp

Cranford PJ, Kamermans P, Krause G, Mazurié J, Buck BH, Dolmer P, Fraser D, Van Nieuwenhove K, Beirn FXO, Sanchez-Mata A, Thorarinsdóttir GG, Strand Ø (2012) An ecosystem-based approach and management framework for the integrated evaluation of bivalve aquaculture impacts. Aquac Environ Interact 2(3):193–213

Daniel TC, Muhar A, Arnberger A, Aznar O, Boyd JW, Chan KMA, Costanza R, Elmqvist T, Flint CG, Gobster PH, Grêt-Regamey A, Lave R, Muhar S, Penker M, Ribe RG, Schauppenlehner T, Sikor T, Soloviy I, Spierenburg M, Taczanowska K, Tam J, von der Dunk A (2012) Contributions of cultural services to the ecosystem services agenda. Proc Natl Acad Sci 109(23):8812–8819

Duncan PF, Ghys A (2019) Shells as collectors items. In: Smaal et al (eds) Goods and services of marine bivalves. Springer, Cham, pp 381–411

FAO (1985) Socio-economic aspects of shellfish aquaculture in Languedoc-Roussillon, France. Seminar socio-economic aspects of aquaculture development in the Mediterranean countries, Mediterranean Regional Aquaculture Project, FAO, Rome

FAO (2016) The state of world fisheries and aquaculture 2016. Contributing to food security and nutrition for all, Rome, 200 pp

Filgueira R, Comeau LA, Guyondet T, McKindsey CW, Byron CJ (2015) Modelling carrying capacity of bivalve aquaculture: a review of definitions and methods. In: Meyers RA (ed) Encyclopedia of Sustainability Science and Technology. Springer, New York, pp 1–33

Glaser M, Krause G, Halliday A, Glaeser B (2012) Towards global sustainability analysis in the Anthropocene. In: Human-nature interaction in the anthropocene: potentials of social-ecological systems analysis. Routledge, New York, pp 193–222

Guillotreau P, Bundy A, Perry I (eds) (2018a) Societal and governing responses to global change in marine systems. In: Routledge studies in environment, culture, and society. Taylor & Francis, New York

Guillotreau P, Le Bihan V, Pardo S (2018b) Mass mortalities of farmed oysters in France: bad responses and good results. In: Societal and governing responses to global change in marine systems. Routledge studies in environment, culture, and society. Taylor & Francis, New York

Héral M, Deslous-Paoli J (1991) Oyster culture in European countries. In: Menzel W (ed) Estuarine and marine bivalve mollusk culture. CRC Press Inc, Boca Raton, pp 153–190

ICES (2009) Report of the working group on marine shellfish culture (WGMASC), 7–9 April 2009, Bremerhaven, Germany. ICES CM 2009/MCC: 02, 91 pp

Kluger LC, Filgueira R, Wolff M (2017) Integrating the concept of resilience into an ecosystem approach to bivalve aquaculture management. Ecosystems:1–19

Krause G, Brugere C, Diedrich A, Ebeling MW, Ferse SCA, Mikkelsen E, Pérez Agúndez JA, Stead SM, Stybel N, Troell M (2015) A revolution without people? Closing the people–policy gap in aquaculture development. Aquaculture 447:44–55

Millennium Ecosystem Assessment (2005) Ecosystems and human well-being: synthesis. Island Press, Washington, DC

Muehlbauer F, Fraser D, Brenner M, Van Nieuwenhove K, Buck BH, Strand O, Mazurié J, Thorarinsdottir G, Dolmer P, O'Beirn F, Sanchez-Mata A, Flimlin G, Kamermans P (2014) Bivalve aquaculture transfers in Atlantic Europe. Part A: transfer activities and legal framework. Ocean Coast Manag 89(Supplement C):127–138

Petersen JK, Holmer M, Termansen M, Hassler B (2019) Nutrient extraction through bivalves. In: Smaal et al (eds) Goods and services of marine bivalves. Springer, Cham, pp 179–208

SAPEA, Science Advice for Policy by European Academies (2017) Food from the oceans: how can more food and biomass be obtained from the oceans in a way that does not deprive future generations of their benefits? SAPEA evidence review report no. 1, Berlin, 160 pp

Seki T (2018) Societal and governing responses to global change in marine systems. In: Guillotreau P, Bundy A, Perry I (eds) Routledge studies in environment, culture, and society. Taylor & Francis, New York, pp 23–39
Shumway et al (2003) Shellfish aquaculture – In praise of sustainable economies and environments. World Aquaculture 34(4):15–17 (Guest Editorial)
Smaal AC, van Duren L (2019) Bivalve aquaculture carrying capacity: concepts and assessment tools. In: Smaal et al (eds) Goods and services of marine bivalves. Springer, Cham, pp 451–483
Whitmarsh D, Wattage P (2006) Public attitudes towards the environmental impact of salmon aquaculture in Scotland. Eur Environ 16(2):108–121

Chapter 18
A Variety of Approaches for Incorporating Community Outreach and Education in Oyster Reef Restoration Projects: Examples from the United States

Bryan DeAngelis, Anne Birch, Peter Malinowski, Stephan Abel, Jeff DeQuattro, Betsy Peabody, and Paul Dinnel

Abstract There is a growing body of science to suggest that there is a mutualistic relationship between habitat restoration projects and community volunteers and participation. Restoration projects and programs benefit from community participation via an added labor force and by fostering community investment and support, which is critical for project success and future restoration investments. Community participants gain physically and psychologically rewarding experiences from being a part of restoration projects, while fostering an environmental ethos. Oyster restoration serves as particularly ideal opportunities for engaging community volunteers

B. DeAngelis (✉)
The Nature Conservancy, North America Oceans and Coasts Program, URI Graduate School of Oceanography, Narragansett, RI, USA
e-mail: bdeangelis@tnc.org

A. Birch
The Nature Conservancy, Florida Chapter, Maitland, FL, USA
e-mail: abirch@tnc.org

P. Malinowski
Billion Oyster Project, New York Harbor Foundation, Brooklyn, NY, USA
e-mail: pmalinowski@nyharbor.org

S. Abel
Oyster Recovery Partnership, Annapolis, MD, USA
e-mail: sabel@oysterrecovery.org

J. DeQuattro
The Nature Conservancy, Gulf of Mexico Program, Mobile, AL, USA
e-mail: jdequattro@tnc.org

B. Peabody
Puget Sound Restoration Fund, Bainbridge Island, WA, USA
e-mail: betsy@restorationfund.org

P. Dinnel
Skagit County Marine Resources Committee, Mount Vernon, WA, USA

A. C. Smaal et al. (eds.), *Goods and Services of Marine Bivalves*,
https://doi.org/10.1007/978-3-319-96776-9_18

and participation. These additional values provided to a community where oyster restoration is taking place is an important additive benefit that oyster restoration provides. The nature by which many oyster restoration projects are implemented offers satisfying opportunities for community members to participate in physically rewarding, hands-on work. Many oyster restoration programs are also ideal for incorporating student or citizen science, or broad-scale education and outreach. Despite the growing science to support the value of volunteer and community participation, coupled with increased oyster restoration, there is a paucity of information for project managers to turn-to for guidance as to how community participation can be built into oyster restoration projects and programs. This chapter presents five cases from the United States to demonstrate the broad, and often unique, opportunities to incorporate community and volunteer participation into oyster restoration.

Abstract in Chinese 摘要:越来越多的证据表明,生物栖息地恢复项目与社区相应活动和志愿者参与之间存在互惠关系。栖息地恢复项目得益于社区劳动力和资金支持,这对项目的成功和未来的栖息地恢复项目投资至关重要。参与者从这类项目中获得了生理和心理上的愉悦以及相应的项目经验,同时也培养了环境保护意识 。牡蛎礁恢复为社区和志愿者们参与活动提供了机会,由此带来的社会价值比恢复的生物量所带来的经济价值重要的多。许多牡蛎礁恢复项目的实施为社区成员提供了令人满意的收益,包括物质奖励和实践工作经验的积累。牡蛎礁恢复项目也是兼顾学生参与互动、公民科学素养培养、全民教育和拓展的良好方式。尽管越来越多的科学项目肯定了志愿者和社区参与越来越多牡蛎礁恢复的价值,但是项目管理人员却缺乏如何将社区参与纳入牡蛎礁恢复项目的经验和信息。本章介绍了五个来自美国的典型案例,介绍了如何将社区和志愿者与牡蛎礁恢复项目进行有机结合。

Keywords Oyster restoration · Volunteers · Community participation · Education

关键词 牡蛎礁恢复 • 志愿者 • 社区参与 • 教育

18.1 Introduction

Restoration practitioners, ecologists and researchers tend to cite the value of shellfish restoration projects in terms of the ecosystem services they provide (Coen et al. 1999; 2004, 2007; Coen and Luckenbach 2000; Brumbaugh et al. 2006; Grabowski and Peterson 2007; Beck et al. 2011; Grabowski et al. 2012). These ecological services are often cited as the primary motivation behind a restoration project. Rarely cited is the ability for shellfish, particularly oyster, restoration projects to serve as ideal opportunities for education and outreach to a variety of groups of citizens of all ages and abilities. Community involvement in restoration has been suggested to be a mechanism for reconnecting communities with their landscape, empowering citizenry, and fostering an environmental ethos (Leigh 2005; Lee and Hancock 2011), while providing the volunteers with a psychologically rewarding experience (Miles et al. 2000). Thus, these additional values provided to a community where restoration is taking place is an important additive benefit that restoration can, and does, provide. Volunteers of ecological restoration projects become advocates for environmental restoration and their participation is motivated by a desire to learn more about nature, while engaging in a fun, social experience (Grese et al. 2000; Ryan et al. 2001).

In the United States, shellfish restoration occurs in every coastal state (e.g. www.projects.tnc.org/coastal and https://restoration.atlas.noaa.gov/src/html/index.html). In the United States, incorporating volunteers in shellfish restoration projects is often standard practice. For example, since 1995 the Office of Habitat Conservation, Restoration Center of the National Oceanic and Atmospheric Administration have implemented over 600 shellfish restoration projects, and documented 68,792 volunteers and over 393,000 volunteer hours through their Community-based Restoration Program (NOAA Restoration Center, personal communication). At these rates of volunteer engagement per restoration project, it's clear that shellfish restoration can change the landscape by returning lost ecological services, but maybe more importantly, shellfish restoration has the ability to transform, through education and hands-on participation in rebuilding nature, how local communities value and perceive the ecological landscape in their own backyards.

The method of how shellfish restoration projects are typically implemented makes them inherently interesting opportunities for volunteers, and serve as education and outreach opportunities. Many shellfish restoration projects tend to utilize implementation techniques that rely on a relatively substantial human workforce. For example, projects often involve collecting shell, bagging shell, or building reefs from consolidated or unconsolidated material that require many hands to lighten the work. These unique volunteer-labor opportunities are also efficient education opportunities since those involved in the labor will be eager to learn how their efforts are contributing to restoring habitat, and what the role of that habitat is in nature (Schroeder 2000; Ryan et al. 2001). Oyster shell collection efforts, perhaps especially, have a wide educational reach, often involving education and collection points at the locations at which the public widely interact with the oysters in their everyday life, such as at restaurants. That said, every restoration activity involving volunteers provides an opportunity to educate the public regarding the benefits of shellfish restoration, and has the added benefit of also fostering a strong sense of ownership and engagement in the volunteers which is valuable in its own right. Furthermore, because a large workforce is required in some instances, using volunteers is often incorporated into projects out of financial necessity (Propst et al. 2003). According to the Independent Sector (https://www.independentsector.org/volunteer_time) the value of volunteer time in 2015 was $23.56 per hour. Using a theoretical example of 35 volunteers and an average of 6 h per volunteer, a project could not only save approximately $5000 in labor, they can also designate these hours towards matching funds that many federal restoration grants require. The restoration project benefits from the volunteer workforce, while the project serves as an ideal and extremely effective platform for engaging and educating local citizens, of all ages, by offering a highly unique opportunity to literally get their hands dirty and become a part of restoring nature.

Experiencing nature through physical interactions encourages humans to understand and connect to the natural world, which is the foundation of environmental stewardship (Van der Werff et al. 2014). As our world is increasingly urbanized, combined with a continued increase in indoor technological activities, there is a growing subset of the human population, especially youth, whose health is at risk from the increasing separation from nature (Sandercock et al. 2010, 2011, 2012; Aggio et al. 2012). It is clear, however, from observing shellfish restoration projects in the U.S, that there is no shortage of volunteers, or willingness to participate. For example, in 2011 The Nature Conservancy organized 600 volunteers over a two-day event to construct the Helen Wood Park oyster restoration project in Mobile, Alabama (Fig. 18.1). In other words, the willingness of volunteers to participate in shellfish restoration demonstrates that the projects themselves serve as a superb opportunity for community engagement. While bivalve populations, particularly oysters, have been highly degraded in the U.S. and nearly extirpated in some locations (Beck et al. 2011; zu Ermgassen et al. 2012), a strong cultural connection still exists to these animals in most coastal communities. This cultural connection elicits a high degree of excitement and the desire to become a part of the restoration project in their community. The implementation method of most shellfish restoration projects, as mentioned above, typically involve out-door, in-water, hands-on work, which can lead to improved self-confidence, teamwork, and a sense of satisfaction from doing "important work" (Miles et al. 1998, 2000). Furthermore, a successful restoration project requires more than just ecological knowledge. Because most shellfish restoration projects happen in public waters, and often with public monies, community investment, and community support are critical to the overall success of a restoration project, as has been documented in other types of ecological restoration (Geist and

Fig. 18.1 Over 600 volunteers were organized in Mobile Alabama for the Helen Wood Park restoration project. (Photo: © Erika Nortemann/The Nature Conservancy)

Galatowitsch 1999). The community at large: individuals, local governments, businesses – all have a stake in the health of their local ecosystem. Involving them directly into the project through community outreach and education will generate more interest in and support for continuing additional restoration investments.

While there is a growing body of literature that evaluates the impacts and benefits of using volunteers in ecological restoration, there is a paucity of published information available to restoration practitioners and project managers describing the breadth of methods of how to incorporate community participation into shellfish projects. To serve that need, here we present five very different cases in the United States to provide examples of the variety of approaches that different shellfish restoration projects and programs have used to incorporate volunteer or community participation into their projects. This is clearly not meant to be an exhaustive list, as there are many noteworthy examples in the U.S. not mentioned here. These examples were chosen based on their uniquely different method of incorporating community participation. We are not attempting to comment on the quantitative or qualitative success or restoration output in terms of goods and services provided from the restoration, but rather describing how the projects each successfully incorporated community participation that provided benefits to both the restoration project, and the participants.

18.1.1 Case I – A Community Gives Back: The Role of Community in Restoring Oyster Habitat in the Charlotte Harbor Estuary, Punta Gorda Florida

Over the past century, the health of the oyster habitat in Florida's bays and estuaries has dramatically declined (Beck et al. 2011). An estimated 85–90% of oyster reefs have been lost in the Charlotte Harbor Estuary, and with it a corresponding loss in the ecosystem services the habitat once provided (Geselbracht et al. 2013; CHNEP 2013). The Nature Conservancy's (TNC) experience working on oyster restoration in Florida shows that, without exception, once a community understands the plight of oysters and the benefits oysters and their habitat offer, they are eager to be involved and help their community thrive. This is the case for the TNC-led restoration project -Trabue Harborwalk Oyster Habitat Project (Trabue) in the City of Punta Gorda – where the success of this project, and future restoration, depends on supportive, active, and committed volunteers throughout the Charlotte Harbor community.

The Trabue project blends science-based restoration with community engagement. The goal of the project is to test three different intertidal oyster restoration methods and purposefully engage volunteers in all phases of the restoration project to stimulate widespread community support to advance future oyster habitat restoration. Two of these methods, oyster mats and oyster bags, provide excellent opportunities for volunteer involvement, both in the construction and deployment in the

Fig. 18.2 Volunteers unload oyster bags in preparation for constructing an experimental reef in Charlotte Harbor, FL. (Photo: Ann Birch)

water (Fig. 18.2). Oyster bags are created by filling tube-shaped plastic netting with fossilized shell material and tied at both ends. The bags are approximately 40–60 cm in length and 4.5 km in weight, which makes the bags easy to lift by adults. Oyster mats are created using 36 individual oyster shells, drilled with one hole near the hinge, and secured to the mat with a cable tie. Each mat, made of aquaculture grade plastic 'mesh' material, is 40.64 cm^2 in size. This size is easy to construct, transport, and arrange during construction of reef beds. Oyster mats have been used in other Florida estuaries for intertidal, low profile reef restoration and have been shown to be successful in providing substrate for oyster larvae to attach and grow, but have also proven to be an ideal technique for volunteer participation, regardless of age or ability of the volunteer. When using oyster mats in restoration, volunteer help is indispensable to complete every stage of mat construction: from cutting the mat material into squares, bundling cable ties, drilling shells, attaching shells to the mat, and deploying the mats in the water to create a reef bed. Even very young children can help when guided by an adult. It's an opportunity for children and adults alike, who may never have seen an oyster, to handle the shell and learn about the oyster's role in an estuary. Making mats in the classroom serves as a quasi 'field trip' for teachers and their students, especially when budgets for travel are sparse. The teachers can weave the activity into their science lessons. Counting out shells and cable ties needed to attach the shell to the mat uses simple math. Attaching the shells to the mat in a random pattern teaches visual skills. And the activity is dirty enough for the kids to have fun but clean enough for the classroom.

Over the course of 2 years (2014–2015), community groups and individuals volunteered their time and expertise in constructing both the materials and the reefs. TNC contracted with the Charlotte Harbor Aquatic Preserve, a state agency, to hire a part-time volunteer coordinator. The coordinator recruited volunteers of all ages from the local community to take part in the project. These included students from kindergarten to college, Girl and Boy Scouts, local fishing and boating clubs, Big Brothers and Big Sisters, non-profit organizations, a local CrossFit business, and realtor and construction companies. Local businesses offered their services that included the use of their forklifts and backhoes, transporting the material and delivery of shell material, and helping promote the project to the community. Private donors and foundations support this project financially, and oyster restoration in general, particularly because of the high level of community involvement. The coordinator either traveled to the groups or volunteers came to the location to construct the bags and mats. An educational presentation introducing the project, its partners, the importance of oysters, and what each of the participants can do to help was provided to each new group of volunteers as an orientation to the project. The presentation was followed by detailed instructions on how to properly construct an oyster mat or bag. In all, an estimated 1300 community volunteers contributed more than 3000 h over the course of 24 months. A total of 900 mats and 1600 oyster bags were constructed during 18 mat-making events and 11 oyster bagging events and deployed over the course of a few weeks.

Conducting science-based pre-and post-reef construction monitoring to determine if the project objectives are met is an essential part of any restoration project. Yet funding to cover monitoring costs can be difficult to find. Likewise, once a project is constructed there is typically little to no opportunity for continued volunteer involvement, even though most community volunteers have been 'hooked' by their experience with oyster restoration and want to stay involved. The Volunteer Oyster Habitat Monitoring Program (VOHMP) in the Charlotte Harbor estuary was established to fill these gaps. The Friends of the Charlotte Harbor Aquatic Preserve received a public grant in 2015 to start-up the VOHMP. A hired volunteer coordinator is responsible for training and organizing a cadre of citizen scientist volunteers to learn science techniques for monitoring the success of the restored reefs. TNC provides oversight of the program, engaging the volunteers ready to help monitor future oyster projects already being planned in the estuary by TNC and partners. In this way, the VOHMP provides a valuable service in an engaging way that keeps the community involved in the project over the long-term, and thus maintains continued investment into their local estuary and ecosystem.

The Trabue project has been embraced and adopted by the City of Punta Gorda and Charlotte Harbor community at large. Working with a community involves not only engaging volunteer citizens but also cultivating relationships and partnerships with the community's decision makers, government agencies, community organizations and businesses; these are the entities that know and care deeply about their community and invest in its quality of life. Investing the time to connect with people, foster trust, show the value a restoration project offers to the community, and to thank them for the opportunity to be part of their community's vision are invaluable and essential ingredients of any project.

It's a rarity for citizens to have easily accessible and inexpensive opportunities to work with marine scientists and actively participate in restoring a marine species and habitat. Shallow water oyster restoration offers both; people of any age or ability can be involved with no other investment but their time, getting their hands dirty or feet wet alongside scientists. The Trabue project is a prime example of how citizens from all walks of life and varied interests joined together for a common cause to help make a difference in their community. TNC is committed to involving communities in restoration activities. Our oyster restoration projects generate a sense of ownership with many volunteers, which is exactly what we hope for; community support and stewardship for their project in their estuary.

To learn more about this project visit https://www.nature.org/ourinitiatives/regions/northamerica/unitedstates/florida/explore/floridas-oyster-reef-restoration-program.xml.

18.1.2 Case II – Billion Oyster Project: Oyster Restoration Through Public Education in New York Harbor

The Billion Oyster Project (BOP) is based on the belief that direct engagement and interaction with wild animals and functioning ecosystems has a transformative effect on young people. As our world is increasingly urbanized there is a growing subset of our human population that is coming of age separate from nature. Simultaneously, efforts abound aimed at increasing student engagement in school to improve outcomes for millions of young people. Too often, these interventions exist in the vacuum of school without the real-world, hands-on implementation that leads to improved self-confidence, authentic problem solving, teamwork, and the belief that anyone and everyone has the power to effect change.

BOP is an attempt to brings these too often separate issues together. It has grown from the belief that if we are to continue living, working, teaching and learning on this planet we must fundamentally change how humans learn about and interact with nature. Our solution began in a high school Aquaculture class and has grown into a region wide initiative involving 70 restaurants, 65 schools, thousands of students, millions of oysters and a dozen active restoration and research sites.

New York Harbor is a massively degraded natural system, oysters are functionally extinct, and every time it rains billions of gallons of untreated household wastewater enter the system. The visibility is very low, often less than a foot. Currents are strong and commercial traffic is constant. To overcome these challenges, it is essential to engage the entire metropolitan community in the work of growing and restoring oysters. Community engagement has become central to the work of Billion Oyster Project. This work is executed through four core programs: Shell Collection, Reef Construction and Monitoring, Schools and Citizen Science and Public

Engagement. Each of these programs is designed to advance the work of growing and restoring oysters while simultaneously building a community of environmental stewards and advocates who will no longer stand for a polluted harbor that lacks its native keystone species.

The Project began at the New York Harbor School, founded in 2003 by Murray Fisher and a small group of passionate educators. The Harbor School aims to prepare students for college and careers by immersing them in New York City's maritime experience. Students at Harbor School first began interacting with oysters as part of an oyster gardening program led by the New York/New Jersey Baykeeper. For its first 7 years, Harbor School was located in Bushwick, Brooklyn, New York's most land locked neighborhood. It was not until 2010 that the school relocated to Governors Island, a stone's throw from lower Manhattan and right in the center of New York Harbor. This move allowed for the development of six Career and Technical Education Programs. Through these programs students have learned to SCUBA dive safely, raise oyster larvae, operate and maintain vessels, build and maintain commercial-scaled oyster nurseries, design underwater monitoring equipment and conduct long-term authentic research projects all in the murky, contaminated, fast moving waters of one of the busiest ports in the country (Fig. 18.3). For these students, Billion Oyster Project provides a complex problem that requires them to practice the skills they are learning and collaborate with their peers from other disciplines. These students are the primary workforce for the Reef Construction and Monitoring Program. Students in individual programs work to produce the raw materials of restoration and research. Together, they plan and execute complex installation and monitoring missions throughout the Harbor. These activities would

Fig. 18.3 Students in the New York Harbor School Aquaculture Program monitor oyster growth and survival at the Billion Oyster Project Community Reef site in Brooklyn Bridge Park. (Photo: Vonwong)

not be possible without the diverse expertise of students in various programs. They are joined by a growing group of industry professionals, divers, captains, welders, advocates, scientists and marine technicians. These BOP Professionals work alongside Harbor School teachers to facilitate the participation of students in all aspects of Reef Construction and Monitoring.

Harbor School students are now joined by students at 65 public middle and high schools and dozens of citizen scientists throughout the five boroughs of New York City. The work of the BOP Schools and Citizen Science Program is built around Oyster Restoration Stations. These small wire cages hold live oysters, settlement tiles and a trap for mobile invertebrates. These components are monitored separately to assess species diversity, succession, and oyster growth and survival. Partner schools contribute by monitoring these stations and supporting breeding colonies at various locations around the Harbor that add to the reproductive potential of the Harbor each spring. These Oyster Restoration Stations also serve as access points that bring math and science classes out of their buildings and down to the water's edge. Through this oyster restoration and research, students learn the science of the estuary and the math of aquaculture and ecosystem restoration. In this way, young people become active stewards of the Harbor. The data collected by these school groups forms a Harbor-wide oyster growth and survival study and a growing water quality data set that together help inform future restoration work. Each year 5000 new students participate in these programs.

A primary challenge of engaging communities and volunteers in the work of oyster restoration in New York Harbor is the physical lack of access to the water. Walk to the water's edge and most often you will be met by fences and steep or vertical bulkheads. There are, however, a few places where access is possible. The Public Engagement Program takes advantage of these access points and is now working with community groups, schools and volunteers in collaboration with the Reef Construction and Monitoring Program to build reefs in communities. These new reefs, for the first time, allow for volunteers and schools to regularly enter the water to participate directly in oyster restoration.

All the above programs require a consistent source of cured oyster shells. Because the oyster industry on the East Coast is dominated almost entirely by the restaurant half shell market, there is no available source of oyster shells besides those that are generated by restaurants. In New York City, a full 35 tons of oyster shells are generated every week. The vast majority of these are, unfortunately, landfilled. The Shell Collection Program currently operates at 70 restaurants, 5 days per week and averages four tons of shell per week. These shells are transferred to a location on Staten Island where they spend a year out of water before they can be returned to the Harbor.

To date, through the implantation of these four programs, BOP has collected over 180 tonnes (400,000 pounds) of shells, engaged over 600 volunteers on Governors Island and at community reef sites and worked with over 10,000 students. All of the 20 million oysters restored to date have been grown and installed by high school students. We are just at the beginning of our journey towards a recovered New York Harbor, and still a long way from understanding what the best strategies are for scaling up our restoration efforts. However, if we are able to restore a sustainable oyster population and build a program that allows teachers and students to be successful in their work of restoring the natural environment, then we will have created a model that is replicable in any city in the world that happens to exist on or near a degraded natural system.

18.1.3 Case III – Building an Engaged Community Program Through Shell Recycling: Creating a Win-Win-Win Strategy

There is widespread interest by lawmakers, environmental groups, commercial growers and the oyster-eating public to have more oysters in the Chesapeake Bay. In the late 1800s, Maryland used to supply the nation with oysters, however, due to historical overfishing, disease, silt and sediments and poor water quality, the oyster population has been reduced to a fraction of its historical peak.

Interest in rebuilding the Maryland's iconic shellfish industry began in the early 1990s through a state-sponsored Oyster Roundtable and the formation of a Maryland-based non-profit, the Oyster Recovery Partnership (ORP) that was dedicated to the implementation of reef restoration. An Environmental Impact Statement completed in 2008 further evaluated oyster restoration alternatives and together with strategies recommended by a state-mandated Oyster Advisory Commission culminated in the large-scale recovery and aquaculture efforts underway a decade later. Scientific advancements and increased production capacity are demonstrating that oyster reefs can be successfully restored on a large scale. Harris Creek, a tributary on Maryland's Eastern shore, was the first of ten tributaries to be restored in the Chesapeake Bay with 350 acres of new oyster reefs.

Oyster shell was found to be most effective and most accepted material for reef recovery efforts – whether to harden bottom or used as substrate for natural or hatchery produced larvae to attach to. Due to a limited availability of shell from traditional sources and rising acquisition costs, the Oyster Recovery Partnership was forced to explore other solutions for its restoration efforts. At the suggestion from local Baltimore oyster shuckers, ORP created the Shell Recycling Alliance and one of the first large-scale, urban-based shell recycling programs in the country. The program launched with two dozen participating restaurants and collected a few thousand bushels in its first year. Six years later, the program has collected more than 100,000 bushels (3500 tons), enough shell to plant 450 million oyster spat (Fig. 18.4). The program now accounts for 25% of the organization's annual shell needs, has grown to 300 active restaurants in the mid-Atlantic region at a cheaper cost that procuring and transporting shell through traditional sources. The average cost to recycle a bushel of shell is $2.70 as compared to $3.50 to $4.75 being spent to purchase and deliver shucked shell from processors to the organization's primary shell processing facility in Cambridge, MD.

Fig. 18.4 Freshly shucked oyster shells from local mid-Atlantic restaurants are collected by the Oyster Recovery Partnership's shell recycling staff and taken to a nearby landfill in Grasonville, MD where it is aged for 1 year before being used for restoration. (Photo: Stephan Abel)

When ORP first began the recycling program, they had no idea as to how popular the program would become. They found that while the general public cares about the environment and health of the Chesapeake Bay, many do not have the time to volunteer or support a specific cause. The value of this program is therefore that it is easy to participate in, and a win for everyone. The public can engage simply by eating oysters, the restaurants reduce waste costs while supporting a local sustainable fishing industry and the non-profit benefits by getting much needed shell coupled with increased public awareness.

Initially, ORP treated the shell recycling program as a logistics business, much like a waste management business, and were fortunate in that they were able to secure private and government grant funds to operate the program. They experimented with various strategies to optimize the city collection route and hauling methodologies. The organization assumed that restaurants would opt to participate in the program in order to reduce their waste removal costs while benefiting from a State-sponsored shell recycling tax credit. However, the tax credit which offers $5 for every bushel recycled has been utilized by only a few dozen restaurants and individuals. While cost savings do play a role in adopting the service, the primary motivator appears to be that the restaurants themselves are eager to do their part by becoming environmental stewards and it often ties into their menu of serving locally sourced, fresh food. When the program first started, ORP staff would regularly go to restaurants and meet with their manager or chef to encourage them to recycle their shell. Now that the program has matured, restaurants proactively contact ORP directly to become members.

When restaurants become Shell Recycling Alliance members, they are provided with a "welcome aboard" package that includes various marketing materials to educate their patrons on why oyster reefs are important and why oyster shell is needed. ORP also approached local County landfills and waste stations and placed dumpsters and trash cans at 70 locations around the state so that residents could also contribute by recycling their used shell.

Over time the restaurants have become more engaged, and many now regularly promote their efforts via social media and/or proactively host an oyster fest 'fundraiser' with proceeds going to ORP. This has proven to be an unanticipated added value of the program. Other partners, like Flying Dog Brewery, now produce an oyster stout and provide a percentage of proceeds back to the Oyster Recovery Partnership for its oyster restoration programs. As the program has gained traction, the media has in turn started to cover the effort more, resulting in the public becoming more educated and engaged. Today, when ORP supports a community event, adults and students ask how oyster restoration is doing, offer to solicit new restaurants for the recycling program, inquire how they can personally recycle their shell or offer to volunteer. This has clearly proven to be a fantastic way to both further the restoration of oysters in the Chesapeake Bay, and to engage and educate the general public about these very efforts.

For a summary of other shell recycling projects around the Country, visit http://oysterrecovery.org/

18.1.4 Case IV – Conservations Corps and Community Engagement: Creating Conservationists with Jobs

Franklin D. Roosevelt created the Civilian Conservation Corps in 1933 to provide jobs for unskilled laborers, who were put to work conserving the land and natural resources owned by Federal, State and Local Governments. While Job Corps and other Government work programs still exist today, many conservation and workforce development corps are managed by non-profit organizations and range in size from thousands of Corpsmembers to one small corps of 12 members or less. Corps crews take on a variety of project types that range from disaster relief, to ecosystem restoration, to community engagement, and much more.

Many Conservation Corps (CC) pay their crewmembers stipends at or near minimum wage and will often offer full or partial tuition for secondary education opportunities after the member completes a timed service period. While the wages are low, joining a CC often leads members to higher positions within the CC, or better outside job opportunities since CC's typically require members to participate in regular training programs that cover soft and hard skills. Of the many societal sectors that they operate, CC's are well-adapted and critical tools for community engagement. Most CC programs are developed locally, by diverse stakeholders who have identified a population of at-risk, young adults, and/or minority groups that are willing and able to work, yet have little economic resources. Many crewmembers originate from the low-income communities in which their CC's operate, and often have never had the opportunity to experience nature and conservation in a meaningful way.

Additionally, CC's engage communities through the projects they undertake. Project types range widely and many have CC's working great distances from human habitation and camping in remote wilderness areas for extended periods of time. Many CC projects, however, have crewmembers working alongside community members and volunteers to build oyster reefs along the coast (Fig. 18.5), providing aid to a disaster-struck community, or going door-to-door to educate residents about risks and how to respond to them. CC's are limber and flexible enough to drop what they are doing and provide rapid assistance as they are needed. Disaster relief, for example gives CC's opportunities for crews to not only do the heavy lifting, including clearing roads and using chainsaws on fallen trees, but also to cook and provide fresh water for displaced residents, or educate them about disaster preparedness or how to receive disaster assistance.

Conservation Corps have been used in oyster restoration efforts on the Atlantic Coast, and in particularly, the Chesapeake Bay area for many years. CC's that

Fig. 18.5 The Conservation Corps of the Forgotten Coast in Apalachicola, Florida bagging oyster shells to be used in a living shoreline. (Photo: Holden Foley)

operate in the Gulf of Mexico, however are just beginning to work on oyster projects as funding from the Deepwater Horizon Oil Spill is released for restoration efforts. An example is a Conservation Corps 2015–2017 initiative in Apalachicola, FL where a partnership with the Corps Network, a local workforce training NGO, and The Nature Conservancy (TNC), worked together to help the Conservation Corps of the Forgotten Coast hire and sustain Corpsmembers for 2 years.

All the Corpsmembers were recruited from areas around Apalachicola, where most families fall into low-to-moderate income levels. The crewmembers on this CC grew up in Apalachicola, and have had a deep understanding of the role that oysters play in their lives and in the health of the Bay for most of their lives, but in many cases this was the first experience that many had with oyster restoration and regulation.

In 2016, the Conservation Corps of the Forgotten Coast, located in Apalachicola, Florida, helped with the construction of a 450-foot long oyster reef living shoreline and have worked with the Florida Fish and Wildlife at oyster monitoring stations to track harvest volumes and oyster sizes. For the living shoreline project, the 8-member crew worked in the Apalachicola National Estuary Research Reserve to improve oyster habitat by placing a lime rock foundation of 50 tons of riprap, which was topped with several layers of bagged oyster shell. Over 900 labor hours were invested in the project. Crew members plan to return to this site in 2017 and plant marsh grass behind the shoreline structure. This CC also worked for the State of Florida's Fish and Wildlife at their oyster monitoring stations where they helped track harvest volumes and ensured that the product meets the minimum of 3 inches in size. The CC is on schedule to have served 4500 h by the completion of the project.

The CC crew in Apalachicola has a unique opportunity to not only conduct oyster restoration in the community that they live for the purpose of ecosystem restoration, but they are also engaged in projects that give them exposure to commercial oyster fishers and the regulatory agencies that manage the oyster resources.

To learn more about a Conservation Corps on the Gulf of Mexico Coast, visit http://www.nature.org/ourinitiatives/regions/northamerica/areas/gulfofmexico/restoration/gulf-of-mexico-stream-assessments.xml.

18.1.5 Case V – Olympia Oyster Restoration in Fidalgo Bay, Washington: How a Single Phone Call Catalyzed the Growth of Community-Based Oyster Restoration in Puget Sound, WA

Olympia oyster restoration in Fidalgo Bay, Washington is an example of how place-based restoration builds community, in unexpected and ever-expanding ways. From its inception, the project has been both an outgrowth of and a catalyzer of community outreach, with community engagement growing in perfect step with a resurgence of the native oyster population.

Olympia oyster restoration in Fidalgo Bay began with a phone call from a local community member. In 2001, an article in the Seattle Post-Intelligencer (*"Rare Shellfish is Sought for Spawning,"* May 30, 2001) reported on Puget Sound Restoration Fund's search for Olympia oysters (the West Coast's only native oyster) in Samish Bay, WA. The article prompted a reader, and local community member, to call and report an Olympia oyster sighting in Fidalgo Bay. The caller provided very specific instructions as to where the native oyster (previously unreported in Fidalgo Bay) could be found. At the time, Puget Sound Restoration Fund (PSRF) was trying to identify restoration sites to help the state of Washington implement an Olympia oyster stock rebuilding plan. PSRF enlisted the help of Bill Taylor, a local business owner of Taylor Shellfish whose family had been growing Olympia oysters for generations, to verify the oyster sighting. Until the call from the community member, though, Fidalgo Bay was not recognized as a potential location to find existing Olympia oysters, or even as a location to support rebuilding efforts. Unexpectedly, Bill spotted perfect habitat conditions for Olympia oysters (but no oysters) during the 2001 trip to Fidalgo Bay and this provided the basis for Olympia oyster seeding efforts that began in Fidalgo Bay in 2002. PSRF partnered with a variety of private, tribal, municipal and nonprofit partners to move forward with

Fig. 18.6 Community volunteers counting and measuring Olympia oyster seed prior to planting in north Fidalgo Bay, Washington. (Photo: Paul Dinnel)

Olympia oyster restoration in Fidalgo Bay. In the beginning, there was just a small nucleus of people involved in the project. One of the first partners to jump on board was the Skagit County Marine Resources Committee (MRC) (a volunteer-based Committee). The MRC recruited volunteers to implement a range of restoration actions, including spreading oyster seed, monitoring the growing population, and assessing natural recruitment to the site (Fig. 18.6). State and private business cooperation was essential from the outset. Washington Department of Fish & Wildlife guided efforts to ensure consistency with their state stock rebuilding plan, and Taylor Shellfish produced seed oysters required for rebuilding oyster stocks. Since then, the effort has attracted a growing galaxy of volunteers, researchers, Tribes, nonprofits, and government agencies. As of 2016, the effort has resulted in an estimated 3.1 million Olympia oysters covering approximately 3.5 acres of habitat. In addition, the Fidalgo Bay population serves as an important broodstock source, enabling the production of hatchery seed for other areas of North Puget Sound where Olympia oyster populations have plummeted from thousands of acres of oysters in the early 1900s to just a few small remnant populations.

From 2001 to 2016, over 25 organizations and over 130 volunteers have participated in various aspects of Olympia oyster research and restoration in Fidalgo Bay. Activities span genetic analysis, tideland authorization, broodstock collection, seed production and spreading, field monitoring, shell spreading, chemical fingerprint analysis, and all of the enabling funding that makes this work possible. The role of the community, particularly groups like Skagit County MRC, has helped Olympia oyster restoration efforts spread to many other locations across Puget Sound and the state of WA.

The story of Fidalgo Bay is a perfect example of building community around resources. The story began with a single phone call from an interested, knowledgeable resident – which led to burgeoning interest and a growing collaboration of non-governmental, state governmental, private industry and community volunteers. As Olympia oysters have been recruiting to Fidalgo Bay over the last 15 years, so too, people have been recruiting to Fidalgo Bay to study and monitor the growing population and habitat, and strengthen our human connections to shellfish resources. And although we couldn't have expected any of this, we shouldn't be at all surprised. Human communities have been attaching themselves to coastal resources for longer than any of us can remember.

To learn more about PSRF, visit: http://restorationfund.org

Information about Skagit MRC and their Olympia oyster work can be found at: http://www.skagitmrc.org/about-us/ (see "Projects").

18.2 Conclusion

Community outreach and education through shellfish restoration can come in many different varieties and forms – and there is no one-size-fits-all formula. The method of engaging citizens and offering outreach and education opportunities should be designed on a case-by-case basis to serve the needs of the individual project, as well as those citizens willing to participate in the project.

There is a growing body scientific literature that documents the motivations of volunteers and the physical and psychological benefits they derive from their participation in ecological restoration (e.g. Miles et al. 1998, 2000; Grese et al. 2000; Schroeder 2000; Ryan et al. 2001; Leigh 2005; Clewell and Aronson 2006; Lee and Hancock 2011; Jacobson et al. 2012). As well as evidence that suggests restoration projects benefit from community participation. It has been suggested that scientific knowledge alone cannot ensure success of an ecological restoration project, and ongoing human participation and commitment are critical to ensuring the long-term success and sustainability of restoration projects (Geist and Galatowitsch 1999). Others suggest that without public support and participation, governments may be unable to generate the political support to undertake programmatic restoration (Clewell and Aronson 2006).

Restoring oyster habitat has become an accepted practice along U.S. coasts, with projects increasing both in number and in scale. As the practice of oyster restoration matures in the U.S., so too does the complexity of projects. To date, there is no evidence to suggest this has reduced the rate of community participation in oyster restoration projects. For example, the currently largest oyster restoration project in the world, Harris Creek, Virginia restored 350 acres of habitat and seeded over 2 billion oysters, while utilizing volunteers in shell planting, shell recycling and other aspects of the project. As projects increase in cost, size and scale and incorporate more engineers and professional contractors, it is unclear whether this mutualistic relationship between the project and community will remain. However, one thing remains clear from observing decades of oyster restoration projects and volunteers in the United States: If you build it, they will come.

Acknowledgements The authors are grateful to two anonymous referees who made valuable comments on the manuscript.

References

Aggio D, Ogunleye AA, Voss C, Sandercock GRH (2012) Temporal relationships between screen-time and physical activity with cardiorespiratory fitness in English schoolchildren: a 2-year longitudinal study. Prev Med 55:37–39

Beck MW, Brumbaugh RD, Airoldi L, Caranza A, Coen LD, Crawford C, Defeo O, Edgar GJ, Hancock B, Kay M, Lenihan H, Luckenbach MW, Toropova CL, Zhang G (2011) Oyster reefs at risk and recommendations for conservation, restoration and management. Bioscience 61(2):107–116

Brumbaugh RD, Beck MW, Coen LD, Craig L, Hicks P (2006) A Practitioners' guide to the design and monitoring of shellfish restoration projects: an ecosystem services approach. The Nature Conservancy, Arlington, VA, MRD educational report no. 22, 28 pp

Charlotte Harbor National Estuary Program (2013) Comprehensive conservation management plan. http://www.chnep.org/CCMP/CCMP2013.pdf

Clewell AF, Aronson J (2006) Motivations for the restoration of ecosystems. Conserv Biol 20(2):420–428
Coen LD, Luckenbach MW (2000) Developing success criteria and goals for evaluating oyster reef restoration: ecological function or resource exploitation? Ecol Eng 15:323–343
Coen LD, Knott DM, Wenner EL, Hadley NH, Ringwood AH (1999) Intertidal oyster reef studies in South Carolina: design, sampling and experimental focus for evaluating habitat value and function. In: Luckenbach MW, Mann R, Wesson JA (eds) Oyster reef habitat restoration: a synopsis and synthesis of approaches. VIMS Press, Gloucester Point, pp 131–156
Coen L, Walters K, Wilber D, Hadley N (2004) A SC Sea grant report of a 2004 workshop to examine and evaluate oyster restoration metrics to assess ecological function, sustainability and success results and related information. Sea Grant Publication, 27 pp
Coen LD, Brumbaugh RD, Bushek D, Grizzle R, Luckenbach MW, Posey MH, Powers SP, Tolley G (2007) AS WE SEE IT. A broader view of ecosystem services related to oyster restoration. Mar Ecol Prog Ser 341:303–307
Geist C, Galatowitsch SM (1999) Reciprocal model for meeting ecological and human needs in restoration projects. Conserv Biol 13(5):970–979
Geselbracht L, Freeman K, Kelly E, Gordon D, Birch A (2013) Retrospective analysis and sea level rise modeling of coastal habitat change in Charlotte Harbor to identify restoration ad adaptation priorities. Fla Sci 76(2):328–355
Grabowski JH, Peterson CH (2007) Restoring oyster reefs to recover ecosystem services. In: Cuddington K, Byers JE, Wilson WG, Hastings A (eds) Ecosystem engineers: concepts, theory and applications. Elsevier Academic Press, Amsterdam, pp 281–298
Grabowski JH, Brumbaugh RD, Conrad RF, Keeler AG, Opaluch JJ, Peterson CH, Piehler MF, Powers SP, Smyth AR (2012) Economic valuation of ecosystem services provided by oyster reefs. Bioscience 62:900–909
Grese RE, Kaplan R, Ryan RL, Buxton R (2000) Psychological benefits of volunteering in stewardship programs. In: Gobster PH, Hull B (eds) Restoring nature: perspectives from the social sciences and humanities. Island Press, Washington, DC, pp 265–280
Jacobson SK, Carlton SJ, Monroe MC (2012) Motivation and satisfaction of volunteers at a Florida natural resource agency. J Park Recreat Adm 30(1):51–67
Lee M, Hancock P (2011) Restoration and stewardship volunteerism. In: Egan D, Hjerpe EE, Abrams J (eds) Human dimensions of ecological restoration. Island Press, Washington, DC, pp 23–38
Leigh P (2005) The ecological crisis, the human condition, and community-based restoration as an instrument for its cure. Eth Sci Environ Polit (ESEP) 2005:3–15
Miles I, Sullivan WC, Kuo FE (1998) Ecological restoration volunteers: the benefits of participation. Urban Ecosys 2(1):27–41
Miles I, Sullivan WC, Kuo FE (2000) The psychological benefits of volunteering for restoration projects. Ecol Restor 18:218–227
Ogunleye AA, Voss C, Sandercock GR (2011) Prevalence of high screen time in English youth: association with deprivation and physical activity. J Public Health 34(1):46–53
Propst DB, Jackson DL, McDonough MH (2003) Public participation, volunteerism, and resource-based recreation management in the U.S.: What do citizens expect? Soc Leisure 26(2):389–415
Ryan RL, Kaplan R, Grese RE (2001) Predicting volunteer commitment in environmental stewardship programmes. J Environ Plan Manag 44:629–648
Sandercock G, Angus C, Barton J (2010) Physical activity levels of children living in different built environments. Prev Med 50:193–198
Sandercock GRH, Ogunleye A, Voss C (2012) Screen time and physical activity in youth: theif of time or lifestyle choice? J Phys Act Health 9:977–984
Schroeder HW (2000) The restoration experience: volunteers' motives, values, and concepts of nature. In: Gobster PH, Hull B (eds) Restoring nature: perspectives from the social sciences and humanities. Island Press, Washington, DC, pp 265–280

Van der Werff E, Steg L, Keizer K (2014) I am what I am, by looking past to present:The Influence of Biospheric Values and Past Behavior on Environmental Self-Identity. Environ Behav 46(5):626–657

zu Ermgassen PSE, Spalding MD, Blake B, Coen LD, Dumbauld B, Geiger S, Grabowski J, Grizzle R, Luckenbach M, McGraw K, Rodney W, Ruesink J, Powers P, Brumbaugh R (2012) Historical ecology with real numbers: past and present extent and biomass of an imperiled estuarine habitat. Proc R Soc B 279:3393–3400

Chapter 19
Bivalve Gardening

C. Saurel, D. P. Taylor, and K. Tetrault

Abstract From an increasing awareness of sustainable food production, the promise of the "blue revolution" and campaigns to ameliorate the marine environment, seafood gardening has emerged from motivated local citizenry as a local food production phenomenon. Bivalve gardening, primarily manifested as oyster gardening, is a relatively new concept, slowly gaining traction worldwide. Terrestrial and marine gardening share the same principles of cultivating organisms and providing ecosystem goods and services. The main differences concern the growing medium – and legislation regarding use and access to gardens. Bivalves appear to be an ideal group of marine organisms for local production, they are low maintenance and do not require external food supplies as they feed directly by filtrating their surrounding growing medium. However, the cultural services provided by bivalve gardening range from social organisation to sustainable engagement; and require certain pillars such as clear objectives, support from the local community and government, dedicated volunteers, native bivalve seed availability, training, and realistic objectives. Moreover, the development of new gardens raises fundamental issues including food safety, regulation, and marine spatial planning. We use two case studies to illustrate different approaches to bivalve gardening: (1) in the U.S. several bivalve gardening initiatives are taking place, it is often referred as oyster gardening and initiated as a bivalve habitat recovery efforts, (2) in Denmark in Europe, several projects have started directly as bivalve gardens for food provisioning and are managed by local associations.

C. Saurel (✉) · D. P. Taylor
Danish Shellfish Centre, DTU Aqua, Nykøbing M, Denmark
e-mail: csau@aqua.dtu.dk; dtay@aqua.dtu.dk

K. Tetrault
Cornell Cooperative Extension, Suffolk County, Riverhead, NY, USA
e-mail: kwt4@cornell.edu

A. C. Smaal et al. (eds.), *Goods and Services of Marine Bivalves*,
https://doi.org/10.1007/978-3-319-96776-9_19

Abstract in Chinese 摘要:在人们对可持续食物产出、“蓝色革命”美好愿景、海洋环境改善越来越关注的背景下,海洋生物资源恢复已经成为广大民众比较接受的食物生产方式。双壳贝类,尤其是牡蛎的种群资源恢复,作为相对较新的概念已经在世界范围内逐渐普及开来。陆基和海水养殖的基本原则都是进行生物培育并且提供生态产品和服务,其主要区别在于养殖媒介以及养殖许可的审批和立法的过程。由于独特的滤食特性且日常维护成本较低,双壳贝类是众多海洋生物中食物供给功能较强的理想物种,。然而,双壳贝类种群资源恢复活动所提供的文化服务功能涵盖社会组织到公众可持续的参与,这需要当地社区和政府的支持、热心的志愿者参与、足量本地苗种的供应、相关技术培训和实际的实施方案制定等作为有效支撑。。此外,新模式的发展也进一步激起了人们对于食品安全、法规和海洋空间规划等基础性问题的讨论。我们用两个案例研究来说明不同的双壳贝类养殖的方法和提供的生态服务:1)在美国,已经在多处开展了双壳贝类(主要是牡蛎)种群资源恢复行动,旨在重建双壳贝类的栖息地; 2)在丹麦,多个双壳贝类种群资源恢复项目由当地协会进行管理,主要目的是提供食物供给功能。

Keywords Bivalve gardening · Cultural services · Oyster · Mussel · Non-commercial aquaculture · Community

关键词 双壳贝类种群资源恢复 • 养殖服务 • 牡蛎 • 贻贝 • 非商业性水产养殖 • 社 区

19.1 Introduction

19.1.1 The Bivalve Garden

Bivalve gardening is a non-commercial activity where bivalves such as mussels and oysters are grown for personal consumption. It is often perceived as a novel activity or concept as there is scarce tradition of private production with a physical garden of marine bivalves for personal consumption; there is rather a more established tradition for hand picking and gathering in the wild. Bivalve production is regarded as one of the most sustainable forms of seafood production, as bivalves extract organic matter from their surrounding environment, mainly by filtering phytoplankton, and thus do not require external food sources. Presently there are few examples of bivalve gardens, mainly based on community/association gardening principles using licenced grounds or individually operated in privately owned coastlines.

In a general sense, gardens are multifunctional and provide many cultural services in addition to the provision of food for personal consumption. While bivalve gardens share attributes with terrestrial gardens/allotments, typically hobby-scale with little infrastructure, the marine medium adds an altogether new dimension to food production with many new challenges.

19.1.2 *History of Bivalve Gardening*

19.1.2.1 From Gathering to Gardening

The development of bivalve gardening in contemporary history follows the development of the paradigm of securing sources of marine animal and vegetable proteins. Marine food production has shifted from gathering to farming at a much slower pace than terrestrial products. Terrestrial farming emerged in the Neolithic Era, ca. 10,000 years ago, through the domestication of terrestrial plants and animals; fundamentally changing human feeding habits and the structure of human life. By comparison, the domestication of aquatic foods has largely developed in recent times (Teletchea 2015). More than 90% of aquatic food domestication took place in the twentieth century while 97% of terrestrial domestication developed more than 2000 years ago (Duarte et al. 2007). Hunting and gathering has almost vanished from a commercial perspective for terrestrial products, while nearly half of global marine products are still extracted rather than cultivated (FAO 2016) from both commercial and recreational fisheries; but also from licenced/regulated hand picking and illegal poaching. Aquaculture has existed for thousands of years, mainly focused on finfish and seaweed, as for instance in China (Rabanal 1988). There are early records of bivalve gardening during the late Holocene in British Columbia where a first nation tribe maintained a garden of butter and littleneck clams (*Saxidomus gigantean, Leukoma staminea*) (Lepofsky and Caldwell 2013). These early gardeners were modifying and transforming the shoreline in order to increase clam production (Groesbeck et al. 2014). More recently, since the seventeenth to eighteenth centuries, oyster ponds on the Atlantic coast of France have been used for family production of oysters.

In recent times, community citizen gardening in the U.S. has developed for marine food consumption, from a movement that originated within bivalve habitat restoration programs in degraded estuarine systems on the East coast. In the Puget Sound shoreline landowners are growing their own bivalves on privately owned beaches or docks, sourcing their bivalve 'seed' and material from commercial bivalve growers (Chase 2017). In France, an activity called "aquaculture de loisir" (recreational aquaculture) could be interpreted to mean bivalve gardening. On the Atlantic coast of France, marshes have been modified to ponds and hillocks as far back as the seventeenth and eighteenth centuries for agriculture, salt ponds, pisciculture, oyster culture, and recreational culture. There, recreational aquaculture represents a social and cultural heritage where oysters were traditionally cultivated in privately owned saltwater ponds ("claires") for familial consumption; nowadays this is shifting towards shrimp culture (Paticat 2007). More recently in Denmark, bivalve gardening is a phenomenon derived not from bivalve habitat restoration but directly targeting food production; the government and private foundations have facilitated its implementation. In Japan, there are a few examples of seafood gardening, mainly focusing on seaweed, and some bivalve gardens also originating from restoration projects. There, personal seaweed growing is termed as an "ownership system" (e.g in 2005 in Minamata city http://bp.eco-capital.net/bps/read/id/88 or in Hiroshima http://www.haff.city.hiroshima.jp/info/2016/11/8982/). In other parts of the world where bivalve restoration projects exists, such as in Australia, oyster gardening is starting to come to fruition (Simon Branigan, The Nature Conservancy Australia pers. comm.).

19.1.2.2 Food Requirement vs. Sustainable Production

The sea has been historically perceived as a source of inexhaustible resources, either as food or raw material. For centuries bivalves have been extracted for food, often ignoring the ecological consequences such as eutrophication (Jackson et al. 2001). An illustration of the extent of extraction can be seen in shell middens, where in coastal areas around the world, gathered shells were piled up over many generations, covering areas up to 600–700 m long (Andersen 2000) and several meters high (Butler et al. 2019). Most of the coastal areas and estuaries worldwide, where bivalves are endemic, have been affected by direct and indirect anthropogenic impacts of securing food and materials; ranging from overexploitation of bivalves, overfishing, nutrient and toxic substances pollution, introduction of invasive species, climate change, and coastal erosion. These impacts have often lead to devastating ecological consequences such as eutrophication and habitat loss (Beck et al. 2011). Anthropogenic impacts were enhanced from the mid-1900s during the green revolution, through the use of modern agricultural technologies (e.g. genetics, fertilizers) and more efficient use of arable lands to improve food security at high environmental costs (Ausubel 2000). Recently, efforts in dissemination of information and research communication on ecosystem functioning and sustainable production have fomented ocean literacy and citizen consciousness regarding imbalances in coastal ecosystems due to pollution and reduction of stocks from overfishing (Gelcich et al. 2014); as well as the need to provide food for the growing world population. Thus, populations are facing a dilemma between food procurement with current access to a large quantity of very diverse foods at a high environmental price, and sustainable production.

19.1.2.3 Food Culture

Bivalve gardening was also born from the comprehension that food security did not equate sustainable food production. At the end of the twentieth century, some consumers became driven by an interest in understanding the origin of their food (Grunert et al. 2014); with the loss of knowledge in composing a proper diet stemming from an the overabundance in the variety of available food, as described in "The Omnivores Dilemma" (Pollan 2006). Food security in this era also entails a vast amount of exotic and processed foods. A growing proportion of consumers are seeking other choices than those that form their current food environment. Thus, from a perceived loss of food culture, emerged new movements, such as the slow food movement (www.slowfood.com, accessed on 01/09/2018, still working) in the late 1980s, based on "preventing the disappearance of local food cultures and traditions, counteracting the rise of fast life and combatting people's dwindling interest in the food they eat". The movement includes three concepts of food: (i) GOOD: quality and healthy food; (ii) CLEAN: sustainable production; and (iii) FAIR: price moderation for consumers and producers (Petrini et al. 2012). Slow food movements are also associated with a wide range of other terms such as: conscious eater, citizen eater, omnivore consciousness, food consciousness, local food movement, locavores, and ethical eaters. Community organisations and shared gardens have been a way to propel the slow food movement, and this is also the case for bivalve gardens. These organisations are connecting food, people, and community; they have a high level of consciousness and they illustrate the social engagement of citizens. Members of bivalve garden associations interact, learn, and comprehend the systemic origins of marine food production and often engage and empower themselves to participate in the restoration of coastal ecosystems.

19.1.3 Services and Social-ecological Systems

Bivalve gardens provide a wide range of ecosystem services (Haines-Young and Potschin 2011) with similar social and ecosystem factors (Table 19.1) of urban or community gardens (Cabral et al. 2017; Camps-Calvet et al. 2016). The main services reside in provisioning and cultural services (Table 19.1) driven by the aspiration for sustainable production of healthy local food. They are not only driven by the good will to provide supportive and regulative services to ecosystems disturbed by e.g. eutrophication and overexploitation.

Analogous to terrestrial gardens, bivalve gardens carry varied significance to different people, ranging from recreational, spiritual, or an educational framework. Bivalve gardens are comparable with so-called terrestrial wildlife gardens, sustainable gardens, and green gardening as a form of sustainable aquaculture in their participation in the enhancement of biodiversity and support of wildlife. Bivalve gardens also permit the maintenance of local varieties to increase resilience of local food supply (Barthel et al. 2014).

Bivalve gardens contribute to raising the public consciousness on environmental issues and sustainable farming, as well as the involvement of the community in protecting the environment from eutrophication or overexploitation. In the practice of gardening, there is an inherent educational aspect, where citizens can learn about aquaculture processes, observe and understand nature and seasons, and become more aware of the surrounding marine ecosystems (Tidball and Krasny 2010). Bivalve gardeners can vitalize the coastal area, share and transmit knowledge, and educate local communities and schools. By cultivating their own food, active citizens can trace healthy seafood from start to plate. Citizens engage socially for local support and community building by collaborative production of local food and the space for production. In many cases, community terrestrial gardens have been a

Table 19.1 Social and ecological factors for ecosystem good and services provided by shellfish gardens

	Social factors	Ecological factors
Provisioning services		
Food supply	Production of healthy local food, increase resilience of local food supply	Addition of food for predators in the system
Shell material	Use of shell material for various purposes e.g. construction	Increase clean substratum for settlement.
Source of seed	Restoring breeding stocks	Export of larvae to the surrounding environment
Regulating services		
Water clarification	Improvement of water for bathing	Improvement of habitat for seagrass and macroalgae and improved water status
Nutrient extraction	Potential nutrient credit	Improvement of habitat and water status
Biodiversity	Improving biodiversity, increase resilience of ecosystem	Increase of substratum and habitat for local species
Cultural services		
Learning & education	Experimentation with gardening practice. Teaching local communities and school regarding aquaculture, sustainable growth, blue growth.	Natural shellfish growth
Recreation & entertainment	Experimentation of boat activities at sea	Biophysical change
Physical exercise	Physical activities from shellfish spat manipulation from spat to harvestable product. Maintenance of the crop.	Removal and addition of mussel and wildlife biomass via harvesting and maintenance of the structure and crop
Spiritual & nature experience	Experimentation and connection with nature, relaxation. Invitation to dream and reflexion at sea	Decreased degradation of environment due to heightened awareness
Social engagement/ political empowerment	Engagement toward sustainable food and local support, and a cause that is meaningful to the community at large	More investment and service for sustainable production with increased potential for natural recruitment into fishery
Community building	Experiment in social cohesion with local community and carry a project together	Incremental improved water quality through stewardship activities
Localivore	Contribute to low carbon footprint and consume locally	Reduction in pollution from food transportation
Food traceability & health	Follow healthy omega3 rich seafood from start to plate	
Food quality & gourmet	Experimentation with new recipes, try new food, open horizons, increase in demand for shellfish and other seafood	
Art craft, design, creativity	Use of shell for creations, design shellfish garden landscape	

forum for participation in democratic processes (Ghose and Pettygrove 2014). Like their terrestrial counterparts, many bivalve gardens embody similar community- and civic-bound structures. At these early stages of development, bivalve gardens are, however, generally less integrative and driven principally for deriving supplementary food supplies rather than addressing food security.

From a health and wellness perspective, gardeners are invigorated via the recreational and community aspect of the activity and benefit to well-being (Egli et al. 2016). This is realized through: (i) social activity, (ii) physical activity by manipulating the farm units, boats and live products, (iii) spiritual discovery and therapeutic effects from contact with nature, (iv) creative use of bivalve products (i.e. arts and crafts, design from shells and raw materials, culinary quality experience (Table 19.1), and potential aesthetic aspirations comparable with land art, eco-design).

19.1.4 Bivalve Gardening Challenges

There is typically little spatial limitation for citizens establishing their own terrestrial vegetable gardens, which can exist on roof tops, as hanging gardens, floating gardens, pots in a kitchen etc. A bivalve garden should be located along the coastline, which is restrictive and this raises the issues of ownership and competition for shoreline and coastal water use. It is assumed that the first nation clam gardeners owned the gardening area in proximity to their settlement by controlling access (Lepofsky and Caldwell 2013). Nowadays, depending on the geographically relevant legislation, the coastline may be state or individually owned. In Denmark, a licence for establishing a community garden is delivered by the state, while in many states in the U.S., individual shoreline landowners can operate on their own plot while following federal and state regulation regarding its use.

Several practical reasons ranging from physical, social, and biological constraints (described in Table 19.2) must be taken into consideration in order to establish a bivalve garden, as well as the existing legislation regarding bivalve trade, biosecurity and food safety. Bivalve gardens, founded under the aquaculture framework, are also implicated with issues such as invasive species or diseases, as for bivalve aquaculture (e.g. EU Regulation 1143/2014 on Invasive Alien Species). For instance in the US or Europe, species cultivated in the bivalve garden must be native and locally present, otherwise, prohibited; wild seed comes from the same water body that the bivalve will be cultivated to reduce spread of potential disease and invasive species (see Puget Sound species recommendations). Although, seed from local species can be provided by certified disease-free hatcheries.

In the following two case studies, we focus on two different approaches for bivalve gardens. In the first, the U.S. case study exemplifies the provisioning service as a derivative of supportive and regulative initiatives from citizens and is illustrated

Table 19.2 Constraints to establish a shellfish garden

Constraints	Description
Physical	Adapted growing structures, water depth, storms, waves, physical carrying capacity and access to the coastline.
Biological	Food availability and quality for the shellfish, local presence of the shellfish cultured, production carrying capacity, food safety and water quality (e.g. low faecal coliform numbers, low toxin and heavy metal contaminations (e.g. EU shellfish directive, the US National Shellfish Sanitation Program). The origin of the juveniles should be either local or from disease free hatchery to prevent introduction of new species or disease.
Ecological	Ecological carrying capacity, potential competition with other present native species
Legal	Delivery time for a licence, regulations might not yet exist for licencing this type of activity.
Social	Management issues from a marine spatial planning point of view, potential user conflicts with other coastal activities, biological, physical and economical.
	Social beliefs: toxin, virus, bacterial contaminations are often in people's minds when it comes to shellfish, and some people would not take the risk to grow their own shellfish.
	No socio-ecological memory of shellfish gardens: unlike terrestrial gardening where a vast range of information, tools and guides are available to grow a salad or a chicken, citizens might feel alienated from the shellfish growing.

Table 19.3 Basic components of the two case studies

U.S. SPAT model	Denmark Fjord garden model
Intensive and extensive training opportunities	Workshops with training opportunities with professionals
Year-round, weekly activities and availability	Year-round, weekly activities and availability
Membership with direct incentives	Membership with direct incentives
Compartmentalized elements with individual leadership (committee concept)	Committee concept emerging but not yet fully implemented
Goal oriented; working towards a cause that is meaningful to the community at large.	Hobby, social aspect and mainly food oriented with ocean literacy goals
Availability of activities for all user groups	Availability of activities for all user groups, embraced depending on capability e.g. sick at sea but happy to cook

by the SPAT program (Suffolk Project in Aquaculture Training). In the second case, the provisioning services from bivalve gardens in Denmark are the main driving forces of the various projects and are illustrated by the Fjord garden project (Table 19.3).

19.2 Case Studies

19.2.1 The United States Case Study: Culture, Restoration, and Food Provisioning

19.2.1.1 Origins and Current Status

Seafood consumption and the activities associated with harvesting of seafood are coupled to the cultural heritage of many coastal communities in the United States (Griffith 1999). Identity and traditions in communities with a heritage of working waters has been shaped by historically important commercial species, while the loss of this heritage and its associated traditions is often lamented (Chambers 2006). Such traditions are often manifested in cultural tourism and seafood festivals (Claesson et al. 2005), where culinary customs and the 'waterman' are celebrated and romanticized. Non-commercial harvesting and gathering in bivalve grounds is historically significant for many coastal communities. While small-scale harvesting has been practiced by immigrating populations since European colonization, Native American groups have been harvesting and nurturing bivalve grounds for millennia (Cardinal and Fluharty 2012). Historical perceptions of many bivalve species harvested from the wild have transformed from sustenance foods to cultural staples, or even luxury items, concurrently with the shift from gathering to industrial harvest. Fisheries depletions are a relic of this affection for certain species; oysters, on both coasts, were severely overexploited by the early twentieth century and wild fisheries never returned to their peak production.

Contemporary oyster gardening in the United States began with the decline of the eastern oyster (*Crassostrea virginica*) population in the Chesapeake Bay; due to a combination of diseases, over fishing and diminished water quality in the 1960s (Mackenzie 2007). In many of the eastern coastal states, active restoration programs have essentially developed out of oyster stock and habitat improvement policy. Decades of work in breeding programs founded in the development of resistance to commercially important oyster pathogens, as well as towards increasing standing stocks of breeding oysters are generally viewed as successful (Brumbaugh et al. 2000). Bivalve gardening in the US originated in many of these restoration programs, and many continue to operate with broad membership and public participation under a restoration mandate (Rossi-Snook et al. 2010). Momentum in the growth of bivalve gardening as a phenomenon has shifted to cultivation for personal consumption in many coastal regions.

Numerous examples, from both Pacific and Atlantic coastal initiatives supporting gardening at the community-level, emphasize individual agency in both restoration processes and food production. Bivalve gardening associations and programs exist in nearly every coastal state in the US. Terminology is not standardized at either the popular or the institutional level, where gardening can indicate simply growing bivalves for one's own purposes, or an established method of bivalve population restoration leveraging public participation. Bivalve gardening "programs" in the US, are typically driven by conservation and/or restoration initiatives, where individuals become 'members' or otherwise obtain a share in the program (e.g. purchasing a 'starter kit' from a conservation group including culture gear, starter seed, and information on husbandry). Bivalve gardening (or grower) "associations" tend to consist of bivalve consuming enthusiasts whom interact on the basis of growing bivalves for personal consumption. These distinctions are not upheld across the US, nor are they exclusive of each other; many groups host blended membership between restoration and personal consumption motivations, which can fluctuate over time. To a degree, this fluid gradient between motivations for food production and restoration represents many societal contemplations of bivalve aquaculture ecosystem services.

Programs and associations are important interfacing fora for the public and the aquaculture industry. Many gardening initiatives in the US stem from the aquaculture industry's development of hatchery-based production, where high quality seed developed for fast growth and disease resistance provide a readily available source of 'seedlings'. Many state agencies (either aquaculture extension programs or regulatory) maintain directories of hatcheries selling seed to the public, readily available through internet search. Wild seed collection is also practiced for several species, for both infauna and epifauna, particularly mussels and clams. Multiple bivalve species are currently cultivated in gardens around the US, segregated by the Atlantic (and Gulf of Mexico) and Pacific coasts. On the Atlantic coast, the eastern oyster (*Crassostrea virginica*) reigns as the most widespread cultured bivalve in the entire US; from the northeast in Maine to Galveston Bay in Texas. This is largely attributed to emphases in oyster breeding programs, hatchery development, and variety (strain) availability. Other principle gardened species include blue mussels (*Mytilus edulis*), and quahogs/littleneck/hard clams (*Mercenaria. mercenaria*). As hatchery technologies develop for Atlantic scallops (*Placopecten magellanicus*), Bay scallops (*Argopecten irradians*), and surf clams (*Spisula solidissima*) it is anticipated that these species will be future candidates for gardening programs on the eastern seaboard. On the west coast, the geographical focus of gardening has resided in the northwest. Species such as Pacific littleneck clams (*Protothaca staminea*), Manila clams (*Venerupis japonica* or *philippinarum*), butter clams (*Saxidomus gigantean*), horse clams (*Tresus spp.*), cockles (*Clinocardium nuttallii*), geoduck (*Panopea generosa*), Olympia oysters (*Ostreola conchaphila*), Kumamoto oysters (*Crassostrea sikamea*), and Pacific oysters (*Magallana gigas*) can be sourced for gardening (Toba, Nosho, Washington Sea Grant Program 2002). While on both coasts many species are available for gardening, the predominant organisms of interest are oysters sourced from existing public or commercial hatchery programs.

19.2.1.2 Organizational Patterns

While programs and associations are an important component of bivalve gardening in the US, the majority of gardeners in many states do not participate in organized initiatives. A large number of gardeners are motivated to grow bivalves for their own consumption on their own property; private property ownership is an important feature shaping the bivalve gardener demographic. Land and water tenure issues in the US can vary considerably between coastal areas (Dellapenna 2009); often complicated, and in the case of private use of waters, the legal and regulatory framework can be difficult to navigate for the potential gardener. Access to growing waters, and the right to use those waters, may be bound to socioeconomic contexts in a region that could further influence formation and compositional patterns of gardening associations. Much of the land adjacent to accessible bivalve growing areas are privately owned, and as such, the use of those waters is largely restricted to property owners or individuals gaining permission from those owners to work the waters.

Alternatives to the mode of private ownership in gardening are emerging. In addition to the community garden spaces maintained in Suffolk County, NY (described below) there are several examples of functioning community gardens on the west coast (Evergreen Shellfish Club, Henderson Inlet Community Shellfish Farm, Pickleweed Point Community Oyster Farm, Port Madison Community Shellfish Farm) and east coast (Great South Bay Oyster Gardening Program, Three Mile Harbor Shellfish Garden). In most coastal states, oyster restoration programs without an explicit gardening component also provide the means to participate in cultivation practices. A subsequent effect of this model is the provision of access to bivalve gardening to participants without ownership of waterfront property.

19.2.1.3 Training

Many bivalve gardening associations and restoration programs base the process of membership accretion on their educational/training syllabus. In general, educational/training components include biology of the cultured species, ecology of the region and its aquatic realms, restoration principles, aquaculture, water management and quality, and seafood safety (Oesterling and Petrone 2012). Many gardening associations and programs provide a training regimen that is packaged with membership/participation. Participants will typically attend a short lecture series on ecology, aquaculture, and bivalve biology, followed by hands-on training with gardening equipment. Aquaculture extension specialists and marine conservation practitioners generally direct training sessions. In terrestrial horticulture, training programs have been developed to empower engaged gardening leadership through a decentralizing process termed "Master Gardener" (Pittenger and University of California 2015). These "Master Gardeners" are entitled to train and mentor individuals within their locality to cultivate crops in a manner specific to the local ecological conditions with techniques refined to the cultivars. Analogous to the terrestrial mode, groups such as the Tidewater Oyster Gardener's Association, hosts a "Master Oyster Gardener" program that envelops similar mentor-dissemination principles in the aquatic realm; instructing present and potential gardeners in bivalve husbandry techniques. "Master Oyster Gardeners" are then deployed into the community to host workshops and support gardening activities within their community. Similar training frameworks are employed in other bivalve gardening programs around the country, where veteran growers guide practical instruction.

19.2.1.4 Permits and Regulation

Bivalve gardening is regulated by state agencies, generally rooted in bivalve sanitation, coastal zone planning, and species restriction (to prevent introduction of invasive species, disease, parasites, etc.). Regulations can vary considerably between states, and applicability can be dependent of personal property law in a given municipality. In accordance with the National Shellfish Sanitation Program (NSSP) Model Ordinance, most states that recognize bivalve gardening activities, govern these activities through permitting and compliance processes. In Virginia, for example, obtaining permission to garden is relatively straightforward; a potential gardener obtains a simple permit from the Marine Resource Commission (MRC). Non-commercial permits (personal consumption) are cost-free, and require specific use constraints, such as avoidance of Submerged Aquatic Vegetation areas, siting to avoid conflict with watercraft and other configuration-specific considerations (4VAC20-336). Some bivalve gardeners pursue permitting to sell their products or use them in a public setting. Additional permits from the MRC are required for sales, depending on the location and physical garden setup; commercial aquaculture operations follow the same regulations. Across the state border, however, in Maryland bivalve gardeners are prohibited from utilizing their oysters for consumption or sale, and must relay them to a restoration site. Gardening activities in Maryland are permitted by the US Army Corps of Engineers under physical and maritime use conditions, and gardeners must register with the state Department of Natural Resources. In New York, a permit is acquired from the state Department of Environmental Conservation, and while personal consumption is encouraged, sale is prohibited without a specific commercial permit.

From the public health realm, regulation of gardening activities is exercised by similar spatial restrictions applied to commercial growers. Taking Virginia again as an example, bivalve growing areas are defined and regularly monitored for algal toxins and human enteric pathogens. When an area is 'condemned', growers are prohibited from harvesting regardless of their permit status. These condemnations may be seasonal, and may permit the grower to relocate their bivalve to another area for depuration. Sales of fresh product must first undergo operational inspection from the state Department of Health, which manages regulation on bivalve sanitation. Shucking or further processing/handling of tissue requires rigorous inspection, planning, and a permit from the same department (12VAC5-150).

19.2.1.5 Oysters in New York

Cornell Cooperative Extension (CCE) of Suffolk, New York began an oyster gardening program in 2000 at its Southold facility on the north fork of Long Island. Beginning in 1992, The Suffolk County Marine Environmental Learning Center (SCMELC) used a small bivalve hatchery to assist local townships with bivalve seed that would be grown in gardens, and then broadcast for bivalve restoration and stock enhancement purposes. The bivalve stocks in this area had been heavily compromised by a harmful algal blooms referred to as the "brown tide" (*Aureococcus anophagefferens*). The facilities produce an average of 4–6 million bivalve seed from three species and provide education to the community on a year-round basis. Increased awareness within the community of bivalve and their potential for improving water quality spurred the need for a more comprehensive approach to nurturing the seed in order to boost survival rates, as well as to foster a greater sense of environmental stewardship. Hence, the Suffolk Project in Aquaculture Training (SPAT) was developed and launched following an introductory open house in December of 1999. Table 19.4 summarize the elements of starting and keeping a successful bivalve garden based on the SPAT experience. This program is distinct from the Billion Oyster Project (BOP) as described in DeAngelis et al. 2019. The BOP project is focused on educational training in schools, rather than participation of private persons as in the SPAT project.

SPAT offers membership to the public requiring a yearly fee, providing 1000 oyster seed that can be deployed and harvested for personal use. This was modified from the original approach, which provided 2000 seed and required 50% of the survivors to be broadcast into the environment. For members that have their own private access to water, specific rules apply; including the inability to sell their stock, requisite cultivation in waters that are certified as sanitary, and to acquire the necessary permits. For members who do not own private access to water, three community gardens are available for planting, which are overseen by CCE staff (Fig. 19.1). In 2016, active membership comprised 226 families, 68 of which owned private waterfront; over 1000 families have interacted with SPAT through its tenure.

While oyster gardens and the ability of individuals to culture their own stock for personal use are essential features of SPAT, the strength of the program has resided in aquaculture training. The program hosts weekly volunteer work sessions; on average, members have collectively logged over 10,000 volunteer hours annually. During these work sessions, members become involved in all aspects of the program, included construction of numerous systems such as floating and land-based upwelling systems. Since 2002, SPAT has operated its own bivalve hatchery, the "SPAT Shack" which was built and funded by the members. In 2016, members added a second hatchery system, producing ~1 million oyster seed and 1 million clam seed. As an essential component of the hatchery, a full nursery system is maintained to hold stocks until ready for deployment in the environment.

The "SPAT Shack" provides members the autonomy to learn bivalve culture techniques without ulterior demands. This allows members to study the cultivation

Table 19.4 Elements of starting and keeping a successful shellfish garden

How to get started	**The committee concept**
Have a clear goal in mind, however small or large.	Division of labour allows for multiple components of the project to be addressed simultaneously and aggressively.
Rectify immediate obstacles (permits, regulations, and local community acceptance).	Utilizing a dedicated core group of volunteers with specific expertise and commitment to a specific component of the project leads to a higher level of quality results.
Solicit some level of funding, however small.	Monthly meetings of the committee chairpersons (advisory board) lends itself to a high level of coordination through solid communication.
Draft a plan of action to achieve the project goals.	Committee chairs network well with the volunteers at large.
Advertise an informal community open house (make sure you invite press and politicians).	A higher level of commitment is necessary from the program coordinator or project group leader in order to maintain coordination.
Follow up with all interested parties.	Involvement of all members
Calendarize some worthy events/ activities.	Priorities must be kept in order for the group to function as a whole.
Delegate important functions to core group.	
Network and develop partnerships.	
Maintain momentum.	
Training sessions	**Facilities and equipment**
Volunteers will understand the process.	Being a turn-key operation takes many years and depends on the various possible site specific constraints.
Volunteers will be learning about techniques that will be used during various phases of a project.	Seed availability is key for starting a sea garden and an operational bottleneck. Hatcheries are expensive and complex to run. Operators say "I wouldn't wish a shellfish hatchery on my worst enemy...".
Questions will be answered on topics of interest or importance to the individual and the group as a whole.	Another saying is "be careful what you wish for." Programs and projects can fall apart by wanting too much too soon.
Confidence and understanding will be gained by the volunteers on the subject matter.	Partnerships with successful operations are always a plus.
Confidence and understanding will be gained by the trainer on how to convey concepts to the group.	Developing a program is like climbing a ladder, taking it one rung at a time (and not looking down)!
The trainer will get to know the individual volunteers.	Be logical, economical and efficient with budgets.
Expertise will be needed by the trainer in the subject matter.	
Commitment will be needed by the program organizers.	
Enables volunteers to become ambassadors for the program.	

Fig. 19.1 Community garden with 50 growers (SPAT – U.S.). (Photo courtesy: Kim Tetrault)

of bivalve seed in a relatively stress free and non-competitive environment. Accessibility to workshop facilities and institutional staff is meaningful to member participation. A year-round lecture series is hosted which includes 11 two-hour lectures offered twice each month covering all aspects of bivalve aquaculture. These lectures are well attended with an average of 35 members per month.

The most essential element of the SPAT program relates to the organisational structure under which it operates. CCE is a non- profit organisation (US, 501- C3) whose primary mandate is to educate members of the community, and assist them in putting their knowledge to work. The marine division of CCE is staffed by marine professionals, providing SPAT full-time oversight from CCE. This dual management system (from members and staff) simultaneously supports the community ownership of the program while maintaining demonstrable standards in its operations.

19.2.2 The Danish Case Study: Food Provisioning, Well-being, and Environmental Awareness

Unlike the U.S., the concept of bivalve gardening in Denmark emerged with the aspiration to empower citizen stewardship on local seafood production for family consumption. Bivalve gardening also aspires to promote a Danish lifestyle of health and well-being. There is a general interest in healthy and organic food and it is also visible in recent movements such as the "New Nordic Cuisine" where Danish chefs

Fig. 19.2 Map of main shellfish gardens in Denmark. Green dots, Fjord garden project, yellow orange and blue independent projects. Red circle indicate the local population size associated to the garden

and citizens are promoting "slow food" by going back to locally grown, wild, healthy and sustainable Scandinavian food delicacies.

Since 2011, several bivalve gardening initiatives have become functional. The concept is expanding in Denmark, culminating in small rural areas ranging from 4000 inhabitants to large cities such as the capital Copenhagen with more than 1.7 million inhabitants (Fig. 19.2). Cultivated species consist of entirely endemic varieties; mussels (*Mytilus edulis*), flat oysters (*Ostrea edulis*), and macroalgae (*Saccharina latissima, Palmaria palmata*). Invasive species such as the Pacific oyster (*Magallena gigas*) are prohibited to be cultivated, hence also in gardens.

Several factors could explain the reasons bivalve gardening in Denmark is flourishing: most of the described constraints to bivalve gardening in Table 19.2 are met:

- **Physical**: Denmark has a long and sheltered coastline. The ratio of land to coastline in Denmark is extremely small (5.8 km^2/km) in comparison to all other bivalve producing European countries (e.g. Norway = 12.1, UK = 19.5, or France = 132 km^2/km, The World Factbook 2017). This geographical-historical feature provides particular access and relationships with the marine environment. Moreover, most sheltered estuaries are quite shallow and protected from large fetch, thus bivalve production infrastructures do not require elevated investment and is easily accessible with a small boat or from structures directly connected to the shore. The conversion of unused industrial harbours to clean areas for new nautical activities is opening easy access space for bivalve gardens, as long as the sanitation is good for growing bivalves.
- **Biological/Ecological**: Good sanitation conditions and high primary productivity make Danish waters very suitable for the aquaculture of filter feeding bivalves. Moreover, production and ecological carrying capacities on basin scales seem far from being realized (Nielsen et al. 2016). Danish waters are highly impacted by eutrophication due to an excess of terrestrial nutrient loading even though there is a mandated policy of reducing nutrients introduction into waterways to comply with EU Water Framework Directive standards (WFD – "Directive 2000/60/EC"). Most coastal areas around Denmark have high hygienic water (Class A areas) thanks to an increasing number of waste waters treatment plants (WWTP, Carstensen et al. 2006) meaning that bivalves can be harvested for direct consumption under the Shellfish Water Directive (2006/113/EC).

Species A key aspect of bivalve gardening is the free access to seed and fast growth of the cultivated animals, which makes it attractive to the gardeners. There is a high level of natural recruitment for local species and it can take less than a year for mussels to reach commercial size in certain areas, such as the Limfjorden. Provided the high natural recruitment of mussels (*Mytilus edulis*), no hatchery are necessary for cultivation. Mussel seed is collected naturally on spat collectors placed in the garden around May, seeds are then sorted and then socked around September, and the crop is harvestable from April the following year. Regarding the native flat oysters (*Ostrea edulis*), spat can be collected from spat collectors deployed in sheltered areas with an existing population of oysters or small oysters can be hand-picked if allowed by the authorities and kept in the gardens for ongrowing. It takes approximately 3 years to reach commercial size in the Limfjorden. A more secure supply of oyster seed would rely on hatchery production, which is expensive and not yet reliable in Denmark.

- **Legal**: The delivery time for a licence can be relatively short in Denmark but depends on the competent authorities and whether there are objections from stakeholders. It normally takes less than 4–5 months to produce a licence; and authorities are considering an easier procedure for sea gardens as long as the production is not commercial due to food sanitation regulation

- **Social and cultural**: New Nordic Cuisine, the slow food movement and connection to the sea are catalysts for the creation of bivalve gardens. Overall, there is a positive acceptance and enthusiasm for sea garden projects by citizenry.

An interesting aspect of bivalve gardening organisations that have developed in Denmark is their different approaches to constitution: (i) One person or a small group of citizens create an association of bivalve gardeners and run it, (ii) A group of citizens or a local agency promote bivalve gardening and create associations together with partners and then recruit citizens and board members to run the association, Most bivalve gardens in Denmark maintain their own informative website often associated to social media such as Facebook. Associations cover an annual fee between 40 and 70 euros and consist of 30–200 members. Associations are generally composed of various groups with key interests such as gourmet foods, aquaculture techniques, art, demonstration of aquaculture practice, food and workshop events, seaweeds, mussels, and oysters. In 2011, a pioneering group of citizens interested in non-commercial seafood production for personal consumption developed the present gardening concept and by 2013, the "Havhaven Ebeltoft Vig" (Sea Garden Ebeltoft Vig) association had created the first hobby-based bivalve garden in Denmark (Fig. 19.2). Their configuration consisted of a few longlines of mussels and seaweed, which are deployed with both common and individual crops. Based on the same principle another association, "Kerteminde Maritime haver" (Kerteminde Maritime gardens), deployed a bivalve farming structure and the first lines and socks of seaweed and mussels in 2016.

In Copenhagen, a non-profit association called "Maritime Nyttehaver" (translated: Maritime allotments) started in 2012. The garden is situated in the middle of the capital harbour, which is now remediated from past polluting industries but not sanitary enough to provide edible bivalves (Fig. 19.2). The main objective of this association is to establish urban aquaculture and promote ocean literacy. Products such as clothing, courses, culture demonstrations, and mussel culture kits (which includes spat collectors, bivalve gardening guide, ropes etc.) for individual use are marketed to help finance the garden operation. These mussel kits could raise some legislation and management concerns regarding their individual use, the deployment location, the potential impact of the material on other users, and ecosystem impacts in the coastal environment beyond deployment.

As another example, Fjord garden project, a large privately funded project in partnership with four municipalities, called "Fjordhaver i Limfjordens havne" (Estuarine gardens in the Limfjord harbour), was launched in 2015. This project enabled the development of four sea gardens in four different harbour cities of the largest Danish estuary (Limfjorden, Fig. 19.2), with the assistance of professional groups to establish the gardens. The development of the sea gardens varied with different constraints at different locations, but employing the same basic principle to create a local association with a maximum of 150–200 members. In less than a year into the project, two of the gardens became operational. The overall purpose of this "Fjord garden" project is to create opportunities for a "good life" in cities around the Limfjorden. As such, the project is driven by five core beliefs: (1) create life in depopulated or unused port areas and provide space for social activities to interested and committed citizens; (2) bring "blue" into the city; (3) empower non-professionals in relation to seafood production to facilitate understanding of the production process and creating a relationship between the product and the educated consumer; (4) promote healthy meals and lifestyles with roots in the local maritime history; and (5) increase ocean literacy to further develop roots in the sea and raise local awareness of the goods and services provided by organic and inorganic extractive aquaculture species (bivalve and macroalgae respectively) as opposed to the fed species (e.g. finfish).

Several workshops have been conducted throughout the Fjord garden project, gathering all members of the 4 associations and including professionals in order to train members and transfer knowledge for the sustainability and legacy of the bivalve gardens beyond the project life (end 2017). Workshops address training on mussel, oyster and seaweed production, food safety, ecology of the fjord, biology and growth cycle for year-round production, garden setup and production material (e.g. knots, buoys, socks, longline setup, boat and platform operation), and finally how to handle the harvested product including preparation. Food safety and culinary workshops have been evaluated has very important by the association members. Project members have participated in the establishment of gardens (Fig. 19.3) on two different types of production structures: long-line configurations accessible by boats, and rafts connected to land (Fig. 19.4). In some associations, subgroups have emerged to specialise in the cultivated species, or boat and structure maintenance. In less than 2 years of their creation, the associations and gardens have already garnered more than 200 members in total, sharing a common interest. Thus far, the relative success of the bivalve sea gardens can be attributed to positive dissemination from local media, distribution of promotional leaflets, professional and financial support from the private fund Nordea and most importantly, motivated and committed members. Challenges however may arise as the project ends in regards to the professional and financial support for the legacy in training of new members and maintenance of facilities and equipment. To stimulate project legacy, a sea garden guide with key information gathered during the project will be produced.

There are several additional challenges to the future development of bivalve gardens in Denmark. In certain areas, predators such as Eider ducks (*Somateria mollissima*) can feed on and eliminate an entire production unit of hanging mussels. In terms of food safety, analyses for water quality are extremely expensive and not carried out by the authorities, but rather by the users. The Danish Veterinary and Food administration from the Ministry of Food and Environment of Denmark has some clear guidelines for private harvesting of mussels and oysters regarding bivalve sanitation: there is no imposed restriction in consuming bivalves from bivalve gardens as long as it is for private use and not sold (update from the 01/09/2018 https://www.foedevarestyrelsen.dk/Selvbetjening/Guides/Kend_kemien/Sider/Indsamling-af-muslinger-og-oesters-mm-hvad-skal-man-vaere-opmaerksom-paa.aspx). Danish coastal waters are divided into production areas, which are by default closed for commercial bivalve harvesting. In order to open an area for harvest, professionals, either from fishery or aquaculture must conduct

Fig. 19.3 Shellfish garden members participate in the line preparation (**a**), sorting (b), socking (**c**) and hanging of mussel socks on both from the Fjord garden project longline and raft (**d** and **e** respectively) (Denmark). (Photo courtesy Carsten Fomsgaard (**a**–**d**), Rikke Frandsen (**e**))

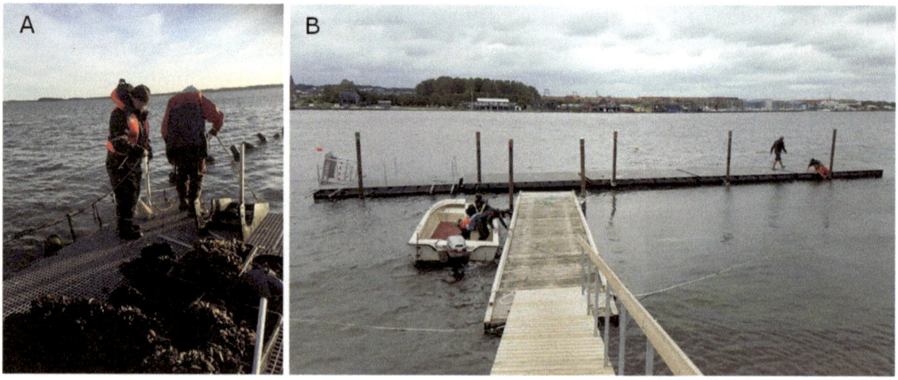

Fig. 19.4 Two different type of production structure: (**a**) longlines area delimited with yellow corner buoys with access by boat and raft used in the Fjord garden project maintenance, and (**b**) raft connected to land (Denmark). (Photo courtesy of Lola Thomsen (**a**) and Esge Hansen (**b**))

microbiological, toxicological (algae) and chemical contaminants analysis on fresh bivalves and water samples from the area. Responsible consumption is therefore advised for both hand-picked bivalves, and bivalve gardeners, whom can benefit from the analyses conducted by professional producers in the same areas.

Although the issue of competition between bivalve gardeners and commercial operators can be problematic, it is marginal due to: (i) the small volume of bivalves produced by e gardens, (ii) the non-commercial definition of bivalve gardens, (iii) the limited domestic market for bivalves in Denmark and (iv) the interest in bivalve consumption by gardeners could be beneficial on the long term to commercial entities. Presently the cohabitation of gardeners and commercial entities is peaceful as there is no conflict of interest for the use of the coastal area or resources. However, commercial entities are concerned that irresponsible consumption by bivalve gardeners regarding food sanitation could tarnish their image.

In terms of marine spatial planning and social acceptance, overall, bivalve gardens in Denmark as well as bivalve aquaculture are relatively well accepted. The visual pollution from buoys is relatively discrete, but complaints of nuisance linked to buoys washed ashore or other conflicts over use of the space, emerge and delay licencing. Notwithstanding, the keen enthusiasm and interest of Danish citizens in bivalve gardening has resulted in a boom of applications for garden licences around the country; at a rate that the aquaculture administration authorities might have problems to handle, not least in view of the emerging mixed forms of sea gardens, e.g. gardens linked to local restaurants thereby compromising both food safety regulations and undue competition with the professional. This could create, in the long run, the occurrence of unregulated individual gardens with a loss of community and social acceptance and a potential danger for other nautical activities (i.e., if culture equipment is placed in coastal areas illegally). Other issues and potential risks exist for future gardeners and the ecosystem. Bivalve gardens are not protected from potential invasive species or translocation of disease from contaminated seeds, if poor practices are exercised in surrounding waters. Pollution, climate change, and low spatfall are also risks to be taken into consideration. The next logical step, as for bivalve aquaculture, is to monitor the environmental impacts of gardens.

19.3 Successful Bivalve Gardens and Future Challenges

As a phenomenon emerging from socially and culturally instructed objectives in sustainable food production and ecology, bivalve gardening is principally an activity that enhances personal experience and community engagement. The provision of seafood for coastal communities is more profound than the fulfilment of nutrition requirements. Gardening gives agency to individuals and communities to play a role in shaping their food system. While seafood can be a very important source of particular fatty acids and minerals (McManus and Newton 2011), bivalve gardening tends not to be an exercise in securing food; it is an embodiment of cultural-ecological principles that align with contemporary ideology of sustainable food systems (Turner 2011). Gardens do not compete with commercial food production; they are an expression of ambitions for greater control of food systems, increased variety in the access to nutritious and sustainable foods.

The case studies describe multiple approaches to bivalve gardening, which can manifest at the individual or group-level (Table 19.3). Bivalve gardeners tend to be motivated by interests in preserving or celebrating cultural and place-based heritage, exercising the motions of self-sustenance and self-determination, as well as altruistic dispositions to improve the environment (Krasny et al. 2014). These motivations can be grounded in historical contexts, such that gardening can provide the means to revitalize tradition, the shift from fishing/hand picking tradition to controlled and sustainable hobby aquaculture, and thus form continuity within a community, and across generations. Ecological consciousness and a drive to take part in processes governing the state of physical surroundings and daily life can develop simultaneous self and collective inclinations of discovery. While the impetus for gardening may originate from different sources between case studies, there are common driving factors for bivalve gardeners that are worth describing.

From a culinary perspective, gardening for consumption exhibits its own discernible motives, accompanied by a deeper appreciation for quality aspects of seafood. In some contexts, bivalve gardening can provide easier access to high-value species that may not be a part of the general food culture. As a hobby, there is emphasis on uniqueness of the product and process, where place and husbandry evoke distinct organoleptic characteristics and values attune to being 'home grown'. Bivalve harvests from the garden for personal consumption are often associated with social gathering. To these effects, bivalve gardening serves an interesting purpose in the expression of cultural and societal values.

In practically every example of bivalve gardening, the desire to learn and educate others about marine ecology, food production, and bivalve biology is strongly expressed. As a physical activity in the outdoors, many participants are drawn to the essence of a structured endeavour that contributes to personal well-being and adds meaning to daily life.

With growing awareness of anthropogenic impacts on aquatic ecosystems and recognition of cultural heritage related to those ecosystems, wide scale interest in mechanisms to positively influence the coastal environment gives rise to organized gardening programs. Restoration of bivalve populations and habitat attracts proponents of environmental stewardship; bivalve gardening provides space for individual ownership of this process (Torres et al. 2017).

The process of establishing bivalve gardens is strongly dependent on sociocultural contexts, such as seafood consumption patterns, environmental awareness, historical perceptions of seafood, social capital, and motivated organizers. Institutional and regulatory frameworks also shape the environment which gardening modes may materialize, such as land tenure regimes and regulation (or the lack thereof). For example, as in the U.S. and Danish cases, the right to deploy a garden in coastal waters can be restricted to shoreline owners or gained through the aquaculture administration authorities and a public arbitration process which may delay licencing. As such, community-based organization is advantageous in acquiring rights to use coastal waters. Bivalve community gardens could learn from well-established terrestrial community gardens and the various toolkits already in place. One of the most important core beliefs in this paradigm is the grassroots approach, where citizens are engaged from the instigation through the operation of the community garden and empowered through stewardship of the food production (Abi-Nader et al. 2001).

As participation in bivalve gardening initiatives expands, present and future challenges will be confronted. Presently, food safety is the immediate concern for public health and the seafood industry. Years of refinement in harvesting techniques and food safety practice (i.e. HACCP) have contributed to the growth of the bivalve industry and subsequent growth in demand for fresh bivalves. In addition to the direct public health impact, disease outbreaks related to unsanitary practices can be damaging to gardening programs and the bivalve industry. With the emerging issues of climate change, invasive species and connectivity between bodies of water (i.e. translocation of seed, ballast), vectors for HABs and pathogens are increasing (Tirado et al. 2010). The regular monitoring for water quality and bivalve contamination is a necessity to guaranty public health. However, certified analyses remain costly and slow and often not affordable for gardeners when the garden does not belong to an area monitored by the state or the industry. The development of new technology (e.g. molecular, genetic) for quicker and more affordable test kits seems inevitable for bivalve gardening.

Analogous to terrestrial gardens, bivalve gardens require space, which can be limited and contentious; particularly in coastal waters where Marine Spatial Planning is confronted with the dilemma of achieving potentially antagonistic goals between good ecological status and economic development of blue growth (Jones et al. 2016). There are a number of common issues that arise counter to the development of coastal aquaculture, including habitat manipulation from fixed structures or working equipment, interference with other recreational uses of the same waters, and aesthetic impacts. Gardening efforts should balance stakeholder perceptions and remain receptive to the community, which hosts the garden.

While bivalve gardening is ordinarily described as an environmentally and socially positive activity, there are circumstances in which gardening could impart negative consequences. The marine environment, as a medium for the growth of gardened species, contrasts with terrestrial gardens in the risk potential for spread of pathogens and invasive species. Indeed, while common stewardship of our land resources in this light should be a broad objective, the marine environment cannot be discretized in a controlled manner. Poor gardening practices, such as haphazard seed sourcing and transfer of organisms between distinct coastal waters, can be catastrophic to local ecology and industry. Conforming to the roots of gardening, strong mentorship is exceedingly important to realize gardening goals and ensure sustainability of its practices. Although in areas such as Denmark where gardens are localised in eutrophic areas where phytoplankton is in excess, the cultivated bivalves do not compete with native wild species, ecological carrying capacity should be considered in the establishment criteria of a garden (Table 19.2).

Within the regulatory framework of many regions, countries, municipalities, and towns, bivalve gardening is often a novel concept with peculiar aspects that may present difficulties in formulating effective regulation. Guiding sustainable gardening practices through regulation should reflect the aforementioned potential problems with gardening, however, doing so may culminate in unwieldy rules that overburden gardening programs; especially while balancing other stakeholder interests. Leaders in the gardening community should reach out to regulators and policy makers to help advise the formulation of regulations.

The longevity and legacy of gardening initiatives can often be overlooked. Accretion of younger generations participating with similar levels of enthusiasm can prove to be very difficult for some groups, particularly in restoration programs. As many gardening efforts are founded in cultural heritage and the motivation to revitalize that heritage, gardeners working in community-based programs should carefully contemplate and plan for conceptual inheritance of bivalve gardening.

Acknowledgements The authors would like to particularly thank Jens Kjerulf Petersen for its constructive comments on the manuscript and various discussions on bivalve gardening. Thanks to Carsten Fomsgaard, Pascal Barreau, Vic Spain, Ramón Filgueira, and Masaki Moro for their assistance regarding bivalve gardens in Denmark, France, U.S. and sea gardens in Japan. We are grateful to Boze Hancock and Vivian Huse as reviewers, for their comments of an earlier version of this paper.

References

4VAC20-336. Virginia Administrative Code: General Permit No. 3 Pertaining to Non commercial Riparian Shellfish Growing Activities

12VAC5-150. Virginia Administrative Code: Regulations for the Sanitary Control of Storing, Processing, Packing or Repacking of Oysters, Clams and Other Shellfish

Abi-Nader J, Dunnigan K, Markley K, Buckley D (2001) Growing communities curriculum. The American Community Gardening Association, Philadelphia

Andersen SH (2000) 'Køkkenmøddinger'(Shell Middens) in Denmark: a survey. Proc Prehist Soc 66:361–384. Cambridge University Press

Ausubel JH (2000) Great reversal: nature's chance to restore land and sea. Technol Soc 22:289–301

Barthel S, Parker J, Folke C, Colding J (2014) Urban gardens: pockets of social-ecological memory. In: Tidball K, Krasny M (eds) Greening in the red zone. Springer, Dordrecht

Beck MW, Brumbaugh RD, Airoldi L, Carranza A, Coen LD, Crawford C, Defeo O, Edgar GJ, Hancock B, Kay M, Lenihan HS, Luckenbach MW, Toropova CL, Zhang G, Guo X (2011) Oyster reefs at risk and recommendations for conservation, restoration and management. Bioscience 61:107–116

Brumbaugh RD, Sorabella LA, Garcia CO, Goldsborough WJ, Wesson JA (2000) Making a case for community-based oyster restoration: an example from Hampton Roads, Virginia, USA. J Shellfish Res 19(1):467–472

Butler PG, Freitas PS, Burchell M, Chauvaud L (2019) Archaeology and sclerochronology of marine bivalves. In: Smaal A, Ferreira JG, Grant J, Petersen JK, Strand O (eds) Goods and services of marine bivalves. Springer, Cham, pp 413–444

Cabral I, Keim J, Engelmann R, Kraemer R, Siebert J, Bonn A (2017) Ecosystem services of allotment and community gardens: a Leipzig, Germany case study. Urban For Urban Green 23(4):44–53

Camps-Calvet M, Langemeyer J, Calvet-Mir L, Gómez-Baggethun E (2016) Ecosystem services provided by urban gardens in Barcelona, Spain: insights for policy and planning. Environ Sci Policy 62:14–23. https://doi.org/10.1016/j.envsci.2016.01.007

Cardinal K, Fluharty DL (2012) Shellfish aquaculture in Puget sound in light of Washington's coastal marine spatial planning. University of Washington

Carstensen J, Conley DJ, Andersen JH, Ærtebjerg G (2006) Coastal eutrophication and trend reversal: a Danish case study. Limnol Oceanogr 51:398–408

Chambers E (2006) Heritage matters: heritage, culture, history, and Chesapeake Bay. Chesapeake perspectives. College Park: Maryland Sea Grant College, University of Maryland

Chase S (2017) Shellfish harvesting and safety. Shore Stewards News, guidelines and resources for leaving near water, established 2003, Fall 2017. Available at http://shorestewards.cw.wsu.edu/. Consulted 04.11.2017

Claesson S, Robertson RA, Hall-Arber M (2005) Fishing heritage festivals, tourism, and community development in the Gulf of Maine. Proceedings of the 2005 Northeastern Recreation Research Symposium, 341(GoMOOS 2003), pp 420–428

DeAngelis B, Birch A, Malinowski P, Abel S, DeQuattro J, Peabody B, Dinnel P (2019) A variety of approaches for incorporating community outreach and education in oyster reef restoration projects: examples from the United States. In: Smaal A, Ferreira JG, Grant J, Petersen JK, Strand O (eds) Goods and services of marine bivalves. Springer, Cham, pp 335–354

Dellapenna JW (2009) United States: the allocation of surface waters. In: Dellapenna JW, Gupta J (eds) The evolution of the law and politics of water. Springer, Dordrecht, pp 189–204

Duarte CM, Marbà N, Holmer M (2007) Rapid domestication of marine species. Science 316:382

Egli V, Oliver M, Tautolo ES (2016) The development of a model of community garden benefits to wellbeing. Prev Med Rep 3:348–352

FAO (2016) The state of world fisheries and aquaculture 2016. Contributing to food security and nutrition for all, Rome. 200 pp.

Gelcich S, Buckley P, Pinnegar JK, Chilvers J, Lorenzoni I, Terry G, Guerrero M, Castilla JC, Valdebenito A, Duarte CM (2014) Public awareness, concerns, and priorities about anthropogenic impacts on marine environments. Proc Natl Acad Sci 111(42):15042–15047

Ghose R, Pettygrove M (2014) Urban community gardens as spaces of citizenship. Antipode 46:1092–1112. https://doi.org/10.1111/anti.12077

Griffith D (1999) The estuary's gift: an Atlantic Coast cultural biography. Pennsylvania State University Press, University Park

Groesbeck AS, Rowell K, Lepofsky D, Salomon AK (2014) Ancient clam gardens increased shellfish production: adaptive strategies from the past can inform food security today. PLoS One 9(3):e91235. https://doi.org/10.1371/journal.pone.0091235

Grunert KG, Hieke S, Wills J (2014) Sustainability labels on food products: consumer motivation, understanding and use. Food Policy 44:177–189

Haines-Young R, Potschin M (2011) Common international classification of ecosystem services (CICES): 2011 Update. Contract No. EEA/BSS/07/007. European Environment Agency, Copenhagen

Jackson JBC, Kirby MX, Berger WH, Bjorndal KA, Botsford LW, Bourque BJ, Bradbury RH, Cooke R, Erlandson J, Estes JA, Hughes TP, Kidwell S, Lange CB, Lenihan HS, Pandolfi JM, Peterson CH, Steneck RS, Tegner MJ, Warner RR (2001) Historical overfishing and the recent collapse of coastal ecosystems. Science 293:629–637

Jones PJS, Lieberknecht LM, Qiu W (2016) Marine spatial planning in reality: Introduction to case studies and discussion of findings. Marine Policy 71:256–264. https://doi.org/10.1016/j.marpol.2016.04.026

Krasny ME, Crestol SR, Tidball KG, Stedman RC (2014) New York City's oyster gardeners: memories and meanings as motivations for volunteer environmental stewardship. Landsc Urban Plan 132:16–25. https://doi.org/10.1016/j.landurbplan.2014.08.003, ISSN 0169-2046

Lepofsky D, Caldwell M (2013) Indigenous marine resource management on the northwest coast of North America. Ecol Process 2:1–12

Mackenzie CL (2007) Causes underlying the historical decline in eastern oyster (*Crassostrea virginica* Gmelin, 1791) landings. J Shellfish Res 26:927–938
McManus A, Newton W (2011) Seafood, nutrition and human health: a synopsis of the nutritional benefits of consuming seafood, Curtin University of Technology. Centre of Excellence for Science, Seafood & Health (CoESSH)
Nielsen P, Cranford PJ, Maar M, Petersen JK (2016) Magnitude, spatial scale and optimization of ecosystem services from a nutrient extraction mussel farm in the eutrophic Skive Fjord, Denmark. Aquac Environ Interact 8:311–329
Oesterling M, Petrone C (2012) Non-commercial oyster culture, or oyster gardening. SRAC Publication 4307:1–12
Paticat F (2007) Flux et usages de l'eau de mer dans les marais salés endigués Charentais : Cas du marais salé endigué de l'île de Ré. Géographie. Université de Nantes. (in French)
Petrini C, Bogliotti C, Rava R, Scaffidi C (2012) The central role of food. Slow Food International. Congress Paper 2012–2016. 22pp
Pittenger D, University of California Division of Agriculture Natural Resources (2015) California master gardener handbook, 2nd edn. University of California System. 3382pp
Pollan M (2006) The omnivore's dilemma: a natural history of four meals. Penguin Press, New York
Rabanal HR (1988) History of aquaculture. ASEAN/SF/88/Tech.7. ASEAN/UNDP/FAO Regional Small-Scale Coastal Fisheries Development Project, Manila, Philippines
Rossi-Snook K, Ozbay G, Marenghi F (2010) Oyster (*Crassostrea Virginica*) gardening program for restoration in Delaware's Inland Bays, USA. Aquac Int 18(1):61–67. https://doi.org/10.1007/s10499-009-9271-5
Teletchea F (2015) Domestication of marine fish species: update and perspectives. J Mar Sci Eng 3:1227–1243
The World Factbook 2017 (2017) Central Intelligence Agency, Washington, DC
Tidball KG, Krasny ME (2010) Urban environmental education from a social-ecological perspective: conceptual framework for civic ecology education. Cities Environ 3(1):11. 20 pp
Tirado M, Clarke R, Jaykus L, Mcquatters-Gollop A, Frank J (2010) Climate change and food safety: a review. Food Res Int 43:1745–1765
Toba DR, Nosho TY, Washington Sea Grant Program, Publisher (2002) Small-scale oyster farming for pleasure and profit in Washington. Washington Sea Grant
Torres AC, Nadot S, Prévot A-C (2017) Specificities of French community gardens as environmental stewardships. Ecol Soc 22(3):28
Turner B (2011) Embodied connections: sustainability, food systems and community gardens. Local Environ 16(6):509–522. https://doi.org/10.1080/13549839.2011.569537
Websites: http://bp.eco-capital.net/bps/read/id/88 consulted 07/07/2016; http://www.haff.city.hiroshima.jp/info/2016/11/8982/ consulted 07/07/2016

Chapter 20
Shells as Collector's Items

Peter F. Duncan and Arne Ghys

Abstract Shell collecting, and the more scientific discipline of conchology, have a long history, and the general activity has made significant contributions to art, commerce and science since at least the seventeenth century. Modern shell collecting encompasses a wide range of molluscan families and species, including numerous bivalve taxa, and collections may be developed via a range of methods including self-collection, purchase from specialised dealers, exchange or from older collections. The fundamentals of building and maintaining a scientifically-valid specimen shell collection are discussed, including the role of conchological organisations in promoting shell collecting and increasing awareness of the activity.

The International shell trade can be locally significant, and some trends in shell collecting are presented, with a particular focus on the most popular bivalve families and online specimen-shell sales. The issues of sustainable harvesting, regulation and enforcement are discussed. However, the importance of shell collections and collectors in relation to molluscan taxonomy is also presented, as is their relevance to environmental awareness and potential role in enabling people to better interact with and understand the marine environment.

A number of important and highly collectable bivalve species are presented as examples.

Abstract in Chinese 贝壳收藏有着悠久的历史，从17世纪开始，已经作为一项非常普遍的活动在艺术，商业和科学的发展方面作出了重要贡献，基于贝壳收藏形成了更科学系统化的贝壳学。　现代贝壳收集包括多样的软体动物家族和物种，双壳贝类包含其中。通过个人收集，专业经销商交易，交换或旧品淘取等一系列方法，贝壳的收集和交易得以发展。我们对贝壳标本收藏系统的科学构建和有效维持的基本原则进行了讨论，包括了贝类学相关机构在促进贝壳收藏和提高相关活动影响力方面所能起到的作用。

国际贝壳贸易具有重要的意义，我们将展示较受欢迎的双壳贝类家族和贝壳样品的销售收藏趋势。同时就如何实现交易的可持续性，交易管理和执法

P. F. Duncan (✉)
University of the Sunshine Coast, Maroochydore DC, QLD, Australia

A. Ghys
Engelstraat, Deerlijk, Belgium

A. C. Smaal et al. (eds.), *Goods and Services of Marine Bivalves*,
https://doi.org/10.1007/978-3-319-96776-9_20

问题进行了讨论。 此外,本文介绍了贝壳收藏活动和贝壳藏品在软体动物分类学方面的重要性,旨在提高人们的环境保护意识,使人们能够更好地了解和爱护海洋。

本节,我们将以一些极具收藏价值双壳贝类物种进行举例说明。

Keywords Conchology · Bivalves · Shell collecting · History · Collections · Taxonomy · Trade · Scientific value

关键词 贝类学 • 双壳贝类 • 贝壳收藏 • 历史 • 收集 • 分类学 • 交易 • 科学价值

20.1 A Short Introduction to Conchology

Humans may have been collecting shells throughout their history as a modern species, and there is evidence to indicate that ancestral humans /con-specifics also collected shells for various purposes (d'Errico et al. 2005, 2009; Zilhão et al. 2010; Joordens et al. 2015). As by-products of food foraging, it is easy to see the practical use of some shell collecting, but at some point shells must also have been collected for aesthetic and ornamental purposes. Evidence from archaeological excavations in Africa suggests that modern humans were using sea shells as ornamental jewellery during the middle stone age (middle Paleolithic) in southern Africa and in the Maghreb (north-west Africa) (d'Errico et al. 2005, 2009) between 70,000 and 120,000 years ago. However, at that time, modern humans are not thought to have existed in Europe, with Neanderthals being the dominant hominid group there.

Zilhão et al. (2010) reported pierced and pigmented shells, considered to be ornamental jewellery, from caves in south-eastern Spain dating to 50,000 years, around 10,000 years before modern humans are believed to have inhabited the area. Therefore, the discovery that Neanderthal hominids also used a range of bivalve families for ornamental purposes, including species of Cardidae, Glycymeridae, Spondylidae and Pectinidae, indicates that the collection and abstract use of molluscan shells is not a modern pursuit, but one practised by both our distant ancestors and closely-related hominids.

Baldwin Brown (1932) (in Dance 1986) discussed the non-practical use of shells as jewellery or aesthetic pieces recovered from the Grimaldi caves in north-west Italy, which were dated to the upper Paleolithic of around 30,000 years ago. Various molluscan species have been recorded from excavations of this cave system, some of which must have been traded over long distances, since several came from the Atlantic Ocean and wider Mediterranean Sea.

However, rather than considering all aspects of shell use, this chapter will focus on the collection of shells in their natural state as the aesthetic and scientific pursuit known as conchology. Conchology may be defined as the study of terrestrial and aquatic molluscan shells, and their associated hard parts, including the operculum and radula. The primary distinction between conchology and the wider discipline of malacology lies in its focus on shells, rather than the mollusc as a whole animal (Dance 1986).

There is also some debate about the definitions of conchology versus shell-collecting. Conchology, as a scientific discipline, has a specific focus on the shell as a means to better understanding molluscs as organisms, whereas shell collecting, in its typical form, can be considered to be the acquisition and collection of shells for primarily aesthetic purposes, with limited relationship to a scientific discipline. However, a properly curated collection of shells, accompanied by its collection data, may have very significant scientific value and will be discussed further in this chapter.

To paraphrase an adage, 'not all conchologists are shell collectors, and not all shell collectors are conchologists'. However, as with most disciplines, divisions are somewhat subjective, although the generalities, as outlined above, are perhaps worth defining from the beginning.

It is also important to recognize that while this volume is primarily on the subject of marine bivalves, it is almost impossible to consider conchology, or shell collecting, without reference to the collection of the other major molluscan classes, in particular the Gastropoda. Conchologists and shell collectors are primarily interested in four of the seven classes of mollusc, namely Polyplacophora (chitons), Scaphopoda (tusk shells), Bivalvia (clams etc.) and Gastropoda (snails etc.), due to the presence of shells in these taxa and their general availability.

The total number of extant mollusc species has been estimated to be as high as 200,000, with between 50,000 and 120,000 having been formally described, although there is no common agreement. For example, Bouchet et al. (2016) noted that WoRMS (World Register of Marine Species) listed around 46,000 valid species of marine mollusc. With the addition of terrestrial and freshwater species (see Rosenberg 2014) the total is around 75,000. Bouchet et al. (2016) also reported 82,000 valid molluscan names, which is a little less than Chapman's (2009) estimate of 87,000 described species. We can probably say that the number of described mollusc species is between 75,000 and 87,000, with a likely tendency towards the lower part of the range. Moreover, Mora et al. (2011) and Bouchet et al. (2016) estimate that un-described species may still constitute between 75% and 91% of the true total. Regarding the diversity in the main molluscan classes, Gastropoda are considered to constitute around 80% of all described species (Ponder and Lindberg 2008), and bivalves around 11–14% (Nicol 1969).

In terms of popularity with collectors, these proportions are also broadly indicative, although for bivalves the number of commonly collected species is probably even less than this. Therefore, while bivalves are the central subjects of this chapter, some important gastropods are included where relevant, in part to provide a more complete context for conchology and shell-collecting in general.

20.2 Historical Aspects and the Development of Conchology

A history of recent shell collecting has been well described by Dance (1986), and some useful examples from this work are included here.

One of the earliest 'modern' records of shell collecting and collections dates back to AD 79 where apparent collections of bivalves and gastropod shells, both marine and freshwater, have been found in the excavations of Pompeii (Tiberi 1879). The purpose of this multi-species collection, whether for aesthetic display, decoration or more serious study, remains unknown, but it did include a specimen of the bivalve *Pinctada magaritifera* (the black-lip pearl oyster), which is a striking, nacreous shell found in the Indian and Pacific Oceans, including the Red Sea. Presumably this indicates that attractive shells were valuable enough to warrant transportation and perhaps trading.

Pliny the Elder, who died in the same volcanic eruption of AD 79, wrote extensively on molluscs, and was particularly comprehensive on the commercially-important species and their products, such as pearls and murex dyes. However, the actual collection of shells for aesthetic or scientific purposes has little definitive history until much later, and was primarily associated with the so-called 'golden age' of exploration, colonialism and trading during the seventeenth century. Unsurprisingly, the popularity of shell collecting was greatest in those countries with the strongest interests in overseas expansion, particularly in areas with significant molluscan biodiversity, such as the tropics and the Indo-Western Pacific biogeographic region. As such, the port and capital cities of the Netherlands, Belgium, France and Great Britain developed as the major shell-collecting centres and, with the wealth from colonial commerce, helped established the inherent value and trading of shells that persists today (Fig. 20.1). See Dance (1986) for a more comprehensive account.

The characteristics of shells that appealed to early collectors and artists; aesthetics, exoticism, rarity, commercial value and durable structure have ensured a certain longevity to shells over time, such that specimens have circulated around collections and persisted over time. Indeed, the value of many exotic shells ensured that only the wealthy were able to accumulate important specimens and collections, that same value ensuring their longevity and provenance. Notably, the royal collections of several European countries have proven particularly enduring and important. For example, the collection of Queen Ursula of Sweden provided Linnaeus with several hundred species for the most important editions of the *Systemae Naturae* (1758–1768, 10th and 12th editions). The advent of a more systematic approach to science and collecting, along with useful and value-enhancing collection information, then provided a basis for the development of more scientific endeavours.

Ultimately, wealth, royal patronage and bequests enabled significant scientific institutions, museums and their shell collections to developed in western Europe from this time, providing the basis for the 'gentlemen scientists' of the 18th and 19th centuries who contributed much to the associated disciplines of conchology and malacology.

Fig. 20.1 'Peace and the Arts' by Dutch artist Cornelis van Haarlem (also known as Cornelis Cornelisz). Painted in 1607, it illustrates the early use of tropical and sub-tropical marine shells in art as a direct consequence of wider international trading and exploration by Europeans. Shell specimens include: *Strombus pugilis* (tropical and sub-tropical west Atlantic and Caribbean), *Trochus niloticus* (tropical Indo-west Pacific), *Cymbiola vespertilio* (tropical western Pacific), *Harpa doris* (tropical and sub-tropical eastern Atlantic), *Conus* sp. and *Hippopus hippopus* (Indo-west Pacific). (Identification from Dance 1986) (Image: National Trust)

20.3 Major Bivalve Families for Collectors

People who collect shells probably fall into two general categories, although in practice it may be a continuum, namely; aesthetic shell collectors, and those with a more specialized focus, approaching a true conchologist as the specialization increases. Aesthetic collectors will probably not collect a single family, but rather a range of more attractive species from all over the world, focusing mainly on the most impressive specimens across a wide range of taxa. By contrast, conchologists, or specialists, typically focus on one, or a very few families, as the major subject for collection. More simply, a specialist will typically identify as a collector of, for example, Pectinidae (scallops), and have reasonable representation of shells across a number of genera within that family. This makes logical sense for several reasons; the form and shape within a family are relatively consistent, lending a degree of aesthetic continuity to a collection. In addition, families comprise a number of sub-taxa that make for a challenging, but potentially attainable completion of a series, as well as a number of individual specimens that make for a good display. Finally, the family level also appears to provide an attainable level of intellectual specialization within the group, enabling an individual to become somewhat expert over the lifetime of a collection.

While conchologists generally specialize in particular families, some may also focus on biogeographic regions, thereby collecting a wider range of taxa from a particular locality, for example Australian, island or deep-water shells.

Individuals may further specialize in rare or low-diversity taxa, providing opportunities for achieving complete collections, and probably over time, a very specific expertise in the group. In addition, some opportunistic collectors may acquire specimens of limited interest to themselves, but with a view to subsequent sale or exchange for more relevant shells.

Therefore, while we can generalize about popular families, shell collector's interests and collections may be as diverse as the molluscs themselves.

As noted previously, bivalves are not as widely collected as gastropods amongst conchologists and collectors for several reasons. The lower diversity of taxa, and perhaps shell forms, probably limits their popularity compared with gastropods and, in general, bivalves are less visually spectacular, usually lacking the intensity and range of colours, patterns and architectural forms of gastropods. In addition, the requirement to retain two valves, and fix them appropriately for display or storage, adds to the problems of space and curation effort by comparison to uni-valved species.

Analysis of the catalogues of online shell retailers, i.e. shells available for sale via the internet, can provide some insight into the relative availability, and assumed desirability of the various molluscan classes and families. For example, using probably the two largest online specimen shell suppliers (Conchology Inc.[1] and Femorale[2]) indicates that for the Gastropoda there were 127 families and 98,335 specimens for sale, compared with 21 families and 684 specimens for Bivalvia (based on early 2017 data).

We should be a little cautious about simple comparisons between gastropods and bivalves using actual numbers offered, since there may be many more species (biodiversity) or specimens of some families compared to others, and it seems reasonable to assume that commercial dealers will be offering shells for sale that are the more desirable species and specimens. However, additional data from the largest internet dealer, indicated that about 14% of handled shells are bivalves (12.9% marine, 0.7% freshwater) (G. Poppe, Pers. Comm.), with gastropods making up the vast majority of the remainder.

Nevertheless, in bivalve taxa where the visual and physical disadvantages are naturally overcome, then these are typically the most popular for collecting, often generating equivalent levels of enthusiasm, passion, availability and sometimes the commercial value found in the most popular gastropod families.

The following analysis (Fig. 20.2) again uses two of the largest online specimen shell retailers and provides a breakdown of availability of bivalves by family, noting that multiple specimens may be offered for sale; therefore the numbers presented for each family are total individual specimens. However, we consider that this is probably reflective of demand and therefore an index of both desirability and collectability.

[1] http://www.conchology.be

[2] http://www.femorale.com

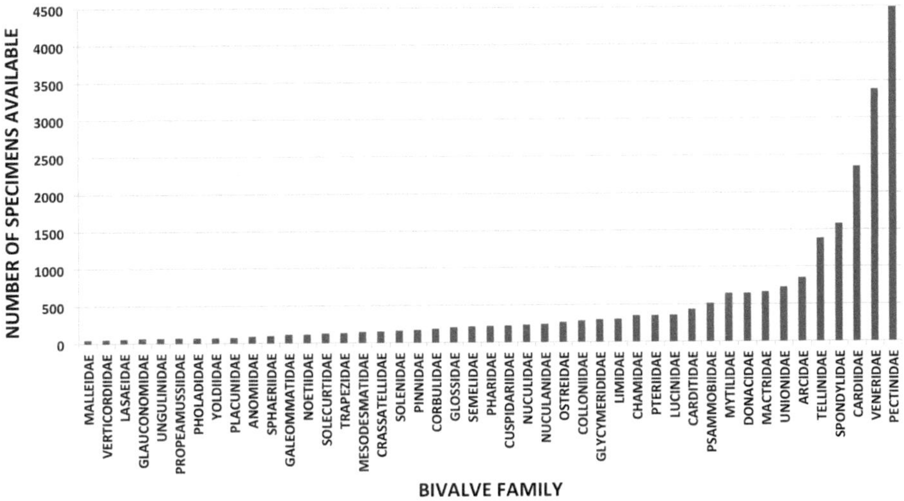

Fig. 20.2 Total number of individual specimen shells by major family (bivalves only) available for sale online (March 2017), from the two largest specimen shell retailers. Only families with 50 or more specimens available are included

This relatively basic analysis indicates that specimens of the following bivalve families are the most available and, by inference, the most popular with collectors. In decreasing order (by number of specimens available), the ten most popular families are; scallops (Pectinidae) (4491 specimens), venerid clams (Veneridae) (3388 specimens), cockles (Cardiidae) (2346 specimens), thorny oyster (Spondylidae) (1579 specimens), tellins (Tellinidae)(1386 specimens), arc shells (Arcidae) (855 specimens), freshwater mussels (Unionidae) (729 specimens), mactrid clams (Mactridae) (667 specimens), donax or wedge clams (Donacidae) (646 specimens), marine mussels (Mytilidae) (645 specimens). All are marine families except for the Unionidae. By comparison, on the same internet sites, the two most popular gastropod families of Conidae and Cypraeaidae constitute more than 15,000 and 12,200 specimens respectively.

In terms of actual sales of specimen shells, anecdotally, it is confirmed that Pectinidae are the biggest sellers, followed by Spondylidae (G. Poppe, Pers. Comm.), which are probably the two most colourful and spectacular bivalve families.

To provide a little more detail on the most collectable bivalve families, Table 20.1 shows a partial taxonomic classification, ordered in terms of their general collectability, and including some key genera and their wider uses and relevance beyond specimen shells. Popularity amongst collectors undoubtedly relates to aesthetic features such as shape and colour, but familiarity, availability and cultural significance may also be important.

Table 20.1 Bivalve taxa of importance to conchological collectors and other goods and services

Superfamily	Family	Genera	Product (goods and services)
Pectinoidea	Pectinidae	Numerous, including *Pecten, Aequipecten, Placopecten, Argopecten, Mizuhopecten, Chlamys* etc.	Food, specimen shells.
	Propeamussidae	Glass scallops, 22 genera including *Cyclopecten, Parvamussium, Similipecten*. Deep water.	Specimen shells.
	Spondylidae	*Spondylus*	Specimen shells, jewellery.
Veneroidea	Veneridae	Numerous, including *Venus, Ruditapes, Mercenaria, Venerupis*	Food, specimen shells.
Cardioidea	Cardiidae	Numerous, including *Acanthocardium, Cerastoderma, Cardium, Fragum*	Food, specimen shells.
	Tridacnidae	*Tridacna*	Food, specimen shells (CITES[a] listed), aquarium.
	Tellinidae	Numerous, including *Tellina, Macoma*	Food, specimen shells.
	Arcidae	Numerous, including *Arca, Anadara*	Food, specimen shells.
	Mactridae	Numerous, including *Mactra, Lutraria, Spisula*	Food, specimen shells.
Pterioidea	Pteriidae	Numerous, including *Pinctada, Pteria, Isognomon*	Pearls, nacre (mother of pearl) products, specimen shells.
	Malleidae	Several including *Malleus*	Specimen shells
Pinnoidea	Pinnidae	Several, including *Pinna, Atrina*	Food, byssus thread (sea silk) fabric etc., pearls, specimen shells.
Ostreoidea	Ostreidae	Several, including *Ostrea, Crassostrea, Saccostrea*	Food, specimen shells.
Anomioidea	Placunidae	Primarily *Placuna placenta*	Shell products and crafts (Philippines), specimen shells.
Unionoidea	Unionidae Margaritiferidae	Numerous, including *Anodonta, Unio, Margaritifera, Hyriopsis*	Jewellery (pearls), nuclei for marine pearl production, specimen shells.

The taxonomy is based on various sources, and may differ between authorities.
[a]CITES: Convention on International Trade in Endangered Species

20.3.1 Some Rare or Highly-Collectible Bivalves

While bivalves are generally not amongst the most desirable or valuable specimen shells, there are a few species that have obtained a high status in both these characteristics over time. As such, when available, they continue to command relatively high prices and interest. Some of these species are included in the work 'Rare Shells' (Dance 1969), which provides details on 50 highly-collectible shells. Here are four bivalve examples;

Fimbria soverbii (Reeve 1842) (Fig. 20.3)
Common name: Common basket lucina
Family: Lucinidae
Locality: Western Pacific from Japan and China to Australia.

Dance (1969) states that "in the second half of the nineteenth century this was one of the few coveted bivalves" with "fine specimens seldom seen in collections." This is probably still true today. At auction in 1865 only *Pholadomya candida* (see below) attracted a higher price. The species is not exceptionally rare, but seems to appear irregularly for sale and still commands a relatively high price of around €70–80, depending on condition, size, appearance and locality.

Fig. 20.3 *Fimbria soverbii* from Australia. (Images courtesy of Marcus Coltro (Femorale))

Pholadomya candida (Sowerby 1823) (Fig. 20.4)
Common name: Caribbean piddock
Family: Pholadomyidae
Locality: South-eastern Caribbean

This is a rather legendary shell, partly due to its general rarity, but mainly because it is considered to be a 'living fossil" (Runnegar 1972). Its closest relatives are from the Miocene period (between 23 and 5 million years ago), and until recently this species was thought to be extinct. However, Diaz and Borrero (1995) reported fresh-dead specimens from Colombia and Venezuela, which were apparently the first traces obtained since the late nineteenth century. More recently, living animals were located in shallow water (3 m) off the Colombian coast and a specimen was obtained, providing opportunities for further study, including DNA analysis for phylogenetic purposes (Dìaz et al. 2009). This re-discovery has provided significant interest for evolutionary biologists and malacologists (Ausubel et al. 2010), since only once before have the soft body parts been studied (Runnegar 1972). The shell itself does, very occasionally, appear for sale, but these have inevitably been recycled from old collections, and the price is high, ranging from €1800 to €2500, making this a very limited market for a specialist type of wealthy collector.

Spondylus regius (Linnaeus 1758) (Fig. 20.5)
Common name: Regal thorny oyster
Family: Spondylidae
Locality: Western Pacific from Japan and China to northern Australia

Spondylus regius is a spectacular shell with a long history of popularity and value. The species was probably known to collectors even before it was described by Linnaeus in 1758, and Dance (1969, 1986) provides a number of anecdotes regarding specimens of this species. Such was the fame and value of some individual shells that their transactions, owners and even prices paid are a matter of historical record. As such, many of these famous shells are still to be found in national museums, providing a physical testament to the provenance. For example, the

Fig. 20.4 *Pholadomya candida*, Colombia. (Image courtesy of Juan M. Dìaz)

Fig. 20.5 *Spondylus regius*. (Image: courtesy of Marcus Coltro (Femorale))

famous conchologist G. B. Sowerby 1st, purchased the famous Tankerville collection in 1824, including a fine specimen of *Spondylus regius*, which was sold the following year for £25 (€30). This is around the same price that a good specimen would cost today, although in 1825 the relative value was the equivalent of over £2500 (€3000). As with many shells, the individual colour, locality and condition, particularly of the prominent spines in this species, determine the value, although recent prices range from as little as €5 up to €50.

Nodipecten magnificus (G.B. Sowerby 1st 1835) (Fig. 20.6)
Common name: Magnificent scallop
Family: Pectinidae
Locality: Ecuador (Galapagos Islands), and possibly Colombia

One of the largest and most striking scallops, this species is considered to be endemic to the Galapagos Islands of Ecuador, although it has supposedly been reported from the Ecuadorian mainland and south-west Colombia (Raines and Poppe 2006). However, its generally accepted centre of distribution as the Galapagos is ecologically relevant. The species can reach 200 mm in height, and its desirability and commercial value, more than €1000 for a large, high-quality specimen, combined with a very limited range makes the species potentially vulnerable to overexploitation and population decline. As such, this species is included on the IUCN (International Union for Conservation of Nature) red list of threatened species as data deficient,[3] acknowledging its vulnerability, but also limited ecological knowledge. This highlights a relatively uncommon, but important aspect of conchological responsibility, that of conservation.

[3] http://www.iucnredlist.org/details/14831/0

Fig. 20.6 *Nodipecten magnificus*. (Image: Arne Ghys, www.pectensite.com)

20.4 Shell Collections

20.4.1 Obtaining Specimens

20.4.1.1 Direct Collection

'Dead' Collecting

This refers to the collection of 'dead' material, i.e. a shell without the presence of the original live animal, although occasionally a shell may be inhabited by another animal, such as a hermit crab. In such cases the shell may be described in accompanying collection data as 'dead-collected' or 'crabbed', although the latter is much more likely for gastropod shells.

Dead material may be of varying quality. For example, a shell collected with original animal remains, perhaps as a result of natural predation, may well be of equivalent quality to live collected. In such cases, where quality is high, they may be referred to as 'fresh dead specimens' in accompanying collection data. However, in most cases some degree of deterioration of the shell will have occurred, ranging from physical damage caused by wave action or predators, natural chemical deterioration due to aerial exposure, or settlement of encrusting organisms, such as barnacles, tube worms or corals. Attempted removal of such organisms may further damage the shell, or leave unsightly residues that diminish the overall appearance and quality of the specimen.

The collection of dead-collected shells, or their inclusion in a collection, may be for several reasons including; the general rarity or difficulty in obtaining a particular species, the inclusion of a 'dead' specimen until a better condition shell can be obtained, or due to an ethical objection to live collecting.

Live Collecting

Although these methods are primarily for the collection of live specimens, 'dead' shells may also occasionally be obtained in the same way.

To make a general point on live-collected shells; since specimens are ultimately intended for a collection, then the best quality examples are those that are specifically collected, either by the collectors themselves, or by professionals for subsequent sale. In either case the desirability and value are significantly affected by condition (see section 20.5). A general rule, often written into shell collecting codes of conduct, is that damaged or unsightly shells, e.g. chipped margins or spines, or with severe growth lines, and also mating, brooding or egg-carrying individuals should not be taken, but instead returned to the environment for breeding.

- Intertidal Collection

  The natural distribution of bivalves ranges from the intertidal zone to the deepest parts of the oceans. The intertidal (or littoral) province represents a highly biodiverse ecological zone containing a very large number of species. It also acts as a receiving environment for offshore areas with specimens being delivered by wave action and storms, and as such it is a profitable area in which to collect shells. Bivalves live in a wide variety of habitats in the intertidal zone, including under rocks, dead coral, and within a range of sediment types from muds to sand and gravels.

  The intertidal zone is also relatively accessible to people, both for food collection and for shells, and care must be taken to limit collection activities and minimize habitat damage. For this reason, national and local rules often apply to intertidal collection. Similarly, many shell clubs have developed codes of conduct/ethics for live shell collecting, which may include bag limits, not taking damaged or breeding individuals, and the replacement of turned boulders, dead coral slabs and other substrate materials.[4]

- Diving (free and scuba)

  Scuba diving and snorkeling offer additional opportunities for shell collecting, particularly for species below the intertidal zone. Scuba in particular provides for relatively extended periods of underwater collection, although shell collecting is also undertaken using surface supplied air (hookah diving). This may be a relatively cheaper, or more convenient option compared with compressed air cylinders, which require refilling, and is often the preferred method in many countries, e.g. the Philippines. However, local regulations may prohibit the use of particular equipment for collecting. In general, diving for shells is not a common activity outside the tropics, and typically requires a lot of experience to develop the skills necessary for finding molluscs underwater.

[4] http://www.conchologistsofamerica.org/conservation/ethics.asp; http://www.sydneyshellclub.net/ethics.html; http://www.malsocaus.org/?page_id=14

- Fishing operations
  Targeted collection of shells by directed fishing, or as secondary bycatch from commercial fishing operations, may also produce commercially valuable shells, and are important sources of specimens, particularly from deeper waters.
  Many different types of fishing can yield shells, and most types of benthic fishing gear, used on the seabed, can provide a variety of species, including infaunal, burrowing species.
  - Benthic trawls and dredges: otter and beam trawls are primarily used for demersal fish, but will also catch infaunal, benthic or swimming bivalves. Toothed and un-toothed dredges are often used for commercial scallop fishing, and also catch other bivalves living in the same habitats.
  - Benthic traps: perhaps surprisingly, baited traps for species such as crabs and lobster can also catch molluscs, although usually they are carnivorous gastropods attracted by baits. However, bivalves, especially swimming species such as scallops may occasionally appear in traps.
  - Nets; tangle or '*lumun-lumun*' nets: directed fishing for specimen shells using specialized nets is most common in the Philippines (Floren, 2003). Such nets, with relatively small mesh sizes, are set overnight in waters down to 100 m and primarily collect gastropods, but also swimming or spiny bivalves. However, this is a labour-intensive method of shell collection with low profit margins, and appears to have become less common in recent years.
  - Remotely-Operated Vehicles (ROVs): in recent years the use of remotely operated vehicles, or ROVs has increased, particularly for high-value species such as Australian *Zoila* cowries.
  - *Ex-pisce*: fishing operations may provide unusual sources of specimen shells. One example, more common for gastropods, but also occasionally for bivalves, is from fish stomachs, or *ex-pisce* as listed on collection data. This source has been particularly important for some rare, deep-water species, such as southern African cowries (Cypraeidae) (Boswell 1964).

20.4.1.2 Indirect Collection

Typically it is not possible to complete a shell collection, or obtain some specimens personally, so shell collectors often use third parties or 'indirect' collection methods.

Most collectors at some point either purchase or exchange specimens with fellow collectors or dealers. Such opportunities arise via shell-collecting clubs, dedicated shell shows (such as Antwerp, Paris, Australian or Conchologists of America), specimen shell shops and, more recently, via dedicated or general online auction sites.

Shell specimens are also sometimes obtainable by the purchase of existing collections, either via disposal through choice, or the death of the owner. Collections may also be acquired by dealers and sold off individually, a practice that has been going on since the seventeenth century and the advent of modern shell collecting (Dance 1986).

20.4.2 *The Shell Collection*

The purpose of a shell collection may be as variable as the collectors or the specimens they contain. It may represent a specific focus and expertise with a narrow range of species, which arguably tends towards a more conchological, or scientific purpose. Other, more eclectic collections may also serve a similar purpose, although diversity, aesthetics or commerce may also be the primary purpose.

Regardless of purpose the utility and value of any shell collection lies in both the specimens themselves, and in the accompanying collection data. Although there is no definitive list of data requirements, the general rule is that more information is better. Conchological publications and organisations often provide advice, but whether for scientific or commercial purposes, specimen shells are significantly more valuable with good data that records some or all of the following;

- Species name and taxonomy: although this may not be the most important information as identification can occur post-collection.
- Location: arguably the most important piece of data, and therefore as much detail as possible.
- Habitat, or ecological data, associated with the specimen; including depth, substrate etc.
- Collection date (and perhaps time of day).
- Specimen dimensions and condition.
- Original collector's name.
- Price paid (if purchased).

Ideally, data should be retained with the specimen itself, to avoid separation. This may be in the form of a paper label retained inside the valve(s), or within an individual compartment in a cabinet. Historically, shell specimens have been marked directly with Indian ink, providing a numerical reference on the shell associated with an accompanying catalogue. More recently computer-based systems have been used for data management and dedicated software is available for shell collectors,[5] along with less specialised database software.

20.4.2.1 Specimen Cleaning, Maintenance and Conservation

Mollusc shells have an organic component to their structure and may also have remains of organic tissue due to incomplete cleaning, especially from live-collected specimens. This presents the risk of specimen deterioration, odours or infestations, and therefore cleaning is important. Initial preparation of bivalves must also consider the hinge and ligament if the specimen is to be preserved with these features intact. Bivalves may be set (air dried) closed, or with the valves separated by breaking the hinge, dependent on eventual specimen purpose. For example, access to the

---

[5] Examples include: http://shellcollections.com/; http://home.global.co.za/~peabrain/software.htm

interior of the shell valves may be important to show features used in classification or ageing, e.g. pallial sinus shape, muscle scars and ligament rings, although separated valves present a greater risk of specimen separation. In this regard, water-soluble glue can provide a useful solution.

Secondary cleaning may also be undertaken, depending on specimen purpose or condition. This may be to remove encrusting organisms, such as tube worms or sponges, and can be achieved via simple brushing, treatment with diluted (e.g. 5–10%) bleach (sodium hypochlorite) or even small rotary power tools, depending on the severity or nature of the fouling. However, as before, it may be important to preserve some delicate components of the shell valves intact, such as the periostracum or shell micro-structure, which can be easily lost by bleach treatment or abrasion. Therefore, cleaning method and extent depends on the specimen, its purpose and the technical experience of the collector.

Emersion and cleaning processes tend to dehydrate shells, resulting in dulling of the surfaces and colour fading, although colour fading generally occurs over time regardless of remedial action. Before depositing in the collection specimens can often benefit aesthetically from a light application of mineral oil, which also helps to reduce desiccation and cracking of the periostracum and other organic components of the shell. Generally, vegetable, or other organic oils, should be avoided as these may provide a substrate for fungal or insect pests.

20.4.2.2 Collection Organization

There is no formal, accepted standard for organisation of a collection, since it depends on many factors such as specimen number, size and collection purpose. In addition to the specimens themselves, space may be a significant factor, since storage containers or cabinets, and the inevitable library of reference material, add additional components to collection management.

However, some fundamental characteristics of shells themselves require basic consideration to maintain their long-term condition.

Firstly, shells are relatively fragile, particularly specimens with delicate marginal lips, spines, projections or naturally thin shell valves, such as deep-water species. As such, they require separation from each other, either in individual, partially air-filled clip-seal polyethylene bags, or in compartmentalized drawers, boxes or inserts.

A variety of materials have been used for specimen shell storage, but exposure to light (which fades colours), air, dust, fungi, insects and potentially corrosive chemicals must be considered. Some of these factors can be controlled by storage in specifically designed cabinets, although the construction material may be important.

Historically shell cabinets were made of wood, particularly tropical hardwoods, which were generally more suitable than woods such as oak. Over time, and particularly in humid, poorly ventilated environments, acidic gases from storage material can significantly and irreversibly damage carbonate structures such as shells. This chemical reaction, one of several related processes, is called Byne's disease, or

Bynesian decay, and is the reaction of acidic vapours on the alkaline shell material, resulting in efflorescence of salts (Byne 1899). Significant literature exists on the subject, see Tennent and Baird (1985) and Cavallari et al. (2014).

For this reason many cabinets and drawers are now made of plastic, laminate or preferably metal.

20.4.2.3 Temporary Exhibition

A shell collection is primarily for the collector, but occasionally it, or components of it, may be exhibited for purposes such as competition or education.

Most shell shows are annual events, typically organized by shell clubs or organisations such as;

- Koninklijke Belgische Vereniging voor Conchyliologie (Royal Belgian Society for Conchology) (www.konbvc.be)
- Association Française de Conchyliologie (French Conchological Association) (www.xenophora.org)
- Conchologists of America (COA) (www.conchologistsofamerica.org)
- British Shell Collectors Club (www.britishshellclub.org)
- Conchological Society of Great Britain and Ireland (www.conchsoc.org)
- The Australian National Shell Show is organized by various state shell clubs on a rotational basis, often in addition to their own club or state shell show.

These shows are an opportunity to buy, exchange and exhibit shells (Fig. 20.7), and often have lectures on conchological subjects. Competitions covering a range of categories, e.g. specific molluscan families, regional or worldwide shells, colour forms, and 'shell of the show', are common and provide an opportunity to display particularly special collections or specimens to other conchologists and the public. Shows are typically open for general admission and well-organised annual shell shows provide advance publicity for the event, including media releases.

Educational displays are typically the preserve of museums, but short-term exhibits, loans or lectures often provide additional opportunities for conchologists to display or discuss their specimens.

Typically, for all such events, a collector will have, or be provided with a small display box which can hold up to 20 specimens and which usually includes a glass or transparent plastic lid to protect and secure shells while on view.

20.4.2.4 Disposal of Collections

A question that collectors have faced since conchology began, is the fate of a collection due to eventual disinterest, death or lack of money.

In simple terms there are probably three options; bequest, sale or donation. Bequest provides continuity for the collection and donation to an educational or conservatorial institution may ensure continued preservation of the collection intact,

Fig. 20.7 International shell shows. Clockwise from top left: dealer tables where specimens are sold, exchanged or displayed, table detail showing specimens and storage containers for sale, prize-winning competition displays showing collections of Spondylidae (thorny oysters) and Pectinidae (scallops). Note the use of standardized display box sizes and specimen collection data for competition entries

provided that the specimens are important enough to be accepted by a museum or equivalent.

Perhaps the most likely fate of a shell collection is purchase by another collector or dealer, in which case it is probable that only some specimens will be retained by the purchaser, with the majority being subsequently offered for individual sale.

20.5 The Value of Conchology

20.5.1 Commercial

Estimating the overall commercial value of collectable mollusc shells alone, as distinct from the rest of the animal as food, may be difficult, although there are some useful indicators. The Philippines is probably the country with the largest contribution to the shell trade and shell collecting, particularly from a specimen supply and trade perspective. The Philippines Bureau of Fisheries and Aquatic Resources collates data on many aquatic products, including shells and shell-byproducts, and have been consistently reported since 1998 (Fig. 20.8) (BFAR 2017). These products are mostly for export, and the vast majority are likely to be shell products, worked shells and display items, rather than specimen shells for collectors. However, some specimen shells are included in the total.

Floren (2003) reported that the peak of Philippines shell and shell-craft exports was in 1988 at 10,000 t, (valued around US$21.45 m) and a low of 1600 t in 2000, although still worth more than US$18 m. Thereafter, as indicated by Laureta (2008),

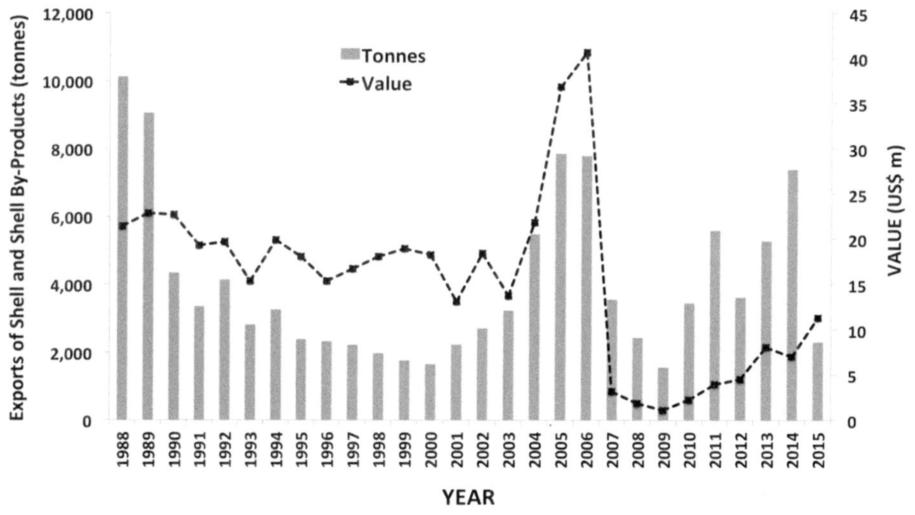

Fig. 20.8 Exports of shell and shell by-products from the Philippines, as reported by BFAR. Exports are in tonnes, values in million US$. (Data source: BFAR http://www.bfar.da.gov.ph/publication)

exports began increasing again to around 2233 t, worth US$13.1 m in 2001. While it seems probable that a relatively small proportion of these volumes are for edible consumption, given high domestic consumption and food regulation requirement in other countries (Duncan et al. 2009), Floren (2003) did note that 300 t of abalone (*H. asinina*) was exported in 1999, which may account for a significant proportion of the value around that time.

More recently BFAR reported that exports of 'shells and by-products' for the years 2005 and 2014 (Fig. 20.8) amounted to 7854 t (worth US$36.8 m), and 7388 t (worth US$7 m), respectively. Therefore, it appears that values and quantities have fluctuated considerably, with particularly low values around 2008–9 of only US$1.1 m. It is unclear whether this was due to global economic conditions, changes in product types, or that statistical data is not directly comparable. Floren (2003) noted discrepancies between national statistics and individual islands, with Cebu alone reporting 19,565 t of shell and shell craft sales in 1999, which was apparently not included in official figures. However, it may also be simply that high-value components, such as abalone, have declined.

The Philippines is also the largest producer of *kapis*, or window-pane shell, (*Placuna placenta*)(Placunidae). The shell itself is not particularly collectible, but it is used for shell craft in a wide variety of applications from lamp shades to windows. Gallardo et al. (1995) reported that exports between 1986 and 1991 were worth around US$36 m, although they also noted that significant depletion of beds was occurring due to overfishing (see also Adan 2000), suggesting an equivalent decline in its commercial value in recent years.

20.5.1.1 Individual Shells

The range of an individual shell's worth varies considerably and essentially depends on rarity and demand. Bivalves do not command the highest prices amongst shell collectors, which tend to be for the rare species of gastropod, but as might be expected, rarer, larger or more colourful specimens of the most collectable bivalve taxa, e.g. Pectinidae or Spondylidae, are not insignificant in terms of individual commercial value.

In shell collecting, specific consideration of rarity, would include the following;

- Numerical scarcity: the actual number known to exist, or the ecological rarity of the species. This can change over time; increasing, with additional specimens becoming available as their habitat becomes better known or explored, or decreasing, if a habitat is damaged or natural mortality increases (e.g. due to disease or pollution), or a species is exploited towards local extinction.
- Collection difficulty: some specimens are simply hard to collect, reflected in numerical availability. However, if a new collection method becomes available then supply may increase and value decrease. Similarly, if existing collecting methods disappear or change, then the opposite may be true.

- Specimen locality, e.g. geographic, depth, protected area etc.; the difficulty of collecting specimens from inhospitable or rare habitats may be reflected in value. Similarly, an area or habitat that becomes protected may increase the value of previously collected specimens, since many shells are 'recycled' from old collections, and a knowledgeable dealer/collector will price according to current conditions. There are also examples of rare shells that existed in low numbers because their actual habitat or location was erroneously reported or unknown. If it subsequently becomes known, then supply may increase and price fall, for example, the scallop *Annachlamys reevei* from the Philippines.
- Specimen size; both unusually large or small shells of a particular species, often command a premium and there are lists of record sizes for different species available with which to gauge such characteristics.
- Specimen condition; is typically very important, unless a poor quality example of a rare species is acceptable until a better replacement is found. Damage, chips, erosion or marine fouling/boring generally detract from value.
- Specimen aesthetics; this characteristic is subjective, but unusual or particularly strong colours or patterns, well-developed or preserved structures, such as spines, scales or ribs can all enhance value. There is also a small, niche market for abnormal specimens, which often command higher prices than typical specimens. Such abnormalities may include unusual shapes, growth patterns, absence or presence of shell characteristics such as ribs or spines, shell deformation due to damage to shell-secreting tissues, melanistic or iron-oxide (rusty) specimens, unusual colours or partial colourations, or even odd associations such as shells with barnacles or other marine growths.
- Provenance, including the quality of accompanying collection data; a designated taxonomic type (usually paratypes, rather than holotypes which are typically museum deposited), illustration in media or part of a famous collection may increase value. Given that provenance can influence value, there is a well-recognised potential for collection data, and even specimens to be forged, although this is common in many human pursuits where significant commercial value is involved.

However, rarity in itself does not make a specimen valuable if it is not from a collectible family. Therefore, as with most objects of value, rarity and desirability, their availability and demand combine in different ways to determine the commercial value of any individual shell.

In terms of the actual range of shell values, it is often said that a good example of any shell with collection locality data is worth at least a single major international currency unit, i.e. €1, £1, $1 etc.

For upper limits, high quality, rare bivalve specimens, from the families Pectinidae or Spondylidae, can be worth up to several hundred Euro/US dollars/Pounds each, but very few exceed €/US$1000 or more. For example, tridacnid clams (Tridacnidae), which are protected under CITES, an international treaty which regulates trade in endangered species, should restrict availability. As such supply is usually limited to older specimens, and particularly large shells may command very high prices, although few other bivalves are anywhere near as valuable.

By comparison, some geographically localized, deep water, rare or particularly spectacular Cypraeidae (Gastropoda) specimens are offered in the tens of thousands of €/US$/£ range. In fact, high prices for particularly rare shells are not new since, relatively speaking, even higher prices have been paid from the beginning of commercial shell trade in the seventeenth century (Dance 1986). However, these are unusual exceptions, and the vast majority of individual specimen shells, even within the most-highly collectible families, can be acquired for less than €/US$/£100.

20.5.2 Scientific

Bouchet et al. (2016) estimated that in recent years, around 443 new species of mollusc have been described annually. This output is based on relatively few individual taxonomic specialists, perhaps only 500 people, and half of all new mollusc species were described by only 34 individuals. Importantly, based on first authors, 57% of new species were actually described by amateur, or 'citizen' scientists, compared to academic scientists. Of the top seven first authors, in terms of new species described, five were 'citizen' scientists and two were academics. Bouchet et al.'s study highlights several important points. Firstly, that amateurs continue to make an important contribution to molluscan science, certainly in relation to taxonomy, as has been the case since the nineteenth century at least (Dance 1986). Secondly, the general paucity of research funding from government and universities for relatively non-applied, baseline subjects such as taxonomy, means that the work is increasingly carried out by self-funded individuals using their own time and resources to help increase the knowledge of molluscan biodiversity. Thirdly, and one specifically noted by Bouchet et al. (2016), '*the factor limiting the description of more new species of marine molluscs is the availability of taxonomists – not the availability of new species to be described – and the difficulties of sampling them.*' As such, it will be increasingly important to encourage and recruit existing and new shell collectors to become knowledgeable and interested in taxonomy to ensure that this biodiversity resource remains available and active. Amateur and academic conchologists must cooperate better to achieve this, rather than pursuing divisive practices of separating shell collectors and scientists. This will be particularly important if the estimated 150,000 un-described molluscan species, including bivalves, are to be named in less than the 300 years that Bouchet et al. (2016) have estimated, based on current description rates.

While species descriptions are certainly significant contributions that amateur conchologists can make, all collectors can improve the scientific value of their collection by maximizing the accuracy and scope of collection data. Over time, a collection focused on a particular taxonomic group or geographic distribution, may approach completeness, enhancing its scientific value. As such, any well-curated collection has a potential long-term scientific contribution.

Some amateur collections can include type material, i.e. those specimens used to first describe a species, particularly given the high representation of amateurs in new species designations. By convention the holotype of any new species should be deposited in a named public institution, which provides for general access to it. However, since many species are described using multiple specimens, there may be multiple paratypes, often quite different from the holotype, and these are often retained in personal collections. While paratypes, and other type specimens, are noted in the original description, including their location, it is important that such specimens are accurately recorded, available for study on request, and not lost to science when collections are eventually dispersed or disposed of.

Amateur conchologists may also rarely own specimens from famous historical collections. Dance (1986) provides an overview of these, generally nineteenth century, collections, and many have remained intact and been ultimately bequeathed to national museums. For example; the Melvill-Tomlin collection in the National Museum of Wales,[6] Linnaeus' collection in the The Linnean Society of London, Darwin and Sowerby's in the British Museum of Natural History, and the Kohn Collection (Conidae) in the Field Museum, Chicago.

While many important collections have remained intact, it is also the case that some have not, or that important specimens were separated over time. If such specimens can be identified, then they may have important scientific, as well as historical and commercial value, but their identification is an important first step, again highlighting the importance of good data, and the retention of original data records.

The active collection of shells by conchologists, either individually or part of an organisation requires the acquisition of knowledge regarding species' habitat preferences and behaviour, as well as general ecological field-work skills. Even biodiverse areas such as tropical reefs and intertidal habitats can appear devoid of molluscan life to inexperienced observers, but a knowledge of tidal state, species' activity patterns and ecological associations can enable much more efficient collection. The ability of experienced shell collectors to find their subjects is best appreciated at first hand, but a good example of such skills has been demonstrated in the case of conotoxin research, which began in the 1990s (Adams et al. 1999). Biochemists, interested in the pharmaceutical application of *Conus* venoms, made use of shell collectors, both amateur and professional, to source a wide range of species for bio-prospecting and research (e.g. Safavi-Hemami et al. 2011; Robinson et al. 2014, 2016). Such was the popularity of the research field, and the effectiveness of collectors, that concerns arose regarding the sustainability of wild harvesting, since researchers, drug companies and shell collectors were searching for a wide range of *Conus* spp. at the same time. At one point it was even suggested that the genus should be CITES listed (Chivian et al. 2003) to protect species from over-collection. While efforts have subsequently been made to harvest cone venoms less destructively or produce them artificially, and the initial research urgency has declined, two points are important. Firstly, experienced shell collector's skills in

[6] http://naturalhistory.museumwales.ac.uk/molluscatypes/Collection.php

finding and identifying molluscs are well-developed and effective, with useful applications in field biology and other research endeavours. For various reasons such local ecological field experience and knowledge is becoming rarer, particularly in developed countries, but its usefulness should be recognised before it is lost to future generations. Secondly, and as with all wild harvesting, care and consideration needs to be taken in relation to sustainable levels of collection.

20.5.3 Education

The collection of shells is not necessarily an individual occupation, and many organisations exist to provide opportunities for collecting trips, discussion, exchange, display and education. Annual shell shows in particular offer a regular and public-facing opportunity for awareness raising of both conchology and wider ecological and biodiversity topics.

Conchological organisations typically provide guidance to members and non-members regarding the ethical or sustainable collection of specimens from the wild, with most making this information available on host websites (see section 20.4 for examples), and many produce regular newsletters or magazines, some of which have been published for many decades (Table 20.2). Bouchet et al. (2016) provide a detailed list of conchological publications, noting that many are now available online, for example http://www.conchology.be/?t=405.

These endeavours by conchological organisations help to ensure responsible behaviour amongst members, provide a focal point for newcomers to begin responsible shell collecting, and provide continuity of essential molluscan knowledge and skills. The encouragement of sustainable practices, but also the promotion of fundamental interests in biodiversity, ecology and the natural world is an important role played by conchological organisations in society.

Table 20.2 Selected list of conchological/malacological organisations and publications

Organisation	Publication	Published Since
Royal Belgian Society for Conchology	*Gloriamaris*	1961
Belgian Society for Malacology	*Novapex*	2000
French Conchological Association	*Xenophora*	1981
Conchologists of America	*American Conchologist*	1972
Hawaiian Malacological Society	*Hawaiian Shell News*	1960 (New Series)–2011
British Shell Collectors Club	*Pallidula*	1970
Conchological Society of Great Britain and Ireland	*Journal of Conchology*	1879
Malacological Society of Australasia	*MSA Newsletter*	1953
	Molluscan Research	1957

20.6 Conchology: Environmental Threat or Conservation Benefit?

Conchology, at least the acquisition of specimens, is essentially a hunter/gatherer activity, and such behaviours are becoming relatively rare in modern agricultural societies, with probably commercial and subsistence fishing as the only remaining large-scale examples.

There are many reason for this behavioural change, but one of the more recent considerations has been the potential damage that wild harvesting can have on natural biological populations, particularly as the human population has grown so rapidly in recent decades. It may be increasingly unacceptable for people in developed countries to kill wild animals for sport, food or leisure when more efficient food production systems exist. Similarly, the sustainability arguments are obvious with the human population expected to reach 8.6 billion by 2030 (United Nations 2017).

Unregulated fishing effort and practices have resulted in many commercial fishery species becoming depleted to dangerous levels through overfishing (Pauly 2008; Jacquet 2009), and, although we are unaware of any specific examples of a mollusc species becoming extinct due to harvesting for conchology, instances of over-harvesting of shellfish for food and shell byproducts have been widely reported. Examples of such depletions include several scallop species (Blake and Shumway 2006; Duncan et al. 2016), *Pholas orientalis* (pacific angel wing) (Laureta and Marasigan 2000; Ronquillo and McKinley 2006), *Lobatus gigas* (formerly *Strombus*) (queen conch) (NOAA 2015) and tridacnid clams (Van Wynsberge et al. 2013), to the extent that the latter two are CITES Schedule II listed, indicating a now-regulated international trade. In all cases the shell has been a byproduct of a food fishery, but habitat damage, overfishing and the effects of pollution have all contributed to the declines, making remaining populations more vulnerable to additional exploitation pressures, whatever their form.

Over-collection of *Conus* spp. due to demand from research and enhanced shell trade, has been discussed with earlier, but another bivalve mollusc, *Placuna placenta*, used primarily for shell craft rather than specimens, has also been subject to over-exploitation in the Pacific and Indian Oceans (Gallardo et al. 1995; Laxmilatha 2015). As a result, restrictions on harvesting this species have been introduced, been implemented, and around the world several management authorities have introduced bag limits or complete collection bans for some molluscs that are collected for food, angling bait, or for their shells.

Typically, such restrictions apply near high-density human populations, e.g. the Sydney area of Australia,[7] but also in various types of marine protected areas, e.g. zoned areas of the Great Barrier Reef Marine Park[8] (Day 2002). Importantly however, zoned management strategies can allow for protection of the most important areas, but also for the continuation of a wide range of managed activities in other

[7] Collecting around Sydney, Australia, http://bit.ly/2dK8JC7

[8] www.gbrmpa.gov.au/zoning-permits-and-plans/zoning/about-zoning

zones, including shell collection and scientific research. Elsewhere in Australia, which arguably has the most developed management for specimen shell collecting, several important initiatives have been introduced, including specimen-shell fishery assessments[9] (Australian Government 2004, 2005), management plans (Queensland Government 2008, 2009) and professional shell-collector licencing.[10] Additional information on regulatory measures and catch statistics (2002–2003) from the specimen-shell fishery in Queensland can be found online.[11]

Australia is certainly not the only jurisdiction to regulate and manage specimen-shell fisheries and collecting, with the Philippines[12] (Philippines Government 2001) and USA, particularly Florida, having a variety of harvest control measures. However, despite regulation and protection, there are ongoing issues with enforcement, highlighted by several authors, e.g. Dolorosa et al. 2013 and Nijman et al. 2015. Nijman et al. (2015) reported on the shell trade in twelve, supposedly protected mollusc species from Indonesia. Research based on market surveys and customs seizures indicated open and significant trade in the shells of seven clam (Tridacnidae) species (CITES listed), four gastropods and one cephalopod, and while the individual values were not particularly high, the volumes were significant, and the trade was only possible due to very limited enforcement.

Legislation is, however, an 'end point', and is only as effective as its enforcement; therefore many responsible conchological organisations and shell clubs have attempted to pre-empt such measures by developing and adopting codes of conduct for responsible collection, as well as awareness raising and education of environmental and sustainability issues via shell shows and publications.

While human activities can have very detrimental effects on wild animal populations through over-harvesting, and conchological collection is no exception, it may be important here to differentiate between specimen shells for collectors (conchologists) and display shells; those particularly large and impressive species, which are collected and sold as souvenir or decorative items, rather than as part of a curated collection. There is some evidence that for particular species, display collecting and trade can be locally detrimental. Trade in bivalve species such as tridacnid clams, and gastropods, such as *Charonia tritonis* (trumpet shell), *Cassis cornuta* (horned helmet shell) and nautilus shells may be locally, and even internationally significant, with 42,000 specimens confiscated by Indonesian authorities between 2005–2013. Similarly, Gössling et al. (2004), estimated annual numbers of these species for sale in Zanzibar between 700 and 1500 per annum. Giant clams (*Tridacna* and *Hippopus* spp.) appear to be particularly vulnerable to over-harvesting, due to slow growth rates and their exploitation for both food and ornaments, and demand has led to local population extinctions (Floren 2003), hence their listing on CITES schedule II.

---

[9] www.fish.wa.gov.au/Documents/sofar/status_reports_of_the_fisheries_2011-12_statewide.pdf

[10] www.daf.qld.gov.au/fisheries/commercial-fisheries/queenslands-commercial-fisheries/harvest-fisheries

[11] https://www.environment.gov.au/system/files/pages/7f4f21c5-7fe2-4bcf-b594-2f17319cc48e/files/exceptional-circumstance-submission.pdf

[12] A useful summary of Philippines shell regulation can be found at: www.conchology.be/?t=1000

However, it is important to differentiate between shell-trade sectors and maintain perspective. UN FAO statistics (FAO 2016) indicate that in 2014 around 740,000 t of scallops (Pectinidae) were harvested from the wild. If we convert that into numbers of shells (at 10 per kg), then this equates to around 7.4 billion shells per annum. By contrast, an annual estimate of pectinid specimen shells traded to collectors (based on comments from online dealers, shell show sales etc.) would amount to a maximum of 5000 shells, and bearing in mind that this is the most popular family of collectible bivalves, and that many shells are 'recycled' through collections.

Nevertheless, all wild capture activities, including shell collecting, require a fundamental appreciation of ecology and sustainability, while the moral issues associated with killing for specimen collections are an individual matter. However, as we have seen, shell collecting is not only about 'killing molluscs'; specimens can be obtained via various other means (see Sect. 20.4), and over time it becomes apparent to the majority of collectors that their interest requires the protection of the environments and habitats that support the animals.

The Audubon Medal recipient, Richard Louv, in his work on "nature-deficit disorder' said, "*We cannot protect something we do not love, we cannot love what we do not know, and we cannot know what we do not see. And touch. And hear.*" (Louv 2005).

If society alienates people from an appreciation of biodiversity and the natural world in general, and collecting in particular, then there is potentially an even greater risk that we will cease to relate to natural environments and the animals and plants that live in them. Unregulated or careless shell collecting certainly has the potential to cause environmental and biodiversity damage, but appropriately managed, well-informed and responsible wild harvest, as part of a wider collecting and curation process, may actually be a greater contribution to environmental good than harm.

20.7 Future Prospects

The inherent value of shells as collector's items has ranged throughout human history from aesthetic pieces of jewellery, exotic curiosities from a new world, subjects of artistic expression and design, scientific specimens, and also as commercial items for sale by specialist dealers.

These different values of shells, and hence shell collections, indeed the existence and persistence of the physical collection itself, have been critical for the development of natural history, taxonomy, systematics and evolutionary biology. Since the curiosity cabinets of seventeenth century Europe, scientists have been inspired to investigate and publish work on shells, enabling the gradual progression and accumulation of ideas and information that is at the core of scientific investigation. So, perhaps more than any other biological collectible, the rich diversity of shells has provided the material and opportunity for both amateurs and professionals to provide insights into the wider natural world.

However, throughout the conchological community there is concern that the demographic is ageing and that relatively few younger people are continuing the long tradition of the discipline. While the value of responsible specimen collecting and accessible institutional natural history collections are in little doubt (Rocha et al. 2014; Bradley et al. 2014), the future development of natural sciences also requires a next generation of interested, passionate and educated individuals to pursue it. Therefore, it is perhaps important to ensure that future generations are given opportunities to develop the interest and practical skills that come with hobbies such as shell collecting. Several high-profile naturalists, such as Sir David Attenborough,[13] have raised concern about the danger of wholesale restrictions on natural history collecting, and the reduced opportunities for younger people to explore their natural environment and develop the fundamental interests that lead onto significant scientific contributions, whether as professionals or amateurs.

There is no question that if we continue to collect live mollusc shells from the wild then we must do so sustainably, and responsible conchological organisations have recognized this for many decades. However, enabling people to pursue their interest in natural history, and better understand the fundamental links between biodiversity and the supporting environment, can help to encourage the next generation of environmental conservationists. The world's seas, particularly near-shore areas, face numerous threats from anthropogenic pollution, habitat loss and fragmentation, and overfishing, which are arguably much greater threats to mollusc populations than shell collecting.

If people retain an interest in these habitats and species we can better understand and protect them, although the converse is equally true. This interest requires appropriately managed access to at least some collecting areas, for both amateurs and professional scientists. In this way, we, as communities, can maintain local familiarity, expertise and relevant data series, as well as recognizing emerging environmental threats and deterioration.

Robert Louis Stevenson (1911) famously wrote; '*It is perhaps a more fortunate destiny to have a taste for collecting shells than to be born a millionaire*'; which, while part of a discussion about finding an interest in life beyond the simple acquisition of wealth, uses an example which continues to generate significant passion, even obsession, amongst its devotees. As such, it seems an appropriate conclusion to this chapter.

Acknowledgements The authors are grateful to two anonymous reviewers for their constructive comments.

[13] http://www.telegraph.co.uk/news/earth/wildlife/9657545/David-Attenborough-I-would-never-have-been-a-naturalist-under-todays-fossil-laws.html

References

Adams DJ, Alewood PF, Craik DJ, Drinkwater RD, Lewis RJ (1999) Conotoxins and their potential pharmaceutical applications. Drug Dev Res 46(3–4):219–234

Adan R (2000) The window-pane (*kapis* shell) industry. SEAFDEC. Asian aquaculture Vol. XXII no. 4 (July–August), 2pp

Australian Government (2004) Assessment of the Queensland marine specimen shell collection fishery. Report for the Department of Environment and Heritage, November 2004, 22 pp. http://bit.ly/2rM3PPV

Australian Government (2005) Assessment of the Western Australian specimen shell managed fishery. Report for the Department of Environment and Heritage, May 2005, 25 pp. http://bit.ly/2s4hukX

Ausubel JH, Crist DT, Waggoner PE (2010) First census of marine life 2010 highlights of a decade of discovery. ISBN: 978-1-4507-3102-7. A publication of the Census of Marine Life. www.coml.org

Baldwin Brown G (1932) The art of the cave dweller: a study of the earliest artistic activities of man. John Murray, London, 280 pp

Blake NJ, Shumway SE (2006) Bay scallop and calico scallop fisheries, culture and enhancement in eastern North America In: Scallops: biology, ecology and aquaculture. Developments in Aquaculture and Fisheries Science, 2nd edn, vol 35, pp 945–964

BFAR (Bureau of Fisheries and Aquatic Resources) (2017) Philippine fisheries profiles 1998–2015, Manila, Philippines. http://www.bfar.da.gov.ph/publication

Boswell H (1964) South Africa's *ex-pisce* shells (removed from fish stomachs). Hawaiian Shell News X11(5). www.internethawaiishellnews.org/HSN/1964/6403.pdf

Bouchet P, Bary S, Héros V, Marani G (2016) How many species of molluscs are there in the world's oceans, and who is going to describe them? In: Héros V, Strong E, Bouchet P (eds) Tropical Deep-Sea Benthos 29. (Mémoires du Muséum National d'Histoire Naturelle; 208). Muséum National d'Histoire Naturelle, Paris, pp 9–24. ISBN: 978-2-85653-774-9

Bradley RD, Bradley LC, Garner HJ, Baker RJ (2014) Assessing the value of natural history collections and addressing issues regarding long-term growth and care. Bioscience 64(12):1150–1158

Byne LSG (1899) The corrosion of shells in cabinets. J Conchol 9(6):172–178

Cavallari DC, Salvador RB, Cunha BR (2014) Dangers to malacological collections: Bynesian decay and pyrite decay. Collection Forum 28(1–2):35–46

Chapman AD (2009) Numbers of living species in Australia and the World, 2nd edn. A report for the Australian Biological Resources Study, Canberra. Retrieved May 2017 ISBN 978-0-642-56861-8 (online)

Chivian E, Roberts CM, Bernstein AS (2003) The threat to cone snails. Science 302(5644):391

Dance SP (1969) Rare shells. Faber and Faber, London, 128 pp

Dance SP (1986) A history of shell collecting. E.J. Brill, Leiden, 265 pp

Day JC (2002) Zoning – lessons from the Great Barrier Reef Marine Park. Ocean Coast Manag 45:139–156

Diaz JM, Borrero FJ (1995) On the occurrence of *Pholadomya candida* Sowerby, 1823 (Bivalvia: Anomalodesmata) on the Caribbean coast of Colombia. J Molluscan Stud 61(3):407–408

Diaz JM, Gast F, Torres DC (2009) Rediscovery of a Caribbean living fossil: *Pholadomya candida,* GB Sowerby I, 1823 (Bivalvia: Anomalodesmata: Pholadomyoidea). Nautilus 123:19–20

d'Errico F, Henshilwood C, Vanhaeren M, van Niekerk K (2005) *Nassarius kraussianus* shell beads from Blombos cave: evidence for symbolic behaviour in the middle stone age. J Hum Evol 48:3–24

d'Errico F, Vanhaeren M, Barton N, Bouzouggar A, Mienis H, Richter D, Hublin J-J, McPherron SP, Lozouet P (2009) Additional evidence on the use of personal ornaments in the middle paleolithic of North Africa. Proc Natl Acad Sci U S A 106:16051–16056

Dolorosa RG, Conales SF, Bundal NA (2013) Status of horned helmet, *Cassis cornuta,* in Tubbataha Reefs Natural Park, and its trade in Puerto Princesa City, Philippines. Atoll research bulletin no. 595, 20 pp

Duncan PF, Andalecio MN, Peralta E, Laureta LV, Hidalgo AR, Napata R (2009) Evaluation of production technology, product quality and market potential for the development of bivalve mollusc aquaculture in the Philippines. ACIAR report FR2009–41 (project FIS/2007/045). Canberra, 193 pp

Duncan PF, Brand AR, Strand Ø, Foucher E (2016) The European scallop fisheries for *Pecten maximus, Aequipecten opercularis, Chlamys islandica*, and *Mimachlamys varia*. In: Ehumway SE, Jay Parsons G (eds) Scallops, 3E. Elsevier Science, Oxford, pp 781–858

FAO (2016) FAO yearbook. Fishery and aquaculture statistics 2014. Rome, 105 pp

Floren AS (2003) The Philippine shell industry with a focus on Mactan, Cebu. Coastal Resource Management Project of the Department of Environment and Natural Resources, 50 pp

Gallardo WG, Siar SV, Encena V II (1995) Exploitation of the window-pane shell *Placuna placenta* in the Philippines. Biol Conserv 73:33–38

Gössling S, Kunkel T, Schumacher K, Zilger M (2004) Use of molluscs, fish, and other marine taxa by tourism in Zanzibar, Tanzania. Biodivers Conserv 13:2623–2639

Jacquet J (2009) Silent water: a brief examination of the marine fisheries crisis. Environ Dev Sustain 11:255–263

Joordens JCA et al (2015) *Homo erectus* at Trinil on Java used shells for tool production and engraving. Nature 518:228–231

Laxmilatha P (2015) Status and conservation issues of window pane oyster *Placuna placenta* (Linnaeus 1758) in Kakinada Bay, Andhra Pradesh. India J Mar Biol Assoc India 57(1):92–95

Laureta LV, Marasigan ET (2000) Habitat and reproductive biology of angelwings *Pholas orientalis* (Gmelin). J Shellfish Res 19:19–22

Laureta LV (2008) Compendium of economically important seashells in Panay, Philippines. University of the Philippines Press, Quezon City, 162 pp

Louv R (2005) Last child in the woods: saving our children from nature-deficit disorder. Algonquin Books, Chapel Hill, 390 pp

Mora C, Tittensor DP, Adl S, Simpson AGB, Worm B (2011) How many species are there on Earth and in the ocean? PLoS Biol 9(8):e1001127. https://doi.org/10.1371/journal.pbio.1001127

Nicol D (1969) The number of living species of molluscs. Syst Zool 18(2):251–254

Nijman V, Spaan D, Nekaris KA-I (2015) Large-scale trade in legally protected marine mollusc shells from Java and Bali, Indonesia. PLoS One 10(12):e0140593. https://doi.org/10.1371/journal.pone.0140593

NOAA (2015) http://www.fisheries.noaa.gov/pr/species/invertebrates/queen-conch.html

Pauly D (2008) Global fisheries: a brief review. J Biol Res (Thessaloniki) 9:3–9

Philippines Government (2001) Fisheries Administrative (Order no. 208, Series of 2001). Conservation of rare, threatened and endangered fishery species. http://www.bfar.da.gov.ph/bfar/download/fao/FAO208.pdf

Ponder WF, Lindberg DR (eds) (2008) Phylogeny and evolution of the Mollusca. University of California Press, Berkeley, 481 pp

Queensland Government (2008) Performance measurement system Queensland marine specimen shell collection fishery, 15 pp

Queensland Government (2009) Queensland marine specimen shell collection fishery 2009. Report on progress against DEWHA conditions and recommendations. Brisbane, Australia, 6 pp. http://bit.ly/2reA8Cl

Raines KR, Poppe GT (2006) A conchological iconography: the family Pectinidae. Conchbooks, Hackenheim 722 pp

Robinson SD, Safavi-Hemami H, Mcintosh LD, Purcell AW, Norton RS, Papenfuss AT (2014) Diversity of conotoxin gene superfamilies in the venomous snail, *Conus victoriae*. PLoS One 9(2):e87648. https://doi.org/10.1371/journal.pone.0087648

Robinson SD et al (2016) A naturally occurring peptide with an elementary single disulfide-directed b-hairpin fold. Structure 24(2):293–299

Rocha LA et al (2014) Specimen collection: an essential tool. Science 344:814–815

Ronquillo J, McKinley RS (2006) Developmental stages and potential mariculture for coastal rehabilitation of endangered Pacific angelwing clam, *Pholas orientalis*. Aquaculture 256(1):180–191

Rosenberg G (2014) A new critical estimate of named species-level diversity of the recent Mollusca. Am Malacol Bull 32(2):308–322

Runnegar B (1972) Anatomy of *Pholadomya candida* (Bivalvia) and the origin of the Pholadomyidae. Proc Malac Soc Lond 40:45

Safavi-Hemami H, Siero WA, Gorasia DG, Young ND, MacMillan D, Williamson NA, Purcell AW (2011) Specialisation of the venom gland proteome in predatory cone snails reveals functional diversification of the cono-toxin biosynthetic pathway. J Proteome Res 10(9):3904–3919

Stevenson RL (1911) Lay morals (chapter IV). In: Lay morals and other papers. Chatto and Windus, London, 320 pp

Tennent NH, Baird T (1985) The deterioration of mollusca collections: identification of shell efflorescence. Studies in conservation. International Institute for Conservation of Historic and Artistic Works (IIC) 30(2):73–85

Tiberi N (1879) Le Conchiglie Pompeiane. Bull Soc Malac Ital 5:139–151

United Nations (2017) World population prospects: the 2017 revision. Published June 2017. https://esa.un.org/unpd/wpp/

Van Wynsberge S, Andréfouët S, Gilbert A, Stein A, Remoissenet G (2013) Best management strategies for sustainable giant clam fishery in French Polynesia islands: answers from a spatial modeling approach. PLoS One 8(5):e64641. https://doi.org/10.1371/journal.pone.0064641

Zilhão J, Angelucci DE, Badal-García E, d'Errico F, Daniel F, Dayet L, Douka K, Higham TFG, Martínez-Sánchez MJ, Montes-Bernárdez R, Murcia-Mascarós S, Pérez-Sirvent C, Roldán-García C, Vanhaeren M, Villaverde V, Wood R, Zapata J (2010) Symbolic use of marine shells and mineral pigments by Iberian Neandertals. PNAS 107(3):1023–1028

Chapter 21
Archaeology and Sclerochronology of Marine Bivalves

Paul G. Butler, Pedro S. Freitas, Meghan Burchell, and Laurent Chauvaud

Abstract In a rapidly changing world, maintenance of the good health of the marine environment requires a detailed understanding of its mechanisms of change, and the ability to detect early signals of a shift away from the equilibrium state that we assume characterized it before there was any significant human impact. Given that instrumental measurements of the oceans go back no further than a few decades, the only way in which we can assess the long-term baseline variability that characterizes the pre-perturbation equilibrium state of the marine environment is by the use of proxy records contained in stratified or layered natural archives such as corals, fish otoliths and bivalve mollusc shells.

In this chapter we will look at the ways in which the environmental signals recorded in the shells of bivalve molluscs can be used to shed light on marine variability both in the present and over past centuries and millennia, and specifically how they can be used to study marine climate, the marine environment and the economic and cultural history of the relationship between humans and the oceans.

The chapter is divided into two parts: section one describes the morphological, geochemical and crystallographic techniques that are used to obtain information from the shells, while section two covers the use of bivalve shells in a wide range of

P. G. Butler (✉)
College of Life and Environmental Sciences, University of Exeter, Penryn Campus, Cornwall, UK
e-mail: p.butler@exeter.ac.uk

P. S. Freitas
Danish Shellfish Centre, Technical University of Denmark, Nykøbing M, Denmark
e-mail: psfr@aqua.dtu.dk

M. Burchell
Department of Archaeology, Faculty of Humanities & Social Sciences, Memorial University, St. John's, NL, Canada
e-mail: mburchell@mun.ca

L. Chauvaud
IUEM-UBO, UMR CNRS 6539, Technopôle Brest-Iroise, Plouzané, France
e-mail: Laurent.Chauvaud@univ-brest.fr

A. C. Smaal et al. (eds.), *Goods and Services of Marine Bivalves*,
https://doi.org/10.1007/978-3-319-96776-9_21

applications, including ecosystem services, environmental monitoring, archaeology, climate reconstruction, and climate modeling.

Abstract in Chinese 摘要：在瞬息万变的世界中，为了维护良好的海洋环境，我们需要对其变化机制有一个详细的了解，以便能够及时获取和辨识由人为影响造成的海洋生态平衡状态改变的早期信号。近几十年来，，我们评估海洋环境平衡扰动的长期基线变化的唯一途径是记录在珊瑚礁，鱼耳石和贝类贝壳内不同年代的环境变化留下的信号。在本章中，我们将着眼于研究当下和过去的数百乃至数千年中双壳贝类壳中记录的环境信号，并基于这些信号来揭示长久以来海洋的变化情况，包括如何利用这些信息来进行海洋气候研究，海洋环境研究以及人类与海洋相互作用在经济与文化方面情况。

本章分为两部分：第一部分描述如何通过形态学，地球化学和晶体学技术从贝壳中获取信息；第二部分介绍了双壳类贝壳在生态系统服务，环境监测，考古学，气候状态重构和气候模拟等研究中的应用。

Keywords Environmental monitoring · Mollusc · Archaeology · Marine climate · Ecosystems

关键词 环境监测 • 软体动物 • 考古学 • 海洋气候 • 生态系统

21.1 Physical and Geochemical Proxies

Everything that is known about past environmental and climatic conditions in the Earth's history prior to the appearance of historical written records and the use of instrumental measurements is based on the identification and interpretation of proxies preserved in biological or geological structures. Proxies are measurable physical or chemical properties of biogenic or abiogenic structures (e.g. shells, coral skeletons, trees, sediments, rocks) that can be interpreted as a signal of one or more environmental variables at the time during which the structures were formed. In addition, proxies enable monitoring of present day environmental conditions in locations where instrumental or historical observations are absent.

The major challenge when using bivalve shell material as a proxy archive (this is common to all proxy archives) is to establish the causal link between the wider environment in which the animal was living and the form, or configuration, with which the proxy manifests itself in the carbonate shell material. This is necessary in order to isolate the influence of the large-scale environment on the proxy from the effects of biomineralization or micro-environments. Complicating factors include vital effects, fractionation, multiple drivers in the environment, diagenesis, temporal lags, determination of the season of growth, and variable growth rates (throughout ontogeny and within each year) (Schöne 2008). While these sources of uncertainty can never be fully eliminated, they can be partially compensated through greater replication of chronologies in space (as the real environmental signal emerges from the background noise) and through mathematical modelling (Mueller et al. 2015;

Goodwin et al. 2009; De Ridder et al. 2004) or forward modeling of the processes of shell growth (Tolwinski-Ward et al. 2011).

The main proxies used in bivalve sclerochronology are: variations in periodic shell growth (usually in the form of daily, tidally or annually deposited increments); stable oxygen, carbon and nitrogen isotopes and elemental composition of the shell; and changes in the shell crystal microstructure.

21.1.1 Shell Growth

Shell growth reflects the complex interactions of biological clocks and physiological processes with recurrent environmental pacemakers such as light/dark cycles, tidal exposure and diurnal or seasonal temperature variations. Interruption or reduction of shell growth results in the formation of distinct lines or bands (see Fig. 21.1), which delimit periodic growth increments at a range of temporal scales from sub-daily to annually.

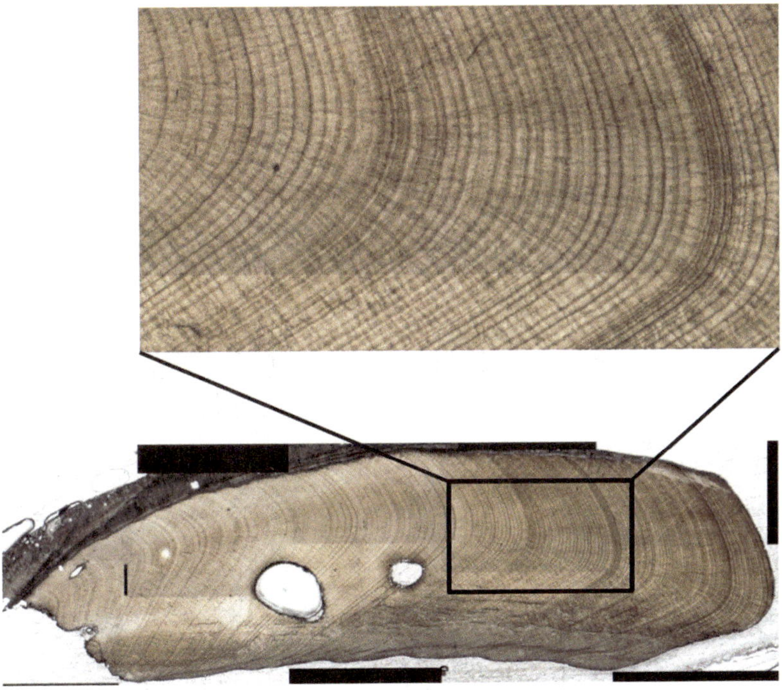

Fig. 21.1 Annually-resolved growth increments imaged in the umbone (hinge) region of a specimen of *Glycymeris glycymeris*. Each increment (the wide lighter bands between the thin dark lines) consists of material laid down during the growth season (usually between 6 and 9 months). (Photo: Pedro Freitas)

While it is a challenging task to disentangle the signals of multiple environmental or climatic drivers in time-series of bivalve growth increments, growth increment series have been interpreted as a response to climate patterns in the Arctic (Ambrose et al. 2006) and north Atlantic (Reynolds et al. 2017, Swingedouw et al. 2015, Schöne et al. 2003,), west African monsoon activity (Azzoug et al. 2012), sea surface temperature (Brocas et al. 2013; Reynolds et al. 2013; Black et al. 2009; Butler et al. 2010) and palaeo-productivity (Wanamaker et al. 2009; Witbaard 1996).

21.1.2 Stable Isotopes

The ability to use stable isotopes as geochemical proxies relies on the fractionation (i.e. the relative preference) between the lighter and heavier isotopes of an element during chemical reactions (e.g. carbonate precipitation or respiration) and the preservation of the resultant stable isotope ratio in the shell material. Stable isotope ratios of oxygen and carbon are commonly used in bivalve shells, while the use of stable isotope ratios of other elements (e.g. magnesium, boron, nitrogen, sulphur or strontium) is less common (e.g. Levin et al. 2015; Liu et al. 2015; Carmichael et al. 2008; Holmden and Hudson 2003), as is the use of clumped isotopes (Eagle et al. 2013).

21.1.2.1 Stable Oxygen Isotopes

The stable oxygen isotope ratio ($\delta^{18}O_{shell}$) of shell carbonate depends on both the ambient temperature and the isotopic composition of the water, the latter being influenced by precipitation-evaporation dynamics and water mass mixing, thus being correlated (in marine environments) with salinity (Carmichael et al. 2008; Epstein et al. 1953; Urey 1947). Empirical palaeotemperature equations have been developed to reconstruct temperature from $\delta^{18}O_{shell}$ (e.g. Kim and O'Neil 1997; Grossman and Ku 1986), although these assume that $\delta^{18}O_{water}$ is known or can be estimated. Bivalves usually precipitate their shell calcite and aragonite in or close to oxygen isotopic equilibrium (e.g. Wefer and Berger 1991) and palaeotemperature equations have been produced for several bivalve species, including *Pecten maximus* (Chauvaud et al. 2005), *Mytilus edulis* (Wanamaker et al. 2007), *Glycymeris glycymeris*(Royer et al. 2013), *Tridacna gigas* (Aharon 1983) and *Tridacna maxima* (Duprey et al. 2015). The effect of seasonally variable growth rates must be taken into account, particularly in annually resolved records, since this causes variable time averaging and bias in $\delta^{18}O_{shell}$ records towards the season of highest growth and may inhibit the preservation of the full seasonal temperature amplitude (Schöne 2008; Goodwin et al. 2003).

21.1.2.2 Stable Carbon Isotopes

The stable carbon isotope composition of bivalve shell carbonate ($\delta^{13}C_{shell}$) has been proposed as a proxy for $\delta^{13}C$ of dissolved inorganic carbon ($\delta^{13}C_{DIC}$), and the processes that control it: salinity, the marine $\delta^{13}C$ Suess effect (Butler et al. 2009), and productivity and respiration (e.g. Schöne et al. 2011; Arthur et al. 1983; Killingley and Berger 1979; Mook and Vogel 1968). However, shell carbon does not originate only from DIC, but also includes a proportion of metabolic carbon with highly depleted $\delta^{13}C$ values and it can also be affected by kinetic isotopic disequilibrium (e.g. Gillikin et al. 2007; Kennedy et al. 2001; Mcconnaughey et al. 1997; Klein et al. 1996; Tanaka et al. 1986). Nevertheless, $\delta^{13}C_{shell}$ can provide valuable information on environmental conditions in species with a stable metabolic influence or where the $\delta^{13}C_{DIC}$ signal is large enough to be preserved in $\delta^{13}C_{shell}$ (Butler et al. 2011; Schöne et al. 2011; Khim et al. 2003).

21.1.2.3 Stable Isotopes in the Shell Organic Matrix

Bivalve shells contain an organic matrix comprising up to 5% of the shell material (Marin et al. 2012), which can be analysed for $\delta^{13}C$, $\delta^{15}N$ and $\delta^{34}S$ (e.g. Carmichael et al. 2008), albeit at a lower temporal resolution than is possible for the inorganic fraction. $\delta^{13}C$, $\delta^{15}N$ and $\delta^{34}S$ in the organic matrix depend, as with bivalve soft tissues, on the isotopic composition of food sources and on fractionations associated with metabolic processes (Vander Zanden and Rasmussen 2001), providing information on primary consumer food sources, ecosystem trophic structure (Graniero et al. 2016; Ellis et al. 2014; Dreier et al. 2012; Versteegh et al. 2011; Mae et al. 2007; O'Donnell et al. 2003, 2007), and anthropogenic nitrogen inputs (Black et al. 2017; Kovacs et al. 2010; Watanabe et al. 2009; Carmichael et al. 2008).

21.1.3 Elemental Composition of Shell Carbonates

The elemental composition of bivalve shells (expressed as normalized E/Ca ratios) has – at least in theory – potential for palaeoceanographic reconstruction and environmental monitoring, this being related to the control of element incorporation by environmental variables such as temperature, or ambient element concentration. However, the complexities of bivalve shell biomineralization lead to strong physiological and kinetic effects related to metabolism, growth rates, ontogenetic age, shell mineralogy, crystal structure and the organic matrix (e.g. Freitas et al. 2008, 2009, 2016; Shirai et al. 2014; Lazareth et al. 2013; Schöne et al. 2013; Carré et al. 2006; Klein et al. 1996; Lorens and Bender 1977). Minor and trace elements can be incorporated in shell carbonate by various processes, including substitution of calcium in the carbonate crystal lattice (Soldati et al. 2016; Lingard et al. 1992); differential adsorption to heterogeneous crystal surfaces (Schöne et al. 2013); binding to organics (Takesue et al. 2008); and co-precipitation as separate mineral phases (Fritz et al. 1990). As a result, the effective use of bivalve shell E/Ca ratios as environmental proxies has been limited, often to species-specific applications or applications restricted to particular environmental settings (e.g. Bougeois et al. 2014; Elliot et al. 2009).

21.1.4 Microstructure

While shell microstructure has commonly been used in phylogenetic studies, only more recently has it been found that crystal fabrics at the micrometre scale might preserve information on environmental conditions at the time of shell formation: these include pH in *Mytilus edulis* shells (Milano et al. 2016; Hahn et al. 2014) and temperature in *Trachycardium procerum* (Perez-Huerta et al. 2013) and *Cerastoderma edule* shells (Milano et al. 2017; Nishida et al. 2012).

21.2 Goods and Services of Bivalve Sclerochronology

21.2.1 Ecosystem Services

21.2.1.1 Introduction

The marine system has been estimated to supply about two-thirds of all ecosystem services provided by the natural environment (Gesamp 2001). These include, but are not limited to, fisheries, aquaculture, carbon sequestration, water quality, energy production, aggregate extraction and biodiversity. However, the present day marine system is challenged by the combined impact of climate change and industrial scale fisheries, and the definition of a "natural" baseline ecosystem or any kind of ecosystem equilibrium is problematic and challenging. While regime shifts in response to natural climate variability undoubtedly occurred before the industrial era (Hare and Mantua 2000; Minobe 1997), these did not take place in the context of steep trends in ocean temperature, ocean pCO_2 and selective harvesting of key species (Rocha et al. 2015). To maintain ecosystem resilience under such conditions is a challenging task for ocean management, and a key part of the process will be to assess the degree of ecosystem variability that characterizes a resilient system (Steinhardt et al. 2016; Willis et al. 2010). It is this degree of variability that can be assessed with the help of biochronologies drawn from multiple sources (eg bivalve molluscs, fish otoliths, corals).

The term "ecosystem variability" in this context includes population dynamics of single species, predator-prey relationships, trophic chains and host-pathogen relationships. Indicators of ecosystem variability that can be identified in bivalve shell archives include growth rate, population dynamics, environmental DNA and stable isotope ratios (see Sects. 21.2.1.1, 21.2.1.2, 21.2.1.3 and 21.2.1.4 below). Although these are inherently limited as ecosystem proxies, being based on single species, their usefulness can be substantially enhanced by comparing them directly with other precisely dated archives, such as tree rings, corals, coralline algae and fish otoliths (e.g. Black 2009; Black et al. 2009) or by characterizing bivalve growth patterns using mixed effects models (Mazloumi et al. 2017; Morrongiello et al. 2012). In this way it is possible to develop detailed timelines of ecosystem variability, including leads and lags at annual and seasonal resolutions between different ecosystems and different parts of the same ecosystem.

With the use of modern statistical techniques such as principal components analysis and mixed effects models to isolate the causes and effects of interacting environmental drivers on multiple proxy archives, it is now possible to reconstruct ecosystem dynamics over many centuries. Long chronologies can be used in tandem with shorter archival records and instrumental data (Black et al. 2014), so that the dynamics of ecosystem regime shifts can be modeled and extended back in time with the application of mixed effects models (Morrongiello and Thresher 2015) to the long proxy archive. The use of networks of bivalve chronologies can add a spatial element to the extended archive (Reynolds et al. 2017).

21.2.1.2 Bivalve Growth Rates

To a first approximation, variability in bivalve growth rates can be assumed to be a response to variations in food supply to the benthos (Witbaard 1996). However, this apparently straightforward assumption is complicated by predator-prey relationships in the upper part of the water column, so that food supply to the bottom-dwelling bivalves can sometimes be anticorrelated with primary production at the surface (Witbaard et al. 2003). A further level of complexity is introduced by the position of the bivalve population above or below the seasonal thermocline, so that more complex ecosystem variability in shallow surface waters appears to result in rather low growth synchrony between animals in the same population (Marali and Schöne 2015), or the reversal of the correlation between growth and seawater temperature (compare Mette et al. 2016, with Brocas et al. 2013, and Butler et al. 2010). Bivalve growth rates therefore seem to reflect an emergent outcome of a complex web of ecosystem relationships in the overlying water column. This complexity can be deconvolved using multivariate analysis techniques such as multiple linear regression (Mette et al. 2016), Bayesian hierarchical modeling (Helser et al. 2012), principle component analysis (Tao et al. 2015), or mixed effects models (Izzo et al. 2017).

21.2.1.3 Population Dynamics of Bivalve Fisheries

The use of long absolutely-dated chronologies adds enormous value to studies of population dynamics and the management of commercial bivalve fisheries (Ridgway et al. 2012; Harding et al. 2008; Kilada et al. 2007). The ability to determine precise dates of settlement over long time frames enables changes of population structure over time to be determined. These changes can be related to information about climate, regime shifts, hydrography and predator-prey relationships, allowing aspects of the underlying ecosystem variability to be inferred (Ridgway et al. 2012; Witbaard and Bergman 2003; Witbaard et al. 1997).

21.2.1.4 Environmental DNA

A number of recent studies have shown that ancient DNA (aDNA) can be recovered from fossil material (e.g. Pruefer et al. 2014; Orlando et al. 2013), raising the possibility that changes in environmental DNA over time could be reconstructed by extracting aDNA from precisely dated fossil shell material. Snippets of aDNA recovered from the organic fraction of the shell matrix may characterize not only the genome of the bivalve itself, but also other species in the environment (from both inside and outside the shell). Metagenomic analysis of modern and fossil shells (Der Sarkassian et al. 2017) has shown that the shell biominerals (depending on their condition) may contain a range of microbial DNA from the marine environment, as well as the DNA of the host organism itself and its pathogens. This work indicates that there is potential to use environmental DNA and ancient DNA in shells to monitor the evolutionary history of bivalve species, their associated microbial communities and their relationship with pathogens.

21.2.1.5 Stable Isotope Ratios

The position of an animal in the surrounding food web is a useful indicator of predator-prey interactions or trophic chains, and one that can be approached through the analysis of stable isotope ratios of carbon and nitrogen in the organic fraction of the shell (e.g. Gillikin et al. 2017). Stable isotopes can also indicate disruption of the ecosystem as a result of anthropogenic inputs such as agricultural runoff or wastewater input (Versteegh et al. 2011). For more information on the use of stable isotopes in the shell organic matrix, see Sect. 21.1.2.3.

21.2.2 *Environmental Services*

21.2.2.1 Introduction

Reliable monitoring of past and present environmental conditions is essential if we are to accurately assess the impacts of anthropogenic and natural changes on the marine environment. Bivalve shells can provide a tool for present and retrospective monitoring, establishing pre-impact environmental baselines, and allowing the reconstruction of marine and freshwater environments that range from estuaries to the deep-sea (e.g. Schöne and Krause 2016; Steinhardt et al. 2016; Fortunato 2014; Richardson 2001; Jones 1983). While the soft tissues of bivalves are commonly used in monitoring projects such as the well-known Mussel Watch program (Schöne and Krause 2016), the use of bivalve shells presents several distinct advantages: (1) shells are usually not affected by post-deposition alterations, while soft tissue decomposes rapidly; (2) there is potential to obtain both high-resolution (circa-daily to annual) records with accurately-dated banded shell material, and lower resolution time-averaged records from whole shells or fractions of shells; (3) temporal snapshots can be obtained from individual specimens; (4) where the bivalves are sufficiently long-lived, the proxy record can be extended into the past beyond the lifetime of single individuals through replicated cross-matched chronologies (see Sect. 21.2.4.1.); (5) shells can provide proxy records for times and locations where instrumental networks and records are absent. However, the use of shells for environmental monitoring and reconstruction is still limited, due to analytical limitations (e.g. stable isotopes in the organic matrix), unknown pathways of incorporation into the shell (e.g. hydrocarbons and other organic pollutants), significant inter-shell variation, or the complex control of most proxies by multiple environmental and biological variables. Consequently, most studies of bivalve shell environmental proxies have focused on the evaluation and validation of environmental and physiological controls for individual species at specific sites. Nevertheless, with the added value provided by crossdating and replication, bivalve shells can provide baseline monitoring and reconstruction services for a range of environmental characteristics (see Schöne and Krause 2016 and Steinhardt et al. 2016 for recent reviews), including contamination events, temperature, salinity and river discharge.

21.2.2.2 Pollution Events

Minor and trace elements, due to their role in biogeochemical processes and their potentially hazardous impact on the environment, have been of particular interest in studies of the capacity of bivalve shells to record natural and anthropogenic changes in ambient chemistry, including pollution events. Most studies have compared whole shells or fractions of shells from contaminated and non-contaminated sites (see Schöne and Krause 2016 for a recent review). Elevated levels of metals (e.g. Mn, Fe, Cu, Zn, Cd, Pb and U) in the shells of several species (e.g. *Arctica islandica*, *Crassostrea gigas*, *Crassostrea virginica*, *Ensis siliqua*, *Modiolus modiolus*, *Mercenaria mercenaria*, *Mya arenaria*, *Mya truncata*, *Mytilus edulis*, *Mytilus californianus*, *Mytilus galloprovinciallis*, *Perna perna*, *Perna viridis* and *Pinctada imbricata*,) have been interpreted as an indication of elevated metal levels in the ambient seawater or sediment (Cariou et al. 2017; Holland et al. 2014; Krause-Nehring et al. 2012; Dunca et al. 2009; Klunder et al. 2008; Protasowicki et al. 2008; Bellotto and Miekeley 2007; Macfarlane et al. 2006; Pearce and Mann 2006; Gillikin et al. 2005a; Liehr et al. 2005; Nicholson and Szefer 2003; Yap et al. 2003; Richardson 2001; Almeida et al. 1998; Puente et al. 1996; Raith et al. 1996; Pitts and Wallace 1994; Fuge et al. 1993; Bourgoin 1990; Koide et al. 1982; Bourgoin and Risk 1987; Chow et al. 1976). However, most of this research is based on snapshots in time, and it rarely involves the use of a truly sclerochronological approach to produce time-series of metal levels in shells (e.g. Vander Putten et al. 2000; Price and Pearce 1997; Carriker et al. 1980). Recent studies have produced decadal to centennial records of environmental heavy metal variability using long-lived species, such as Pb and Fe in *Arctica islandica* (Holland et al. 2014; Krause-Nehring et al. 2012) or Pb in *Mercenaria mercenaria* (Gillikin et al. 2005b). Metal levels in the shells of freshwater bivalves (e.g. Mn, Co, Ni, Cu, Zn, Cd, and Pb) have also been demonstrated to record contamination from industrial or mining activities (Markich et al. 2002; Schettler and Pearce 1996; Anderson 1977) or even, using Na shell content, contamination from road-salt (O'Neil and Gillikin 2014). Elemental proxies have also been proposed as archives of changes in pelagic primary production, e.g. Mo/Ca (Barats et al. 2010; Thébault et al. 2009a), Ba/Ca (Barats et al. 2009) or Li/Ca (Thébault and Chauvaud 2013; Thébault et al. 2009b). However, in general, trace element ratios to calcium (Mg/Ca, Sr/Ca, Li/Ca, Mn/Ca and Ba/Ca) are more difficult to interpret in bivalves and seem to be very sensitive to vital effects, especially growth rate (Carré et al. 2006; Gillikin et al. 2005c; Takesue and Van Geen 2004). In addition to minor and trace elements, other proxies in bivalve shells have been used to record anthropogenic contamination. For instance, the $\delta^{15}N$ composition of the shell organic matrix can provide information on nitrogen anthropogenic wastewater inputs to estuarine ecosystems (e.g. Gillikin et al. 2017; Kovacs et al. 2010; Watanabe et al. 2009; Carmichael et al. 2008).

21.2.2.3 Temperature, Salinity and the Stable Oxygen Isotope Proxy

Bivalve $\delta^{18}O_{shell}$ values obtained at daily, annual or decadal resolutions reflect a wide range of habitats and species from deep-sea oysters *Neopycnodonte zibrowii* (Wisshak et al. 2009) to coastal (or estuarine) and freshwater bivalves, such as mussels (e.g. *Unionidae* (Dettman et al. 1999), *Mytilus trossulus* (Klein et al. 1996), *Pinna nobilis* (Kennedy et al. 2001), scallops (*Pecten maximus* – Chauvaud et al. 2005) and oysters (*Crassostrea gigas* – Ullmann et al. 2010); the geographical spread ranges from tropical (e.g. *Tridacna gigas* (Elliot et al. 2009), *Hippopus hippopus* (Aubert et al. 2009), *Comptopallium radula* (Thébault et al. 2007)) and temperate waters (e.g. *Glycymeris glycymeris* – Royer et al. 2013) to sub-polar (e.g.

Arctica islandica – Schöne et al. 2004, Marsh et al. 1999) and polar waters (Carroll et al. 2009; Tada et al. 2006; Simstich et al. 2005)(e.g. *Astarte borealis* (Simstich et al. 2005) *Laternula elliptica* (Tada et al. 2006), *Serripes groenlandicus* (Carroll et al. 2009)). They also perform an important analytical function, since diurnal or seasonal variation in $\delta^{18}O_{shell}$ effectively validates the periodicity of the growth patterns (e.g. Schöne and Giere 2005; Goodwin et al. 2001; Brey and Mackensen 1997; Jones and Quitmyer 1996; Witbaard et al. 1994; Krantz et al. 1984).

However, the most powerful application of $\delta^{18}O_{shell}$ has been to provide information about the oceanographic and climatic processes that control seawater temperature and salinity. For instance, bivalve $\delta^{18}O_{shell}$ has been used to determine changes in: seasonality (e.g. Beierlein et al. 2015; Wanamaker et al. 2011; Schöne and Fiebig 2009; Schöne et al. 2005b); ocean circulation and atmospheric forcing dynamics (e.g. Reynolds et al. 2017; Wanamaker et al. 2008); shelf and coastal seas hydrography, reflecting changes in circulation (e.g. Torres et al. 2011), river discharge (e.g. Muller-Lupp and Bauch 2005; Simstich et al. 2005; Dettman et al. 2004; Schöne et al. 2003; Surge et al. 2003; Khim et al. 2003; Mueller-Lupp et al. 2003; Ekwurzel et al. 2001) or glacial ice-melt runoff (e.g. Versteegh et al. 2012; Tada et al. 2006; Ekwurzel et al. 2001; Azetsu-Scott and Tan 1997); ENSO variability (e.g. Welsh et al. 2011; Carré et al. 2005); West African Monsoon variability (e.g. Azzoug et al. 2012); and coastal upwelling (e.g. Jolivet et al. 2015). However, within-shell trends in isotopic amplitudes and averages may also reflect decreases in growth rate rather than environmental fluctuations. Therefore, particular care should be taken when interpreting inter-annual isotope profiles from long-lived species (Goodwin et al. 2003).

In addition to $\delta^{18}O_{shell}$, other proxies have been proposed to record changes in salinity and river discharge of coastal and estuarine waters, e.g. Sr isotopes (Widerlund and Andersson 2006) and Ba/Ca ratios (Poulain et al. 2015; Carroll et al. 2009; Gillikin et al. 2006, 2008).

21.2.3 *Cultural Services*

21.2.3.1 Introduction

Human interactions with the intertidal zone, including shellfish collection and in particular harvesting of bivalves have been part of human, and some non-human, cultures for over a hundred thousand years. Bivalve shells have become an increasingly valuable resource for archaeological studies of food habits, patterns of seasonal site occupation, migration, tool use, ornamentation, and the dating of archaeological sites (Thomas 2015a, b; Andrus 2011; Andrus and Crowe 2000; Claassen 1998). Interdisciplinary approaches, combining archaeology, biology and geochemistry have significantly contributed to increased understanding and interpretations of the long-term contributions of bivalves to human culture. The application of bivalve sclerochronology in archaeology is expanding the range of questions archaeologists can ask about past human-environmental interactions.

The earliest evidence of the intentional gathering of bivalves by humans is found at Terra Amata, France (Claassen 1998; Stein 1992). Bivalves were used as a staple and supplementary food source, and empty shells were used for tools and ornamentation by humans and even by our non-human ancestors (Duncan and Ghys 2019). Some of the earliest evidence for understanding human cognition and symbolism comes from the preservation of shell artefacts. Individuals were crafting marine shells into beads to be worn as ornamentation in Israel and Algeria 75,000 years ago (Vanhaeren et al. 2006), and as far back as 82,000 years ago, shell beads decorated with red ochre were left behind in human occupied caves in North Africa (Bouzouggar et al. 2007). Our extinct cousins, the Neanderthals, crafted adornments from the shells of the marine bivalves *Pecten, Glycymeris, Spondylus and Acanthocardia* at Cueva de los Avoines, Iberia, in the Middle Palaeolithic 50,000 years ago (Zilhao et al. 2010). Freshwater shells first appear in the archaeological record in southern Egypt 24,000 years ago. In hunter-fisher-gatherer societies, both past and present, shell tools are used as part of everyday tasks and shells have been valued as ornamentation and symbolic objects; in many early communities, thousands of shell disc beads were used to adorn the dead to prepare them for the afterlife and commemorate their status within society; for example, burials dating between 4000–3500 Cal. BP from the Salish Sea in British Columbia in Canada contain individuals who were buried with up to 350, 000 individual stone and shell beads (Coupland et al. 2017).

21.2.3.2 Shell Middens

The most abundant bivalve remains in the archaeological record appear in the context of shell middens and shell mounds (Roksandic et al. 2014) (Fig. 21.2). Cumulative everyday acts of bivalve collection over decades, centuries or millennia resulted in the formation of shell middens that can be found along almost all of the world's coastlines. Shell middens are deposits that consist primarily of shell, although their micro-constituents can vary with site formation, duration of occupation, population size and purpose (Alvarez et al. 2011; Claassen 1998; Stein 1992). The size and shape vary, from small mounds of finely crushed shells, to extended mounds stretching over tens of kilometres of coastline and over 10 m in height (or depth). For example, modern-day shell middens, and archaeological shell middens from the Saloum Deltal in Senegal are over 15 m in height (Hardy et al. 2016). Shell middens are frequently regarded as homogenous 'garbage heaps' that can represent periods of shell acquisition and disposal that range from a few days of intensive harvesting to continuous harvesting over millennia. In other contexts, they can take the form of monumental architecture, such as those in the south-eastern United States, specifically the shell rings in Georgia and South Carolina (Marquardt 2010; Thompson and Andrus 2011). Our understanding of variability in the nature and extent of shell middens and mounds has been enriched by work on notable shell

Fig. 21.2 (top) Shell midden excavation in progress from the Salish Sea, southern British Columbia. (Photo: Terence Clark); (bottom) Intact sediment block from a shell midden embedded with fiberglass resin, north Calvert Island, British Columbia, Canada. (Photo: Meghan Burchell)

midden regions including the Jomon middens of Japan (e.g. Habu et al. 2011), Sambaquis of Brazil (Okumara and Eggers 2014), the Pacific Northwest Coast (Moss 2011), as well as the kitchen middens ("*køkkenmøddinger*") on the coasts of Scandinavia (Anderson 2008), to name just a few studies among many.

21.2.3.3 Bivalves as a Food Source

Contemporary studies of indigenous populations attest to the importance of shellfish, particularly bivalves, as a food source, especially on the coasts of Australia, Chile, Papua New Guinea, Mozambique and South Africa (Bird and Bird 1997; Kyle et al. 1997), and demonstrate that bivalve gathering is not always a random activity, but is often governed by social and environmental circumstance. Bivalves in cultural contexts go beyond being simply a food source – their shape, colours, and sounds have influenced human cultural activities for millennia, and shells still hold a prominent place in many origin myths and rituals within indigenous societies today. Through the analysis of bivalves in various contexts, interpretations about long-term human environmental interactions and human interaction with the supernatural world can be interpreted.

Collection of oysters and limpets is first recorded at the open-air site of Terra Amata in France, and further evidence suggest gathering of bivalves began elsewhere in Europe as long ago as 450,000 years (Bailey and Milner 2008). The intensification of gathering of bivalves and other marine molluscs as a food source has been observed at approximately 9000 years ago at Cantabria, northern Spain (Waselkov 1987); overall, the increased visibility of bivalve collection has been associated with human population growth, economic intensification, and changes in sea level (Bailey and Craighead 2003).

Bivalves and other intertidal resources have, for the most part, been considered an insignificant, or 'fall back' resource at coastal sites, especially when compared to other food sources such as fish or marine and terrestrial mammals (Eerkens et al. 2016; Erlandson 2001). In previous archaeological studies, the presence of bivalves has often been little more than acknowledged (Fitzhugh 1995); however, as new methods emerge for studying the season and intensity of gathering, archaeologists are becoming better able to understand the role of bivalves in coastal economies of the past, especially regarding seasonal patterns of resource acquisition and by proxy, site occupation.

21.2.3.4 Bivalve Sclerochronology and Seasonality of Human Occupation

Seasonality plays a critical role in hunter-fisher-gatherer societies, particularly in temperate locations, since it influences the availability of food resources and structures the organization of activities and the timing of events. Seasonal changes are therefore integrated into all social, economic and settlement activities. Seasonal subsistence practices are scheduled to optimize the acquisition of resources that vary in quantity, availability and abundance. The importance of seasonality also varies by location and is enhanced in areas with a 'hungry season', where food resources need to be stored to ensure a supply throughout the year (De Garine and Harrison 1988). Bivalves have been identified as a seasonally critical food source and a required source of carbohydrates and proteins during 'lean seasons'. In response to seasonal changes, hunter-gatherers can practice resource management, including season-specific bivalve harvesting (Smith and Wishnie 2000) and drying and storage of their meat (Henshilwood et al. 1994). Using stable oxygen isotope analysis of shell carbonate, archaeologists are able to identify the season, or seasons of bivalve collection (Jew et al. 2013; Hallmann et al. 2009; Deith 1986; Killingley 1981; Shackleton 1973) and interpret long- and short-term settlement patterns (Prendergast and Schöne 2017; Burchell et al. 2013). Seasonality can also be determined through the analysis of sub-annual and annual growth patterns by measuring the distance between seasonally deposited lines (Carré et al. 2009; Milner 2001; Lightfoot et al. 1993). However, the methods used to identify seasonality are contingent on species, shell growth and locality. Some bivalves produce multiple 'annual lines' in their shells, and with these species, seasonality can only be resolved with high-resolution stable oxygen isotope analysis. This has been a critical advance in understanding how hunter-fisher-gatherers co-ordinated movements between sites and developed permanently settled villages. For example, by combining season-of-harvest determined from stable oxygen isotope analysis of the bivalve *Saxidomus gigantea* with sclerochronology, radiocarbon dating and ancillary lines of archaeological evidence, year-round occupation of the village site at Namu for at least 4500 years has been confirmed (Burchell et al. 2013; Cannon and Burchell 2017), predating previous ideas about when hunter-fisher-gatherers established permanent villages on the Pacific Northwest Coast of North America.

21.2.3.5 Bivalve Sclerochronology and Accurate Radiocarbon Dating

The marine radiocarbon reservoir effect (i.e. the uncertainty in radiocarbon dating of marine samples because the measured radiocarbon has spent an unknown period in the marine system before being taken up into the sample) is a longstanding challenge for archaeology in coastal sites, that can be usefully approached using bivalve sclerochronology. If the regional marine reservoir can be independently determined by radiocarbon analysis of an absolutely dated bivalve chronology (Wanamaker et al. 2012; Butler et al. 2009), a more accurate radiocarbon calibration can be applied to midden shells from the same region. Conversely, if there is an independent assessment of the date of occupation, radiocarbon dating of midden shells can be used to determine the regional marine reservoir (Ascough et al. 2006). It is also possible to further constrain radiocarbon dating of coastal sites by the construction of crossmatched floating chronologies using shells found at different levels in middens (Helama and Hood 2011).

21.2.4 Climate Services

21.2.4.1 Introduction

The key characteristics that make this archive so powerful are: (a) that the animals deposit periodic (daily, fortnightly or annual) well-defined increments in the shell; (b) that growth is synchronous within populations; (c) that individuals of certain species can live for hundreds of years (*Arctica islandica* (Butler et al. 2013, Schöne et al. 2005a) and *Glycymeris glycymeris* (Reynolds et al. 2013)); (d) that most species precipitate calcium carbonate in isotopic equilibrium with seawater; (e) that natural or anthropogenic deposits of bivalve shells are widespread and are found at all latitudes. Synchronous growth patterns provide *prima facie* evidence that the shells are recording a common environmental signal, while annual banding allows the precise calendar year of each band to be determined (as long as the year of the most recent band is known). In addition, where species have extended lifespans, the years of fossil shells can be precisely determined by comparing their banding patterns with those from live collected shells. In this way, crossdated and replicated timelines (chronologies) of shell material can be built that go back much further in time than the lifetimes of any live collected shell. For example, specimens of *A. islandica* off the north coast of Iceland regularly live for more than 300 years (Schöne et al. 2005a), and one specimen collected there in 2006 is (at 507 years; Fig. 21.3) the longest-lived non colonial animal known to science whose age can be precisely determined (Butler et al. 2013).

While the multicentennial length of long chronologies adds value to proxy-based reconstructions derived from them, these are only available for certain regions (in particular the temperate and boreal North Atlantic Ocean). In low latitudes, bivalves with much shorter lifespans can be used to reconstruct paleoclimate, albeit in shorter and less precisely dated windows. These include studies of seasonality in the Eocene in central Asia (Bougeois et al. 2014) and in the Miocene in the Amazon (Kaandorp et al. 2005; Vonhof et al. 1998), ENSO variability in the eastern (Carré et al. 2014) and western (Driscoll et al. 2014) tropical Pacific, and Holocene climate variability in the southwest Pacific (Duprey et al. 2012, 2014).

21.2.4.2 The Use of Proxy Archives in Climate Modelling

With the atmospheric concentration of CO_2 passing the 400 ppm threshold in 2016 and unlikely to fall back below it for the foreseeable future (Betts et al. 2016) and emissions continuing to increase at a rate equivalent to business as usual (Boden et al. 2015), the need for climate scientists to generate useful projections to inform mitigation and adaptation policy is more acute than ever. Impacts of climate change on the marine system include species range shifts, loss of ecosystems and biodiversity and impacts on coastal livelihoods (IPCC 2014). Accurate projections of regional change in the short to medium term (which are of most interest to policy-makers) require sufficiently high resolution in climate models, and this in turn

Fig. 21.3 Shell valves of a 507-year old specimen of *Arctica islandica* collected near Grimsey island, north of Iceland in 2006, with processing notes. (Photo: Bangor University)

depends on access to similarly high resolution instrumental and proxy data for assimilation (Fang and Li 2016; Phipps et al. 2013). Suitable methods are already being used for modeling of terrestrial systems using proxies from tree-rings (Breitenmoser et al. 2014; Loader et al. 2013) and speleothems (Baker et al. 2012) and for tropical marine systems using proxies from corals (Evans et al. 1998), and more recently it has been demonstrated that reconstructions of marine climate using bivalve shells can provide high resolution real world data for the temperate and boreal oceans that can be used to test and constrain coupled climate models (Pyrina et al. 2017; Emile-Geay et al. 2016; Swingedouw et al. 2015).

21.2.4.3 Marine Climate of the North Atlantic Ocean

Arctica islandica is a particularly important proxy in this respect because of its distribution in the shelf seas surrounding the North Atlantic Ocean (Schöne 2013; Dahlgren et al. 2000). The North Atlantic is a highly sensitive sentinel of change in the climate system. Heat is transferred from the tropical to the boreal latitudes in the Gulf Stream/North Atlantic Current system. As the water gives up its heat by evaporation at high latitudes in the Labrador and Nordic Seas it becomes dense and sinks, a mechanism (the Atlantic Meridional Overturning Circulation (AMOC)) that plays an important part in driving the global ocean circulation system. Model experiments indicate that the AMOC will weaken during the twenty-first century (Liu et al. 2017; Weaver et al. 2012), and there are some indications that this is already happening (Rahmstorf et al. 2015).

Recent research using the 1357-year *A. islandica* time series for the North Icelandic Shelf (NIS) (Butler et al. 2013) illustrate some of the climate services that can be obtained from this proxy. By measuring the radiocarbon age of shell material that has been independently dated using sclerochronology, it is possible to determine the radiocarbon age of the water mass in which the shell was deposited (i.e. the length of time since the water was last ventilated at the ocean surface). In the case of the NIS, this has enabled researchers to map changes in the relative strength of water masses with Arctic and Atlantic origins and gain a unique insight into the mechanisms driving the marine system in the North Atlantic (including the AMOC) during the past millennium (Wanamaker et al. 2012).

The NIS *A. islandica* series has also been used to validate models of the response of the ocean to large volcanic eruptions. For example, the modeled effect of a particular class of large tropical eruption on bidecadal North Atlantic Ocean circulation variability appears to be mirrored in growth variations in *A. islandica* from the NIS (Swingedouw et al. 2015).

Most recently, the first 1000-year annually-resolved stable oxygen isotope ($\delta^{18}O$) series for the marine environment has been obtained by sampling carbonate from individual increments in shells used in the NIS *A. islandica* chronology (Reynolds et al. 2017). By comparing the shell record with tree ring records, the researchers demonstrated a significant change in the lead-lag relationship between the marine and atmospheric systems. Before the industrial period (AD 1000–1800), changes in the marine system (forced by solar and volcanic variability and internal dynamics of ocean circulation) led changes in Northern hemisphere surface air temperatures (SATs), whereas after ~1800 the relationship was reversed, with changes in SATs leading changes in marine variability. This suggests that the climate effect of rapid increases in atmospheric greenhouse gases has masked the effects of natural external forcing and internal variability.

21.2.4.4 History of Carbon Cycling

The ocean currently acts as a buffer against rapidly increasing concentrations of CO_2 in the atmosphere, taking up between 26% and 34% of the net anthropogenic emissions (Sabine et al. 2004). Because CO_2 derived from fossil fuels is depleted in the heavy isotope ^{13}C, measurement of the stable carbon isotope ratio ($\delta^{13}C$) in dated marine shells from different parts of the ocean can be used to determine spatio-temporal variability in the activity of the ocean as a sink for atmospheric CO_2. The temporal trend in atmospheric $\delta^{13}C$ (^{13}C Suess effect (Francey et al. 1999)) is an indicator of the increasing presence in the atmosphere of CO_2 derived from fossil fuels. The coeval trend in oceanic dissolved inorganic carbon ($\delta^{13}C_{DIC}$), which can be determined from time series of $\delta^{13}C$ measured in absolutely dated bivalve shell material (Schöne et al. 2011; Butler et al. 2009), varies according to water depth and the age of the local water mass, and indicates the spatial distribution of the rate at which the ocean has been acting as a sink for excess atmospheric CO_2.

In addition, radiocarbon in the shell can be used in water mass detection, since the age of the ambient water can be determined by measuring radiocarbon in shells of known age (Scourse et al. 2012; Wanamaker et al. 2012; Fontugne et al. 2004; Ingram and Southon 1996; Southon et al. 1995), and is also related to sea-air CO_2 exchange (Carré et al. 2016).

21.3 Conclusion

During their lifetimes, bivalve molluscs deposit carbonate material to form their shells. This material constitutes a physical archive, which may be time-delimited (usually with daily, tidal or annual periodicity) by well-defined banding patterns in the shells. This archive contains multiple morphological, structural and geochemical records which can be related to the environment in which the shell material was deposited, and which can be analysed as well-ordered and periodically-constrained time series by reference to the banding patterns. The ubiquity and durability of the shells enhances the power of the archives, so that the records contained within them can be used as environmental proxies for a wide range (both in space and in time) of marine and coastal settings.

In this paper, we have described and assessed some of the most notable proven applications of bivalve sclerochronology in ecosystem, environmental, cultural, and climate services. However, it has been necessary to address a vast amount of research in a limited number of words, and the examples described here do not by any means constitute an exhaustive selection. With analytical techniques continually being refined and updated, and new ones being developed (e.g. clumped isotopes and Raman spectroscopy), there is very significant potential in the coming decades for new applications and improved reliability of existing applications. For as long as human society values the environment within which it is constrained to exist, it will find useful tools in the insights into past environments provided by the shells of animals that actually lived in those settings.

Acknowledgements We are grateful to two referees for valuable comments on the manuscript.

References

Aharon P (1983) 140,000-Yr isotope climatic record from raised coral reefs in New Guinea. Nature 304:720–723

Almeida MJ, Moura G, Pinheiro T, Machado J, Coimbra J (1998) Modifications in *Crassostrea gigas* shell composition exposed to high concentrations of lead. Aquat Toxicol 40:323–334

Alvarez M, Briz Godino I, Balbo A, Madella M (2011) Shell middens as archives of past environments, human dispersal and specialized resource management. Quat Int 239:1–7

Ambrose WG Jr, Carroll ML, Greenacre M, Thorrold SR, Mcmahon KW (2006) Variation in *Serripes groenlandicus* (Bivalvia) growth in a Norwegian high-Arctic fjord: evidence for local- and large-scale climatic forcing. Glob Chang Biol 12:1595–1607

Anderson RV (1977) Concentration of cadmium, copper, lead, and zinc in 6 species of freshwater clams. Bull Environ Contam Toxicol 18:492–496

Anderson S (2008) Shell middens ("køkkenmøddinger"): the Danish evidence. In: Antczak A, Cipriani R (eds) Early human impacts on megamollusks (British Archaeological Series). Archaeopress, Oxford

Andrus CFT (2011) Shell midden sclerochronology. Quat Sci Rev 30:2892–2905

Andrus CFT, Crowe DE (2000) Geochemical analysis of *Crassostrea virginica* as a method to determine season of capture. J Archaeol Sci 27:33–42

Arthur MA, Williams DF, Jones DS (1983) Seasonal temperature-salinity changes and thermocline development in the mid-Atlantic bight as recorded by the isotopic composition of bivalves. Geology 11:655–659

Ascough PL, Cook GT, Church MJ, Dugmore AJ, Arge SV, Mcgovern TH (2006) Variability in North Atlantic marine radiocarbon reservoir effects at c. AD 1000. The Holocene 16:131–136

Aubert A, Lazareth CE, Cabioch G, Boucher H, Yamada T, Iryu Y, Farman R (2009) The tropical giant clam *Hippopus hippopus* shell, a new archive of environmental conditions as revealed by sclerochronological and $\delta^{18}O$ profiles. Coral Reefs 28:989–998

Azetsu-Scott K, Tan FC (1997) Oxygen isotope studies from Iceland to an East Greenland Fjord: behaviour of glacial meltwater plume. Mar Chem 56:239–251

Azzoug M, Carré M, Schauer AJ (2012) Reconstructing the duration of the West African Monsoon season from growth patterns and isotopic signals of shells of *Anadara senilis* (Saloum Delta, Senegal). Palaeogeogr Palaeoclimatol Palaeoecol 346:145–152

Bailey GN, Craighead AS (2003) Late Pleistocene and Holocene coastal palaeoeconomies: a reconsideration of the molluscan evidence from northern Spain. Geoarchaeology 18:175–204

Bailey GN, Milner N (2008) Molluscan archives from European prehistory. In: Antczak A, Cipriani R (eds) Early human impact on megamolluscs (British archaeological series). Archaeopress, Oxford

Baker A, Bradley C, Phipps SJ, Fischer M, Fairchild IJ, Fuller L, Spoetl C, Azcurra C (2012) Millennial-length forward models and pseudoproxies of stalagmite $\delta^{18}O$: an example from NW Scotland. Clim Past 8:1153–1167

Barats A, Amouroux D, Chauvaud L, Pecheyran C, Lorrain A, Thébault J, Church TM, Donard OFX (2009) High frequency barium profiles in shells of the Great Scallop *Pecten maximus*: a methodical long-term and multi-site survey in Western Europe. Biogeosciences 6:157–170

Barats A, Amouroux D, Pecheyran C, Chauvaud L, Thébault J, Donard OFX (2010) Spring molybdenum enrichment in scallop shells: a potential tracer of diatom productivity in temperate coastal environments (Brittany, NW France). Biogeosciences 7:233–245

Beierlein L, Salvigsen O, Schöne BR, Mackensen A, Brey T (2015) The seasonal water temperature cycle in the Arctic Dicksonfjord (Svalbard) during the Holocene climate optimum derived from subfossil *Arctica islandica* shells. The Holocene 25:1197–1207

Bellotto VR, Miekeley N (2007) Trace metals in mussel shells and corresponding soft tissue samples: a validation experiment for the use of *Perna perna* shells in pollution monitoring. Anal Bioanal Chem 389:769–776

Betts RA, Jones CD, Knight JR, Keeling RF, Kennedy JJ (2016) El Nino and a record CO2 rise. Nat Clim Chang 6:806–810

Bird DW, Bird RLB (1997) Contemporary shellfish gathering strategies among the Meriam of the Torres Strait islands, Australia: testing predictions of a central place foraging model. J Archaeol Sci 24:39–63

Black BA (2009) Climate-driven synchrony across tree, bivalve, and rockfish growth-increment chronologies of the Northeast Pacific. Mar Ecol Prog Ser 378:37–46

Black BA, Copenheaver CA, Frank DC, Stuckey MJ, Kormanyos RE (2009) Multi-proxy reconstructions of northeastern Pacific Sea surface temperature data from trees and Pacific geoduck. Palaeogeogr Palaeoclimatol Palaeoecol 278:40–47

Black BA, Sydeman WJ, Frank DC, Griffin D, Stahle DW, Garcia-Reyes M, Rykaczewski RR, Bograd SJ, Peterson WT (2014) Six centuries of variability and extremes in a coupled marine-terrestrial ecosystem. Science 345:1498–1502

Black H, Andrus C, Lambert W, Rick T, Gillikin D (2017) $\delta^{15}N$ values in *Crassostrea virginica* shells provides early direct evidence for nitrogen loading to Chesapeake Bay. Sci Rep 7

Boden TA, Marland G, Andres RJ (2015) Global, regional, and national fossil-fuel CO2 emissions. Carbon Dioxide Information Analysis Center, Oak Ridge National Laboratory, U.S. Department of Energy, Oak Ridge

Bougeois L, De Rafelis M, Reichart G-J, De Nooijer LJ, Nicollin F, Dupont-Nivet G (2014) A high resolution study of trace elements and stable isotopes in oyster shells to estimate Central Asian Middle Eocene seasonality. Chem Geol 363:200–212

Bourgoin BP (1990) *Mytilus-Edulis* shell as a bioindicator of lead pollution – considerations on bioavailability and variability. Mar Ecol Prog Ser 61:253–262

Bourgoin B, Risk M (1987) Historical changes in lead in the Eastern Canadian Arctic, determined from fossil and modern *Mya truncata* shells. Sci Total Environ 67:287–291

Bouzouggar A, Barton N, Vanhaeren M, D'Errico F, Collcutt S, Higham T, Hodge E, Parfitt S, Rhodes E, Schwenninger J-L, Stringer C, Turner E, Ward S, Moutmir A, Stambouli A (2007) 82,000-year-old shell beads from North Africa and implications for the origins of modern human behavior. Proc Nat Acad Sci U S A 104:9964–9969

Breitenmoser P, Broennimann S, Frank D (2014) Forward modelling of tree-ring width and comparison with a global network of tree-ring chronologies. Clim Past 10:437–449

Brey T, Mackensen A (1997) Stable isotopes prove shell growth bands in the Antarctic bivalve *Laternula elliptica* to be formed annually. Polar Biol 17:465–468

Brocas WM, Reynolds DJ, Butler PG, Richardson CA, Scourse JD, Ridgway ID, Ramsay K (2013) The dog cockle, *Glycymeris glycymeris* (L.), a new annually-resolved sclerochronological archive for the Irish Sea. Palaeogeogr Palaeoclimatol Palaeoecol 373:133–140

Burchell M, Cannon A, Hallmann N, Schwarcz H, Schöne B (2013) Inter-site variability in the season of shellfish collection on the central coast of British Columbia. J Archaeol Sci 40:626–636

Butler PG, Scourse JD, Richardson CA, Wanamaker AD Jr, Bryant CL, Bennell JD (2009) Continuous marine radiocarbon reservoir calibration and the ^{13}C Suess effect in the Irish Sea: results from the first multi-centennial shell-based marine master chronology. Earth Planet Sci Lett 279:230–241

Butler PG, Richardson CA, Scourse JD, Wanamaker AD Jr, Shammon TM, Bennell JD (2010) Marine climate in the Irish Sea: analysis of a 489-year marine master chronology derived from growth increments in the shell of the clam *Arctica islandica*. Quat Sci Rev 29:1614–1632

Butler PG, Wanamaker AD Jr, Scourse JD, Richardson CA, Reynolds DJ (2011) Long-term stability of $\delta^{13}C$ with respect to biological age in the aragonite shell of mature specimens of the bivalve mollusk *Arctica islandica*. Palaeogeogr Palaeoclimatol Palaeoecol 302:21–30

Butler PG, Wanamaker AD Jr, Scourse JD, Richardson CA, Reynolds DJ (2013) Variability of marine climate on the North Icelandic shelf in a 1357-year proxy archive based on growth increments in the bivalve *Arctica islandica*. Palaeogeogr Palaeoclimatol Palaeoecol 373:141–151

Cannon A, Burchell M (2017) Reconciling oxygen isotope sclerochronology with interpretations of millennia of seasonal shellfish collection on the Pacific Northwest Coast. Quat Int 427:184–191

Cariou E, Guivel C, La C, Lenta L, Elliot M (2017) Lead accumulation in oyster shells, a potential tool for environmental monitoring. Mar Pollut Bull 125:19–29

Carmichael RH, Hattenrath T, Valiela I, Michener RH (2008) Nitrogen stable isotopes in the shell of *Mercenaria mercenaria* trace wastewater inputs from watersheds to estuarine ecosystems. Aquat Biol 4:99–111

Carré M, Bentaleb I, Blamart D, Ogle N, Cardenas F, Zevallos S, Kalin RM, Ortlieb L, Fontugne M (2005) Stable isotopes and sclerochronology of the bivalve *Mesodesma donacium*: potential application to Peruvian paleoceanographic reconstructions. Palaeogeogr Palaeoclimatol Palaeoecol 228:4–25

Carré M, Bentaleb I, Bruguier O, Ordinola E, Barrett NT, Fontugne M (2006) Calcification rate influence on trace element concentrations in aragonitic bivalve shells: evidences and mechanisms. Geochim Cosmochim Acta 70:4906–4920

Carré M, Klaric L, Lavallee D, Julien M, Bentaleb I, Fontugne M, Kawka O (2009) Insights into early Holocene hunter-gatherer mobility on the Peruvian Southern Coast from mollusk gathering seasonality. J Archaeol Sci 36:1173–1178

Carré M, Sachs JP, Purca S, Schauer AJ, Braconnot P, Falcon RA, Julien M, Lavallee D (2014) Holocene history of ENSO variance and asymmetry in the eastern tropical Pacific. Science 345:1045–1048

Carré M, Jackson D, Maldonado A, Chase BM, Sachs JP (2016) Variability of C-14 reservoir age and air-sea flux of CO2 in the Peru-Chile upwelling region during the past 12,000 years. Quat Res 85:87–93

Carriker MR, Palmer RE, Sick LV, Johnson CC (1980) Interaction of mineral elements in seawater and shell of oysters (*Crassostrea virginica* (Gmelin)) cultured in controlled and natural systems. J Exp Mar Biol Ecol 46:279–296

Carroll ML, Johnson BJ, Henkes GA, Mcmahon KW, Voronkov A, Ambrose WG Jr, Denisenko SG (2009) Bivalves as indicators of environmental variation and potential anthropogenic impacts in the southern Barents Sea. Mar Pollut Bull 59:193–206

Chauvaud L, Lorrain A, Dunbar RB, Paulet YM, Thouzeau G, Jean F, Guarini JM, Mucciarone D (2005) Shell of the great scallop *Pecten maximus* as a high-frequency archive of paleoenvironmental changes. Geochemistry Geophysics Geosystems 6:Q08001

Chow TJ, Snyder HG, Snyder CB (1976) Mussels (mytilus sp.) as an indicator of lead pollution. Sci Total Environ 6:55–63

Claassen C (1998) Shells. Cambridge University Press, Cambridge

Coupland G, Bilton D, Clark T, Cybulski JS, Frederick G, Holland A, Letham B, Williams BG (2017) A wealth of beads: evidence for material wealth-based inequality in the Salish Sea region, 400-3500 Cal. BP. Am Antiq 81:294–315

Dahlgren TG, Weinberg JR, Halanych KM (2000) Phylogeography of the ocean quahog (*Arctica islandica*): influences of paleoclimate on genetic diversity and species range. Mar Biol 137:487–495

De Garine I, Harrison GA (1988) Coping with uncertainty in food supply. Clarendon Press, Oxford

De Ridder F, Pintelon R, Schoukens J, Gillikin D, Andre L, Baeyens W, De Brauwere A, Dehairs F (2004) Decoding nonlinear growth rates in biogenic environmental archives. Geochemistry Geophysics Geosystems 5:Q12015

Deith MR (1986) Subsistence strategies at a Mesolithic camp site – evidence from stable isotope analyses of shells. J Archaeol Sci 13:61–78

Der Sarkassian C, Picheraeu V, Dupont C, Ilsøe PC, Perrigault M, Butler PG, Chauvaud L, Eiriksson J, Scourse JD, Paillard C, Orlando L (2017) Ancient DNA analysis identifies marine mollusc shells as new metagenomic archives of the past. Mol Ecol Resour 17:835–853

Dettman DL, Reische AK, Lohmann KC (1999) Controls on the stable isotope composition of seasonal growth bands in aragonitic fresh-water bivalves (unionidae). Geochim Cosmochim Acta 63:1049–1057

Dettman DL, Flessa KW, Roopnarine PD, Schöne BR, Goodwin DH (2004) The use of oxygen isotope variation in shells of estuarine mollusks as a quantitative record of seasonal and annual Colorado River discharge. Geochim Cosmochim Acta 68:1253–1263

Dreier A, Stannek L, Blumenberg M, Taviani M, Sigovini M, Wrede C, Thiel V, Hoppert M (2012) The fingerprint of chemosymbiosis: origin and preservation of isotopic biosignatures in the nonseep bivalve *Loripes lacteus* compared with Venerupis aurea. FEMS Microbiol Ecol 81:480–493

Driscoll R, Elliot M, Russon T, Welsh K, Yokoyama Y, Tudhope A (2014) ENSO reconstructions over the past 60 ka using giant clams (Tridacna sp.) from Papua New Guinea. Geophys Res Lett 41:6819–6825

Dunca E, Mutvei H, Goransson P, Morth C-M, Schöne BR, Whitehouse MJ, Elfman M, Baden SP (2009) Using ocean quahog (*Arctica islandica*) shells to reconstruct palaeoenvironment in A-resund, Kattegat and Skagerrak, Sweden. Int J Earth Sci 98:3–17

Duncan PF, Ghys A (2019) Shells as collectors' items. In: Smaal A et al (eds) Goods and services of marine bivalves. Springer, Cham, pp 381–411

Duprey N, Lazareth CE, Correge T, Le Cornec F, Maes C, Pujol N, Madeng-Yogo M, Caquineau S, Derome CS, Cabioch G (2012) Early mid-Holocene SST variability and surface-ocean water balance in the southwest Pacific. Paleoceanography 27:12

Duprey N, Galipaud JC, Cabioch G, Lazareth CE (2014) Isotopic records from archeological giant clams reveal a variable climate during the southwestern Pacific colonization ca. 3.0 ka BP. Palaeogeogr Palaeoclimatol Palaeoecol 404:97–108

Duprey N, Lazareth CE, Dupouy C, Butscher J, Farman R, Maes C, Cabioch G (2015) Calibration of seawater temperature and $\delta^{18}O$ (seawater) signals in Tridacna maxima's $\delta^{18}O$ (shell) record based on in situ data. Coral Reefs 34:437–450

Eagle R, Eiler J, Tripati A, Ries J, Freitas P, Hiebenthal C, Wanamaker A, Taviani M, Elliot M, Marenssi S, Nakamura K, Ramirez P, Roy K (2013) The influence of temperature and seawater carbonate saturation state on ^{13}C-^{18}O bond ordering in bivalve mollusks. Biogeosciences 10:4591–4606

Eerkens JW, Schwitalla AW, Spero HJ, Nesbit R (2016) Staple, feasting, or fallback food? Mussel harvesting among hunter-gatherers in interior central California. J Ethnobiol 36:476–492

Ekwurzel B, Schlosser P, Mortlock RA, Fairbanks RG, Swift JH (2001) River runoff, sea ice meltwater, and Pacific water distribution and mean residence times in the Arctic Ocean. Journal of Geophysical Research-Oceans 106:9075–9092

Elliot M, Welsh K, Chilcott C, McCulloch M, Chappell J, Ayling B (2009) Profiles of trace elements and stable isotopes derived from giant long-lived *Tridacna gigas* bivalves: potential applications in paleoclimate studies. Palaeogeogr Palaeoclimatol Palaeoecol 280:132–142

Ellis GS, Herbert G, Hollander D (2014) Reconstructing carbon sources in a dynamic estuarine ecosystem using oyster amino acid $\delta^{13}C$ values from shell and tissue. J Shellfish Res 33:217–225

Emile-Geay J, Cobb KM, Carré M, Braconnot P, Leloup J, Zhou Y, Harrison SP, Correge T, Mcgregor HV, Collins M, Driscoll R, Elliot M, Schneider B, Tudhope A (2016) Links between tropical Pacific seasonal, interannual and orbital variability during the Holocene. Nat Geosci 9:168–173

Epstein S, Buchsbaum R, Lowenstam HA, Urey HC (1953) Revised carbonate-water isotopic temperature scale. Bull Geol Soc Am 64:1315–1326

Erlandson JM (2001) The archaeology of aquatic adaptations: paradigms for a new millennium. J Archaeol Res 9:287–350

Evans MN, Kaplan A, Cane MA (1998) Optimal sites for coral-based reconstruction of global sea surface temperature. Paleoceanography 13:502–516

Fang M, Li X (2016) Paleoclimate data assimilation: its motivation, progress and prospects. Sci Chin-Earth Sci 59:1817–1826

Fitzhugh B (1995) Clams and the Kachemak: seasonal shellfish use on Kodiak Island, Alaska (1200–800 BP). Res Econ Anthropol 16:129–176

Fontugne M, Carré M, Bentaleb I, Julien M, Lavallee D (2004) Radiocarbon reservoir age variations in the south Peruvian upwelling during the holocene. Radiocarbon 46:531–537

Fortunato H (2014) Mollusks: tools in environmental and climate research. Am Malacol Bull 33:310–324

Francey RJ, Allison CE, Etheridge DM, Trudinger CM, Enting IG, Leuenberger M, Langenfelds RL, Michel E, Steele LP (1999) A 1000-year high precision record of $\delta^{13}C$ in atmospheric CO_2. Tellus Ser B-Chem Phys Meteorol 51:170–193

Freitas PS, Clarke LJ, Kennedy HA, Richardson CA (2008) Inter- and intra-specimen variability masks reliable temperature control on shell Mg/Ca ratios in laboratory- and field-cultured *Mytilus edulis* and *Pecten maximus* (bivalvia). Biogeosciences 5:1245–1258

Freitas PS, Clarke LJ, Kennedy H, Richardson CA (2009) Ion microprobe assessment of the heterogeneity of Mg/Ca, Sr/Ca and Mn/Ca ratios in *Pecten maximus* and *Mytilus edulis* (bivalvia) shell calcite precipitated at constant temperature. Biogeosciences 6:1209–1227

Freitas PS, Clarke LJ, Kennedy H, Richardson CA (2016) Manganese in the shell of the bivalve *Mytilus edulis*: seawater Mn or physiological control? Geochim Cosmochim Acta 194:266–278

Fritz LW, Ragone LM, Lutz RA, Swapp S (1990) Biomineralization of barite in the shell of the freshwater Asiatic clam *Corbicula fluminea* (Mollusca, Bivalvia). Limnol Oceanogr 35:756–762

Fuge R, Palmer TJ, Pearce NJG, Perkins WT (1993) Minor and trace-element chemistry of modern shells – a laser ablation inductively coupled plasma mass-spectrometry study. Appl Geochem Suppl 2:111–116

Gesamp (2001) A sea of troubles. (IMO/FAO/UNESCO-IOC/WMO/WHGESAMP (IMO/ FAO/ UNESCO-IOC/WMO/WHO/IAEA/UN/UNEP Joint Group of Experts on the Scientific Aspects of Marine Environmental Protection) and Advisory Committee on Protection of the Sea

Gillikin DP, De Ridder F, Ulens H, Elskens M, Keppens E, Baeyens W, Dehairs F (2005a) Assessing the reproducibility and reliability of estuarine bivalve shells (*Saxidomus giganteus*) for sea surface temperature reconstruction: implications for paleoclimate studies. Palaeogeogr Palaeoclimatol Palaeoecol 228:70–85

Gillikin DP, Dehairs F, Baeyens W, Navez J, Lorrain A, Andre L (2005b) Inter- and intra-annual variations of Pb/ca ratios in clam shells (*Mercenaria mercenaria*): a record of anthropogenic lead pollution? Mar Pollut Bull 50:1530–1540

Gillikin DP, Lorrain A, Navez J, Taylor JW, Keppens E, Baeyens W, Dehairs F (2005c) Strong biological controls on Sr/Ca ratios in aragonitic marine bivalve shells. Geochem Geophys Geosyst 6:Q05009–Q05009

Gillikin DP, Lorrain A, Bouillon S, Willenz P, Dehairs F (2006) Stable carbon isotopic composition of *Mytilus edulis* shells: relation to metabolism, salinity, $\delta^{13}C$ (DIC) and phytoplankton. Org Geochem 37:1371–1382

Gillikin DP, Lorrain A, Meng L, Dehairs F (2007) A large metabolic carbon contribution to the $\delta^{13}C$ record in marine aragonitic bivalve shells. Geochim Cosmochim Acta 71:2936–2946

Gillikin DP, Lorrain A, Paulet Y-M, Andre L, Dehairs F (2008) Synchronous barium peaks in high-resolution profiles of calcite and aragonite marine bivalve shells. Geo-Mar Lett 28:351–358

Gillikin DP, Lorrain A, Jolivet A, Kelemen Z, Chauvaud L, Bouillon S (2017) High-resolution nitrogen stable isotope sclerochronology of bivalve shell carbonate-bound organics. Geochim Cosmochim Acta 200:55–66

Goodwin DH, Flessa KW, Schöne BR, Dettman DL (2001) Cross-calibration of daily growth increments, stable isotope variation, and temperature in the Gulf of California bivalve mollusk *Chione cortezi*: implications for paleoenvironmental analysis. PALAIOS 16:387–398

Goodwin DH, Schöne BR, Dettman DL (2003) Resolution and fidelity of oxygen isotopes as paleotemperature proxies in bivalve mollusk shells: models and observations. PALAIOS 18:110–125

Goodwin D, Paul P, Wissink C (2009) MoGroFunGen: a numerical model for reconstructing intra-annual growth rates of bivalve molluscs. Palaeogeogr Palaeoclimatol Palaeoecol 276:47–55

Graniero LE, Grossman EL, O'Dea A (2016) Stable isotopes in bivalves as indicators of nutrient source in coastal waters in the Bocas del Toro Archipelago, Panama. Peerj 4:e2278–e2278

Grossman EL, Ku TL (1986) Oxygen and carbon isotope fractionation in biogenic aragonite – temperature effects. Chem Geol 59:59–74

Habu J, Matsui A, Yamamoto N, Kanno T (2011) Shell midden archaeology in Japan: aquatic food acquisition and long-term change in the Jomon culture. Quat Int 239:19–27

Hahn S, Griesshaber E, Schmahl WW, Neuser RD, Ritter A-C, Hoffmann R, Buhl D, Niedermayr A, Geske A, Immenhauser A (2014) Exploring aberrant bivalve shell ultrastructure and geochemistry as proxies for past sea water acidification. Sedimentology 61:1625–1658

Hallmann N, Burchell M, Schöne BR, Irvine GV, Maxwell D (2009) High-resolution sclerochronological analysis of the bivalve mollusk *Saxidomus gigantea* from Alaska and British Columbia:

techniques for revealing environmental archives and archaeological seasonality. J Archaeol Sci 36:2353–2364
Harding JM, King SE, Powell EN, Mann R (2008) Decadal trends in age structure and recruitment patterns of ocean quahogs *Arctica islandica* from the Mid-Atlantic Bight in relation to water temperature. J Shellfish Res 27:667–690
Hardy K, Camara A, Pique R, Dioh E, Gueye M, Diadhiou HD, Faye M, Carré M (2016) Shellfishing and shell midden construction in the Saloum Delta, Senegal. J Anthropol Archaeol 41:19–32
Hare SR, Mantua NJ (2000) Empirical evidence for North Pacific regime shifts in 1977 and 1989. Prog Oceanogr 47:103–145
Helama S, Hood BC (2011) Stone Age midden deposition assessed by bivalve sclerochronology and radiocarbon wiggle-matching of *Arctica islandica* shell increments. J Archaeol Sci 38:452–460
Helser TE, Lai H-L, Black BA (2012) Bayesian hierarchical modeling of Pacific geoduck growth increment data and climate indices. Ecol Model 247:210–220
Henshilwood C, Nilssen P, Parkington J (1994) Mussel drying and food storage in the late Holocene, SW Cape, South-Africa. J Field Archaeol 21:103–109
Holland H, Schöne B, Marali S, Jochum K (2014) History of bioavailable lead and iron in the Greater North Sea and Iceland during the last millennium – a bivalve sclerochronological reconstruction. Mar Pollut Bull 87:104–116
Holmden C, Hudson J (2003) ^{87}Sr-^{86}Sr and Sr/Ca investigation of Jurassic mollusks from Scotland: implications for paleosalinities and the Sr/Ca ratio of seawater. Geol Soc Am Bull 115:1249–1264
Ingram BL, Southon JR (1996) Reservoir ages in eastern Pacific coastal and estuarine waters. Radiocarbon 38:573–582
IPCC (2014) Summary for policymakers. In: Field CB, Barros VR, Dokken DJ, Mach KJ, Mastrandrea MD, Bilir TE, Chatterjee M, Ebi KL, Estrada YO, Genova RC, Girma B, Kissel ES, Levy AN, MacCracken S, Mastrandrea PR, White LL (eds) Climate change 2014: impacts, adaptation, and vulnerability. Part A: global and sectoral aspects. Contribution of Working Group II to the Fifth Assessment Report of the Intergovernmental Panel on Climate Change. Cambridge University Press, Cambridge/New York, pp 1–32
Izzo C, Doubleday ZA, Grammer GL, Disspain MCF, Ye Q, Gillanders BM (2017) Seasonally resolved environmental reconstructions using fish otoliths. Can J Fish Aquat Sci 74:23–31
Jew NP, Erlandson JM, Watts J, White FJ (2013) Shellfish, seasonality, and stable isotope sampling: ^{18}O analysis of mussel shells from an 8,800-year-old shell midden on California's Channel Islands. J Island Coast Archaeol 8:170–189
Jolivet A, Asplin L, Strand O, Thébault J, Chauvaud L (2015) Coastal upwelling in Norway recorded in great scallop shells. Limnol Oceanogr 60:1265–1275
Jones DS (1983) Sclerochronology – reading the record of the molluscan Shell. Am Sci 71:384–391
Jones DS, Quitmyer IR (1996) Marking time with bivalve shells: oxygen isotopes and season of annual increment formation. PALAIOS 11:340–346
Kaandorp RJG, Vonhof HB, Wesselingh FP, Pittman LR, Kroon D, Van Hinte JE (2005) Seasonal Amazonian rainfall variation in the Miocene climate optimum. Palaeogeogr Palaeoclimatol Palaeoecol 221:1–6
Kennedy H, Richardson CA, Duarte CM, Kennedy DP (2001) Oxygen and carbon stable isotopic profiles of the fan mussel, *Pinna nobilis*, and reconstruction of sea surface temperatures in the Mediterranean. Mar Biol 139:1115–1124
Khim BK, Krantz DE, Cooper LW, Grebmeier JM (2003) Seasonal discharge of estuarine freshwater to the western Chukchi Sea shelf identified in stable isotope profiles of mollusk shells. J Geophys Res Oceans 108:3300–3300
Kilada RW, Roddick D, Mombourquette K (2007) Age determination, validation, growth and minimum size of sexual maturity of the Greenland smooth cockle (*Serripes groenlandicus*, Bruguiere, 1789) in eastern Canada. J Shellfish Res 26:443–450

Killingley JS (1981) Seasonality of mollusk collecting determined from [18]Oprofiles of midden shells. Am Antiq 46:152–158
Killingley JS, Berger WH (1979) Stable isotopes in a mollusk shell – detection of upwelling events. Science 205:186–188
Kim ST, O'Neil JR (1997) Equilibrium and nonequilibrium oxygen isotope effects in synthetic carbonates. Geochim Cosmochim Acta 61:3461–3475
Klein RT, Lohmann KC, Thayer CW (1996) Sr/Ca and C-13/C-12 ratios in skeletal calcite of *Mytilus trossulus*: covariation with metabolic rate, salinity, and carbon isotopic composition of seawater. Geochim Cosmochim Acta 60:4207–4221
Klunder M, Hippler D, Witbaard R, Frei D (2008) Laser ablation analysis of bivalve shells – archives of environmental information. Geol Surv Denmark Greenland Bull:89–92
Koide M, Lee DS, Goldberg ED (1982) Metal and transuranic records in Mussel shells, byssal threads and tissues. Estuar Coast Shelf Sci 15:679–695
Kovacs CJ, Daskin JH, Patterson H, Carmichael RH (2010) *Crassostrea virginica* shells record local variation in wastewater inputs to a coastal estuary. Aquat Biol 9:77–84
Krantz DE, Jones DS, Williams DF (1984) Growth-rates of the sea scallop, *Placopecten-Magellanicus*, determined from the $\delta^{18}O$ record in shell calcite. Biol Bull 167:186–199
Krause-Nehring J, Brey T, Thorrold SR (2012) Centennial records of lead contamination in northern Atlantic bivalves (*Arctica islandica*). Mar Pollut Bull 64:233–240
Kyle R, Pearson B, Fielding PJ, Robertson WD, Birnie SL (1997) Subsistence shellfish harvesting in the Maputaland marine reserve in northern KwaZulu-Natal, South Africa: rocky shore organisms. Biol Conserv 82:183–192
Lazareth CE, Le Cornec F, Candaudap F, Freydier R (2013) Trace element heterogeneity along isochronous growth layers in bivalve shell: consequences for environmental reconstruction. Palaeogeogr Palaeoclimatol Palaeoecol 373:39–49
Levin LA, Hoenisch B, Frieder CA (2015) Geochemical proxies for estimating faunal exposure to ocean acidification. Oceanography 28:62–73
Liehr GA, Zettler ML, Leipe T, Witt G (2005) The ocean quahog *Arctica islandica* L.: a bioindicator for contaminated sediments. Mar Biol 147:671–679
Lightfoot KG, Cerrato RM, Wallace HVE (1993) Prehistoric shellfish-harvesting strategies – implications from the growth-patterns of soft-shell clams (*Mya arenaria*). Antiquity 67:358–369
Lingard SM, Evans RD, Bourgoin BP (1992) Method for the estimation of organic-bound and crystal-bound metal concentrations in bivalve shells. Bull Environ Contam Toxicol 48:179–184
Liu W, Xie S-P, Liu Z, Zhu J (2017) Overlooked possibility of a collapsed Atlantic meridional overturning circulation in warming climate. Sci Adv 3:e1601666
Liu YW, Aciego SM, Wanamaker AD Jr (2015) Environmental controls on the boron and strontium isotopic composition of aragonite shell material of cultured *Arctica islandica*. Biogeosciences 12:3351–3368
Loader NJ, Young GHF, Grudd H, McCarroll D (2013) Stable carbon isotopes from Tornetrask, northern Sweden provide a millennial length reconstruction of summer sunshine and its relationship to Arctic circulation. Quat Sci Rev 62:97–113
Lorens RB, Bender ML (1977) Physiological exclusion of magnesium from Mytilus-edulis calcite. Nature 269:793–794
Macfarlane GR, Markich SJ, Linz K, Gifford S, Dunstan RH, O'connor W, Russell RA (2006) The Akoya pearl oyster shell as an archival monitor of lead exposure. Environ Pollut 143:166–173
Mae A, Yamanaka T, Shimoyama S (2007) Stable isotope evidence for identification of chemosynthesis-based fossil bivalves associated with cold-seepages. Palaeogeogr Palaeoclimatol Palaeoecol 245:411–420
Marali S, Schöne B (2015) Oceanographic control on shell growth of *Arctica islandica* (Bivalvia) in surface waters of Northeast Iceland – implications for paleoclimate reconstructions. Palaeogeogr Palaeoclimatol Palaeoecol 420:138–149
Marin F, Le Roy N, Marie B (2012) The formation and mineralization of mollusk shell. Front Biosci S4:1099–1125

Markich SJ, Jeffree RA, Burke PT (2002) Freshwater bivalve shells as archival indicators of metal pollution from a copper-uranium mine in tropical northern Australia. Environ Sci Technol 36:821–832

Marquardt WH (2010) Shell mounds in the southeast: middens, monuments, temple mounds, rings, or works? Am Antiq 75:551–570

Marsh R, Petrie B, Weidman CR, Dickson RR, Loder JW, Hannah CG, Frank K, Drinkwater K (1999) The 1882 tilefish kill – a cold event in shelf waters off the North-Eastern United States? Fish Oceanogr 8:39–49

Mazloumi N, Burch P, Fowler AJ, Doubleday ZA, Gillanders BM (2017) Determining climate-growth relationships in a temperate fish: a sclerochronological approach. Fish Res 186:319–327

Mcconnaughey TA, Burdett J, Whelan JF, Paull CK (1997) Carbon isotopes in biological carbonates: respiration and photosynthesis. Geochim Cosmochim Acta 61:611–622

Mette MJ, Wanamaker AD Jr, Carroll ML, Ambrose WG Jr, Retelle MJ (2016) Linking large-scale climate variability with *Arctica islandica* shell growth and geochemistry in northern Norway. Limnol Oceanogr 61:748–764

Milano S, Schöne BR, Wang S, Mueller WE (2016) Impact of high pCO(2) on shell structure of the bivalve Cerastoderma edule. Mar Environ Res 119:144–155

Milano S, Schšne BR, Witbaard R (2017) Changes of shell microstructural characteristics of *Cerastoderma edule* (Bivalvia) – a novel proxy for water temperature. Palaeogeogr Palaeoclimatol Palaeoecol 465:395–406

Milner N (2001) At the cutting edge: using thin sectioning to determine season of death of the European oyster, Ostrea edulis. J Archaeol Sci 28:861–873

Minobe S (1997) A 50–70 year climatic oscillation over the North Pacific and North America. Geophys Res Lett 24:683–686

Mook WG, Vogel JC (1968) Isotopic equilibrium betweenshells and their environment. Science 159:874–875

Morrongiello JR, Thresher RE (2015) A statistical framework to explore ontogenetic growth variation among individuals and populations: a marine fish example. Ecol Monogr 85:93–115

Morrongiello JR, Thresher RE, Smith DC (2012) Aquatic biochronologies and climate change. Nat Clim Chang 2:849–857

Moss M (2011) Northwest coast: archaeology as deep history. SAA Press, Washington, DC

Mueller P, Taylor MH, Klicpera A, Wu HC, Michel J, Westphal H (2015) Food for thought: mathematical approaches for the conversion of high-resolution sclerochronological oxygen isotope records into sub-annually resolved time series. Palaeogeogr Palaeoclimatol Palaeoecol 440:763–776

Mueller-Lupp T, Erlenkeuser H, Bauch HA (2003) Seasonal and interannual variability of Siberian river discharge in the Laptev Sea inferred from stable isotopes in modern bivalves. Boreas 32:292–303

Muller-Lupp T, Bauch H (2005) Linkage of Arctic atmospheric circulation and Siberian shelf hydrography: a proxy validation using $\delta^{18}O$ records of bivalve shells. Glob Planet Chang 48:175–186

Nicholson S, Szefer P (2003) Accumulation of metals in the soft tissues, byssus and shell of the mytilid mussel Perna viridis (Bivalvia : Mytilidae) from polluted and uncontaminated locations in Hong Kong coastal waters. Mar Pollut Bull 46:1039–1043

Nishida K, Ishimura T, Suzuki A, Sasaki T (2012) Seasonal changes in the shell microstructure of the bloody clam, Scapharca broughtonii (Mollusca: Bivalvia: Arcidae). Palaeogeogr Palaeoclimatol Palaeoecol 363:99–108

O'Donnell TH, Macko SA, Chou J, Davis-Hartten KL, Wehmiller JF (2003) Analysis of $\delta^{13}C$, $\delta^{15}N$, and $\delta^{34}S$ in organic matter from the biominerals of modern and fossil Mercenaria spp. Org Geochem 34:165–183

O'Donnell TH, Macko SA, Wehmiller JF (2007) Stable carbon isotope composition of amino acids in modern and fossil Mercenaria. Org Geochem 38:485–498

O'Neil DD, Gillikin DP (2014) Do freshwater mussel shells record road-salt pollution? Sci Rep 4:7168–7168

Okumara M, Eggers S (2014) The cultural dynamics of shell-matrix sites. In: Roksandic M, De Souza SM, Eggers S, Burchell M, Klokler D (eds) The cultural dynamics of shell-matrix sites. University of New Mexico Press, Albuquerque

Orlando L, Ginolhac A, Zhang G, Froese D, Albrechtsen A, Stiller M, Schubert M, Cappellini E, Petersen B, Moltke I, Johnson PLF, Fumagalli M, Vilstrup JT, Raghavan M, Korneliussen T, Malaspinas A-S, Vogt J, Szklarczyk D, Kelstrup CD, Vinther J, Dolocan A, Stenderup J, Velazquez AMV, Cahill J, Rasmussen M, Wang X, Min J, Zazula GD, Seguin-Orlando A, Mortensen C, Magnussen K, Thompson JF, Weinstock J, Gregersen K, Roed KH, Eisenmann V, Rubin CJ, Miller DC, Antczak DF, Bertelsen MF, Brunak S, Al-Rasheid KAS, Ryder O, Andersson L, Mundy J, Krogh A, Gilbert MTP, Kjaer K, Sicheritz-Ponten T, Jensen LJ, Olsen JV, Hofreiter M, Nielsen R, Shapiro B, Wang J, Willerslev E (2013) Recalibrating Equus evolution using the genome sequence of an early middle Pleistocene horse. Nature 499:74–78

Pearce NJG, Mann VL (2006) Trace metal variations in the shells of Ensis siliqua record pollution and environmental conditions in the sea to the west of mainland Britain. Mar Pollut Bull 52:739–755

Perez-Huerta A, Etayo-Cadavid M, Andrus C, Jeffries T, Watkins C, Street S, Sandweiss D (2013) El Nino impact on mollusk biomineralization-implications for trace element proxy reconstructions and the paleo-archeological record. PLoS One 8:e54274

Phipps SJ, Mcgregor HV, Gergis J, Gallant AJE, Neukom R, Stevenson S, Ackerley D, Brown JR, Fischer MJ, Van Ommen TD (2013) Paleoclimate data-model comparison and the role of climate forcings over the past 1500 years. J Clim 26:6915–6936

Pitts LC, Wallace GT (1994) Lead deposition in the shell of the bivalve, Mya-arenaria – an indicator of dissolved lead in seawater. Estuarine Coastal and Shelf Science 39:93–104

Poulain C, Gillikin DP, Thébault J, Munaron JM, Bohn M, Robert R, Paulet YM, Lorrain A (2015) An evaluation of Mg/Ca, Sr/Ca, and Ba/Ca ratios as environmental proxies in aragonite bivalve shells. Chem Geol 396:42–50

Prendergast A, Schöne B (2017) Oxygen isotopes from limpet shells: implications for palaeothermometry and seasonal shellfish foraging studies in the Mediterranean. Palaeogeogr Palaeoclimatol Palaeoecol 484:33–47

Price GD, Pearce NJG (1997) Biomonitoring of pollution by Cerastoderma edule from the British isles: a laser ablation ICP-MS study. Mar Pollut Bull 34:1025–1031

Protasowicki M, Dural M, Jaremek J (2008) Trace metals in the shells of blue mussels (Mytilus edulis) from the Poland coast of Baltic Sea. Environ Monit Assess 141:329–337

Pruefer K, Racimo F, Patterson N, Jay F, Sankararaman S, Sawyer S, Heinze A, Renaud G, Sudmant PH, De Filippo C, Li H, Mallick S, Dannemann M, Fu Q, Kircher M, Kuhlwilm M, Lachmann M, Meyer M, Ongyerth M, Siebauer M, Theunert C, Tandon A, Moorjani P, Pickrell J, Mullikin JC, Vohr SH, Green RE, Hellmann I, Johnson PLF, Blanche H, Cann H, Kitzman JO, Shendure J, Eichler EE, Lein ES, Bakken TE, Golovanova LV, Doronichev VB, Shunkov MV, Derevianko AP, Viola B, Slatkin M, Reich D, Kelso J, Paeaebo S (2014) The complete genome sequence of a Neanderthal from the Altai Mountains. Nature 505:43–49

Puente X, Villares R, Carral E, Carballeira A (1996) Nacreous shell of Mytilus galloprovincialis as a biomonitor of heavy metal pollution in Galiza (NW Spain). Sci Total Environ 183:205–211

Pyrina M, Wagner S, Zorita E (2017) Pseudo-proxy evaluation of climate field reconstruction methods of North Atlantic climate based on an annually resolved marine proxy network. Clim Past 13:1339–1354

Rahmstorf S, Box JE, Feulner G, Mann ME, Robinson A, Rutherford S, Schaffernicht EJ (2015) Exceptional twentieth-century slowdown in Atlantic Ocean overturning circulation. Nat Clim Chang 5:475–480

Raith A, Perkins WT, Pearce NJG, Jeffries TE (1996) Environmental monitoring on shellfish using UV laser ablation ICP-MS. Fresen J Anal Chem 355:789–792

Reynolds DJ, Butler PG, Williams SM, Scourse JD, Richardson CA, Wanamaker AD Jr, Austin WEN, Cage AG, Sayer MDJ (2013) A multiproxy reconstruction of Hebridean (NW Scotland)

spring sea surface temperatures between AD 1805 and 2010. Palaeogeogr Palaeoclimatol Palaeoecol 386:275–285

Reynolds DJ, Richardson CA, Scourse JD, Butler PG, Hollyman P, Roman-Gonzalez A, Hall IR (2017) Reconstructing North Atlantic marine climate variability using an absolutely-dated sclerochronological network. New Res Methods Appl Sclerochronol 465(Part B):333–346

Richardson CA (2001) Molluscs as archives of environmental change. Oceanogr Mar Biol 39:103–164

Ridgway ID, Richardson CA, Scourse JD, Butler PG, Reynolds DJ (2012) The population structure and biology of the ocean quahog, *Arctica islandica*, in Belfast Lough, Northern Ireland. J Mar Biol Assoc U K 92:539–546

Rocha J, Yletyinen J, Biggs R, Blenckner T, Peterson G (2015) Marine regime shifts: drivers and impacts on ecosystems services. Philosophical Transactions of the Royal Society B-Biological Sciences 370:20130273–20130273

Roksandic M, Mendonca De Souze S, Klokler D, Eggers S, Burchell M (2014) Cultural dynamics of shell-matrix sites: diverse perspectives on biological remains from shell mounds and shell middens. University of New Mexico Press, Alburquerque

Royer C, Thébault J, Chauvaud L, Olivier F (2013) Structural analysis and paleoenvironmental potential of dog cockle shells (*Glycymeris glycymeris*) in Brittany, Northwest France. Palaeogeogr Palaeoclimatol Palaeoecol 373:123–132

Sabine CL, Feely RA, Gruber N, Key RM, Lee K, Bullister JL, Wanninkhof R, Wong CS, Wallace DWR, Tilbrook B, Millero FJ, Peng TH, Kozyr A, Ono T, Rios AF (2004) The oceanic sink for anthropogenic CO2. Science 305:367–371

Schettler G, Pearce NJG (1996) Metal pollution recorded in extinct Dreissena polymorpha communities, Lake Breitling, Havel Lakes system, Germany: a laser ablation inductively coupled plasma mass spectrometry study. Hydrobiologia 317:1–11

Schöne B (2008) The curse of physiology – challenges and opportunities in the interpretation of geochemical data from mollusk shells. Geo-Mar Lett 28:269–285

Schöne B (2013) *Arctica islandica* (Bivalvia): a unique paleoenvironmental archive of the northern North Atlantic Ocean. Glob Planet Chang 111:199–225

Schöne B, Giere O (2005) Growth increments and stable isotope variation in shells of the deep-sea hydrothermal vent bivalve mollusk *Bathymodiolus brevior* from the North Fiji Basin, Pacific Ocean. Deep-Sea Research Part I-Oceanographic Research Papers 52:1896–1910

Schöne B, Fiebig J (2009) Seasonality in the North Sea during the Allerod and late medieval climate optimum using bivalve sclerochronology. Int J Earth Sci 98:83–98

Schöne B, Krause R (2016) Retrospective environmental biomonitoring – Mussel Watch expanded. Glob Planet Chang 144:228–251

Schöne B, Oschmann W, Rossler J, Castro A, Houk S, Kroncke I, Dreyer W, Janssen R, Rumohr H, Dunca E (2003) North Atlantic oscillation dynamics recorded in shells of a long-lived bivalve mollusk. Geology 31:1037–1040

Schöne B, Castro A, Fiebig J, Houk S, Oschmann W, Kroncke I (2004) Sea surface water temperatures over the period 1884-1983 reconstructed from oxygen isotope ratios of a bivalve mollusk shell (*Arctica islandica*, southern North Sea). Palaeogeogr Palaeoclimatol Palaeoecol 212:215–232

Schöne B, Fiebig J, Pfeiffer M, Gless R, Hickson J, Johnson A, Dreyer W, Oschmann W (2005a) Climate records from a bivalved methuselah (*Arctica islandica*, Mollusca; Iceland). Palaeogeogr Palaeoclimatol Palaeoecol 228:130–148

Schöne B, Pfeiffer M, Pohlmann T, Siegismund F (2005b) A seasonally resolved bottom-water temperature record for the period ad 1866-2002 based on shells of *Arctica islandica* (Mollusca, North Sea). Int J Climatol 25:947–962

Schöne B, Wanamaker A, Fiebig J, Thébault J, Kreutz K (2011) Annually resolved $\delta^{13}C$ (shell) chronologies of long-lived bivalve mollusks (*Arctica islandica*) reveal oceanic carbon dynamics in the temperate North Atlantic during recent centuries. Palaeogeogr Palaeoclimatol Palaeoecol 302:31–42

Scourse JD, Wanamaker AD Jr, Weidman C, Heinemeier J, Reimer PJ, Butler PG, Witbaard R, Richardson CA (2012) The marine radiocarbon bomb pulse across the temperate North Atlantic: a compilation of Delta C-14 time histories from *Arctica islandica* growth increments. Radiocarbon 54:165–186

Schöne B, Radermacher P, Zhang Z, Jacob D (2013) Crystal fabrics and element impurities (Sr/Ca, Mg/Ca, and Ba/Ca) in shells of *Arctica islandica*-implications for paleoclimate reconstructions. Palaeogeogr Palaeoclimatol Palaeoecol 373:50–59

Shackleton NJ (1973) Oxygen isotope analysis as a means of determining season of occupation of prehistoric midden sites. Archaeometry 15:133–141

Shirai K, Schöne BR, Miyaji T, Radarmacher P, Krause RA Jr, Tanabe K (2014) Assessment of the mechanism of elemental incorporation into bivalve shells (*Arctica islandica*) based on elemental distribution at the microstructural scale. Geochim Cosmochim Acta 126:307–320

Simstich J, Harms I, Karcher MJ, Erlenkeuser H, Stanovoy V, Kodina L, Bauch D, Spielhagen RF (2005) Recent freshening in the Kara Sea (Siberia) recorded by stable isotopes in Arctic bivalve shells. J Geophys Res Oceans 110:C08006–C08006

Smith EA, Wishnie M (2000) Conservation and subsistence in small-scale societies. Annu Rev Anthropol 29:493–524

Soldati AL, Jacob DE, Glatzel P, Swarbrick JC, Geck J (2016) Element substitution by living organisms: the case of manganese in mollusc shell aragonite. Sci Rep 6:22514–22514

Southon JR, Rodman AO, True D (1995) A comparison of marine and terrestrial radiocarbon ages from northern Chile. Radiocarbon 37:389–393

Stein J (1992) Deciphering a shell midden. Academic, San Diego

Steinhardt J, Butler PG, Carroll ML, Hartley J (2016) The application of long-lived bivalve sclerochronology in environmental baseline monitoring. Front Mar Sci 3:176

Surge DM, Lohmann KC, Goodfriend GA (2003) Reconstructing estuarine conditions: oyster shells as recorders of environmental change, Southwest Florida. Estuar Coast Shelf Sci 57:737–756

Swingedouw D, Ortega P, Mignot J, Guilyardi E, Masson-Delmotte V, Butler P, Khodri M, Seferian R (2015) Bidecadal North Atlantic ocean circulation variability controlled by timing of volcanic eruptions. Nat Commun 6:6545

Tada Y, Wada H, Miura H (2006) Seasonal stable oxygen isotope cycles in an Antarctic bivalve shell (Laternula elliptica): a quantitative archive of ice melt runoff. Antarct Sci 18:111–115

Takesue R, Van Geen A (2004) Mg/Ca, Sr/Ca, and stable isotopes in modern and holocene *Protothaca staminea* shells from a northern California coastal upwelling region. Geochim Cosmochim Acta 68:3845–3861

Takesue RK, Bacon CR, Thompson JK (2008) Influences of organic matter and calcification rate on trace elements in aragonitic estuarine bivalve shells. Geochim Cosmochim Acta 72:5431–5445

Tanaka N, Monaghan MC, Rye DM (1986) Contribution of metabolic carbon to mollusk and barnacle shell carbonate. Nature 320:520–523

Tao J, Chen Y, He D, Ding C (2015) Relationships between climate and growth of Gymnocypris selincuoensis in the Tibetan Plateau. Ecol Evol 5:1693–1701

Thébault J, Chauvaud L (2013) Li/ca enrichments in great scallop shells (*Pecten maximus*) and their relationship with phytoplankton blooms. Palaeogeogr Palaeoclimatol Palaeoecol 373:108–122

Thébault J, Chauvaud L, Clavier J, Guarini J, Dunbar RB, Fichez R, Mucciarone DA, Morize E (2007) Reconstruction of seasonal temperature variability in the tropical Pacific Ocean from the shell of the scallop, Comptopallium radula. Geochim Cosmochim Acta 71:918–928

Thébault J, Chauvaud L, L'Helguen S, Clavier J, Barats A, Jacquet S, Pecheyran C, Amouroux D (2009a) Barium and molybdenum records in bivalve shells: geochemical proxies for phytoplankton dynamics in coastal environments? Limnol Oceanogr 54:1002–1014

Thébault J, Schöne BR, Hallmann N, Barth M, Nunn EV (2009b) Investigation of Li/Ca variations in aragonitic shells of the ocean quahog *Arctica islandica*, northeast Iceland. Geochem Geophys Geosyst 10:Q12008–Q12008

Thomas KD (2015a) Molluscs emergent, part II: themes and trends in the scientific investigation of molluscs and their shells as past human resources. J Archaeol Sci 56:159–167

Thomas KD (2015b) Molluscs emergent, part I: themes and trends in the scientific investigation of mollusc shells as resources for archaeological research. J Archaeol Sci 56:133–140
Thompson VD, Andrus CFT (2011) Evaluating mobility, monumentality, and feasting at the Sapelo Island Shell ring complex. Am Antiq 76:315–343
Tolwinski-Ward S, Evans M, Hughes M, Anchukaitis K (2011) An efficient forward model of the climate controls on interannual variation in tree-ring width. Clim Dyn 36:2419–2439
Torres ME, Zima D, Falkner KK, MacDonald RW, O'Brien M, Schöne BR, Siferd T (2011) Hydrographic changes in Nares Strait (Canadian Arctic archipelago) in recent decades based on $\delta^{18}O$ profiles of bivalve shells. Arctic 64:45–58
Ullmann CV, Wiechert U, Korte C (2010) Oxygen isotope fluctuations in a modern North Sea oyster (*Crassostrea gigas*) compared with annual variations in seawater temperature: implications for palaeoclimate studies. Chem Geol 277:160–166
Urey HC (1947) The thermodynamic properties of isotopic substances. J Chem Soc:562–581
Vander Putten E, Dehairs F, Keppens E, Baeyens W (2000) High resolution distribution of trace elements in the calcite shell layer of modern *Mytilus edulis*: environmental and biological controls. Geochim Cosmochim Acta 64:997–1011
Vander Zanden MJ, Rasmussen JB (2001) Variation in $\delta^{15}N$ and $\delta^{13}C$ trophic fractionation: implications for aquatic food web studies. Limnol Oceanogr 46:2061–2066
Vanhaeren M, D'Errico F, Stringer C, James SL, Todd JA, Mienis HK (2006) Middle Paleolithic shell beads in Israel and Algeria. Science 312:1785–1788
Versteegh EAA, Gillikin DP, Dehairs F (2011) Analysis of $\delta^{15}N$ values in mollusk shell organic matrix by elemental analysis/isotope ratio mass spectrometry without acidification: an evaluation and effects of long-term preservation. Rapid Commun Mass Spectrom 25:675–680
Versteegh EAA, Blicher ME, Mortensen J, Rysgaard S, Als TD, Wanamaker AD Jr (2012) Oxygen isotope ratios in the shell of *Mytilus edulis*: archives of glacier meltwater in Greenland? Biogeosciences 9:5231–5241
Vonhof HB, Wesselingh FP, Ganssen GM (1998) Reconstruction of the Miocene western Amazonian aquatic system using molluscan isotopic signatures. Palaeogeogr Palaeoclimatol Palaeoecol 141:85–93
Wanamaker AD Jr, Kreutz KJ, Borns HW, Introne DS, Feindel S, Funder S, Rawson PD, Barber BJ (2007) Experimental determination of salinity, temperature, growth, and metabolic effects on shell isotope chemistry of *Mytilus edulis* collected from Maine and Greenland. Paleoceanography 22:PA2217
Wanamaker AD Jr, Kreutz KJ, Schöne BR, Pettigrew N, Borns HW, Introne DS, Belknap D, Maasch KA, Feindel S (2008) Coupled North Atlantic slope water forcing on gulf of Maine temperatures over the past millennium. Clim Dyn 31:183–194
Wanamaker AD Jr, Kreutz KJ, Schöne BR, Maasch KA, Pershing AJ, Borns HW, Introne DS, Feindel S (2009) A late Holocene paleo-productivity record in the western gulf of Maine, USA, inferred from growth histories of the long-lived ocean quahog (*Arctica islandica*). Int J Earth Sci 98:19–29
Wanamaker AD Jr, Kreutz KJ, Schöne BR, Introne DS (2011) Gulf of Maine shells reveal changes in seawater temperature seasonality during the Medieval Climate Anomaly and the Little Ice Age. Palaeogeogr Palaeoclimatol Palaeoecol 302:43–51
Wanamaker AD Jr, Butler PG, Scourse JD, Heinemeier J, Eiriksson J, Knudsen KL, Richardson CA (2012) Surface changes in the North Atlantic meridional overturning circulation during the last millennium. Nat Commun 3:899–899
Waselkov GA (1987) Shellfish gathering and shell midden archaeology. In: Schiffer MB (ed) Advances in archaeological method and theory. Academic, New York
Watanabe S, Kodama M, Fukuda M (2009) Nitrogen stable isotope ratio in the manila clam, *Ruditapes philippinarum*, reflects eutrophication levels in tidal flats. Mar Pollut Bull 58:1447–1453
Weaver AJ, Sedlacek J, Eby M, Alexander K, Crespin E, Fichefet T, Philippon-Berthier G, Joos F, Kawamiya M, Matsumoto K, Steinacher M, Tachiiri K, Tokos K, Yoshimori M, Zickfeld K

(2012) Stability of the Atlantic meridional overturning circulation: a model intercomparison. Geophys Res Lett 39:L20709–L20709
Wefer G, Berger WH (1991) Isotope paleontology – growth and composition of extant calcareous species. Mar Geol 100:207–248
Welsh K, Elliot M, Tudhope A, Ayling B, Chappell J (2011) Giant bivalves (Tridacna gigas) as recorders of ENSO variability. Earth Planet Sci Lett 307:266–270
Widerlund A, Andersson PS (2006) Strontium isotopic composition of modem and Holocene mollusc shells as a palaeosalinity indicator for the Baltic Sea. Chem Geol 232:54–66
Willis KJ, Bailey RM, Bhagwat SA, Birks HJB (2010) Biodiversity baselines, thresholds and resilience: testing predictions and assumptions using palaeoecological data. Trends Ecol Evol 25:583–591
Wisshak M, Lopez Correa M, Gofas S, Salas C, Taviani M, Jakobsen J, Freiwald A (2009) Shell architecture, element composition, and stable isotope signature of the giant deep-sea oyster *Neopycnodonte zibrowii* sp n. from the NE Atlantic. Deep-Sea Res Part I-Oceanogr Res Pap 56:374–407
Witbaard R (1996) Growth variations in *Arctica islandica* L (Mollusca): a reflection of hydrography-related food supply. ICES J Mar Sci 53:981–987
Witbaard R, Bergman MJN (2003) The distribution and population structure of the bivalve *Arctica islandica* L. in the North Sea: what possible factors are involved? J Sea Res 50:11–25
Witbaard R, Jenness MI, Vanderborg K, Ganssen G (1994) Verification of annual growth increments in *Arctica islandica* L from the North-Sea by means of oxygen and carbon isotopes. Neth J Sea Res 33:91–101
Witbaard R, Duineveld GCA, Dewilde P (1997) A long-term growth record derived from *Arctica islandica* (Mollusca, Bivalvia) from the Fladen Ground (northern North Sea). J Mar Biol Assoc U K 77:801–816
Witbaard R, Jansma E, Klaassen US (2003) Copepods link quahog growth to climate. J Sea Res 50:77–83
Yap CK, Ismail A, Tan SG, Rahim IA (2003) Can the shell of the green-lipped mussel Perna viridis from the west coast of Peninsular Malaysia be a potential biomonitoring material for Cd, Pb and Zn? Estuar Coast Shelf Sci 57:623–630
Zilhao J, Angelucci DE, Badal-Garcia E, D'Errico F, Daniel F, Dayet L, Douka K, Higham TFG, Jose Martinez-Sanchez M, Montes-Bernardez R, Murcia-Mascaros S, Perez-Sirvent C, Roldan-Garcia C, Vanhaeren M, Villaverde V, Wood R, Zapata J (2010) Symbolic use of marine shells and mineral pigments by Iberian Neandertals. Proc Natl Acad Sci U S A 107:1023–1028

Part IV
Assessment of Services

Chapter 22
Introduction of Assessments

Joao G. Ferreira and Jens K. Petersen

Abstract The quantitative assessment and evaluation of services is a complex topic, and there is controversy over what currency to use for comparisons. These issues are particularly challenging for e.g. regulatory and cultural services, whereas for provisioning, market mechanisms furnish the price. However, for a complete picture, an integrated evaluation of the the different services is needed. In this section, various case studies are presented to exemplify the application of different types of decision-making tools used to assess and evaluate services.

Keywords Modelling · Valorisation · Nutrient credit trading · Indices

The assessment of services provided by bivalve shellfish is a complex and controversial theme. The complexity arises from the range of services supplied, as discussed in the preceding sections of this book, the currencies that may be employed for evaluation, and from over-arching questions such as whether such an assessment should be applied only to cultivation or for the full range of services.

The controversy is associated for instance to different visions of the role of aquaculture. In the European Union the two key legislative instruments governing water—the Water Framework Directive (WFD – 2000/60/EC) and Marine Strategy Framework Directive (MSFD – 2008/56/EC) classify aquaculture only as a pressure; this ignores the fact that 56% of aquaculture in the EU (Ferreira and Bricker 2016) is extractive in nature as no feed is added to the production. Likewise, in the United States, farmed bivalves make up 50% of aquaculture.

Despite these numbers, and the consequent role of bivalve aquaculture in top-down control of eutrophication, there is a frequent misconception that bivalves pol-

J. G. Ferreira (✉)
DCEA, FCT, New University of Lisbon, Monte de Caparica, Portugal
e-mail: joao@hoomi.com

J. K. Petersen
Danish Shellfish Centre, Institute of Aquatic Resources, Danish Technical University, Nykoebing Mors, Denmark
e-mail: jekjp@aqua.dtu.dk

A. C. Smaal et al. (eds.), *Goods and Services of Marine Bivalves*,
https://doi.org/10.1007/978-3-319-96776-9_22

lute the environment rather than cleaning it up. It must therefore be emphasized that organisms that sequester organic particles in order to grow must by definition cause a net removal of both phytoplankton and detrital organic matter from the water column. The only persuasive approach for demonstrating this thermodynamic axiom is through the quantitative assessment of bivalve services.

Assessment is therefore defined herein as the quantitative estimate of the value of such services—this requires that robust methods may be applied for its determination. The valuation of services can potentially be represented by various metrics (e.g. mass of nitrogen removed, area of habitat created) but ultimately a comparative assessment is by definition monetary, since this is the standard indicator for comparison of trade-offs (Costanza et al. 1997).

While for provisioning services this is straightforward, since it relies on financial and landings data, for other types of services valuation is more complex. This may be further complicated by considerations such as non-use value, e.g. when considering landscape aspects, biodiversity or uniqueness.

The final section of this book consists of five chapters outlining the current knowledge on the assessment of bivalve services. Mathematical models, of various types and differing complexity, are at the core of this assessment. By definition they combine techniques developed within the natural and social sciences, and deal with a range of aspects:

(i) the value of bivalves in the context of wider ecosystem functioning. Models and indices that deal with carrying capacity at an ecosystem scale are particularly important as managers consider plans for aquaculture expansion—the value of current production is relatively easy to calculate, but expected increases in spatial occupation must be assessed in the light of food depletion and co-use of marine space;

(ii) the various ways in which the role of bivalves in nutrient removal can be priced. It is recognized that comparative pricing of competing technologies yield estimates orders of magnitude apart, and that the use of downstream approaches such as top-down control may be an important complement to source control, particularly when dealing with diffuse nutrient loading;

(iii) which ecological and economic instruments can be used to execute an informed assessment of the mass balance of substances of interest, and create a market in which benefits of both cultivation and restoration may be traded.

At a time when Europe, the United States, and Canada, are committed to expanding marine cultivation, and doing so in a sustainable manner, an integrated assessment of the value of bivalve shellfish will help to improve social acceptance, promote food security, economic growth, and employment.

References

Costanza R, d'Arge R, de Groot RS, Farber S, Grasso M, Hannon B, Limburg K, Naeem S, O'Neill RV, Paruelo J, Raskin RG, Sutton P, van den Belt M (1997) The value of the world's ecosystem services and natural capital. Nature 387:253–260

Ferreira JG, Bricker SB (2016) Goods and services of extensive aquaculture: shellfish culture and nutrient trading. Aquac Int 24(3):803–825

Chapter 23
Bivalve Aquaculture Carrying Capacity: Concepts and Assessment Tools

Aad C. Smaal and L. A. van Duren

Abstract The carrying capacity concept for bivalve aquaculture is used to assess production potential of culture areas, and to address possible effects of the culture for the environment and for other users. Production potential is depending on physical and production carrying capacity of the ecosystem, while ecological and social carrying capacity determine to what extent the production capacity can be realized. According to current definitions, the ecological carrying capacity is the stocking or farm density of the exploited population above which unacceptable environmental impacts become apparent, and the social capacity is the level of farm development above which unacceptable social impacts are manifested. It can be disputed to what extent social and ecological capacities differ, as *unacceptable impacts* are social constructs. In the approach of carrying capacity, focus is often on avoiding adverse impacts of bivalve aquaculture. However, bivalve populations also have positive impacts on the ecosystem, such as stimulation of primary production through filtration and nutrient regeneration. These ecosystem services deserve more attention in proper estimation of carrying capacity and therefore we focus on both positive and negative feedbacks by the bivalves on the ecosystem. We review tools that are available to quantify carrying capacity. This varies from simple indices to complex models. We present case studies of the use of clearance and grazing ratio's as simple carrying capacity indices. Applications depend on specific management questions in the respective areas, the availability of data and the type of decisions that need to be made.

For making decisions on bivalve aquaculture, standards, threshold values or levels of acceptable change (LAC) are used. The FAO framework for aquaculture is formulated as The Ecosystem Approach to Aquaculture. It implies stakeholder

A. C. Smaal (✉)
Wageningen UR – Wageningen Marine Research (WMR), Yerseke, The Netherlands

Department of Aquaculture and Fisheries, Wageningen University, Wageningen, The Netherlands
e-mail: aad.smaal@wur.nl

L. A. van Duren
Deltares, Delft, The Netherlands
e-mail: luca.vanduren@deltares.nl

A. C. Smaal et al. (eds.), *Goods and Services of Marine Bivalves*,
https://doi.org/10.1007/978-3-319-96776-9_23

involvement, and a carrying capacity management where commercial stocks attribute in a balanced way to production, ecological and social goals. Simulation models are being developed as tools to predict the integrated effect of various levels of bivalve aquaculture for specific management goals, such as improved ecosystem resilience. In practice, bivalve aquaculture management is confronted with different competing stocks of cultured, wild, restoration and invasive origin. Scenario models have been reviewed that are used for finding the balance between maximizing production capacity and optimizing ecological carrying capacity in areas with bivalve aquaculture.

Abstract in Chinese 双壳贝类水产养殖容量的概念往往用于养殖区生产潜力的评估,并确定养殖对环境和其他区域使用者的潜在影响。生产潜力取决于生态系统的物理状态和养殖容量,而生态和社会容量则决定生产能力的实现程度。根据目前的定义,生态容量是指对养殖水域生态系统产生不良生态影响的最小养殖密度,社会容量是指不引起负面社会影响的最大养殖密度或规模。由于这种不可接受影响的评定往往基于社会主观因素,对社会容量和生态容量认知的差异程度因人而异。在容量评估方法方面,我们通常更关注如何避免贝类养殖对环境造成不可逆的负面影响。但是双壳贝类群体对生态系统同样存在正面的影响,它们可以通过滤食过程和营养物质释放来提高生态系统的初级生产力。双壳贝类的生态系统服务功能使我们在进行容量评估时需要同时考虑正负两方面的效应。本篇总结了一些容量评估的量化工具,其中包括简单的参数指标和复杂的生态模型。我们列举了一些利用滤水率和摄食率作为简单容量估算指标的案例研究。容量评估工具的应用取决于养殖区域面临的具体问题,具体环境,数据的可用性以及需要作出的规划类型。

为了制定双壳贝类关于养殖标准,养殖对环境影响限值或可接受程度的相关规划(LAC),联合国粮农组织制定了水产养殖框架(The Ecosystem Approach to Aquaculture, 水产养殖的生态系统方法),这意味着在制定决策时需要考虑生产效益,生态效益和社会效益三者兼顾的容量管理。各种生态模型正在发展为预测不同水平的贝类养殖对既定管理目标(如改善生态系统弹性等)综合影响的评估工具。

在实际生产过程中,双壳贝类的养殖管理面临着不同的养殖种群,野生种群,恢复种群和入侵种竞争的局面。针对上述的不同场景,可以应用生态系统模型对双壳贝类养殖的最大生产力和生态容量之间平衡点的进行研究。

Keywords Production carrying capacity · Ecological carrying capacity · Social carrying capacity · Indicators · Indices · Models

关键词 生产容量 • 生态容量 • 社会容量 • 指示物 • 指标 • 模型

23.1 Introduction

Marine bivalves are usually cultivated under natural conditions in open water systems and depend on feed, seed and space available in the natural ecosystem. Hence it is an extractive form of aquaculture, using resources supplied by the local ecosystem, and closely linked to natural processes. There are many interactions between the bivalves and their environment. High density bivalve populations filter large quantities of water, take up phytoplankton, reduce turbidity, excrete dissolved nutrients, and produce biodeposits. Under nutrient limited conditions, nutrient regeneration may stimulate primary production, providing a positive feedback on phytoplankton availability for the bivalves. If bivalve stocks are too large, filtration may be larger than the total system can sustain (i.e. the rate of primary production plus the rate of import of food into the system). This incurs phytoplankton depletion, being a negative feedback on food availability. As shown in Jansen et al. (2019, this volume), the balance of positive and negative feedbacks between the bivalves and their food determines the provisioning services of bivalve aquaculture. This also depends on the interaction between cultivated and wild bivalve stocks (Newell 2004). If expansion of bivalve aquaculture stimulates nutrient regeneration and primary production, the impact is positive for all filter feeder stocks. If bivalve culture expansion implies a total stock size that leads to phytoplankton depletion, it has negative impacts for the ecosystem as well as for bivalve aquaculture. So, knowledge of the feedback processes is needed as a basis for addressing sustainable bivalve aquaculture production, and to establish the optimum cultivated stock size. Recent reviews have been published on bivalve carrying capacity studies (McKindsey 2013; Filgueira et al. 2015). These reviews analyse carrying capacity in the framework of the ecosystem approach to aquaculture (EAA) that means: (i) to be developed in the context of ecosystem functions and services with no degradation beyond the resilience capacity, (ii) to improve human well-being and equity for all relevant stakeholders, and (iii) to be developed in the context of other relevant sectors. Critical in this approach is the involvement of stakeholders (Soto et al. 2008; see also Byron et al. 2011a).

McKindsey 2013, presents an overview of the various impacts of bivalve aquaculture, as a basis for addressing the different types of carrying capacity, and how these can be used in decision making processes. Filgueira et al. 2015, also review the bivalve aquaculture impacts, and tools like models and indices to address the links between the different carrying capacity types, with a focus on ecological and social carrying capacities.

Our paper is based on these reviews, as a basis for addressing adverse ecosystem impacts as well as ecosystem services provided by bivalve aquaculture culture. The approach in this review is focused on a scale that is larger than the farm scale. The effects of bivalve aquaculture, including positive and negative feedbacks on the ecosystem's carrying capacity, require integration of farm scale impacts on the surrounding environment (watershed). This review therefore focusses on the scale of an entire bay/watershed.

We have analysed case studies where the role of various filter feeder stocks have been taken into account: culture stocks, wild stocks, and stocks of introduced invasive populations, as a basis for understanding factors that determine bivalve carrying capacity and for developing tools for ecosystem based management.

23.1.1 Concepts

The carrying capacity concept originally comes from the logistic population growth curve, that reaches the asymptote K when the population size is at maximum. This growth curve shows maximum growth rate at half the carrying capacity. So, maximum yield, either in fisheries or in aquaculture, is achieved at a population size that corresponds with half the K value in the logistic function (Odum 1953; Kashiwai 1995). This shows the difference between the carrying capacity concept for aquaculture versus carrying capacity for natural populations: in contrast to natural populations, carrying capacity for exploitation is maximized at a population size that is typically not at maximum size (Smaal et al. 1998). Rather than a population parameter, the carrying capacity concept can also be considered as a characteristic of the ecosystem: Dame and Prins 1998 define bivalve carrying capacity as the total bivalve biomass supported by a given ecosystem as a function of the water residence time, primary production time and bivalve clearance time. They show that carrying capacity for bivalve exploitation depends primarily on the availability of food – through transport and primary production - in relation to filter feeding capacities.

Inglis et al. 2000, proposed a distinction in physical, production, ecological and social carrying capacity. Physical carrying capacity defines the total area of farms that can be accommodated in a given space; the production capacity is defined as the standing stock at which the annual production of the marketable cohort is maximised. The ecological carrying capacity is the stocking or farm density of the exploited population above which unacceptable environmental impacts become apparent, and the social capacity is the level of farm development above which unacceptable social impacts are manifested. As pointed out by Gibbs 2009, this approach to ecological capacity is a social construct, encapsulated by the social carrying capacity. Gibbs defines ecological carrying capacity as the yield that can be produced without leading to significant changes to ecological processes, species, populations or communities. However, the assumption that aquaculture can produce yield without any significant ecological change is far from realistic. Moreover, the question remains what level of ecological change is acceptable for society, which is similar to the social carrying capacity.

In the approach of Inglis and others, ecological carrying capacity is not established as an intrinsic feature of the ecosystem like in the original concept as being the maximum population size supported by the ecosystem. Rather, ecological carrying capacity is now defined by what society considers acceptable, hence there is a circular argument: social carrying capacity is determined by what stakeholders

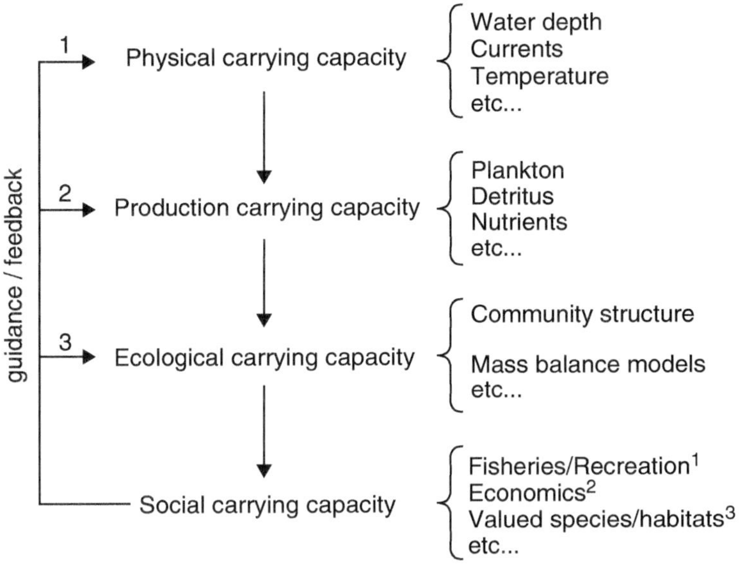

Fig. 23.1 Hierarchical structure of the different types of bivalve carrying capacity. Social carrying capacity provides guidance to choosing pertinent response variables and on establishing limits for these. Superscripts indicate examples of the type of information that informs the selection of response variables for other carrying capacity categories. (McKindsey 2013)

consider acceptable effects on ecological carrying capacity, and the ecological carrying capacity is defined by what stakeholders consider acceptable (Fig. 23.1).

McKindsey et al. 2006 and McKindsey 2013, following the definitions of Inglis et al. 2000, acknowledge the complexity of social and ecological carrying capacities. They propose a hierarchical approach, with physical capacity as a boundary condition for bivalve aquaculture, given the characteristics of the area and the needs of the farmers. Production capacity primarily depends on food availability i.e. primary production, transport of food through water movement and size of competing filter feeders stocks. Ecological carrying capacity is the level of change at which ecological impacts of bivalve aquaculture is considered *acceptable* for society. The question is how and by whom "acceptable" needs to be defined, as being a social concept. Indeed, McKindsey et al. 2006, 2013 consider social carrying capacity as the outcome of a process where the interests of all stakeholders are addressed. So in their approach, social capacity is defined in terms of the decision making process that leads to agreement on the level of ecological impacts that are considered acceptable.

Meanwhile there is a close link between production and ecological capacity. *Firstly*, establishing the cultured bivalve stock size that would give maximum yield, requires information about the size and activity of competing wild filter feeders stocks as they all depend on the same resources. Introduced filter feeder stocks, either as a side effect of bivalve aquaculture or by other causes, need to be included as well (Ruesink et al. 2005; Cugier et al. 2008). The same holds for bivalve stocks that are enhanced by bivalve restoration projects, as they also take their piece of the cake. The point that introducing filter feeders for bivalve aquaculture implies that other filter feeders will face food competition is explicitly addressed by Gibbs 2007. He points to the effects on the foodweb, for zooplankton in particular. This is substantiated by the model simulations as published in Byron et al. 2011b, where they calculated the impact of expansion of oyster culture on zooplankton and fish.

Secondly, cultured stocks can have positive and negative effects on the ecosystem, through various feedback processes. Addressing the different roles of cultured stocks in the ecosystem, specifically the ecosystem services they provide, not only for production but also for ecological response variables, is a prerequisite in understanding and managing the ecological carrying capacity for bivalve aquaculture.

23.1.2 Impacts and Services

The ecological impacts of bivalve aquaculture are a function of size of the culture and scale at which processes operate (farm scale / bay scale). This can be analysed for the pelagic habitat, the benthic habitat, and for the ecosystem functions (McKindsey 2013; see also the review by Filgueira et al. 2015). For the *pelagic habitat*, filtration of the water column by cultivated and wild filter feeders, in relation to food production and transport, is often used as an index for the carrying capacity at bay scale (Dame and Prins 1998; Gibbs 2009; Cranford et al. 2012).

For the *benthic habitat*, biodeposition and subsequent accumulation of organic material and potential oxygen depletion and sulphite release, are being used as impact parameters, particularly for suspended cultures (McKindsey 2013). Bivalve filtration and biodeposition enhance benthic-pelagic coupling, facilitating nutrient regeneration and denitrification (Cranford et al. 2007; McKindsey et al. 2011).

For *ecosystem functions*, the interactions between bivalve aquaculture and the ecosystem need to be addressed, as it concerns complex processes that provide ecosystem services. This includes filtration, biodeposition, nutrient regeneration (Jansen et al. 2019, this volume), selective retention of phytoplankton size classes impacting the pelagic food web structure (Cranford et al. 2009), interaction with higher trophic levels (Byron et al. 2011b; Aguera et al. 2015; Kluger et al. 2016a, b) as well as habitat modification and impacts on local biodiversity (Filgueira et al. 2015; Craeymeersch and Jansen 2019, this volume). These processes have an effect at farm scale as well as at bay scale. As mentioned by McKindsey 2013, these effects can act "positive" and "negative" on the carrying capacity.

Table 23.1 gives a summary of the main feedbacks that apply for bivalve aquaculture in the respective environments.

Various authors have proposed schemes for quantifying impacts as a basis for setting standards for ecological carrying capacity. McKindsey et al. 2006, proposed a conceptual scheme where the impact is plotted as a function of production level. This allows quantifying the maximum production level that gives an "acceptable" impact (Fig. 23.2).

Table 23.1 Positive and negative feedbacks between bivalve aquaculture and the environment, based on McKindsey 2013, Filgueira et al. 2015 and Jansen et al. 2019, this volume

Environment	Feedbacks on	Positive feedbacks	Negative feedbacks
Pelagic environment	Food production	Nutrient regeneration	Phytoplankton depletion
		Turbidity reduction	Zooplankton depletion
		Dentrification	
Benthic environment	Habitat availability	Habitat creation	Degradation
		Increased niche complexity	Resuspension
		Coastal protection	
Ecosystem functions	Food web	Predators	Pathogens, parasites
		Benthic fauna	Invasive species
		Fouling species	

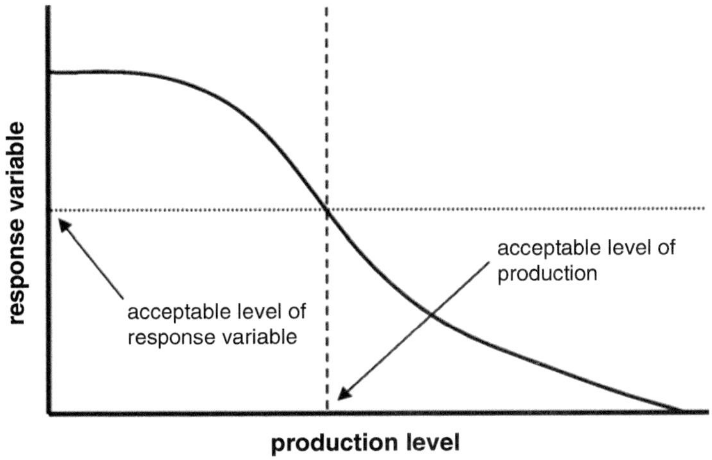

Fig. 23.2 Hypothetical response curve of an environmental variable as a function of bivalve production levels. The acceptable level of response gives the corresponding maximum production level. (McKindsey et al. 2006, see also Tett et al. 2011)

Although Cranford et al. 2012 and McKindsey 2013 refer to positive effects of bivalve aquaculture, the schemes do not explicitly address the bivalve services to the ecosystem. Therefore the hypothetical response curve can be extended to production levels of bivalve aquaculture that stimulate particulate response parameters, as reviewed by Jansen et al. (2019, this volume): Fig. 23.3.

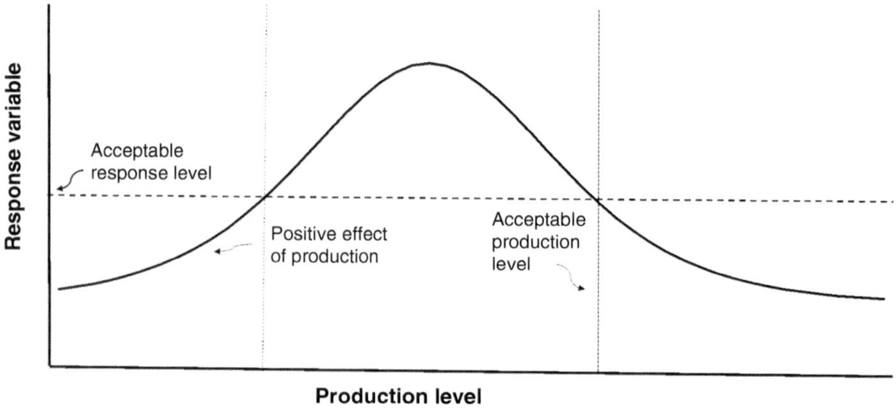

Fig. 23.3 Adapted hypothetical response curve of Fig. 23.2, showing that increase in bivalve production can stimulate a response variable from sub-optimal to an optimal level. Further increase of production leads to adverse impacts, below a level that is considered acceptable

23.2 Approaches

23.2.1 Social Carrying Capacity

Techniques for inferring social carrying capacity (SCC) are still being developed. It is considered as the most complex carrying capacity to determine, as it depends on various groups of stakeholders with different interests. Following the approach of McKindsey 2013, SCC is defined in terms of the decision making process that leads to agreement on the level of ecological impacts that are acceptable. Approaches to SCC deal with various decision making techniques: market, public regulation, multi-stakeholder agreements, and self-regulation (Table 23.2). These techniques are of a generic nature, not specific for bivalve aquaculture. For aquaculture this is based on criteria for the Ecosystem Approach to Aquaculture.

23.2.1.1 Market

It can be argued that economic carrying capacity is part of the social domain. Filgueira et al. 2015 explore social capacity from an economic perspective through the concept of "willingness to pay" that may be used to quantify consumer preferences. However, social capacity is generally considered in a much broader sense than can be quantified by economic approaches.

Table 23.2 Various approaches to carrying capacity, the activity involved, required tools and management options (see below), and the actors

Carrying capacity	Approach	Activity	Tools	Management options	Actors
Social	Market economy	"Willingness to pay" surveys	Free market	Secure level playing field	Consumers, government, research
	Regulation	Directives, licensing	Standards	Licenses	Government
	Multi-stakeholder approach	SES, LAC, Convenant partnerships	Agreements	Process facilitation	Industry, consumers, government,
	Self-organisation	Labelling, best practice	Protocols and assessments	Certification	Industry, consumers, consultants
Ecological	Apply standards	Monitoring and enforcement	Standards	Licenses	Government
	Site-specific measures	Appropriate assessement for licensing	Standards	Licenses	Government, industry, consultants
	Knowledge based aquaculture	Monitoring and assessment	Models, indices	Scenarios	Researchers
	Innovation	R&D; developing integrated models	Integrated models	Scenarios	Researchers
Production	Trial and error	Yield and stock monitoring	Production indices	Learning by doing	Industry
	Site selection	Collect data, run scenarios	Models, indices	Licenses	Industry, consultants, government
	Knowledge based aquaculture, innovation	R&D; monitoring and assessment	Models	Scenarios	Industry, researchers
	Self-organisation	BMP, BEP	Integrated models	Scenarios	Industry, consultants
Physical	Identify boundary conditions	Collect data	Monitoring	Licenses	Industry, researchers, government
	Site selection	Collect data, run scenarios	Monitoring	Licenses	Industry, researchers, government
	Knowledge based aquaculture	R&D; monitoring and assessment	Monitoring	Scenarios	Industry, researchers

23.2.1.2 Public Regulation

Given limitation of the free market, governmental regulations are often required to handle conflicts about common resource exploitation. With respect to regulations, Sequeira et al., 2007 present an overview of legislation and worldwide policy instruments on the protection of the marine environment. It shows the various types of regulation that are implemented in China, USA, Europe and Australia/New Zealand. At global scale there is an apparent lack of UN regulations, although in the framework of the Convention on Biodiversity a global policy has been achieved for environmental protection (www.cbd.int).

23.2.1.3 Multi-stakeholder Agreements

Stakeholder support for regulation can often be a more effective approach than top-down rules. Ostrom 2009 evaluated management strategies of social-ecological systems (SES, Walker et al. 2003) – that can also be used for analysing social carrying capacity – and showed that proper stakeholder involvement delivered a more sustainable approach than top-down rules (McKindsey 2013). The concept of "limits of acceptable change" (LAC) is about indicator selection through a collaborative approach, rather than setting limits as such. Application in the case of New Zealand bivalve aquaculture showed that this approach provides a management framework to prevent negative effects of the activity, supported by stakeholders (Zeldis 2005). There are many ways to organise multi-stakeholder involvement, depending on site specific social, economic and cultural factors. The concept of social-ecological systems stress the idea that society is part of the ecosystem, hence it is logical to include relevant stakeholders, as is also acknowledged in the ecosystem approach to aquaculture.

23.2.1.4 Self-Organisation

Self-organisation by the industry aims to achieve self-imposed goals either through Good Agricultural Practices (GAP), Codes of Practice and Codes of Conduct, implemented through Best Managing Practices (BMP) (Hargreaves 2011). Labelling is also part of self-organisation. This is done for aquaculture and fisheries – including extensive bivalve cultures as enhanced fisheries – through ASC and MSC certification respectively. Self-organisation is generally inferred through external pressure, like from retailers and consumer interests.

For all approaches to infer social carrying capacity, it is a challenge to focus on both sides of bivalve aquaculture: ecological impacts as well as ecological benefits.

23.2.2 *Ecological Carrying Capacity*

Ecological carrying capacity (ECC) is about the various positive and negative feedbacks between bivalve aquaculture and the ecosystem (Table 23.1), and about what is decided at the level of social capacity with regard to management aims. According to the Ecosystem Approach to Aquaculture, this should not only focus on avoiding unacceptable impacts, but it should take into account bivalve ecosystem services that reinforce management aims, such as, for example, ecosystem resilience. In the current approaches, however, focus is merely on avoiding adverse impacts.

23.2.2.1 Standards

The general approach to environmental impacts is based on standards that define the acceptable impact level of a given activity for a given environment. Application of standards is based on the idea that there are threshold values for various parameters that should not be exceeded. Standards are straightforward, relatively simple and relatively easy to implement and to reinforce. However, once defined they are static and any adjustment is complicated. Moreover, positive feedbacks are generally not taken into account. Therefore more dynamic approaches are being developed, like, for example, thresholds of potential concern (TPC). TPCs are a set of operational goals along a continuum of change in selected environmental indicators. TPCs are being continually adjusted in response to the emergence of new ecological information or changing management goals (Cranford et al. 2012). The previously mentioned LAC approach (Limits of Acceptable Change) is also more dynamic as it is used as a basis to achieve consensus in multi-stakeholder groups.

23.2.2.2 Site-Specific Measures

Site-specific measures have the potential to address ecosystem characteristics of the bivalve aquaculture environment. In many cases this is the natural ecosystem in relatively undisturbed areas. Bivalve aquaculture requires relatively unpolluted water, in many countries regulated by water quality standards for bivalve production. It turns out that bivalve production waters are now recognised as areas of high natural value. Hence they attract attention to be protected in the framework of nature conservation. In the European Union, for example, this is regulated in the framework of the Natura 2000 policy. Natura 2000 is a network of core breeding and resting sites for rare and threatened species, and some rare natural habitat types which are protected in their own right. The aim of the network is to ensure the long-term survival of Europe's most valuable and threatened species and habitats listed under the Birds and Habitats Directives, approved by national laws (http://ec.europa.eu/environment/nature/natura2000). In these areas an activity such as bivalve aquaculture is only allowed through a permit of the government. To achieve a permit the farmer has to prove that there is no negative impact of the activity on the maintenance goals that are set for the given area. So the standard is in this case the absence of negative impacts on a given set of parameters, like numbers of protected bird species. In principle this approach could take positive impacts into account. In practice, however, the standards are based on the absence of negative impacts.

23.2.2.3 Knowledge Based Aquaculture

In this approach available knowledge about the interactions between bivalve aquaculture and the ecosystem is mobilized to evaluate positive and negative feedbacks. This requires empirical data of key ecosystem parameters and processes. As this is rather complex, in many cases mathematical models are used to integrate data. With these models processes can be simulated in order to address the optimum level of bivalve aquaculture giving the maximum level of ecological response (Fig. 23.3). A formal scheme to address the impact was applied by Tett et al. 2011 following the driver-pressure-status-impact-response (DPSIR) approach; see also Nobre 2009 for dynamic DPSIR application in decision making. If not many data are available, indices like presented by Dame and Prins 1998 can be used to identify the relation between bivalve stock size, water renewal and food production (see case study below).

23.2.2.4 Innovation

An innovative approach to ecological carrying capacity would require information of the various goods and services that a specific bivalve aquaculture activity provides for the ecosystem. This includes interactions in the water column, in the benthic system and in the ecosystem functions. Given the complexity of the interactions, models are needed to integrate data and processes (Smaal et al. 1998). As reviewed by Filgueira et al. 2015 various types of models are being used for estimating carrying capacity for bivalve aquaculture. This includes farm models, spatial models, food web models, benthic models and habitat models. See also Newell et al. (2019), Grant and Pastres (2019), Ferreira et al. (2019) and Bacher et al. (2019) this volume. Yet a generic integrated approach where the various ecosystem services of bivalve aquaculture, like habitat provisioning, facilitation and nutrient control, are included in an integrated model, is still to be developed.

23.2.3 Production Carrying Capacity

23.2.3.1 Trial and Error

Although production carrying capacity (PCC) of the ecosystem is the most studied type of bivalve aquaculture carrying capacity (McKindsey 2013) the approach most applied in practice, seems to be based on trial and error. This sounds logical as the farmers are well equipped to estimate ups and downs in the yields and should be able to make the link with variation in the size of the stock that they are controlling. However, this can be obscured by dynamics of other filter feeder stocks, either natural or commercial. Other factors like elevated predation, will also influence yields. So, data on stock size as a basis for PCC management are indispensable. This not only concerns total commercial filter feeder stocks in the culture area, but also natural, restored and introduced stocks of filter feeders.

23.2.3.2 Site Selection

PCC depends on local conditions like stability of the sediment for bottom culture, water flow for food supply and waste dispersal, and hydrodynamic forces for suspended culture. So, at farm scale level data is needed of these variables to select optimal sites.

23.2.3.3 Innovation

For PCC, as for ECC, innovation concerns the development of tools and models that take positive feedbacks into account, as a basis for ecosystem services. Given the state-of-the-art, it can be considered innovative if the industry would apply models that allow the calculation of the optimum stock size to provide maximum yield of their cultures. This not only requires proper calculations but also cooperation in managing the stock size. As individual farmers use the common pool as a resource they have to organise themselves to overcome the "tragedy of the commons".

23.2.3.4 Self-Organisation

In addition to self-organisation as mentioned under SCC, approaches like Best Management Practice (BMP) are described for bivalve aquaculture in the framework of sustainable aquaculture (Hargreaves 2011). As pointed out by Hargreaves 2011, BMP is an approach at farm scale that has limitations for the wider scale, although within producer organisations collective approaches for BMP are applied. Yet it remains a voluntary activity that often lacks sufficient assessment and monitoring.

23.2.4 Physical Carrying Capacity

Physical carrying capacity defines the boundary conditions in physical terms for the extent of bivalve culture in a certain area. This depends on hydrodynamics (currents, waves, wind forcing), bathymetry, water quality and available space. The approach is to collect data of the key parameters and identify proper sites. Generally new sites will need pilots to test the proper conditions at the local scale in practice.

23.2.5 Integrated Carrying Capacity

There is a direct link between social, ecological and production carrying capacity. Sequeira et al., 2007 show that partitioning of food for example, between wild and cultivated stocks differs considerably between culture areas in China and Europe, with a much greater portion for bivalve aquaculture than for wild stocks in the Chinese cases, in comparison with the European cases. They clearly show that the proportion of ECC vs PCC can be quite different in different societies, reflecting differences in SCC. In countries where food production is acknowledged as a high priority issue by society, PCC is dominant over ECC. In countries with a large environmental concern in society, this seems to be the reverse. However, the Ecosystem Approach to Aquaculture, that aims to address goods and services of aquaculture and to inform and involve stakeholders, might provide a common framework for an integrated approach of carrying capacity (Byron et al. 2011b; Filgueira et al. 2015). In this approach, the challenge is to achieve a carrying capacity for bivalve aquaculture where commercial stocks attribute in a balanced way to production, ecological and social goals (cf Triple P, see also Cranford et al. 2012).

23.3 Tools

For managing bivalve aquaculture in the natural environment, tools are needed to estimate the different types of carrying capacity and to identify the optimal production level in relation to management goals of a given area. Therefore we will discuss indicators and models as tools to address the interaction between bivalve aquaculture and the ecosystem, and their relevance for decisions on how to manage carrying capacity.

23.3.1 Indicators and Indices

Bivalve carrying capacity indicators should address the positive and negative impacts of bivalve aquaculture for production and ecological carrying capacity. We use an indicator as a parameter to establish the value of a variable or a set of parameters with a specific meaning, often called an index; an index is a calculation based on a set of variables, that can be used for comparative analysis.

In their review on tools for sustainable management, Cranford et al. 2012, discussed pelagic, benthic, production performance and socio-economic indicators of bivalve aquaculture. *Pelagic indicators* address the influence of bivalve suspended and bottom culture farms in the water column; for the pelagic system the interaction between the bivalves and their food is the most important determinant of both PCC and ECC. *Benthic indicators* address the impact of suspended culture on the benthic habitat, comprising of the effect of organic enrichment of the sediment, and the consequences for the benthic community. Benthic indicators predominantly describe impacts at farm scale rather than bay scale, while pelagic indicators tend to be more relevant at bay scale. *Production indicators*, like bivalve condition indices and yield, address the effectiveness of the culture practice in the given environment and provide a tool to evaluate culture measures. For *socio-economic indicators*, Cranford et al. (2012) refer to the social acceptability of the bivalve culture, the supply availability to the market, the livelihood security for the local communities, and the economic efficiency of bivalve culture operations. These indices and indicators can be used to address the effects of bivalve aquaculture on the social carrying capacity (Table 23.3).

Case Study: Pelagic Indices

For the various pelagic indicators as shown in Table 23.3, the ones that address the interaction between the bivalves and their food – phytoplankton biomass, depletion index - directly link to carrying capacity. As shown by Filgueira et al. 2015, the use of "depletion" in the literature to address the uptake of particles by the bivalves, may suggest exhausting of algal cells. This would indicate a rather extreme case of overgrazing, while filtration in combination with nutrient regeneration can also stimulate food availability. Rather than depletion, indicators for the interaction between the bivalves and the pelagic processes should take into account in how far particle filtration is compensated by renewal of the particle stock by import from outside and local primary production. This has been worked out by Dame and others (Smaal and Prins 1993; Dame 1996; Dame and Prins 1998; Prins et al. 1998). They describe the impact of the bivalves on processes in the water column in relation to the water residence time and the primary production in the Bay. Water residence time (RT) is the time it takes to renew the water body by exchange water from a defined area with the adjacent ecosystem. Primary production time (PT) is the time it takes to renew the phytoplankton stock in a given area. Clearance time (CT) is the time it takes for the bivalves to filter the water body in a given area. See Dame and Prins 1998 for the calculations. We define CT/RT as the *clearance ratio*, and CT/PT as the *grazing ratio*. If the clearance ratio (CT/RT) >1, then water renewal time is shorter than bivalve clearance time, hence the system is relatively open and the bivalves have little control over the ecosystem. At a clearance ratio < 1, bivalves filter the water column faster than this water is renewed, hence the bivalves potentially control pelagic processes through their grazing activity. In this case the internal primary production determines the carrying capacity, and this is expressed as PT. At a grazing ratio (CT/PT) > 1, primary production exceeds bivalve filtration capacity, hence food is produced faster than consumed. At a grazing ratio < 1 the system will collapse as food is depleted. Actually if the grazing ratio is just above 1, the system will be unstable as depletion due to daily variation in primary production may occur.

Table 23.3 Potential indicators of bivalve aquaculture impacts in the framework of sustainable management (see Cranford et al. 2012 for descriptions of the indicators)

Pelagic indicators	Sediment indicators	Benthic community indicators	Production indicators	Socio-economic indicators
Nutrient concentration	Sedimentation rate	Biodiversity metrics	Bivalve growth rate	Profitability
Dissolved oxygen	Biodeposition rate	Indicator species	Conditon index	Total employment
Bacterial abundance	Sediment texture	Trophic indices	Meat yield	Gross value added/ employment
Phytoplankton biomass	Onanie enrichment	Benthic similarity	Stocking density	Tax revenues
Depletion index	N and P enrichment		Production time series	Social acceptability
Phytoplankton size	Sediment quality			Conflict assessment
Trophic heterogeneity	Redox potential			
	Total free sulfides			
	Water content			
	Dissolved oxygen			
	Benthic/pelagic flux			
	Pigments			
	Visual observations			
	Benthic Enrichment Index (BEI)			
	Benthic Habitat Quality Index (BHQ)			

Therefore in practice, a buffer capacity is required for a stable solution. This indicator tool has been used in various studies to estimate carrying capacity (Gibbs 2007;Thompson 2005; Filgueira et al. 2015). In his paper of 2007, Gibbs used the term clearance efficiency for the ratio between CT and RT. He defined Filtration Pressure being the food uptake rate as a fraction of primary production; this resembles the grazing ratio. The Regulation Ratio is defined as the Clearance Rate relative to the water mass (= 1/CT), as a fraction of the phytoplankton turnover rate (=1/PT). So RR = (1/CT)/(1/PT) which is similar to the inverse of the grazing ratio.

The clearance and grazing ratios have the elegance of simplicity, but limitations are the lack of spatial and temporal differentiation and assumptions about mixing of the water body and various eco-physiological processes. In cases where not many data and no models are available, these indicators can be used to characterize the potential of an area for bivalve aquaculture. It also has been used to make a comparative analysis of different culture areas, or changes over time. It should be noticed that clearance time by the cultured stocks does not represent total clearance time that includes wild, invasive and restored stocks.

We analysed 20 areas that are used for bivalve aquaculture and we made a comparison on the basis of existing literature values. Only for a limited number of areas distinction can be made between different stocks, as shown in Annex I.

Figure 23.4 shows the log transformed clearance ratio, showing areas with positive values that have a clearance time larger than the water residence time in the right panel, and in the left panel the areas with a clearance time shorter than the residence time. In these areas the bivalves potentially control pelagic processes.

Figure 23.5 shows the Grazing Ratio in areas where clearance time is shorter than water residence time. This means that in these areas the local primary production is the main factor determining carrying capacity. The grazing ratio in these areas ranges from below 1 to over 11. The graph shows changes over time for The Oosterschelde (SW Netherlands), as grazing ratio was 3.4 in 1996, then went down to 2.5 in 2009. This is consistent with the expansion of the invasive Pacific Oyster stocks, as shown in Smaal et al. 2013. Also for the Western Wadden Sea the grazing

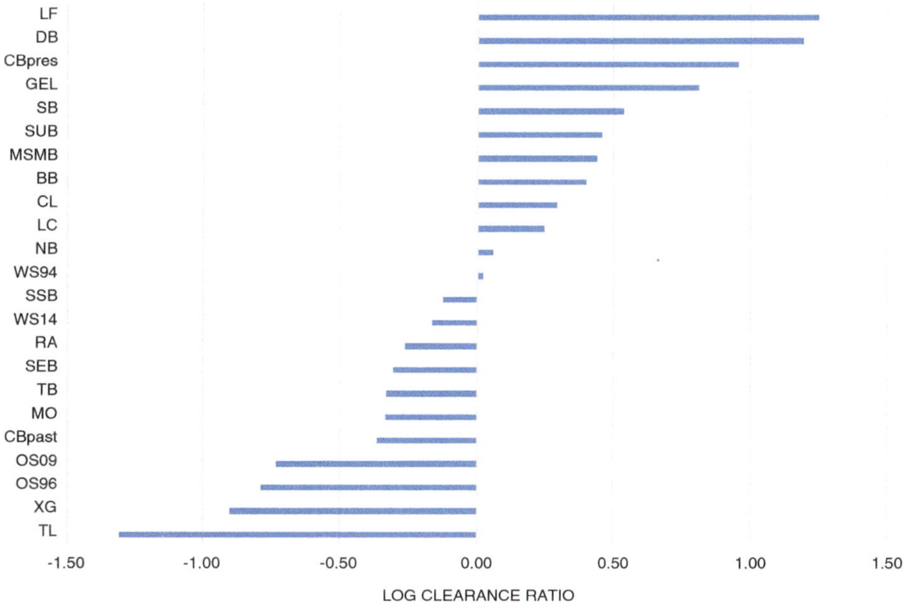

Fig. 23.4 Log Clearance Ratio (CT/RT) of areas with bivalve aquaculture. Values >0 show areas with faster water renewal than water potentially cleared by the bivalves. This holds for Lysefjord (LF, Norway), Delaware Bay (DB, USA), Chesapeake Bay present (CBpres, USA), Grand Entree Lagoon (GEL, Canada), Saldanha Bay (SAB, S-Africa), Sungo Bay (SB, China), Mont St Michel Bay (MSMB, France), Beatrix Bay (BB, N-Zealand), Loch Creran (LC, UK), Narragansett Bay (NB, USA) and Wadden Sea 1994 (WS94) (see Annex I for details and references)
Values <0 show areas where bivalve filtration potentially regulates water column processes as clearance time is shorter than residence time; this is the case for South San Francisco Bay (SSB, USA), Western Wadden Sea 2014 (WS14, The Netherlands), Ria de Arosa (RA, Spain), Sechura Bay (SEB, Peru), Tracadie Bay (TB, Canada), Marennes-Oleron Bay (MO, France), Chesapeake Bay past (CBpast, USA), The Oosterschelde in 1996 and 2009 (OS, The Netherlands), Xiangang Bay (XG, China) and Thau Lagoon (TL, France)

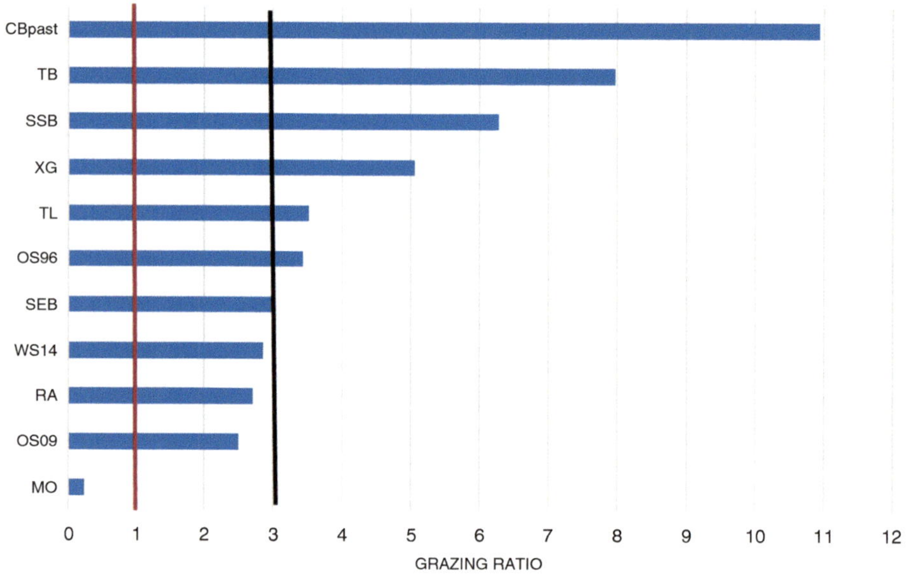

Fig. 23.5 Grazing ratio in areas with a Clearance Ratio < 1, abbreviations in legend of Fig. I.1. Red line shows the theoretical minimum grazing ratio below which the system collapses; the black line gives the threshold of potential concern (TPC) value as set by the ASC; between 1 and 3 the system has the risk of overgrazing

ratio is now 2.86, while in 1994 it was 5.9 (Annex I). This is ascribed to the expansion of the invasive species *Crassostrea gigas* and *Ensis americanus*.

As mentioned before, areas with a grazing ratio around or smaller than 1 are considered unstable, as in these areas, the short clearance time will exhaust primary production. For Marennes-Oleron Bay it is known that microphytobenthos is a major food source for the bivalves that is not represented in primary production data (Héral et al. 1988). The grazing ratio should in practice be above 1. The ASC has set a limit of 3 as a minimum value, a Threshold of Potential Concern. The graph shows that areas such as the Oosterschelde, Ria de Arosa and the Western Wadden Sea have values between 2 and 3. It should be noted that this regards the grazing ratio of the combined cultivated, wild and invasive stocks. Expansion of bivalve aquaculture in these areas requires decisions on the production versus the ecological carrying capacity, and on the fate of invasive stocks.

23.3.2 Models

There are several modelling tools available that are specifically designed to assess carrying capacity, and there are more generic ecosystem models that can also be used to assess system limits in terms of carrying capacity. The basic ones are generally relatively simple box models (Dowd 2005; Grant et al. 2007), the more sophisticated ones are fully coupled physical – ecological models that are spatially explicit.

Models have to be fit for purpose. Some questions do not require very elaborate modelling. To get a first order impression of whether a system is approaching carrying capacity for bivalve, relatively simple models that include data on grazing rates, primary production rates and retention times will suffice. Other questions regarding optimal locations and optimal spacing of bivalve farms need much more explicit detail on transport of nutrients, algae and other constituents and therefore at the very least need a well validated hydrodynamic model as a basis.

23.3.2.1 Physical Carrying Capacity

Physical carrying capacity is basically determined by the availability of suitable habitat. For benthic bivalve beds (natural as well as cultivated) this is determined by the combination of bivalve bed composition (hard or soft substrate); fluid dynamics (high bed shear stress on soft sediment limits settlement possibilities) and parameters such as oxygen concentration. Areas with regular stratification and extended periods of oxygen depletion near the bed will not have bivalve beds, even if there is plenty of food supply (but see Petersen et al. 2013). Habitat models can be used to ascertain the physical carrying capacity for bivalves (and other species) in ecosystems (Cozzoli et al. 2014; Brinkman et al. 2002). They rely on relationship of the bivalves for each parameter in the model and spatial information on the range of values of these parameters in each model. The approach by e.g. Cozzoli et al. (2014) is based on non-linear quantile regression analysis, a powerful statistical method that (given enough data) can be used to set up a predictive model for the distribution of species, related to environmental characteristics such as grain size. Other models calculate a habitat suitability index based on the upper and lower limits of characteristics that bivalves and the optimum range of these limits where they occur (e.g. Barnes et al. 2007).

In systems with benthic cultures, physical carrying capacity may be a limiting factor, particularly in systems with a very large exchange rate (i.e. small retention time) where food supply is large and suitable space limited. In systems with rope cultures the physical carrying capacity is artificially increased by providing suitable physical habitat (generally also outside oxygen depleted layers) and generally not an issue.

23.3.2.2 Production and Ecological Carrying Capacity

Production carrying capacity and ecological carrying capacity are determined by food supply and models need to capture the relationship between local (system) primary production, import of algal biomass from external sources, and cultivated bivalves and natural grazing stocks.

Models can act at different scales. At the scale of a farm local depletion is very relevant, determining local stocking density (Cranford et al. 2014; Newell et al. 2019, this volume). These models give some information on production carrying capacity at very local scales. In terms of carrying capacity of ecosystems for bivalves, farm-scale models are of limited interest. Even if strong depletion occurs within a farm, this does not imply that there are problems with carrying capacity (either production or ecological) in the system as a whole. These models are therefore not considered further here, we concentrate on regional or ecosystem-scale models.

Particularly when considering ecological carrying capacity, it is imperative to consider not only the growth of cultivated bivalves, but also the transfer of carbon (or energy) to other trophic levels. Foodweb models such as Ecopath with Ecosim (EwE – www.ecopath.org) have proved very valuable to assess energy flows between different trophic levels between different species (Wolff et al. 2000). The EwE approach has been used in the Mississippi Delta plain to assess the effect of large scale river diversions on landings of bivalve and fish (Mutsert et al. 2017), using the Delft 3D hydrodynamic modelling suite to provide boundary conditions on transport and primary production. This approach works very well for impact studies where the major changes in the system are physical and not strongly influenced by feedback processes mediated by the bivalves themselves.

For 'true' carrying capacity studies on bivalves, these feedback mechanisms are crucial to consider. In most systems primary productivity is determined by nutrient and light availability. In bivalve dominated systems, the grazing activity of the bivalves strongly influences both these factors. Filtration activity lowers the algal concentration in the water as well as removes fine sediment particles from the water column. This reduces light attenuation and increases productivity in light limited systems. The remineralisation of nutrients increases nutrient availability in systems that are nutrient limited (Van Broekhoven et al. 2014; Jansen et al. 2019, this volume). In bivalve dominated systems (i.e. many systems with large scale bivalve cultures) it is not possible to accurately model primary production without taking these feedback loops into account. Therefore it is in such systems also not possible to accurately calculate bivalve or primary production without a dynamic coupling between bivalve growth and primary production (Filgueira et al. 2015).

In recent years, major advances have been made with fully integrated online coupling of ecosystem hydrodynamics models with nutrient dynamics, light attenuation, primary production and bivalve growth. Bivalve growth is sometimes parameterised as Scope for Growth (SfG), but more often bivalves are modelled using Dynamic Energy Budget modelling (Kooijman 2010; Troost et al. 2010; Guyondet et al. 2010). In this modelling concept, energy entering an organism can be used for either reproduction, or for maintenance, or it can be stored. The model relies on accurate parameterisation of functional responses to environmental variables (food availability, food quality temperature, particulate matter concentrations etc.) that have to be determined for each species, for each system. Depending on the spatial organisation of the model, this approach gives the opportunity to assess ecological carrying capacity and production carrying capacity for bivalve culture at different spatial scales (Guyondet 2010). With this approach (and assuming all parameters are available) it is in principle possible to have multiple bivalve stocks (natural and cultivated) competing for phytoplankton in the same system (Troost 2011 – in Dutch). This allows the assessment of bivalve cultivation on natural bivalve stocks. The study by Troost (2011) yielded response curves for increasing levels of bivalve culture very similar to the conceptual picture in Fig. 23.3. While the unidirectional models have proven their value in indicating the limits of carrying capacity and the potential damage of bivalve aquaculture to the system, the lack of including feedback mechanisms limits their ability to quantify potential benefits.

At present there are no modeling tools available that include the carbon or energy fluxes in a system as well as habitat characteristics. For a comprehensive assessment of ecological carrying capacity it is important that not only the trophic interactions are taken into account, but that also other ecological functions are assessed. For example, bivalves are ecosystem engineers (Passarelli et al. 2014; Ysebaert et al. 2019, this volume). Epibenthic bivalves can transform soft sediment into hard substrate, altering the physical state of the environment and providing habitat for a different range of biota. This is true for benthic assemblages, but also rope cultures can become hotspots of associated fauna and change the species composition in ecosystems. Sometimes these effects of bivalves are seen as very positive. The loss of large scale reef structures of flat oysters from the North Sea has decreased its biodiversity and led to the North Sea being considered an impoverished system. However, very large assemblages of cultivated bivalves may also lead to a switch in species composition due to these ecosystem engineering effects. E.g. in Sechura Bay in Peru, an increased cultivation of the Peruvian scallop, *Pacoplecten purpuratus*, led to large-scale shifts in the species composition in the bay. This significant change in benthic community composition, together with an increase in the predator biomass, paralleled by a decrease in the biomass of their competitors; a change in species diversity and maturity; a system increase in size (in terms of biomass and total flows) and a decrease in energy cycling, led to the conclusion that the ecological carrying capacity of the system had been transgressed (Kluger 2016b). The energy cycling was investigated in this study with an EwE model, the other effects were addressed deriving indicators from datasets.

23.3.2.3 Social Carrying Capacity

There are also models available to calculate social carrying capacity, however at present these do not seem to have been applied to carrying capacity for bivalve cultivation. Most applications of model calculations on social carrying capacity have been applied to tourism and to the adaptive management of e.g. national parks (e.g. Lawson et al. 2003) or recreational areas (e.g. Tarrant and English 1996). These are tools that consider a relatively limited set of parameters, generally based on interviews with stakeholders, assessing their tolerance levels for particular activities or number of tourists in the area. Social carrying capacity is influenced by economic arguments (so production carrying capacity and communication surrounding the maximum profitability of bivalves) and by arguments as well as regulations regarding ecological carrying capacity (in many areas there is legislation (national or in the form of EU directives) determining boundaries of what society sees as acceptable impacts). Byron et al. (2015) produced one of the first modelling approaches to integrate ecological and socio-economical aspects of bivalve aquaculture. Such modelling tools can be very valuable in communication about the costs and benefits of aquaculture to society and perhaps increasing the social carrying capacity for aquaculture. Although, these models do not assess social carrying capacity in itself, they may become increasingly important in adaptive management of systems.

23.4 Carrying Capacity Management

Management of bivalve aquaculture involves the bivalve industry, governmental organisations and other stakeholders such as consumers, environmental NGO's and other users of the culture areas like fishermen and recreationists. Decisions on how to manage carrying capacity, deal with the level of production that is commercially feasible, as well as considered acceptable by society. Hence stakeholders need to decide on what is acceptable, and governmental organisations need to implement regulations and management systems. Tools to support this process need to provide the required information on possible impacts, *what if* scenarios and target values of the management aims, such as standards, thresholds of potential concern (TPC), risk assessments or development scenarios. In Table 23.2 tools and options for managing the various approaches of bivalve carrying capacity have been summarized.

Bivalve aquaculture usually requires a license to operate a farm in a certain area. To acquire a license it has to be proven that the impact of the culture is within acceptable limits. The limits are often implemented as standards or thresholds. As an example, the impact of suspended culture for the quality of the benthic environment can be evaluated by using the level of free sulphite as an indicator. A threshold value of 1500 μM S^{2-} is set as a TPC in the Aquaculture Stewardship Certification (WWF 2010). The ASC also has set a threshold for the pelagic impacts in the form of a grazing ratio of 3 at minimum (WWF 2010). Other carrying capacity thresholds are still under development. In Natura 2000 areas in Europe, TPCs are being developed to protect birds that feed on bivalves that may be in competition with bivalve aquaculture. As this should not have a negative impact on the availability of intertidal bivalves as feed for birds, a monitoring and modelling program is carried out to quantify the impact of seed mussel collectors in Dutch coastal waters on intertidal cockle populations that are a prerequisite for protected birds like the Oystercatcher (Kamermans and Capelle 2019, this volume).

As mentioned before, the Ecosystem Approach to Aquaculture aims to achieve a carrying capacity management for bivalve aquaculture where commercial stocks attribute in a balanced way to production, ecological and social goals. This approach, based on the ecosystem services concept, asks for a management system that is based on knowledge of the complex interactions between bivalve aquaculture and the ecosystem, rather than whether standards are met or thresholds are not surpassed. This is the domain of scenario analysis that can make use of advanced modelling to quantify the effects of different management decisions. Examples in literature deal with *what if* scenarios that calculate the impact of expansion of aquaculture stocks to a level that generates adverse effects, as shown in the model paragraph.

A fundamental question arises when the various categories of competing filter feeders are taken into account. Expansion of bivalve aquaculture will involve increased competition for food with other plankton consumers, varying from other bivalves to predatory ciliates. This is based on the idea that production and consumption are in balance, hence expanding bivalve aquaculture means decrease in other consumer groups. In a bivalve filter feeder dominated ecosystem in a moderate climate, the year-round appearance of bivalve filter feeding in contrast to the seasonal cycle of zooplankton, prevents a dominance of zooplankton (Herman and Scholten 1990). If we disregard the role of heterotrophic micro-organisms, bivalve aquaculture competes for food with the following dominant categories: wild bivalves stocks, comprised of native stocks, invasive stocks and in some cases restored stocks; in addition there are stocks of tunicates, sponges and other epibenthic filter feeders, including fouling organisms on the aquaculture structures that depend on phytoplankton. Hence the basic management question is about the partitioning of available food between these categories. A critical analysis of the role of invasive stocks has led to management measures in Europe to keep them under control (EU 2017). This is of practical relevance for managing bivalve stocks in the Oosterschelde (NL), for example. Competition of wild stocks of the Pacific oyster *Crassostrea gigas* with commercial stocks of the Pacific oyster has provoked efforts to remove wild oyster beds by oyster farmers to maintain the production capacity for their culture (Smaal et al. 2009, 2013). Also in areas such as the Mont St Michel Bay, management of the slipper limpet is investigated as this invasive species has the largest biomass of all bivalves in the Bay and is considered a threat for bivalve aquaculture (Cugier et al. 2008).

Management of competing stocks with different roles in the ecosystem requires an integrated approach, based on knowledge of their ecosystem services, and the relevance for various stakeholders. The Ecosystem Approach to Aquaculture provides a framework for bivalve aquaculture management that takes ecosystem services into account as well as stakeholder interests.

23.5 Conclusions

1. The carrying capacity concept for bivalve aquaculture has been applied to assess production potential of culture areas, and to address possible effects of the culture for the environment and for other users. Production potential is depending on physical and production carrying capacity of the ecosystem, while ecological and social carrying capacity determine in how far the production capacity can be

realized. This is embedded in the Ecosystem Approach to Aquaculture that says that ecosystem functions, services and resilience have to be taken into account as well as stakeholders interests. For bivalve aquaculture it means that attention is given to both positive and negative feedbacks to the ecosystem. In general, in the literature and in management approaches, attention has focused on avoiding adverse effects; the large potential of bivalve aquaculture for providing ecosystem services is generally underestimated. It is therefore concluded that analyses of ecosystem services and feedback mechanisms that attribute to the ecological carrying capacity need more attention.

2. The approach for bivalve aquaculture carrying capacity follows a hierarchical structure. Social carrying capacity is determined by what stakeholders consider to be acceptable effects of bivalve aquaculture. Hence social CC deals with decision making mechanisms, varying from market to self-organisation. These mechanisms are not different from other decision making processes, but in this case they are based on knowledge about ecological carrying capacity. However, as long as ecological carrying capacity is not established as an intrinsic feature of the ecosystem, and is defined by what society considers acceptable, ecological carrying capacity is part of social carrying capacity.
3. The key issue in bivalve aquaculture carrying capacity is the relation between production and ecological capacity. It requires detailed information on the positive and negative effects of bivalve aquaculture on the ecosystem to evaluate the relation between production and ecological carrying capacity, as well as reference values to establish thresholds and standards. Tools that are available to address this relationship, consist of simple indices to complex models. Applications depend on specific management questions in the respective areas.
4. Management of bivalve aquaculture capacity can make use of standards and threshold values to avoid adverse effects. Given the role of bivalves in the ecosystem including the various feedback types, a more advanced approach is to make use of simulation models that predict the integrated effect of various levels of bivalve aquaculture for specific management goals, such as improved ecosystem resilience. In practice, bivalve aquaculture management is confronted with different competing stocks of cultured, wild, restoration and invasive origin. Scenario models can help in finding the balance between maximizing production capacity and optimizing ecological carrying capacity in areas with bivalve aquaculture.

Acknowledgements The authors are grateful to dr Henrice Jansen and dr Jon Grant for their constructive comments.

Annex I

Overview of ecosystem characteristics of bivalve culture areas related to indices for carrying capacity estimation (Tables 23.4, 23.5, 23.6 and Fig. 23.6).

Table 23.4 shows selected ecosystems, dominant species, divided in culture, wild and invasive species, and the area (km^2), the average depth (m) and the volume (10^6 m^3)

System	Code	Country	Bivalve culture	Dominant wild spec	Invasive spec	Area	Depht	Volume
						km^2	m	10^6 m^3
Beatrix Bay	BB	NZL	Perna canaliculus			20	35	696
Carlingford lough	CL	N-IE	C gigas/M edulis	Modiolus modiolus		49	5	460
Chesapeake Bay past	CBpast	USA	C virginica			11,500	7	27,300
Chesapeake Bay present	CBpres	USA	C virginica			11,500	7	27,300
Delaware Bay	DB	USA	C virginica			1942	10	19,420
Great Entry Lagoon	GE	CAN	M edulis			58	3	117
Loch Creran	LC	UK	C gigas/M edulis	Corbula gibba		15	20	240
Lysefjord	LF	NO	M edulis			44	14	880
Marennes-Oleron	MO	FR	C gigas/M edulis			135.7	5	675
Mont St Michel Bay	MSMB	FR	C gigas/M edulis	C edule	Crepidula fomicata	240	0-10	2400
Narragansett Bay	NB	USA	C virginica	Mercenaria mercenaria		328	8.3	2724
Oosterschelde 1996	OS96	NL	C gigas/M edulis	C edule	C gigas	351	7.83	2750
Oosterscschelde 2009	OS09	NL	C gigas/M edulis	C edule	C gigas/E americanus	351	7.83	2750
Ria de Arosa	RA	SP	M galloprovincialis			228	19	4335
Saldanha bay	SB	SA	M galloprovincialis			132	10	596
Sanggou bay	SUB	China	various			154	10	1486
Sechura Bay	SEB	Peru	Argopecten purpura tus			400	15	6000
South San Fransisco Bay	SSB	USA	C virgvitca			490	5.1	2500
Thau lagoon	TL	FR	C gigas/M galloprovincialis			75	4	300
Tracadie Bay	TB	CAN	M edulis			19.4	2.5	41
Western Wadden Sea 1994	WS94	NL	M edulis	C edule		1386	2.9	4020
Western Wadden Sea 2014	WS14	NL	M edulis	C edule; M arenaria	E americanus/C gigas	1386	2.9	4020
Xiangshan Gang	XG	China	various			365	10.4	3803

Table 23.5 shows the parameters chlorophyll concentration, total phytoplankton stock, annual primary production, total ecosystem production, the biomass of the various stocks, the average clearance rate per day and the total clearance rate of the bivalves

System	Code	Chl-a	Phyto stock	PP	System PP	Total bivalve biomass	Cultured stocks	Wild stocks	Invasive stocks	Clearance rate per gram	Clearance rate total
		mg/m^3	10^6 g C	$gC/m^2/year$	10^6 gC	10^6 g ADW				l/g ADW/d	mln m^3
Beatrix Bay	BB	1.25	34.8	120	6.6	254	254			84	21.3
Carlingford lough	CL	2.34	43.1	56	7.5	244	152	92		48	11.7
Chesapeake Bay past	CBpast	10	10,920.0	400	12,602.7	40000	40,000			72	2880.0
Chesapeake Bay present	CBpres	6.9	7534.8	191	6017.8	1900	1900			72	136.8
Delaware Bay	DB	9.9	7690.3	146	776.8	178	178			72	12.8
Great Entry Lagoon	GE	1.8	8.4	135	21.5	15	15			48	0.7
Loch Creran	LC	0.8	7.7	18.6	0.8	284	24	260		48	13.6
Lysefjord	LF	1.25	44.0	120	14.5	94	94			48	4.5
Marennes-Oleron	MO	11	297.0	60	22.3	2850	2850			72	205.2
Mont St Michel Bay	MSMB	5	480.0	100	65.8	12,100	1900	5200	5000	72	871.2
Narragansett Bay	NB	3	326.9	270	242.6	1267				72	91.2
Oosterschelde 1996	OS96	6.3	693.0	380	365.4	8800	7000	1250	550	48	422.4
Oosterschelde 2009	OS09	4	440.0	155	149.1	5471	1561	1250	2660	68	372.0
Ria de Arosa	RA	4.6	797.6	277	173.0	4809	4809			72	346.2
Saldanha bay	SB	8.6	205.0	1240	448.4	300	300			72	21.6
Sanggou bay	SUB	1.7	101.0	32.6	13.8	4801	3420	1381		72	345.7
Sechura Bay	SEB	3.9	936.0	797	873.4	25,603				72	1843.4
South San Fransisco Bay	SSB	2.6	260.0	146	196.0	6255				48	300.2
Thau lagoon	TL	10	120.0	400	82.2	810	810			72	58.3
Tracadie Bay	TB	2.9	4.8	218	11.6	261	261			48	12.5
Western Wadden Sea 1994	WS94	17.5	2810.6	412	1564.5	14,700				26	382.2
Western Wadden Sea 2014	WS14	8.5	1366.8	150	569.6	12,200	3700	7466	1033	48	585.6
Xïangshan Gang	XG	5.7	867.1	500	500.0	4475	2343	2132		72	434.0

Table 23.6 shows the calculated water residence time, the clearance time, the primary production time, the clearance ratio, the log clearance ratio and the grázing ratio

System	Code	RT d	PT d	CT d	CT/RT	logCT/RT	CT/PT	log CT/PT	References
Beatrix Bay	BB	13	5.29	32.62	2.51	0.40	6.16	0.79	Gibbs (2007)
Carlingford lough	CL	20	5.73	39.28	1.96	0.29	6.86	0.84	Ferreira et al. (2008) and Sequeira et al. (2008)
Chesapeake Bay past	CBpast	22	0.87	9.48	0.43	−0.37	10.94	1.04	Harding et al. (1986)
Chesapeake Bay present	CBpres	22	1.25	199.56	9.07	0.96	159.38	2.20	Newell (1988)
Delaware Bay	OB	97	9.90	1515.29	15.62	1.19	153.06	2.18	Biggs and Howell (1984)
Great Entry Lagoon	GEL	25	0.39	162.50	6.50	0.81	413.81	2.62	Trottet et al. (2008)
Loch Creran	LC	10	10.05	17.61	1.76	0.25	1.75	0.24	Sequeira et al. (2008)
Lysefjord	LF	11	3.04	195.04	17.73	1.25	64.12	1.81	Aure et al. (2007)
Marennes-Oleron	MO	7.1	13.31	3.29	0.46	−0.33	0.25	−0.61	Heral et al. (1988), Bacher et al. (1998)
Mont St Michel Bay	MSMB	1	7.30	2.75	2.75	0.44	0.38	−0.42	Cugier et al. (2008)
Narragansett Bay	NB	26	1.35	29.86	1.15	0.06	22.16	1.35	Pilson (1985)
Oosterschelde 1996	OS96	40	1.90	6.51	0.16	−0.79	3.43	0.54	Smaal et al. (2001)
Oosterschelde 2009	OS09	40	2.95	7.39	0.18	−0.73	2.50	0.40	Smaal et al. (2013)
Ria de Arosa	RA	23	4.61	12.52	0.54	−0.26	2.72	0.43	Tenore et al. (1982) and Filgueira et al. (2010)
Saldanha bay	SB	8	0.46	27.59	3.45	0.54	60.35	1.78	Pitcher and Calder (1998) and Stenton-Dozey et al. (2001)
Sanggou bay	SUB	1.5	7.35	4.30	2.87	0.46	0.59	−0.23	Sequeira et al. (2008)
Sechura Bay	SEB	6.6	1.07	3.25	0.49	−0.31	3.04	0.48	Kluger et al. (2016)

(continued)

Table 23.6 (continued)

System	Code	RT	PT	CT	CT/RT	logCT/RT	CT/PT	log CT/PT	References
		d	d	d					
South San Fransisco Bay	SSB	11.1	1.33	8.33	0.75	–0.12	6.28	0.80	Cloem (1982)
Thau lagoon	TL	105	1.46	5.14	0.05	–1.31	3.52	0.55	Plus et al. (2006)
Tracadie Bay	TB	7	0.41	3.27	0.47	–0.33	7.97	0.90	Cranford et al. (2007)
Western Wadden Sea 1994	WS94	10	1.80	10.52	1.05	0.02	5.85	0.77	Dame et al. (1991)
Western Wadden Sea 2014	WS14	10	2.40	6.86	0.69	–0.16	2.86	0.46	WMR data
Xiangsnan Gang	XG	70	1.73	8.76	0.13	–0.90	5.05	0.70	Sequeira et al. (2008)

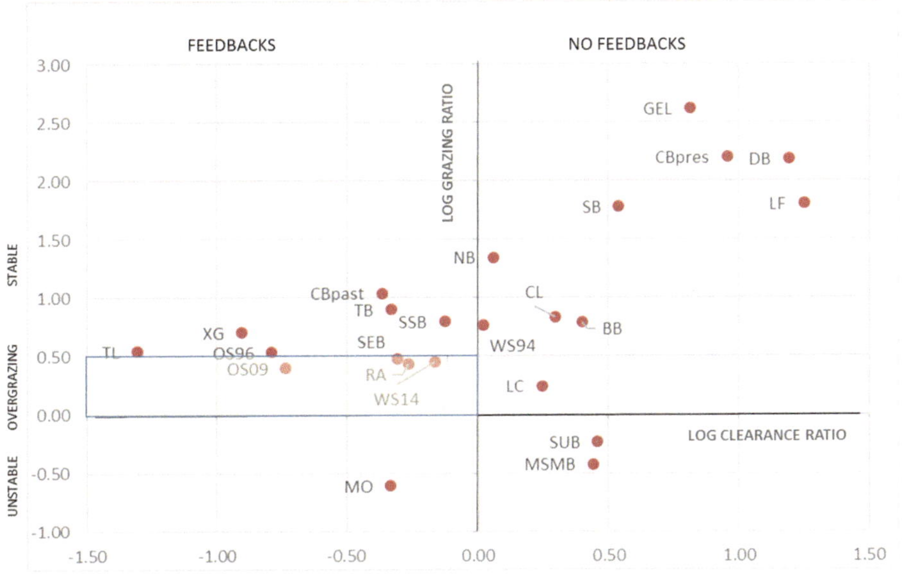

Fig. 23.6 Log Grazing Ratio as a function of Log Clearance Ratio with the right panel showing the areas with a shorter residence time than clearance time, hence less regulation or feedbacks by the bivalves on the ecosystem. The left panel shows the areas where the bivalves potentially have significant feedback effects to ecosystem processes. The index also shows a risk of overgrazing at Log GR values <0.5. This seems to be the case for Wadden Sea 2014, Ria de Arosa and Oosterschelde 2009. As mentioned previously, the bivalves in Marennes-Oleron (MO) rely on microphytobenthos rather than phytoplankton; in Sungo Bay (SUB) and Mt. Saint Michel Bay (MSMB) the grazing ratio is low but the residence time is so short that bivalves rely on food import rather than local production, as log CR > 0

References

Aguera A, van de Koppel J, Jansen JM, Smaal AC, Bouma TJ (2015) Beyond food: a foundation species facilitates its own predator. Oikos 124:1–7

Aure J, Strohmeier T, Strand Ø (2007) Modelling current speed and carrying capacity in long-line blue mussel (*Mytilus edulis*) farms. Aquac Res 38:304–312

Bacher C, Duarte P, Ferreira JG, Héral M, Raillard O (1998) Assessment and comparison of the Marennes-Oléron Bay (France) and Carlingford Lough (Ireland) carrying capacity with ecosystem models. Aquat Ecol 31:379–394

Bacher C et al (2019) Spatial, ecological and social dimensions of assessments for bivalve farm-ing management. In: Smaal A, Ferreira JG, Grant J, Petersen JK, Strand O (eds) Goods and services of marine bivalves. Springer, Cham, pp 527–549

Barnes TK, Volety AK, Chartier K, Mazzotti FJ, Pearlstine I (2007) A habitat suitability index model for the eastern oyster (Crassostrea virginica), a tool for restoration of the C aloosa-hatchee estuary, Florida. J Shellfish Res *26*(4):949–959

Biggs RB, Howell BA (1984) The estuary as a sediment trap: alternate approaches to estimating filter efficiencies. In: Kennedy VS (ed) The estuary as a filter. Academic, New York, pp 107–129

Byron C, Bengtson D, Costa-Pierce B, Calanni J (2011a) Integrating science into management: ecological carrying capacity of bivalve shellfish aquaculture. Mar Policy 35:363–370

Byron C, Link J, Costa-Pierce B, Bengtson D (2011b) Calculating ecological carrying capacity of shellfish aquaculture using mass-balance modelling: Narragansett Bay, Rhode Island. Ecol Model 222:1743–1755
Byron CJ, Jin D, Dalton TM (2015) An integrated ecological-economic modeling framework for the sustainable management of oyster farming. Aquaculture 447:15–22
Cloern JE (1982) Does the benthos control phytoplankton biomass in south San Francisco Bay? Mar Ecol Prog Ser 9:191–202
Craeymeersch J, Jansen H (2019) Bivalve assemblages as hotspots for biodiversity. In: Smaal A, Ferreira JG, Grant J, Petersen JK, Strand O (eds) Goods and services of marine bivalves. Springer, Cham, pp 275–294
Cranford PJ, Strain PM, Dowd M, Hargrave BT, Grant J, Archambault MC (2007) Influence of mussel aquaculture on nitrogen dynamics in a nutrient enriched coastal embayment. Mar Ecol Prog Ser 347:61–78
Cranford PJ, Hargrave B, Li W (2009) No mussel is an island. ICES Insight 46:44–49
Cranford PJ, Kamermans P, Krause G, Mazurié J, Buck BH, Dolmer P, Fraser D, Van Nieuwenhove K, O'Beirn FX, Sanchez-Mata A, Thorarinsdótir GG, Strand Ø (2012) An ecosystem-based approach and management framework for the integrated evaluation of bivalve aquaculture impacts. Aquac Environ Interact 2:193–213
Cugier P, Struski C, Blanchard M, Mazurie J, Pouvreau J, Óliver F (2008) Studying the carrying capacity of Mont Saint Michel Bay (France): respective role of the main filter feeders communities, ICES online at www.ices.dk/products/CMdocs/CM-2008/H/H0108.pdf. Nov 2010
Dame RF (1996) Ecology of marine bivalves: an ecosystem approach. CRC Mar Sci Ser 272 pp
Dame RF, Prins TC (1998) Bivalve carrying capacity in coastal ecosystems. Aquat Ecol 31:409–421
Dame R, Dankers N, Prins T, Jongsma H, Smaal A (1991) The influence of mussel beds on nutrients in the Western Wadden Sea and Eastern Scheldt Estuaries. Estuaries 14:130–138
Dowd M (2005) A bio-physical coastal ecosystem model for assessing environmental effects of marine bivalve aquaculture. Ecol Model 183:323–346
European Union, 2017. Regulation (EU) No 1143/2014 of the European Parliament and of the Council of 22 October 2014 on the prevention and management of the introduction and spread of invasive alien species
Ferreira JG, AJS H, Monteiro P, Moore H, Service M, Pascoe PL, Ramos L, Sequeira A (2008) Integrated assessment of ecosystem-scale carrying capacity in shellfish growing areas. Aquaculture 275:138–151
Ferreira JG et al (2019) Assessment of nutrient trading services from bivalve farming. In: Smaal A, Ferreira JG, Grant J, Petersen JK, Strand O (eds) Goods and services of marine bivalves. Springer, Cham, pp 551–584
Filgueira R, Comeau LA, Guyondet T, McKindsey CW, Byron CJ (2015) Modelling carrying capacity of bivalve aquaculture: a review of definitions and methods. Encyclopedia of sustainability science and technology. Springer, New York
Gibbs MT (2007) Sustainability performance indicators for suspended bivalve aquaculture activities. Ecol Indic 7:94–107
Gibbs MT (2009) Implementation barriers to establishing a sustainable coastal aquaculture sector. Mar Policy 33:83–89
Grant J, Pastres R (2019) Ecosystem models of bivalve aquaculture: implications for supporting goods and services. In: Smaal A, Ferreira JG, Grant J, Petersen JK, Strand O (eds) Goods and services of marine bivalves. Springer, Cham, pp 507–525
Grant J, Curran KJ, Guyondet TL, Tita G, Bacher C, Koutitonsky V, Dowd M (2007) A box model of carrying capacity for suspended mussel aquaculture in Lagune de la Grande-Entrée, Iles-de-la-Madeleine, Québec. Ecol Model 200:193–206
Harding LW, Meeson BW, Fisher TR (1986) Phytoplankton production in two east coast estuaries: photosynthesis light functions and patterns of carbon assimilation in Chesapeake and Delaware Bays. Est Coast Shelf Sci 23:773–806

Hargreaves JA (2011) Molluscan shellfish aquaculture and best practices management. In: Shumway SE (ed) Shellfish aquaculture and the environment. Wiley-Blackwell, Ames, pp 51–80

Héral M, Deslous-Paoli J-M, Prou J (1988) Approche de la capacité trophique d'un écosystème conchilicole. J Cons Int Explor Mer, Cm 1988/K 22 p

Herman PMJ, Scholten H (1990) Can suspension-feeders stabilize estuarine ecosystems? In: Barnes M, Gibson RN (eds) Trophic relationships in the marine environment. Proceedings of the 24th European marine biology symposium. Aberdeen. University Press, Aberdeen, pp 104–116

Inglis GJ, Hayden BJ, Ross AH (2000) An overview of factors affecting the carrying capacity of coastal embayments for mussel culture. NIWA, Christchurch. Client Report CHC00/69: vi + 31 p

Jansen Henrice M, Strand Ø, van Broekhoven W, Strohmeier T, Verdegem MC, Smaal AC (2019) Feedbacks from filter feeders: review on the role of mussels in cycling and storage of nutrients in oligo- meso- and eutrophic cultivation areas. In: Smaal A, Ferreira JG, Grant J, Petersen JK, Strand O (eds) Goods and services of marine bivalves. Springer, Cham, pp 143–177

Kamermans P, Capelle J (2019) Provisioning of mussel seed and its efficient use in culture. In: Smaal A, Ferreira JG, Grant J, Petersen JK, Strand O (eds) Goods and services of marine bivalves. Springer, Cham, pp 27–49

Kashiwai M (1995) History of carrying capacity concept as an index of ecosystem productivity (Review). Bull Hokkaido Natl Fish Res Inst 59:81–101

Kluger LC, Taylor MH, Mendo J, Tam J, Wolff M (2016a) Carrying capacity simulations as a tool for ecosystem-based management of a scallop aquaculture system. Ecol Model 331:44–55

Kluger LC, Taylor MH, Rivera EB, Silva ET, Wolff M (2016b) Assessing the ecosystem impact of scallop bottom culture through a community analysis and trophic modelling approach. MEPS 547:121–135

Kluger LC, Filgueira R, Wolff M (2017) Integrating the concept of resilience into an ecosystem approach to bivalve aquaculture management. Ecosystems. https://doi.org/10.1007/s10021-017-0118-z

Lawson SR, Manning RE, Valliere WA, Wang B (2003) Proactive monitoring and adaptive management of social carrying capacity in Arches National Park: an application of computer simulation modeling. J Environ Manag *68*(3):305–313

McKindsey CW (2013) Carrying capacity for sustainable bivalve aquaculture. In: Christou P, Savin R, Costa-Pierce B, Misztal I, Whitelaw B (eds) Sustainable food production. Springer, New York, pp 449–466. https://doi.org/10.1007/978-1-4614-5797-8

McKindsey CW, Thetmeyer H, Landry T, Silvert W (2006) Review of recent carrying capacity models for bivalve culture and recommendations for research and management. Aquaculture 261(2):451–462

McKindsey CW, Archambault P, Callier MD, Olivier F (2011) Influence of suspended and off-bottom mussel culture on the sea bottom and benthic habitats: a review. Can J Zool 89:622–646

Newell RIE (1988) Ecological changes in Chesapeake Bay: are they the result of overharvesting the American oyster, *Crassostrea virginica*? Understanding the estuary: advances in Chesapeake Bay Research. Chesapeake Research Consortium Publication, Baltimore

Newell RIE (2004) Ecosystem influences of natural and cultivated populations of suspension-feeding bivalve molluscs: a review. J Shellfish Res 23:51–61

Newell CR, Brady D, Richardson J (2019) Farm-scale production models. In: Smaal A, Ferreira JG, Grant J, Petersen JK, Strand O (eds) Goods and services of marine bivalves. Springer, Cham, pp 485–506

Nobre AM (2009) An Ecological and Economic Assessment Methodology for Coastal Ecosystem Management. Environ Manag 44:185–204

Odum EP (1953) Fundamentals of ecology. Saunders, Philadelphia. pp 574

Ostrom E (2009) A general framework for analyzing sustainability of social-ecological systems. Science 325:419–422

Passarelli C, Olivier F, Paterson DM, Meziane T, Hubas C (2014) Organisms as cooperative ecosystem engineers in intertidal flats. J Sea Res 92:92–101
Petersen JK, Maar M, Ysebaert T, Herman PMJ (2013) Near-bed gradients in particles and nutrients above a mussel bed in the Limfjorden: influence of physical mixing and mussel filtration. MEPS 490:137–146
Pilson MQ (1985) On the residence time of water in Narragansett Bay. Estuaries 8:2–14
Pitcher GC, Calder D (1998) Shellfish mariculture in the Benguela system: phytoplankton and the availability of food for commercial mussel farms in Saldanha Bay, South Africa. J Shellfish Res 17:15–24
Plus M, La Jeunesse I, Bouraoui F, Zaldivar JM, Chapelle A, Lazure P (2006) Modelling water discharges and nitrogen inputs into a Mediterranean lagoon – impact on the primary production. Ecol Model 193:69–89
Prins TC, Smaal AC, Dame RF (1998) A review of the feedbacks between bivalve grazing and ecosystem processes. Aquat Ecol 31:349–359
Ruesink JL, Lenihan HS, Trimble AC, Heiman KW, Micheli F, Byers JE, Kay MC (2005) Introduction of non-native oysters: ecosystem effects and restoration implications. Annu Rev Ecol Syst 36:643–689
Sequeira A, Ferreira JG, Hawkins AJS, Nobre A, Lourenco P, Zhang XL, Yan X, Nickell T (2008) Trade-offs between shellfish aquaculture and benthic biodiversity: a modelling approach for sustainable management. Aquaculture 274:313–328
Smaal AC, Prins TC (1993) The uptake of organic matter and the release of inorganic nutrients by bivalve suspension feeder beds. In: Dame RF (ed) Bivalve filter feeders in estuarine and coastal ecosystem processes. Springer, Berlin, pp 271–298
Smaal AC, Prins TC, Dankers N, Ball B (1998) Minimum requirements for modelling bivalve carrying capacity. Aquat Ecol 31:423–428
Smaal A, van Stralen M, Schuiling E (2001) The interaction between shellfish culture and ecosystem processes. Can J Fish Aquat Sci 58:991–1002
Smaal AC, Schellekens T, van Stralen MR, Kromkamp JC (2013) Decrease of the carrying capacity of the Oosterschelde estuary (SW Delta, NL) for bivalve filter feeders due to overgrazing? Aquaculture 404–405:28–34
Soto D, Aguilar-Manjarrez J, Brugère C, Angel D, Bailey C, Black K, Edwards P, Costa-Pierce B, Chopin T, Deudero S, Freeman S, Hambrey J, Hishamunda N, Knowler D, Silvert W, Marba N, Mathe S, Norambuena R, Simard F, Tett P, Troell M, Wainberg A (2008) Applying an ecosystem based approach to aquaculture: principles, scales and some management measures. In: Soto D, Aguilar-Manjarrez J, Hishamunda N (eds) Building an ecosystem approach to aquaculture. FAO/Universitat de les Illes Balears Expert Workshop. 7–11 May 2007, Palma de Mallorca, Spain, FAO fisheries and aquaculture proceedings no 14. Rome, FAO, pp 15–35
Stenton-Dozey J, Probyn T, Busby A (2001) Impact of mussel (*Mytilus galloprovincialis*) raft culture on benthic macrofauna, in situ oxygen uptake, and nutrient fluxes in Saldanha Bay, South Africa. Can J Fish Aquat 58:1021–1031
Tarrant MA, English DBK (1996) A crowding-based model of social carrying capacity: applications for white-water boating use. J Leis Res 28(3):155–168
Tenore KR, Boyer LF, Cal RM, Corral J, Garcia-Fernandez C, Gonzalez N, Gonzalez-Gurriaran E, Hanson RB, Iglesias J, Krom M, Lopez-Jamar E, McClain J, Pamatmat MM, Perez A, Rhoads DC, de Santiago G, Tietjen J, Westrich J, Windom HL (1982) Coastal upwelling in the Rias Bajas, NW Spain: contrasting the benthic regimes of the Rias de Arosa and de Muros. J Mar Res 40:701–772
Tett P, Portilla E, Gillibrand P, Inall M (2011) Carrying and assimilative capacities: the ACExR-LESV model for sea-loch aquaculture. Aquac Res 42:51–67
Thompson JK (2005) One estuary, one invasion, two responses: phytoplankton and benthic community dynamics determine the effect of an estuarine invasive suspension feeder. In: Dame RF, Olenin S (eds) The comparative roles of suspension feeders in ecosystems, Nato science series IV, vol 47, pp 291–316

Trottet A, Roy S, Tamigneaux E, Lovejoy C, Tremblay R (2008) Influence of suspended mussel farming on planktonic communities in Grande-Entrée Lagoon, Magdalen Islands (Québec, Can- ada). Aquaculture 276:91–102

Walker B, Carpenter S, Anderies J, Abel N, Cumming G, Janssen M, Lebel L, Norberg J, Peterson GD, Pritchard R (2003) Resilience management in social-ecological systems: a working hypothesis for a participatory approach. Conserv Ecol 6:1–14

WWF (2010) Bivalve aquaculture dialogue standards. www.worldwildlife.org/what/globalmarkets/aquaculture/WWFBinaryitem17872.pdf

Ysebaert T, Walles B, Haner J, Hancock B (2019) Habitat modification and coastal protection by ecosystem-engineering reef-building bivalves. In: Smaal A, Ferreira JG, Grant J, Petersen JK, Strand O (eds) Goods and services of marine bivalves. Springer, Cham, pp 253–273

Zeldis J (2005) Magnitudes of natural and mussel farm-derived fluxes of carbon and nitrogen in the Firth of Thames, NIWA Client Report: CHC2005-048

Chapter 24
Farm-Scale Production Models

Carter R. Newell, Damian C. Brady, and John Richardson

Abstract Farm-scale production models of bivalves have been used for site selection, optimization of culture practices, and the estimation of ecosystem goods and services. While all farm models require physical forcing through hydrodynamic models, the input of measured or modelled bivalve growth drivers, and a bioenergetic growth model which predicts individual growth and farm yield as a function of husbandry practices, some models are also embedded in a GIS system to allow for a "point and click" ability to test different locations and production strategies at various locations within the modeled domain. More generic Web-based models such as the Farm Aquaculture Resource Management are relatively simple to use, provide a link to larger ecosystem models, and provide direct estimates of ecosystem services. More detailed models, such as *ShellGIS,* may be more data intensive and require detailed bathymetry, spatial velocity fields, information about boundary layer and aquaculture structure hydrodynamics and particle depletion. However, these models provide the detailed spatial and temporal results that can optimize farm productivity and assess benthic impacts. New approaches using high resolution remote sensing satellites and powerful physical-biogeochemical models using unstructured grids to link farm scale models with ecosystem models in a GIS platform have potential to provide improvements in the utility of farm scale models for the estimation of bivalve aquaculture ecosystem goods and services.

Abstract in Chinese 摘要:养殖场规模的双壳贝类产量评估模型已经被广泛应用于养殖选址,养殖配置优化以及生态系统产品和服务评估。大部分的养殖场规模模型都需要通过水动力模型提供驱动,使用实测或模拟的贝类生长数据作为初始和驱动条件,并通过个体生长预测模型及产量评估模型进行结

C. R. Newell (✉)
Maine Shellfish R+D, Damariscotta, ME, USA

D. C. Brady
School of Marine Sciences, Darling Marine Center, University of Maine, Walpole, ME, USA
e-mail: damian.brady@maine.edu

J. Richardson
Blue Hill Hydraulics, Blue Hill, ME, USA
e-mail: jrichardson@bluehillhydraulics.com

A. C. Smaal et al. (eds.), *Goods and Services of Marine Bivalves*,
https://doi.org/10.1007/978-3-319-96776-9_24

果的验证和应用。一些模型还可以嵌入到GIS系统中，允许通过“点选”来测试在模拟区域内，对不同位置进行不同生产策略的组合所产生的效果。一些基于网络的通用模型（比如养殖水域资源管理系统）的使用相对简单，这些系统可以通过网络链接到位于服务器上的大型生态系统模型，通过调用模型结果从而对生态系统服务进行直接评估。一些更加具体的模型（如ShellGIS），可能需要更多的数据（如详细的地形， 边界层信息）来进行获取相应的结果（如养殖设施周围流场结构和示踪粒子扩散分布情况）。虽然这些比较复杂且要求数据较多，但是他们提供了养殖水域内更详细的物理环境状况，模型结果可以用于优化养殖布局以及进行底质环境的评估。利用高分辨率遥感卫星和非结构化网格的物理-生物地球化学耦合模型可以将养殖场尺度模型与GIS平台中的生态系统模型联系起来，这种新技术有助于推动养殖场尺度模型在双壳贝类养殖生态系统优势和生态服务评估方面发挥重要作用。

Keywords Farm-scale bivalve production models FARM · Geographical Information Systems · Particle depletion · Computational fluid dynamics · ShellGIS · Bivalve growth

关键词 养殖场尺度贝类生产模型（FARM） • 地理信息系统（GIS） • 粒子扩散 • 计算流体力学 • ShellGIS • 双壳贝类生长

24.1 Introduction

The role of bivalve farms in the provision of ecosystem goods and services has been reviewed by Ferreira et al. 2011, using the Farm Aquaculture Resource Management (FARM) model (Ferreira et al. 2007), and stressing the importance of bivalve farms in mitigating the consequences of nutrient loading. The application of dynamic biogeochemical, bivalve ecophysiological, and physical oceanographic models to predict bivalve growth have been reviewed by Grant and Filgueira (2011). In this chapter, we concentrate on bioenergetic and mechanistic mass balance models which predict farm production and regulating services, as opposed to statistical models which utilize hypothesized or measured relationships among variables in a specific data set such as the statistical relationships between clam growth, the flux of seston and bottom characteristics (Grizzle and Lutz 1989).

Production models for marine bivalve farms may be used for site selection for new farms, to determine the production capacity and optimal seeding density for farms, as well as to predict benthic and water column interactions and regulating services. While there have been numerous studies of bivalve carrying capacity and bay scale production capacity, relatively few have dealt with smaller scale aquaculture farm models. One of the difficulties in assessing farm-scale production models is that they rely on environmental drivers of production related to larger, bay-scale and ecosystem-scale processes. These large scale dynamics are connected to farm-scale models simulating local oceanographic conditions and culture practices that affect food availability, feeding and growth of the bivalves on the farms. Some approaches also use coupled physical-biogeochemical models and animal growth models to investigate farm and ecosystem interactions (Ferreira et al. 2008; Guyondet et al. 2010; Filgueira et al. 2014), and the effects of husbandry practices on both local and system scales (Smaal et al. 1997; Saurel et al. 2014). Reviews of the types of models available for aquaculture site selection and carrying capacity are presented in McKindsey et al. (2006), Ross et al. (2013), and Filgueira et al. (2015).

The ecological and biogeochemical models influencing bivalve food production (phytoplankton and detritus) such as the Simulation Model for the Oosterschelde (SMOES; Scholten and van der Tol 1994), ECOWIN (Ferreira 1995), RMA (King 2003), Row-Column Advanced Ecological Systems Modeling Program (RCA; Testa et al. 2014), General Aquaculture Model for Bivalve Equilibrium Yield (GAMBEY; Nunes et al. 2003) and others (Grant et al. 2008) are not usually considered in farm scale models due to a mismatch between the spatial scale of the typical application (i.e., bay or ecosystem scale) and the scale of an aquaculture farm (10–100's of meters). Filgueira et al. (2014) used an ecosystem model, a biogeochemical model, and a bivalve growth model to determine oyster carrying capacity in subareas of a region of complex geomorphology in New Brunswick, Canada, but not individual farms within each of these subareas. Since the consumption of food by bivalves occurs at relatively small spatial scales, the processes are important when coupling farm scale production models to the surrounding ecosystem, since local particle flux and consumption can influence bivalve growth and regulating services. This is especially true when food availability for bivalves on the farm is affected by particle depletion as occurs within dense populations of bivalves. Particle depletion is also affected by farming practices, or husbandry. Ultimately, the inclusion of husbandry is an important distinguishing feature of farm-scale models. For example, there models can account for the time of year seeded, the seed genetic origin and size, stocking density, animal biomass (Ferreira et al. 2007), nursery and grow-out gear type, and placement, spacing and orientation of gear on the farm (Bacher et al. 1997, 2003; Campbell and Newell 1998; Comeau et al. 2008; Drapeau et al. 2006; Duarte et al. 2008; Rosland et al. 2011; Newell and Richardson 2014).

The purpose of this chapter is to compare the advantages and disadvantages of different farm-scale production models, and to highlight promising new approaches which can improve the predictive value of farm-scale production models for ecosystem goods and services, and to suggest areas for future research.

24.2 Farm-Scale Models

There are two fairly well established farm-scale models in use today: *FARM* and *ShellGIS*. The *FARM* model (Ferreira et al. 2007) allows for the input of farm dimensions, species, density, cultivation period, temperature, water velocity, chlorophyll-a, particulate organic matter (POM), total particulate matter (TPM) and dissolved oxygen (DO) to calculate growth and harvestable biomass, using a growth model such as *AquaShell*™ (Silva et al. 2011; Saurel et al. 2014). *ShellGIS* (Newell et al. 2013) is a GIS system which also calculates growth and harvestable biomass, using the shellfish growth model *ShellSIM* (Hawkins et al. 2013a, b), and georeferenced growth driver data (temperature, salinity, chlorophyll-a, POM, TPM, DO) as well as water velocity from a flow model. Both shellfish growth models are a mechanistic function of the concentrations and quality of food particles either forced by monitoring or model data (no feedbacks) or embedded in a farm scale model with feedbacks on the environment. The *FARM* model scale is entered by the user, and consists of the farm dimensions (length, width, and depth). In *ShellGIS*, the scale of the flow model (50 m) defines the smallest scale (one cell), but results from multiple grids can be selected by the user by drawing a rectangle in the system. In the FARM model, water velocity and water quality data are entered for each farm. In ShellGIS, the program uses water velocity generated from each grid point and water quality data from georeferenced water samples, water quality model output, or interpolated values from nearby locations.

The *FARM* model produces growth rates, total harvest biomass, biomass seed to harvest ratio, profit, nitrogen credits, and an Assessment of Estuarine Trophic State (ASSETS) eutrophication score. *FARM* outputs are customized to provide direct estimates of bivalve ecosystem goods and services.

ShellGIS has been customized primarily to optimize farm production. *ShellGIS* walks the user through a list of frequently asked questions that determine target species, culture type (bottom or suspended), density (for bottom culture), angle of flow (degrees), stocking density (for suspended), start date, and period to run, including the following:

- What space is available to grow shellfish? (site selection)
- How does water flow through the system? (site selection)
- How do temperature and salinity vary through the system? (site selection)
- How fast will shellfish grow? (production)
- How to minimize time to market? (production and husbandry)
- How to maximize yield and profit? (production)
- How do variations between growing years affect yield? (production)
- What is the hydraulic zone of influence around the farm? (regulation)
- Results from farm or entire embayment (site selection)

While *ShellGIS* does not provide summary information for ecosystem services, *ShellSIM* can be used to plot not only growth but also physiological rates such as oxygen consumption, clearance rate, ammonium excretion and biodeposition rates at any location or time. These data could be integrated into a calculation of ecosystem services through improvements in the software if regulators or growers demonstrated interest. A comparison among *FARM* and *ShellGIS* is presented in Table 24.1.

Table 24.1 Comparisons between the FARM and ShellGIS farm scale models using the criteria of Nath et al. 2000

	FARM	ShellGIS
Objectives	Site selection, optimization of culture practice, ecological effects	Site selection, optimization of culture practices
Target audience	Regulators, growers	Growers
Analytical methods	Simple mass balance model Model forcing is constant Does not include turbulent mixing Shellfish growth model Economic model ASSETS score	GIS-based interactive model Model forcing varies spatially and temporally Includes turbulent mixing in benthic boundary layer and structure models Shellfish growth model Economic model
Analytical methods and results	Bioenergetic growth model Economic model Eutrophication assessment	Bioenergetic growth model GIS layers of bivalve growth drivers Aquaculture structure model Economic model
Geographic area and scale	Farm dimensions and embayment	Embayment and 50 m in GIS framework
Actual use	Extensive	Limited
Comments	Available on the web	Requires running high resolution flow model and collection of bivalve growth drivers
Scale of physical forcing	Single velocity	Spatial velocity fields

24.3 Geographic Information Systems (GIS)

The utility of farm models for decision-making is facilitated by a GIS platform, where specific areas of the farm and culture structures are georeferenced. While farm models have not extensively been embedded in GIS frameworks, this section is included because the future development of bivalve ecosystem goods and services models will be improved by advances in remote sensing and GIS.

Nath et al. (2000) listed GIS platforms and compared the use of GIS systems in aquaculture decision-making, in a number of case studies according to the following criteria:

- Objectives
- Target decision support audience
- Geographic area and scale of analysis
- Analytical methods and results
- Actual use for decision making

GIS systems range from large to fine scale for a variety of purposes including site selection, environmental impacts, and farm productivity estimates for both finfish and shellfish. On a large scale (km), Buitrago et al. (2005) used 20 variables to choose optimum oyster raft sites in Venezuela in a Multi-Criteria Evaluation (MCE). In the intertidal zone, Congleton et al. (1999) used a GIS system to combine intertidal height and water velocity for soft clam mariculture siting, and Arnold et al. (2000) used multiple water quality and benthic habitat criteria to identify sites for hard clam aquaculture. Radiarta et al. (2008) used satellite imagery of chlorophyll-a and temperature, a weighted bio-physical, social-infrastructural and constraint criteria and model builder in ArcGIS to identify the best sites for scallop grow-out. Thomas et al. (2011)) also utilized satellite imagery to predict mussel growth in Mont St.-Michel Bay in France using a Dynamic Energy Budget (DEB) model, with 1 km resolution. Longdill et al. (2008) used a GIS approach to identify Aquaculture Management Areas (AMA's) in New Zealand, which combined residual water velocity, benthic habitat, primary productivity, marine protected areas and constraints and conflicting uses using MCE techniques. Tissot et al. (2012) integrated multiple spatial and temporal environmental data into a GIS based productivity model for oyster farms. Improvements in web-based and GIS-based tools, are likely to improve appropriate siting of farms, but smaller scale models utilizing fine scale hydrodynamics, biomass, and density distribution within the farms are required to provide greater insight into optimization within the farms themselves. In all of these GIS approaches, the presentation of data and model results are only as good as the resolution of the underlying data. Higher resolution ($\leq$100 m) satellites such as Landsat 8 have promise in providing data on temperature, chlorophyll-*a*, and turbidity for site selection and growth drivers for farm scale models (Snyder et al. 2017), especially if they are combined with bivalve growth and aquaculture structure models. If satellite data can be made available at a high enough frequency and resolution, it could be eventually used to provide environmental drivers of production for farm-scale models.

24.4 Farm Model Components

Farm models all have components which simulate water movement (hydrodynamic models) and bivalve growth (growth models) using forcing functions for growth (environmental growth drivers like food and temperature), a description of husbandry practices, and a process for accounting for food particle depletion within the farm. Depletion (reduction in the particulate phytoplankton and detritus, or seston) is a function of food supply (concentration × flow rate) and food demand (filtration by the bivalves). Biogeochemical feedbacks and nutrient cycling on the farms also are important in relation to bivalve biodeposition and excretion, regulating services of the farms. Specifically, bivalves do not simply clear the water of suspended particulates (Newell et al. 1989), but rather they participate in nutrient recycling and are involved in benthic/pelagic coupling.

24.5 Physical Models

Physical models are essentially advection-diffusion equations that predict water velocity at the farm-scale (*i.e.*, 50–100 m) or in the embayment or ecosystem (100–1000 m). The models require data on bathymetry, as well as the meteorological (e.g., wind and precipitation), river, and tidal drivers of circulation. Although the following list is far from exhaustive, some of the more common models are Delft-3D (Delft Hydraulics 2006), the Regional Ocean Modeling System (ROMS; Wilkin et al. 2005), MIKE 21 (Warren and Bach 1992), and the Finite Volume Community Ocean Model (FVCOM; Chen 2012). Some models (*i.e.*, Computational Fluid Dynamic (CFD) models, Hirt and Nichols 1988), although more computationally intensive, can also include aquaculture structure hydrodynamics (*e.g.*, suspended longlines, rafts, trays, racks and bags) as well as benthic boundary layer flow and benthic-pelagic coupling. Hydrodynamic-structure interaction modeling requires detailed physical representations of the aquaculture structures used on the farms (e.g. Plew et al. 2005; Stevens et al. 2008; Delaux et al. 2011; Newell and Richardson 2014; Tseung et al. 2016). The effects of the farms themselves on circulation can be estimated using farm drag coefficients (Grant and Bacher 2001; Pilditch et al. 2001; Plew 2011). A simple representation of water velocity as uniform throughout the site, both in the benthic boundary layer, and in and around aquaculture structures, is not representative of conditions on the farms. The ease of use of a simpler model (*e.g., FARM*) is a trade-off with greater complexities in a more detailed flow model approaches (*e.g., ShellGIS*). In *ShellGIS*, the choice of the type of husbandry (e.g., bottom culture, floating cages, rafts) activates a physical model which incorporates the aquaculture structure hydrodynamics to model food supply and seston depletion (below) more effectively. Finally, wave models such as the Simulating Waves Nearshore model (SWAN; Booij et al. 1997) can be used to model the height of waves which can disrupt bivalve feeding, growth, and farm yield, especially for suspended cultures, due to oscillating high velocities (Dewhurst 2016) and bottom resuspension that can inhibit feeding (Newell et al. 1989).

24.6 Organism Growth Models

Biological models predict organism growth as a mechanistic function of the concentrations and quality of food particles either forced by monitoring data or model data (no feed-backs) or embedded in a farm scale model with feed-backs. Bioenergetic growth models such as ShellSim (Hawkins et al. 2013a, b), Ecophysiological Model of *Mytilus edulis* (EMMY; Scholten and Smaal 1998), MUSMOD (Campbell and Newell 1998), Oyster-DEB (Pouvreau et al. 2006) and AquaShell™ (Silva et al. 2011) are used in farm models to predict bivalve production, including shell growth and tissue growth, and they can also be used to estimate regulating services of farms (carbon sequestration and nitrogen removal on harvest). Recently, Filgueira et al.

(2011) and Larsen et al. (2014) compared bioenergetic scope for growth (SFG, Bayne and Newell 1983) and dynamic energy budget (DEB, Van Haren and Kooijman 1993) bivalve growth models. While both models do a reasonable job of predicting bivalve growth, there are limitations related to the use of chemical proxies of food availability (chlorophyll-*a*, particulate organic matter (POM) and particulate organic carbon and nitrogen (POC, PON) as growth drivers for bivalves since the consumption and absorption of seston is related to its biochemical composition and digestibility.

In these models, growth is represented as a function of the concentration and quality of food particles. However, in farm scenarios, populations of bivalves may locally deplete food concentrations within the culture structures, within the farms, and within the embayments (see Sect. 24.8 below). While the removal of suspended particulates by the bivalves is considered an ecosystem service, it impacts the growth rates of bivalves if there is not a careful consideration of food supply and demand on the farm. Bacher et al. (2003) gave a more realistic approach to modeling scallop growth in Sungo Bay, China over a range of stocking densities when they included particle depletion by the bivalves and the influence of the structures on water velocity.

Some models also may include density dependent growth rates within the "culture units" such as a suspended ropes, pegged ropes, trays, bags, or bottom "patches" of bivalves (Campbell and Newell 1998). Organism growth may also be influenced by space as well as food (Frechette and Lefaivre 1990).

24.7 Environmental Growth Drivers

Bivalve growth drivers include the environmental conditions, water quality parameters and the live phytoplankton, detritus, and other suspended particles which contribute to bivalve feeding and growth. For each species, there are different requirements for these growth drivers, based on the feeding behavior and particle size retention efficiency of the ctenidia, and sensitivity to parameters such as temperature and salinity, including direct effects of water velocity on feeding (Wildish and Miyares 1990; Newell et al. 2001), red tides (Shumway 1990), hypoxia and pollutants. Growth models often use chlorophyll-*a* and POM as bivalve food, but POM can vary in quality (Newell et al. 1998) so the deterministic mussel growth model MUSMOD (Campbell and Newell 1998) utilized phytoplankton biovolume to carbon conversions and detrital carbon (and the detrital N/C ratio) to characterize bivalve food and model mussel growth (Campbell and Newell 1998). ShellSIM uses a chlorophyll-*a* to carbon conversion to quantify food quality as *selected organic matter* (SELORG; the phytoplankton-based carbon), and *remaining organic matter* (REMORG; Hawkins et al. 2013a, b).

Even within a farm, food quality can vary depending on the location of the bivalves. Muschenheim and Newell (1992) found that mussels on the edge of a bottom bed had enhanced access to benthic diatoms and organic detritus in water from

0–10 cm off the bottom, whereas mussels further in the farm relied on water being mixed from surface. It is known that bivalves grow faster on some kinds of algae than others (Epifanio 1979), so chemical measures of food concentration (*i.e.,* chlorophyll-a) have their limitations. Recent (unpublished) data has shown that American oysters (*Crassostrea virginica*) have increased clearance rates and absorption efficiencies with a natural diet dominated by large ciliates than with chain-forming diatoms a few weeks later.

24.8 Depletion Models

Depletion models are used to predict water column effects of bivalve farms, both in modifying available food within the farms, and also quantifying ecosystem services of reducing turbidity, grazing down phytoplankton blooms and benthic pelagic coupling. Simple models of food supply and demand in marine bivalve aquaculture systems are based on mass balances. For example, Incze and Lutz (1980) investigated the flux and consumption of particles for a mussel long-line system based on a predicted filtration rate of 2.4 l h^{-1} per mussel. Rosenberg and Loo (1983) used the approach of Incze and Lutz (1980) to estimate the food ration and flow speed necessary to supply food to a variable number of longlines in Sweden. Both studies concluded that food supply was a function of both current velocity and ambient particle concentration. Carver and Mallet (1990) examined the carrying capacity of a Nova Scotia inlet for mussel longline culture in a similar way, as did Ferreira et al. (2007) with the FARM model. Simple box models generally cannot incorporate the effects of different aquaculture structures on flow, benthic boundary layer dynamics for bottom culture, or density dependent growth rates within culture units.

More complex models of seston depletion consider the characteristics of aquaculture structures and benthic boundary layer dynamics, where reduced velocities are observed from bottom friction or physical drag (Plew 2005). Depletion models of food resources in rafts, longline cages, and benthic free planted bivalves can provide a more realistic picture of localized food concentration than the simple models described above. The following test cases illustrate the importance of aquaculture structure hydrodynamics and particle depletion models in determining the food availability at local scales to bivalves. Ultimately, these dynamics control bivalve growth rate on the farm scale.

24.8.1 Boundary Layer Depletion Models: Free-Planted Mussels

An advection-diffusion equation using water depth, initial phytoplankton concentration, vertical eddy diffusivity, phytoplankton uptake by mussels (*Mytilus edulis*) of known biomass or "filtration velocity" (w_{filt}) and downstream distance in the

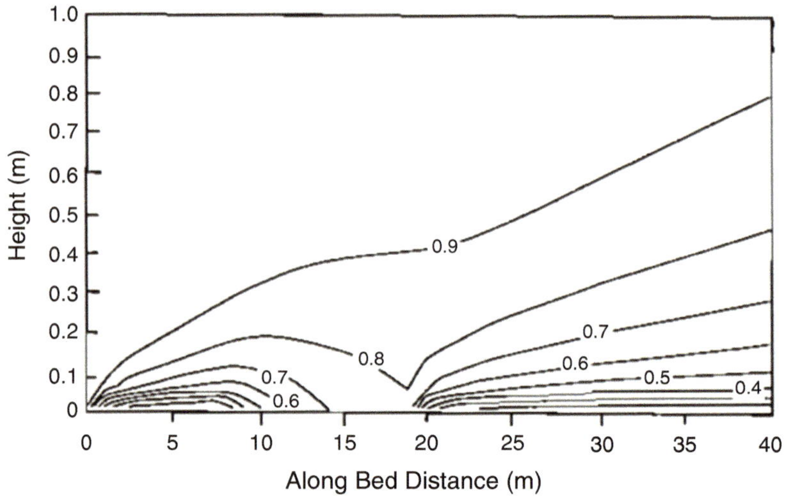

Fig. 24.1a Phytoplankton concentration along a patch of bottom cultured mussels due to boundary layer depletion. (From Newell and Shumway 1993, Fig. 26)

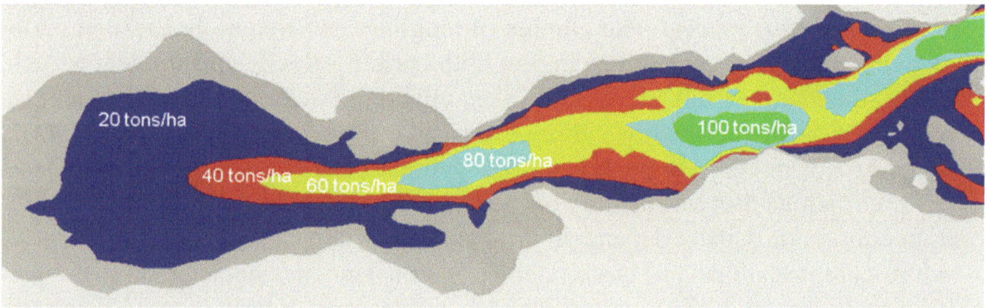

Fig. 24.1b Recommended seeding densities for Carlingford Lough, Ireland using MUSMOD. (Campbell and Newell 1998)

mussel patch, was used to model phytoplankton concentration at the height of ingestion by mussels (Fig. 24.1a), based on flume studies (Butman et al. 1994) and field work (Muschenheim and Newell 1992). In this case, the mussel beds are placed in two patches, from 0–10 m and from 20–40 m from the edge of a farm, and rapid particle depletion is observed in the benthic boundary layer just meters from the edges of both of these high density aggregations. This figure illustrates the importance of minimizing bottom patch size when spreading seed on the bottom in bivalve bottom culture.

The mussel growth model, MUSMOD, was developed for mussel bottom culture (Campbell and Newell 1998) in Maine in order to optimize mussel growth based on seeding density, the bottom shear velocity U* (Campbell and Newell 1998, Table 6), and boundary layer physics to estimate the food supply to bottom cultured mussels.

Using measured water velocities and a 2-d flow model, water samples and buoy temperature, salinity and chlorophyll-a data from Carlingford Lough, Ireland, MUSMOD estimated the best seeding densities for mussel farms (Fig. 24.1b) to maximize seed to harvest yields, mussel growth rates, and meat yields, assuming the mussels were spread well to eliminate density dependent "patch" effects (Newell 1990). The model simulations showed that areas of higher local velocity could support higher seeding densities, as long as the mussels were spread evenly and not in concentrated "patches". Using this type of approach, bottom culture farms could be optimized for production and ecosystem services instead of "trial and error" aquaculture which often results in lower meat yields, farm productivity, and seed survival.

24.8.2 Boundary Layer Depletion Models: Free-Planted Oysters

In this example, a boundary-layer algorithm was used to calculate food concentration in the middle of a benthic planting of 24 g live weight seed oysters (*Crassostrea virginica*) within a hypothetical "patch" of 100 × 100 m over a range of bottom densities from 50 to 1000 m^{-2}, based on water depth, oyster biomass, filtration rate, and free stream velocity from the 50 m resolution flow model ShellGIS (Newell et al. 2013). A reduction in food concentration in the benthic boundary layer was used in conjunction with ShellSIM to estimate the per cent reduction in oyster live weight in the middle of the hypothetical patch of mature oysters in the Damariscotta River, Maine, U.S.A. at two planting densities (100 oysters m^{-2} Fig. 24.2, left, and 500 oysters m^{-2}, Fig. 24.2, right). The color in each 50 × 50 m grid point shows the

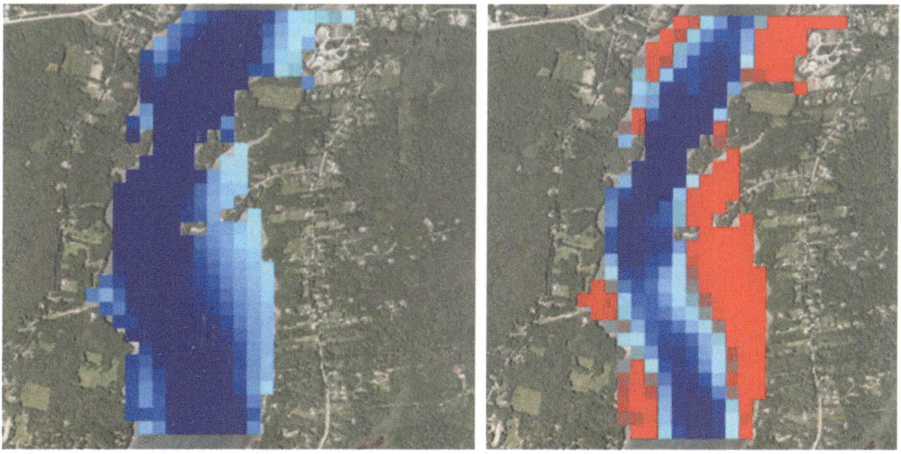

Fig. 24.2 Percent reduction of total fresh weight of oysters in the middle of a 100 × 100 m bottom planting after a year at 100 oysters m^{-2} (light blue is 25% reduction, left) and 500 m^{-2} (red is 50% reduction, right) using ShellGIS. (Newell et al. 2013)

results of utilizing the conditions at that grid point (water velocity, temperature, salinity, chl a, POM) on how growth would be reduced in the middle of a 100 × 100 m patch with the center at that grid point, and does not consider interactions with other grid points (*i.e.,* uses a single 100 × 100 m farm).

The differences in growth rates are a function of different tidally driven water velocities in the system and reduced food in the middle of the oyster patches. The areas in red and light blue have lower production than the higher current dark blue areas, but they may be managed effectively using lower seeding densities. Output of the ShellGIS farm model results shown in Fig. 24.2 are facilitated by the ability to choose, by a point and click method, the results of any farm location within the model domain, and determine production at any point during the growing season.

24.8.3 Computational Fluid Dynamics (CFD) Depletion Models

Aquaculture structure models use CFD and a technique known as the Fractional-Area-Volume-Obstacle-Representation (FAVOR) method to represent the structures within a rectangular grid (Hirt and Sicilian 1985). The FAVOR method uses partial control volumes to provide the advantages of a body-fitted grid but retains the construction simplicity of ordinary rectangular grids. The method also allows for the calculation of flow through "porous" media. It is also used to model seston depletion if there is uptake (filtration) of particles by bivalves at any location within the model domain. While CFD methods are more complicated and require a detailed physical representation of the system and more high performance computing, they can provide great insight into not only the hydrodynamic characteristics of a culture system but also ways to optimize farm productivity. Generally, we observe that chlorophyll-depletion in aquaculture structures increases with shellfish biomass in those structures which matches the patterns in the flow modeling results. Including structure porosity in the models can be a valuable proxy for understanding the impact of biofouling on farm structures and demonstrates the value and optimal timing for farmers to reduce biofouling.

24.8.4 Husbandry Practices

One of the major utilities of farm scale models is the ability to assess different husbandry practices and determine how they might be optimized on a shellfish farm. Factors under control of the farmer include the time of year seeded, the seed genetic origin and size, animal stocking density and biomass (Ferreira et al. 2007), nursery and grow-out gear type, and the placement, spacing and orientation of gear on the farm for scallop longlines (Bacher et al. 1997, 2003), mussel rafts (Duarte et al.

2008; Newell and Richardson 2014), mussel bottom culture (Campbell and Newell 1998); and mussel longline culture (Comeau et al. 2008; Drapeau et al. 2006; Rosland et al. 2011).

24.8.4.1 CFD: Oyster Bags on Racks

As part of the Understanding Irish Shellfish Culture Environments (UISCE) project (Dallaghan 2009), we performed CFD analyses of flow and predicted chlorophyll-*a* concentration in pacific oyster (*Crassostrea gigas*) bags on bottom trestles in Dungarvan, County Waterford, Ireland. Water flow through bags and trestle systems show significant depletion (blue areas) when the orientation of the trestles to flow direction was 0 degrees (Fig. 24.3, left), but flow is significantly improved when the angle is from the side, effectively increasing the cross-sectional area of the bags (Fig. 24.3, right). We found similar patterns in chlorophyll-a depletion.

24.8.4.2 CFD: Mussel Rafts

Newell and Richardson (2014) used a CFD model of flow and chlorophyll-*a* depletion for individual and multiple mussel rafts in Maine to determine that the mean velocity going through a mussel raft was only about 20% of the ambient flow (*i.e.*, 30 cm s^{-1} outside raft = 6 cm s^{-1} inside) (Fig. 24.4), and that particle depletion could be minimized by changing raft orientation to flow direction. These simulations provided guidance for site selection for locations which would minimize depletion in rafts (*i.e.*, outside raft velocities over 25 cm s^{-1}), and allow for adjustments to food concentration experienced by mussels in the rafts as input to production models.

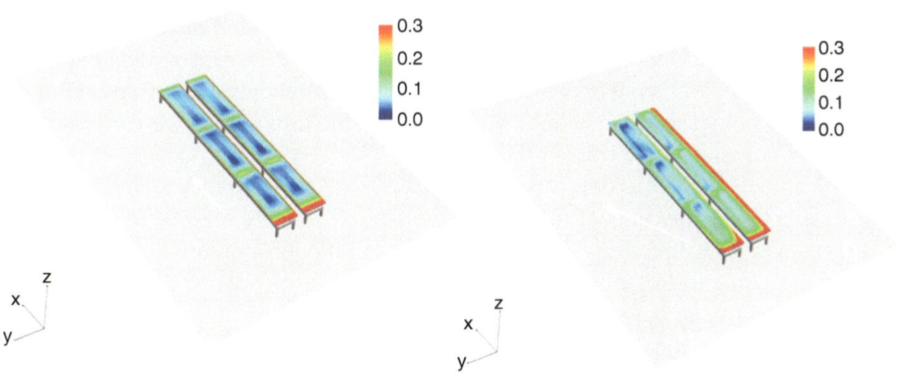

Fig. 24.3 Water velocity (m s^{-1}) in oyster bags on trestles with orientation at 0 degrees to flow direction (**a**) or 15 degrees to flow direction (**b**). (Richardson and Newell 2008)

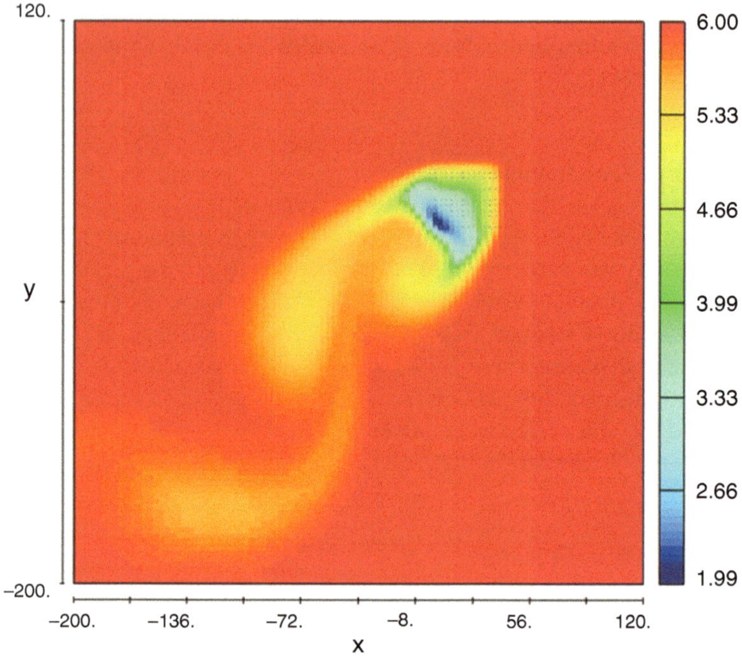

Fig. 24.4 Chlorophyll depletion at 15 cm s^{-1} flow in a Maine mussel raft with a 45° orientation to flow direction. Y-axis is chl-a μg l^{-1}. (Newell and Richardson 2014)

24.8.4.3 CFD: Oyster Rafts, Trays and Mussel Longlines

Oyster stick or tray rafts were modeled in Gorge Harbor, Cortes Island, British Columbia, Canada using Computer Aided Design (CAD) drawings of oyster trays and the CFD FAVOR method described above. Plan view and side view velocity models, combined with consumption estimates, were used to recommend spacing between multiple raft systems (4–5 raft diameters), and indicated that upwelling of deeper, chlorophyll-a waters would occur between the rafts if they were arranged in rows perpendicular to the current direction.

CFD models were also developed for OysterGro™ trays floating in longline systems, using CAD drawings of the trays and modeled velocity relative to orientation of the trays to flow direction (Fig. 24.5). A series of model simulations were used to predict the depletion of phytoplankton in arrays of suspended trays. At the high velocity areas (mean velocity over 20 cm s^{-1}), depletion was only significant in the oyster cages when the long-line system had a 0° orientation to the flow direction (Fig. 24.6).

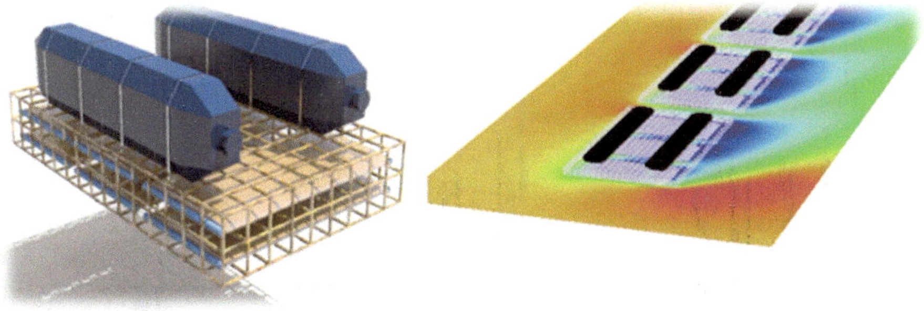

Fig. 24.5 Oystergro™ trays (left) and modeled velocity (right) with orientation of trays 45 degrees to flow direction in the Damariscotta River, Maine (ShellGIS oyster structure model)

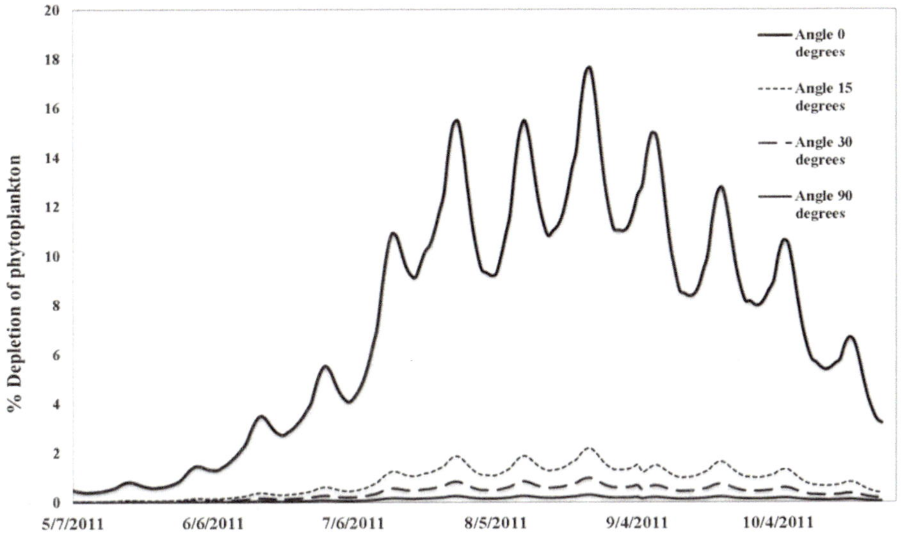

Fig. 24.6 Percent depletion of phytoplankton in surface Oystergro™ cages as a function of time of year and orientation of longline system to flow direction (ShellGIS oyster structure model)

As part of the UISCE project (Dallaghan 2009), we modeled depletion of chlorophyll-a through mussel longlines in Killary Harbor, Ireland, where field measurements and CAD drawings were used to develop a CFD depletion model for single dropper longline systems and an expert system for optimizing longline configuration. Longlines have also been modeled using CFD techniques by Delaux et al. 2011. Again, the depletion was very sensitive to the angle of orientation of the longlines, with little depletion when angles were more than 15 degrees from current direction (Richardson and Newell 2008).

24.8.5 Benthic Impacts

While some models have attempted to simulate benthic effects of bivalve organic matter deposition (Weise et al. 2009), it is often the site-specific balance between sedimentation of biodeposits, resuspension, burial and decay that results in benthic impacts (Testa et al. 2015), especially in shallow water. For example, using a mass-balance approach, Testa et al. (2015) calculated that wave induced resuspension within a Maryland, U.S.A. farm allowed tidal currents outside the farm to export the majority of nitrogen deposition (Fig. 24.7). Understanding the interplay between transport and biogeochemistry may represent the future of sustainable siting. Export of organic matter from the farm was dependent on estimated shear stress from tides and waves which in turn caused resuspension of the oyster biodeposits. Importantly, these dynamics were sensitive to local bathymetry.

The release of nutrients from biodeposits on shellfish farms may balance consumption indirectly (Asmus and Asmus 1991; Testa et al. 2015) by stimulating phytoplankton growth. In this case, the farms themselves may influence bay scale productivity. The flux of dissolved inorganic nitrogen, phosphate, and silica from the decomposition of bivalve biodeposits, and its stimulation of localized phytoplankton blooms is poorly understood, but may be important in maintaining phytoplankton concentrations near the farms when estuarine productivity is low. Organism growth models can also predict individual rates of water filtration and particle consumption, oxygen consumption, ammonium excretion, and biodeposit production,

Fig. 24.7 Erosional area from tides (light green) and waves (dark green) with the sedimentation rates indicated on an oyster farm, channel and control areas in Maryland, U.S.A. (Testa et al. 2015)

or be used in combination with hydrodynamics to predict deposition, resuspension and benthic impacts (Grant et al. 2005).

24.9 Conclusions

While there have been numerous studies on bivalve carrying capacity and bay scale production capacity, relatively few have dealt with farm models, and even fewer capture important small scale effects related to local bathymetry, aquaculture structures, and their orientation. A comparison between the FARM and ShellGIS models is presented in Table 24.1, using the criteria of Nath et al. 2000.

Perhaps one of the more interesting implications of the co-advances in both hydrodynamic-biogeochemical models and aquaculture structure models is the potential to more easily link these modeling platforms in the near future. For example, the Chesapeake Water Quality and Sediment Transport Model has been under continuous development since 1984 (Cerco and Noel 2013) and has moved from spatial resolutions on the order of kilometers to meters. Unstructured grids, such as FVCOM and SCHISM (Ye et al. 2016), are allowing for resolution at the farm scale in nearshore environments.

Better parameterization of the food supply of bivalves, both in terms of the concentration and quality of detritus, and the food value of different species of phytoplankton, will improve the fit of growth models with field data. In addition, understanding the time scales of nutrient cycling by bivalves related to phytoplankton growth and residence time of the water will help shed light on farm scale productivity and aquaculture-environmental feedbacks.

Improvements in web-based and GIS-based tools and advances in remote sensing are likely to improve appropriate siting of farms, but smaller scale models such as ShellGIS, utilizing fine scale hydrodynamics, biomass, and density distribution within the farms, and animal growth models are required to provide greater insight into optimization within the farms themselves or provide better production estimates for ecosystem models. While simplified web-based modeling tools such as FARM can provide quick insights into the results of different farm management scenarios, it is the specific conditions on the farms (bathymetry, localized water velocity, food resources, placement and arrangement of gear, local rope/bag/raft density and biomass, and structure hydrodynamics) which ultimately control the farm productivity and ecosystem services.

Applications that add aquaculture specific biogeochemical parameters, such as SPM, POM, and chlorophyll-*a* to new high resolution grids are also increasingly available (Xia and Jiang 2016; Testa et al. 2014). A challenge for coupling these models with aquaculture structure models with GIS capability, like ShellGIS, will be creating a data framework capable of transitioning model output into formats appropriate for estimating ecosystem goods and services and well as farm siting, and production modeling, based on the ability to nest the higher resolution farm scale models into a bay scale and ecosystem framework like the FARM model currently accomplishes.

Acknowledgements This chapter was supported in part by the National Science Foundation EPSCoR Cooperative Agreement IIA-1355457 to the University of Maine. Any opinions, findings, and conclusions or recommendations expressed in this material are those of the authors and do not necessarily reflect the views of the National Science Foundation. It was also supported by research funded by the Bord Iascaigh Mhara U.I.S.C.E project (Republic of Ireland), the USDA Northeast Regional Aquaculture Center, the Maine Aquaculture Innovation Center, the NOAA National Sea Grant Program and the USDA SBIR program. We acknowledge the constructive remarks of two referees.

References

Arnold WS, White MW, Norris HA, Berrigan ME (2000) Hard clam (Mercenaria spp.) aquaculture in Florida, USA: geographic information system applications to lease site selection. Aquac Eng 23(1):203–231

Asmus RM, Asmus H (1991) Mussel beds: limiting or promoting phytoplankton? J Exp Mar Biol Ecol 148(2):215–232

Bacher C, Duarte P, Ferreira JG, Héral M, Raillard O (1997) Assessment and comparison of the Marennes-Oléron Bay (France) and Carlingford Lough (Ireland) carrying capacity with ecosystem models. Aquat Ecol 31(4):379–394

Bacher C, Grant J, Fang J, Zhu M, Besnard M (2003) Modelling the effect of food depletion on scallop growth in Sungo Bay (China). Aquat Living Resour 16(1):10–24

Bayne BL, Newell RC (1983) Physiological energetics of marine molluscs. In: Saleuddin ASM, Wilbur KM (eds) The Mollusca, vol 4. Academic, London, pp 407–515

Booij N, Holthuijsen LH, Ris RC (1997) The "SWAN" wave model for shallow water. Coast Eng 1996:668–676

Buitrago J, Rada M, Hernández H, Buitrago E (2005) A single-use site selection technique, using GIS, for aquaculture planning: choosing locations for mangrove oyster raft culture in Margarita Island, Venezuela. Environ Manag 35(5):544–556

Butman CA, Frechette M, Geyer WR, Starczak VR (1994) Flume experiments on food supply to the blue mussel Mytilus edulis L. as a function of boundary-layerflow. Limnol Oceanogr 39(7):1755–1768

Campbell DE, Newell CR (1998) MUSMOD©, a production model for bottom culture of the blue mussel, Mytilus edulis L. J Exp Mar Biol Ecol 219(1):171–203

Carver CEA, Mallet AL (1990) Estimating the carrying capacity of a coastal inlet for mussel culture. Aquaculture 88(1):39–53

Cerco CF, Noel MR (2013) Twenty-one-year simulation of Chesapeake bay water quality using the CE-QUAL-ICM eutrophication model. J Am Water Resour Assoc 49(5):1119–1133

Chen C (2012) An unstructured-grid, finite-volume Community Ocean model: FVCOM user manual, Sea Grant College Program. Massachusetts Institute of Technology, Cambridge, MA

Comeau LA, Drapeau A, Landry T, Davidson J (2008) Development of longline mussel farming and the influence of sleeve spacing in Prince Edward Island, Canada. Aquaculture 281(1):56–62

Congleton WR, Pearce BR, Parker MR, Beal BF (1999) Mariculture siting: a GIS description of intertidal areas. Ecol Model 116(1):63–75

Dallaghan B (2009) UISCE project-virtual aquaculture. Aquaculture Ireland

Delaux S, Stevens CL, Popinet S (2011) High-resolution computational fluid dynamics modelling of suspended shellfish structures. Environ Fluid Mech 11(4):405–425

Dewhurst T (2016) Dynamics of a submersible mussel raft (Doctoral dissertation, University of New Hampshire)
Drapeau A, Comeau LA, Landry T, Stryhn H, Davidson J (2006) Association between longline design and mussel productivity in Prince Edward Island, Canada. Aquaculture 261(3):879–889
Duarte P, Labarta U, Fernández-Reiriz MJ (2008) Modelling local food depletion effects in mussel rafts of Galician Rias. Aquaculture 274(2):300–312
Epifanio CE (1979) Growth in bivalve molluscs: nutritional effects of two or more species of algae in diets fed to the American oyster Crassostrea virginica (Gmelin) and the hard clam Mercenaria mercenaria (L.). Aquaculture 18(1):1–12
Ferreira JG (1995) ECOWIN—an object-oriented ecological model for aquatic ecosystems. Ecol Model 79(1–3):21–34
Ferreira JG, Hawkins AJS, Bricker SB (2007) Management of productivity, environmental effects and profitability of shellfish aquaculture—the Farm Aquaculture Resource Management (FARM) model. Aquaculture 264(1):160–174
Ferreira JG, Hawkins AJS, Monteiro P, Moore H, Service M, Pascoe PL, Ramos L, Sequeira A (2008) Integrated assessment of ecosystem-scale carrying capacity in shellfish growing areas. Aquaculture 275(1):138–151
Ferreira JG, Hawkins AJ, Bricker SB (2011) The role of shellfish farms in provision of ecosystem goods and services. In: Shumway SE (ed) Shellfish aquaculture and the environment. Wiley, Singapore, pp 3–31
Filgueira R, Rosland R, Grant J (2011) A comparison of scope for growth (SFG) and dynamic energy budget (DEB) models applied to the blue mussel (Mytilus edulis). J Sea Res 66(4):403–404
Filgueira R, Guyondet T, Comeau LA, Grant J (2014) A fully-spatial ecosystem-DEB model of oyster (Crassostrea virginica) carrying capacity in the Richibucto Estuary, Eastern Canada. J Mar Syst 136:42–54
Filgueira R, Comeau LA, Guyondeta T, McKindseyb CW, Byronc CJ (2015) Modelling carrying capacity of bivalve aquaculture: a review of definitions and methods, vol 2. DFO Canadian Fisheries Advisory Secretariat Research Document
Frechette M, Lefaivre D (1990) Discriminating between food and space limitation in benthic suspension feeders using self-thinning relationships. Mar Ecol Prog Ser 65(0):15–23
Grant J, Bacher C (2001) A numerical model of flow modification induced by suspended aquaculture in a Chinese bay. Can J Fish Aquat Sci 58(5):1003–1011
Grant J, Cranford P, Hargrave B, Carreau M, Schofield B, Armsworthy S, Burdett-Coutts V, Ibarra D (2005) A model of aquaculture biodeposition for multiple estuaries and field validation at blue mussel (Mytilus edulis) culture sites in eastern Canada. Can J Fish Aquat Sci 62(6):1271–1285
Grant J, Bacher C, Cranford PJ, Guyondet T, Carreau M (2008) A spatially explicit ecosystem model of seston depletion in dense mussel culture. J Mar Syst 73(1):155–168
Grant J, Filgueira R (2011) The application of dynamic modeling to prediction of production carrying capacity in shellfish farming. In: Shumway SE (ed) Shellfish aquaculture and the environment. Wiley, Singapore, pp 135–154
Grizzle RE, Lutz RA (1989) A statistical model relating horizontal seston fluxes and bottom sediment characteristics to growth of Mercenaria mercenaria. Mar Biol 102(1):95–105
Guyondet T, Roy S, Koutitonsky VG, Grant J, Tita G (2010) Integrating multiple spatial scales in the carrying capacity assessment of a coastal ecosystem for bivalve aquaculture. J Sea Res 64(3):341–359
Hawkins AJS, Pascoe PL, Parry H, Brinsley M, Black KD, McGonigle C, Moore H, Newell CR, O'Boyle N, Ocarroll T, O'Loan B (2013a) Shellsim: a generic model of growth and environmental effects validated across contrasting habitats in bivalve shellfish. J Shellfish Res 32(2):237–253
Hawkins AJS, Pascoe PL, Parry H, Brinsley M, Cacciatore F, Black KD, Fang JG, Jiao H, Mcgonigle C, Moore H, O'boyle N (2013b) Comparative feeding on chlorophyll-rich versus remaining organic matter in bivalve shellfish. J Shellfish Res 32(3):883–897

Hirt CW, Sicilian JM (1985) A porosity technique for the definition of obstacles in rectangular cell meshes. In: Proceeding of the 4th International Conf. Ship Hydropower. National Academy of Science, Washington, DC
Hirt CW, Nichols B (1988) Flow-3D User's manual. Flow Science Inc.
Hydraulics D (2006) Delft 3D-FLOW user manual. Delft, the Netherlands
Incze LS, Lutz RA (1980) Mussel culture: an east coast perspective. In: Lutz RA (ed) Mussel culture and harvest, a North American perspective, Developments in aquaculture and fisheries science, vol 7. Elsevier Press, Amsterdam, pp 99, 350 pp–140
King IP (2003) RMA-11—a three dimensional finite element model for water quality in estuaries and streams. Resource Modelling Associates, Sydney
Larsen PS, Filgueira R, Riisgård HU (2014) Somatic growth of mussels Mytilus edulis in field studies compared to predictions using BEG, DEB, and SFG models. J Sea Res 88:100–108
Longdill PC, Healy TR, Black KP (2008) An integrated GIS approach for sustainable aquaculture management area site selection. Ocean Coast Manage 51(8–9):612–624
McKindsey CW, Thetmeyer H, Landry T, Silvert W (2006) Review of recent carrying capacity models for bivalve culture and recommendations for research and management. Aquaculture 261(2):451–462
Muschenheim DK, Newell CR (1992) Utilization of seston flux over a mussel bed. Marine ecology progress series. Oldendorf, 85(1), pp 131–136
Nath SS, Bolte JP, Ross LG, Aguilar-Manjarrez J (2000) Applications of geographical information systems (GIS) for spatial decision support in aquaculture. Aquac Eng 23(1):233–278
Newell CR (1990) The effects of mussel (*Mytilus edulis*, Linnaeus, 1758) position in seeded bottom patches on growth at subtidal lease sites in Maine. J Shellfish Res 9:113–118
Newell CR, Shumway SE (1993) Grazing of natural particulates by bivalve molluscs: a spatial and temporal perspective. In: Bivalve filter feeders. Springer, Berlin/Heidelberg, pp 85–148
Newell CR, Shumway SE, Cucci TL, Selvin R (1989) The effects of natural seston particle size and type on feeding rates, feeding selectivity and food resource availability for the mussel Mytilus edulis Linnaeus, 1758 at bottom culture sites in Maine. J Shellfish Res 8(1):187–196
Newell CR, Campbell DE, Gallagher SM (1998) Development of the mussel aquaculture lease site model MUSMOD©: a field program to calibrate model formulations. J Exp Mar Biol Ecol 219(1–2):143–169
Newell CR, Wildish DJ, Mac Donald BA (2001) The effects of velocity and seston concentration on the exhalant siphon area, valve gape and filtration rate of the mussel *Mytilus edulis*. J Exp Mar Biol Ecol 262:91–111
Newell CR, Hawkins AJS, Morris K, Richardson J, Davis C, Getchis T (2013) ShellGIS: a dynamic tool for shellfish farm site selection. World Aquacult 44:50–53
Newell CR, Richardson J (2014) The effects of ambient and aquaculture structure hydrodynamics on the food supply and demand of mussel rafts. J Shellfish Res 32:257–272
Nunes JP, Ferreira JG, Gazeau F, Lencart-Silva J, Zhang XL, Zhu MY, Fang JG (2003) A model for sustainable management of shellfish polyculture in coastal bays. Aquaculture 219(1):257–277
Pilditch CA, Grant J, Bryan KR (2001) Seston supply to sea scallops (Placopecten magellanicus) in suspended culture. Can J Fish Aquat Sci 58(2):241–253
Plew DR (2005) The hydrodynamic effects of long-line mussel farms. Ph. D. Thesis, University of Canterbury, 330 pp
Plew DR, Stevens CL, Spigel RH, Hartstein ND (2005) Hydrodynamic implications of large offshore mussel farms. IEEE J Ocean Eng 30(1):95–108
Plew DR (2011) Shellfish farm-induced changes to tidal circulation in an embayment, and implications for seston depletion. Aquac Environ Interact 1(3):201–214
Pouvreau S, Bourles Y, Lefebvre S, Gangnery A, Alunno-Bruscia M (2006) Application of a dynamic energy budget model to the Pacific oyster, Crassostrea gigas, reared under various environmental conditions. J Sea Res 56(2):156–167
Radiarta IN, Saitoh SI, Miyazono A (2008) GIS-based multi-criteria evaluation models for identifying suitable sites for Japanese scallop (Mizuhopecten yessoensis) aquaculture in Funka Bay, southwestern Hokkaido, Japan. Aquaculture 284(1):127–135

Richardson J, Newell C (2008) Physical and numerical modeling of flow through aquaculture gear. Presentation. Irish Sea Fisheries Board "Understanding Irish shellfish culture environments", Westport, Ireland

Rosenberg R, Loo LO (1983) Energy-flow in a Mytilus edulis culture in western Sweden. Aquaculture 35:151–161

Rosland R, Bacher C, Strand Ø, Aure J, Strohmeier T (2011) Modelling growth variability in long-line mussel farms as a function of stocking density and farm design. J Sea Res 66(4):318–330

Ross LG, Telfer TC, Falconer L, Soto D, Aguilar-Manjarrez J, Asmah R, Bermúdez J, Beveridge MCM, Byron CJ, Clément A, Corner R (2013) Carrying capacities and site selection within the ecosystem approach to aquaculture. In: Ross LG, Telfer TC, Falconer L, Soto D, Aguilar-Majarrez J (eds) Site selection and carrying capacities for inland and coastal aquaculture, pp 19–46 FAO

Saurel C, Ferreira JG, Cheney D, Suhrbier A, Dewey B, Davis J, Cordell J (2014) Ecosystem goods and services from Manila clam culture in Puget Sound: a modelling analysis. Aquac Environ Interact 5:255–270

Scholten H, van der Tol MW (1994) SMOES: a simulation model for the Oosterschelde ecosystem. Hydrobiologia 282(1):453–474

Scholten H, Smaal AC (1998) Responses of Mytilus edulis L. to varying food concentrations: testing EMMY, an ecophysiological model. J Exp Mar Biol Ecol 219(1):217–239

Shumway SE (1990) A review of the effects of algal blooms on shellfish and aquaculture. J World Aquacult Soc 21(2):65–104

Silva C, Ferreira JG, Bricker SB, Delvalls TA, Martín-Díaz ML, Yáñez E (2011) Site selection for shellfish aquaculture by means of GIS and farm-scale models, with an emphasis on data-poor environments. Aquaculture 318(3):444–445

Smaal AC, Prins TC, Dankers NMJA, Ball B (1997) Minimum requirements for modelling bivalve carrying capacity. Aquat Ecol 31(4):423–428

Snyder J, Boss E, Weatherbee R, Thomas AC, Brady D, Newell C (2017) Oyster aquaculture site selection using Landsat 8-Derived Sea surface temperature, turbidity, and chlorophyll a. Front Mar Sci 4:190

Stevens C, Plew D, Hartstein N, Fredriksson D (2008) The physics of open-water shellfish aquaculture. Aquac Eng 38(3):145–160

Testa JM, Li Y, Lee Y, Li M, Brady DC, Di Toro DM, Kemp WM (2014) Quantifying the effects of nutrient loading on dissolved O_2 cycling and hypoxia in Chesapeake Bay using a coupled hydrodynamic-biogeochemical model. J Mar Syst 139:139–158

Testa JM, Brady DC, Cornwell JC, Owens MS, Sanford LP, Newell CR, Suttles SE, Newell RI (2015) Modeling the impact of floating oyster (Crassostrea virginica) aquaculture on sediment-water nutrient and oxygen fluxes. Aquac Environ Interact 7(3):205–222

Thomas Y, Mazurié J, Alunno-Bruscia M, Bacher C, Bouget JF, Gohin F, Pouvreau S, Struski C (2011) Modelling spatio-temporal variability of Mytilus edulis (L.) growth by forcing a dynamic energy budget model with satellite-derived environmental data. J Sea Res 66(4):308–317

Tissot C, Brosset D, Barillé L, Le Grel L, Tillier I, Rouan M, Le Tixerant M (2012) Modeling oyster farming activities in coastal areas: a generic framework and preliminary application to a case study. Coast Manag 40(5):484–500

Tseung HL, Kikkert GA, Plew D (2016) Hydrodynamics of suspended canopies with limited length and width. Environ Fluid Mech 16(1):145–166

Van Haren RJF, Kooijman SALM (1993) Application of a dynamic energy budget model to Mytilus edulis (L.). Neth J Sea Res 31(2):119–133

Warren IR, Bach H (1992) MIKE 21: a modelling system for estuaries, coastal waters and seas. Environ Softw 7(4):229–240

Weise AM, Cromey CJ, Callier MD, Archambault P, Chamberlain J, McKindsey CW (2009) Shellfish-DEPOMOD: modelling the biodeposition from suspended shellfish aquaculture and assessing benthic effects. Aquaculture 288(3):239–253

Wildish DJ, Miyares MP (1990) Filtration rate of blue mussels as a function of flow velocity: preliminary experiments. J Exp Mar Biol Ecol 142(3):213–219

Wilkin JL, Arango HG, Haidvogel DB, Lichtenwalner C, Glenn SM, Hedström KS (2005) A regional ocean modeling system for the long-term ecosystem observatory. J Geophys Res Oceans 110(C6)
Xia M, Jiang L (2016) Application of an unstructured grid-based water quality model to Chesapeake Bay and its adjacent Coastal Ocean. J Marine Sci Eng *4*(3):52
Ye F, Zhang YJ, Friedrichs MA, Wang HV, Irby ID, Shen J, Wang Z (2016) A 3D, cross-scale, baroclinic model with implicit vertical transport for the Upper Chesapeake Bay and its tributaries. Ocean Model 107:82–96

Chapter 25
Ecosystem Models of Bivalve Aquaculture: Implications for Supporting Goods and Services

Jon Grant and Roberto Pastres

Abstract In this paper we focus on the role of ecosystem models in improving our understanding of the complex relationships between bivalve farming and the dynamics of lower trophic levels. To this aim, we review spatially explicit models of phytoplankton impacted by bivalve grazing and discuss the results of three case studies concerning an estuary (Baie des Veys, France), a bay, (Tracadie Bay, Prince Edward Island, Canada) and an open coastal area (Adriatic Sea, Emilia-Romagna coastal area, Italy). These models are intended to provide insight for aquaculture management, but their results also shed light on the spatial distribution of phytoplankton and environmental forcings of primary production. Even though new remote sensing technologies and remotely operated in situ sensors are likely to provide relevant data for assessing some the impacts of bivalve farming at an ecosystem scale, the results here summarized indicate that ecosystem modelling will remain the main tool for assessing ecological carrying capacity and providing management scenarios in the context of global drivers, such as climate change.

Abstract in Chinese 本文重点关注通过生态系统模型的方法去理解双壳贝类养殖活动与低营养级种群动力学之间的复杂关系。为此，我们回顾了受双壳贝类摄食影响的浮游植物空间显式模型，并讨论了三个有关的实例，包括河口(Baie des Veys，法国)，海湾(特拉卡迪湾，加拿大爱德华王子岛)和海岸带开放海域(亚得里亚海，艾米利亚-罗马涅沿岸，意大利)。这些模型旨在为水产养殖管理提供更深层次的信息，但其结果也包含了浮游植物的空间分布和环境压力下的初级生产力情况。我们或许可以通过新的遥感技术和原位传感器远程传输的相关数据来评估生态系统规模的双壳贝类养殖造成的影响，但所有结果表明，生态系统模型仍将是评估生态容量的主要工具，并在气候变化等全球范围影响的背景下提供养殖管理方面的参考信息。

J. Grant (✉)
Department of Oceanography, Dalhousie University, Halifax, NS, Canada
e-mail: jon.grant@dal.ca

R. Pastres
Dipartimento di Scienze Ambientali, Informatica e Statistica, Mestre, VE, Italy
e-mail: pastres@unive.it

A. C. Smaal et al. (eds.), *Goods and Services of Marine Bivalves*,
https://doi.org/10.1007/978-3-319-96776-9_25

Keywords Ecosystem modelling · Low trophic levels · Bivalve farming · Phytoplankton depletion · Aquaculture

关键词 生态系统建模 • 低营养级 • 双壳贝类养殖 • 浮游植物消耗 • 水产养殖

25.1 Introduction

The culture of marine suspension-feeding bivalves involves farming extensive coastal areas at high biomass. The ability of these animals to influence ecosystem processes is a central theme of this book. Ecosystem goods and services, such as provision of harvested protein, require that energy or matter flow be directed through cultured populations, and potentially diverted from other pathways (e.g. wild species requirements). The concept of carrying capacity has been subdivided to reflect this definition (McKindsey et al. 2006). For example, ecological carrying capacity would apply to an environmental threshold beyond which the ecological integrity of the ecosystem would be considered compromised. This approach requires assessment of inputs and outputs of matter and energy to coastal systems, and ecosystem modeling has frequently been utilized for this purpose (Grant and Filgueira 2011).

Ecosystem models applied to shellfish culture may be categorized in two ways:

1. Mitigation models, which seek to address the role of bivalves in reducing 'excess' phytoplankton arising from eutrophication. This topic is addressed explicitly in Petersen et al. (2019).
2. Carrying capacity models, which seek to determine food limitation of cultured bivalves. This chapter is addressed from a provisioning point of view in Smaal and van Duren (2019).

In both cases, the concentration of phytoplankton biomass, usually quantified as photopigments, primarily chlorophyll, has been the focus of ecosystem models. Phytoplankton are at the base of all marine food webs, and may be characterized as the most important part of marine ecosystems, and certainly the most important of supporting services (Richardson and Shoeman 2004). Regulation of phytoplankton biomass classically occurs through either bottom up (nutrients) or top down (grazing) processes. The biomass of natural bivalve populations is equilibrated with its food supply, so excessive grazing would not be an ongoing feature of the ecosystem. However, bivalves stocked in culture could easily overgraze their food supply, the essence of carrying capacity. Several consequences would ensue, including reduced growth or increased mortality of farmed animals, and competition with other grazers such as wild bivalves and zooplankton. In order to preserve the supporting service of phytoplankton, criteria have been established based on the abundance of phytoplankton that should be 'left over' once bivalve nutrition is satisfied. Grant and Filgueira (2011) argued that the extent of depletion should not exceed the natural spatiotemporal variation of phytoplankton in a given culture system, effectively parameterizing a sustainability criterion.

Although this value can be expressed as an average, the spatial distribution of phytoplankton can be very complex, as well as the biological and physical processes that lead to its renewal. In bays and estuaries, exchange with the coastal ocean has a large influence on phytoplankton, as does grazing and sinking (Cloern 1996). Moreover, watershed-derived nutrients are a key factor in phytoplankton production, as occurs in eutrophication (Cloern 2001). Despite numerous studies of phytoplankton in estuaries, there are few which attempt to map their spatial distribution.

Sampling to create those maps is difficult due to temporal variation at very small spatial scales. Although quantities such as phytoplankton may be expressed as chlorophyll and observed through satellite remote sensing, there are several drawbacks to this approach. First, coastal bays are not ideal for ocean colour measurements since pixel resolution may be coarse, and many pixels are masked due to land proximity, water depth, and turbidity. Despite this limitation, Radiarta and Saitoh (2009) were able to detect both spatial and temporal patterns of chlorophyll and turbidity in Funka Bay (Japan), although the 1 km resolution was appropriate to the ~300 km scale of the bay. Moreover, the impact of bivalves on chlorophyll cannot be easily observed, despite one example from a high resolution CASI image (Grant et al. 2007). Some of these limitations may be overcome in the near future as the spatial and temporal resolution of satellite data increase. Furthermore, at local scales it may become possible to use underwater or aerial autonomous vehicles equipped with ocean colour sensors for detecting phytoplankton depletion due to the presence of shellfish farms (Ludvigsen and Sorensen 2016). Quantification of local depletion has been accomplished with towed sensors (Nielsen et al. 2016).

Modelling is perhaps the only way to address these processes at larger scales and produce maps of chlorophyll simulated in the presence and absence of aquaculture, as well as in alternative management scenarios, e.g. relocation of farms, changes in stocking density, and introduction of new species. Modelling is also the only option for exploring the consequences of climate change on shellfish production, as shown in Canu et al. (2010) and Guyondet et al. (2015). Although model simulations can create detailed spatial maps, they are difficult to validate, especially the null scenario in the absence of shellfish at an established aquaculture site. In fact, in many cases, the consequences of siting shellfish leases in terms of chlorophyll are retrospective – extensive bivalve aquaculture is already in place. There are few examples where aquaculture site planning has been carried out on the basis of predicted phytoplankton spatial distribution (Filgueira et al. 2015). We suggest that the necessity of understanding food limitation in cultured bivalves has advanced an understanding of phytoplankton distribution in general, as well as models to elaborate this occurrence.

Based on these considerations, we review spatially explicit models of phytoplankton impacted by bivalve grazing and pose the following questions:

- How has ecosystem modelling been used to map chlorophyll in the presence and absence of bivalve culture?
- Are phytoplankton submodels used for this purpose adequate?
- Are these maps representative of ecosystem-scale properties?

25.2 The Structure of Ecosystem-Wide Depletion Models

Most models of the interaction between suspension feeders and phytoplankton are classical PNZ models with varying degrees of complexity in trophic structure (see review in Grant and Filgueira 2011). These range from simplified models where there are no other grazers except cultured bivalves, to more fully configured pelagic food chains. Although these ecosystem models can be simulated over an annual cycle, they are more commonly used to represent spring and summer for purposes of emphasizing spatial over temporal changes. This occurs because the focus is primarily on explicit aquaculture locations and the local or regional spatial impacts of grazing on phytoplankton. This focus is opportune because the case of 'no grazers' must inevitably act as reference point which yields insight into the dynamics of coastal phytoplankton.

Because interannual variation in seasonal forcing such as precipitation and river flow have such large impacts on phytoplankton production, it is possible to understand longer term trends in food supplies which might influence bivalve production (Grangeré et al. 2009; Thomas et al. 2011). However, depending on the importance of top down regulation, models which neglect other grazers such as zooplankton might be expected to perform poorly in simulating annual phytoplankton cycles. Nonetheless, if suspension feeding bivalves pre-empt zooplankton grazing pressure, annual phytoplankton cycles will reflect predation by shellfish since aquaculture is persistently in place and forces temporal changes based on harvest and stocking. These changes in phytoplankton production have been observed in San Francisco Bay due to an invasive suspension feeding clam (Cloern 1982). Regardless, the dominance of bivalves in controlling phytoplankton is also dependent on the spatial extent of aquaculture in the system. Below, we present case studies where bivalve culture is spread throughout a bay (Canada), where it is localized in a semi-enclosed system (France), and where it occurs along a stretch of open coast (Adriatic Sea, Italy).

25.3 Phytoplankton in Estuaries – Distribution

Due to the importance of eutrophication and of phytoplankton in estuarine food chains, there is a substantial general literature on this topic. However, fewer studies deal with spatial distribution of phytoplankton, and as indicated above, remote sensing of chlorophyll is difficult in these environments. Moreover, there are few models of phytoplankton that attempt to simulate relatively small-scale spatial detail, including the effects of river, tide, wind, bathymetry, etc. However, we reiterate that due to the importance of microalgal distribution for aquaculture success, models of seston depletion by shellfish have provided general insight into the topic of phytoplankton ecology.

Information on spatiotemporal variation in phytoplankton is significant in being able to characterize the 'normal' range of biomass or chlorophyll, so that grazer perturbations due to aquaculture may be gauged. The spatial distribution of phytoplankton is partially a balance between primary production and advection; these dynamics are explored in the case studies below. Although photosynthesis allows phytoplankton biomass to accumulate, advection may either contribute to this buildup as in convergence zones, or act to disperse cell populations and reduce local biomass (Cloern and Nichols 1985; Lucas et al. 1999a, b).

Photosynthesis is a result of both nutrient supply and the light field, both of which are highly variable in coastal systems. The role of rivers in supplying nutrients to estuaries has been extensively studied due to the prevalence of eutrophication (Cloern 2001). The impact of bivalve aquaculture in modulating these processes through grazing has also been long recognized (Meeuwig 1999) and utilized in bioremediation programs (see Cranford 2019; Petersen et al. 2019). Turbidity may impose serious limits on photosynthesis (May et al. 2003) and typically occurs in bivalve culture areas which are shallow, dominated by soft sediments, and thus subject to resuspension.

25.4 Phytoplankton in Estuaries – Composition

While we have emphasized in this chapter the effects of bivalves on chlorophyll as a bulk biological water property, this is clearly an over-simplification. Phytoplankton undergo a seasonal succession of species composition described by Cloern (1996) as follows: 'A common annual cycle begins with large winter-spring diatom blooms followed by summer blooms of small flagellates, dinoflagellates, and diatoms and then autumn blooms dominated by dinoflagellates'. The selectivity of bivalve grazers for certain classes of microalgae is well known and selection not only removes bulk chlorophyll but creates a preponderance of small cells referred to as picoplankton (>2–3 μm) (Smaal et al. 2013; Zhao et al. 2016). Size-selective feeding in bivalve culture can influence the entire phytoplankton size spectrum in coastal waters as documented in multiple studies (Cranford et al. 2011). It has been further suggested that this alteration may be an indicator of carrying capacity for bivalve aquaculture (Cranford et al. 2008; Jiang et al. 2016). Cultured bivalves can, however, derive significant nutrition from picoplankton (Sonier et al. 2016), so their indicator value is not straightforward.

From an aquaculture modelling perspective, multiple classes of phytoplankton are less commonly implemented in favour of the more tractable unimodal phytoplankton component, expressed solely as chlorophyll, with non-specific size and species composition. In the example below, Grangeré et al. 2010 focus on diatoms as the dominant phytoplankton class in Baie des Veys but found that modelled chlorophyll was underestimated due to *Phaeocystis* blooms which were not part of the simulation. However, because this genus is colony forming, its value as oyster food is variable, thus impacting model chlorophyll predictions but not necessarily bivalve feeding. It is feasible that high levels of chlorophyll could occur in ecosystems controlled by bivalve grazing where there are an abundance of picoplankton with a size refuge from suspension feeders (e.g. Comeau et al. 2015). Although there are bivalve ecosystem models with several phytoplankton size/composition classes (Cugier et al. 2010; Brigolin et al. 2011; Guyondet et al. 2015), this is more often formulated for temporal succession of phytoplankton rather than spatial distribution of size classes. The topic of harmful algal blooms (HAB) is essential in any discussion of coastal phytoplankton composition and shellfish culture. It is beyond the scope of this chapter and is covered in Wijsman et al. 2019.

25.5 Case Studies

We utilize only case studies where a map of modelled chlorophyll is depicted, rather than a change map (i.e. % depletion), since the 'no bivalves' case requires these units. However, it is recognized that even with this restriction, there are many more examples than can be covered herein.

25.5.1 Baie des Veys

A series of studies based on the Normandy Coast of France represent among the most comprehensive examples of chlorophyll models applied to bivalve culture. We discuss Grangeré et al. (2010), conducted in the Baie des Veys. This is a funnel shaped sub-estuary in the eastern Baie de Seine including the entrance of four rivers dominated by the Vire River. Due to macrotidal conditions, there are extensive tidal flats which include wild cockle populations. The primary culture species is Pacific oyster, *Crassostrea gigas,* grown on intertidal oyster tables, but *Mytilus edulis* is also farmed.

A significant part of this study is the extent to which the phytoplankton submodel was calibrated, largely by comparing modelled and measured primary production (Grangeré et al. 2009). Specifically, a variety of photosynthesis-intensity (PI) curves were generated for the Baie des Veys and compared to field observations of primary production (^{14}C method) and light. Consideration of both C:Chl and nutrient limitation formulations was also made. Because the calibrated phytoplankton submodel was used in Grangeré et al. (2010), this study represents a comprehensive examination of their spatial and temporal distribution.

Model results indicate that phytoplankton production is stimulated in the spring by river nutrient input, first appearing on the western side of the bay, and proceeding until the head of the bay has enhanced chlorophyll (Fig. 25.1b). Pigment levels

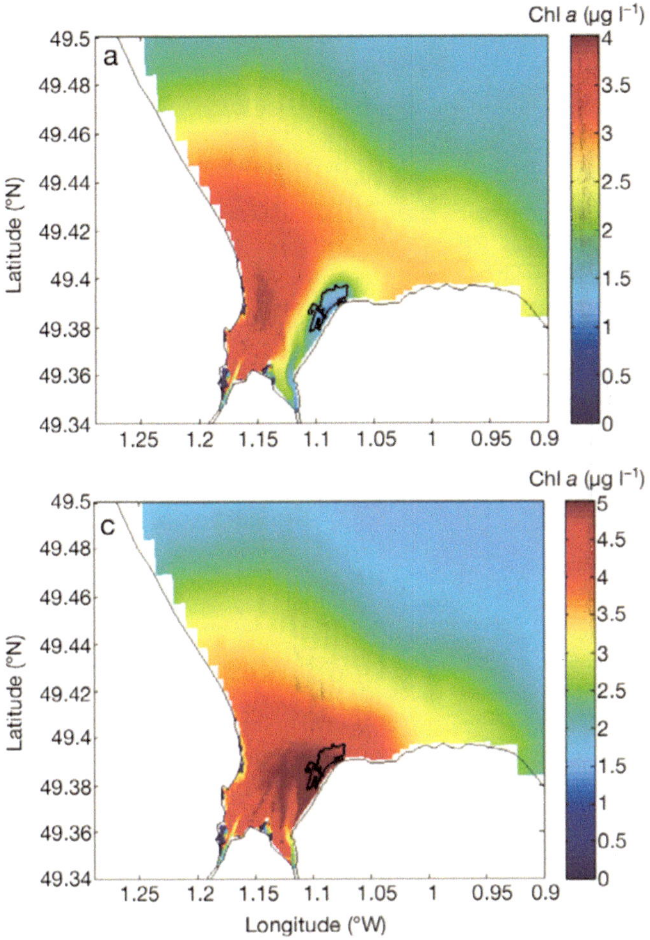

Fig. 25.1 A comparison of ecosystem scale chlorophyll distribution in the Baie de Veys, Normandy in the presence (**a**) and absence (**b**) of suspension feeding benthos (cultured and wild). The oyster farming area is shown in black outline

gradually attenuate offshore presumably through mixing. This appears to be a somewhat classical bottom-up scenario for a phytoplankton bloom. The simulation that includes oyster aquaculture shows a strong influence of bivalve grazing, with chlorophyll depleted by a factor of about threefold in the culture area (Fig. 25.1a). Finer scale views of the oyster culture areas revealed additional structure, with oysters at the northern limit of the culture area achieving superior growth due to better advective renewal of seston. As Grangeré et al. (2009) state "Top-down effects of oysters on phytoplankton at local scales were revealed, whereas bottom-up effects drove primary productivity at the whole bay scale. In general we conclude that spatial modelling is particularly appropriate to reveal spatial properties which would be difficult to observe directly."

It is important to emphasize that their model was applied to the seasonal dynamics of oyster growth, and not an instantaneous or averaged assessment of carrying capacity. The results of their studies indicate several principles for resolving shellfish-food chain interactions at the ecosystem scale:

1. An ecosystem model with sufficient spatial scale and appropriate structure to account for processes forced by an offshore boundary as well as a land-based source of nutrients, i.e. rivers
2. Validation of the phytoplankton model parameters and groundtruthing of chlorophyll via water samples.
3. The ability to distinguish between classes of phytoplankton, including those that are rejected by suspension feeding bivalves for either size or composition.
4. Clear delineation of bivalve culture areas, and the importance of their spatial extent in forcing localized versus far field chlorophyll distribution.

25.5.2 Tracadie Bay, Prince Edward Island, Canada

Prince Edward Island (PEI) has the largest mussel aquaculture industry in North America, producing ~19,000 tonnes annually. Much of the province is characterized by shallow sandy river estuaries, ideal for shellfish farming. Multiple studies have been conducted on the North Shore of the Island in Tracadie Bay (see references in Filgueira et al. 2015), but we highlight the spatial model in Grant et al. (2008). The bay is characterized by a barrier island at the entrance to a small inlet with a complex tidal delta, and a gradual narrowing over its 5 km length. The Winter River enters into a small side bay. Mussels (*Mytilus edulis*) are cultured on longlines in most of the bay, excluding the inlet region with a large intertidal zone. The ecosystem model in Grant et al. (2008) included dissolved nutrients, phytoplankton, detritus, and mussels coupled to a 2D circulation model with 606 nodes. The phytoplankton submodel is forced by light and nutrients. The daily and annual light fields were also modelled with respect to latitude according to Grant et al. (1993) Nutrient fields were available from a sampling program on the Winter River described in Cranford et al. (2007).

The example output (Fig. 25.2) demonstrates both the dynamics of chlorophyll (expressed as carbon equivalents) as well as the effects of grazer control. In this summer example without mussels, exchange at the inlet locally dilutes chlorophyll, but a gyre region behind the barrier island allows phytoplankton to accumulate with the benefit of a continual offshore nutrient renewal. However, this effect gradually tapers off through the interior region as reduced flushing causes nutrient limitation and reduced biomass. About midway along the length, entrance of Winter River nutrients causes localized increases in chlorophyll, an effect that is more pronounced in spring (not shown) when river nutrients are higher. Tracadie Bay is surrounded by

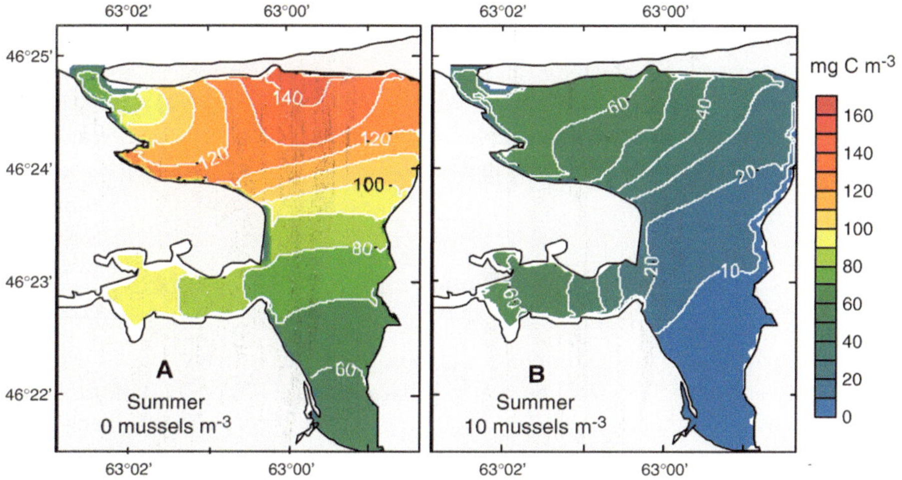

Fig. 25.2 Modelled chlorophyll carbon maps of Tracadie Bay, Nova Scotia in the presence and absence of cultured mussels. Units are carbon equivalents of PEI biomass converted with C:Chl = 50

agriculture (as are many estuaries in PEI), and the potential eutrophication is only offset by mussel grazing (Cranford et al. 2007; Guyondet et al. 2015; Meeuwig 1999).

The effects of mussel grazing on this system are dramatic. There is a reduction in chlorophyll of 2–6x. The effect of the gyre behind the barrier as a chlorophyll sink is eliminated and the sharp landward reduction in chlorophyll is even more pronounced. Although the local inlet area is not subject to depletion due to tidal exchange, the rest of the bay has chlorophyll levels that are less than any location in the absence of mussels.

This study is unique in that a towed fluorometer (Acrobat) was available to groundtruth model results. While Acrobat data basically validated model predictions, it also demonstrated a tidal signal in depletion as Winter River emptied its high biomass to the larger bay at low tide. These field results illustrate the contrast that observations, including sampling and remote sensing, are snapshots whereas models are averaged, in this case daily.

The ultimate field experiment in Tracadie Bay was conducted in December 2009 when a winter storm opened a new tidal inlet along the barrier island (Filgueira et al. 2014). Water renewal time for the whole bay was reduced by 1/3 or more. As a result, cultured harvest increased by about 1/3 even with the same mussel stocking density. The alleviation of seston depletion by flushing was clearly demonstrated. Moreover, the effects of climate change on coastal geomorphology were expressed through increased estuarine productivity.

Model outcomes provide the following generalities:

1. The ability of shellfish aquaculture to dominate chlorophyll spatial distribution in a small bay.
2. The importance of flushing and renewal as a mitigation against seston depletion.
3. The use of sophisticated spatial survey methods to groundtruth model results
4. The success of a one-class phytoplankton model

25.5.3 *Adriatic Sea, Emilia-Romagna Coastal Area, Italy*

The two previous case studies emphasize pelagic dynamics with bivalves as primary consumers. However, the shallow waters characteristic of bivalve culture areas invariably involve tight benthic-pelagic coupling. Benthic processes have been studied extensively in both the Baie des Veys and Tracadie Bay (e.g. Cranford et al. 2009; Ubertini et al. 2012), but we use an example from Adriatic Italy to bring together benthic and pelagic dynamics as they relate to suspended bivalve culture.

Shellfish culture is an important activity along the Adriatic and Ionian Italian coasts. The two main products are: (i) Manila clams *(Tapes philippinarum)*, which are farmed in the Northern Adriatic lagoons, such as those of Marano, Venice, Goro and Scardovari, and (ii) Mediterranean mussel (*Mytilus galloprovincialis*), which

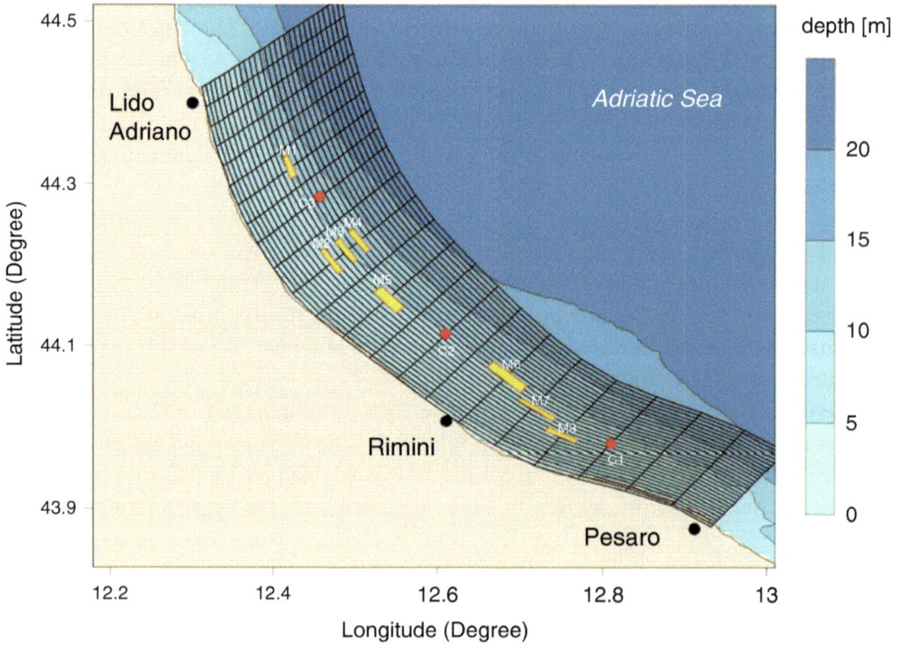

Fig. 25.3 Map of the coastal area investigated in the case study. Water quality monitoring stations are shown in red, mussel farms in yellow

are farmed mainly off-shore on longlines from the Gulf of Trieste in the North to the Gulf of Taranto in the South. This case study, presented in detail in Brigolin et al. (2008), was focused on investigating the impact of mussel farming on lower trophic levels and on the biogeochemistry of surface sediments along the coastal area of Emilia-Romagna (Fig. 25.3), which in 2013 produced about 22,000 tonnes of Mediterranean mussels, i.e. about one third of the Italian production. This study differs from the previous cases as it deals with an open, though shallow, coastal area where processes driven by the North-South WACC (Western Adriatic Coastal Current) are effectively transported along the coast, mixing dissolved compounds and suspended particles.

The above issues were investigated using an integrated model (Brigolin et al. 2008), which included: (1) a 2D transport module, (2) a pelagic biogeochemical module, (3) a farmed mussel population dynamics module, (4) a module for the simulation of early diagenesis processes in surface sediments. The model was designed to simulate the population dynamics of farmed mussels and their impact on the pelagic environment, as well as on the fluxes of oxygen and nitrogen due to the remineralization of mussel faeces and pseudo-faeces in surface sediment. Therefore, it can be used for estimating the biomass yield and quantifying the effect of seston depletion due to mussel filtration as in Dowd (2005), Grant et al. (2005), and Ferreira et al. (2007). Furthermore, the explicit inclusion of early diagenetic processes allows assessment of the influence of mussel farming on the overall C and N biogeochemical cycles. This context expands the more pelagic focus of the Canadian and French case studies, as well as expanding the community composition of the phytoplankton submodel.

The first module solves the advection-diffusion equation. Input data for water velocities and elevation were provided by a 2D finite difference hydrodynamic model, which was previously calibrated to simulate the hydrodynamic circulation in the NW Adriatic Sea under realistic forcings induced by tides and meteorological fields for the year 2004 (Lovato et al. 2010). The hydrodynamic model was applied to the whole Adriatic, including the Lagoon of Venice, using a curvilinear boundary-conforming grid, composed of 287,363 nodes, with mesh sizes varying from approximately 12 km to 50 m.

The pelagic biogeochemical module, described in detail in Brigolin et al. (2011), included 14 state variables in order to simulate the dynamics of carbon, nitrogen (nitrate, ammonia), phosphorus and silica, and to mimic the main features of the observed seasonal succession of the phytoplankton community. Therefore, besides the concentrations of the above inorganic nutrients and dissolved oxygen, the module simulates the evolution of three phytoplankton functional types: winter diatoms, summer diatoms and flagellates. The set of state variables also includes four pools of dissolved organic detritus, one for each macronutrient. Beside allowing closure of biogeochemical cycles, the carbon detritus represents an additional source of energy for farmed mussels, which in some instances can compensate for the lack of phytoplankton (Brigolin et al. 2009). Diatoms were divided into winter and summer types as winter diatom blooms are mainly accounted for by *Skeletonema marinoi*, while autumn peaks are related to the presence of *Chaetoceros socialis* and other

Chaetoceros spp. The flagellate functional type was meant to model various classes (Prasinophycea, Haptophycea, Chlorophycea, Cryptophycea, and Chrysophycea). Zooplankton variables were defined according to size, in order to take into account the role of micro- and meso-zooplankton in controlling phytoplanktonic biomass. The above biotic variables were expressed as carbon content of planktonic tissue. Elemental fluxes of N, P and Si through the ecosystem are quantified by assuming a fixed C:N:P:Si ratio.

The third module was based on the individual bioenergetic model described in detail in Brigolin et al. (2009), which simulates the evolution of dry weight, and through correlation, wet weight and length of an average Mediterranean mussel (*Mytilus galloprovincialis*) individual. The population dynamics of cohorts of farmed mussels at each farming site were simulated by following the evolution of an ensemble of individuals by means of a Monte Carlo approach. The parameterization of each individual was slightly different in order to mimic the observed variability of the output variables. In particular, the maximum clearance rate *CRmax* and the maximum respiration rate *Rmax,* were treated as Gaussian stochastic variables and randomly assigned to each individual. Mortality rate was assumed to be constant throughout the grow-out phase. The daily release of mussel bio-deposits from a given mussel farm was subsequently estimated on the basis of individual emissions and stocking density. Mussel biodeposits were transported using a Lagrangian particle tracking module, which was originally developed for investigating the impact of fish farming on the benthic community and tested at Mediterranean fish farms (Brigolin et al. 2014). The module was recently updated and employed for mapping the environmental impact of shellfish farms, as part of a systematic procedure for assessing the suitability of this coastal area for oyster and mussel farming, applied in the context of the Maritime Spatial Planning EU Directive (Brigolin et al. 2017).

Mussel biodeposits and organic detritus derived from the decomposition of phyto and zooplankton eventually settle on the seabed; this flow of organic matter to surface sediments represents the input for the early diagenesis module, which enables estimation of the steady-state vertical profile of ammonia, nitrate and reactive phosphorus in a sediment core. Early diagenesis processes are presented in detail in Brigolin et al. (2011); they include the oxic degradation of organic matter, as well as the main anoxic pathways, in which microbial communities use nitrate, sulfate, and oxidized forms of iron and manganese as electron acceptors. Re-oxidation processes of reduced products are also taken into account, since they contribute to depletion of dissolved oxygen concentration in the upper sediment layers.

Setting boundary conditions for an open coastal area is not easy and there are always sources of uncertainty. Boundary conditions for the pelagic model were estimated on the basis of a year long time series of sea surface temperature and concentrations of ammonium, nitrate, dissolved inorganic phosphorus, dissolved oxygen, reactive silica and chlorophyll collected approximately every 2 weeks at 6 monitoring stations close to the boundary of the computational domain.

The results of this study indicate:

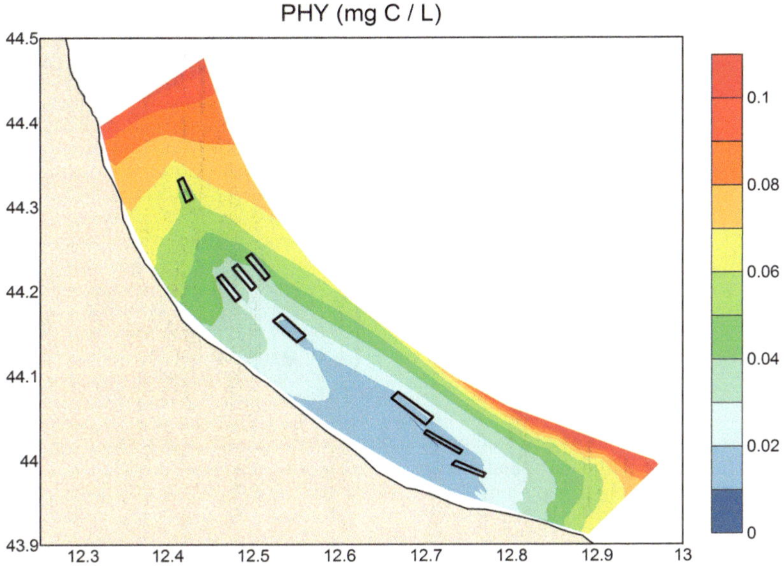

Fig. 25.4 Spatial distribution of chlorophyll in the Emilia-Romagna study area (Italy) in Spring 2004. Units are mg C l^{-1}

1. As shown in Fig. 25.4, the model predicts different mussel biomass yields caused by a north-south chlorophyll gradient (expressed as mg C l^{-1}) related to the nutrient enriched waters discharged by the Po River.
2. Local depletion of chlorophyll by cultured mussels could not be unequivocally related to the presence of the farms, even though Fig. 25.4 suggests that mussel filtration in spring could locally reduce phytoplankton density, particularly in the northern part of the study area. In this regard, increasing the spatial resolution of the model could help in improving the description of transport and mixing processes and their role in ecosystem dynamics.
3. The effects of mussel farms on phytoplankton biomass and C and N cycles are more clearly revealed in Fig. 25.5, which compares the deposition of organic particles per unit surface beneath mussel farms with those estimated at control sites (Fig. 25.3). The fluxes of organic carbon at the sediment-water interface are similar beneath farms M1-M6 and about 8 times higher than those at control sites. Even though these deposition rates are much lower in comparison with those originated in sea-cage fish farming, the overall impact of mussel farming on the C and N cycles may be more significant at a regional scale, because of the much larger extent of leased areas. Furthermore, fluxes of organic carbon are significantly lower beneath farm M7, which is located in between farm M6 and M8, in the southern part of study area. Since these fluxes are correlated with the amount of phytoplankton and non-living organic particles cleared by mussels, this result provides indirect evidence of phytoplankton depletion, due to the cumulative effect of the adjacent farms in clearing suspended particles.

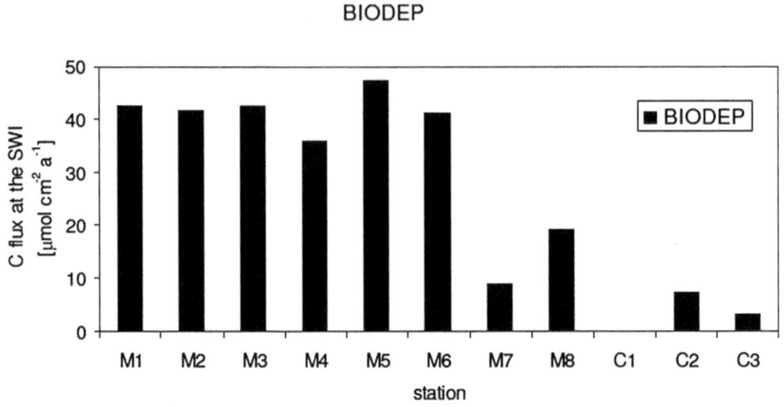

Fig. 25.5 Annual fluxes of organic carbon in surface sediment beneath the mussel farms located in the Emilia-Romagna study area and at the three monitoring stations shown in Fig. 25.3, which can be taken as control sites

4. The model can be used for assessing the overall "ecological carrying capacity" of the coastal zone with respect to mussel farming and is thus a useful tool for managing this activity within an Integrated Coastal Zone Management approach. Patterns and levels of biodeposition provide evidence of the spatial pattern of bivalve grazing in a way that would not be indicated by chlorophyll depletion.
5. The value of both benthic and pelagic processes in modelling bivalve-phytoplankton interactions is clearly shown, as the feedback to nutrient regeneration has implications for phytoplankton production, including favouring certain cell types (Zhao et al. 2016).

25.6 Management and Husbandry Considerations

The implications of spatial scale for shellfish culture are immediately obvious. In the case of Baie de Veys, culture density is limited to the intertidal, excluding huge areas of the bay. For this reason, it would be difficult for aquaculture to cause ecosystem-wide depletion (see also Dabrowski et al. 2013). However, this does not preclude the significance of more localized depletion, which is obvious from model results. Moreover, this depletion causes local variation in oyster growth including depressed growth in the chlorophyll depletion zone. In contrast, the case study of Tracadie Bay indicates system-wide seston depletion. In this example, mussel culture on suspended longlines is practiced throughout the bay, and thus culture occupies much of the surface area. Consequently, baywide seston depletion occurs, and diminishes mussel growth in the upper parts of the bay where renewal of depleted water is reduced (Waite et al. 2005).

The results presented in the Adriatic case suggest that even in open coastal areas the cumulative impact of shellfish farms on phytoplankton dynamics should be taken into account when estimating both the ecological and production carrying capacities. However, these findings should be interpreted with care due to: (i) the rather coarse resolution of the model, (ii) the uncertainty in ocean boundary conditions, whose effect on the results is more relevant than in the other two case studies here presented. Those issues are to some extent related, and they can be tackled in future studies by nesting coastal models within models developed for operational oceanography and using higher resolution ocean colour products. In this regard, the Copernicus Marine Environment Monitoring Service (http://marine.copernicus.eu/) is already providing reanalysis and real-time data concerning both water temperature and biogeochemical variables.

Given that some culture scenarios cause system reduction of chlorophyll, questions of standards and thresholds quickly arise. For this reason, many results are reported as the spatial distribution of % depletion (Filgueira et al. 2015). This format is important because it specifies a map of depletion and its degree of localization. Even when large-scale aquaculture scenarios are compared with models (Filgueira et al. 2013), it is obvious that some culture densities are beyond carrying capacity as defined by depletion thresholds.

The utility of chlorophyll maps in aquaculture planning has recently emerged in modelling studies. Filgueira et al. (2010) used a simplified mussel-phytoplankton model in a Norwegian fjord where an upweller was used to stimulate phytoplankton production via nutrient diffusion. They deployed optimization to indicate culture locations which minimized advective loss of enhanced chlorophyll. Subsequently, Filgueira et al. (2015), answered a planning question regarding the severity of depletion under proposed additions of mussel culture longlines into a bay with existing aquaculture leases.

We subscribe to an ecosystem approach to aquaculture (EAA) as promulgated by Aguilar-Manjarrez et al. (2017). This is consistent with our stated goal of maintaining ecosystem services within their natural limits (Grant and Filgueira 2011). These limits expressed as temporal variation is site dependent (e.g. Thomas et al. 2011; Cloern and Jassby 2010) and a discussion of intra-annual variation is beyond the scope of this chapter. When aquaculture is discretely located, i.e. single farm sites, effects produced including nutrient release or biodeposition tend to be near-field. In this case, scaling of the magnitude of these effects relative to the size of the ecosystem is essential, even though this is rarely considered. In the case of shellfish culture, effects such as depletion of chlorophyll tend to be pervasive, even impacting large scale phytoplankton spatial distribution as in Tracadie Bay. As shown in the Adriatic case, spatial patterns of biodeposition and subsequent diagenesis are also apparent. Even though new remote sensing technologies and remotely operated in situ sensors are likely to provide relevant data for assessing some of these impacts, we emphasize that ecosystem modelling will remain the main tool for interpreting these processes, assessing ecological carrying capacity and providing management scenarios in the context of global drivers, such as climate change. The Tracadie Bay case of a new inlet is a graphic example (Filgueira et al. 2014).

Conservation of primary production by microalgae, the most important of supporting services, can thus be managed with respect to aquaculture development. Modelling has been underutilized in marine spatial planning applied to aquaculture, but has huge scope for furthering the EAA approach, as in Brigolin et al. (2017) and Filgueira et al. (2015). Although we highlight the value of these models in their ability to elucidate the spatial dynamics of phytoplankton, several questions arise as to why the models work so well in terms of simulating both bivalve growth and chlorophyll distribution.

For example, size composition of phytoplankton seems to be an ecosystem-wide response of differential grazing. However, spatial variation in phytoplankton communities may be difficult to characterize with sampling (see Zhao et al. 2016). Although remote sensing has been used to distinguish taxonomic makeup of autotrophs including size classes (Brewin et al. 2011) the problems of ocean colour detection in the coastal zone persist.

In conclusion, we pose a few questions in the context of how they might impact model structure and predictive power, with a suggestion for the direction of answers, recognizing that these are very much topics for future research.

What are the consequences to other grazers for changes in phytoplankton size distribution? Comeau et al. (2015) examined partitioning of particle sizes between cultured bivalves and tunicates, and assessed the extent to which tunicates 'removed' carrying capacity through grazing competition.

What is the role of aggregation in masking apparent size classes? Feeding experiments demonstrate that small cells are readily ingested when embedded in mucus aggregates (Kach and Ward 2008; Cranford et al. 2011), but aggregate size structure, their incorporation of phytoplankton, and implications for bivalve food models are poorly known.

Why isn't resuspension more important in these shallow systems and thus necessary in models? There is an extensive literature showing positive, negative, or neutral effects of resuspension in bivalve growth (e.g. Grant et al. 1990; Kang et al. 2006; Ubertini et al. 2012). This likely occurs because, despite the potential for entrainment of benthic microalgae into suspension, excess suspended load dilutes seston quality. Although similar questions have been posed for detritus as a supplemental food source (e.g. macrophyte debris), the bivalve-phytoplankton trophic link is unquestionably central to food limitation and carrying capacity for aquaculture species.

Acknowledgements The authors are grateful to the referees for their valuable comments.

References

Aguilar-Manjarrez J, Soto D, Brummett R (2017) Aquaculture zoning, site selection and area management under the ecosystem approach to aquaculture. Full document. Report ACS113536. Rome, FAO, and World Bank Group, Washington, DC, 395 pp

Brewin RJW, Hardman-Mountford NJ, Lavender SJ, Raitsos DE, Hirata T, Uitz J, Devred E, Bricaud A, Ciotti A, Gentili B (2011) An intercomparison of bio-optical techniques for detecting dominant phytoplankton size class from satellite remote sensing. Remote Sens Environ 115:325–339

Brigolin D, Lovato T, Ciavatta S, Pastres R (2008) The impact of mussel farming on the biogeochemistry of the northern Adriatic coastal ecosystem: preliminary results from a modelling study. ICES CM Documents 2008 of the ICES 2008 annual science conference, Halifax, Canada, Document CM 2008/L:12

Brigolin D, Dal Maschio G, Rampazzo F, Giani M, Pastres R (2009) An individual-based population dynamic model for estimating biomass yield and nutrient fluxes through an off-shore mussel (*Mytilus galloprovincialis*) farm. Estuar Coast Shelf Sci 82:365–376

Brigolin D, Lovato T, Rubino A, Pastres R (2011) Coupling early-diagenesis and pelagic biogeochemical models for estimating the seasonal variability of N and P fluxes at the sediment–water interface: application to the northwestern Adriatic coastal zone. J Mar Syst 87:239–255

Brigolin D, Meccia VL, Venier C, Tomassetti P, Porrello S, Pastres R (2014) Modelling biogeochemical fluxes across a Mediterranean fish cage farm. Aquacult Environ Interact 5:71–88

Brigolin D, Porporato EMD, Prioli G, Pastres R (2017) Making space for shellfish farming along the Adriatic coast. ICES J Mar Sci 74:1540–1551

Canu DM, Solidoro C, Cossarini GF (2010) Effect of global change on bivalve rearing activity and the need for adaptive management. Clim Res 42:13–26

Cloern JE (1982) Does the benthos control phytoplankton biomass in South San Francisco Bay? Mar Ecol Prog Ser 9:191–202

Cloern JE (1996) Phytoplankton bloom dynamics in coastal ecosystems: a review with some general lessons from sustained investigation of San Francisco Bay, California. Rev Geophys 34:127–168

Cloern JE (2001) Our evolving conceptual model of the coastal eutrophication problem. Mar Ecol Prog Ser 210:223–253

Cloern JE, Jassby AD (2010) Patterns and scales of phytoplankton variability in estuarine–coastal ecosystems. Estuar Coasts 33:230–241

Cloern JE, Nichols FH (1985) Time scales and mechanisms of estuarine variability, a synthesis from studies of San Francisco Bay. Hydrobiologia 129:229–237

Comeau LA, Filgueira R, Guyondet T, Sonier R (2015) The impact of invasive tunicates on the demand for phytoplankton in longline mussel farms. Aquaculture 441:95–105

Cranford P (2019) Magnitude and extent of water clarification services provided by bivalve suspension feeding. In: Smaal et al (eds) Goods and services of marine bivalves, Springer, Cham, pp 119–141

Cranford PJ, Strain PM, Dowd M, Hargrave BT, Grant J, Archambault M (2007) Influence of mussel aquaculture on nitrogen dynamics in a nutrient enriched coastal embayment. Mar Ecol Prog Ser 347:61–78

Cranford PJ, Li W, Strand Ø, Strohmeier T (2008) Phytoplankton depletion by mussel aquaculture: high resolution mapping, ecosystem modeling and potential indicators of ecological carrying capacity. ICES CM Document 2008/H:12. 5p. www.ices.dk/products/CMdocs/CM-2008/H/H1208.pdf

Cranford PJ, Hargrave BT, Doucette LI (2009) Benthic organic enrichment from suspended mussel (Mytilus edulis) culture in Prince Edward Island, Canada. Aquaculture 292:189–196

Cranford P, Ward JE, Shumway SE (2011) Bivalve filter feeding: variability and the limits of the aquaculture biofilter. In: Shumway SE (ed) Aquaculture and the environment. Wiley, New York

Cugier P, Struski C, Blanchard M, Mazurié J, Pouvreau S, Olivier F, Trigui JR, Thiébaut E (2010) Assessing the role of benthic filter feeders on phytoplankton production in a shellfish farming site: Mont Saint Michel Bay, France. J Mar Syst 82:21–34

Dabrowski T, Lyons K, Curé M, Berry A, Nolan G (2013) Numerical modelling of spatio-temporal variability of growth of *Mytilus edulis* (L.) and influence of its cultivation on ecosystem functioning. J Sea Res 76:5–21

Dowd M (2005) A biophysical coastal ecosystem model for assessing environmental effects of marine bivalve aquaculture. Ecol Model 183:323–346

Ferreira JG, Hawkins AJS, Bricker SB (2007) Management of productivity, environmental effects and profitability of shellfish aquaculture—the Farm Aquaculture Resource Management (FARM) model. Aquaculture 264:160–174
Filgueira R, Grant J, Strand Ø, Asplin L, Aure J (2010) A simulation model of carrying capacity for mussel culture in a Norwegian fjord: role of induced upwelling. Aquaculture 308:20–27
Filgueira R, Grant J, Stuart R, Brown M (2013) Ecosystem modelling for ecosystem-based management of bivalve aquaculture sites in data-poor environments. Aquacult Environ Interact 4:117–133
Filgueira R, Guyondet T, Comeau LA, Grant J (2014) Storm-induced changes in coastal geomorphology control estuarine secondary productivity. Earth's Futur 2(1):1–6
Filgueira R, Guyondet T, Bacher C, Comeau LA (2015) Informing marine spatial planning (MSP) with numerical modelling: a case-study on shellfish aquaculture in Malpeque Bay (Eastern Canada). Mar Pollut Bull 100:200–216
Grangeré K, Lefebvre S, Ménesguen A, Jouenne F (2009) On the interest of using field primary production data to calibrate phytoplankton rate processes in ecosystem models. Estuar Coast Shelf Sci 81:169–178
Grangeré K, Lefebvre S, Bacher C, Cugier P, Ménesguen A (2010) Modelling the spatial heterogeneity of ecological processes in an intertidal estuarine bay: dynamic interactions between bivalves and phytoplankton. Mar Ecol Prog Ser 415:141–158
Grant J, Filgueira R (2011) The application of dynamic modelling to prediction of production carrying capacity in shellfish farming. In: Shumway SE (ed) Shellfish culture and the environment. Wiley-Blackwell, New York
Grant J, Enright CT, Griswold A (1990) Resuspension and growth of *Ostrea edulis*: a field experiment. Mar Biol 104:51–59
Grant J, Dowd M, Thompson K, Emerson C, Hatcher A (1993) Perspectives on field studies and related biological models of bivalve growth and carrying capacity. In: Dame RF (ed) Bivalve filter feeders. NATO ASI series (Series G: ecological sciences), vol 33. Springer, Berlin/Heidelberg
Grant J, Cranford P, Hargrave B, Carreau M, Schofield B, Armsworthy S, Burdett-Coutts V, Ibarra D (2005) A model of aquaculture biodeposition for multiple estuaries and field validation at blue mussel (*Mytilus edulis*) culture sites in eastern Canada. Can J Fish Aquat Sci 62:1271–1285
Grant J, Bugden G, Horne E, Archambault M-C, Carreau M (2007) Remote sensing of particle depletion by coastal suspension-feeders. Can J Fish Aquat Sci 64:387–390
Grant J, Bacher C, Cranford PJ, Guyondet T, Carreau M (2008) A spatially explicit ecosystem model of seston depletion in dense mussel culture. J Mar Syst 73:155–168
Guyondet T, Comeau LA, Bacher C, Grant J, Rosland R, Sonier R, Filgueira R (2015) Climate change influences carrying capacity in a coastal embayment dedicated to shellfish aquaculture. Estuar Coasts 38:1593–1618
Jiang T, Chen F, Yu Z, Lu L, Wang Z (2016) Size-dependent depletion and community disturbance of phytoplankton under intensive oyster mariculture based on HPLC pigment analysis in Daya Bay, South China Sea. Environ Pollut 219:804–814
Kach DJ, Ward JE (2008) The role of marine aggregates in the ingestion of picoplankton-size particles by suspension-feeding molluscs. Mar Biol 153:797–805
Kang CK, Lee YW, Choy EJ, Shin JK, Seo IS (2006) Microphytobenthos seasonality determines growth and reproduction in intertidal bivalves. Mar. Ecol Prog Ser 315:113–127
Lovato T, Androsov A, Romanenkov D, Rubino A (2010) The tidal and wind induced hydrodynamics of the composite system Adriatic Sea/Lagoon of Venice. Cont. Shelf Res 30:692–706
Lucas LV, Koseff JR, Monismith SG, Cloern JE, Thompson JK (1999a) Processes governing phytoplankton blooms in estuaries. II: the role of horizontal transport. Mar Ecol Prog Ser 187:17–30
Lucas LV, Koseff JR, Cloern JE, Monismith SG (1999b) Processes governing phytoplankton blooms in estuaries. I: the local production-loss balance. Mar Ecol Prog Ser 187:1–15
Ludvigsen M, Sørensen A (2016) Towards integrated autonomous underwater operations for ocean mapping and monitoring. Annu Rev Control 42:145–157
May CL, Koseff JR, Lucas LV, Cloern JE, Schoellhamer DH (2003) Effects of spatial and temporal variability of turbidity on phytoplankton blooms. Mar Ecol Prog Ser 254:111–128

McKindsey CW, Thetmeyer H, Landry T, Silvert W (2006) Review of recent carrying capacity models for bivalve culture and recommendations for research and management. Aquaculture 261:451–462

Meeuwig JJ (1999) Predicting coastal eutrophication from land-use: an empirical approach to small non-stratified estuaries. Mar Ecol Prog Ser 176:231–241

Nielsen P, Cranford PJ, Maar M, Petersen JK (2016) Magnitude, spatial scale and optimization of ecosystem services from a nutrient extraction mussel farm in the eutrophic Skive Fjord, Denmark. Aquacult Environ Interact 8:311–329

Petersen JK, Holmer M, Termansen M, Hassler B (2019) Nutrient extraction through bivalves. In: Smaal et al (eds) Goods and services of marine bivalves, Springer, Cham, pp 179–208

Radiarta IN, Saitoh S-I (2009) Biophysical models for Japanese scallop, *Mizuhopecten yessoensis*, aquaculture site selection in Funka Bay, Hokkaido, Japan, using remotely sensed data and geographic information system. Aquac Int 17:403–419

Richardson AJ, Schoeman DS (2004) Climate impact on plankton ecosystems in the Northeast Atlantic. Science 305:1609–1612

Smaal AC, van Duren L (2019) Bivalve aquaculture carrying capacity: concepts and assessment tools. In: Smaal et al (eds) Goods and services of marine bivalves, Springer, Cham, pp 451–483

Smaal AC, Schellekens T, van Stralen MR, Kromkamp JC (2013) Decrease of the carrying capacity of the Oosterschelde estuary (SW Delta, NL) for bivalve filter feeders due to overgrazing? Aquaculture 404–405:28–34

Sonier R, Filgueira R, Guyondet T, Tremblay R, Olivier F, Meziane T, Starr M, LeBlanc AR, Comeau LA (2016) Picophytoplankton contribution to *Mytilus edulis* growth in an intensive culture environment. Mar Biol 163:73

Thomas Y, Mazurié J, Alunno-Bruscia M, Bacher C, Bouget J-F, Gohin F, Pouvreau S, Struski C (2011) Modelling spatio-temporal variability of *Mytilus edulis* (L.) growth by forcing a dynamic energy budget model with satellite-derived environmental data. J. Sea Res 66:308–317

Ubertini M, Lefebvre S, Gangnery A, Grangeré K (2012) Spatial variability of benthic-pelagic coupling in an estuary ecosystem: consequences for microphytobenthos resuspension phenomenon. PLoS One 7

Waite L, Grant J, Davidson J (2005) Bay-scale spatial variation of mussels *Mytilus edulis* in suspended culture, Prince Edward Island, Canada. Mar Ecol Prog Ser 297:157–167

Wijsman JWM, Troost K, Fang J, Roncarati A (2019) Global production of marine bivalves. Trends and challenges for the future. In: Smaal et al (eds) Goods and services of marine bivalves, Springer, Cham, pp 7–26

Zhao L, ZhaoY XJ, Zhang W, Huang L, Jiang Z, Fang J, Xiao T (2016) Distribution and seasonal variation of picoplankton in Sanggou Bay, China. Aquacult Environ Interact 8:261–271

Chapter 26
Spatial, Ecological and Social Dimensions of Assessments for Bivalve Farming Management

C. Bacher, A. Gangnery, P. Cugier, R. Mongruel, Øivind Strand, and K. Frangoudes

Abstract The general purpose of assessment is to provide decision-makers with the best valuable data, information, and predictions with which management decisions will be supported. Using case studies taken from four scientific projects and dealing with the management of marine bivalve resources, lessons learned allowed identifying some issues regarding assessment approaches. The selected projects also introduced methodological or institutional frameworks: ecosystem approach to aquaculture (EAA), system approach framework (SAF), marine spatial planning (MSP), and valuation of ecosystem services (ES).

The study on ecosystem services linked ES to marine habitats and identified ES availability and vulnerability to pressures. The results were displayed as maps of resulting potential services with qualitative metrics. The vulnerability value is an alternative to monetary valuation and, in addition to identifying the most suitable

C. Bacher (✉) · P. Cugier
IFREMER, Centre de Bretagne, DYNECO-LEBCO, Plouzané, France
e-mail: cedric.bacher@ifremer.fr; philippe.cugier@ifremer.fr

R. Mongruel
IFREMER, Centre de Bretagne, Ifremer, UMR 6308, AMURE, Plouzané, France
e-mail: remi.mongruel@ifremer.fr

A. Gangnery
IFREMER, Station de Port en Bessin, Port en Bessin, France
e-mail: aline.gangnery@ifremer.fr

Ø. Strand
Institute of Marine Research, Bergen, Norway
e-mail: oivind.strand@imr.no

K. Frangoudes
Univ Brest, IFREMER, CNRS, UMR 6308, AMURE, IUEM, Plouzané, France
e-mail: katia.frangoudes@univ-brest.fr

A. C. Smaal et al. (eds.), *Goods and Services of Marine Bivalves*,
https://doi.org/10.1007/978-3-319-96776-9_26

areas for each type of ES, this metric allows identifying the management strategies that will most probably maintain or affect each individual ES.

The MSP example focused on bivalve farming activity and accounted for several criteria: habitat suitability, growth performance, environmental and regulation constraints and presence of other activities. The ultimate endpoint of such an approach is a map with qualitative values stating whether a location is suitable or not, depending on the weight given to each criterion.

In the EAA case study, the indicator was defined by the growth performance of cultivated bivalves in different locations. This indicator is affected by distant factors – e.g. populations of marine organisms competing for the same food resource, nutrient inputs from rivers, time to renew water bodies under the action of tidal currents. The role and interactions of these factors were assessed with a dynamical ecosystem model.

Examples illustrate that the assessment is often multi-dimensional, and that multiple variables would interact and affect the response to management options. Therefore, the existence of trade-offs, the definition of the appropriate spatial scale and resolution, the temporal dynamics and the distant effects of factors are keys to a policy-relevant assessment. EA and SAF examples show the interest of developing models relating response to input variables and testing scenarios. Dynamic models would be preferred when the relationship between input and output variables may be masked by non-linear effects, delay of responses or differences of scales.

When decision-making requires economic methods, monetary values are often of poor significance, especially for those ecosystem services whose loss could mean the end of life, and appear to be a comfortable oversimplification of reality of socio-ecological systems which cannot be summarized in single numbers. Alternative methods, such as the ones proposed in the SAF and ES examples, would preferably consider institutional analysis or multicriteria assessment rather than single monetary values.

Case studies also highlighted that credibility of assessment tools benefit from the association of stakeholders at different stages, among which: identification of the most critical policy issues; definition of system characteristics including ecological, economical and regulation dimensions; definition of modelling scenarios to sort out the most effective management options; assessment of models and indicators outputs.

Abstract in Chinese 摘要:总体而言,进行评价的目的是为决策者提供最有效的数据,信息和预测,从而支持管理决策的制定。通过四个关于海水双壳贝类资源管理案例研究,我们明确了关于评估中存在的问题。 这些案例研究包含了方法学和一些制度框架,其中包括:水产养殖生态系统方法(EA),系统管理方法(SA),海洋空间规划(MSP)和生态系统服务价值评估(ES)。

在生态系统服务价值评估(ES)的案例研究中,我们将其与海洋生境联系起来,确认了生态系统价值评估的有效性和应对不同压力时的脆弱性。 评估将定性结果在地图上进行相应的展示。 脆弱值是经济估值的替代参考,除了为每种类型的生态服务确定最适区域外,脆弱性指标可以帮助我们确认管理策略是否会对不同生态系统服务功能造成影响。

海洋空间规划的案例研究侧重于双壳贝类养殖活动并考虑了几个条件:适宜性评价,生长情况,环境和法规限制以及其他相关活动。海洋空间规划的最终目的是绘制具有定性值的示意图,根据标准的权重来确定地点是否适合进行养殖活动。

在水产养殖生态系统方法案例研究中,指标由不同地点养殖的贝类生长情况来确定。 这类指标同样受其他因素的影响,例如与贝类进行食物竞争的其他生物种群,来自河流的营养物质输入,潮汐作用下水体的更新时间等。 我们可以利用生态系统动态模型来评估这些因素的相互作用。

这些系统往往是多维的,多个变量会相互影响,并影响管理系统的选择。因此,各方面因素的协调与权衡,适当的模型空间尺度和分辨率,动态的时序模拟和一些外部驱动因素是进行养殖规划和策略评估的关键。 我们可以通过模型开发将输入数据与结果相应联系起来。 当输入和输出变量之间的联系是非线性、或存在响应延迟以及尺度差异时,动态模型的建立成为有效的解决方法。

当决策制定需要进行经济效益评估时,单纯的经济核算通常意义不大,因为生态系统服务功能难以通过简单的货币化进行全面概括,某些生态功能的缺失将意味着生态系统的崩溃。其它可供选择的评估方法,例如在系统管理方法和生态系统评估案例中的方法,也不应该仅仅考虑货币价值,应当进行制度分析或者多因素综合分析 。

评估工具的可靠性可以从不同阶段的利益相关群体获得提升, 其中包括:确定最关键的政策问题;确定包括生态、经济以及政策法规等方面的系统特征; 定义具体的模型场景以采取最有效的管理手段;评估模型和指标的产出成果。

Keywords Ecosystem services · Marine spatial planning · Ecosystem approach to aquaculture · System approach · Modeling · Stakeholder involvement

关键词 生态系统服务 • 海域空间规划 • 水产养殖生态系统方法 • 系统管理方法 • 模型模拟 • 利益相关者的参与

26.1 Introduction

Valuation of Ecosystem Services (ES) is one among several management frameworks, concepts and approaches that support the implementation of several legislative tools (Lonsdale et al. 2015). These frameworks have different scopes which have been extensively described in handbooks, and discussed and compared in international working groups (e.g. FAO, ICES). They all aim at improving the management of natural resources and refer more or less explicitly to the need for long-term actions to make the use of resources sustainable. In a position paper published by the Marine Board, Rice et al. (2010) also highlighted the multiple dimensions of science-policy integration for decision-making with respect to management of marine resources. Recommendations included the links between ecosystem services and management policies, and the need for science support to strategic environmental assessments, including socioeconomic factors.

The objective of this chapter is to show some specific examples taken from scientific projects and introduce some issues regarding the integration of knowledge and assessment tools, rather than to review existing literature. The term assessment is taken in a broad sense as a "formal effort to assemble selected knowledge with a view to making it publicly available in a form intended to be useful for decision-making" (Rice et al. 2010). Our selection of examples will highlight the spatial, ecological and social dimensions that have been addressed through assessments of bivalve-related activities within several management frameworks: the ecosystem approach to aquaculture, marine spatial planning, system approach, and ecosystem services. We first review below the general definitions of these frameworks.

Historically, Integrated Coastal Zone Management (ICZM) may be considered as one of the first frameworks which dealt with the difficulty to manage coastal human activities competing for the of use natural resources (including space). Quoting Pinot (1998), Cormier et al. (2013) defined the objective of ICZM as "the disposition of each coastal segment to the most appropriate business, according to decisions taken by the public authorities in light of scientific knowledge, thanks to which we can ensure consistency in the use (avoiding the adverse effects that would result in sterilization of the rich shores), and harnessing the energy of nature to serve our needs rather than abruptly counter the natural system". This definition introduces key concepts, which are overarching across all other frameworks: role of stakeholders, resolution of conflicts, sustainable use of the coastal zone, use of scientific knowledge, interdependency of activities, multiple social and biophysical dimensions – which are highlighted in the following definitions for each framework.

Aguilar-Manjarrez et al. (2010) have defined Ecosystem Approach to Aquaculture (EAA) as "a strategy for the integration of the activity within the wider ecosystem in such a way that it promotes sustainable development, equity, and resilience of interlinked social and ecological systems". They stated three main principles for aquaculture development: (1) no degradation of ecosystem functions and services beyond their resilience capacity; (2) improvement of human wellbeing and equity for stakeholders; and (3) consideration for other relevant sectors. They also emphasized that EAA applies at different scales: the farm, the waterbody and its watershed/aquaculture zone, and the global, market-trade scale.

Marine Spatial Planning (MSP) is a "process of analyzing and allocating parts of three-dimensional marine space to specific uses, to achieve ecological, economic, and social objectives that are usually specified through the political process; the MSP process usually results in a comprehensive plan or vision for a marine region" (Aguilar-Manjarrez et al. 2010). MSP is generally defined as a means to "create and establish a more rational organization of the use of marine space and the interactions between its uses, to balance demands for development with the need to protect the environment, and to achieve social and economic objectives in an open and planned way" (Douvere 2008). This applies to aquaculture development where planning is an important process, which is expected to stimulate and guide the evolution of the sector by providing incentives and safeguards, attracting investments and boosting development, while ensuring its long-term sustainability to ultimately contribute to economic growth and poverty alleviation (Brugère et al. 2010).

The System Approach Framework (SAF) builds upon the systems science and aims to incorporate the ecological, social, and economic dimensions of coastal systems and integrate knowledge, to support decision-making (Tett et al. 2011). Dynamic models have been developed and used to explore alternative policy options following a problem-oriented and scenario-based approach. This approach involves several steps: consultation of stakeholders to prioritize one management issue; definition of the natural, social, and economic dimensions of the coastal system; building a mathematical model of the ecological and social processes likely to explain the dynamics of the system; defining scenarios and indicators to analyze model outputs with stakeholders.

Following MEA (2005), Ecosystem Services (ES) are defined as the benefits people obtain from ecosystems and include provisioning services such as food, water, and raw materials; regulating services such as climate regulation, protection from floods and storms, water quality and waste bioremediation; cultural services that provide recreational, aesthetic, and spiritual benefits; and supporting services such as biologically mediated habitats and nutrient cycling (Liquete et al. 2013).

To allow comparisons and discussions, the case studies described below are presented with a similar structure: issue identification, system definition, assessment principles, main results and lessons learned. On this basis, we review some of the key issues and features of the assessment: system boundaries, stakeholder involvement, tools availability, contribution to the decision-making process. All these examples deal with spatial aspects but more generally deal with multiple dimensions in relation to the questions raised and the framework used.

26.2 Ecosystem Approach to Aquaculture – Bay of Mont Saint Michel Case

26.2.1 Issue Identification

Mont Saint Michel Bay is a place of major bivalve farming in the northwestern part of France with an annual production of around 10,000 tonnes of the blue mussel *Mytilus edulis*, 5000 tonnes of the Pacific oyster *Magallana gigas* and 1000 tonnes of the European flat oyster *Ostrea edulis*. At the beginning of the 2000s, farmers requested a spatial extension of their concessions to avoid excessive siltation (in oyster areas) or growth limitation (in mussel areas). Furthermore, the industry has been responsible for the introduction of the invasive gastropod, *Crepidula fornicata*, the slipper limpet, through the importation of *M. gigas* during the 1970s (Blanchard 1997). Since then, the slipper limpet proliferated in the subtidal area of the bay to reach a biomass of ca. 150,000 tonnes in 2004 (Blanchard 2009), corresponding to the highest biomass of filter feeders in the bay. The carrying capacity of the bay has thus become an important question for scientists and stakeholders, especially farmers. These issues have been addressed through the IPRAC[1] national project.

26.2.2 System Definition

Mont Saint Michel Bay is a sandy and muddy bay of 500 km^2 with a high tidal range (up to 15 m) and a large intertidal zone reaching half of the total surface area. The ecosystems and landscapes of the bay represent a remarkable natural and cultural heritage, subject to numerous protection measures, but also support a wide variety of human activities: bivalve farming, professional and recreational fishing, hunting, tourism, sheep farms on salt meadows or intensive farming on polders. An ICZM approach has been initiated by local authorities to build a shared vision of the bay and to define common and future management objectives.

26.2.3 System Assessment

From an ecological perspective, primary productivity, ecosystem carrying capacity and trophic interactions between natural and cultivated filter feeders have been investigated as well as the economic drivers of the aquaculture activity.

A numerical ecosystem model of the Mont Saint Michel bay has been developed, which couples a 3D hydrodynamic model to a primary production model and to a benthic model (Cugier et al. 2010a). The primary production model allows a realistic simulation of phytoplankton dynamics; the benthic model takes into account the main filter feeders present in the bay and the interactions between primary production and the ecophysiology of cultivated oysters and mussels.

This model was used as a tool to better understand the functioning of the bay, mediate stakeholder interactions, and co-construct scenarios of future changes. The stakeholders involved were local administrations, watershed managers, farmers, environmental non-governmental organizations, and recreational fishing representatives. A participatory approach was implemented to achieve this objective. A first series of meetings were organized to inform stakeholders about the scientific consortium, the modelling tool, its possibilities and limits, and get the perception of stakeholders regarding the trophic resource availability and sharing in the bay. Following that, a second series of meetings based on focus groups allowed the definition of a list of scenarios. These groups highlighted three categories of questions corresponding to more than 30 scenarios:

[1] IPRAC – Impact of environmental factors and shellfish culture practices on the ecosystem of Mont Saint Michel Bay and shellfish production. Study through modelling scenarios (2007–2010).

- The link between watersheds and the bay through nutrient inputs to test potential effects on primary resource availability and bivalve production. Various scenarios have been proposed e.g. reduction in nutrient inputs, linked to EU Directive 2000/60/EC (European Community 2000), also known as Water Framework directive (WFD), or national regulations; increase of nutrient inputs based on the hypothesis of agriculture development.
- The proliferation of *C. fornicata* to investigate the potential trophic competition with other wild and cultivated filter feeders. Related scenarios explored further proliferation and control measures to limit it.
- Evolution of the shellfish farming practices (changes in standing stocks and/or variations in cultivated areas) and their possible effect on trophic resource availability.

Results of modelling scenarios were interpreted in terms of growth performance of cultivated species.

26.2.4 Main Results

Apart from scenarios dealing with aquaculture management, scenarios concerning proliferation of *C. fornicata* appear to have the most potential impacts in terms of trophic competition (Fig. 26.1; Cugier et al. 2010b). Objectives for reduction of nitrogen inputs from watersheds, as stated by the WFD or national directives, have a moderate impact on primary production in the bay and thus on bivalve production. In the first meetings with stakeholders, this reduction was not necessarily viewed as "a good thing", especially by bivalve farmers who expected a potential risk of

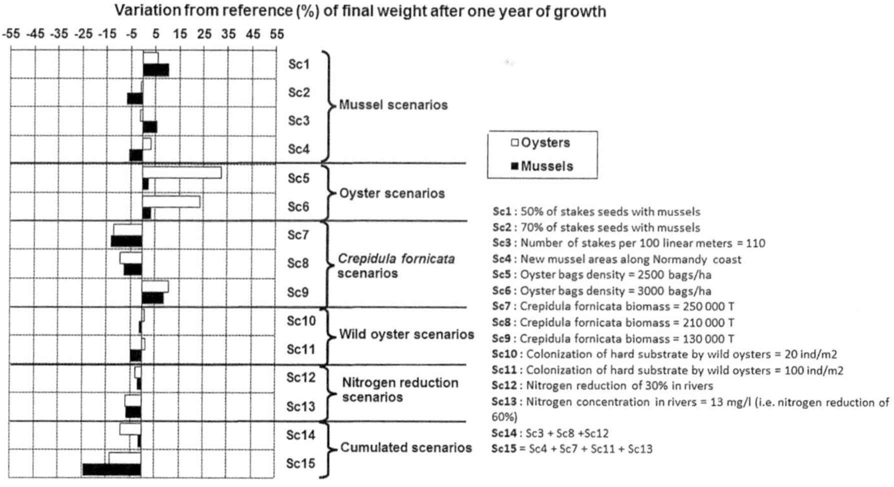

Fig. 26.1 Results of scenarios tested with the model of Bay of Mont Saint Michel. (Cugier et al. 2010b)

growth performance decrease. Results show that this risk remains very limited, far behind the one related to *C. fornicata* proliferation. Finally, scenarios exploring farming practices modifications show a potential significant impact on food availability and bivalve growth. This result can be interpreted as a potential control lever to compensate for the negative effect due to *C. fornicata*.

26.2.5 Lessons Learned

The tested scenarios are neither predictive (they do not state the ecological future of the bay) nor normative (they should not be considered as real wishes concerning the evolution of the bay and what should be done or not). They are exploratory and designed to understand system dynamics and responses to more or less strong variations of its forcing functions. In this context, stakeholders perceive the model as a powerful tool with which "everything is possible". Stakeholders would use scenario results to prioritize the various forcing functions and their variations according to their impact on the trophic resource and the management objectives of this resource as set by themselves. They could be included into ongoing consultation processes at the bay and watershed levels.

26.3 Marine Spatial Planning – Normandy Case

26.3.1 Issue Identification

Normandy is located in the northwestern part of France and includes the bay of Mont Saint-Michel and also Cancale Bay (the latter does not belong to the Normandy Region, but lies in the same ecoregion). In response to the EU Directive on Marine Spatial Planning (MSP), the French government has set up a management plan at the scale of the four coastal regions (East Channel – North Sea for Normandy and North Atlantic – West Channel for Cancale Bay). The aquaculture (essentially bivalve) sector is also driven by specific spatial planning policy.

Linking social demand and scientific progress to develop operational tools for decision-makers and stakeholders has recently been identified as an important issue (Byron et al. 2011). On the one hand, MSP should be based on an ecosystem approach and must rely on the best scientific knowledge, research and innovation. One the other hand, scientists have developed expertise in Geographic Information Systems (GIS), remote sensing data and numerical modelling, which are well recognized as powerful tools to assist the development and management of sea use and the sustainable management of living resources. This issue was addressed through the AquaSpace[2] project.

[2] Making space for increased aquaculture production – http://www.aquaspace-h2020.eu (2015–2018).

26.3.2 System Definition

The whole length of the case study coastline is about 450 km for a total area of 20,000 km^2 (including terrestrial and marine zones). The case study belongs to two administrative regions but also contains two biological entities: a part of the Gulf Normand Breton located on the Western part of the Cotentin peninsula and the Bay of Seine located on the Eastern part. Normandy represents a series of economic, cultural and environmental issues due to a large range of activities (e.g. bivalve farming, tourism, commercial and recreational fishing, agriculture, nuclear power plant, fuel processing industries, sand and gravel extraction and, in the future, offshore renewable energy) and a complex governance system based upon several administrations (AAMP 2009). In this area, bivalve aquaculture largely dominates aquaculture and is usually located in sheltered and intertidal areas (bays, estuaries). Two species are cultivated: the oyster *M. gigas*, and the mussel *M. edulis* with annual productions around 34,000 and 29,000 tonnes, respectively.

26.3.3 System Assessment

A web based dynamic GIS tool named AkvaVis (www.akvavis.no), developed by the Institute of Marine Research, Christian Michelsen Research and Hordaland County Council in Norway, has been deployed and adapted to the Normandy case. It allows the integration of data, model outputs, regulatory frameworks, and expert knowledge by applying a web-based dialogue where the user would receive instant response from the tool to any choice requested from the tool. Under the name of SISAQUA,[3] Akvavis is mainly targeting the development of bivalve aquaculture in Normandy. Through the definition of spatial indicators of aquaculture suitability, it aims at helping end-users and decision-makers to optimize aquaculture performance (e.g. maximize individual growth, control water quality, rearrange existing bivalve culture areas) and to develop aquaculture activities (e.g. selecting new potential sites).

A working group has been set up to associate the main stakeholders (e.g. national, regional and local authorities, aquaculture industry and representatives, technical centers, non-governmental organizations, public institutions). The objectives were to engage a consultation and conciliation process around the issues related to aquaculture development in Normandy and the data needed to implement aquaculture MSP. This group is also being used to test and improve the SISAQUA tool (Fig. 26.2).

[3] Spatial Information System for Aquaculture in Normandy.

Fig. 26.2 Snapshot of SISAQUA showing maps of predicted mussel shell length and areas with specific protection measures. (http://sextant.ifremer.fr/fr/web/sisaqua)

26.3.4 Main Results

Group work allowed the identification and ranking of different types of issues regarding: policy and management, economic and market, aquaculture-environmental issues, data and demand for tools.

With respect to indicators of site suitability for bivalve aquaculture, SISAQUA displays and combines spatial information related to:

- Physical and biological characteristics of the site based on observed data and outputs of hydrobiological models;
- Potential bivalve growth performance based the assimilation of remote sensing data in ecophysiological model;
- Public information on various regulations and other marine activities.

26.3.5 Lessons Learned

Based on the AkvaVis development in Norway over several years, and the ongoing work with SISAQUA in Normandy, several concerns have been raised:

(i) Data quality and integration. The user output of these systems fully relies on the quality of data used and the integration with other information from the regulatory framework, industry practice etc. Information outputs are characterized by a spatial and sometimes also temporal dimension, which requires data with a certain level of continuity, often from modelling. One issue is to be able to match data of different characteristics with, e.g. regulatory information, to make them assessable and significant for user application and analysis. For instance, physical data integrated in SISAQUA (such as waves or currents) are constrained by the extent of the hydrobiological model used, which does not presently include all the Normandy region. The quality of data provided relates also to the question how the update should be taken into account, often needed in cases with dynamic processes like rapid development of the aquaculture industry in extent or structure.

(ii) Stakeholder interactions. The development of GIS tools like AkvaVis/SISAQUA is based on a demand for helping in analysis and decision-making. The experience showed that various stakeholders need to be involved in different stages of the process, to establish a dialogue and maximize chances to avoid possible conflicts. Stakeholders needed at an early stage of the process might not be of relevance for later stages. Strong stakeholder consultations through the development stages may also provide information needed for an efficient evaluation, for instance by user inquiry at completion. The process also highlights the necessity to avoid sectoral approaches when applying marine spatial planning. In AquaSpace, a stakeholder group was set up since the beginning of the project to create a framework for discussions and improve the tool development.

(iii) Tool limitations. AkvaVis, like SISAQUA, was constructed for the purpose of aquaculture development. Beyond limitations related to purely technical aspects, experience gained through AquaSpace showed that stakeholder demands are diverse, and that tools taking into account the links and interactions between sectors are needed.

26.4 System Approach Framework – Pertuis Charentais Case

26.4.1 Issue Identification

The Pertuis Charentais area is located on the French Atlantic coast. This site is characterized by the vulnerability of the continuum between the freshwater from the Charente catchment, a flat hydrological basin with a pluvial regime, and the coastal waters, which are subject to varying salinity gradients. Much of the human activities in the area require freshwater: availability of drinking water for households and tourists; good ecological status of the coastal ecosystems (rivers, saltmarshes, nurseries, coastal water productivity); agriculture (irrigation during summer for crop); shellfish farming (freshwater supports spat production and river nutrients support oyster growth). The local governance system implements regulations and management measures to maintain freshwater quality and sustainable levels of extractive use, while giving priority to the availability of freshwater for natural habitat protection and for consumption of drinking water. Nevertheless, the Charente watersheds frequently experience an acute summer freshwater deficit due to low rainfall and excessive irrigation. The research project SPICOSA[4] addressed these freshwater management issues through the development of a System Approach Framework (SAF) for coastal zone management using virtual and simulation models in order to provide integrated assessments of the coastal zone (Tett et al. 2011).

26.4.2 System Definition

A focus group of local administrators involved in the Charente catchment management worked with the scientists in order to refine the definition of the issue at stake. According to the current management plan, Reachable Discharge Thresholds (RDT), which are supposed to guarantee the first two priorities (good ecological status of natural habitats and availability of drinking water for households), have been defined at different control points in the river catchment. The operational objective of the management plan ensures that the system can reach the RDTs during the summer in at least eight years out of ten. The stakeholder group's main expectations concerned the options available for achieving the already fixed objectives of this management system. The project thus focused on the quantitative management of the freshwater in the Charente catchment.

The Ecosystem Services (ES) approach was then used for depicting the user conflicts generated by the scarcity of freshwater in the Charente catchment. Four main conflicts are generated by the competing uses of the freshwater services in the catchment, the last two being classic cases of common-pool resource rivalries: (1) conflict between the two extractive uses of freshwater (irrigation and drinking water); (2) conflict between extractive uses (provisioning services) and other services (support, regulatory, and cultural) provided by freshwater; (3) rivalry among land farmers, who are direct users of freshwater; and (4) rivalry among farmers, who are indirect users of nutrients supplied by the river to the coastal waters. A model was built to simulate the impact of governance scenarios on the availability of freshwater for all uses (Mongruel et al. 2011).

26.4.3 System Assessment

A model of the social-ecological system has been set up in three tiers, which are interconnected through endogenous processes: resources and ecological functions (Charente hydrology and coastal water productivity), uses (agriculture, household drinking water consumption and bivalve farming) and governance mechanisms (water discharge thresholds and water use restrictions) (Fig. 26.3). The Charente river dynamics is represented by the equations of the hydrological model, which is

[4] Science and Policy Integration for Coastal Systems Assessment (2007–2011).

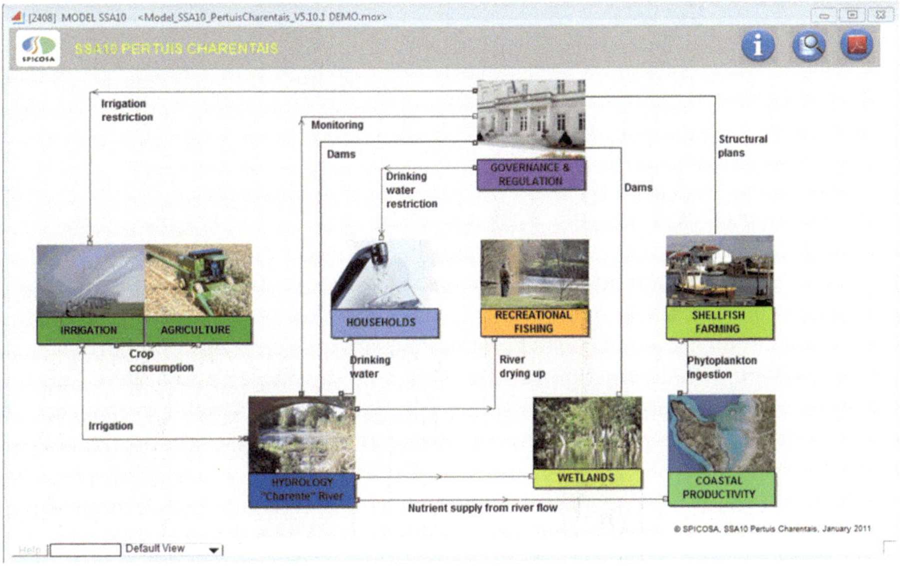

Fig. 26.3 System modelling framework as shown in the model user interface. (From Mongruel et al. 2011)

used by water managers to monitor the daily flow levels of the Charente and restrict irrigation during droughts. The agriculture module is connected to the hydrological sub-model and simulates crop water consumption under various irrigation strategies. These strategies depend on the institutional arrangements chosen in the governance module, which also simulates restriction rules triggered by critical discharge levels at monitoring stations. The model was able to estimate several indicators of the level of the main ecosystem services according to various climate and governance scenarios.

The assessment focused on the institutional arrangements regarding the freshwater use-rights of the land farmers. Downstream farmers have access to their whole annual use-right at any time. The restrictions imposed by water shortages apply to this annual use-right: farmers are likely to adopt short-sighted irrigation strategies because they have no incentive to anticipate future reductions of their permitted volumes, which are far higher than their actual needs. Farmer strategies in six of the upstream sub-basins are based on a planned schedule of irrigation needs that distributes annual use-rights over segmented periods of the irrigation season (periodic strategy). Some upstream farmers have adopted collaborative irrigation strategies for severe drought situations by taking turns to pump water in some locations (collaborative strategy). The simulations have explored the gradual harmonization of the irrigation schemes at river catchment scale, under various climate conditions.

26.4.4 Main Results

The results indicate that any attempt to preserve coastal ecosystems through irrigation practices that consume less water would also probably mean productivity losses for farming of arable land. However, when achieved through "soft" institutional change, significant positive effects on the environment (expressed in terms of crisis event reduction) would generate fairly reasonable decreases in irrigation consumption. Coastal productivity is much more sensitive to inter-annual changes in precipitation than to the institutional arrangements regarding freshwater use. Intermediate production (half-grown oysters) is much more sensitive to the availability of primary production than the harvested production, and may decrease by 24% during a dry year. As improved irrigation strategies have no positive effect on their production during normal years, this may explain why oyster farmers prefer to concentrate their demands on the possibility of obtaining freshwater releases during severe droughts.

Protecting the ecosystems that depend on the Charente has been defined as the primary objective of water governance, an objective considered to be achieved when crisis situations due to an unbalanced water budget are avoided eight years out of ten. This "zero crisis" criterion is much more likely to be met during normal years than during dry years. The results of the simulation model suggest two directions for improved freshwater governance: (i) implementing planned individual strategies on the downstream area is a necessary condition for avoiding crisis events during normal years, and (ii) the most efficient institutional scheme for all climatic conditions would be to implement collaborative strategies in the entire river catchment.

26.4.5 Lessons Learned

Governance scenarios for coastal system assessments should pay attention to the complexity of institutional change. Most of the models for coastal system assessment simulate the introduction of a new management measure without considering the impact of existing measures and their evolution. The SPICOSA experiment in the Charente river catchment addressed "soft" institutional change, in which improved operational agreements, based on local collective organization, are taken into consideration, rather than more drastic change through top-down decisions.

The outputs of the simulations are expressed in terms of ES' physical availability and of production yield (for provisioning services), which is a first step toward estimates of costs, benefits, and their distribution. It is worth noting that, for collaborative institution analysis, transaction costs should also be taken into account, since these costs may discourage the emergence of effective partnerships (Lubell et al. 2002).

From a broader perspective, when the sustainability of a complex common-pool resource is at stake, some users may develop adaptive strategies by searching for alternative resources in external areas: this is already true for oyster farmers of the Charente region, who carry out the early stages of the growth cycle in other production basins. Such strategies may indicate decreasing robustness of the social-ecological system, since adaptive behaviours prefer solutions other than collective action against resource overexploitation (Anderies et al. 2004).

26.5 Valuation of Ecosystem Services – Normand-Breton Gulf Case

26.5.1 *Issue Identification*

The Normand-Breton Gulf (NBG) encompasses a variety of natural habitats and marine ecosystems, which make it a candidate for the creation of Marine Protected Area (MPA). For this kind of large ecosystems with multiple issues (see also EA and MSP case studies), the French Administration has created a conservation and management tool called 'Marine Natural Park' which combines a series of objectives regarding knowledge improvement, habitat and species conservation, preservation of environmental quality, sustainable development of economic activities and cultural identity of the territory. VALMER[5] was a French-British project which aimed at developing approaches for marine ES assessment in support of marine ecosystem management. In this project, the objective of the ES assessment was to provide an initial diagnosis of the area, in order to help future MPA managers to elaborate their management plan.

26.5.2 *System Definition*

The size of the whole area includes the Normandy and Brittany coasts, and covers a number of marine habitats and islands, totaling ca. 9971 km^2. It is characterized by a landscape of various habitats: tidal flats, rock plates, and subtidal regions with a depth of up to 80 m (Cabral et al. 2015). Bivalve farming is one activity among many others (see MSP case study). These activities contribute to the economy of the region, which is populated by ca. 600,000 inhabitants who live at a distance less than 3 km from the coastline. Cabral et al. (2015) aimed to estimate the vulnerability of marine habitats as a proxy of their potential to deliver ES according to different management scenarios. The approach relies upon the assumption that the increase of vulnerability is likely to decrease the supply of ecosystem services.

[5] Valuing ecosystem services in the Western Channel (2012–2015).

26.5.3 System Assessment

The concept of vulnerability considers the degree of exposure to environmental changes, the sensitivity of the natural system to these changes and the adaptive capacity of human systems. The assessment was built upon a combination of spatial information dealing with marine habitat characteristics and pressures related to the main human activities. Expert knowledge and habitat maps were combined to build a risk map using the InVEST habitat risk assessment model (HRA). The model output is a map of relative values showing the habitats, which are more or less sensitive to changes. This step was completed with a matrix of the potential contribution of each habitat to the main ecosystem services. Again, the output can be displayed as map of more or less high availability of a given ES. The final step of this method considers that a high level of ES availability makes a given habitat less vulnerable to a given risk. ES availability is then defined as a proxy of adaptive capacity. The ratio between habitat risk and ES availability yields a spatial vulnerability index, which is a qualitative measure of the habitat's ability to deliver ES, taking into account the level of pressure on the habitat. On this basis, theoretical management scenarios considered the increase or decrease of human pressure (e.g. conservation, development) and the vulnerability index was recalculated for each ES category. The map of vulnerability index changes for each ES type highlighted which areas are most sensitive to management actions, with respect to ES. Results can also be summarized as an average percentage of ES delivery change for a given habitat.

26.5.4 Main Results

In this study, bivalve shellfish farming is one among provisioning services and its importance is balanced by the amount of other services measured as the percentage of occupied habitat. The Habitat Risk Assessment model identified which habitats are the most important for the supply of ES, considering the usual classification of ES (Fig. 26.4). Simple theoretical scenarios were used to help understanding of potential degradation or improvement resulting from management options. For instance, results showed that the near shore areas exhibit higher risk values to various ES, which means that these habitats are more exposed to pressures unlike the habitats in the offshore areas (Cabral et al. 2015).

26.5.5 Lessons Learned

The HRA model was chosen because it allowed using available data and expert knowledge to define proxies of ES availability and relation to pressures. Model outputs are qualitative and can be used to identify the direction of relative changes

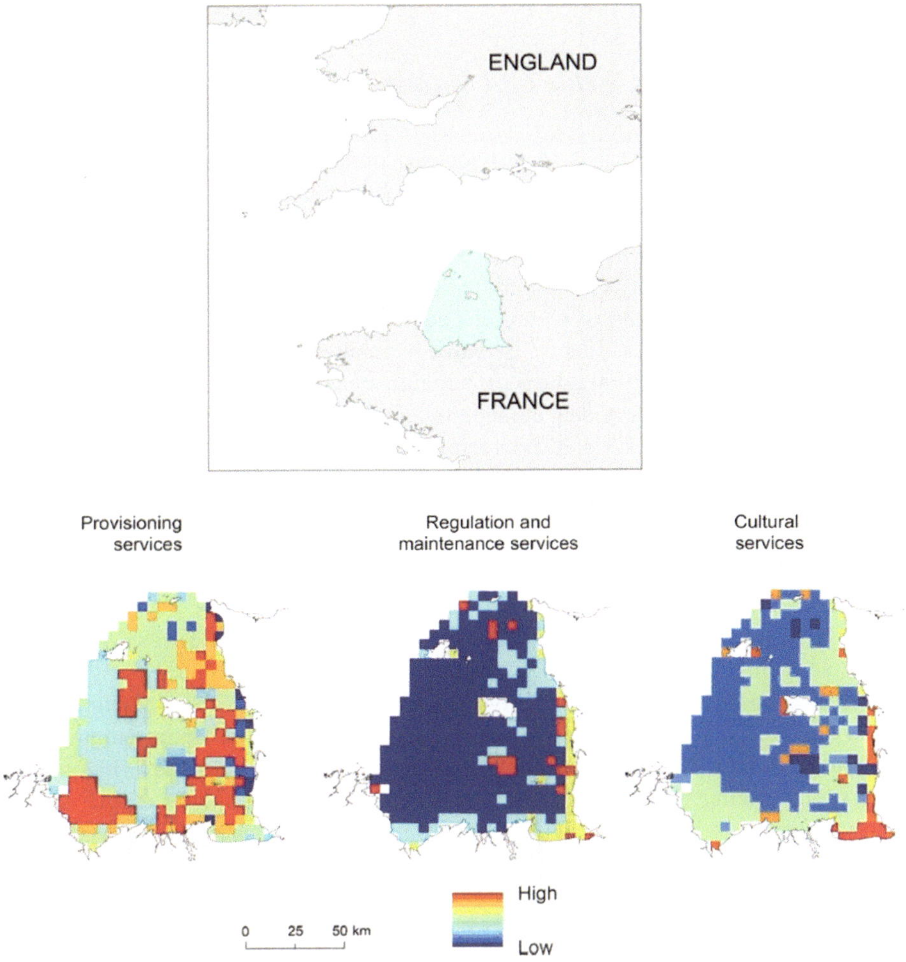

Fig. 26.4 Map of ES availability for different types of service. (From Cabral et al. 2015)

in ES supply when management scenarios are defined. As such, the results were only published to demonstrate and evaluate the applicability of the HRA method in the NGB context. The utility for decision-making has still to be tested and discussed with stakeholders.

Cabral et al. (2015) highlighted several difficulties and limitations. Among the most important ones, the quality and availability of reliable data is critical. Though a large effort was accomplished to gather and integrate existing information, input data were not always accurate and no validation of the results was possible. The approach was still not efficient for estimating regulating services, which cannot rely upon the classification of benthic habitats and would require more relevant information (e.g. pelagic ecosystem component).

26.6 Discussion

26.6.1 Assessment, Scale and Management

The purpose of assessment is to provide decision-makers with the best valuable data, information, and predictions with which management decisions will be supported. In our examples, assessment deals with bivalves in different frameworks. One single study dealt with Ecosystem Services strictly speaking, and linked ES to marine habitats to identify ES availability and vulnerability to pressures. The results can be displayed as maps of resulting potential services with qualitative metrics (from low to high). The vulnerability value is an alternative to monetary valuation and, in addition to identifying the most suitable areas for each type of ES, this metric allows identifying the management strategies that will most probably maintain or affect each individual ES. In this example, bivalve farming and fishery are viewed as activities which take place in locations only defined by pressures and marine habitat features using a standard classification. Though qualitative, the indicators of ES availability and vulnerability can be used as a metric to compare ES and to assess how management strategies would affect the habitat's vulnerability for delivering ES.

It is interesting to note that this framework links to MSP, as illustrated in another example within approximately the same geographical region. In our example, MSP is focusing on bivalve farming activity and accounts for several criteria: habitat suitability, growth performance, environmental and regulation constraints and presence of other activities. The ultimate endpoint of such an approach is a map with qualitative values stating whether a location is suitable or not, depending on the weight given to each criterion. Though the output is limited to bivalve activity, it could be extended to all other activities and feed the ES framework as described in the previous example. Besides, this MSP approach only depicts site suitability (which relates to ES availability seen above) but does not directly give clues on the effect of management strategies.

However, none of these approaches accounts for more complex features related to system functioning. The two other examples clearly illustrate how interactions between some components of a system allow building of indicators of the sensitivity to changes. In the EA case study, the indicator is defined by the growth performance of cultivated bivalves in different locations. This indicator does not only depend on local conditions but is affected by distant factors – e.g. populations of marine organisms competing for the same food resource, nutrient inputs from rivers, time to renew water bodies under the action of tidal currents. These interactions portray a system made of variables, biological and physical processes and management scenarios. As for the previous examples, spatial distribution of these components is a key characteristic in the building of the indicator. In addition, the temporal dynamics and distant effects of factors are central to the assessment of the indicator responses to environmental changes or management decisions. In this example, management deals with measures regarding ecosystem health and aquaculture development – e.g. aquaculture extension, restoration of river quality, control of pest invasion. All these factors relate to the dynamics of a single primary resource (e.g. phytoplankton), which is shared by the main system components, resulting in trade-offs.

The existence of trade-offs and the definition of the appropriate spatial scale and resolution are keys to a policy-relevant assessment, which has been clearly exemplified by Nelson et al. (2009) in their modelling of multiple ecosystem services. In the example dealing with SAF, the scope shifted to account for the management of the freshwater resource and to integrate coastal zone and catchment areas. Since freshwater supply is the resource at stake for several activities (ecosystem preservation, drinking water, agriculture, bivalve culture), freshwater management units determines the system domain and spatial resolution. Regarding bivalves farming activity, the coastal zone is given as a single spatial unit for the sake of simplicity, and the endpoint is total annual production. It is worth noting that management rules are part of the system description and that, as for the EA case study, the connections between the components of the system drive the dynamics of the whole system in a set of cause-effect relationships.

26.6.2 Methods and Tools

Spatial data are essential for any assessment method. Tempera et al. (2016) provided an assessment of the spatial distribution of marine ecosystem service capacity in the European seas using habitat maps based on a EUNIS typology of marine habitats. Maes et al. (2012) reviewed current mapping methods, and identified current knowledge gaps to assess ES at European scales. Two of the gaps usually identified are the data quality and the capacity of data to inform on ecosystem functions. Both examples on MSP and ES illustrate how habitat mapping provides the first data layer to the assessment of ES, and how the combination of environmental data and information on human activities can inform MSP. Despite the gaps which have been outlined, more and more spatial data are being used to assess, directly or indirectly (e.g. proxies), biodiversity and ecosystem functions (Walters and Scholes 2016). In the field of aquaculture and fisheries, GIS and remote sensing data have been reviewed and promoted by Meaden and Aguilar-Manjarrez (2013). Therefore, the use of spatial data, combined with the tools and methods dealing with spatial and landscape ecology, and assessment tools such as InVEST, are promising.

Models have been central in some of our examples. Vulnerability assessment in the case of NBG relies on the calculation of indicators based on multidimensional input data regarding marine habitats, availability of ES and pressures due to human activities. In two other examples, bay of Mont Saint Michel and Pertuis Charentais, mathematical models have been set up to quantify the interactions between system components. In the first case, the model accounts for the ecological interactions between bivalve farming, primary productivity and wild benthic populations which are competing for the primary resource. In the second case, the primary resource was defined as the freshwater flow, which supports agriculture and primary production in the coastal zone. The model therefore accounts for several uses and also considers regulation rules. In both examples, the mathematical model allows the simulation of the dynamics of bivalve production, and scenarios have been setup with stakeholders to explore the response of bivalve yield to management options.

Models relate response to input variables. The choice between dynamic and statistical models depends on available data and system properties. Dynamic models must be preferred when the relationship between input and output variables may be masked by non-linear effects, delay of responses or differences of scales. The system is often multi-dimensional, and multiple variables would interact and affect the response of the system. It is also clear that spatial interactions must be taken into account and that spatial resolution depends on the issues addressed by the modelling approach. In the Pertuis Charentais example, spatial dimensions include the catchment area of the main river, and the coastal zone is defined as a single spatial entity. On the other hand, a high-resolution spatial model was set up in the Bay of Mont Saint Michel example to account for the competition between cultivated bivalves and wild populations of filter-feeders. Most important is the non-linear dynamics of most natural and human systems. Koch et al. (2009) noticed that most valuation processes assume that a quantity of an ecosystem function varies linearly with forcing variables. They suggest that understanding and quantifying non-linearities in ecosystem functions would provide more realistic ES values. This is the reason why several approaches of ES assessment may be complementary and would need to be combined.

There is a debate regarding the economic methods that are most appropriate to support decision-making with respect to ecosystem management. The European Union Biodiversity Strategy to 2020 aims at assessing the economic value of ES, and promotes the integration of these values into accounting and reporting systems at EU and national levels by 2020 (Mongruel et al., in Cormier et al. 2013). However Balvanera et al. (2016, in Walters and Scholes 2016) stated that there are strong biases for economic values, "which are the product of markets and incentives, and do not necessarily account for the marginal contribution of ecosystems to food production through primary productivity, water for irrigation, soil fertility, pollination, or pest regulation, relative to those contributed by society. Also, these values do not include the negative impacts of agricultural intensification and expansion, nor that of industrial fisheries, on biodiversity conservation and the degradation of supporting and regulating". Monetary values are often of poor significance, especially for those ecosystem services whose loss could mean the end of life, and appear to be a comfortable oversimplification of reality of socio-ecological systems which cannot be summarized in single numbers (Mongruel et al., ibid.). Alternative methods would therefore consider institutional analysis or multicriteria assessment rather than only monetary values of ES.

26.6.3 Participatory Approach

In most of our examples, stakeholders have been associated to the development of the assessment tools. In Normandy, stakeholder groups have been setup to identify the most critical issues in terms of aquaculture spatial planning. Lessons learned with MSP showed that stakeholders should be involved beforehand in the process to establish a dialogue, minimize conflicts of uses and avoid sectoral approaches. In the Bay of Mont Saint Michel study, the ecosystem model has been proposed as a tool to understand the main ecological interactions and to identify the main drivers of ecosystem carrying capacity. Stakeholders were consulted to define modelling scenarios, which were run to sort out the most effective management options with respect to bivalve biological production. Stakeholder involvement was also at the core of the System Approach Framework (SAF) in the Pertuis Charentais case study. Decision-makers contributed to the definition of the policy issue, the system characteristics, and regulation procedures. The development of the mathematical model resulted from this consultation process and the assessment of the model outputs was conducted with decision-makers.

In all our examples, the assessment tool could not be transferred to some administration, private company or national agency for further use. This poses some limitation on the maturity of the assessment approach, rather than the methods themselves. In a paper on carrying capacity assessment, Byron et al. (2011) outlined that a major gap in effective decision-making is due to poor communication between scientists and stakeholders. This fact has been turned into general principles (Byron et al. 2011): several categories of stakeholders must be involved (e.g. end-users, decision-makers, etc.); the stakeholder process should be conducted in an independent and unbiased way; stakeholders should be involved early in the process, should have an opportunity for input, have influence over the final decision; the stakeholder process and objectives should be transparent. One lesson learned from the SAF case study presented here is that collective thinking allows identification of issues, building system representation, and evaluation of the outputs of the assessment, and ensures the engagement of the stakeholders, the credibility and the acceptance of the assessment. This is specifically true when the assessment is based on models (Voinov et al. 2016).

Acknowledgements The work presented here has been partially supported by EU Horizon 2020 FPR 'Aquaspace' grant No. 633476. We are grateful for the comments of two referees on the manuscript.

References

AAMP (2009) Analyse des enjeux et propositions pour une stratégie d'aires marines protégées Bretagne Nord /Ouest Cotentin. Agence des Aires Marines Protégées. 38 pp + annexes

Aguilar-Manjarrez J, Kapetsky JM, Soto D (2010) The potential of spatial planning tools to support the ecosystem approach to aquaculture. Expert. FAO Fisheries and Aquaculture Proceedings. No. 17, 176 pp

Anderies JM, Janssen MA, Ostrom E (2004) A framework to analyze the robustness of social-ecological systems from an institutional perspective. Ecol Soc 9(1):18

Blanchard M (1997) Spread of the slipper limpet (Crepidula fornicata) in Europe. Current state and consequences. Sci Mar 61:109–118

Blanchard M (2009) Recent expansion of the slipper limpet population (Crepidula fornicata) in the Bay of Mont-Saint-Michel (Western Channel, France). Aquatic Living Ressour 22:11–19

Brugère C, Ridler N, Haylor G, Macfadyen G, Hishamunda N (2010) Aquaculture planning: policy formulation and implementation for sustainable development. FAO Fisheries and Aquaculture Technical Paper No. 542, FAO, Rome, 70p

Byron C, Bengtson D, Cosat-Pierce B, Calanni J (2011) Integrating science into management: ecological carrying capacity of bivalve shellfish aquaculture. Mar Policy 35:363–370

Cabral P, Levrel H, Schoenn J, Thiébaut E, Le Mao P, Mongruel R, Rollet C, Dedieu K, Carrier S, Morisseau F, Daures F (2015) Marine habitats ecosystem service potential: a vulnerability approach in the Normand-Breton (Saint Malo) Gulf, France. Ecosyst Serv 16:306–318

Cormier R, Davies I, Kannen A (eds) (2013) Integrated coastal-zone risk management ICES Cooperative Research Report 320. 145 pp

Cugier P, Struski K, Blanchard M, Mazurié J, Pouvreau S, Olivier F, Trigui J, Thiébaut E (2010a) Assessing the role of benthic filter feeders on phytoplankton production in a shellfish farming site: Mont Saint Michel Bay, France. J Mar Syst 82:21–34

Cugier P, Frangoudes K, Blanchard M, Mongruel R, Pérez Agúndez JA, Le Mao P, Robin T, Fontenelle G, Mazurié J, Cayocca F, Pouvreau S, Olivier F (2010b) Impact des facteurs environnementaux et des pratiques conchylicoles sur la baie du Mont Saint-Michel et la production conchylicole. Etude de scenarii par modélisation. Programme Liteau 3. Rapport Final, 176 p. http://archimer.ifremer.fr/doc/00026/13707/

Douvere F (2008) The importance of marine spatial planning in advancing ecosystem based sea use management. Mar Policy 32:762–771

European Community (2000) Directive of the European Parliament and of the council 2000/60/EC, establishing a framework for community action in the field of water policy. 62 pp

Koch E, Barbier E, Silliman B, Reed D, Perillo G, Hacker S, Granek E, Primavera J, Muthiga N, Polasky S, Halpern B, Kennedy C, Kappel C, Wolanski E (2009) Non-linearity in ecosystem services: temporal and spatial variability in coastal protection. Front Ecol Environ 7:29–37

Liquete C, Piroddi C, Drakou EG, Gurney L, Katsanevakis S, Charef A, Egoh B (2013) Current status and future prospects for the assessment of marine and coastal ecosystem services: a systematic review. PLoS One 8(7):e67737. https://doi.org/10.1371/journal.pone.0067737

Lonsdale JA, Weston K, Barnard S, Boyes S, Elliott M (2015) Integrating management tools and concepts to develop an estuarine planning support system: a case study of the Humber Estuary, Eastern England. Mar Pollut Bull 100:393–405

Lubell M, Schneider M, Scholz JT, Mete M (2002) Watershed partnerships and the emergence of collective action institutions. Am J Polit Sci 46(1):148–163

Maes J, Egoh B, Willemen L, Liquete C, Vihervaara P, Schägner JP, Grizzetti B, Drakou EG, Notte AL, Zulian G, Bouraoui F, Luisa Paracchini M, Braat L, Bidoglio G (2012) Mapping ecosystem services for policy support and decision making in the European Union. Ecosyst Serv 1:31–39

Meaden GJ, Aguilar-Manjarrez J (eds) (2013) Advances in geographic information systems and remote sensing for fisheries and aquaculture, FAO fisheries and aquaculture technical paper no. 552. Rome, FAO, 425 pp

MEA (2005) Ecosystem and human well being: synthesis, Millenium ecosystem assessment. Island Press, Washington, DC, 137 pp

Mongruel R, Prou J, Ballé-Béganton J, Lample M, Vanhoutte-Brunier A, Réthoret H, Pérez Agúndez J, Vernier F, Bordenave P, Bacher C (2011) Modeling soft institutional change and the improvement of freshwater governance in the coastal zone. Ecol Soc 16(4):15. https://doi.org/10.5751/ES-04294-160415

Nelson E, Mendoza G, Regetz J, Polasky S, Tallis H, Cameron R, Chan K, Daily G, Glodstein J, Kareiva P, Lonsdorf E, Naidoo R, Richetts T, Shaw M (2009) Bodiversity conservation, commodity production, and tradeoffs at landscape scales. Front Ecol Environ 7(1):4–11

Pinot JP (1998) L'outil par excellence de l'aménagement intégré du littoral: le SMVM, voeux pieux et réalités. In: Miossec A, Perron F (eds) Analyse et gestion intégrée des zones côtières, Seminaire de l'UMR 6554. CNRS, Nantes, pp 33–39, cited in Cormier et al. 2013

Rice J et al (2010) Science dimensions of an Ecosystem Approach to Management of Biotic Ocean Resources (SEAMBOR). Marine Board Position Paper 14

Tempera F, Liquete C, Cardoso AC (2016) Spatial distribution of marine ecosystem service capacity in the European seas. In: EUR 27843. Publications Office of the European Union, Luxembourg

Tett P, Sandberg A, Mette A (2011) Sustaining coastal zone systems. Academic Press, Dunedin, 173 pp

Voinov A, Kolagani N, Mccall M, Glynn P, Kragt M, Osermann F, Pierce S, Ramu P (2016) Modelling with stakeholders – next generation. Environ Model Softw 7:196–220

Walters M, Scholes R (eds) (2016) The GEO handbook on biodiversity observation networks. Springer Open, Cham, 326 pp

Chapter 27
Assessment of Nutrient Trading Services from Bivalve Farming

J. G. Ferreira and S. B. Bricker

Abstract This review examines key aspects of bivalve services, with a dual emphasis on commercial production and eutrophication control, and explores how the two can be combined by means of market instruments. Our focus is on regulatory trading services, in particular on ways in which nutrient credits can be traded for improved water quality management and better food security. We provide budgets for nutrient loading in Europe, North America, and China, factoring in point and non-point loading, and assess the contribution of finfish aquaculture. We then review the role of commercially cultivated bivalves for the same geographic areas, to assess the scope of combining farmed bivalves and top-down control of symptoms of nutrient enrichment. Water quality trading has existed as a concept for the past 40 years, but it can claim few success stories; we examine some of the challenges and potential solutions, as well as practical implementations, with a focus on non-point trading, for mitigation of diffuse nutrient loading. Finally, we discuss options for different indicators, and provide examples of how an assessment can be made, including the valuation of regulatory services provided by commercially grown bivalves. We conclude that the role of bivalves in nutrient credit trading programmes should form an integral part of ecosystem-based management. From the perspective of aquaculture enhancement, which is fundamental for improved food security, this is a triple-win, providing competitiveness of agriculture, eco-intensification of aquaculture, and greater consumer safety.

Abstract in Chinese 摘要：本文综述了双壳贝类服务价值的主要方面，重点强调了贝类在商业化生产和富营养化控制方面的作用，并探讨了如何通过市场手段将两者结合起来。我们的关注点在于调节类的配额贸易服务，特别是如何利用"营养盐排放配额"的方式来促进水质改善管理和粮食安全保障。我们举例说明欧洲，北美和中国一些水域的营养负荷收支情况，分别从点源以及

J. G. Ferreira (✉)
DCEA, FCT, New University of Lisbon, Monte de Caparica, Portugal
e-mail: joao@hoomi.com

S. B. Bricker
NOAA – National Ocean Service, NCCOS, Silver Spring, MD, USA
e-mail: Suzanne.Bricker@noaa.gov

A. C. Smaal et al. (eds.), *Goods and Services of Marine Bivalves*,
https://doi.org/10.1007/978-3-319-96776-9_27

非点源输入两方面评估了鱼类养殖的贡献份额。然后，我们总结分析了在相同地理区域开展商业化双壳贝类养殖对富营养化的下行控制作用。水质配额贸易的概念已经存在四十多年,但目前并没有什么成功的运用案例。本文研讨了水质配额贸易推行存在的一些挑战和潜在的解决方案，并且着重以非点源输入配额贸易为例探讨了减轻扩散性营养物输入的实施方案。最后,我们对评估指标的的选择进行了讨论,并提供了一些评估实例,包括对商业规模双壳贝类养殖的生态调节功能的评估。双壳贝类在“营养盐配额贸易”项目中的作用应该被视为生态系统管理的一部分。从加强水产养殖的角度考量，这对改善粮食安全可起到根本作用，因此可以三赢：一是提高了农业竞争力，二是有助于实现水产养殖生态集约化,三是更进一步的保障消费者需求。

Keywords Bivalves · Eutrophication · Regulatory services · Nutrient credit trading · Trading mechanisms · Indicators and assessment

关键词 双壳贝类 • 富营养化 • 调节服务 • “营养盐配额贸易” • 贸易机制 • 指标和评估

27.1 Introduction and Scope

Nutrient discharge to coastal waters is a major driver in the development of eutrophication symptoms (Bricker et al. 2003; Borja et al. 2008; Diaz and Rosenberg 2008). The conceptual relationship for these primary and secondary symptoms, also called direct and indirect effects (OSPAR 2010), is illustrated in Fig. 27.1.

Eutrophication has been defined in several different ways (e.g. Anonymous 1991a, b; Nixon 1995; Cloern 2001; Andersen et al. 2006); for this review, we have adopted the European Union (EU) Marine Strategy Framework Directive (MSFD, 2008/56/EC) definition, since our emphasis is on the trading potential of nutrient abatement services. The MSFD defines eutrophication (Ferreira et al. 2011) as 'a process driven by enrichment of water by nutrients, especially compounds of nitrogen and/or phosphorus, leading to: increased growth, primary production and biomass of algae; changes in the balance of organisms, and water quality degradation. The consequences of eutrophication are undesirable if they appreciably degrade ecosystem health and/or the sustainable provision of goods and services.'

Nutrient pressures on estuarine and marine areas have intensified in many parts of the world as populations are increasingly drawn to coastal zones. Nevertheless, efforts to control loading have been mostly successful in the reduction of point-source discharges, particularly in the Western world, but diffuse inputs from agriculture are far less easy to reduce (e.g. Gunningham and Sinclair 2005; Collins and McGonigle 2008).

In the West, some agricultural outputs (diffuse sources) are used as fertilizer for land-based crops; in other parts of the world, particularly Asia, where nutrient sup-

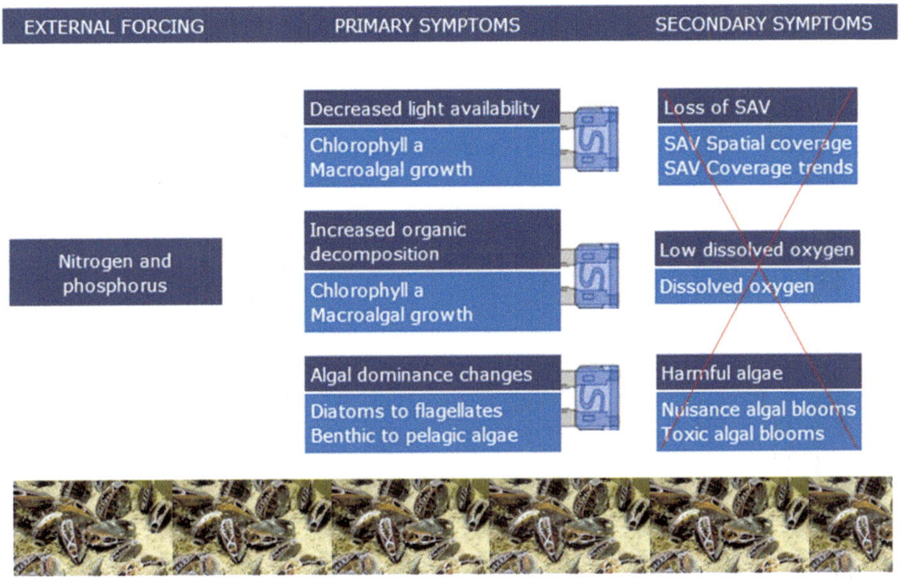

Fig. 27.1 General conceptual scheme of eutrophication, including top-down control by filter-feeding bivalves. The boxes for primary and secondary symptoms (identical to direct and indirect effects), show the symptom name (e.g. Decreased light availability), and below it the indicators for assessment. Bivalves act as a circuit-breaker (marked S), interrupting the organic decomposition cycle (secondary symptoms), which are thus (as a group) marked with an X (*SAV* Submerged Aquatic Vegetation, normally considered to mean seagrasses rather than macroalgae)

ply is a key limiting factor for food production, re-use takes place both on land and in water. In the latter case, nutrients may be taken up directly in inorganic extractive aquaculture, e.g. for seaweeds such as Nori (*Porphyra yezoensis*), and other plants such as water spinach (*Ipomoea aquatica*), but also indirectly through organic extraction.

The *indirect* re-use of dissolved nitrogen and phosphorus, after conversion into particulate organic forms through primary production, is a key step in the removal of these compounds from coastal ecosystems; this is largely mediated by filter-feeding bivalves (Gerritsen et al. 1994; Higgins et al. 2011; Petersen et al. 2014; Ferreira and Bricker 2016).

The world's annual aquaculture production in 2014 was estimated to be 73.8×10^6 tonnes (FAO 2016), of which 50% corresponds to non-fed, i.e. extractive, aquaculture. World bivalve production for 2014 was 16×10^6 tonnes (FAO 2016), of which 1.7% takes place inland (all in Asia). Overall, bivalve aquaculture accounts for 21.6% of the total production, or about two-fifths of total extractive aquaculture.

Farmed bivalve production shows a strong regional imbalance: Asia grows 94.2% of all molluscs, while the Americas and Europe account for 1.6 and 4.2% respectively. In Europe, practically all production takes place in the European Union, where bivalves account for 44% of total aquaculture (Ferreira and Bricker

2015; European Commission 2016). The inclusion of Norway brings this figure down to 20%, which is more in line with the world average (Ferreira and Bricker 2015).

Thirty-five years ago, two seminal papers (Cloern 1982; Officer et al. 1982) described the role of benthic filter-feeders in top-down control of eutrophication—both authors cite Mann and Ryther (1977), who discussed extractive organic aquaculture. Together, these publications are at the core of subsequent work on nutrient-related bivalve ecosystem services (e.g. Lindahl et al. 2005; Xiao et al. 2007; Kellogg et al. 2014; Saurel et al. 2014; Rose et al. 2015). In recent years, this has gained attention as a promising nutrient management practice to complement traditional land-based measures (Rose et al. 2014, 2015; Petersen et al. 2014).

As integrated coastal zone management (ICZM) evolved, and legislative instruments (e.g. the EU Water Framework Directive: WFD, 2000/60/EC), and policy guidance documents (e.g. USEPA 2008a) became available, options for nutrient abatement were reviewed in detail. In particular, cost-benefit (Nunneri et al. 2007) and cost-effectiveness (Gren et al. 2008; Lancelot et al. 2011) analysis was used as a tool, and the potential role of nutrient credit trading was considered, especially on the eastern seaboard of the United States (Virginia DEQ 2008; CT-DEP 2010).

Despite clear evidence that filter-feeding bivalves play an important role in nutrient management, or more specifically in management of nutrient-related *issues* (e.g. water clarity), policy-makers have been slow to embrace the fact that top-down eutrophication control mechanisms associated with commercial bivalve farming should be part of any integrated watershed-level management strategy.[1]

In this review, we examine (i) nutrient loading and the role of commercially cultivated bivalves; (ii) nutrient credit trading mechanisms and indicator selection; and (iii) potential assessment methodologies and their application.

27.2 Nutrient Loading and the Role of Cultivated Bivalves

An assessment of the potential role of filter-feeding bivalves in offsetting eutrophication symptoms requires an evaluation of the magnitude of both the inputs and outputs, i.e. land-based nutrient loading (to which nutrient emissions from finfish cage culture could be added), and bivalve production and nitrogen removal. Management emphasis is typically placed on nitrogen rather than phosphorus, since the former is considered to be the limiting nutrient for primary production in estuarine and coastal systems (Ryther and Dunstan 1971; Boynton et al. 1982; Nixon and

[1]Virginia is an exception: House Bill N°176 (2012) includes in Article 1.B.1: '*...incineration or management of manures, land use conversion, stream or wetlands restoration, bivalve aquaculture, algal harvesting, and other established or innovative methods of nutrient control or removal.*' More recently the Chesapeake Bay Partnership has approved the use of harvested oyster tissue as a nutrient best management practice (BMP) whereby MD and VA jurisdictions are allowed to use nutrient credits from oyster tissue to count toward fulfilment of nutrient reduction goals (Oyster BMP Expert Panel 2016).

Pilson 1983; NRC 1993). It is worth noting, however, that this is not universally accepted (see Howarth 1988, for a review).

From the standpoint of land-based emissions control, the distinction is probably irrelevant, since wastewater treatment facilities (WWTF) remove both nitrogen and phosphorus (USEPA 2004a), and fertilizer reduction measures for agriculture do likewise. From the perspective of top-down control by bivalves, this is probably also a moot point, because filter-feeders remove both elements. Where the question *may* become relevant is in the valuation of a specific nutrient, but this can be overcome either by (i) using population-equivalent (PEQ) coefficients for both N and P (e.g. Ferreira et al. 2007a), thereby dealing with avoided costs; or (ii) using an indicator associated with the reduction of symptoms rather than causative factors (see section on indicator suitability).

Recent work in the United States (Oyster BMP Expert Panel 2016) already takes both nitrogen and phosphorus into account when considering regulatory services from bivalve aquaculture.

27.2.1 Nutrient Loading to the Coastal Zone

The nutrient loading and eutrophication status of some European waters were reviewed in Ferreira and Bricker (2016); source-apportionment of nutrient loads is key for policy decisions, but in many regions this is not fully available. In this review, we have expanded and improved the European data set (see Table 2 in Ferreira and Bricker 2016) to include loading data for major parts of the world's coastal ocean (Table 27.1); where possible, we have discriminated the nutrient sources by combining data from various authors, including: (i) Ærtebjerg et al. (2001) for Europe; (ii) NRC (2000), and Wise and Johnson (2011) for North America; and (iii) Tong et al. (2015) and the China Fishery Statistical Yearbook (2016) for China.

Where point-source and diffuse inputs can be assessed separately, the latter are typically 70–80% of the total loading. This is reflected in the ratio between calculated population-equivalents (PEQ) for load estimates and population data. These ratios are 1.7 for Europe, 3.3 for the US, and 6.4 for Canada. The ratio for China is below one, which suggests that a significant component of the total load to the coastal area is not included—coastal diffuse source loads per unit area tend to be higher than those from major rivers, since intensive agriculture is often concentrated close to the coast, and smaller rivers draining these areas would have a higher load than major rivers that also drain inland areas with natural land uses (Nunes, pers. com.)

China's freshwater finfish aquaculture (27.2×10^6 t in 2015, China Fishery Statistical Yearbook 2016) far exceeds that of marine finfish (Table 27.1); carp (grass, silver, and bighead) account for about half the freshwater production, but, although the last two are planktivores, and also feed on particulate organic detritus, their role in reducing loading is questionable, because they are mainly cultivated

Table 27.1 Nitrogen and phosphorus loading to marine waters (10^3 tonnes, percent total in parentheses where applicable) for major areas of the world

	Total nitrogen	Total phosphorus	Redfield ratio	Notes
Europe				
Norwegian Sea[a]	28.4 (30.7)	1.5	18.9	
Barents Sea[a]	5.4 (5.8)	0.3	18.0	
Sub-total direct loading	33.8 (36.6)	1.8	18.8	
Finfish aquaculture (Norway) [b, c]	55.9 (60.5)	14.5	3.9	
Finfish aquaculture (Faroe Islands) [d, e]	2.7 (2.9)	0.7	3.9	
Sub-total finfish aquaculture	58.6 (63.4)	15.2	3.9	
Sub-total Arctic waters	92.4	17.0	5.4	
Baltic Sea [f]				
Point sources	243.0 (29.0)	12.0	20.2	EEA ratio but 2010 HELCOM figure
Diffuse sources	592.0 (70.7)	29.3	20.2	EEA ratio but 2010 HELCOM figure
Finfish aquaculture [b]	2.5 (0.3)	0.6	3.9	N/P ratios calculated for salmon
Sub-total Baltic Sea	837.5	42.0	20.0	
North Sea, Celtic Sea, Bay of Biscay [g]				
Point sources	368.0 (30.5)	24.0	15.4	EEA ratio but 2010 OSPAR data
Diffuse sources	837.0 (69.4)	54.5	15.4	EEA ratio but 2010 OSPAR data
Finfish aquaculture	0.3 (<0.1)	0.1	3.9	
Sub-total North Sea, Celtic Sea, Biscay	1205.3	78.5	15.3	
Mediterranean Sea [a]				
Nutrient hotspots (S. Europe and N. Africa)	259.7 (12.9)	75.2	3.5	
Potential diffuse sources	1747.7 (87.1)	126.7	13.8	Estimated: Total-hotspots-aquaculture
Finfish aquaculture (gilthead bream) [b]	4.3 (0.2)	1.1	3.9	
Finfish aquaculture (European seabass) [b]	3.1 (0.2)	0.8	3.9	
Sub-total finfish aquaculture	7.4 (0.4)	1.9	3.9	
Sub-total Mediterranean Sea	2007.4	201.9	9.9	Includes N. African discharge to Med

(continued)

Table 27.1 (continued)

	Total nitrogen	Total phosphorus	Redfield ratio	Notes
Total Europe	**4142.6**	**339.4**		1255 million PEQ[h]; tot. pop. 726 million[i]
United States				NRC (2000)
NE coast[j]				
Rivers and estuaries	270.0 (40.3)	17.6	15.4	
Atmospheric	210.0 (31.3)	13.7		
SE coast[j]				
Rivers and estuaries	130.0 (19.4)	8.5	15.4	
Atmospheric	60.0 (9.0)	3.9		
Sub-total US east coast	670.0	43.6		
Gulf of Mexico[j]				
Rivers and estuaries	2100.0 (88.2)	136.8	15.4	
Atmospheric	280.0 (11.8)	18.2		
Sub-total US Gulf of Mexico	2380.0	155.0		
Pacific Northwest[k]				
Point sources	100.9 (21.8)	6.6	15.4	
Diffuse sources	362.2 (78.2)	23.6	15.4	
Sub-total US Pacific NW	463.2	30.2		1% load from watersheds in Western Canada
Marine finfish aquaculture[l]	0.9	0.2	3.9	≈0% total loading
Total United States	**3514**	**229**		1065 million PEQ[h]; tot. pop. 319 million
Canada				
NE Canada[j]				
Rivers and estuaries	160.0 (21.9)	10.4	15.4	
Atmospheric	100.0 (13.7)	6.5		
St. Lawrence watershed[j]				
Rivers and estuaries	340.0 (46.6)	22.1	15.4	
Atmospheric	130.0 (17.8)	8.5		
Finfish aquaculture[c,m]	1.2 (0.2)	0.3	3.9	
Sub-total Canadian east coast	731.2	47.8		
Western Canada				
Finfish aquaculture[c,m]	2.2	0.6	3.9	
Sub-total Canadian west coast	–	–		4000 t $year^{-1}$ into US west coast from Canada
Total Canada	**733.3**	**48.4**		222 million PEQ[h]; tot. Pop. 35 million

(continued)

Table 27.1 (continued)

	Total nitrogen	Total phosphorus	Redfield ratio	Notes
China				
Major rivers[n]				
Yangtze	1690.0 (62.5)	168.0	10.1	
Huanghe	16.5 (0.6)	0.8	20.5	
Liaohe	3.8 (0.1)	0.3	11.1	
Haihe	4.4 (0.2)	0.2	22.4	
Huaihe	38.2 (1.4)	2.6	14.6	
Qiantangjiang	47.3 (1.7)	1.7	28.2	
Minjiang	87.0 (3.2)	3.2	27.4	
Zhujiang	785.9 (29.0)	30.6	25.7	
Sub-total river loading	2673.2 (98.8)	207.5	12.9	
Coastal finfish aquaculture [o,p]	32.8 (1.2)	8.5	3.9	
Total China	**2706.0**	**215.9**		820 million PEQ[h]; total pop of 1.4 billion
Total Europe, North America, and China	**11095.9**	**832.8**		3.36 billion PEQ[h,q];

[a]Ærtebjerg et al. (2001)
[b]Ferreira and Bricker (2016)
[c]Feed and faeces N/P ratios from Wang et al. (2013); N/P ratios for excretory products from Wang et al. (2014)
[d]Production data: http://www.salmon-from-the-faroe-islands.com/
[e]Weight conversion coefficients recalculated from Acharya (2011)
[f]HELCOM (2010)
[g]OSPAR (2010)
[h]1 PEQ = 3.3 kg N ind^{-1} year^{-1} (Ferreira and Bricker 2016)
[i]Population: 508 × 10^6 for EU; 118 × 10^6 for other European countries; 100 × 10^6 for North African Maghreb
[j]NRC (2000)
[k]Wise and Johnson (2011)
[l]Data for 2012, FAO FishStatJ; http://www.fao.org/fishery/statistics/software/fishstatj/en
[m]Canadian Department of Fisheries and Oceans (DFO), 2016; reported live weight production of 30,266 t year^{-1} (East Coast) and 56,276 t year^{-1} (West Coast) http://www.dfo-mpo.gc.ca/stats/aqua/aqua14-eng.htm
[n]Tong et al. (2015)
[o]Marine finfish live weight production = 1.31 × 10^6 t year^{-1} (China Fishery Statistical Yearbook 2016) as compared to 27.15 × 10^6 t year^{-1} in freshwater (China Fishery Statistical Yearbook 2016)
[p]Loading calculated using data from Ferreira and Bricker (2016), assuming a cultivation period of 500 days and 0.5 kg biomass per fish
[q]Calculations were based on a PEQ equivalent for treated domestic effluent. If a coefficient of 4.4 kg N PEQ^{-1} year^{-1} (untreated effluent) is used, the equivalent population is reduced to 2.52 billion (all PEQ values will be lower by 25%)

with grass carp, and re-use waste feed and other side-streams of fed aquaculture. Equally, it is unclear how much nitrogen and phosphorus these 27 million tonnes of farmed fish might *add* to the overall load, because of the practice of carp polyculture, where organically extractive species offset pellet-fed ones, and because of the widespread use of Integrated Multi-Trophic Aquaculture (IMTA).

Table 27.1 also includes loading from (marine) finfish aquaculture for Europe, USA, Canada, and China, which together account for 1% of the overall nitrogen discharge (3.3% of the phosphorus, because of lower N:P ratios in finfish emissions). Arctic waters are the only area where the proportion of N load due to finfish is significant (63.4%, mainly due to Norwegian salmon and trout production), but the *total* contribution of this region to the European[2] budget is only 2%.

For all the areas considered, with the possible exception of Canada, eutrophication has been identified as an issue (see e.g. for Europe: HELCOM 2009, 2014; OSPAR 2010; Ferreira and Bricker 2016; US: Howarth et al. 2002; Bricker et al. 2008; China: Xiao et al. 2007; SOA 2016). In the US, a large part of the NE seaboard and Gulf of Mexico are impacted (Bricker et al. 2008), and in China, 9.8×10^4 km^2 were affected in 2012 (Tong et al. 2015). Secondary symptoms of eutrophication (*sensu* Bricker et al. 2003) such as hypoxia and nuisance and toxic blooms (HAB) typically occur as a consequence of excessive primary production—in China, 73 offshore HAB events were reported in 2012, affecting an area of almost 8000 km^2 (Tong et al. 2015). Eutrophication-related hypoxia has been documented in Europe (Diaz and Rosenberg 2008), the US (Bricker et al. 2008), and China (Tong et al. 2015), leading in extreme cases to the development of 'dead zones' (e.g. Rabalais et al. 2002).

Cultivation of bivalve species is spatially ubiquitous in the parts of the world covered in Table 27.1, although the stocking density varies widely, as do the main species farmed. All this production shares a common ecosystem service by exerting top-down control on primary symptoms of eutrophication, and acts as a circuit-breaker in the eutrophication cycle (Ferreira and Bricker 2016), as illustrated in Fig. 27.1.

27.2.2 *Bivalve Production*

A detailed breakdown of national bivalve production for Europe is given in Ferreira and Bricker (2015). The global European production is shown in Fig. 27.2, with the bivalve bivalve producing nations highlighted.

The production analysis for Europe has been extended herein to match the nutrient loading data shown in Table 27.1: total numbers for Europe, the United States, Canada, and China are given in Table 27.2. In total, almost 13×10^6 t year^{-1} are produced in the areas considered, about 79% of the estimated world production

[2] Includes North African Maghreb for estimates of loading to the Mediterranean Sea.

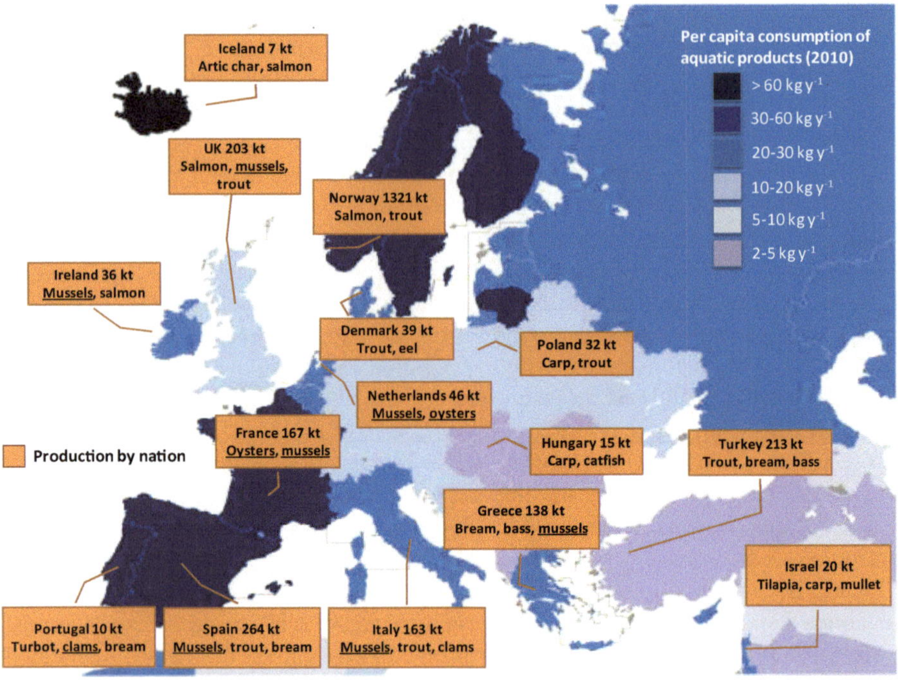

Fig. 27.2 Aquaculture production in Europe (bivalves underlined), illustrating the wide distribution of bivalve aquaculture, and its spatial relevance to top-down nutrient control. The *per capita* consumption of aquatic products is also shown

(FAO 2016). However, the global numbers include other molluscs such as abalone, snails, limpets, and octopi, not considered here because they are not filter-feeders—China alone produces over 90 kt year^{-1} of abalone, and almost 112 kt year^{-1} of the freshwater mystery snail (*Bellamia chinensis*), which together practically equal all the North American bivalve production.

The coefficients used in Ferreira and Bricker (2016), obtained through the application of the FARM model to the main cultivated bivalve species, were used to calculate the potential net nitrogen removal for the world production of filter-feeding bivalves listed in Table 27.2. In total, about 635 kt N may be removed annually (Table 27.3), a regulatory service that is unaccounted for but corresponds to almost 192 million population-equivalents. Within a nutrient credit trading framework, this would correspond to a potential minimum value of 7.7 billion USD.

A comparison of nutrient loading and nitrogen offsets by farmed bivalves is given in Table 27.4, broken down by world areas. Aquaculture is a very small contributor to nutrient budgets in the West, both as a source (fed aquaculture, mainly finfish) and a sink (bivalves), due to social licence constraints to expansion. The contribution of marine aquaculture to the total nitrogen loading to the coastal zone ranges from trivial (1.2% in China) to insignificant (0.02% in the United States). However, in both cases, there is a significant input of nitrogen from land-based

Table 27.2 Bivalve production for major areas of the world (tonnes live weight year^{-1})

Group, genus, or species	European Union[a]	United States[b]	Canada[c]	China[b]	Total
Oysters	92,620		12,604		105,224
Cupped oysters (*Crassostrea sp.*)	89,870	131,849		3,948,817	4,170,536
Flat oyster (*O. edulis*)	2750	4			2754
Mussels (*M. edulis, M. galloprovincialis*)	405,195	3127	25,464	764,395	1,198,181
Scallops	56		114	1,419,956	1,420,126
Clams, cockles, arkshells	34,438		1626		36,064
Cockles	4431	1		278,058	282,490
Clams	29,766				29,766
Soft clam	14	683			697
Good clam (*V. decussatus*)	5628				5628
Carpet shell (*V. pullastra*)	339				339
Manila clam	23,779	4126		3,735,484	3,763,389
Razor clam (*Solen sp., Sinonovacula sp.*)	5			720,466	720,471
Quahog (*M. mercenaria*)	1	27,704			27,705
Geoduck clam		534			534
Pen shells (*Pinnidae*)				15,061	15,061
Other		51	119	897,116	897,286
Freshwater molluscs				147,040	147,040
Total	532310[d]	168,079	39,927	11,926,393	12,666,709
Percentage of total (%)	4.2	1.3	0.3	94.2	100

[a]Data for 2013, see Ferreira and Bricker (2015) for data sources and national breakdown
[b]Data for 2012, FAO FishStatJ; http://www.fao.org/fishery/statistics/software/fishstatj/en
[c]Data for 2014, Canadian Department of Fisheries and Oceans (DFO), 2016; http://www.dfo-mpo.gc.ca/stats/aqua/aqua14-eng.htm
[d]Data shown for 2013, updated from the 2011 Eurostat dataset given in Ferreira and Bricker (2016). The major change from 2011 to 2013 was that significant blue mussel production volumes were moved from aquaculture to fisheries. As an example, Eurostat reported Danish blue mussel aquaculture in 2011 as 47,907 t, and reduced it to 560 t in 2013. Though less extreme, reductions were also made to estimates for Germany, The Netherlands, and Ireland. These were not reductions in capacity, but a reclassification. The number given herein agrees well with the European Commission (2016) Common Fisheries Policy report, which gives a total EU aquaculture production of 1,211,259 t for 2013, of which 43.6% (520,841 t) are molluscs and crustaceans

freshwater fish farming (see footnote, Table 27.4) which is not shown here, since we are only considering coastal systems where a clear link between different loading sources and bivalve aquaculture can be established.

Table 27.4 shows that for Europe, on a mass balance basis, bivalves offset over half the total fed aquaculture nitrogen load, and in Canada, they offset almost 90% of the N load from finfish culture. In both the USA and China, the *relative* role of bivalves in removing the nutrients discharged by finfish culture is far more relevant than in the other areas considered, but the differences in scale of production must be

Table 27.3 Bivalve nitrogen removal calculated with the FARM model for major areas of the world (tonnes N year^{-1})

Group, genus, or species	European Union	United States	Canada	China	Total
Oysters[a]	9461		1287		10,749
Cupped oysters (*Crassostrea sp.*)	3439	5045		151,110	159,595
Flat oyster (*O. edulis*)	105	0.2			105
Mussels (*M. edulis, M. galloprovincialis*) [b]	25,341	196	1593	47,805	74,933
Scallops[c]	2		4	54,338	54,344
Clams, cockles, arkshells[d]	2418		114		2532
Cockles	311	0.1		19,524	19,835
Clams	2090				2090
Soft clam	1	48			49
Good clam (*V. decussatus*)	395				395
Carpet shell (*V. pullastra*)	24				24
Manila clam	1670	290		262,293	264,252
Razor clam (*Solen sp., Sinonovacula sp.*)	0.4			50,589	50,589
Quahog (*M. mercenaria*)	0.1	1945			1945
Geoduck clam		37			37
Pen shells				1058	1058
Total	37,222	7562	2999	586,716	634499[e]
Percentage of total (%)	5.9	1.2	0.5	92.5	100

[a]Calculated using an N removal of 38.2 kg N t FW^{-1} year^{-1}, from FARM model outputs for Pacific oyster (Ferreira and Bricker 2016)
[b]Calculated using an average N removal of 62.5 kg N t FW^{-1} year^{-1}, by combining FARM model outputs for blue mussel and Mediterranean mussel (Ferreira and Bricker 2016).
[c]Calculated using Pacific oyster N removal (Ferreira and Bricker 2016), no scallop model available.
[d]Calculated using an N removal of 70.2 kg N t FW^{-1} year^{-1}, from FARM model outputs for Manila clam (Ferreira and Bricker 2016).
[e]Corresponds to 192,272,352 PEQ year^{-1}, which would have a potential value of 7690.89 million USD year^{-1}, using a PEQ conversion factor for land-based removal from Lindahl et al. (2005)

taken into account—the 870% offset of finfish culture in the US is more due to the very low finfish production than to a significant bivalve production. By contrast, in China, the removal of 587 × 10^3 t N year^{-1} undoubtedly plays a role in mitigating coastal eutrophication.

Although, by definition, in organically extractive aquaculture there is a net removal of particulate organic matter (POM), bivalve culture at a high stocking density in suspended structures such as rafts or longlines may locally impact the bottom in a similar way to finfish cage culture (Grant, pers. com.). Particle consolidation by bivalves into pseudofaeces and faeces might result in faster settling, and therefore part of the phytoplankton nitrogen which might be otherwise be flushed out of an estuary or embayment could be retained within an estuary or bay.

Table 27.4 Nitrogen loading and offsets for major areas of the world

	Europe	USA	Canada	China	Total
Total N load (10^3 t N year^{-1})	4142.6	3514.0	733.3	2706.0	11095.9
Fed aquaculture N load (10^3 t N year^{-1})	68.8	0.9[a]	3.3	32.8[b]	105.8
Organic extractive N removal (10^3 t N year^{-1})	37.2	7.6	3.0	586.7	634.5
Proportion of total N load due to fed aquaculture (%)	1.7	0.02	0.5	1.2	
Proportion of fed aquaculture N load offset by bivalves (%)	54.1	870.2	89.6	1790.8	
Proportion of total N load offset by bivalves (%)	0.9	0.2	0.4	21.7	

[a]Only marine aquaculture, mainly salmonids; excludes 229 × 10^3 t live weight year^{-1} freshwater production, of which 67% are channel catfish
[b]Only marine aquaculture; excludes 27,150 × 10^3 t live weight year^{-1} freshwater production, of which 49% are grass carp, silver carp, and bighead carp

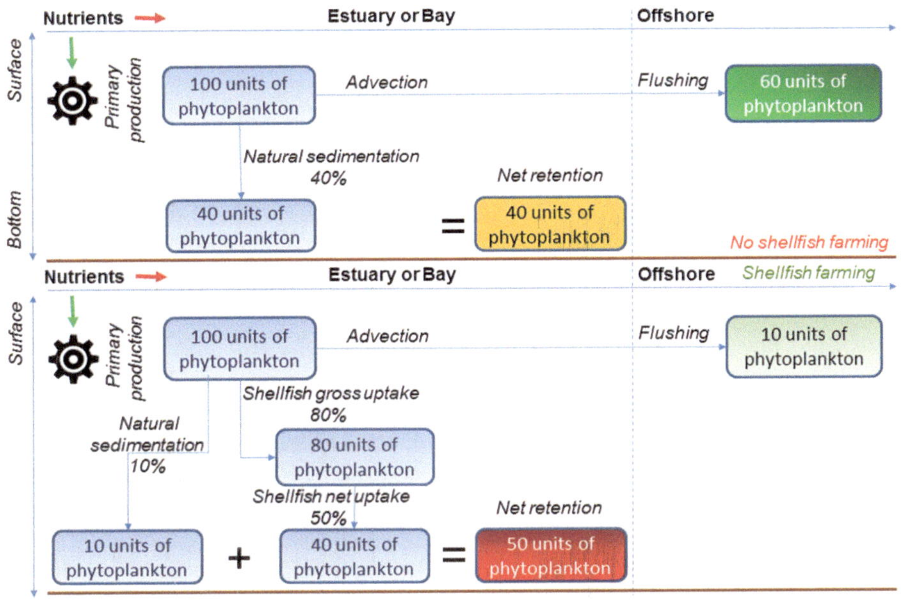

Fig. 27.3 Potential net phytoplanton nutrient retention in an estuary or bay where large-scale bivalve farming is practised

Figure 27.3 shows the *conceptual representation* for a system where, with no top-down control by bivalves (upper pane), there would be a net export of 60 phytoplankton 'units' from the system to offshore waters, and a retention of 40 units due to sedimentation. In the lower pane, bivalves would remove 80 units through gross uptake, of which 50% (40 units) are lost to the sediment through pseudofaeces and faeces, both of which sediment rapidly within the system. A further 10 units are lost through natural sedimentation of phytoplankton and remain within the estuary or bay, and therefore 10 units are exported offshore. As a consequence, although there would be a net removal (top-down control of phytoplankton) of 40 units of POM, which are used for bivalve growth, 50 units of POM are nevertheless retained in the estuarine sediment, i.e. 25% more than in the non-bivalve model. We emphasize only that this *may* occur, and underscore that it should not be seen as a typical situation. However, this conceptual example helps to illustrate that the use of bivalve aquaculture in nutrient management is a complex issue, and must be carefully considered.

27.3 Trading Mechanisms

27.3.1 How Does a Trading Program Work?

Sixty-five percent of US estuaries and many in the EU and elsewhere are impacted by nutrient loads and do not meet established water quality standards (e.g. Bricker et al. 2007; HELCOM 2010, 2014). Legislation such as the EU WFD and US Clean Water Act establish a basis for regulating pollutants from both point and non-point sources. Despite these regulations and attempts to reduce nutrient discharges, many waterbodies remain impaired. This situation has created increasing interest in the concept of nutrient credit trading as a means of achieving water quality goals in a timely and cost-effective manner (USEPA 2004b). In the US, a Total Maximum Daily Load (TMDL; USEPA 2017) analysis is conducted on a waterbody that does not meet water quality standards to determine the maximum amount of a pollutant (nutrients) that can be discharged to the waterbody and still meet water quality goals. That maximum, or cap, is used to allocate maximum allowable loads from regulated point sources (e.g. WWTF) discharging to the waterbody.

A nutrient trading program provides the opportunity for point-source dischargers who reduce their nutrient loads below those allocated target levels to sell their surplus reductions or nutrient 'credits' to other dischargers in the same watershed who are unable or face higher-cost nutrient reduction options. A credit is the difference between the discharge allowance for a point source and the measured discharge from that source. In the case of unregulated non-point sources, a credit or offset is a nutrient reduction by that source that must be certified by a regulatory agency and is referred to as a Best Management Practice (BMP). Non-point source BMPs include agricultural nutrient management practices (e.g. cover crops, riparian buffers), wetland construction, and urban stormwater controls. Trading programs are designed to establish a market-based approach to nutrient management by providing economic incentives for achieving nutrient load reductions (Lindahl et al. 2005; Jones et al. 2010; Lal 2010; Stephenson et al. 2010). The overall goal of trading programs is to meet regulatory requirements at lower overall costs, but they can potentially generate greater environmental benefits than would be achieved under traditional regulation, and may also address and raise awareness of other sources contributing to water quality degradation.

Nutrient credit trading programs are already a reality in parts of the US (Lal 2010; Branosky et al. 2011; Ferreira et al. 2011; STAC 2013). The Connecticut Nitrogen Credit Exchange (CNCE) is a nutrient trading program created in 2002 to address nutrient-related hypoxia conditions in Long Island Sound (LIS), where the state acts as broker and price setter. This is one of the few mature and successful examples of water quality credit trading. The program provided an alternative compliance mechanism for 79 WWTFs throughout the state, with 15.5 million nitrogen credits bought and sold during 2002–2009, representing a value of $45.9 million US. The cost savings of the exchange's credit trading were estimated at $300–$400 million (CT DEP 2010), compared to improving nitrogen removal technologies.

As more facilities successfully attained their final waste load allocations, the number of buyers of nitrogen credits decreased, though there are still some buyers and the program continues (M. Tedesco, Long Island Sound Study, pers. com.). It is important to note that the CNCE includes only point sources, though a mechanism for including non-point sources to meet more stringent future allocations is being discussed.

27.3.2 Non-point Source Trading Challenges

The inclusion of non-point sources in credit trading programs is intended to increase flexibility and provide additional options for regulated sources to achieve reductions through trades with unregulated non-point sources. The U.S. Environmental Protection Agency's (EPA) 2008 national water quality trading policy supports creation of non-point source water quality trading credits through agricultural BMPs, creation and restoration of wetlands, stormwater control construction, and more recently have included nutrient assimilation offsets that remove nutrients directly from the water, such as bivalve aquaculture (USEPA 2008a).

Because trading programs must ensure that water quality goals are met, regulators must be certain that the off-site non-point source load reduction will yield similar or superior water quality conditions. Trading programs must ensure equivalent outcomes when controls take place at different nutrient sources and locations in the watershed (Stephenson and Shabman 2017a). Unlike the success demonstrated by the point-to-point trading in LIS, very few trades have been made in the many non-point source trading programs developed, due to the high costs of assuring equivalence between point and non-point sources (Stephenson and Shabman 2017a; STAC 2013, Ribaudo and Gottlieb 2011).

Several regulatory requirements to assure equivalence contribute to the high costs of purchasing agricultural non-point credits, may hinder establishment of trading programs. These include: trading ratios, setting of baselines, and quantification and verification of non-point source control effectiveness (Ribaudo and Gottlieb 2011; Stephenson and Shabman 2017b).

Transaction costs for generating credits, and for monitoring and enforcement of crediting projects, may increase production costs (DeBoe and Stephenson 2016). Costs associated with generating credits range from \$1865–\$8705 per three-year project, depending on the complexity of the contract (DeBoe and Stephenson 2016). The cost of monitoring and verification of reduction performance, once controls are implemented, varies depending on the type of monitoring and verification (onsite vs remote), as well as the project duration (permanent or term credits) and frequency of required verification (i.e. 5 year vs 1 year verification). DeBoe and Stephenson (2016) describe potential 82–96% reductions in monitoring and verification costs with self-reporting and remote monitoring. Comparison of annualized transaction costs for projects generating permanent credits (\$257), 10 year fixed term credits (\$534–\$864) and 3 year fixed term credits (\$1801–\$4144) are considered modest due to the type of activity being credited (mostly land conversion), though working land BMPs may cost more. Thus, costs are currently not seen as a barrier to trade (DeBoe and Stephenson 2016).

A further analysis was done to evaluate other reasons for the lack of non-point source trading using three well-developed Virginia nutrient trading programs (Stephenson and Shabman 2017b). The analysis included industrial and municipal WWTF, municipal stormwater programs, and land development programs and showed that obstacles to nutrient credit trading are regulatory:

1. Regulatory requirements and trade restrictions where on-site nutrient reductions at permitted sources (called sequencing) are preferred. Permittees are required to operate installed capital equipment to design capability to meet mandatory effluent concentrations regardless of possible cost advantages trading with other sources, the state prioritizes point-to-point trading, and land development requires 75% of nutrient control to occur on-site.
2. Overlapping regulatory requirements where (i) WWTF have non-transferable requirements for the total volume of stormwater runoff from a site, and those lower discharges result in nutrient reductions; (ii) proposed wastewater re-use; and (iii) aquifer recharge, will further reduce nutrient loads. Water quality improvement grants pay 30–90% of WWTF upgrade costs reducing the need for off-site credits.
3. Compliance preference of regulated dischargers where: regulated sources prefer to achieve compliance with on-site technologies and control practices where the risk of non-compliance is under their direct control.

Issues that suppress point-source demand for non-point source credits in Virginia are representative of conditions found elsewhere in the US. For decades, federal and state programs have provided farmers with financial assistance ('cost-share') to implement specific agricultural practices that reduce pollutant loads. These programs pay farmers to implement practices, rather than paying directly for pollutant load reductions. Recent efforts to boost the supply of non-point source load reduction credits for trading demonstrates that non-point source practices can be quantified and certified into estimated load reductions. If governments would apply these non-point source crediting tools and methods along with competitive bid processes to identify low cost non-point source options with public non-point source funding, non-point source trading would thrive (Stephenson and Shabman 2017b).

27.3.3 Inclusion of Bivalves in Credit Trading Programs

As shown above, there is compelling evidence in support of the use of bivalves as a nutrient removal BMP for inclusion in nutrient credit trading. The documented nutrient removal capacity shown in multiple studies (e.g. Lindahl et al. 2005; Kellogg et al. 2014; Petersen et al. 2014; Rose et al. 2015) is as effective as BMPs that have already been approved for use in trading programs. Table 27.5 shows that annualized nitrogen removal by bivalve farms compared favourably to removal by stormwater control measures, based on two lines of evidence: (i) the nitrogen removal per unit area was highest for bivalve and gravel wetlands, all other

Table 27.5 Annual nitrogen removal (kg ha^{-1}) by different types of stormwater control measures, installed at the University of New Hampshire Stormwater Center, and by agricultural best management practices in the Chesapeake Bay watershed, as approved by the Virginia Department of Environmental Quality

Management practice	Annual nitrogen removal (kg ha^{-1})[a]	Cost (€ kg^{-1} N)[a]
Bivalve farms	118–1520 (819)	11–278 (145)
Stormwater control measures (modified from Houle et al. 2013)		56–6720 (3388)
Vegetated swale	0	
Wet pond	293	
Dry pond	222	
Sand filter	0	
Gravel wetland	1111	1.1–396 (199)
Porous asphalt	0	
Approved agricultural BMP (modified from Stephenson et al. 2010) minimum–maximum		0.2–870 (435)
Early cover crop	0.04–1.23 (0.63)	
15% N reduction	1.24–4.72 (2.98)	
Continuous no-till	0.80–2.01 (1.41)	
15% N reduction + continuous no-till	1.85–5.62 (3.74)	
Crop to forestland conversion	4.16–12.98 (8.57)	
Wastewater treatment upgrades		0.9–14,093 (7047)
Other		5.2–404 (205)

The final column provides data on reported costs for six categories of non-point-source nitrogen removal strategies. Each strategy includes a range of subcategories. Reported costs have been converted to € kg^{-1} N (adapted from Rose et al. 2015). Mean values are given in brackets where applicable

[a]For a breakdown of detail for ranges provided in this column, please see online supplementary material in Rose et al. (2015)

stormwater control measures were far less effective; (ii) the implementation cost per unit nitrogen is lowest for bivalves, followed by wetlands (37% higher). Taken together, bivalve aquaculture and wetlands are the most promising BMPs in terms of both competitive cost and nitrogen removal per unit area.

In general, nutrient removal by bivalve farms, and by specific stormwater control measures such as wetlands and ponds, was far higher than the removal reported for agricultural BMPs (Rose et al. 2015); in addition, the unit cost for those options was less than half that of agricultural BMPs. This comparison suggests that both stormwater control measures and bivalve aquaculture would be more desirable for non-point-source credit trading than agricultural practices.

Rose et al. (2015) expanded the analysis to evaluate comparative costs for nitrogen removal strategies (Table 27.5). The last column in the table shows that non-point source credits produced by cultivated bivalves are similar to those produced by agricultural non-point nutrient management strategies and both are more cost-effective than urban stormwater strategies and wastewater treatment upgrades. This analysis of removal efficiencies and cost-effectiveness confirms that bivalves are a promising nutrient removal strategy that could potentially be successfully used in a credit trading program.

A recent analysis of agricultural and assimilative service BMPs further supports the potential successful use of bivalves in trading programs. Stephenson and Shabman (2017a) evaluated approved agricultural BMPs (structural i.e. riparian buffers, grass filter strips; management i.e. cover crops, tillage practices, nutrient management, and land conversion) and aquatic plant biomass creation and harvest, bivalve aquaculture, stream restoration, and wetland restoration and creation. Five water quality criteria were evaluated, including quantification certainty, temporal matching, additionality, and leakage. Table 27.6 provides results of the assessment and shows that assimilation reduction strategies such as biomass harvest and bivalve aquaculture provide more assurances of equivalence than agricultural non-point sources.

There is high certainty in quantification with the nutrient harvest technologies, as well as better temporal matching, and lower non-additionality and leakage risks than in agricultural non-point source projects (Stephenson and Shabman 2017a). Agricultural BMPs present challenges to equivalency due to uncertain quantification of nutrient reduction performance, temporal mismatching of loads, and leakage. Osmond et al. (2012) also note the uncertainties associated with quantification of agricultural BMPs due to deficiencies in existing modelling tools, and suggest that due either to problems with modelling or water quality data, or both, the models grossly overestimate the effectiveness of conservation practices. It must be noted, however, that there are potential uncertainties associated with bivalve culture, and therefore with its potential role in nutrient management. Examples include HAB- or disease-related mortalities, and loss of gear and stock in extreme weather, which can result in an increase of nutrients in the water column.

Nutrient assimilation credits have the potential to increase both the quantity and the quality of credits used by regulated point sources to achieve compliance. If they can provide more certain water quality outcomes, then a strong case can be made for

Table 27.6 Summary of water quality equivalence of nutrient credit trading options

	Quantification of outcome[a]	Temporal matching[b]	Spatial redistribution[c]	Leakage[d]
Non-point source credits				
Structural agricultural BMPs	Observed behaviours: Modelled outcomes	Stochastic loads, load averaging across time	Requires delivery attenuation estimates	Some leakage potential
Management agricultural BMPs	Observed or reported behaviors: Modelled outcomes	Stochastic loads, load averaging across time	Requires delivery attenuation estimates	Some leakage potential
Land conversion	Observed behaviors: Modelled outcomes	Stochastic loads, load averaging across time	Requires delivery attenuation estimates	Some leakage potential
Nutrient assimilation wetlands	Measured or Modelled outcomes	Potentially stochastic loads, load averaging across time	Requires delivery attenuation estimates	Minimal
Bivalve aquaculture	Biomass harvest: Measure burial/ denitrification: model	Temporal matching of load reductions with buyers	Requires delivery attenuation estimates	Some leakage potential
Algal harvest	Measure outcomes	Temporal matching of load reductions with buyers	May require delivery attenuation estimates	Minimal
Seaweed and aquatic plant harvest	Measure outcomes	Temporal matching of load reductions with buyers	May require delivery attenuation estimates	Minimal
Stream restoration	Model outcomes	Potentially stochastic loads, load averaging across time	Requires delivery attenuation estimates	Minimal

Adapted from Stephenson and Shabman (2017a)

A glossary of terms is provided in the notes for this table

[a]Quantification of nutrient reduction credits should be estimated with a similar level of certainty. A nutrient credit is defined as a nutrient load reduction, relative to a baseline, over a specific period of time (e.g., kg of nitrogen per year). For example, point sources typically quantify nutrient loads by direct measurement of flow and sampling of effluent concentrations. Non-point source credits are more difficult to quantify

[b]Temporal matching means that the timing of the load reduction from the credit is the same as the timing of the point-source load being offset. When the timing is the same, there is no risk of an adverse effect on water quality conditions as a result of the trade

[c]Spatial redistribution requires trading programs to define a specific geographic location in the watershed where water quality outcomes will be compared and evaluated for equivalency. Nutrient credit trading spatially redistributes nutrient loads actions across a watershed. For example, a trading program may allow a point source to buy credits from a non-point source regardless of the location as long as 'delivered loads' to the watershed impairment point (such as a downstream estuary) is the same between buyer and seller. A point source could buy credits from a downstream non-point source, thereby increasing nutrient loads in the watershed between the point and non-point source, but producing equivalent delivered loads below the non-point source. Note that trading program provisions explicitly prohibit transactions that would impair local water quality along the delivery route

[d]Leakage occurs when a nutrient credit trade produces another form of unaccounted increase in nutrient loads

their inclusion in nutrient trading programs. Nutrient assimilation credits, relative to agricultural non-point source load reductions, can offer greater assurances of equivalence for trades with regulated point sources.

27.3.4 Oyster BMP in Chesapeake Bay

Inclusion of bivalves in trading programs is viewed as a positive addition to nutrient trading programs (Stephenson et al. 2010; Rose et al. 2014; Stephenson and Shabman 2017a) but until recently they were not an approved BMP and thus could not be included. Recently, the Chesapeake Bay Program Oyster BMP Expert Panel evaluated and approved nutrient removal reduction by cultured oysters and developed a framework for crediting and verification for application of an oyster BMP (Oyster BMP Expert Panel 2016). The BMP is for harvested tissue only (Table 27.7); recommendations for development of a BMP for oyster shell, denitrification, and burial are anticipated in 2017.

Recommended default estimates for nutrient credits production by harvested oyster tissue were derived from oyster growth (shell height to dry tissue weight regressions) and tissue nutrient concentration data from several Chesapeake Bay locations (Oyster BMP Panel 2016). Differences in biomass between diploid and triploid oysters warranted the use of separate regression equations. The 50th quantile was used to conservatively account for differences in culture method and type (off-bottom/on-bottom, hatchery-produced/wild). The final default recommendations for average nutrient content of 8.2% nitrogen and 0.9% phosphorus, based on dry tissue, are applied regardless of location or ploidy to avoid biases (i.e. site specific, variability in time). However, the framework allows for development of site-

Table 27.7 Recommendations for crediting of nitrogen and phosphorus removal by harvested oyster tissue in Chesapeake Bay

Best Management Practice (BMP) Name	lbs N reduced per 10^6 oysters harvested	lbs P reduced per 10^6 oysters harvested
Diploid Oyster Aquaculture 2.25 Inches	110	22
Diploid Oyster Aquaculture 3.0 Inches	198	22
Diploid Oyster Aquaculture 4.0 Inches	331	44
Diploid Oyster Aquaculture 5.0 Inches	485	44
Diploid Oyster Aquaculture ≥ 5.5 Inches	683	66
Triploid Oyster Aquaculture 2.25 Inches	132	22
Triploid Oyster Aquaculture 3.0 Inches	287	22
Triploid Oyster Aquaculture 4.0 Inches	573	66
Triploid Oyster Aquaculture 5.0 Inches	970	110
Triploid Oyster Aquaculture ≥ 5.5 Inches	1,477	154
Site-Specific Monitored Oyster Aquaculture	N/A	N/A

Adapted from Oyster BMP Panel (2016)

specific removal rates by interested growers, in conjunction with the state and the Chesapeake Bay Partnership with costs assumed by the grower.

The importance of these recommendations is that states can now legally use nitrogen and phosphorus removed in harvested oyster tissue as a BMP in trading programs within the Chesapeake Bay Watershed, with potential use by other states that support oyster growth. At present, credits earned would count toward nutrient reductions required by the Chesapeake Bay TMDL, i.e. nutrient pollutant clean-up plan—although full inclusion in trading programs requires additional discussion, this is an encouraging step, and the use of bivalve BMPs for regulatory compliance should be encouraged.

27.3.5 Bivalve Aquaculture for Water Quality Improvement in Massachusetts

The successful inclusion of bivalves for nutrient water quality compliance has been demonstrated in the town of Mashpee, Massachusetts. As part of a Comprehensive Watershed Nitrogen Management Plan (CWNMP) the plan incorporates several traditional nitrogen reduction approaches and the harvest of cultivated Eastern oysters (*Crassostrea virginica*) and hard clams (quahogs; *Mercenaria mercenaria*) to meet TMDL water quality goals and restore bivalve resources (Town of Mashpee Sewer Commission 2015). Comparison of estimated costs (Net Present Value – NPV) for bivalve implementation ($22 million NPV) and sewer mains, pumping stations, and road construction for collection systems ($80 million NPV) shows significant savings are expected from inclusion of aquaculture. Other advantages are that the bivalves remove nitrogen from the water column by filtering organic particulates, the capital costs are lower ($180 million for Phase 1 with bivalves, $360 million without), it helps restore bivalve resources and has the potential to generate other positive impacts related to habitat. Some disadvantages are that only watersheds with appropriate habitat can be targeted, long-term performance is unknown, predators and diseases may impact performance, long-term maintenance is unknown, annual seeding of bivalve beds may be required, and bacterial pathogens from septic system effluents may not be addressed.

Mashpee's GIS Department mapped the bivalve habitat based on GPS data collected from the estuaries, and determined that there is sufficient habitat to support the proposed densities of bivalves. In samples from Mashpee harvest areas, both clams and oysters were found to have 0.5% nitrogen. Thus, a 3.5-inch 100 g harvest size oyster would represent removal of 0.5 g N, and a 60 g harvest size clam would remove 0.3 g N (Reitsma et al. 2016). The plan targets harvest of 9 million oysters and 26.5 million clams to remove a total of 12.6 metric tons of nitrogen, 73% of the nitrogen reduction required by the TMDL. This is considered a conservative estimate since it does not include potentially significant losses from denitrification or burial (Kellogg et al. 2013). It will be a challenge to maintain annual bivalve harvest at these levels, but the plan accommodates annual seeding if necessary to produce the harvest necessary to include aquaculture as a nutrient management option.

Other more traditional management measures like WWTFs will be much slower to come online and will only reach full build-out if aquaculture fails. After implementation, performance will be evaluated every 5 years. Bivalve aquaculture holds great promise in helping to reach water quality goals affordably and in compliance with the state 208[3] water quality management plan requirements. The Mashpee CWNMP is an example of how to include bivalves in comprehensive management plans. Through programs such as this, water quality compliance will be successfully achieved with the added benefit of supporting domestic production of seafood.

27.3.6 Indicators and Assessment Methodologies

Water quality trading (WQT) mechanisms were first proposed by Dales (1968), and gained traction in the US during the 1980s and 1990s, as water authorities reviewed management options for meeting TMDLs (Shortle 2013). At the beginning of this century, the US EPA began to support WQT both technically and financially, and nutrient credit trading developed as a concept (e.g. Stephenson et al. 2010), and has subsequently been implemented to some degree, largely in the United States. The Connecticut Department of Environmental Protection began its participation in a watershed-scale trading programme in 2002, largely because of concerns related to eutrophication in LIS: this is a chronic issue in LIS, attributed to excessive nutrient loading, and manifests itself e.g. through low dissolved oxygen—a secondary, or well developed, indicator of eutrophication (Fig. 27.4). For the western area of LIS, hypoxia, i.e. dissolved oxygen values lower than 3.5 mg L^{-1}, has been a problem for 90–100% of the period between 1991 and 2008, and has been recognized as a serious water quality impairment since long before that.

Rice and Stewart (2013) report spring chlorophyll peaks in LIS averaging 8.9 μg L^{-1} for the period 1995–2010, down from 25.3 μg L^{-1} in previous decades, which suggests that nutrient source control has been effective in reducing primary symptoms of eutrophication. Nevertheless, as seen in Fig. 27.4, this appears to be insufficient to reverse hypoxia, although the spatial extent has been reduced from 800 km^2 in 1987 to 330 km^2 in 2002 (Ferreira et al. 2007b).

The inclusion of filter-feeding bivalves in nutrient credit trading programmes is at best incipient, and has only been examined as a management tool in the United States (e.g. Stephenson et al. 2010; STAC 2013; Oyster BMP Expert Panel 2016). As discussed in the previous section, the emphasis has been on the removal of nitrogen from the receiving water by bivalves, with a possible extension to phosphorus, should P be relevant as a limiting nutrient.

The premise is that source control of N or P loading will lead to a reduction in eutrophication symptoms, e.g. lower concentration maxima of phytoplankton

[3] The 208 programme is a state of Massachusetts water quality management plan.

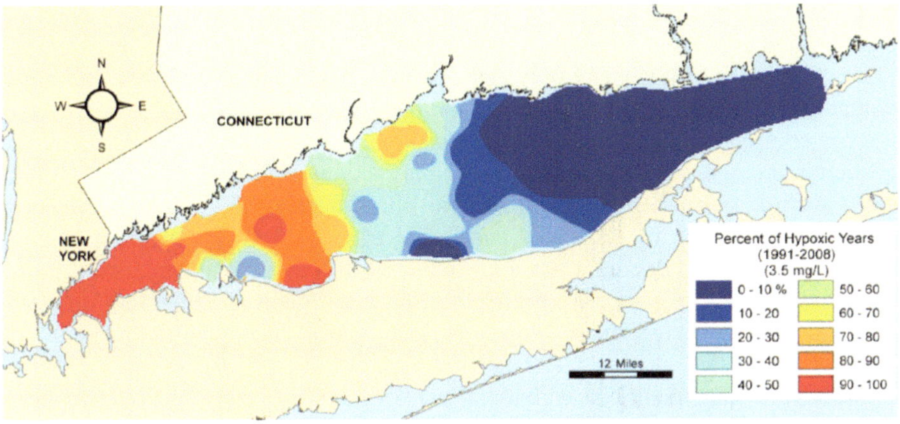

Fig. 27.4 Hypoxia in Long Island Sound: a motivation for nutrient trading schemes for eutrophication management

blooms, and smaller spatial and temporal extent of impairment. However, the key effect of bivalve filter-feeders is to attenuate *direct* symptoms, rather than to reduce nutrient load— this attenuation can be (qualitatively) evaluated by means of indicators such as water clarity (e.g. Cranford, this volume).

Although at present nitrogen removal is used as a currency to assess the regulatory ecosystem services of bivalves with respect to nutrient control, emphasis could instead be placed on how source control of emissions compares with top-down control, in terms of the reduction of *symptoms*. From the perspective of eutrophication management, the relevant indicator is not the change in the causative factor, i.e. the nutrient load (and its associated valuation or cost) but the change in the relevant target variables, such as chlorophyll and dissolved oxygen. If we select chlorophyll (α) as a management indicator in an estuary or bay, as is the case in the US (Bricker et al. 2008; USEPA 2008b), EU (WFD, see Ferreira et al. 2006; MSFD, see Ferreira et al. 2011), and elsewhere (see reviews in Borja et al. 2008; Zaldivar et al. 2008), an objective function for chlorophyll reduction α could be written as:

$$\alpha = \min\{f(\lambda,,,\mu,,,\rho,,,\phi)\} \qquad (27.1)$$

where λ is the nutrient loading, μ is the physical exchange (advection and diffusion), ρ is primary production, and ϕ is bivalve filtration. These variables (and others) have an effect on chlorophyll concentration, but some, such as physical exchange, are not amenable to management measures—however, μ may strongly condition the value of α, particularly in high energy systems, because it is a key determinant of system susceptibility, influencing both water turnover and light climate (the latter particularly when there is strong benthic-pelagic coupling).

An analysis of 1100 chlorophyll and Total Particulate Matter (TPM) measurements in LIS for the period 2000–2002, including surface, mid-water, and bottom samples (data supplied by J. Rose, NOAA) shows that phytoplankton, normalised as POM and expressed as percentage of TPM,[4] averages 7.8%, with a high coefficient of variation (147%). Not only is the chlorophyll signal often masked by other components of TPM, i.e. detrital POM and particulate inorganic matter (PIM), but there is no way to connect the measurements with the fluxes that generate them, i.e. advection and dispersion, sediment-water interactions, and biological sources and sinks such as primary production, bivalve filtration (Eq. 27.1), and zooplankton grazing. A reduction of suspended particulate matter (TPM) in the water column can therefore be considered a *potential* indicator of lower phytoplankton biomass, but there is typically a very low signal to noise ratio (e.g. in LIS the two variables show a very poor correlation, with $r = 0.19$), and source apportionment is not possible.

Although water quality measurements cannot be used to assess the relative influence of emissions control and bivalve drawdown on chlorophyll concentrations in the receiving water, ecosystem models allow a comparison to be made, provided that such models (i) explicitly simulate the relevant state variables and processes; (ii) simulate nutrient discharge from the catchment as part of the modelling framework, allowing different source-control scenarios to be compared with changes in bivalve stocking density.

A modelling framework of this type (Ferreira et al. 2016) typically includes the elements shown in Fig. 27.5, and simulates nutrient loading from the catchment, water circulation and exchange with the ocean, pelagic and benthic primary production, bivalve growth by means of some form of individual-based modelling (IBM),

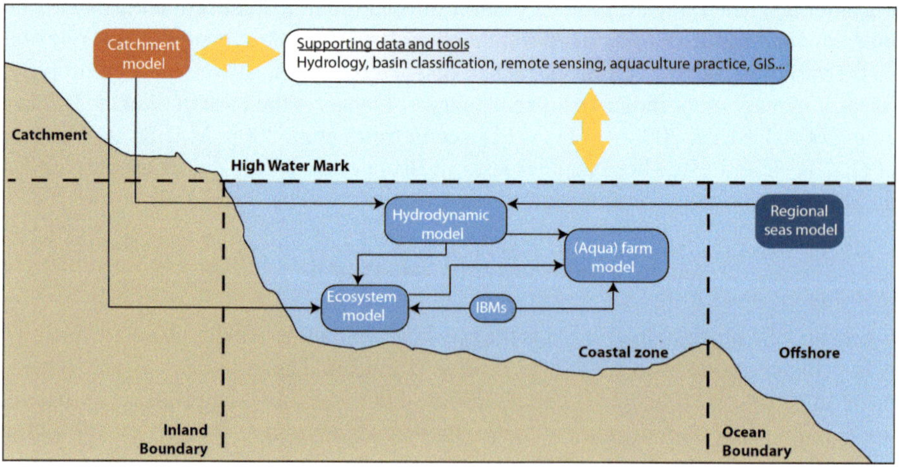

Fig. 27.5 Multi-model simulation framework applied for coastal systems analysis

[4]Converted to POM (mg L^{-1}) using a C:chl ratio of 50, and a POC:POM ratio of 0.38.

and bivalve population dynamics, including harvesting of the marketable cohort (see Ferreira et al. 2008, and Nobre et al. 2010, for examples from Europe and Asia).

Figure 27.6 illustrates the application of this modelling framework to Lough Foyle, a large (179 km^2) estuary that forms the northern border between Northern Ireland (UK) and Ireland. Twenty percent of the lough is intertidal, and there is a substantial and centuries-old production of bivalves, including the blue mussel *Mytilus edulis*, the European oyster *Ostrea edulis*, and more recently the Pacific oyster *Crassostrea gigas*.

The modelling framework is typically run for a decadal period, allowing the integration of multiple culture cycles (typically of the order of 2–3 years), and the effect on chlorophyll concentrations in different parts of the Foyle of 'switching' bivalve cultivation on or off is shown in Fig. 27.6. The model results suggest a strong top-down control of phytoplankton blooms, with typical draw-down of 2–8 $\mu g\ L^{-1}$ during the spring-summer bloom periods; however, at the head of the estuary, this effect may be substantially greater, reaching 16 $\mu g\ L^{-1}$ during the spring. Lough Foyle is particularly interesting from a regulatory perspective, because over 98% of the nitrogen loading to the estuary is derived from diffuse sources within the catchment (Nunes and Ferreira 2016). Phytoplankton growth thus depends little on urban nutrient sources, which means that excessive algal blooms cannot easily be controlled at source by nutrient removal, since that would require substantial changes to agricultural practices such as fertilizer application—these are both costly and socially unpopular.

In this example, bivalves therefore provide an important contribution to nutrient management and legal compliance, based on WFD biological quality elements

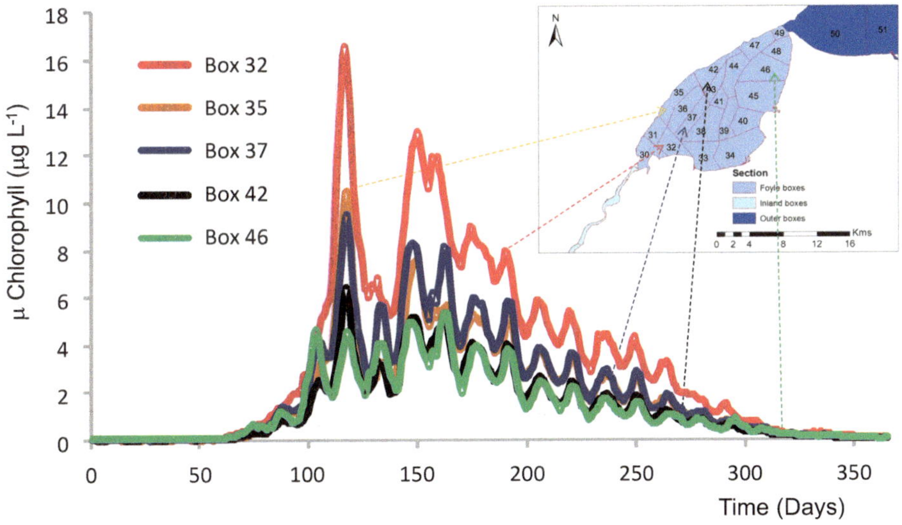

Fig. 27.6 Phytoplankton drawdown in Lough Foyle simulated with the EcoWin.NET system-scale model

(BQE) such as chlorophyll concentration. The removal of algae (primary symptoms of eutrophication) before the organic decomposition stage (secondary symptoms) also acts to reduce hypoxia, since it greatly lowers the availability of particulate organics, but it should be noted that the role of bivalve filtration in regulating chlorophyll concentration is obviously dependent on various factors, including bivalve stocking density and areal coverage of cultivation, and physical aspects such as flushing time.

This modelling framework was also applied (Fig. 27.7) to analyse various nutrient loading scenarios, and their effect on chlorophyll concentration. The percentile 90 value was chosen as the appropriate indicator, for consistency with the ASSETS model for eutrophication assessment (Bricker et al. 2003), and mean values are shown for all the modelling domain.

The lower line considers the standard nutrient loading and varying stocking densities for bivalves, and the upper line represents the effect of source-control on primary production, without any cultivated bivalves in the system. Under natural conditions (no agricultural activity or urban areas), simulated using the Soil and Water Assessment Tool (SWAT) hydrological model (e.g. Gassman et al. 2007), the chlorophyll P_{90} is about 6 μg L^{-1}, increasing to 9 μg L^{-1} in the present day (without bivalves).

Bivalves, under standard (present-day) nutrient loading conditions, lower the P_{90} to 4 μg L^{-1}, i.e. (in the model) bivalve filter feeders are considerably more successful in mitigating elevated chlorophyll concentrations in Lough Foyle than nitrogen source control.

From a management perspective, it is interesting to analyse the comparative effect of source control and bivalve regulatory services in economic terms. One approach for valuation is shown in Fig. 27.8, which provides cost estimates for both

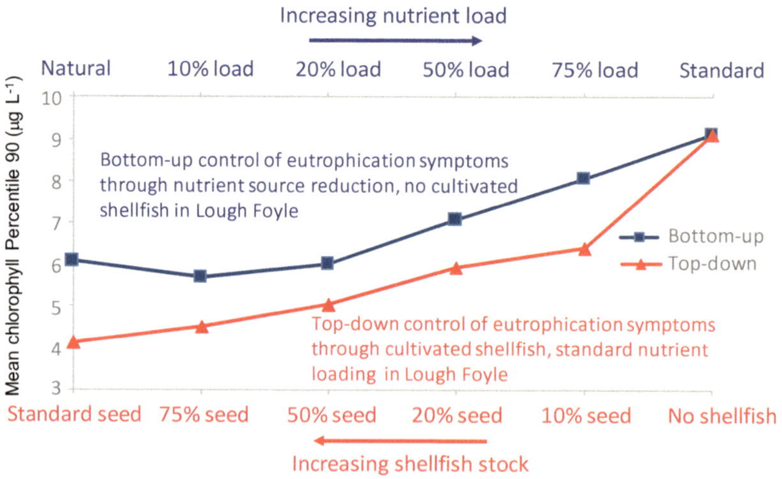

Fig. 27.7 Chlorophyll drawdown with bottom-up and top-down control simulated with the EcoWin.NET system-scale model

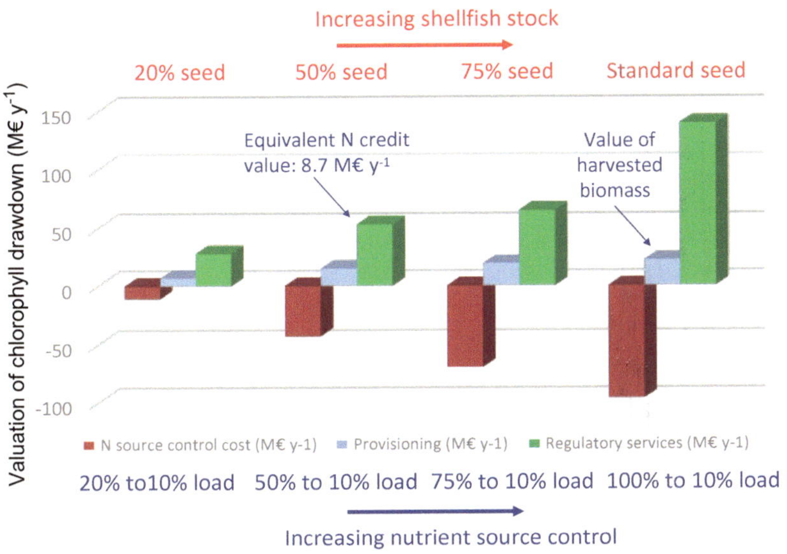

Fig. 27.8 Valuation of bivalve ecosystem services in Lough Foyle calculated using the EcoWin. NET system-scale model; negative values (red bars) are the cost of nutrient source control

types of management measures, i.e. nutrient source control (in red), and bivalve regulatory services (in green). The figure also shows (in blue) the provisioning service from bivalves, i.e. the value of harvested biomass—in the ecosystem model, the biomass of cultivated animals above a user-defined weight threshold is removed from the Lough during the period of harvest, and accrued. The value of the total harvested biomass is then estimated based on the farmgate price of the product.

The calculations are made separately for the two types of measures, but common sense dictates that combined solutions should be the preferred option, not least because of the danger of moral hazard in exempting agriculture from better management practices.

The decrease in N load, ΔL (t year^{-1}), was correlated with the corresponding reduction in chlorophyll P_{90}, Δα (μg L^{-1}). The cost of reducing emissions at source was determined by considering a unit cost 10.8 € kg^{-1} N, converted from a value of 12.4 USD kg^{-1} N, estimated by Lindahl et al. (2005) for 47 small stabilization ponds (lagoons) in Sweden, and multiplying by the load reduction ΔL. Load reduction can thus be expressed in monetary units C (M€ year^{-1}), and regression analysis yields Eq. 27.2, with a correlation coefficient r = 0.999 ($p_{<0.01}$).

$$C = 27.2\,\Delta\alpha + 3.37 \tag{27.2}$$

Equation 27.2 states that for Lough Foyle, a reduction of 1 μg L^{-1} for chlorophyll P_{90} costs 30.57 M€ year^{-1} in terms of source control. Furthermore, the cost per kg applied is low when compared with data for non-point mitigation (Table 27.5) proposed by Stephenson et al. (2010). Equation 27.2 was used to determine the alternative cost of the regulatory service provided by bivalves in Lough Foyle, by calculating the value associated with the chlorophyll P_{90} decrease for four scenarios, 20%, 50%, 75%, and 100% present bivalve stocking density, when compared to no bivalves in the lough. In parallel, equivalent source-control costs are shown for 4 N loading reduction scenarios, relative to 10% of the present-day load.

Apart from the systematically higher offset provided by bivalves in each scenario when compared to source control, the most striking observation is the difference in the value of the regulatory service provided by bivalves calculated using the different approaches, i.e. nutrient removal (N), and chlorophyll abatement ($\Delta\alpha$). The ratio of symptom value (chlorophyll) /causative factor value (N) for the four scenarios varies between 5.8 and 13.7, for the lowest to highest stocking densities (20%, 50%, 75%, and 100%, see Fig. 27.8).

This appears to be the first comparative analysis that focuses on eutrophication indicators, and suggests that for this particular system, the value of regulatory ecosystem services supplied by bivalves—in this case including three different bivalve species—will be underestimated by an order of magnitude if the approach is based on an equivalence of source control.

The degree to which such an approach can be generalised, without development of a complex suite of models for different estuaries and bays, is a question that requires further analysis. In particular, variations in water residence time and underwater light climate will undoubtedly affect the ratio above, since it is well established since the 1950s (Ketchum 1954), that physical conditions strongly constrain phytoplankton bloom development.

Our aim in relating regulation services provided by bivalves to other nutrient management options should be to establish which indicators provide the best metrics for assessment, which methodologies can be used for comparative analysis and valuation, and to develop those outcomes into tools for practical ecosystem management.

27.4 Conclusions

Nutrient management in coastal waters requires a holistic approach, and the role of bivalves in nutrient credit trading programmes should form an integral part of ecosystem-based management. This can only be achieved if it is recognized that bivalve farmers should play an active part in market-based control strategies. From the perspective of aquaculture enhancement, which is fundamental for improved food security, this is a triple-win, providing competitiveness of agriculture, eco-intensification of aquaculture, and consumer safety.

The food safety issue is particularly relevant because organically extractive aquaculture relies on local environmental conditions, and bivalve filter-feeders can enhance negative aspects, including heavy metals and organic micropollutants, through bioaccumulation and bioamplification. This underscores the need for improved traceability, which is required for any credit trading scheme. An improved understanding of husbandry, and better stock control, brings several other practical benefits, including certification, consumer confidence, and access to insurance markets.

From a food security perspective, since the European Union currently imports 71% of the aquatic products it consumes (European Commission 2016), and the United States imports 86% (Tiller et al. 2013), any mechanism that can reduce this trade deficit is welcome. The enhancement of bivalve production in European and North American bays and estuaries, where aspects such as xenobiotics are far better regulated than in other parts of the world, will also promote branding (e.g. Made in Europe, Born in the USA), which can drive exports to markets where confidence in internal product safety is weak.

The challenge of sector growth in the West is mainly linked to social licence, which limits spatial expansion. However, if market instruments such as nutrient credit trading expand to accommodate bivalve producers, then existing sites will eco-intensify, boosting yield, improving profitability, and creating jobs.

At present, discussions of valuation such as were presented herein are relevant because they review current knowledge and promote the implementation of integrated management, but from an economic point of view, an ecosystem service is worthless if there is no market for it. It is clear from our analysis of trading mechanisms that the US is by far the most advanced nation in the field of WQT, although HELCOM produced a framework document in 2008 for the Baltic (Green Stream Network 2008), which does not, however, make any reference to bivalves.

In order to promote a European context for involvement of the bivalve aquaculture industry in nutrient credit trading frameworks, it is worth speculating on why the US is considerably more advanced in this area. Potential reasons are: (i) differences in legislation and policy instruments; (ii) concerns that a reduced focus on source control may detract from efforts to reduce land-based nutrient discharge; and (iii) uncertainties about effectiveness as a management tool. While a full discussion of these issues is beyond the scope of this work, we believe that all these aspects need a detailed analysis, if Europe is to move towards integrated nutrient management measures, which insofar as possible internalise the mechanisms used at the basin scale.

Two key differences between the US and Europe can be readily identified: (i) Europe has to deal with enclosed seas such as the Baltic, Black Sea, and the Mediterranean basins, whereas the US marine systems are open; and (ii) as discussed earlier, two of the key EU legal instruments for water policy, the WFD and MSFD, attempted to provide a complete framework for management, but left out aquaculture. Europe is moving toward a much better integration of those instruments with the policies for aquaculture eco-intensification, but there is still some way to go. The US approach of analysing BMPs and enabling approval by regulators of specific aspects such as bivalve grower participation, is a promising approach.

We envisage that nutrient credit trading, and the integration of aquaculture stakeholders, including both finfish producers as emitters and bivalve growers as offset providers, as well as land-based non-point dischargers, will grow substantially over the next decades. More appropriate indicators of ecosystem health will be used, models will play an increasingly important role in assessment and valuation, and communities and coastal management alike will benefit from greater cost internalisation, better traceability, and a closer connection between natural and social systems.

Acknowledgements The authors wish to thank J.P. Nunes and C.B. Zhu for contributions to the compilation of nutrient loading data, and J. Rose for supplying a water quality dataset for Long Island Sound. We are grateful to K. Stephenson for thoughtful discussion of non-point source nutrient trading issues, and to T. O'Higgins for re-working the modelling framework shown in Fig. 27.5. We would also like to thank the EASE project team (listed in Ferreira et al. 2016) for the work executed for Lough Foyle. Finally, we are grateful to three reviewers whose comments substantially improved an earlier draft of this work.

References

Acharya D (2011) Fillet quality and yield of farmed Atlantic salmon (*Salmo salar* L.): variation between families, gender differences and the importance of maturation. M.Sc. Thesis, Norwegian University of Life Sciences, 64 pp. https://brage.bibsys.no/

Ærtebjerg G, Carstensen J, Dahl K, Hansen J, Nygaard K, Rygg B, Sørensen K, Severinsen G, Casartelli S, Schrimpf W, Schiller C, Druon JN, (2001) Eutrophication in Europe's coastal waters. European Environment Agency, Topic report 7/2001, 86 pp. http://www.eea.europa.eu/

Andersen JH, Schlüter L, Ærtebjerg G (2006) Coastal eutrophication: recent developments in definitions and implications for monitoring strategies. J Plankton Res 28(7):621–628

Anonymous (1991a) Council Directive of 21 May 1991 concerning urban waste water treatment (91/271/EEC). Off J L135

Anonymous (1991b) Council Directive 91/676/EEC of 12 December 1991 concerning the protection of waters against pollution caused by nitrates from agricultural sources. Off J L375

Borja A, Bricker SB, Dauer DM, Demetriades NT, Ferreira JG, Forbes AT, Hutchings P, Jia X, Kenchington R, Marques JC, Zhu CB (2008) Overview of integrative tools and methods in assessing ecological integrity in estuarine and coastal systems worldwide. Mar Pollut Bull 56:1519–1537

Boynton WR, Kemp WM, Keefe CW (1982) A comparative analysis of nutrients and other factors influencing estuarine phytoplankton production. In: Kennedy VS (ed) Estuarine comparisons. Academic, New York, pp 69–90

Bricker SB, Ferreira JG, Simas T (2003) An integrated methodology for assessment of estuarine trophic status. Ecol Model 169(1):39–60

Bricker SB, Longstaff B, Dennison W, Jones A, Boicourt K, Wicks C, Woerner J (2007) Effects of nutrient enrichment in the Nation's estuaries: a decade of change, National estuarine eutrophication assessment update. NOAA Coastal Ocean Program Decision Analysis Series No. 26. National Centers for Coastal Ocean Science, Silver Spring, MD. 322 pp

Bricker SB, Longstaff B, Dennison W, Jones A, Boicourt K, Wicks C, Woerner J (2008) Effects of nutrient enrichment in the nation's estuaries: a decade of change. Special issue of Harmful Algae 8:21–32

Branosky E, Jones C, Selman M (2011) Compareison tables of state nutrient trading programs in the Chesapeake Bay watershed. Fact sheet. World Resources Institute, Washington, DC

China Fishery Statistical Yearbook (2016) Bureau of Fisheries, Ministry of Agriculture of the People's Republic of China, 2016. China Agricultural Press, Beijing. ISBN: 978-7-109-21691-4

Cloern JE (1982) Does the benthos control phytoplankton biomass in South San Francisco Bay? Mar Ecol Prog Ser 9:191–202

Cloern J (2001) Our evolving conceptual model of the coastal eutrophication problem. Mar Ecol Prog Ser 210:223–253

Collins AL, McGonigle DF (2008) Monitoring and modelling diffuse pollution from agriculture for policy support: UK and European experience. Environ Sci Pol 11(2):97–101

Connecticut Department of Environmental Protection (CT DEP) (2010) An incentive-based water quality trading program. The Connecticut Department of Environmental Protection. Bureau of Water Protection and Land Reuse, Hartford, 10pp

Dales JH (1968) Pollution, property & prices: an essay in policy-making and economics. University of Toronto Press, Toronto, pp VII–111S

DeBoe G, Stephenson K (2016) Transactions costs of expanding nutrient trading to agricultural working lands: a Virginia case study. Ecol Econ 130:176–185

Diaz RJ, Rosenberg R (2008) Spreading dead zones and consequences for marine ecosystems. Science 321:926–929

European Commission (2016) Facts and figures on the common fisheries policy. European Commission, 56 pp

FAO (Food and Agriculture Organization of the United Nations) (2016) The state of world fisheries and aquaculture (SOFIA). FAO, Rome, 204 pp

Ferreira JG, Nobre AM, Simas TC, Silva MC, Newton A, Bricker SB, Wolff WJ, Stacey PE, Sequeira A (2006) A methodology for defining homogeneous water bodies in estuaries – application to the transitional systems of the EU Water Framework Directive. Estuar Coast Shelf Sci 66(3/4):468–482

Ferreira JG, Hawkins AJS, Bricker SB (2007a) Management of productivity, environmental effects and profitability of bivalve aquaculture – the Farm Aqua-culture Resource Management (FARM) model. Aquaculture 264:160–174

Ferreira JG, Bricker SB, Simas TC (2007b) Application and sensitivity testing of an eutrophication assessment method on coastal systems in the United States and European Union. J Environ Manag 82:433–445

Ferreira JG, Hawkins AJS, Monteiro P, Moore H, M. Service, Pascoe PL, Ramos L, Sequeira A (2008) Integrated assessment of ecosystem-scale carrying capacity in shellfish growing areas. Aquaculture 275:138–151

Ferreira JG, Andersen JH, Borja A, Bricker SB, Camp J, Cardoso da Silva M, Garcés E, Heiskanen A-S, Christoph H, Ignatiades L, Lancelot C, Menesguen A, Tett P, Hoepffner N, Claussen U (2011) Indicators of human-induced eutrophication to assess the environmental status within the European Marine Strategy Framework Directive. Estuar Coast Shelf Sci 93:117–131

Ferreira JG, Bricker SB (2015) Nitrogen remediation through shellfish aquaculture – a model analysis. AE2015, Rotterdam, European Aquaculture Society, http://www.ecowin.org/eas2015

Ferreira JG, Bricker SB (2016) Goods and services of extensive aquaculture: shellfish culture and nutrient trading. Aquac Int 24(3):803–825

Ferreira JG, Moore H, Boylan P, Jordan C, Lencart-Silva JD, McGonigle C, McLean S, Nunes JP, Service M, Zhu CB (2016) Application of a multi-model framework for integrated ecosystem management in Lough Foyle. In: EAAP 2016, 67th annual meeting of the European Federation of Animal Science, Belfast, UK, 29/8-2/9 2016

Gassman PW, Reyes MR, Green CH, Arnold JG (2007) The soil and water assessment tool: historical development, applications, and future research directions. Am Soc Agric Biol Eng 50(4):1211–1250

Gerritsen J, Holland AF, Irvine DE (1994) Suspension-feeding bivalves and the fate of primary production: an estuarine model applied to Chesapeake Bay. Estuaries 17(2):403–416

Green Stream Network (2008) Framework for a nutrient quota and credits trading system for the contracting parties of HELCOM in order to reduce eutrophication of the Baltic. Final report to the Nordic Environment Finance Organization. Green Stream Network, Helsinki

Gren IM, Jonzon Y, Lindqvist M (2008) Cost of nutrient reductions to the Baltic Sea – technical report. SLU, Institutionen för ekonomi 2008:1, 65 pp

Gunningham N, Sinclair D (2005) Policy instrument choice and diffuse source pollution. J Environ Law 17(1):51–81

HELCOM (2009) Andersen JH, Laamanen M (eds) Eutrophication in the Baltic Sea – an integrated thematic assessment of the effects of nutrient enrichment and eutrophication in the Baltic Sea region.Baltic Sea Environment Proceedings No. 115B. 148pp

HELCOM (2010) Ecosystem health of the Baltic Sea 2003–2007: HELCOM initial holistic assessment.Baltic Sea Environment Proceedings No. 122

HELCOM (2014) Eutrophication status of the Baltic Sea 2007–2011 - a concise thematic assessment. Baltic Sea Environment Proceedings No. 143.

Higgins CB, Stephenson K, Brown BL (2011) Nutrient bioassimilation capacity of aquacultured oysters: quantification of an ecosystem service. J Environ Qual 40(1):271–277

Houle J, Roseen R, Ballestero T, Puls T, Sherrard J (2013) A comparison of maintenance cost, labor demands, and system performance for LID and conventional stormwater management. J Environ Eng 139:932–938

Howarth RW (1988) Nutrient limitation of net primary production in marine ecosystems. Ann Rev Ecol 19:89–110

Howarth RW, Sharpley A, Walker D (2002) Sources of nutrient pollution to coastal waters in the United States: implications for achieving coastal water quality goals. Estuaries 25(4b):656–676

Jones C, Branosky E, Selman M, Perez M (2010) How nutrient trading could help restore the Chesapeake Bay. World Resources Institute, Washington, DC. http://wwwwriorg/publication/how-nutrienttrading-could-help-restore-chesapeake-bay Accessed 12 Feb 2017

Kellogg ML, Cornwell JC, Owens MS, Paynter KT (2013) Denitrification and nutrient assimilation on a restored oyster reef. Mar Ecol Prog Ser 480:1–19

Kellogg ML, Smyth AR, Luckenbach MW, Carmichael RH, Brown BL, Cornwell JC, Piehler MF, Owens MS, Dalrymple DJ, Higgins CB (2014) Use of oysters to mitigate eutrophication in coastal waters. Estuar Coast Shelf Sci 151:156–168

Ketchum BH (1954) Relation between circulation and planktonic populations in estuaries. Ecology 35:191–200

Lancelot C, Thieu V, Polard A, Garnier J, Billen G, Hecq W, Gypens N (2011) Cost assessment and ecological effectiveness of nutrient reduction options for mitigating Phaeocystis colony blooms in the Southern North Sea: an integrated modeling approach. Sci Total Environ 409(11):2179–2191

Lal H (2010) Nutrient credit trading- a market-based approach for improving water quality. In: Delgado JA, Follett RF (eds) Advances in nitrogen management for water quality. Soil and Water Conservation Society, Ankeny, pp 344–361

Lindahl O, Hart R, Hernroth B, Kollberg S, Loo L, Olrog L, Rehnstam-Holm A (2005) Improving marine water quality by mussel farming: a profitable solution for Swedish Society. Ambio 34(2):131–138

Mann R, Ryther JH (1977) Growth of six species of bivalve molluscs in a waste recycling-aquaculture system. Aquaculture 11:231–245

National Research Council (NRC) (1993) Managing wastewater in coastal urban areas. National Academy Press, Washington, DC

National Research Council (NRC) (2000) Clean coastal waters: understanding and reducing the effects of nutrient pollution. National Academy Press, Washington, DC

Nixon SW, Pilson MEQ (1983) In: Carpenter EJ, Capone DG (eds) Nitrogen in the marine environment. Academic Press, New York, pp 565–648

Nixon SW (1995) Coastal marine eutrophication: a definition, social causes, and future concerns. Ophelia 41:199–219

Nobre AM, Ferreira JG, Nunes JP, Yan X, Bricker S, Corner R, Groom S, Gu H, Hawkins A, Hutson R, Lan D, Lencart e Silva JD, Pascoe P, Telfer T, Zhang X, Zhu M (2010) Assessment of coastal management options by means of multilayered ecosystem models. Estuar Coast Shelf Sci 87:43–62

Nunes JP, Ferreira JG (2016) Comparing agricultural and urban nutrient loads to coastal systems. EAAP 2016, 67th annual meeting of the European Federation of Animal Science, Belfast, UK, 29/8-2/9 2016

Nunneri C, Windhorst W, Kerry Turner R, Hermann Lenhart H (2007) Nutrient emission reduction scenarios in the North Sea: an abatement cost and ecosystem integrity analysis. Ecol Indic 7(4):776–792

Officer CB, Smayda TJ, Mann R (1982) Benthic filter feeding: a natural eutrophication control. Mar Ecol Prog Ser 9:191–202

Osmond D, Meals D, Hoag D, Arabi M, Luloff A, Jennings G, Mcfarland M, Spooner J, Sharpley A, Line D (2012) Improving conservation practices programming to protect water quality in agricultural watersheds: lessons learned from the National Institute of Food and Agriculture–conservation effects assessment project. J Soil Water Conserv 67(5):122A–127A. https://doi.org/10.2489/jswc.67.5.122A

OSPAR (2010) Moffat C, Emmerson R, Weiss A, Symon C, Dicks L (eds) Quality status report 2010. OSPAR Commission, London. 176pp

Oyster BMP Expert Panel (2016) Panel recommendations on the Oyster BMP nutrient and suspended sediment reduction effectiveness determination decision framework and nitrogen and phosphorus assimilation in oyster tissue reduction effectiveness for oyster aquaculture practices. Report submitted to the Chesapeake Bay partnership water quality goal implementation team, September 22, 2016. www.chesapeakebay.net/calendar/event/24330/.

Petersen JK, Hasler B, Timmermann K, Nielsen P, Tørring DB, Larsen MM, Holmer M (2014) Mussels as a tool for mitigation of nutrients in the marine environment. Mar Pollut Bull 82:137–143

Rabalais N, Turner RE, Wiseman WJ Jr (2002) Gulf of Mexico Hypoxia, A.K.A. "The Dead Zone". Annu Rev Ecol Syst 33:235–263

Rose JM, Bricker SB, Ferreira JG (2015) Comparative analysis of modeled nitrogen removal by shellfish farms. Mar Pollut Bull 91(1):185–190

Reitsma J, Murphy DC, Archer AF, York RH (2016) Nitrogen extraction potential of wild and cultured bivalves harvested from nearshore waters of Cape Cod, USA. https://doi.org/10.1016/j.marpolbul.2016.12.072

Ribaudo MO, Gottlieb J (2011) Point-nonpoint trading-can it work? J Am Water Resour Assoc 47(1):5–14

Rice E, Stewart G (2013) Analysis of interdecadal trends in chlorophyll and temperature in the Central Basin of Long Island Sound. Estuar Coast Shelf Sci 128:64–75

Rose JM, Bricker SB, Tedesco MA, Wikfors GH (2014) A role for shellfish aquaculture in coastal nitrogen management environ. Sci Technol 2014(48):2519–2525. https://doi.org/10.1021/es4041336

Ryther JH, Dunstan WM (1971) Nitrogen, phosphorus, and eutrophication in the coastal marine environment. Science 171:1008–1013

Saurel C, Ferreira JG, Cheney D, Suhrbier A, Dewey B, Davis J, Cordell J (2014) Ecosystem goods and services from Manila clam culture in Puget Sound: a modelling analysis. Aquac Environ Interact 5:255–270

Shortle J (2013) Economics and environmental markets: lessons from water-quality trading. Agric Resour Econ Rev 42(1):57–74

STAC (Chesapeake Bay Program Scientific and Technical Advisory Committee) (2013) Evaluation of the use of shellfish as a method of nutrient reduction in the Chesapeake Bay. STAC Publ. #13-005, Edgewater. 65 pp

State Oceanic Administration (SOA) of China (2016) Bulletin of Marine Environmental Status of China for the year of 2015. SOA, 2016. http://www.coi.gov.cn/gongbao/nrhuanjing/nr2015/

Stephenson K, Aultman S, Metcalfe T, Miller A (2010) An evaluation of nutrient nonpoint offset trading in Virginia: a role for agricultural nonpoint sources? Water Resour Res 46:W04519

Stephenson K, Shabman L (2017a) Nutrient Assimilation Services for water quality credit trading programs: a comparative analysis with nonpoint source credits. Coast Manag 45(1):1–20. https://doi.org/10.1080/08920753.2017.1237240

Stephenson K, Shabman L (2017b) Where did the agricultural nonpoint source trades go? Lessons from Virginia water quality trading programs. J Am Water Resour Assoc 53(5):1178–1194

Tiller R, Gentry R, Richards R (2013) Stakeholder driven future scenarios as an element of interdisciplinary management tools; the case of future offshore aquaculture development and the potential effects on fishermen in Santa Barbara, California. Ocean Coast Manag 73:127–135

Town of Mashpee Sewer Commission (2015) Final recommended plan/final environmental impact report. Comprehensive wastewater management plan, Town of Mashpee. Prepared by GHD, Inc., Hyannis, Massachusetts. 349 pp

Tong Y, Zhao Y, Zhen G, Chi J, Liu X, Lu Y, Wang X, Yao R, Chen J, Zhang W (2015) Nutrient loads flowing into coastal waters from the main rivers of China (2006–2012). Sci Rep 5:16678. https://doi.org/10.1038/srep16678
United States Environmental Protection Agency (USEPA) (2004a) Primer for municipal wastewater treatment systems. EPA 832-R-04-001, U.S. Environmental Protection Agency. Office of Water, Washington, DC, 30 pp
United States Environmental Protection Agency (USEPA) (2004b) Water quality trading assessment handbook: can water quality trading advance your watershed's goals? EPA 841-B-04-001. Office of Water, Washington, DC
United States Environmental Protection Agency (USEPA) (2008a) Handbook for developing watershed plans to restore and protect our waters. EPA 841-B-08-002, U.S. Environmental Protection Agency. Office of Water, Washington, DC, 400 pp
United States Environmental Protection Agency (USEPA) (2008b) EPA's 2008 Report on the Environment. National Center for Environmental Assessment, Washington, DC; EPA/600/R-07/045F. http://www.epa.gov/roe, http://cfpub.epa.gov/ncea/cfm/recordisplay.cfm?deid¼190806
United States Environmental Protection Agency (USEPA) (2017) Implementing Clean Water Act Section 303(d): impaired waters and Total Maximum Daily Loads (TMDLs). https://www.epa.gov/tmdl. Accessed 9 Feb 2017
Virginia DEQ (Department of Environmental Quality) (2008) Trading nutrient reductions from nonpoint source best management practices in the Chesapeake Bay Watershed: Guidance for Agricultural Landowners and Your Potential Trading Partners. http://www.deq.virginia.gov/Portals/0/DEQ/Water/PollutionDischargeElimination/VANPSTradingManual_2-5-08.pdf. 40 pp
Xiao Y, Ferreira JG, Bricker SB, Nunes JP, Zhu M, Zhang X (2007) Trophic assessment in Chinese coastal systems—review of methods and application to the Changjiang (Yangtze) Estuary and Jiaozhou Bay. Estuaries & Coasts 30(6):901–918
Wang X, Andresen K, Handå A, Jensen B, Reitan KI, Olsen Y (2013) Chemical composition and release rate of waste discharge from an Atlantic salmon farm with an evaluation of IMTA feasibility. Aquacult Environ Interact 4:147–162
Wang X, Broch OB, Forbord S, Handå A, Skjermo J, Reitan KI, Vadstein O, Olsen Y (2014) Assimilation of inorganic nutrients from salmon (*Salmo salar*) farming by the macroalgae (*Saccharina latissima*) in an exposed coastal environment: implications for integrated multi-trophic aquaculture. J Appl Phycol 26:1869–1878
Wise DR, Johnson HM (2011) Surface-water nutrient conditions and sources in the United States Pacific Northwest. J Am Water Resour Assoc 47(5):1110–1135 https://www.ncbi.nlm.nih.gov/pmc/articles/PMC3307616/
Zaldivar JM, Cardoso AC, Viaroli P, Newton A, de Wit R, Ibanez C, Reizopoulou S, Somma F, Razinkovas A, Basset A, Holmer M, Murray N (2008) Eutrophication in transitional waters: an overview. Transitional Waters Monogr 1:1–78

Correction to: Goods and Services of Marine Bivalves

Aad C. Smaal, Joao G. Ferreira, Jon Grant, Jens K. Petersen, and Øivind Strand

Correction to:
A. C. Smaal et al. (eds.), *Goods and Services of Marine Bivalves*, https://doi.org/10.1007/978-3-319-96776-9

This book was inadvertently published with an incorrect copyright year '2018' within the book references. This has now been amended throughout the book to the correct copyright year '2019'.

The updated versions of the book can be found at
https://doi.org/10.1007/978-3-319-96776-9

A. C. Smaal et al. (eds.), *Goods and Services of Marine Bivalves*,
https://doi.org/10.1007/978-3-319-96776-9_28

Epilogue

In this book various functions of marine bivalves in the ecosystem were addressed, as a basis for analysing their goods and services, defined as the direct and indirect benefits people obtain from marine bivalves (adapted after Beaumont et al. 2007). Following the structure of the goods and services concept, we described the goods provided by bivalve aquaculture, the functions of bivalves that act as regulating services in the ecosystem, and the cultural services provided by bivalves to various human communities. In the final part, tools have been reviewed to assess bivalve services.

Provisioning Services

The most obvious benefit people obtain from bivalves are from bivalve aquaculture and fisheries. FAO data on global aquaculture and fishery production of marine bivalves for human consumption show a steady increase of 6.5% per year over the last 20 years, up to 16 million tons in 2015. It now comprises about 13% of total marine production in the world from wild catch and aquaculture, in addition to other molluscs (5%), seaweed (22%), crustaceans (7%) and fish (53%). In contrast to marine fish that comes for ca 90% from wild catch, bivalve production derives almost 90% from aquaculture. This illustrates the context of bivalve aquaculture in provisioning human nutrition. Wild fisheries in general are confronted with a plateau in yield. It is widely recognised that sea harvest can only increase through controlled production i.e. aquaculture. Marine bivalve production is now based on the domestication of more than 70 species (FAO data). It should be mentioned however, that the provisioning of seed in many cases occurs through fishery of wild juvenile stocks, and only a minor part comes from hatchery/nursery production. Products from bivalve aquaculture and fisheries not only comprise human nutrition. An issue that deserves more attention is the immense production of shell material that results from bivalve harvest. On average meat content is ca 30%, so annually

A. C. Smaal et al. (eds.), *Goods and Services of Marine Bivalves*,
https://doi.org/10.1007/978-3-319-96776-9

70% of 18 million tons production, ca 13 million tons, is shell that has no or low value application so far. Given the need to efficiently use resources, as exemplified by a circular economy approach, it is a challenge to make more use of the various types of shell material (see the review by Morris et al. 2018).

The global challenge is to secure sustainable food production for an increasing human population. Marine bivalves are a renowned resource of protein and healthy fatty acids. Bivalve aquaculture has the potential to contribute to future food requirements, and meanwhile fulfil relevant ecological functions. Given the need to feed the growing world population with healthy food that also supports ecological functions, the question is what contribution can be realized by the use of marine bivalve aquaculture. In a recent report SAPEA (2017) analysed the potential food production from the oceans, and they conclude that "*Basically, there is only one way to obtain significantly more food and biomass from the ocean and that is to harvest seafood that on average is from a lower trophic level than today*". Expansion of bivalve aquaculture production from 18 million tons at present to 100 million tons in ca 20 years is one of the options proposed by SAPEA. Present limitations are competing spatial claims in the coastal zone, water quality requirements, lack of offshore technology, episodic mortalities and concerns about invasive species and interactions with wild stocks. Ocean acidification is also considered as a threat for the bivalves. Yet, coupling with seaweed farming may counteract this process (SAPEA 2017). The report is directed to the European Commission. This is relevant as particularly in Europe, bivalve production has decreased since 2000, while prices go up and demands are increasingly covered by imports. Space for bivalve aquaculture is considered as the main limitation, due to maximum use of production carrying capacity in traditional culture areas, together with enhanced environmental regulation i.e. limiting ecological capacity, and competing claims from other stakeholders, i.e. limiting social carrying capacity.

As for aquaculture in general, the leading bivalve producer is China, accounting for 85% of the world production, and still growing. Production and ecological carrying capacity are in many areas the main factors that determine the actual potential for bivalve aquaculture in China. The limitations for further growth are now recognised and when new policies, legislation and management measures are successively implemented, growth in China will face limits.

New developments would be needed to benefit from the provisioning services of the marine bivalves like advanced marine spatial planning (see Gentry et al. 2017), further expansion of hatchery production, application of seed collectors and trials with offshore aquaculture (see Buck and Langan 2017). Indeed, if the expected world population of 10 billion people in 2050 would consume 250 g of bivalve meat per week, the required annual global production would be 500 million tons of bivalves including shell. This is ca 30 times the actual production.

Regulating Services

Since bivalve aquaculture is extensive, the cultured bivalves form an integral part of the ecosystem. This is the key in understanding other functions the bivalves have in the ecosystem, including both wild and cultured bivalve stocks. Yet, cultured stocks are under human control, providing the opportunity to make use of bivalve regulating services. In this book various regulating services that are attributed to bivalves have been reviewed.

By their functional role as filter feeders, bivalves regulate water quality, primary production and nutrient dynamics. In their shell and tissues they accumulate nutrients and carbon. These functions make them instrumental in mitigating eutrophication, wastewater discharges, fish farming impacts and sequestration of carbon dioxide.

By their capacity to form structures, epibenthic bivalve reefs (oysters and mussels) are considered eco-engineers, that modify the physical environment with effects on the ecological features. If proper substrate is provided at selected locations, reefs can develop as self-sustaining systems. They are used to enhance coastal protection as they hamper wave action and increase sedimentation beyond the reefs. As a consequence, other communities develop in the 'shadow' of the reef, like seagrass as well as benthic infauna that is beneficial for i.e. shorebirds.

Oyster reefs and mussel beds are biodiversity hot spots for epifauna, and a refuge for mobile fauna including fish. The reefs can also act as an additional food source for local human populations. The eco-engineering capacities are applied in shellfish restoration, in that oyster reef substrate is deployed in areas where original stocks have disappeared and conditions can be created to restore them. This is a major challenge as bivalve reefs are worldwide under pressure (Beck et al. 2011).

The functional and structural features of bivalves are not only provided by wild stocks; similar functions can be achieved by bivalve aquaculture.

Cultural Services

As reviewed by Daniel et al. (2012), cultural services are widely recognised as important but they are often considered as subjective, difficult to quantify and dependent on social context. Although this also holds for the other services, as by definition it is about benefits people obtain, the benefits of cultural services generally depend on the specific social context. For example in the US, private initiatives for shellfish restoration occur on a large scale in different communities, while in Europe this is typically the domain of the public sector. As addressed in Part III, in Europe, private initiatives with bivalves are more related to leisure activities such as cultivation in a social context for culinary use. Yet, preferences may easily change in the future. Some private initiatives for bivalve restoration are already in place in Europe (www.ark.eu/en/projects/shellfish-reefs).

A prominent example of cultural services of bivalves and bivalve shells considers the use in arts and decoration. This is a wide topic that has not been addressed in this book. However it is extensively covered in scientific and popular literature.

Cultural services are in a way multi-dimensional in comparison with the direct and indirect material services of marine bivalves. To acknowledge the beauty of shells, for example, is a service that goes beyond quantification and valorisation.

Assessment Tools and Valorisation

In Part V, various approaches are reviewed on how to assess the goods and services of marine bivalves. As bivalves are part of the ecosystem, assessment is complex because the functional role depends on ecosystem processes. An integrated approach is followed that includes the processes relevant for the purpose of the modelling. For farm scale bivalve production estimations, for example, the model requirements are different from spatial planning models. Yet, for all types of models localized data are needed for calibration and validation. If data are scarce, a first estimate can be based on simple indicators, and when this leads to further steps, more detailed information can be collected to support decision making.

The aim of this book is to show that knowledge and inclusion of the relevant goods and services of cultivated and wild bivalves in decision making will give more integrated solutions; the model tools can be used to achieve this broader picture. An important issue is how model scenarios can be translated into decision making. Attempts are made to express the goods and services in monetary terms, see for example Grabowski et al. (2012), in order to include them in trade market based decisions. Valorisation of certain services raises the critical issue of how prices for goods and services are established. These are the product of markets and incentives, and do not necessarily account for the contribution of the marine bivalves to non-provisioning services. In specific cases, nitrogen can be converted into credits because nitrogen input and output can be quantified in a uniform way and compared with other costs related with nitrogen management. This tool for nutrient trading may simplify management decisions. However, this is difficult to achieve for more complex decisions that concern marine spatial planning where multiple stakeholders play a role. For these types of decision making, models based on multi-criteria analysis may be a more suitable decision support instrument.

Yet, many initiatives are underway to improve quantitative knowledge of ecosystem goods and services. For example in the TEEB framework (the economics of ecosystems and biodiversity: www.teebweb.org) various studies are being carried out on quantification and valorization of ecosystem services, focusing on development of methodologies as well as tests in case studies. For the marine bivalves further studies are needed in order to improve the tools as reviewed in Part IV and apply these in decision making.

Integration

This book is about the direct and indirect benefits people obtain from marine bivalves. This issue is much broader than generally acknowledged, as is apparent when the various goods and services are reviewed. The question remains as to how to achieve synergy in the use of these goods and services. The urgency to do so not only comes from the challenge to contribute to human nutrition for an expanding world population – the predicted food requirement by 2050 may ask for a 30-fold increase in production, but also to restore threatened bivalve communities, as many bivalve reefs have disappeared over the last decades.

The key factor in providing bivalves as food is to find space where the carrying capacity can be exploited in a sustainable way. This means that physical and production capacities need to be sufficient for the aquaculture of bivalves. It also means that people accept the aquaculture production in the given area, i.e. the social carrying capacity needs to be addressed concomitantly. This can only be realized when the ecological carrying capacity is not exceeded. A net consideration of the role of the bivalves in the ecosystem need to be analysed including possible positive and negative impacts.

This book shows that a major benefit of this type of extensive aquaculture can be found in the other services provided by the bivalves. Their role as filter feeders that stimulate nutrient cycling and phytoplankton turnover is beneficial for water quality management. The extraction of nitrogen and phosphate through harvesting mitigates adverse eutrophication effects. Enhanced water transparency due to filtration promotes submerged aquatic vegetation, while biodeposition may enhance denitrification. Bivalve beds including cultivated bivalves, facilitate a rich community of epifauna and mobile fauna, including fish. Their role as eco-engineers implies structural changes in the direct environment of the beds, promoting sedimentation, infauna development, enhancement of bird food and seagrass extension, hence secure shoreline resilience. Integration of bivalve aquaculture and bivalve restoration to achieve optimal use of bivalve goods and services is a substantial challenge.

We hope this book stimulates the further development of the bivalve goods and services concept, as well as the implementation of an integrated approach for restoration and sustainable exploitation of marine bivalves.

Yerseke, The Netherlands — Aad C. Smaal
Monte de Caparica, Portugal — Joao G. Ferreira
Halifax, NS, Canada — Jon Grant
Nykøbing Mors, Denmark — Jens K. Petersen
Bergen, Norway — Øivind Strand

References

Beaumont NJ, Austen MC, Atkins JP, Burdon D, Degraer S, Dentinho TP, Derous S, Holm P, Horton T, van Ierland E, Marboe AH, Starkey DJ, Townsend M, Zarzycki T (2007) Identification, definition and quantification of goods and services provided by marine biodiversity: implications for the ecosystem approach. Mar Pollut Bull 54:253–265

Beck MW, Robert D, Brumbaugh LA, Carranza A, Coen LD, Crawford C, Defeo O, Edgar GJ, Hancock B, Kay MC, Lenihan HS, Luckenbach MW, Toropova CL, Zhang G, Guo X (2011) Oyster reefs at risk and recommendations for conservation, restoration, and management. Bioscience 61:107–116

Buck BH, Langan R (eds) (2017) Aquaculture perspective for multi-use sites in the open ocean – the untapped potential for marine resources in the Anthropocene. Springer Open, 402 p

Daniel TC, Muhar A, Arnberger A, Aznar O, Boyd JW, Chan KMA, Costanza R, Elmqvist T, Flint CG, Gobster PH, Grêt-Regamey A, Lave R, Muhar S, Penker M, Ribe RG, Schauppenlehner T, Sikor T, Soloviy I, Spierenburg M, Taczanowska K, Tam J, von der Dunk A (2012) Contributions of cultural services to the ecosystem services agenda. PNAS 109(23):8812–8819

Gentry RR, Froehlich HE, Grimm D, Kareiva P, Parke M, Rust M, Halpern BS (2017) Mapping the global potential for marine aquaculture. Nat Ecol Evol 1:1317–1324

Grabowksi JH, Brumbaugh RD, Conrad RF, Keeler AG, Opaluch JJ, Peterson CH, Piehler MF, Powers SP, Smyth AR (2012) Economic valuation of ecosystem services provided by oyster reefs. Bioscience 62:900–909

Morris JP, Backeljau T, Chapelle G (2018) Shells from aquaculture: a valuable biomaterial, not a nuisance waste product. Rev Aquac 11:1–6

SAPEA, Science Advice for Policy by European Academies (2017) Food from the oceans: how can more food and biomass be obtained from the oceans in a way that does not deprive future generations of their benefits? SAPEA, Berlin

A. C. Smaal et al. (eds.), *Goods and Services of Marine Bivalves*,
https://doi.org/10.1007/978-3-319-96776-9